Practical and Theoretical Geoarchaeology

Practical and Theoretical Geoarchaeology

Second Edition

Paul Goldberg
and
Richard I. Macphail

With Chapter Contributions by Chris Carey and Yijie Zhuang

WILEY Blackwell

Registered Offices
John Wiley & Sons, Inc., 111 River Street, Hoboken, NJ 07030, USA
John Wiley & Sons Ltd, The Atrium, Southern Gate, Chichester, West Sussex, PO19 8SQ, UK

Editorial Office
9600 Garsington Road, Oxford, OX4 2DQ, UK

For details of our global editorial offices, customer services, and more information about Wiley products visit us at www.wiley.com.

Wiley also publishes its books in a variety of electronic formats and by print-on-demand. Some content that appears in standard print versions of this book may not be available in other formats.

Library of Congress Cataloging-in-Publication Data

Names: Goldberg, Paul, author. | Macphail, Richard, author.
Title: Practical and theoretical geoarchaeology / Paul Goldberg and Richard
  I. Macphail.
Description: Second edition. | Hoboken, NJ : Wiley-Blackwell, 2022. |
  Includes bibliographical references and index.
Identifiers: LCCN 2022024105 (print) | LCCN 2022024106 (ebook) | ISBN
  9781119413196 (paperback) | ISBN 9781119413202 (adobe pdf) | ISBN
  9781119413219 (epub)
Subjects: LCSH: Archaeological geology.
Classification: LCC CC77.5 .G65 2022 (print) | LCC CC77.5 (ebook) | DDC
  930.1–dc23/eng/20220630
LC record available at https://lccn.loc.gov/2022024105
LC ebook record available at https://lccn.loc.gov/2022024106

Cover Design: Wiley
Cover Images: © The Authors

Set in 9.5/12.5pt STIXTwoText by Straive, Pondicherry, India

SKY10035602_080522

PG: To all my friends and colleagues who really helped write the book.

RIM: To Jill, Flora, Steve, Marilyn, Pete and Sue; L&D University Hospital and the Exeter and District Ramblers; and a very special thanks to all the archaeological organisations who have contributed to the advances in this book.

CC: To Sam, mum and dad, my friends and colleagues.

YZ: To my mum and family.

Contents

Preface to Revised Edition

It has been more than a decade and a half since the first edition appeared. Over this time, Geoarchaeology as a broad discipline has developed and matured significantly. Although the fundamentals of geological and pedological sciences that are employed during fieldwork and laboratory investigations essentially remain the same, the ways data are gathered and the accuracy of their interpretations have clearly advanced. In addition, we have seen an enormous expansion in the geographical and cultural scope of sites and issues now being addressed as a part of interdisciplinary investigations. The sites, countries, and methods employed by the authors are both worldwide in scope and up to date. This is because the four authors – who are based in three continents – have combined their practical experience and academic study to this end. Whilst several recent works have focused on single-technique approaches, this book has the aim of bringing all geoarchaeological methods to the table. For example, traditional fieldwork is now augmented by remote sensing techniques and three-dimensional modelling through geotechnical approaches. We also show how laboratory studies, using both traditional methods (bulk chemistry and soil micromorphology), can be enhanced by employing various new instrumental techniques. Just as importantly, we demonstrate how data can be fully integrated with other palaeoenviromental findings, and then graphically portrayed employing GIS, for example.

In particular, the past decade or so displayed a burgeoning development of geoarchaeology in Asia and Africa. Reflecting on the great achievements that have been made by colleagues working in these diverse environmental and ecological settings, we have incorporated some recent geoarchaeological studies on these regions in relevant chapters. We believe that more dialogues among geoarchaeologists working in different global regions will stimulate methodological and theoretic innovations in geoarchaeology and we hope that our preliminary effort in this book will start to provide a useful platform for such important scholarly understanding in global geoarchaeology.

After discussions over the years with fellow geoarchaeologists (e.g., Sarah Sherwood, Sewanee, USA), as well as archaeologists and palaeoenvironmentalists who equally need to be able to appreciate what Geoarchaeology can achieve for their site and projects, we have retained a similar thematic structure. In the first part, fundamental aspects of Sediments, Stratigraphy, and Soils are presented, which, with numerous case studies and examples, should prove interesting to both novices and professionals who want to brush up their skills and renew their acquaintance with these subjects. We then examine a broad series of depositional environments and their differing roles in use of the landscape and potential for archaeological site preservation. We have grouped the first three into topics associated with the hydrological cycle. Slopes, for example, includes issues of both erosion and colluviation; Rivers examines the physics of flow, sediment transport, and deposition, such as alluviation and its effects on landscapes given over to rice production in south-east Asia, for example; Lakes are important for wetland resources, and for their sediments that preserve records of early human activity. Other special environments involve wind as an agency of deposition of both sands and silts, which provide both background environments as well some typical site formation processes (Aeolian). Similarly, while Marine Coasts can be characterized by both recent and ancient dune formation, an understanding of intertidal archaeology and sea level changes can be key to interpreting site analyses and associated formation processes. Caves and Rockshelters are very special depositional environments where human use dates from Early Palaeolithic hunting and gathering to recent animal stabling; some of the earliest recognized cognitive activity is recorded in them.

After this discussion of sedimentary environments, we present five thematic chapters demonstrating the many facets of Geoarchaeology. Human Impact mainly deals with some important activities of people, such as woodland clearance and cultivation, the latter including a large variety of cultivation and manuring methods from across the globe; the effects of mining and water management in arid environments are also considered. This is followed by a chapter on The Human Use of Materials, which involves both the use of natural soils for construction and the manufacture of lime-based materials, and the investigation of residues from ferrous and non-ferrous metal working. This section then leads to Anthropogenic Deposits, which begins with a review of how these can be modelled for better understanding site formation processes and use of space, for example. In addition, we investigate mounds and mound-like deposits such as tells, utilizing examples from the New World, Eastern Europe, and the Near East. This chapter also encompasses urban and settlement archaeology, and formation of specific occupation deposits. Special topics (e.g. *terra preta*, Dark Earth, and pit houses) are also headlined. An important chapter explains how Geoarchaeologists arrive at sound interpretations of past sites and landscapes. This is presented in Experimental and Ethno-Geoarchaeology, where we discuss experiments ranging from clearance and cultivation methods, to monitoring changes to buried soils through time and how these findings have been applied to sites. In addition, we discuss experimental aspects of inundation of coastal sites, creation of reference materials, animal management, and observations of deposits that are relicts of house burning and decay. The last chapter in this thematic series, Geoarchaeology of Forensic Science and Mortuary Practices, was developed because of the increasing interest in both forensic science and the allied study of various funerary practices of cultural importance that need to be differentiated.

The last part of the book is given over to methods and the presentation of geoarchaeological findings. Fieldwork (Field Methods) can now involve coring, remote sensing, use of drones, 3D modelling, as well as careful logging of profiles and scientific sampling protocols. The chapter on Laboratory Methods discusses the use of traditional physical and chemical analyses employing instrumental techniques on bulk samples to map and characterize patterns of occupation, for example; parallel thin section studies have the advantage of contextualizing and closely linking analytical results to the microstratigraphy. For the latter, we briefly give examples of the use of associated SEM/EDS and micro-FTIR techniques to produce the kind of hard data that was only dreamt about in the past. Lastly, no matter how good the field and laboratory data are, they are not useful if they are not presented properly. In Reporting and Publishing we not only provide instances of how a variety of site findings can be documented and illustrated for the client and wider scientific audience, but we also show that the very act of preparing an article can improve the interpretation of sites when all data are integrated.

xiii

Acknowledgments

This revision took a lot more effort than we had budgeted, and we would like to thank a number of people who helped us along the way.

We benefited from numerous discussions and collaborations and retrieved a number of valued insights from the late Ofer Bar-Yosef, Takis Karkanas, Sarah Sherwood, Carolina Mallol, the late Harold Dibble, Shannon McPherron, Dennis Sandgathe, Chris Miller, Vera Aldeias, Susan Mentzer, as well as Mike Allen, Jan Bill, John Crowther, Gill Cruise, Henri Galinié, Lars Erik Gjerpe, Joachim Henning, Johan Linderholm, the late Peter Reynolds, Pat Wiltshire and Seonbok Yi. Amalia Pérez-Juez looked over a number of chapters and made several constructive and valuable comments. The staff at Wiley-Blackwell has been always there, and Andrew Harrison and Rosie Hayden, and Antony Sami have been very helpful and understanding throughout the course of the revisions.

The University of Tours, and UCL, Institute of Archaeology, and members of the Archaeological Soil Micromorphology Working Group served as test subjects for various parts of this book over the last few decades. Longtime colleagues at HU (Na'ama Goren-Inbar, Nigel Goring-Morris, Anna Belfer-Cohen, and Erella Hovers) were involved in much of the geoarchaeological work in Israel, and Nigel's work provided several examples of key sites there.

Over the years, the interaction with many geoarchaeologists similarly shaped our thinking. At the outset, the late Henri Laville served as an inspiration for cave sediments. Marie-Agnès Courty has carried on this tradition of scholarship and friendship and has set the bar for geoarchaeological standards. Our first collaboration with her was invigorating and subsequent geoarchaeological interactions have helped us all develop and profit. We continue to be indebted to her.

Rolfe Mandel, and Vance Holliday, are among the foremost geoarchaeology researchers, particularly in the New World, and discussions with them delivered knowledge and insights over many years. Vance Holliday in particular, devoted a lot of his time to furnish us with timely, constructive comments on the first edition and they have helped hone our message and make it suitable to readers. Over the years, Takis Karkanas has been a fountain of wisdom, and many and long conversations with PG in Athens and in the field. We also would like to thank all reviewers of the 1st edition. Over the years, we have been stimulated by many other people and organizations who also supplied us with help, support, and information: Bud Albee, John Allan, Nick Barton, Jan Bill, Martin Bell, Grethe Bjørkan Bukkemoen, Sandra Bond, David Bone, Quentin Borderie, David Bowsher, Mike Bridges, Brian Byrd, Nick Conard, Marie Agnès Courty, Gill Cruise, Kim Darmark, Yannick Devos, Michael Derrick, Søren Diinhoff, Roger Engelmark, John G. Evans, the late Nick Fedoroff, Ben Ford, Erik Daniel Fredh, Charly French, Silje E Fretheim, Henri Galinié, Kasia Gdaniec, Anne Gebhardt, Lars Erik Gjerpe, Liz Graham, Tom Gregory, Ole Grøn, Sue Hamilton,

Joachim Henning, John Hoffecker, Marine Laforge, Bruce Larson, Tom Levy, Ole Christian Lønaas, Elisabeth Lorans, Jessica Leigh McGraw, Stephen Macphail, Curtis Marean, Chris McLees, Roberto Maggi, Nicolas Naudinot, Rebecca Nicholson, Cristiano Nicosia, Kevin Reeves, Mark Roberts, Arlene and Steve Rosen, Thilo Rehren, Philippe Rentzel (and Basel team), Christian Løchsen Rødsrud, Stephen Rowland, Silje Rullestad, Tom Rynsaard, Raymond Sauvage, Alex Smith, Graham Spurr, Julie Stein, Terje Stafseth, the late Georges Stoops, Chris Stringer, Kathryn Stubbs, Ken Thomas, Christine Tøssebro, the late Peter Ucko, Brigitte Vliet-Lanöe, Luc Vrydaghs, Steve Weiner, Emma West, Tim Williams, Liz Wilson, Jamie Woodward, Sarah Wyles, Ingrid Ystaard, the CNRS, Universities of Bergen, Frankfurt, Stavanger, Tours, and Wollongong, and UCL, Institute of Archaeology, The Alexander von Humboldt Foundation, University of Tübingen, Albion Archaeology, Butser Ancient Farm, Cotswold Archaeology, English Heritage, Framework Archaeology, KHM-UiO, Headland Archaeology, MOLA, NiKU, Norfolk Archaeological Unit, NTNU, University, Suffolk Archaeological Unit, Oxford Archaeology, Wessex Archaeology. We are grateful to all of you.

Yijie Zhuang would like to thank Professor Guoping Sun from the Zhejiang Provincial Institute of Cultural Relics and Archaeology for generously providing the unpublished excavation photos from the Tianluoshan site.

Nicholas Crabb is thanked for taking time away from his PhD writing to provide some multi-spectral data sets. Jacky Nowakowaski, Joe Sturgess, Andy Jones and the Royal Institution of Cornwall, Truro are thanked for help with the Gwithian Archive. Mike Allen and Charly French are thanked for providing the North Farm Section. Tony Brown, Lee Bray and Andy Howard are thanked more generally for their geoarchaeological input over the years!

1

Introduction to Practical and Theoretical Geoarchaeology

1.1 Introduction

People were doing geoarchaeology long before this term was invented for earth sciences applied to archaeology. One of the authors (PG) can remember a lecture in his first year at the University (1961) by Sheldon Judson on stream erosion in Italy (Judson, 1961, 1963). Shortly after that, he discovered others in the Old World who had carried out or summarized what would be considered "modern" geoarchaeology. These were published as major books that include, for example, works by Cornwall (1953), Zeuner (1946, 1958, 1959), and Butzer (1960, 1964; Butzer and Cuerda, 1962). It gained proper name recognition with the publication of an edited volume: *Geoarchaeology: Earth Science and the Past* (Davidson and Shackley, 1976).

Since that time, geoarchaeology has become highly prominent and almost common parlance on archaeological sites. Geoarchaeological investigations, either as independent research or tied to archaeological projects, appear in reports, monographs, books, and journal articles, and they may be either within a specific section of an article or as a stand-alone publication. The namesake geoarchaeological journal is simply *Geoarchaeology* (Wiley), but the discipline receives some attention in other, more broadly science-focused publications, such as *Journal of Archaeological Science* and *Archaeological and Anthropological Science, Quaternary International, Quaternary Science Reviews, Quaternary Research*, and others. Finally, geoarchaeological subjects make it into other publications that touch on more mainstream archaeological, anthropological, or geological subjects: *Journal of Human Evolution, American Antiquity, Journal of Sedimentary Research, Antiquity*, and *Sedimentary Geology*. There have also been inclusions in high-end science journals, *Nature* and *Science*.

In the United States, annual meetings of both the Geological Society of America (GSA) and the Society for American Archaeology (SAA), generally have at least one session or poster session, in addition to society-sponsored symposia on the subject. The GSA has an Archaeological Geology Division, and the SAA has the Geoarchaeology Interest Group. It can be noted here that two of the authors are recipients of the GSA's *Rip Rapp Award for Archaeological Geology*. The Association of American Geographers (AAG) commonly has geoarchaeology sessions at their annual meetings. In Europe, more and more scientific meetings (e.g. Association for Environmental Archaeology, European Association of Archaeologists, European Geosciences Union, UISPP, and International Union of Soil Science (IUSS), Paleopedology Commission) include some aspect of geoarchaeology, including paleopedology, past agricultural practices and other human influences on the landscape, stratigraphy/microstratigraphy, and micromorphology of archaeological soil-sedimentary sequences and living floors.

Practical and Theoretical Geoarchaeology, Second Edition. Paul Goldberg and Richard I. Macphail.
© 2022 John Wiley & Sons Ltd. Published 2022 by John Wiley & Sons Ltd.

Likewise, the most exciting archaeological sites that one reads about today, either in the popular press or professional literature, commonly have a substantial geoarchaeological component. The reader has only to be reminded about the significance of the geoarchaeological aspect of sites that are concerned with major issues relating to human development and culture. Some high profile issues and sites include: the use and evidence of the controlled use of fire (Zhoukoudian, China; Wonderwerk, South Africa; Schöningen, Germany); the sedimentary context and the origin of various hominins (Dmanisi, Republic of Georgia; Denisova, Russia; Liang Bua, Indonesia; Boxgrove, UK; Atapuerca, Spain; Mediterranean and South African caves – Gorham's Cave, Gibraltar; Hayonim Cave, Israel, Blombos Cave, South Africa; Olduvai Gorge, Tanzania); peopling of the New World (Gault/Buttermilk Creek sites, Texas); Asian rice cultivation (Huizui, China); large Eastern European and Near Eastern settlements (Borduşani-Popină, Romania; Çatal Höyük, and Aşıklı Höyük, Turkey; Tel Dor, Israel); early management of domestic animals (L'abri Pendimoun à Castellar, France; Arene Candide, Italy; Negev Desert, Israel); tropical and European Dark Earth (Marco Gonzalez, Belize; London Guildhall, UK); and worldwide settlement morphology and funerary practices (Heimdalsjordet and Gokstad Mound, Norway).

These well-known landmark sites have really drawn attention to the contribution that geoarchaeology can make to, and its necessity in, modern archaeological studies. This situation was not the case only a few decades ago when only a handful of archaeological projects utilized the skills of the geoarchaeologist. Still, the best results have come from highly focused geoarchaeological investigations, which have employed the appropriate techniques, and which have been intimately linked to multidisciplinary studies that provide consensus interpretations.

This totally revised book is about how to approach geoarchaeology and use it effectively in the study of archaeological sites and contexts (see Preface). We shall not enter into any detailed discussion of the origins and etymology of "Geoarchaeology" vs. "Archaeological Geology" (full discussions of this irrelevant debate can be found in Butzer, 1982; Courty et al., 1989; Rapp, 1975; Rapp and Hill, 1998; Waters, 1992). In a prescient, no-frills view of the subject Renfrew (1976: 2) summed it up concisely and provided these insights into the nature of geoarchaeology:

> This discipline employs the skills of the geological scientist, using his concern for soils, sediments, and landforms to focus these upon the archaeological "site", and to investigate the circumstances which governed its location, its formation as a deposit and its subsequent preservation and life history. This new discipline of geoarchaeology is primarily concerned with the context in which archaeological remains are found. And since archaeology, or at least prehistoric archaeology, recovers almost all its basic data by excavation, every archaeological problem starts as a problem in geoarchaeology.

These issues of context, and what today would be called "Site Formation Processes" in its broadest sense, can and should be integrated regionally to assess concerns of site locations and distributions, and geomorphic filters that might have controlled their visibility on the landscape.

Geoarchaeology exists and is performed at different scales (Stein and Linse, 1993). Its usage and practice vary according to the training of the people involved and the goal of their study. For example, geologists and geographers may well emphasize the mapping of large-scale geological and geomorphological features, such as the location of a site within a drainage system or other regional landscape feature, and some may call this the geotechnical approach. This perspective is at a regional scale that exists in three dimensions, with relative relief possibly being measured in 1,000s of meters, especially if working in the Alps and Andes. Much of the geoarchaeological research carried out in North America is focused at this landscape scale, while it is more of a preliminary study approach in Europe. Geologists would also be interested in the overall *stratigraphy* of a site (including sediments, soils, features, etc.) and how these aspects might interrelate with major landforms, such as stream terraces, glacial features, and loess plateaus.

Pedologists, on the other hand, would be more concentrated on the parent materials, the surfaces upon which soils formed, and how both have evolved in conjunction with the landscape; these materials can be buried by subsequent deposition or occur on the present-day surface. In either case, pedologists' focus tends to be on the scale of the soil pit, i.e., on the order of meters.

Archaeologists themselves may want to focus geoarchaeological attention upon microscale, cm-thick occupation deposits: what they are, and how they reflect specific or generalized past human activities, and how they may fit into larger behavioral patterns. In the case of rescue/mitigation archaeology, in the USA commonly termed Cultural Resource Management (CRM), geoarchaeology is tailored to the nature of the "job specifications" proscribed by the developer under the guidelines of salvage operations. In Europe, the whole funding remit is to extract as much geoarchaeological and associated paleoenvironmental information as possible, and specialists have urged the importance of site visits, and advising and training of site staff even before machining and excavation commences. Thus, the geoarchaeologist may well be just one member of an environmental team whose task is to reconstruct the full biotic/geomorphic/pedologic character of a site and its setting, and how these environments were interrelated with past human occupations. All the above approaches can be relevant depending on the research questions involved; holistically they could be subsumed under the term "site formation processes" (Schiffer, 1987).

Archaeologists come from a variety of backgrounds. As stated above, in North America, archaeology is taught predominantly in anthropology programs, although very rarely, some universities (e.g. Simon Fraser University in Canada) actually have archaeology departments; Classical and Near Eastern Archaeology programs are not rare, but these tend to emphasize written sources over excavation. In Europe, archaeology is included within Programs, or in Departments and Institutes, and not necessarily as an extension of anthropology. Many archaeologists there have no science background, and come from History or Art History.

Although in the UK geoarchaeology is taught in a number of archaeological departments, it is not taught in all archaeology degrees and this is the same across Europe as a whole. In France for example, this subject may only be taught to prehistorians and not to classical or medieval archaeologists. Commonly, even in the UK and elsewhere in the world, geoarchaeology is often an "optional module" or is found as an *ad hoc* offshoot of geology and geography. In North America, it is not anchored in any particular department and may be cross listed among Anthropology, Archaeology, Geology, and Geography. Despite good intentions and good training, many geoscientists tend to be naïve in their approach to solving archaeological problems, and therefore they effectively reduce their potential in advancing this application of their science. This situation often diminishes or even negates their contributions to interdisciplinary projects. The opposite situation can be found, where an archaeologist does not know what questions to ask of a deposit sequence or feature (Goldberg, 1988; Thorson, 1990). Recognition of these educational constraints to what geoarchaeology can achieve for both the site and the researcher is the main *raison d'être* for this 2nd edition.

Thus, as Renfrew (1976) so cogently demonstrated, geoarchaeology provides the ultimate context for all aspects of archaeology from understanding the position of a site in a landscape setting to a comprehension of the context of individual finds and features. As such, it serves as the lowest common denominator to all archaeological sites worldwide. Without such knowledge, even the most sophisticated isotope study has limited meaning and interpretability. As banal as it might sound, the adage, "garbage in, garbage out" is wholly pertinent if the geoarchaeological aspects of a site are ignored.

In the past, geoarchaeology was carried out very much by individual innovators. In North America, the names Claude Albritton Jr., Kirk Bryan, E. Antevs, E.H. Sellards, and C. Vance

Haynes immediately come to mind as the early and prominent leaders in incorporating the geosciences into the framework of archaeology (see Holliday (1997) and Mandel (2000a) for details). In fact, Mandel concisely points out that for the Great Plains, geoarchaeology, or at least geological collaboration, locally constituted an active part of archaeological survey for several areas, although it was patchy in space and time. Much of the emphasis was focused on evaluating the context of Paleoindian sites and how these occurrences figured into the peopling of the New World (Mandel, 2000b).

In Europe during the 1930s to the 1950s, Zeuner at the Institute of Archaeology (now part of University College London), developed worldwide expertise in the study of the geological settings of numerous Quaternary and Holocene sites that ranged from India to Gibraltar (Zeuner, 1946; Zeuner, 1953; Zeuner, 1959). After Kubiëna called the world's attention to soil micromorphology (Kubiëna, 1938, 1953; Kubiëna, 1970; Zeuner, 1946, 1953, 1959), Cornwall, also at the Institute of Archaeology, applied this technique to archaeology for the first time (Cornwall, 1953) (see below). At the same time, Dimbleby (and later, J. G. Evans) developed the link between archaeology and environmental studies, and produced one of the first detailed investigations of past vegetation and monument-buried soils for Bronze Age England (Dimbleby, 1962; Evans, 1972). Duchaufour in France also systematically studied environmental change and pedogenesis (Duchaufour, 1982), and some of the earliest paleo-pastoralism rock shelter studies were developed in France (Binder et al., 1993; Brochier, 1983). In mainland Europe, the legendary French prehistorian François Bordes, whose doctorate in geology was concerned with the study of loess, paleosols, and archaeological sites, principally in Northern France (Bordes, 1954), placed the French Paleolithic within its geomorphologic setting. Vita-Finzi, working in the Mediterranean Basin, used archaeological sites to suggest the chronology of Mediterranean valley fills, which he related to both climatic and anthropogenic factors (Vita-Finzi, 1969). Cremaschi (1987) investigated paleosols and prehistoric archaeology in Italy.

Although some geoarchaeological research is funded by granting agencies (NSF, NGS, NERC, CNRS, DFG, ARC), much, if not most, of modern geoarchaeological work, in both the New and Old Worlds, is fostered and sponsored by CRM projects, ultimately related to human development throughout the world. Approaches and job specifications vary according to whether investigations are at one end of the spectrum, short-term one-off studies, or long-term research projects at the other. Geoarchaeological work can be done by single private contractors or by huge international teams, which may well include specialists who also act as private contractors. Nowadays, local authorities, government agencies (e.g. State Departments of Transportation in the USA; National Cultural Heritage Administration, China) and national research funding agencies (e.g. NSF in the USA, AHRC and Historic England in the UK, and AFAN and the CNRS in France; Nara National Institute, Japan; Cultural History Museum, University of Oslo, NiKU and NTNU, Norway) may all be involved in commissioning geoarchaeological investigations. It is currently a very flexible field. It is also one where there is an increasing need for formal training, but where relatively few practitioners have been in receipt of one.

Geoarchaeological work is now often broken up into several phases, with desktop investigations, fieldwork survey, excavation, sample assessment and laboratory study, all being likely precursors to full analysis and final publication. This is all part of modern funding and operational procedures.

Single-job or site-specific studies may be as straightforward as finding out "What is this fill?" On the other hand, problem-based research could involve the gathering of geoarchaeological data on the possible controlled use of fire, as at Zhoukoudian, China (Goldberg et al., 2001a; Weiner, 1998) or origin of salt working coastal "redhills" in England (Biddulph et al., 2012). Sites are investigated at different scales and sometimes, for very different reasons. At one time, "Dark Earth", the dark colored Roman-medieval urban deposits found in urban sites across northern Europe, engaged the particular interest of geoarchaeologists because these enigmatic deposits commonly span the "Dark Ages", and human activities at this time have been poorly understood (Macphail et al., 2003a; Nicosia et al., 2017), while South and meso-American tropical Dark Earth (*terra preta*) recorded pre-conquest settlements in some cases (Arroyo-Kalin, 2017). Analysis of "Dark Earth" therefore, became a research-funded topic for urban development sites (CRM projects in urban areas) across Belgium, France, and the UK, for example, Brussels being a particularly well-studied urban area (Devos et al., 2020a).

On the other hand, attention can be focused on individual middens and midden formation because they provide a wealth of material remains, particularly organic, that are normally poorly preserved and complex to understand and interpret (Stein, 1992). Regional studies of the intertidal zone, for example, may include the investigation of middens as one single component, an early interdisciplinary study being Mesolithic Westward Ho! (UK) (Balaam et al., 1987); more recently, South American middens and an Antarctic seal-hunting site have come under scrutiny (Villagran et al., 2009; Villagran et al., 2013). Submerged sites (Grøn et al., 2021) and deeply stratified deposits can be found, and in some cases accessed through cofferdams or coring (Linderholm et al., Submitted) (Macphail and Goldberg, 2018a: 15–20). Equally, studies of alluvial deposits and associated floodplains (Brown, 1997; French, 2003) have involved the search for buried sites, within the overall realm of evaluating the distribution and history of archaeological sites and past land uses such as herding (Macphail, 2011a); in Norway there can be the added complication of landslides (Macphail et al., 2016). The Po plain of Italy (Cremaschi, 1987; Cremaschi and Nicosia, 2012) and the Yellow River of China (Kidder and Liu, 2017) both feature a series of late prehistoric settlements; water management and wet rice paddy fields are also phenomena of prehistoric east Asia (Lee et al., 2014; Zhuang, 2018). Many of the most significant Paleoindian and Archaic sites in the USA are situated within alluvial sequences (Ferring, 1992; Mandel, 2000a; Mandel, 2008; Mandel et al., 2018).

Modern geoarchaeological research makes use of a vast number of techniques that either have been used in geology and pedology or have been developed or refined for geoarchaeological purposes. Early geoarchaeological research until the latter part of the last century, at least in North America, was predominantly field based and made use of both natural exposures and excavated areas. More recently, field techniques have become more improved and technologically sophisticated. Natural exposures can be supplemented with surface satellite remote sensing data, as well as subsurface data derived from machine-cut backhoe trenches, augering, coring, and advanced geophysical techniques (e.g. magnetometry, electrical resistivity and ground-penetrating radar; see references in Gilbert et al., 2017). Moreover, such data can be assembled and interrogated using Geographic Information Systems (GIS) (Landeschi, 2019; Wheatley and Gillings, 2002) that produce deposit models and which can be used to generate and test hypotheses (Carey et al., 2018).

Laboratory techniques have similarly become more varied and sophisticated. At the outset, many geoarchaeological studies adopted techniques from geology and pedology that were aimed at sediment/soil characterization. Thus traditional techniques characteristically consisted of grain size analysis (granulometry), coupled with other physical attributes (e.g. particle shape, bulk density, bulk mineralogy), as well as basic chemical analyses of organic matter, calcium carbonate content, extractable iron, etc. The analysis of phosphate to elucidate activity areas or demarcate site limits has a longer history spanning over 70 years (Arrhenius, 1931, 1934; Parnell et al., 2001). Conventional techniques with long historical pedigrees, such as x-ray diffraction (XRD; now supplemented with micro-XRD), electron microprobe, x-ray fluorescence (XRF and micro-XRF), and instrumental neutron activation analysis (INAA), atomic absorption (AA) have been enhanced by rapid chemical, elemental, and mineralogical analyses of samples through the use of Fourier transform infrared spectrometry (FTIR and micro-FTIR), Raman spectrometry, and by inductively coupled plasma-atomic emission spectrometry (ICP-AES) or mass spectrometry (ICP-MS) (Artioli, 2010; Gilbert et al., 2017; Weiner, 2010).

In addition, a notable advance in geoarchaeology has been the application of soil micromorphology to illuminate a wide variety of geoarchaeological issues (Courty et al., 1989; French, 2003). These earlier works have been enhanced with new books reflecting the evolution and maturity of the discipline (French, 2015; Karkanas and Goldberg, 2018; Macphail and Goldberg, 2018a; Nicosia and Stoops, 2017; Stoops et al., 2018a). Important topics range from the development of soil and landscape use (French, 2015; Gebhardt et al., 2014; Zhuang et al., 2013), the formation of anthropogenic deposits (Banerjea et al., 2015a; Cammas et al., 1996b; Macphail, 1994a; Macphail et al., 2007a; Matthews et al., 1997), to the evaluation of the first uses of fire (Berna et al., 2012; Goldberg et al., 2001, 2017b; Stahlschmidt et al., 2015b), and the use of experiments and ethnoarchaeology to produce such insights (Banerjea et al., 2015b; Cammas, 2018; Carey et al., 2014; Friesem et al., 2014a; Macphail et al., 2004). The science has also been strengthened by geoarchaeologists standing by their analyses (Goldberg et al., 2009a; Macphail, 1998; Macphail et al., 2006).

Finally, geoarchaeological research has been facilitated by the development of numerous dating techniques just within the past two to three decades. Now, sites within the span beyond the widely accessible limits of radiocarbon are potentially datable with techniques, such as thermoluminescence (TL) (Mercier et al., 2007), optically stimulated luminescence (OSL) (Jacobs et al., 2019; Jankowski et al., 2020), and electron spin resonance (ESR) (Duval et al., 2018; Rink and Schwarcz, 2005).

In this book we aim to present a fundamental, wide-ranging perspective of the essentials of modern geoarchaeology in order to demonstrate the breadth of the approaches and the depth of problems that can be proposed and tackled. Additionally, it is aimed to promote a basic and straightforward line of communication and understanding among all multi-disciplinarians. We cover a variety of topics that discuss thematic issues, as well as practical skills. The former encompasses such broad concepts as stratigraphy, Quaternary and environmental studies, sediments, and soils. We then present a survey of some of the most common geological terrains that provide the natural settings for most archaeological sites, and expanded into chapters on "slopes", "rivers", "lakes, "aeolian settings", "marine coasts", and "caves". These are established geoarchaeological topics into which we have incorporated some new findings. Unlike many books on geoarchaeology, we have dedicated a major portion of the volume to topics that were normally not treated in many geoarchaeological texts. While the first edition (2006) pioneered chapters on "human impact", "occupation deposits", "human materials", and "applications to forensic science", for example, these have been revised into "clearance, cultivation and other soil modifications" (e.g. from mining), "human use of materials", "anthropogenic deposits", "experimental and ethno-geoarchaeology", and "forensic and mortuary geoarchaeology". The topics span the course of human history from early hominins in African caves, major food production developments across south-east Asia to settlement patterns and urban sites across Europe and south-west Asia.

Similarly, it is important also to obtain some insights into practical aspects of geoarchaeology, including how geoarchaeologists should specifically fit in to a project. Similarly, two chapters are devoted to a presentation of pragmatic and theoretical methods currently used in geoarchaeology. These include not only field techniques (e.g. from remote sensing, satellite and drone imagery, coring, to describing a profile and collecting samples), but also those techniques that are used in the laboratory (varying from bulk, microscope to instrumental approaches). Although we summarize the "what" and "how", we also try to emphasize the "why" and provide several example-based caveats for important techniques. A final facet deals with the practical aspects of manipulating and reporting geoarchaeological results (e.g. with GIS) while keeping in mind that material presented in reports differs from that in articles. Reports essentially present the full database and arguments, whereas articles are commonly more thematic and focused, and by necessity are constrained to present results more concisely. Reports, which are seldom published in full, constitute the "gray literature" and make up an important part of the scientific database. They are too commonly overlooked, ignored, or simply are not readily accessible. We suggest electronic archives can be best accessed online, but these websites need to have been built and be under continual maintenance.

As a final point, we maintain that geoarchaeology in its broadest sense, must be made understandable to all players involved, be they archaeologists with strong training in anthropology, or the geophysicist, with minimal exposure to archaeological issues. All participants should have enough of a background to understand what each participant is doing, why they are doing it, and most importantly, what the implications of the geoarchaeological results are for all team members. Too often we hear about the geo-specialist simply turning over results to the archaeologist, essentially being unaware of the archaeological problem(s), both during the planning stages and later, after execution of the project. Hence, they cannot correctly put their results to use. On the other hand, many archaeologists tacitly accept results produced by specialists with few notions on how to evaluate them. This book attempts to level the playing field by providing a cross-disciplinary background to both ends of the spectrum. Such basic material is needed to establish a dialogue among the participants so that problems can be mutually defined, mutually understood, and best interpreted.

2

Sediments

2.1 Introduction

Sediments – and their alteration products, soils (see Chapter 4) – are the backbone of most archaeological sites (see Figure 2.1 for example). They are the glue that hold the rest of the archaeological record (lithics, bones, ceramics, even architecture) in place, and they are what archaeologists "dig". In most sites, archaeological objects are articulated with the sediments, and are commonly transported with the natural/geological components. Furthermore, the site itself and its deposits can be integrated with sediments that occur on the landscape scale, such as a building situated on a floodplain and perhaps buried by stream deposits (Figure 2.1). So, it is important that we provide some of the basic aspects of what sediments are, how we describe them, and how we can use them to extract information about their history and their relationship to archaeological materials, contexts, and past environments.

Geologists describe sediments, loose or indurated (sedimentary rocks), as materials deposited at the earth's surface under low temperatures and pressures (Pettijohn, 1975). They are found all over the globe – typically Pleistocene and younger (<2.5 my). Sedimentology, the study of sediments, also takes into consideration pores, the void spaces between the sediments that help reveal the history of the sediments.

Sediments differ from soils (treated in Chapter 4) in that sediments we observe today are not formed in place. Instead, in the case of clastic/detrital sediments (see below), they consist of individual particles that can greatly vary in size and organization. Particles such as gravel or sand originate from a source, are transported (e.g. by water, wind, or gravity), and then are deposited. Thus, there is a physical displacement from point A to point B that can be reconstructed using specific parameters of the sediment (discussed below) and the attributes of the deposit, or the three-dimensional unit distinguished in the field on the basis of observable changes in some physical properties (e.g. color, sediment size or composition) (discussed in Chapter 3).

In this section we examine some of the basic characteristics of sediments. Since many soils develop from the *in situ* weathering of sedimentary deposits, many of these characteristics are applicable to soils as well (Chapter 4). We have two principal goals. Since sediments largely guide the framework for archaeological site interpretation, it is necessary to have at least a working knowledge of their basic characteristics and the terminology used to describe them. Essentially, these descriptive characteristics constitute a lexicon: the term "sand", for example, corresponds to a defined size range, irrespective of composition. Secondly, and perhaps more important, is that many of the descriptive parameters that we observe in sediments commonly reflect – either

Practical and Theoretical Geoarchaeology, Second Edition. Paul Goldberg and Richard I. Macphail.
© 2022 John Wiley & Sons Ltd. Published 2022 by John Wiley & Sons Ltd.

Figure 2.1 Examples of sediment types associated with archaeological sites. (a) Die Kelders Cave, South Africa, one of the best-studied Middle Stone Age (MSA) sites in the region. The cave (green arrow) is formed along the unconformable contact between Paleozoic quartzite (Table Mountain Sandstone) below and sandy limestone above (arrow); it is covered with a drape of bedded Quaternary calcareous sandstone of aeolian origin (aeolianite) (Marean, et al., 2000). (b) Die Kelders Cave interior that shows profile of bedded sand, ashes, and organic-rich occupation material that has been partially redistributed by wind; note the shift from the predominantly geogenic lower part to the increasingly anthropogenic-rich upper part. The profile is 2 m across. (c) Profile of part of the Lower Paleolithic site of Schöningen, Germany, a site known for its wooden spears, butchered horses, and stone tools. Shown here is extensively iron-stained lacustrine marls (calcareous muds) overlain and truncated by organic-rich lake deposits (Stahlschmidt, et al., 2015a, Stahlschmidt, et al., 2015b). Red scale is 20 cm. (d) Uppermost sediment of the cave site of Tabun, Mt. Carmel, Israel, one of the baseline sites for the Lower and Middle Paleolithic of the Near East. Gary Rollefson is standing on diagenetically altered silty sands (of marine and aeolian origin), which grade upward into siltier deposits capped by red and white banded clayey and ashy deposits and ultimately, red clay that accumulated through a chimney hole at the back of the cave. (Jelinek, et al., 1973, Shimelmitz, et al., 2016, Tsatskin, et al., 1992).

individually or collectively – the history of the deposit, including its (1) origin, (2) transport agent, and (3) the types of depositional processes where it was put down, its *environment of deposition* (see Stein, 1987 for a geoarchaeological perspective). Figuring out these aspects of a sediment's history constitutes a mental structure for every sedimentologist, whether they are studying a 100 m-thick sequence of Carboniferous sandstones in Nova Scotia, a 10 cm-thick sandy layer within a Late Pleistocene prehistoric cave in the Mediterranean, or a "destruction layer" from a tell in the Middle East. In sum, by observing and recording the attributes of a sediment we not only provide an objective set of criteria to describe and communicate those attributes, but also a means towards reconstructing its history.

In what follows, we attempt to provide an overview of the essential aspects of sediments themselves. The emphasis here is sediments of essentially geological/natural origin (geogenic). Archaeological sediments (anthropogenic) – those related to human activities – are discussed in other chapters (10, 13) where we provide examples and more specific details about their application and interpretation in geoarchaeological settings.

2.2 Types of Sediments

Sediments can be classified into three basic types, clastic (*or* detrital), chemical, and organic, all of which are pertinent to geoarchaeology. *Clastic* sediments are the most abundant. They are composed of fragments (clasts) of rock or minerals, other sediment, or soil material. Natural clastic sediments that are most relevant to archaeology are terrigenous sediments (*vs.* marine sediments) deposited by agents such as wind (e.g. sand dunes), running water (e.g. streams, beaches), and gravity (e.g. landslides, slumps, colluvium). Anthropogenic sediments (Ch. 10, 12, and 13) are largely composed of clastic material. An example of a typical clastic sediment is quartz sand on a beach in some vacation wilderness like Cape Cod, USA. Intertidal fine sediments, sometimes known as marine alluvium, can both bury marine inundated coastal sites or form terrestrial landscapes when post-glacial uplift has occurred (Chapter 9).

Volcaniclastic sediments (Table 2.1) are those associated with volcanic activity and include volcanic ash, blocks, bombs, and pyroclastic flow debris (Fisher and Schmincke, 2012). These types of sediments appear in geoarchaeological contexts with geologically relatively young volcanic activity where the finer debris can weather into highly fertile soils. Many of these regions have been important areas of human activity in the past (e.g. Polynesia, Central Basin of Mexico, Mediterranean, addressed in Chapter 4) (James et al., 2000). Volcaniclastic sediments have additional importance in archaeological contexts where they have been catastrophically ejected, burying Bronze Age villages (Matarazzo, 2015; Matarazzo et al., 2010) and better known sites such as Pompeii and Herculaneum, Roman towns buried on the Bay of Naples in Southern Italy when Vesuvius erupted

Table 2.1 Types of sediments; consolidated (lithified), rock equivalents for clastic sediments are given in parentheses and *italics*. Note that *bioclastic* limestones, composed of biologically precipitated shell fragments (e.g. coquina), can be thought to be both clastic and biochemical in origin.

Clastic and bioclastic		Non-clastic	
Volcaniclastic	**Terrigenous, marine, and lacustrine**	**Chemical**	**Biological**
lapilli, blocks, bombs	Cobbles, boulder; gravel (*conglomerate*)	Carbonates (*limestones*)	peat (lignite, coal)
ash (*welded tuff*)	sand (*sandstone*)	Evaporates (chlorides, sulfates, silicates; phosphates)	
	silt (*siltstone*)	Travertines and flowstones (cave and karst settings)	Algae, bacteria, diatoms, ostracodes, foraminifera
	clay (*shale*)		
	Bioclastic (carbonate): coarse (*coquina*); fine (*chalk*)		
	Bioclastic (siliceous): *diatomite*		

in AD79. Pompeii is considered the type site for the preservation of instances in archaeological time (the so-called "Pompeii Premise" (Binford, 1981) (further discussion on Pompeii is presented in Chapters 12–13). These sites are covered by meters of volcaniclastic debris (tephra), consisting of pumice, volcanic sand, lapilli (2–64 mm), and ash less than 2 mm (Giuntoli, 1994). Tephra marker horizons provided dating opportunities at Upper Paleolithic Kostenki, Russia and Viking Age Landnám – or first occupation – in Iceland (Holliday et al., 2007; Sigurgeirsson et al., 2013). The site of Ceren in San Salvador represents a similar setting, where structures and agricultural fields were buried under several meters of ash and tephra resulting in a rare snapshot of daily life among rural commoners during the Late Classic Mayan period (Sheets, 2002; Slotten et al., 2020).

Volcaniclastic deposits are widespread in rift valleys and because they are materials that can be dated, they play critical roles in the dating and stratigraphy of Pliocene and Pleistocene deposits from sites in East Africa, the Jordan Rift, and Turkey and Georgia. Rift settings and sites such as Olduvai Gorge, Koobi Fora, Gesher Benot Ya'akov, and Dmanisi figure strongly in the study of human origins, and they are just a few of the sites where archaeological deposits and hominin remains are associated or intercalated with volcanic rocks and tephra (Ashley and Driese, 2000; Blegen et al., 2016; Blumenschine et al., 2012; Ferring et al., 2011; Gabunia et al., 2000; Goren-Inbar et al., 2018; Stern et al., 2002).

Bioclastic sediments can exhibit a wide range of sediment types and sizes from both marine or terrigenous sources. Marine organisms such as mollusks and corals produce shells of calcium carbonate that can remain whole when buried. On the other hand, in cases where these hard body parts are subjected to wave action the shell can be broken into small cm- to mm-size particles, that when cemented result in the formation of a bioclastic limestone, for example (Table 2.2; see Ørlandet, Norway in Chapter 18). Coquina is an example of a coarse bioclastic sediment composed of such particles. Alternatively, a fine-grained calcareous equivalent is chalk, which is composed of silt- and fine sand-size tests of marine organisms (foraminifera); in other instances, organisms with siliceous skeletons, such as diatoms, result in the formation of diatomite. Bioclastic sediments, *per se*, are relatively rare. However, these minute biological remains, such as ostracodes, diatoms, and foraminifera, can be preserved within otherwise mostly mineralogenic deposits, such as in lake sediments (see Figure 18.23b) (Cruise et al., 2009). The moat deposits from the Tower of London, for example, contain numerous diatoms, which point to shallow, turbid water in disturbed sediments

Table 2.2   Common minerals and rock fragments in sediments (modified from Boggs, 2012).

Major minerals

Quartz

Potassium and plagioclase feldspars

Clay minerals

Accessory minerals

Micas: muscovite and biotite

Heavy minerals (those with specific gravity >2.9):

- Zircon, tourmaline, rutile
- Amphiboles, pyroxenes, chlorite, garnet, epidote, olivine
- Iron oxides: Hematite, limonite, magnetite

Rock fragments

Igneous

Metamorphic

Sedimentary

(Keevill, 2004). These conditions were inferred to be a result of inputs into the moat of waste disposal, surface water, and water from the Thames and the City Ditch. Foraminifera analyses were key to reconstructing the early hominin (Middle Pleistocene) coastal environment at Boxgrove and early Holocene inundated coastal sites in Essex, UK (Macphail et al., 2010; Pope et al., 2020).

Chemical sediments are those produced by direct precipitation from solution. Lakes in semi-arid areas with strong evaporation, for example, can exhibit a number of precipitated minerals, such as halite (table salt), gypsum (calcium sulfate), or calcite or aragonite (both forms of calcium carbonate). On occasion, these are visible in the field (Chapter 4) and recognizable under the petrological microscope (Durand et al., 2018; Mees and Tursina, 2018; Poch et al., 2018). In cave environments, chemical sediments are widespread and typically produce sheets of calcium carbonate (e.g. *travertine* or *flowstone*), or a variety of ornamental forms such as stalactites and stalagmites that are usually composed of calcite or aragonite (White and Culver, 2012). In addition, other minerals can form as a result of direct precipitation or transformation of previously present minerals; new minerals can include phosphates, nitrates, or sulfates, for example (Hill and Forti, 1997; Karkanas and Goldberg, 2018a).

The third group, *biological* sediments, is composed mostly of organic materials, typically plant matter. Peat, lignite, or organic-rich clays in swampy areas and depressions are characteristic examples. These type of sediments, peats in particular, are key components of archaeological studies reconstructing changes in water levels, including sea levels (e.g. Smith, 2020), lake levels (e.g. Jackson et al., 2000), and even freshwater springs (e.g. Toffolo et al., 2017) where peat can represent now inundated extant surfaces that can contain evidence of human occupation.

2.2.1 Clastic Sediments

Clastic sediments (Table 2.1) have been the object of study by sedimentologists for decades and so they have a firm basis for understanding them. Clastic sediments present a number of attributes (also studied by pedologists) that can be readily described. When examined individually or together they can lead to often robust interpretations of their source, transport, and environment of deposition. These attributes generally include composition, texture (grain size and shape), fabric, and sedimentary structures.

Composition – Sediments can exhibit a wide variety of composition of mineral and rock types, and normally this is a function of the source of the material (Tables 2.2 and 2.3). As a consequence, geologists have been able to reconstruct geological landscapes (e.g. former landmasses; Table 2.3) that have long since been eroded. In spite of their wide variety, certain rocks and minerals occur repeatedly in

Table 2.3 Heavy mineral associations and related geological sources (modified from Pettijohn et al., 1975).

Mineral association	Typical geological source
Apatite, biotite, brookite, hornblende, monazite, muscovite, rutile, titanite, tourmaline, zircon	Acid igneous rocks
Augite, chromite, diopside, hypersthene, ilmenite, magnetite, olivine	Basic igneous rocks
Andalusite, corundum, garnet, phlogopite, staurolite, topaz, vesuvianite, wollastonite, zoisite	Contact metamorphic rocks
Andalusite, chloritoid, epidote, garnet, glaucophane, kyanite, sillimanite, staurolite, titanite, zoisite-clinozoisite	Dynamothermal metamorphic rocks
Barite, iron ores, leucoxene, rutile, tourmaline (rounded grains), garnet, illmanite, magnetite, zircon (rounded grains)	Reworked sediments

sediments ("major minerals" in Table 2.2). Their relative abundance in a sediment can vary with location and age. In the case of the latter, some minerals (e.g. olivine) are more susceptible to alteration/destruction than others (e.g. quartz), and thus they can be less persistent in older sediments. Furthermore, overall sediment composition can be influenced by secondary processes (e.g. soil formation and *diagenesis*), which result in the precipitation of minerals that either cement the skeletal grains of the sediment, or that precipitate as concentrations within the sedimentary mass (e.g. *nodules and concretions*). Secondary mineralization may involve different chemical groups, including carbonates (e.g. calcite, aragonite), silicates (e.g. opal, microcrystalline quartz/chert); sulfates (e.g. gypsum, barite), phosphates (e.g. apatite, leucophosphite), and iron oxides (e.g. limonite, goethite).

Texture – Texture involves attributes of individual grains or clasts, and these like other traits, have both descriptive and interpretative value. One of the most basic and widespread attributes is that of grain size (Table 2.4), and it is one that both earth scientists (geologists and pedologists) and archaeologists use and intuitively understand: "this deposit is fine-grained, while the one above is coarser and sandier." Clearly, we need to be more precise than this example, and both geologists and pedologists employ formal names and size limits to describe the classes of particle sizes that

Table 2.4 Common grain size scales used in geology and pedology.

Wentworth class (geology)[1]	Size range	Phi (Φ) units[2]	UK soil science class equivalent[3]	Size range	USA soil science class equivalent[4]	Size range
Boulder	>256 mm	−8	Boulders	>600 mm	Boulders	>600 mm
			Very large stones	200–600 mm	Stones	250–600 mm
Cobble	64–256 mm	−6 to −8	Large stones	60–200 mm	Cobbles	76–250 mm
Pebble	4–64 mm	−2 to −6	Medium stones	20–60 mm	Coarse Gravel	20–76 mm
			Small stones	6–20 mm	Medium Gravel	5–20 mm
Granule	2–4 mm	−1 to −2	Very small stones	2–6 mm	Fine Gravel	2–5 mm
Very coarse sand	1–2 mm	0–1			Very coarse sand	1–2 mm
Coarse sand	0.5–1 mm	1–0	Coarse sand	0.6–2 mm	Coarse sand	0.5–1 mm
Medium sand	250–500 µm	2–1	Medium sand	212–600 µm	Medium sand	250–500 µm
Fine sand	125–250 µm	3–2	Fine sand	63–212 µm	Fine sand	100–250 µm
Very fine sand	63–125 µm	4–3			Very fine sand	50–100 µm
Coarse silt	31–63 µm	5–4	Coarse silt	20–63 µm	Coarse silt	20–50 µm
Medium silt	16–31 µm	6–5	Medium silt	6–20 µm	—	
Fine silt	8–16 µm	7–6	Fine silt	2–6 µm	Fine silt	2 to 5 µm
Very fine silt	4–8 µm	8–7				
Clay	<4 µm	>8	Clay	<2 µm	Clay	< 2 µm

[1] After Nichols (1999)
[2] $\Phi = -\log_2 d$ (d = grain diameter)
[3] Avery, 1990; Hodgson, 1997
[4] Schoeneberger, et al., 2012

range from fine, micron-size (1 μm = 0.001 mm) grains of dust, up to large boulders several meters across (Table 2.4).

The *grain size scale* commonly used by geologists in the United States is that developed by (Wentworth, 1922) (Table 2.4); note that this scale is a geometric grade scale with the limits between classes having a constant ratio of 1/2 (Krumbein and Sloss, 1963). Geologists have later modified this by introducing the *phi (Φ) scale*, which is a logarithmic transformation of the grain size to the base 2 (Φ = −log₂d, where d is the grain size in mm). These arithmetic intervals are convenient for the graphical presentation of grain size data, such as histograms (see Chapter 17), and for performing statistical analyses. In any case, as can be seen in Table 2.4, the limit between silt and clay is 3.9 μm in the Wentworth scale, and the limit between silt and sand is 62.5 μm. Furthermore, we highlight that soil scientists (in both the U.S. and UK) employ different limits between sand/silt and silt/clay: for soils in the U.S, silt includes material between 50 μm and 2 μm, whereas in the UK, it is 63 μm to 2 μm (Table 2.4). These differences are particularly significant when evaluating data in reports and maps, and whether the descriptions have been done by a geologist or pedologist: the same sediment can have different percentages of silt or sand, depending if it were analyzed by a geologist or pedologist and the geographic location of the laboratory.

Methods used in grain size analysis are discussed in Chapter 17. Nevertheless, we need to point out here that sediments (and soils) are usually mixtures of different sizes of particles, as for example, sand, silt, and clay for finer materials. Graphic representation of such mixtures are commonly presented in triangular diagrams (Figure 2.2), which have three major end members, each representing 100% sand, silt, or clay. As can be seen, different mixtures are partitioned into grouped and given names to characterize each group, depending on the proportions of the different end members. Note that the boundaries/proportions between groups can vary considerably within geology (Figure 2.2 a, b) and between disciplines (geology vs. pedology; Figure 2.2c). Soil scientists use the additional term, loam, that is a mixture of predominantly sand and silt with some clay (Schoeneberger et al., 2012); a mixture of say, 40% silt, 30% sand and 30% clay would be called a clay loam.

Loam is commonly used in the geoarchaeological literature but it should be remembered that it is a term from pedology with the mindset of soil, and is out of place for describing geological sediments. The point is that, again, when evaluating grain size analyses of a given study, one should be aware of its goals, as well as the background and country of origin of the author, if the nomenclatural system is not presented. Field methods for estimating particle size are given in Chapter 16.

Sorting – Sorting is a term applied to the proportion and number of different size classes comprising the grain populations. In particular, it relates to the statistical distribution of sizes around the mean (the standard deviation; see Tucker (1988) for ways to measure and evaluate it). Sorting can be readily visualized in Figure 2.3. The predominance of one size of particle indicates a well-sorted sediment. Beach and dune sands, for example, are characteristically well sorted, as are windblown dust deposits, known as *loess* (see Chapter 8). A poorly sorted sediment consists of varying particle sizes. Slope deposits, where a mass of sediment has been moved down hill (the process of *colluviation*: see Chapter 5) and glacial till are typically poorly sorted deposits due to the lack of selectivity of the transport agent: glaciers simply pick up, grind, transport, and deposit the substrate material along their paths.

Particle shape is usually considered for gravel, pebbles, and sand-size particles. It is another descriptive parameter and indicator of the history of the sediment. Three related features of shape are generally considered. *Form* refers to the general outline of the grain and ranges from equant grains (with roughly equal length, width, and thickness dimensions approaching the shape of a sphere), to platy or disc-shaped grains, in which the thickness is markedly less than the length or

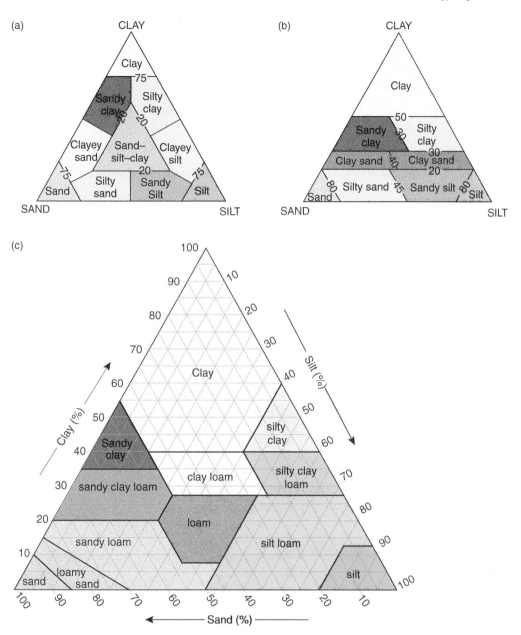

Figure 2.2 Triangular textural diagrams showing classes of mixtures of sand, silt, and clay; the ends of the triangles denote, 100% of sand (left), silt (right), and clay (top). Class boundaries used in geology are shown in (a) and (b), whereas a typical one in pedology is shown in (c) ((a) and (b) after Pettijohn, 1975; (c) Christopher Aragón / Wikimedia Commons / CC BY-SA 4.0.

width (see Boggs, 2009 for methodological details to measure shape). *Roundness* relates to the angularity of a grain and is concerned with the jaggedness of the edges (Figure 2.4b). Finally, *surface texture* refers to the microtopographic features of the grain, such as pitting, etch marks, and micro fractures; grains may vary from pitted to smooth (Figure 2.5). Other traditional morphology

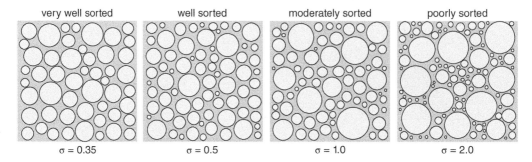

Figure 2.3 A graphical representation of sorting, the distribution of grain sizes in a sediment. Very well sorted sediments have a narrow range of sizes, whereas poorly sorted ones have a variety of sizes (modified from Flügel 2009).

descriptions (e.g. Folk, 1974) also include sphericity, expressed mathematically as how equal the three dimensions are. In the shape system advocated here this is rolled (e.g. through long-term wave activity) into roundness.

The importance of form is not universally agreed upon, and its measurement should be viewed in light of other sedimentary parameters (Boggs, 2014). Form is more pertinent to the coarser, pebble fraction, as the finer sand and silt fractions (generally quartz) seem to be only slightly modified in shape if at all by transport. The most consistent example comes from beach pebbles and cobbles, which tend to be flattened, although the reasons for this are not clear. In fluvial environments, shapes can be associated with ease of transport, so that spherical and prolate shapes tend to be more readily transportable than are blady particles (see Figure 2.4a).

Grain roundness is a function of mineralogy, size, transport, and distance of transport (Boggs, 2014). For sand-size quartz grains, for example, increased roundness is commonly associated with aeolian transport, which is more effective than water in rounding grains. For larger pebbles, composition plays an increasingly important role. Limestone pebbles whose composition is more easily reduced by chemical and physical weathering, for example, are much easier to round in fluvial environments than are those from the far more resistant cherts, which tend to fracture before they become rounded. On the other hand, the presence of well-rounded chert pebbles is commonly indicative of several cycles of reworking and points to relatively greater age. In any case, well-rounded pebbles point to fluvial transport, where rounding tends to take place relatively rapidly after a clast enters the fluvial system.

Studies of surface texture usually involve sand-size particles (Figure 2.5). Signs of fracture and abrasion produced during transport, as well as *diagenetic* changes expressed as etching and secondary precipitation can be observed under the optical (binocular and petrographic) and electronic (scanning electron – SEM) microscopes (see Chapter 17). Study of scores of quartz grains under the SEM from modern environments revealed that different sedimentary environments impart different combinations of surface signatures on the quartz grains (Krinsley and Doornkamp, 2011; Smart and Tovey, 1981, 1982; Vos et al., 2014).

Tabun Cave on Mt Carmel, Israel, is one of the most important Lower (LP) and Middle Paleolithic (MP) sites in south-west Asia (Albert et al., 1999; Goldberg and Bar-Yosef, 2019; Jelinek, 1982; Leshchinskiy et al., 2021; Mercier and Valladas, 2003; Vita-Finzi and Stringer, 2007). The cave contains hominid remains (including a well-preserved Neanderthal burial) and a rich assemblage of

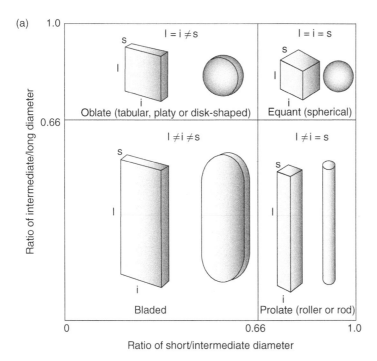

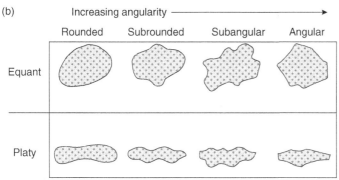

Figure 2.4 Particle shape (form) and roundness. (a) Aspects of particle shape as expressed as the ratio of long (DI), intermediate (Di), and short (Ds) axes of the particle (modified from Tucker, 2011). (b) Two features of shape are depicted here, roundness and sphericity. The columns depict changes from well-rounded (top) to angular grains (bottom). The rows illustrate grains of different sphericity classes: the grains on the left are more elongated than the more equi-dimensional grains on the right (modified from Courty et al., 1989).

lithic tools within an 18 m-thick sedimentary sequence that spans close to 400,000 years and includes several marine isotope stages and sea level fluctuations. SEM surface analyses of quartz grains assisted in understanding the environment of deposition of the sandy components in the cave (Figure 2.5). Bull and Goldberg (1985) found that the LP basal deposits (Layers F and G of excavator Dorothy Garrod – Garrod and Bate, 1937) show aeolian and diagenetic surface characteristics. The middle unit, Layer E (also LP) is principally aeolian but with signs of marine alteration. Quartz grains in Layers D, C, and B (MP) in the upper part, show only aeolian modifications. These

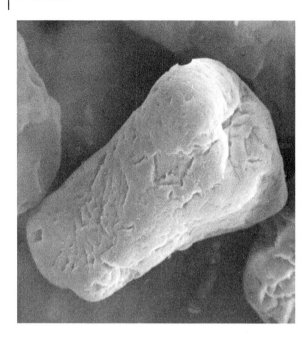

Figure 2.5 Surface texture of grains, illustrated here as an SEM photo of quartz grain from Lower Paleolithic Layer E in Tabun Cave, Israel. Shown here is chemical etching of the grain (Bull and Goldberg, 1985) indicative of strong diagenesis of these deposits. Magnification of grain is 340× (photo courtesy of Darwin Spearing).

results in broad terms point to a regressive trend in sea level over this time frame helping to explain varying use of the cave over time.

Fabric is used by sedimentologists and pedologists in different ways. For sedimentologists (e.g. Boggs, 2014; Tucker, 2009), fabric relates to the orientation and packing of grains, which is commonly a function of flow direction. Elongated pebbles, for example, can be oriented in the same direction, either parallel or perpendicular to the flow, depending on the hydraulic characteristics in the environment of deposition. Sand-size particles will commonly be oriented parallel to the flow direction. Another sedimentological concept is *packing*, which describes the contacts between grains; it is associated with porosity and permeability. Sedimentologists differentiate *clast-supported fabric*, in which grains are in contact with adjacent ones, from *matrix-supported fabric*, where coarser clasts (e.g. sand, pebbles) are enclosed within a finer matrix (Figure 2.6).

Pedologists, on the other hand, (especially those analyzing soil/sediment in thin section, micromorphology, see Chapter 17) have a slightly broader and more nuanced view of fabric (Courty et al., 1989; FitzPatrick, 1993; Macphail and Goldberg, 2018a; Stoops, 2020). The most generalized definition that fits into most geoarchaeological contexts was first stated by Bullock et al. (1985: 17): "Soil fabric deals with the total organization of a soil, expressed by the spatial arrangements of the soil constituents (solid, liquid and gaseous), their shape, size, and frequency, considered from a configurational, functional and genetic viewpoint." This approach to fabric encourages detailed observation of all the components of a deposit, enabling the deconstruction of a soil or sediment into its essential primary (depositional) and secondary (post-depositional) elements (see Chapter 17, lab techniques) (Karkanas and Goldberg, 2018b).

Bedding, bedforms, and sedimentary structures – Bedding is an important sedimentary feature, as it reveals information about the environment of deposition as well as post-depositional changes, such as bioturbation, that may erase original traces of bedding. Although in theory an individual bed accumulated "under constant physical conditions" (Reineck and Singh, 1986: 95), it is commonly difficult or impossible to recognize such individual events and conditions. A single bed is distinguished from adjoining ones by surfaces generally called bedding planes, which mark beds

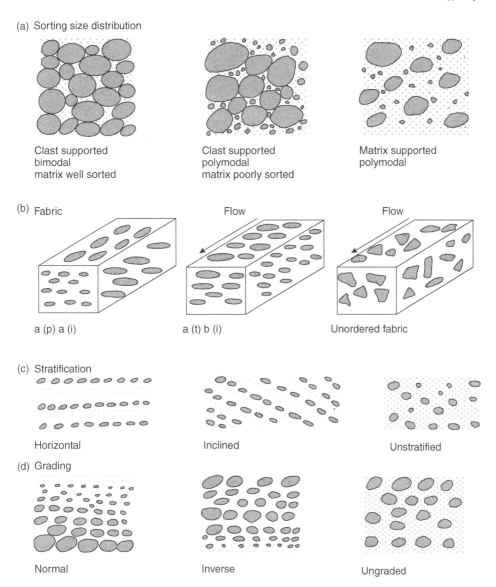

(a) Sorting size distribution

Clast supported
bimodal
matrix well sorted

Clast supported
polymodal
matrix poorly sorted

Matrix supported
polymodal

(b) Fabric

Flow

Flow

a (p) a (i)

a (t) b (i)

Unordered fabric

(c) Stratification

Horizontal

Inclined

Unstratified

(d) Grading

Normal

Inverse

Ungraded

Figure 2.6 Textural and organizational aspects of clasts in fluvial conglomerate, showing features such as sorting and internal organization, including fabric, stratification, and grading. This visualization can be equally applied to other types of deposits (e.g., slope deposits in (2) on the right) (modified from Graham, 1988: fig. 2.1).

of different composition reflected in texture, for example (Figure 2.7). Bedding can be described by a number of criteria, including thickness (Tables 2.5a, b) and shape (Figure 2.8). A lamina is a thin bed, and generally has uniform composition and smaller areal extent than a bed; it results from a minor fluctuation in flow conditions rather than representing constant physical condition; individual laminae are bounded by laminar surfaces (Reineck and Singh, 1986).

Bedforms are surface morphological features that are produced by the interaction of flow (water or wind) and sediment on a bed (Nichols, 2013). Familiar examples are ripple marks as seen on beaches and in streams (Chapter 9), and dunes (Chapter 8). Bedding represents sedimentary

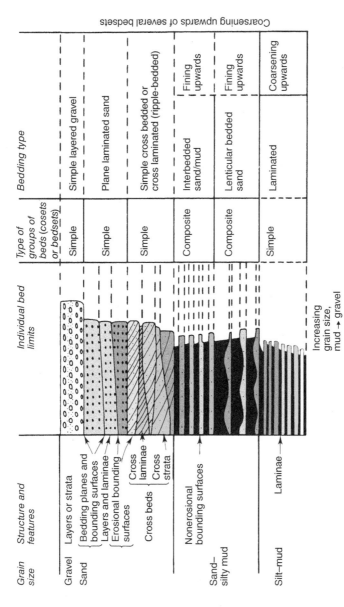

Figure 2.7 Grouping and subdivision of sedimentary beds according to grain size and sedimentary structure (modified from Collinson and Mountney, 1989).

Table 2.5a Nomenclature used to describe bedding types (adapted from Collinson and Mountney, 2019).

Name	Sedimentary unit	Comments
Bed	>1 cm thick	Upper and lower surfaces called bedding planes Subdivided into '*layers*' or '*strata*'
Cosets, bedsets	Groups of beds	*Cross-bedding/cross lamination* = Bedding inclined to depositional surface
Laminae	<1 cm thick	

Table 2.5b Nomenclature applied to describe thickness characteristics of beds and laminae (adapted from Boggs, 2014).

Beds		Laminae	
Thickness (mm)	Name	Thickness (mm)	Name
>1000	very thick	>30	very thick
300–1000	thick	10–30	thick
100–300	medium	3–10	medium
10–100	thin	1–3	thin
<10	very thin	<1	very thin

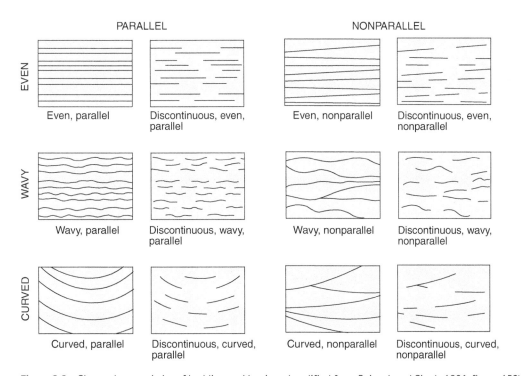

Figure 2.8 Shape characteristics of bedding and laminae (modified from Reineck and Singh, 1986: figure 152).

structures associated with these surface bedforms, denoting higher order arrangements or organizations of groups of particles within a sediment. As with grain size, both the bedforms and associated structures have descriptive value, which furnish information about the environment of deposition, including direction, depth, and intensity of flow. Systematic treatments of sedimentary structures can be found in Boggs (2009) and Collinson and Mountney (2019), with only a few mentioned here.

Current ripple marks, with flow from one direction, are small-scale bedforms, on the order of centimeters. In plan view, they can give different morphologies that include parallel or subparallel, and sinuous to lunate shaped (Figure 2.9). In cross-section, they may be symmetrical or asymmetrical, with crests that range from sharp to flattened. Internally, stratification is expressed by various forms of cross-bedding that are divided into two basic types, tabular and trough cross-bedding as exposed in three-dimensional view (Figure 2.9). Moreover, cross-bedding can be produced in a number of different ways in different types of sedimentary environments, such as channels, point bars in meandering streams (Chapter 6), beaches (Chapter 9), and sand dunes (Chapter 8).

Cross-bedding occurs at a variety of scales that range from millimeters to meters. In small-scale forms, cross-beds range from mm up to ~ 4 cm thick and are usually trough shaped. They are produced by migrating current and wave ripples (Reineck and Singh, 1986). Larger-scale cross-bedding units range from >4 mm up to 1 to 2 m in thickness, and can be both planar and trough shaped. They can be found in different environments (e.g. dunes, longshore bars).

Other structures found within sediments and soils can be of non-depositional origin and produced by physical, chemical or biological agents, some partially synchronous (penecontemporaneous) with the original deposition. They include features produced by freeze-thaw (e.g. ice cracking and frost wedging; Figure 2.10a), and desiccation, bioturbation (e.g. burrowing by rodents and soil insect fauna; Figure 2.10b), and deformation, convolution structures, and load structures (Figures 2.10 c, d) (Karkanas, 2019).

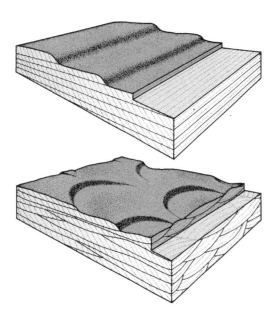

Figure 2.9 Ripple marks. Block diagrams showing two types of surface ripple bedforms. The underlying sediments are organized into cross beds, of which two types are illustrated here. The upper block illustrates tabular cross-bedding, which is produced by the migration of ripples with straight crests; in the lower block, troughs are formed by the migration of sinuous ripples (Tucker, 1981) (modified from Tucker, 1981: fig. 2.21).

Figure 2.10 Examples of sedimentary structures. (a) Ice wedge formed in bedded and iron-stained silts from Normandy, northern France. (b) Burrows produced from small rodents at the site of Hayonim Cave, Israel. In addition, the finely bedded and laminated nature of these silty and clay deposits points to deposition in standing water near the entrance. (c) Convolution structures in the lighter area of Layers 3/4b/4u at Boxgrove, UK in which calcareous pond deposits were deformed while they were still wet and undergoing dewatering. (d) Soft sediment deformation of finely laminated marly lake sediments from the Pleistocene Lisan Lake, Dead Sea Region, Jordan Valley; these deformation features are likely related to earthquakes within the rift valley (Enzel, et al., 2000).

2.2.2 Chemical Sediments (or Non-clastic Sediments)

Chemical sediments are those precipitated from solution, and in certain geoarchaeological contexts (e.g. caves) they can form a significant and critical part of the deposits and geoarchaeological story.

In the open-air context, where rates of evapotranspiration are high, chemical sediments can be related to large-scale features, such as old lake basins and sediments (e.g. Great Salt Lake and other lakes in the Great Basin, USA; and Dead Sea, at the boundary of Jordan and Israel). Many lacustrine sediments (discussed in more detail in Chapter 11) in these arid areas commonly tend to be rich in carbonates (calcite – $CaCO_3$; magnesite – $MgCO_3$, and dolomite – $CaMg(CO_3)_2$), as well as chlorides (halite – NaCl) and sulfates (gypsum – $CaSO_4 \cdot 2H_2O$; anhydrite – $CaSO_4$). The Dead Sea Scrolls, for example, were found in caves developed within Late Pleistocene lacustrine calcareous and gypsiferous marls deposited in Lake Lisan, the precursor to the modern Dead Sea (Bartov et al., 2002; Enzel and Bar-Yosef, 2017) (Figure 2.10d). Similar types of deposits in archaeological contexts are not uncommon, and are known from the Lubbuck Lake on the American High Plains (e.g. Holliday and Johnson, 1989) and Australia's Lake Mungo (Bowler et al., 2003; Fitzsimmons et al., 2014; Jankowski et al., 2020) among others.

Many of these same secondarily precipitated salts and minerals also occur in the soil environment, either on the surface or at shallow depths. Calcite, gypsum, and halite are all common minerals in arid and semi-arid environments, and it is not uncommon to find ceramic sherds or lithic artifacts with a thin carbonate crust on their undersides. The formation of iron and manganese oxides are also widespread in waterlogged environments, such as the Lower Paleolithic site of Boxgrove, on the south coast of England (Macphail, 1999a; Roberts and Pope, 2018; Roberts and Parfitt, 1999). The site of Bilzingsleben in Germany is noted for its detailed Middle Pleistocene sequence, including its collection of *Homo erectus* remains. The site formed as a combined result of lacustrine carbonate deposition and extensive travertine formation, which sealed much of the deposits resulting in preservation of both archaeological remains (Mania and Mania, 2005; Müller and Pasda, 2011) as well as key paleoenvironmental clues including animal bones, pollen, and plant impressions.

Calcareous seeps can produce tufa in rivers and along spring lines, as at Kostenki, Russia where they are seemingly linked to an Upper Paleolithic kill site (Holliday et al., 2007). At Huizui, Henan Province, China, tufa slabs were used to construct Middle Neolithic (Yangshao) floors, an identification that was at first controversial because tufa had not been found locally; tufa slabs had been misidentified as manufactured lime floors (Macphail and Crowther, 2007).

Caves are the focus of an astonishing variety of chemical sediments that comprise most of the major mineral groups, including, oxides, sulfides, nitrates, halides, carbonates, phosphates, and silicates (Ford and Williams, 2007; Ford and Cullingford, 1976; Hill, 1998; White and Culver, 2012) (see Chapter 10). Since the majority of prehistoric and archaeological caves are in karstic terrains, the most common minerals are the carbonates. These tend to occur in a variety of well-known geometric forms, including stalactites, stalagmites, and massive bedded accumulations (*flowstones*) (White and Culver, 2012).

Other minerals, such as phosphates, can be quite frequent in many cave sites, particularly those from Mediterranean climates but less so from temperate areas. These phosphates are derived from bat and bird guano, large vertebrate excreta (e.g. hyaena), bones, vegetal materials, and ashes. The minerals range from simple apatite (calcium phosphate and related minerals) to more complex aluminum, iron, and calcium phosphates (e.g. crandallite, montgomeryite, leucophosphite) produced by the reaction of siliceous clay with guano-derived phosphate (Hill, 1998; Karkanas et al., 2000; Karkanas and Goldberg, 2018a; Karkanas et al., 1999; Weiner, 2010; Weiner et al., 2002).

2.3 Organic Matter and Sediments

Organic matter among sediments is generally localized in subaerial or subaqueous environments. In subaerial environments, such as A horizons in soils, organic matter occurs in various forms, depending on the degree of breakdown (Babel, 1975). Organic matter decomposes into *humus*, a product of plant decomposition aided by microbial activity. This material can be defined as aging plant tissues that undergo "biochemical changes through the synthesis of protease enzymes, the rupture of cell membranes with consequent mixing of cellular constituents, and the auto-oxidation and polymerization of phenolic-type compounds" (White, 2013: 45).

In more poorly drained situations, the presence of water and anaerobic conditions slows the breakdown of organic matter by bacteria and oxidation, and thick accumulations can occur in the form of *peats* (see Chapter 4 for Histosols). Because of the associated acidity and anaerobic conditions, bones tend to decay while plant remains (including pollen) and animal soft tissue preserve well. Some spectacular remains have been discovered in peat bogs including the Lindow Man in England (Stead et al., 1986), Tollund Man in Denmark (Glob, 1969). The Windover Site in Florida had been a pond used as an Archic Period cemetery where bodies were submerged and anchored underwater. This site produced the rare find of intact brain tissue from individuals buried over 7,000 years ago (Doran, 2002).

In addition to these Holocene examples, we note the 400,000-year-old Lower Paleolithic site of Schöningen in Germany (Thieme, 1997) where remarkably well preserved, worked wooden spears occur at the contact between lacustrine marl and an overlying organic-rich lacustrine mud that originally was identified as peat (Stahlschmidt et al., 2015a); until recently, the site had also been thought to have one of the earliest incidences of fire in Europe but these claims have been discredited (Stahlschmidt et al., 2015b).

2.4 Archaeological Sediments

One of the aspects of this text that sets it apart from others in geoarchaeology is an emphasis on archaeological sediments. These deposits – a main subject area in geoarchaeological research – have been dealt with only slightly in geoarchaeological textbooks apart from in Courty et al. (1989). They have recently been the focus of research, at both the field and the micro-scale (Karkanas and Goldberg, 2018b; Nicosia and Stoops, 2017). For this reason and because any sediment that may have a cultural connection can be termed an archaeological sediment, or even in the purist sense, an artifact (Karkanas and Goldberg, 2018b; Miller, 2011), we devote Chapters 12 and 13 to the subject. Dark earth, a generic term for poorly defined archaeological deposits, is given special prominence (Arroyo-Kalin et al., 2009; Nicosia et al., 2017).

2.5 Conclusions: Sediments vs. Soils

In closing, we wish to reiterate the differences between sediments and soils, discussed in Chapter 4 (see also Mandel and Bettis, 2001). A good part of sediments in archaeological contexts are clastic in nature, with the remainder being composed of organic-rich deposits, and in special environments (particularly caves), chemical sediments. As such, these clastic sediments typically embody the concepts of source, transport, and ultimately deposition. As discussed in the next two chapters, soils form under a different set of conditions, those that take place in a single, essentially stable location. Whereas detrital components, such as silt and clay may be *translocated* or moved within the soil profile, and salts may be leached or precipitated in these same horizons, the dynamic is one of vertical movements or displacements generally on the order of centimeters or decimeters. Thus, one should be cautious in the casual use of phrases such as "soil samples", "archaeological soils", or "compacted soil" when actually the term "sediments" should be applied. This point might appear to be fastidious, but as discussed in many places in this book the dynamics of formation of sediments are very different from those of soils, and so are the interpretations.

3

Stratigraphy

3.1 Introduction

Geoarchaeology, like geology and archaeology, is foremost a field-based endeavor that relies on empirical data. No matter how much we might like to theorize in either discipline, the bottom line is that we are constrained by the observations we make first in the field and later in the laboratory. This notion, along with the reality that geology (in its broadest sense) and pedology are the lowest common denominator for virtually all archaeological sites. Together, they provide the ultimate context for the sites themselves *and* their contents. They thus underscore the need for people working in the field to have at least a rudimentary understanding of the materials (e.g. sediments, soils, human constructions, and their products) that they are actually digging, as well as the landscapes that articulate with sites and the archaeological record that they enclose.

These materials commonly epitomize a complex, interwoven concoction of geological and human accumulations that normally have been modified after their deposition by natural soil-forming processes (pedogenesis), or by human activities, such as trampling, digging, dumping, or by building activities (Figure 3.1). In order to carry out accurate geoarchaeological and archaeological research, we must be cognizant of such processes. Moreover, we must be able to record them in a standardized and systematic way that can portray the information objectively and enable others to interpret them in their own way.

The purpose of this chapter is to set out some of the ground rules and guidelines for observing and recording field data related to soils and sediments, both of which have complex and interrelated origins in archaeological settings. We begin with a general discussion of stratigraphy and how it is structured to examine temporal and spatial relationships among sediments and soils. Many previously published geoarchaeological books and articles have emphasized geogenic deposits and their context to the relative exclusion of anthropogenic accumulations and modifications. An extraordinary amount of heretofore untapped information – often of quality similar to or higher than traditional data derived from artifacts or faunal remains – resides in these deposits, and the reader should be aware of their nature and usefulness (Karkanas and Goldberg, 2018; Macphail and Goldberg, 2018a; Miller, 2011; Shahack-Gross et al., 2005; Sherwood and Kidder, 2011).

3.2 Stratigraphy and Stratigraphic Principles

Stratigraphy has an intuitive meaning to many people, even if specific definitions vary among authors and disciplines. In the more classical geological literature, stratigraphy tends to be viewed

Practical and Theoretical Geoarchaeology, Second Edition. Paul Goldberg and Richard I. Macphail.
© 2022 John Wiley & Sons Ltd. Published 2022 by John Wiley & Sons Ltd.

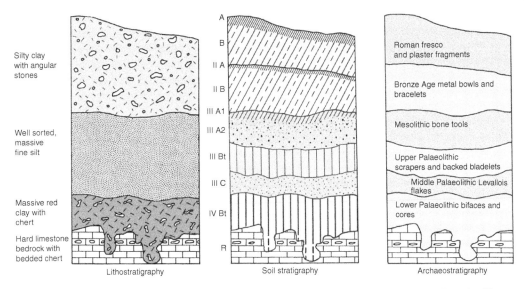

Silty clay
with angular
stones

Well sorted,
massive
fine silt

Massive red
clay with
chert

Hard limestone
bedrock with
bedded chert

A
B
II A
II B
III A1
III A2
III Bt
III C
IV Bt
R

Roman fresco
and plaster fragments

Bronze Age metal bowls and
bracelets

Mesolithic bone tools

Upper Palaeolithic
scrapers and backed bladelets

Middle Palaeolithic Levallois
flakes

Lower Palaeolithic bifaces and
cores

Lithostratigraphy   Soil stratigraphy   Archaeostratigraphy

Figure 3.1 Hypothetical triptych illustrating the complexities involved in archaeological deposits. Shown here is the same profile as viewed from the eyes of the geologist (lithostratigrapher), pedologist (soil stratigrapher), and archaeologist (archaeostratigrapher). In the first panel, the lithostratigraphy is presented on the basis of the lithological characteristics of the deposits, such as composition, texture, and bedding. The second panel illustrates division of the profile on the basis of soil-forming events as expressed by different soil horizons (see Chapter 4); these features represent post-depositional modifications of the original (primary) sediments that took place on stable or semi-stable surfaces. The third panel depicts the profile from the standpoint of the archaeological materials it contains, such as artifacts, features, or architecture. Thus, the stratigraphy from an archaeological site can be seen through different "methodological eyes". Moreover, they can represent the combined effects of depositional and post-depositional processes, coupled with human activities, such as flint knapping and pottery manufacture and discard, as well as construction activities associated with buildings. The trick is to isolate these factors. The first stage to accomplish this is the detailed recording of the stratigraphy in the field (modified from Courty et al., 1989: figure 3.3).

at the regional scale: "The crux of much of stratigraphy is the spatial relationships of rocks over geographic areas" (Schoch, 1989). For F. J. Pettijohn, a renowned sedimentologist in his time, it also encompasses physical characteristics of sediments that include notions of space and time:

> stratigraphy in the broadest sense is the science dealing with strata and could be construed to cover all aspects – including textures, structures, and composition. . . .. Stratigraphers are mainly concerned with the stratigraphic order and the construction of the geologic column. Hence the central problems of stratigraphy are temporal and involve the local succession of beds (order of superposition), the correlation of local sections, and the formulation of a column of worldwide validity. Although these are the objectives of stratigraphy, the measurement of thickness and description of gross lithology are commonly considered a part of the stratigrapher's task
>
> **Pettijohn, 1975: 1**

Archaeologists are used to viewing stratigraphy, deposits, features, and artifacts at the scale of the site and perhaps its immediate surroundings (Stein and Linse, 1993). For them, stratigraphy can be simply, "the natural and cultural layering of the soil at a site" (Feder, 1997) (intuitively correct despite the incorrect use of "soil"). D. H. Thomas comes closer to a more holistic definition of stratigraphy: "An analytical interpretation of the structure produced by the deposition of geological and/or cultural sediments into layers, or strata" (Kelly and Thomas, 2017). He rightly includes the interpretative aspects, "since stratigraphy does not exist by itself". For better or worse, it is something that is recognized somewhat subjectively, and different persons – depending on experience and background (archaeologist, geologist, geographer, pedologist, other environmentalist) might construct different stratigraphic sequences for the same physical stratigraphic profile. The International Commission on Stratigraphy defines stratigraphy as

> . . .the description of all rock bodies forming the Earth's crust and their organization into distinctive, useful, mappable units based on their inherent properties or attributes in order to establish their distribution and relationship in space and their succession in time, and to interpret geologic history.
>
> **Murphy and Salvador, 2020**

In the geoarchaeological sector, Waters takes a traditional geographic/geological, "depositional" viewpoint, emphasizing deposits and soils that are observed principally on a geomorphologic scale: "Stratigraphy is the study of the spatial and temporal relationships between sediments and soils. Stratigraphic sequences are created because depositional environments are dynamic and constantly changing" (Waters, 1992: p. 60). On this regional scale, stratigraphic study is useful to organize sediments and soils into objectively identifiable packages that can be arranged in some kind of chronological order and absolute age on the basis of temporal markers, such as soil formation, or perhaps erosion. "Parcels" of soils/sediments can be physically (spatially) or temporally linked and integrated over an area, whether on the scale of meters to kilometers (correlation; see below).

In this book we take a scale-neutral and broad approach to stratigraphy and *characterize* it as the three-dimensional organization in space and time of geological layers, soils, archaeological features, and artifacts. In a sense, it embodies the total fabric of these entities, whereby stratigraphy and the stratigraphic record represent encoded pages of geoarchaeological history, including the gaps (blank pages) where "apparently" nothing took place. By being able to view these pages within a structure – the stratigraphic framework – we are better able to decode these pages and reconstruct the depositional, pedological, and geomorphological history of the site and its surroundings, *including* the activities of the people who acted within and around the site. It also enables us to evaluate the relative timings of human and geological events which can be extended to, and correlated with, other sites and regions.

During the latter part of the nineteenth century, classical geologists realized that it was necessary to separate the physical characteristics of a sedimentary rock (i.e., color, texture, composition) from the time in which it formed. This came about because they realized from its fossil contents that a geological unit can accumulate over long period of time and can be older in one place than in another (diachronous or "time-transgressive" units: Brown et al., 2013; Krumbein and Sloss, 1963; Vita-Finzi, 1973). Deltaic deposits, for example, become progressively younger seaward, even though the "deltaic lithologies" overall do not change radically horizontally within the area of the delta. To overcome these issues of time and space, they recognized a number of different types of stratigraphic units, which at present are embodied in the *International Stratigraphic Guide* (Murphy and Salvador, 1998).

Table 3.1 summarizes the most common types of stratigraphic units. Note that the two major groups of units are delineated on the basis of content or physical limits (Group I), as opposed to

Table 3.1 Types of stratigraphic units (modified from NASCN, 1983: table 1 and Fritz and Moore, 1988: table 1.3).

Stratigraphic unit	Example of unit name	Real-life geological example
I. Units based on content or physical characteristics		
Lithostratigraphic	• Regional scale: Formation • Site scale: Bed	• Regional scale: 'Ubeidiya Formation, Israel (interbedded layers of clay, marl and gravel with intercalated Lower Paleolithic site (Bar-Yosef and Tchernov, 1972)) • Site scale: band of yellow silt over layer of clay (Figure 3.5)
Biostratigraphic	Biozone	Younger Dryas (pollen zone), NW Europe (Faegri and Iverson, 1989)
Magnetopolarity	Polarity Zone	Dmanisi, Georgian Republic, spans normal polarity of Olduvai subchron to reversed in Upper Matuyama chron (~1.8 Ma) (Ferring et al., 2011); normal polarity magnetozone (Jaramillo (1.0–1.1 Ma) within Reverse zone (Matuyama) in Gran Dolina at Atapuerca, Spain (Parés et al., 2018)
Pedostratigraphic	Geosol	Sangamon Geosol (USA) (Jacobs et al., 2009); Barham Soil (UK) (Kemp et al., 1993); Usselo Soil (Denmark, west Germany, The Netherlands) (Kaiser et al., 2006)
Allostratigraphic	Alloformation	Unit Q2 in Figure 3.4
II. Units based on age time units		
Chronostratigraphic	Period	Quaternary Period, ca. 1.8 to 0 my
Polarity chronologic	Polarity Chron	Brunhes Chron of "normal" polarity (see Table 3.2)

Table 3.2 Main paleomagnetic polarity events for the last 3.5 million years (not all excursions are indicated) (modified from Bradley (1999) and Singer (2014)). Normal polarity in roman, reversed polarity in italic.

Age (my)	Chrons (epochs)	Subchrons (events)	Age range (my)
0 to 0.773	Brunhes		
		Laschamp excursion	*0.041*
0.773	*Matuyama*		
		Jaramillo	1.048–1.008
		Cobb Mt.	1.211–1.189
		Olduvai	1.934–1.787
		Réunion	2.200
2.610			
	Gauss		
		Kaena	*3.04–3.11*
		Mammoth	*3.22–3.33*
3.58			

those units in Group II which are defined solely on the basis of time. Thus, the units in Group I are derived, for example, on the basis of their mineral composition or texture, biological/fossil constituents, soil characteristics, or stratigraphic boundaries or gaps marked by lapses in time (unconformities).

Lithostratigraphic units are the most basic, ubiquitous, and relevant to the majority of geoarchaeological situations. They are denoted on the basis of lithological characteristics, such as color, texture, composition, thickness, upper, and lower boundaries. They do not imply any notions of time, just the descriptive aspects of the sedimentary bodies. In regional-scale geological contexts, the primary lithostratigraphic unit is the formation, which is one that can be mapped over a region. It can be represented by a 20 m-thick accumulation of alternating beds of clay and silt or by a 2 m-thick massive layer of limestone. Two or more formations that form a consistently uniform lithological package can be combined into a group. On the other hand, formations can be subdivided into smaller units, such as a member, and even finer, bed. The latter is smallest unit of the lithostratigraphic units and is lithologically distinct from units above it and below it.

On an archaeological scale, which size-wise can be on the order of tens of square meters (e.g. rockshelter, building) up to thousands (e.g. tell or ancient city), a lithostratigraphic unit can take the form of a cm-thick band of red clay underlying a plastered floor from a Bronze Age house or a meter-thick waste dump. In caves, a layer of rock fall mixed with clay would also be common type of lithostratigraphic unit (see discussion of facies, below).

Biostratigraphic units are delineated on the basis of the fossils that they contain, including their appearance, disappearance, or relative abundance. The fossils can be either plants or animals, marine or terrestrial. For example, the Tertiary Period was subdivided by Lyell based on the abundance of fossil molluskan species that are living today. The Eocene, for example, has only 3.5 % of living forms whereas the Pleistocene has about 90% of modern species (Farrand, 1990).

Magnetic polarity units represent certain remnant magnetic properties of a sediment (e.g. dipole-field pole position, intensity of the magnetic field, etc.) that differ from those above and below it (Murphy and Salvador, 1998). The most often-cited example of magnetic polarity units is based on reversals in the earth's magnetic field, and rocks or sediments could possess either "normal" or "reversed" polarity. At the important site of Dmanisi, Georgian Republic, magnetic polarity determinations show that hominids, animals, and lithics date to 1.85–1.78 Ma (Ferring et al., 2011).

Related to these are *polarity chronologic* units, which refer to the specific *time* interval over which a state of magnetic polarity exists. Both basalts and water-deposited sediments are typically measured, as they possess magnetic polarity properties of the earth's magnetic field at the moment that the basalt crystallized, or the sediment accumulated (Table 2.2). Major polarity periods (*Chrons*) exist on the order of millions of years, whereas shorter-term polarity events take place on the order of 10^4 to 10^5 years (Bradley, 1999). So, for example, the Matuyama Reversed Chron corresponds to a period between 2.581 to 0.78 my ago, and deposits at this time display polarity opposite to today's. Subchrons represent shorter intervals or "events" within longer magnetic polarity periods. Thus, at the Lower Paleolithic site of Olduvai Gorge, Tanzania, basalts dated to about 1.95 to 1.77 my ago (the Olduvai subchron) display normal polarity within the longer Matuyama Reversed Chron (Table 2.2).

Pedostratigraphic units represent whole or part of a buried soil, which exhibits one or more soil horizons (see Chapter 4) that is preserved in a rock or sediment, and is traceable over an extended area (Edwards and Owen, 2005; Schoch, 1989). These types of units, foremost of which is the *geosol*, represent relatively short periods of geological time. Hence, they serve as temporal pegs or "marker horizons" that temporally place the relative ages of deposits and events that overlie and underlie the soil (Holliday, 1990). Geosols also have paleoenvironmental significance (Fedoroff and Goldberg, 1982). The interglacial Sangamon soil in the Midwest USA has been extensively used as a stratigraphic tool (Birkeland, 1999; Jacobs et al., 2009), and similar such soils have been documented in the UK by Kemp et al. (1993). The Barham Soil in eastern England, for example, is a clear stratigraphic marker horizon that denotes a major landscape change of marked climatic degradation (Rose et al., 1985). In Denmark, The Netherlands, and western Germany coversand soils (Arenosols) include the late-glacial Usselo Soil, characterized by concentrations of charcoal variously attributed to human manipulation of the vegetation by fire and/or extraterrestrial impact (Kaiser et al., 2006).

Allostratigraphic units are rock or sediment bodies that are overlain and underlain by temporal discontinuities (unconformities) (Hughes, 2010). A widespread example is stream terrace deposits produced by successive episodes of alluvial deposition and erosion, commonly in the form of downcutting resulting in terrace formation. Allostratigraphic units are a convenient means to map widespread fluvial deposits, for example, and they constitute a basic component in documenting the Holocene fluvial geoarchaeological history of much of the mid-continent USA and the eastern Mediterranean (Brown, 1997; Ferring, 2001; Goldberg, 1986; Holliday, 1990; Mandel, 1995). Such fluvial deposits commonly contain archaeological sites and can be traced over several km within the same allostratigraphic unit (see Boxes 3.1 and 6.2).

Geochronologic units are those which are defined solely on the time interval that they encompass. As such, they do not represent actual specific rocks but are more conceptual, representing specific intervals of time. These units are differentiated on the basis of radiometric dating, such as potassium/argon ($^{40}K/^{40}Ar$) or uranium-series techniques; they can contain mixtures of different types of fossils and lithologies. The Pleistocene Epoch, for example, that includes much of the geoarchaeological record, began 2.58 million years ago, and ended with the Holocene Epoch 11,700 years ago (U.S. Geological Survey Geologic Names Committee, 2018).

The system of stratigraphic units described above, evolved within the geological community over many decades as a practical means to facilitate organizing, describing, and interpreting the rock record. In the 1980s, the archaeological community expressed concerns about the lack of attention being paid to archaeological sediments and stratigraphy, and attempts were made to develop a system of nomenclature geared to the recording of archaeostratigraphy (Gasche and Tunca, 1983; Stein, 1990). The former two authors formally proposed, for example, the term "layer" for the fundamental lithological unit, and "phase" for chronostratigraphic unit.

Both these and similar concepts have since faded away as they appeared to have little practical merit, as the foundations for the objective description of soils and sediments (whether of geological or human origin) exists within these earth-science disciplines, and there is no need to add another level of descriptive complexity. A "4 cm-thick lenticular mass of crumbly organic-rich sand" can be simply described as that. Moreover, the use of a descriptive term in this "archaeostratigraphic" system, such as "layer", adds a clearly genetic flavor to the description. In the Earth Sciences, small-scale stratigraphic subdivisions, such as beds, are expected to represent individual natural depositional episodes. In archaeological settings, the sedimentary dynamic is not always apparent as they can be of natural (geogenic) or human origin (anthropogenic). Thus, it is generally not realistic to attempt to correctly identify a distinct layer – especially in the field – that would appear to indicate a specific "geoarchaeological" event. In addition, the isolation of "layers" may likely be unwarranted and unwanted, since it supposes that such layers are solely anthropogenic in nature. It would be at odds with a situation that involves reworking of a previously deposited anthropogenic accumulation. As we discuss below in Chapter 16, field descriptions must remain as free as possible from genetic overtones. This is especially true because field observations provide only partial insights into the history of a deposit; analytical data from the laboratory are normally needed to determine more fully the complete history of a deposit or stratigraphic sequence.

3.3 Facies and Microfacies

Geologists realized that at a given time, different types of sediments can be accumulating within the different depositional environments. So, for example, at the coast sandy sediments accumulate at the beach zone, but further offshore towards deeper water, the sediments often become increasingly fine-grained and "muddier". Landward, lateral change in lithologies might include *aeolian* (windblown) sand or organic-rich, sandy silty deposits found in coastal marshes. In solely continental (i.e., terrestrial) environments, we might see coarse gravel deposits accumulate within a river channel, which grade laterally into silt and clay deposits of the floodplain (see Chapter 5). These laterally equivalent lithological bodies that accumulate at the same time are called *facies*, and the term permits us to visualize how sediments/rocks of different compositions can be of the same age.

Over decades the term "facies" has been used in a variety of ways (Boggs, 2012). In our view, the most appropriate use is descriptive and includes terms such as "sandy" facies and "clayey organic" facies. Moreover, different sub-lithofacies can be recognized on the basis of lithological characteristics (e.g. massive sand vs. finely bedded sand), or bio-facies base on fossil content (e.g. gastropod–rich clay vs. clay with bivalves) in order to provide more flexibility (Figure 3.2) (Murphy and Salvador, 1998). Less desirable practices involve names that have a genetic flavor such as environment of deposition (e.g. "shallow marine" facies or "intertidal" facies). This practice of using facies names that connote interpretation or genesis should be avoided, as they can be commonly subjective or wrong. In sum, facies must be identified on the basis of *objective* criteria, such as lithology or fossil content.

The facies concept is very valuable and functional approach in geoarchaeology, both at regional scales, as above, and at scales related to individual stratigraphic units exposed in an excavated balk. At the regional level, recognition of different facies enables detection of different parts of former landscapes and geological environments; it also links up with questions, such as site location, site burial and erosion, and resource availability. At the site-specific level where individual deposits are concerned, the concept of "archaeological" *microfacies* has been developed.[1] It involves the recognition of facies at the microscope level using petrographic thin sections or polished slabs of impregnated blocks of sediment. The use of the microfacies concept is critically important in deciphering the formation of archaeological sediments, site-specific activities of past inhabitants, and the integrity of the archaeological record (Courty, 2001). As succinctly stated by Courty (2001, p. 229), ". . .the ultimate goal of archaeological facies analysis is to restore the three-dimensional image of a human-related space at a given time and to describe its evolution." Thus, to be able to recognize penecontemporaneous (approximately the same time) human activities within different areas at a site (e.g. cooking area *vs.* storage area), it is necessary to recognize and be able to trace small-scale "microfacies" across the site and see where one type of microfacies grades into another. So, for example, a layer of ashy silts in an area can be resolved microscopically into individual ash accumulations resulting from *in situ* burning in one area; however, laterally these ashy deposits change into loose, organic-rich ashy silts produced by trampling and of the dumping of cleaned-out hearth materials that is mixed with organic refuse and discard. Applications of this concept are outlined in Chapter 17 and utilized in Chapter 13, in which domestic floor sub-types can be grouped separately from those of stabling floors (Macphail et al., 2007a).

---

1 Courty's (2001) use of the term "microfacies" is different from that used in carbonate sedimentary petrography, where it refers to ". . .the total of all sedimentological and paleontological data which can be described and classified from thin sections, peels, polished slabs or rock samples" (Flügel, 2009: 1) (see Chapter 17 for details).

(a)

(b)

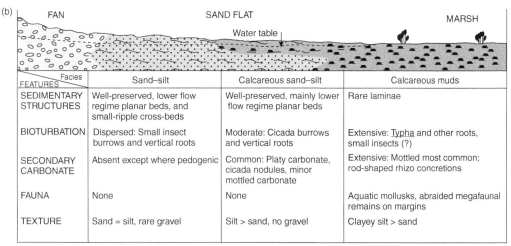

FEATURES \ Facies	Sand–silt	Calcareous sand–silt	Calcareous muds
SEDIMENTARY STRUCTURES	Well-preserved, lower flow regime planar beds, and small-ripple cross-beds	Well-preserved, mainly lower flow regime planar beds	Rare laminae
BIOTURBATION	Dispersed: Small insect burrows and vertical roots	Moderate: Cicada burrows and vertical roots	Extensive: Typha and other roots, small insects (?)
SECONDARY CARBONATE	Absent except where pedogenic	Common: Platy carbonate, cicada nodules, minor mottled carbonate	Extensive: Mottled most common; rod-shaped rhizo concretions
FAUNA	None	None	Aquatic mollusks, abraided megafaunal remains on margins
TEXTURE	Sand = silt, rare gravel	Silt > sand, no gravel	Clayey silt > sand

Figure 3.2 Facies. (a) Photograph of alluvial fan in the Ka Valley of Western Sinai, Egypt showing the flattening of the fan surface from the mountain front as it grades topographically into a plane at the toe of the fan. From the base of the mountain front at the head of the fan down-fan towards the toe of the fan. (b) A stratigraphic cross-section through the lower part of such a fan sequence shows the downward lateral lithostratigraphic changes in facies from gravels to sands and ultimately calcareous muds. Other lithostratigraphic characteristics are also shown. Modified from Ashley, 2001.

3.4 Correlation

A related concept to facies is correlation. As used in an everyday sense, correlation implies a certain degree of bringing together two similar things. In geology, the idea of correlation, like facies, is used in different ways. Two are most prevalent. One slant refers to a demonstration of equivalence of time, so that two bodies of rock were formed at the same time. The alluvial facies example

---

Box 3.1   Facies and stratigraphy: the Paleoindian-Archaic site of Wilson-Leonard, Texas

Facies and microfacies are important facets of geoarchaeology. Understanding lateral and vertical changes in the lithological aspects of deposits over a landscape and within a site is vital for comprehending fully the deposits and stratigraphy, and integrating them into the reconstruction of site settlement and sediment history. The Paleoindian-Archaic site of Wilson-Leonard in central Texas illustrates some of these concepts.

The Wilson-Leonard site contains a rather complete record of occupations in the Southern Plains of the US (Bousman et al., 2002; Collins et al., 1993), spanning Paleoindian through Late Prehistoric times, roughly from about 11,400 radiocarbon years ago up to about 4,000 years ago. The site is located about 33 km north-northwest of Austin, Texas (Figure 3.3) within the valley of Brushy Creek, a tributary of the Brazos River (Q2 in Figure 3.4; Collins and Mear, 1998). This valley displays several Quaternary fluvial terraces, which, following custom in North America, are labeled Q1, Q2, Q3, etc., from youngest to oldest, and usually from lowest to

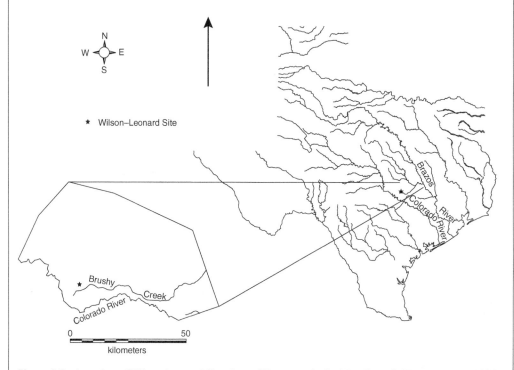

Figure 3.3   Location of Wilson-Leonard Site, Central Texas marked with * (from Collins and Mear, 1998).

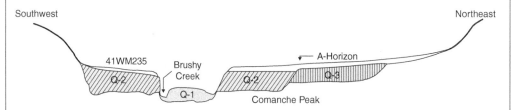

Figure 3.4   Schematic cross-section through Brushy Creek, showing the position of the Wilson-Leonard site in relation to the Q-2 fluvial terrace, which can be seen on both sides of the valley (from Collins and Mear, 1998). Q3 is an older Pleistocene terrace, whereas Q1 is modern gravelly alluvium.

Box 3.1 (Continued)

highest. Excavations in the 1980s and early 1990s in the Q2 terrace fill revealed a sequence of deposits ranging from gravels at the base to organic-rich silts at the top that are marked by the presence of numerous remains of burned rock ovens. The accompanying schematic cross-section and photograph of the Western profile of the site (Figures 3.5a and 3.5b) illustrate a

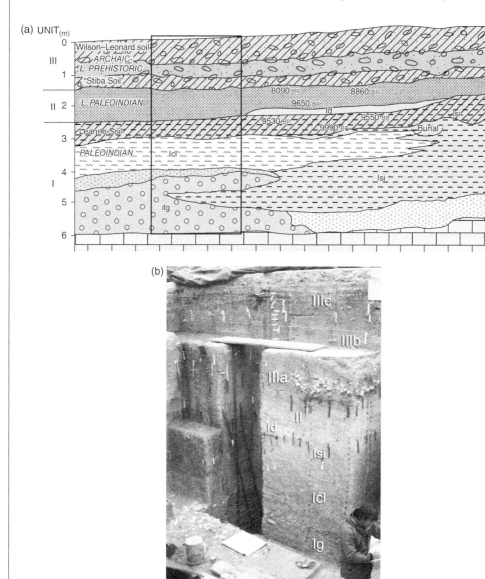

Figure 3.5 (a) Schematic view of major lithostratigraphic units and their lateral and vertical facies variations as shown by the interfingering of some of the units. *Unit Icl*, for example (a clayey *cienega* – depression – deposit) grades laterally into massive alluvial silt of *Unit Isi*. The profile is approximately 8 m wide. (b) Photograph of the area outlined in (a) with some of the labeled stratigraphic units. Geologist T. Stafford is taking notes.

(Continued)

Box 3.1 (Continued)

number of facies that are characteristic of a typical Late Quaternary fluvial setting in this part of Texas (Goldberg and Holliday, 1998), where rivers and creeks drain the Edwards Plateau and flow into the Gulf of Mexico. The following is a description of some of the different types of lithostratigraphic units at the site, which are keyed to a depiction to its historical evolution and its local landscape (Figure 3.6).

Occurring at the base of the section is *Unit Ig*, which consists of medium to coarse, well-rounded gravel that laterally interfingers with *Unit Isi*. The gravels are interpreted to be channel deposits that accumulated when Brushy Creek was flowing at the location of the site (Figure 3.6, #1). The silty sediments of *Unit Isi* that overlie and interfinger with the gravels of

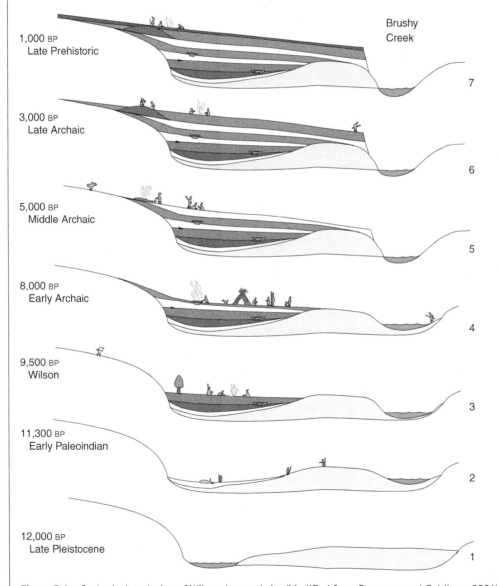

Figure 3.6 Geological evolution of Wilson-Leonard site (Modified from Bousman and Goldberg, 2001).

Box 3.1 (Continued)

Unit Ig are overbank deposits that are partly contemporary with *Unit Ig* as shown by their inter-fingering. *Unit Icl* is an organic silty clayey lens that overlies *Unit Ig* but also interfingers with *Unit Isi. Unit Id* is a thin, discontinuous lens of silt that is draped on the underlying deposits of *Unit Isi. Unit II* consists of fine gravelly silt that overlies *Subunits Isi* and *Id* with a sharp, ero-sional contact, thus representing a different facies and depositional regime. It, in turn, is over-lain by thick, darker brown locally organic-rich accumulations of gravelly silts in the upper part of the profile. Intercalated within these upper deposits of *Unit III* are numerous burned rock features with fire-cracked rock that represent clear anthropogenic features. This uppermost *Unit III* was further subdivided based on color and texture, whose changes reflect the relative contributions and rates of sedimentation from different sources: darker, organic matter-rich components are derived from fire-related anthropogenic activities, whereas the lighter-colored silts indicate relatively more geogenically derived, calcareous alluvium and colluvium, the lat-ter originating from the slopes behind the site. Thus, the darker color of deposits from *Subunits IIIa* and *IIIc* result from high inputs of organic matter, fuel, and plant residues associated with numerous burned rock oven features (Goldberg and Guy, 1996; the base of Unit III in Figure 3.3b).

These descriptive lithostratigraphic units and facies relate to specific geogenic and anthro-pogenic processes acting at the Wilson-Leonard site. They include calcareous fluvial deposits associated with Brushy Creek (e.g. *Units Ig* and *Isi*) and stony colluvial material derived from the slopes behind the site to the West. By being able to recognize these different facies and their lateral and vertical changes, we are able to piece together the evolution of the geological environment of the site during the Late Pleistocene-Holocene (Figure 3.6; numbers below refer to individual within the figure). Accordingly, *Unit Ig* represents a gravel bar that formed within the channel of Brushy Creek over 11,000 years ago (*#1*). When this channel was sud-denly abandoned by avulsion, and the stream began to flow within a new channel (right side of figure), it left behind a closed, elongated depression (*cienega*) that was successively filled in with organic-rich clays (*Unit Icl*; *#2, 3*) that accumulated during flooding events. As this mass of sediment continued to accumulate, or aggrade, the influence of stony colluviation began to be more pronounced with the accumulation of Unit II (*#4–7*); fluvial material originating from Brushy Creek concomitantly became less important, and likely covered the area of occupation only during times of major floods. During Early Archaic times (*#4*), the intensity and area of occupation increased, along with the use of numerous burned rock ovens. These factors led to the addition of anthropogenically derived organic matter-rich fine detritus mixed with burned stones and colluvially derived stony silts (*Units IIIa* and *IIIb*).

This example from the Wilson-Leonard site illustrates that contemporaneous lateral facies shifts in lithologies from gravels to silts can take place even within the boundaries of an exca-vated area of only 8 × 8 m; in fact, the geoarchaeological picture of the site was initially rather confusing during the re-excavation of the site during the 1990s. Thus, understanding these lateral and vertical facies changes is crucial to understanding the overall stratigraphy at the site, the geological context from which the artifacts come, the local environment in which the former inhabitants lived, and how all these changed through time.

discussed above (channel gravels vs. floodplain silts; up-fan vs. down-fan changes), illustrates this concept. Another perspective about correlation is more encompassing: equivalence is not strictly based on time equivalence, but on lithologic, paleontologic, and chronologic grounds (Murphy and Salvador, 1998). Thus, in the case of *lithocorrelation*, relationships are established on the basis of similar lithologies or stratigraphic position. Two clay deposits, for example, may look similar and be "correlatable", but they *may* not necessarily be the same age. Similarly, a gray ashy band in the corner of a Bronze Age building may not be the same ashy deposit on the other end of the building.

Biocorrelation is effected on the basis of fossil make-up or by biostratigraphic comparisons, but it does not necessarily imply equivalence in age (Murphy and Salvador, 1998). On the other hand, in *chronocorrelation*, comparisons are made by means of chronological age. Whereas with older geological periods it is critical to keep these different types of correlation separated, this practice is less critical in younger, geoarchaeological settings because stratigraphic and temporal resolution is greater and lithologic, paleontologic, and chronostratigraphic boundaries tend to overlap, or coincide. In any case, in most (geo)archaeological cases, investigators are probably most interested in knowing whether a deposit (e.g. mudbrick collapse), activity (e.g. excavating pits), or event (e.g. downcutting or flooding/deposition by the river) that took place in one place is the same as that in another part of the site, or between two sites, or over a region.

3.5 Keeping Track: The Harris Matrix

One of the thorny, operational issues that arise during the investigation of archaeological deposits, particularly those from sites with complex stratigraphic sequences, is how to organize all the stratigraphic data (features, stratigraphic units, etc.) and how to correlate stratigraphic units across space. This is less of an issue with most open-air hunter and gatherer sites, where more homogeneous, stratigraphic units can be recognized in the field and which tend to be more laterally and vertically extensive. But when an area is intensely occupied, such as in a tell or ancient city, or in a cave habitation site, layers may extend only a short distance, and there can be crosscutting pits and ditches, and complex architectural features.

A technique that has received much attention is that of the Harris Matrix (Harris, 1979, 1989; www.harrismatrix.com). Essentially, this is a house-keeping strategy that permits one to visualize the stratigraphic relationships among stratigraphic units and features in diagrammatic form. Harris stressed the recognition of not only the stratigraphic units but the contacts between them, which reflect specific human actions (e.g. clearing, pit filling). Whereas there are some controversies associated with the technique and concept (see Barham, 1995; Farrand, 1984; Karkanas and Goldberg, 2018; Stein, 1987) it nevertheless may serve as a convenient form of shorthand, if it can be implemented with caution in the proper context. In many cases, it is quite difficult to produce a realistically accurate and understandable diagram because of complexities and ambiguities of the stratigraphic deposits and their relationships. A computer program (e.g. https://harrismatrixcomposer.com/#/) is available, but it should not be considered as a black box, and considerable attention to details of data input is still necessary, as is a complete understanding of all the stratigraphic relationships at the site.

3.6 Concluding Comments

The above information is meant to familiarize the reader – particularly those not trained in the geosciences – with many facets and complexities of stratigraphy. We cannot overemphasize the importance of stratigraphy, its concepts, and how it is put into practice. The techniques and strategies used to recognize and record stratigraphic information in the field are discussed in Chapter 16 (Field techniques).

A few additional, somewhat general points are worth emphasizing. A principal one is that there is a difference between the sediments we find at a site, for example, and its "stratigraphy". The sediments are what they are, geological and/or archaeological materials (in their broadest meanings) that were deposited at a place (by whatever means natural or human) and were perhaps later modified chemically (e.g. dissolution or precipitation of calcite, phosphate, iron, etc.) or physically (trampling, burning, burrowing, roof fall, whatever).

Stratigraphy on the other hand is largely a construct made by us humans. This is clearly evident if one looks at a profile that may have been described by different people, or by a geologist and a pedologist (see Figure 3.1, for a stylized version; but see also Bruxelles et al., 2014; Collins et al., 1998; Goldberg and Holliday, 1998). In the geological literature, we commonly see revisions of stratigraphic sections and sequences by different authors over time, even though the sediments and/or rocks have not changed for thousands or millions of years. In this light, the point is that we should not make statements such as, "animal burrowing or human construction, *destroyed* or *modified* the stratigraphy". These kinds of activities are part of the stratigraphy itself and never detached from it. In fact, a well thought out stratigraphic description/conception of what is seen in the field, should make it easier for us down the road when we try to interpret the stratigraphy. Ultimately, however, stratigraphy can be simple, complex, interesting, dull, or difficult to interpret. But there is no such thing as "good" or "bad" stratigraphy. It can be correctly conceptualized or poorly studied, but these deficiencies are human facets and not geo/archaeological ones.

4

Soils

> The soil is a natural body of animal, mineral and organic constituents differentiated into horizons of variable depth which differ from the material below in morphology, physical make-up, chemical properties and composition, and biological characteristics.

> **(Joffe, 1949) cited by Bridges (1970)**

4.1   Introduction

This chapter introduces the importance of soils in geoarchaeology, and why it is essential that soils are clearly differentiated from natural and archaeological sediments. In order that both environmental and human influences on soil formation are understood, the concept of the *five soil-forming factors* is introduced. This provides the basis for analysis and reconstruction of past soil landscapes, and shows how soil classification and mapping are essential tools in this endeavor. Basic soil types and the horizons that make up different soil profiles are also given, and some important soil-forming processes are described. Generally, we employ generic soil names (e.g. podzol) and their Soil Survey of England and Wales equivalents (Avery, 1990), as well as international soil classification terminology from the USA and Europe, which are used worldwide (Soil Survey Staff, 2014; WRB, 2015).

An understanding of soils is fundamental to archaeology. Apart from soils being an essential component of any environmental reconstruction and background to any human activity, they provide an absolutely essential resource to humans. For example, woodland type, richness of pasture, and character of agricultural undertaking, are all governed by soil conditions. Moreover, the nature of human impact on soils, through erosion, acidification, desertification, and management practices can contribute information on the nature of contemporary human societies.

Furthermore, an archaeologist has to have at a least a vague notion of what soils are, their characteristics, and the processes under which they form in order to differentiate them from sediments. This holds true not only as a general statement, but also in specific instances such as in the field where one may need to distinguish naturally occurring soils from archaeological and geological sediments. Misidentifications can have significant ramifications in interpreting stratigraphic sequences at the site, regional, and local levels.

At the broad scale, soils were, and in many cases, still are one of the most important factors affecting human occupation and development in the Holocene, whether in Asia, Mesoamerica, or Europe (Beach et al., 2015; Dunning et al., 2012; Gebhardt et al., 2014; Limbrey, 1990;

Practical and Theoretical Geoarchaeology, Second Edition. Paul Goldberg and Richard I. Macphail.
© 2022 John Wiley & Sons Ltd. Published 2022 by John Wiley & Sons Ltd.

Neogi et al., 2020). For example, fertile soils, resulting from regular rainfall, warm temperatures, and amenable geology were often the foci for early arable activity. It was in the area of the "Fertile Crescent" that the earliest arable activity was recorded in cultures that were aceramic (Rosen, 2007).

At the more detailed scale, many sites are situated on soils, with these soils and soil-based sediments normally comprising most of the archaeological deposits and the archaeological stratigraphy (Karkanas and Goldberg, 2018). Thus, soils that make up archaeological deposits also form the substrate upon which occupation takes place, and which may themselves be involved in the preservation of a site as overburden, can all provide crucial information about site history (see Box 4.1; cf. colluvium in Chapter 5).

The study of *paleosols* (here, a generic term for ancient soils) and associated hominid/human remains, both in the New and Old Worlds, have provided clear clues to past environmental conditions (Catt, 1986; Catt and Bronger, 1998; Holliday, 2004, Chapter 8; see review in Cremaschi et al., 2018). Moreover, local soils may have been utilized in the manufacture of ceramics, daub and mud brick (Chapter 12), and in mound construction (Sherwood and Kidder, 2011), or served as the medium for arable activity (Chapter 11). Finally, the site itself may be buried under slope deposits that result from soil erosion (Chapter 5). In the last case, conditions of burial relate directly to soil conditions, such as vegetation cover, aridity, waterlogging, and acid or alkaline influences, which in turn affect conservation of archaeological materials and environmental reconstructions (see Table 4.4; Figure 4.10); this information can be crucially important in forensic and mortuary practice studies (Chapter 15). Of greater importance to ancient populations, however, are soilworkability and soil stability/erodibility, because these factors will govern how a landscape can be exploited for agriculture. Pedogenesis can produce dramatic alteration changes to the geological parent material, by forming soil horizons (Figures 4.1 and 4.9; see below).

The term "soil" means different things to different people. Colloquially, it can be equated with "dirt", something that should be avoided here. To an engineer, it is the loose substrate upon which

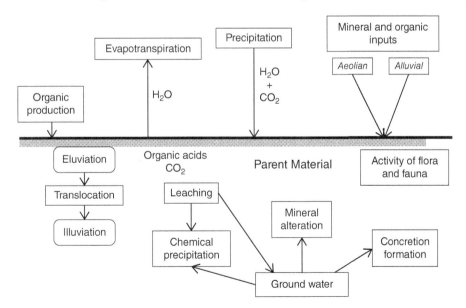

Figure 4.1 Soil-forming processes (Figure 4.9), controls on soil composition (Figures 4.3 and 4.4), and links to climate and plants leading to different horizon formation (Figure 4.2; Tables 4.1 and 4.2) (after a range of authors: Avery, 1990; Brady and Weil, 2008; Duchaufour, 1982).

buildings are constructed, whereas to an agronomist, it is the material upon which crops are grown. For Quaternary scientists and geoarchaeologists, on the other hand, soils are valuable records of past environments and environmental change.

In contrast to sediments, which are laid down by various processes discussed elsewhere in the book (e.g. alluvium, aeolian sand, slope deposits) and usually form the original (parent) material in most geoarchaeological contexts, soils form by *in situ* development on a stable substrate (see below). A good, pragmatic and generalized definition of soil by Joffre (1949) is cited by (Bridges, 1970) (see above).

The specific factors responsible for producing different types of soil horizons are discussed in detail below. However, in general terms, soils and their horizons are formed by combinations of inputs, outputs, and movements of solutions, gases, mineral and organic matter between the atmosphere and the soil/ground surface and subsurface (Figure 4.1). The surface can be augmented by aeolian or alluvial sedimentation, for instance, or with water from precipitation and the incorporation of organic matter; conversely, soil moisture can be depleted by evapotranspiration into the atmosphere. Below the surface, faunal and floral activity can add organic material and move soil particles physically by bioturbation. Similarly, grains of silt and clay, for example, can be moved downward or laterally (translocated) by water percolation either from the surface or by groundwater. Water can also bring about chemical and mineralogical changes with leaching, dissolution, and reprecipitation, resulting in the formation of nodules of calcite, gypsum, or Fe (e.g. FeOOH goethite), for example.

4.2 Soil Horizons

There are several types of soil horizons, which aid in the description/characterization of the soil (Table 4.1; Figure 4.2) and point to its origin. Not all horizons appear in every soil, a result of the various soil-forming processes discussed below. Moreover, soil horizon nomenclature and terminology varies according to the soil classification being used, which in turn is a function of the country and scientific body that classified them (e.g. US = Department of Agriculture, UK = Soil Survey of England and Wales, Europe = FAO-UNESCO, World Reference Base for Soil Resources) (Avery, 1990; Soil Survey Staff, 2014; WRB, 2015). The common soil horizons and the processes associated with their formation are given in Table 4.1.

Regardless of the specific letter used to designate a specific horizon, soil horizons can be grouped according to their position in the soil. From top to bottom, these include (i) superficial organic horizons, (ii) topsoil horizons (Epipedons), (iii) upper subsoil horizons, (iv) subsoils, and (v) parent materials/geological regolith, the material within which the soil is developed (Figure 4.2a). A simplified version of Table 4.1 provides the general types of processes associated with four generalized horizons, A, B, C, and R. Graphically, this is illustrated in Figure 4.2a with the addition of the organic, O horizon. Field examples of shallow and deep soil profiles are also presented (Figures 4.2b and 4.2c).

The various soil-forming processes (e.g. leaching, translocation) normally take place on an extant, exposed substrate. Sediments are generally little affected by the *five soil-forming factors* described below, which transform the parent material (typically sediments or rocks, but also archaeological substrates) into soils (Box 4.1; Figures 4.2–4.4; Box 13.1). During soil formation (*pedogenesis*; see Box 4.2 for upland soil development examples), minerals are subjected to weathering, elements may become hydrated or leached, and biological activity – whether involving bacteria or larger plants and soil animals – mixes organic matter with mineral material; the sum total is the initiation of soil horizon formation (Figures 4.2a–4.2c).

Figure 4.2a Generalized soil profile showing bedrock (R), soil parent material (C horizon) and overlying lower subsoil B, upper subsoil E, topsoil A, and surficial organic horizons (L – litter; F – fermentation; H – humus; see Table 4.1).

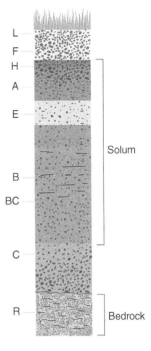

Figure 4.2b Field photo of shallow humic soil formed directly on chalk (Rendol/Rendzina) with the humic mineral topsoil Ah horizon (epipedon) developed under grassland (pasture), on the weathered chalk substrate (Cu layer). Icknield soil series, south-east England (Jarvis et al., 1984: 208–210).

Figure 4.2c Field photo of deep argillic brown earth soil profile (Alfisol/Luvisol), with relatively minerogenic topsoil Ah horizon, Eb (A2) upper subsoil horizon, Bt, and Btg lower subsoil horizons formed in superficial drift including loess, over Chalk. Winchester soil series, south-east England (Jarvis et al., 1984: 97–101, 392).

Table 4.1 Soil horizons – with suggested associated major soil processes and international terms (possible associated USDA diagnostic horizon[1] are also given).

Common horizon term	Europe (FAO/UNESCO)	UK	USA (USDA[1])	Common soil name Europe/UK/USA	Major soil process
Superficial organic horizons					
L (litter)	Litter	L	~Oi	Mull/ O1/cf. Oi	High biological activity; only accumulation of plant fragments (L)
L/F (fermentation)	Fermentation	F	~Oe	Moder/ O2/cf. Oe	Moderately low biological activity; accumulation of excrements of soil fauna and decomposing plant fragments (F) below L.
L/FH (humus)	Humus	H	~Oa	Mor/ O2/cf Oa	Very low biological activity; accumulation of amorphous organic matter termed humus (H) below, L and F.
Peat	Histic H	Peat O horizon	~Oa (Histic epipedon)	Peat/O	Dominant accumulation of organic matter because waterlogging inhibits biological breakdown of organic matter.
Topsoil horizons (Epipedons)					
Humic topsoil	e.g. Mollic A, Umbric A	Ah horizon	A (e.g. Mollic epipedon, Umbric epipedon)	Humic topsoil	Accumulation of organic matter in the mineral soil (along with associated nutrients of N, K and P); focus of biological activity and organic matter turnover (oxidation and alteration) Mollisols include grassland prairie soils and chernozems
Plaggen	Plaggen (Cultosol)	Cultosol (Ap)	Ap (Plaggen epipedon; Anthropic epipedon[2])	Plaggen/ Anthrosol	Over-thickened (0.40–1.0 m) humic topsoil developed through additions of manure, turf, household waste. (e.g. since AD 1000 in Holland)
Ploughsoil	Cultivated A	Ap	Ap (in the USA includes both ploughed soils and pasture)	Ploughsoil/ Arable topsoil	Topsoil, mechanically homogenized to depth of plough share (ca. 0.40 m) or ard (e.g. 0.06 m); liable to loss of organic matter through oxidation; arable soils ameliorated by additions of organic manures, possibly since Neolithic.
Upper subsoil horizons					
Leached/eluviated upper subsoils	Albic E (A2)	Eb horizon	E (Albic)	Argillic/Luvisol/ Alfisol	Eluviation of clay (along with cations, including iron; organic matter and phosphorus)
Leached/eluviated upper subsoils	Albic E (A2)	Ea horizon	E (Albic)	Podzol/Spodosol/ Podzol	Eluviation of iron and aluminum (sesquioxides) (after acid breakdown of clay and mobilization by plant chelates)

Subsoils

Cambic B	Cambic B	Bw horizon	Bw (Cambic)	Cambisol/Brown soil/Cambisol	General weathering of minerals and clay formation.
Argillic Bt	Argillic B	Argillic Bt	Bt (Argillic)	Luvisol/Argillic brown soil/Alfisol	As above, but in some cases, with illuviation of clay from overlying A2 horizon (clay translocation)
Spodic Bs/Bh/Bhs	Spodic Bs/Bh/Bhs	Podzol Bs/Bh/Bhs	Bh/Bs/Bhs (Spodic)	Podzol/podzol/Spodosol	Illuviation of sesquioxides (Fe and Al) often along with humus
Calcic B and K horizon	—	—	Alkaline Bk	Calcisols/Calcic brown soils/Aridisols	Illuviation of alkaline earths, especially calcium carbonate; cementation by calcium carbonate (K horizon/calcrete)
Gypsic horizon	n.a.	Gypsol?	By (Gypsic)	N.a./gypsic/gypsid	Concentration/cementation by gypsum
Gleyed (Ag, Bg, G) horizon	Gleyic G (Bg)	Gleyic G (Bg)	Bg	Gleys/Gleys/Aquents, Aquepts	Gleying or hydromorphic process of reducing iron (Fe^{+++} to Fe^{++}); pale (G) to mottled (Bg) colors
Ironpan	Ironpan Bf	Placic Bf		Ironpan/ironpan/ironpan, plinthic	Cementation by iron (usually under gleyed conditions)
Oxic B	n.a.	Ferrasols	Bo (Oxic)	N.a./ferrasols/Oxisols	Strong (tropical) weathering and formation of iron-rich laterite (hematite and goethite)
Rubefied argillic B	n.a.	Planosols	B (Argillic)	N.a./Acrisols/Ultisols	Extreme (tropical) illuviation of clay from E horizon
Fragipan	Fragipan	Fragipan	Bx (Fragic)	?/Cryosols/Gelisols	Compaction of soil material associated with freezing and thawing
Andic horizon	n.a.	Andic horizon	B (Andic)	n.a./andosols/Andepts	Organic matter accumulation and weathering of volcanic ash (e.g. Tephra) to produce allophanes.
Natric and salic horizons	—	—	Btn (Natric) / Bz (Salic)	Saline soils/Salorthids, Natrustalfs, Natrargids	Concentration of salts, especially sodium salts (halite) in Natric horizons

(Continued)

Table 4.1 (Continued)

Common horizon term	Europe (FAO/UNESCO)	UK	USA (USDA[1])	Common soil name Europe/UK/USA	Major soil process
Parent material					
C horizon	C horizon	C	C (Cox, Cr, Cu)	n.a.	Partial weathering of substrate (poorly consolidated or unconsolidated regolith or surficial sediment)
R horizon	R horizon	R	R	n.a.	Unweathered hard bedrock

[1] In contrast to field identified A, B, and C horizons, diagnostic horizons of the USDA's *Keys to Soil Taxonomy* (2014) are rigidly defined according to criteria measured in the field and laboratory.

[2] Anthropic epipedon requires >250 ppm citrate-extractable P_2O_5 (a mollic epipedon has <250 ppm citrate-extractable P_2O_5).

See also Bridges, 1990; Pape, 1970; WRB, 2015.

Caveat: workers should check their classifications (and possible revised versions) with the appropriate texts for their regions. Some good examples of UK soils classified according to the US classification system can be found in Avery (1990). Also, the soils of Norway were also mapped in the 1980s (Soil Map Norway – Jordbunnskart – ESDAC – European Commission (europa.eu)).

Table 4.2   Soil classification (Soil Survey Staff, 2014; USDA, 2014).

Soil order	Main characteristics
Entisol	No or weak evidence for soil horizon formation
Inceptisol	Humid region soils with soil horizon formation, loss of bases but weatherable minerals still present
Aridisol	Soils formed with little/rare precipitation
Mollisol	"Grassland" (prairie/steppe/pampa) base-rich soil
Alfisol	"Forest" base-rich soil with diagnostic argillic subsoil horizon
Spodosol	Leached soil with diagnostic spodic subsoil horizon, variously enriched in Fe, Al, and humus
Gelisol	Cold climate permafrost soils
Ultisol	Warm climate, low base status soils with diagnostic argillic subsoil horizon
Oxisol	Tropical and subtropical soils often with reddish iron oxide-rich horizons of great age
Andisol	Often warm, humid climate soils formed on volcanic substrates
Vertisol	Humid climate soils, characterized by seasonal swelling and deep cracking
Histosol	Peats

Archaeological deposits themselves may have accumulated through sedimentary processes, some or all of which are purely anthropogenic in origin. Furthermore, they themselves may be affected by pedological or geogenic post-depositional processes (e.g. cryoturbation; see Figures 4.6b and 4.10c; Box 4.3), and they may destroy original layering and transform or completely remove some easily weatherable materials, such as wood ash (Weiner, 2010). Extreme pedological alteration of archaeological deposits results in the formation of *dark earth* as found in Roman/early medieval Europe (Borderie et al., 2014; Devos et al., 2016; Macphail, 2014a) and South- and Mesoamerica where it is more commonly known as *terra preta* (Arroyo-Kalin, 2017; Glaser and Woods, 2004; Graham, 2006; Graham et al., 2016) (see Chapter 13; Box 13.1).

Box 4.1   The five factors of soil formation and the Bronze Age site of Brean Down, UK

Factors of Soil Formation

As discussed in the main text, Jenny (1941) enumerated five factors that affect the formation of soils:

$$s = f'(\text{Cl, O, R, P, and T}), \text{ where}$$

s = soil properties, dependent on the function (f) of soil-forming factors;

Cl = climate (Arctic conditions, for example, result in slight soil formation; high effective precipitation can lead to leaching of soils and peat; high evapotranspiration can lead to secondary precipitation of carbonates, gypsum, halite, and other salts);

O = organisms (both large and small: plants – mosses to mangroves; animals – ants to antelopes);

R = relief (droughty ridges and uplands, eroding slopes, colluvial footslopes and boggy valleys);

P = parent material (hard igneous rocks may produce thin ranker soils, soft clays are easily eroded, but are poorly drained)

T = time (permits the increased development of the soil profile horizonation – A, B, C horizons. Changes relating to time: age of landscape, with older exposed landscapes having more strongly developed horizons) (see also Jenny, 1941 in Bridges, 1970 and Brady and Weil, 2008).

(Continued)

Box 4.1 (Continued)

The Five Soil-forming Factors at Bronze Age Brean Down, UK

These soil-forming factors can be illustrated with the multi-period site of Brean Down (Somerset coast of Bristol Channel). This site is noted for its late Glacial deposits (studied by Ian Cornwall of the Institute of Archaeology, UCL), Beaker soils and land use, and Middle Bronze Age settlement; it is also one of the earliest salt working sites in the UK (Bell, 1990). Natural and excavated exposures revealed Pleistocene slope deposits and Holocene soil formation; Beaker (Early Bronze Age) through Iron Age occupations and colluvium, punctuated by episodes of aeolian reworked beach sand. The local geology is composed of Carboniferous limestone, and the talus slope deposits that constitute the Brean Down headland form the basal deposits at the site. Pleistocene alfisols (argillic brown earths/luvisols) formed on the limestone (and probably on the fine drift cover). Here and on other limestones, such as the Jurassic Oolitic Limestone (e.g. Cotswolds, Gloucestershire), slow weathering of the bedrock produces a red-colored clay-rich β horizon, which formed along the boundary between the overlying *illuvial* Bt horizon and underlying weathered C horizon (limestone substrate) (Catt, 1986) (Figure 4.3a; see also Figures 2b and 2c). This soil profile shows the main results of *time* (well-developed ABC profile forming since the Late Glacial), *parent material* (limestone), *climate* (western humid temperate – inducing leaching and clay translocation), and *organisms* (under woodland).

At Brean Down, however, the soil-forming factors are complex and not straightforward (Figure 4.3b):

- *topography* and *relief* play a role, as the site is located on a slope towards the base of the catena, and soil materials are affected by colluviation;
- multiple *parent materials* exist, including limestone talus, previously formed red β soils (*terra fusca*), "Mendip loess", local Pleistocene sand and silt, and reworked beach sand and estuarine silt;
- *time* is represented by some 5,000 years from the Late Glacial to the Atlantic Period;
- *organisms* here involve not only natural soil fauna, but also human activity, such as uprooting and clearance of trees (Macphail, 1990c) (Figures 4.3c and 4.3d). The results of all these *factors* are revealed in the soil micromorphology (and grain size and chemistry) of the Beaker paleosol (buried by Bronze Age colluvium and blown sand) (Bell, 1990).

These features are manifestations of the following soil-forming factors at the site: *Climate* (mid-Holocene humid western temperate climate), *Organisms* (human-induced plough erosion and colluviation; inclusion of burned and charcoal-rich soils (relict from clearance)), *in situ* cultivation disturbance; *in situ* earthworm working during post-cultivation pasture stage), *Relief* (footslope position resulting in colluvial additions to the soil and over-thickening of the profile), *Parent Material* (as Figure 4.2.), with inclusion of various soil fragments from the relict, terra fusca (β horizon) Holocene profile, and burned soil, which for example affect measurements of soil organic matter, grain size, magnetic susceptibility, and phosphate), and *Time* (here very old, Late Glacial), earlier (clearance, ploughsoil colluvium), and contemporary (Beaker Period pasture topsoil) (from Macphail, 1990c).

Box 4.1   (Continued)

Figure 4.3a   Alfisol formed on limestone under woodland in a humid western temperate climate, long weathering developing a reddish clay-rich β horizon (cf. humic decalcified woodland *terra fusca* and more strongly rubified red "Mediterranean" soils).

Hypothetical forest soil horizonation on limestone

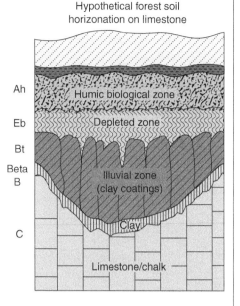

Figure 4.3b   Field photo, Brean Down profile, as Figures 4.3 and 4.4; at the base there is the Early Bronze Age paleosol (EBPS) formed from reddish terra fusca and limestone clasts, which is overlain by the Early Bronze Age (EBA) or "Beaker" colluvial clearance and ploughsoil. The dark Bronze Age (BA) layer recorded two Middle Bronze Age round houses and a Late Bronze Age midden, and some of the earliest evidence of salt working in the UK (Bell, 1990). Above, are Roman and Iron Age layers (R & IA).

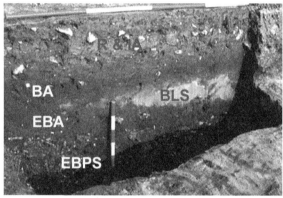

Soil history and micromorphology

(a) Late Glacial/Early Flandrian

(b) Atlantic

(c) Neolithic/Beaker Disturbance

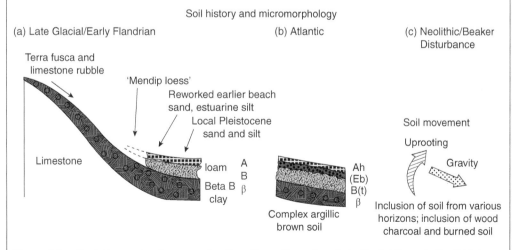

Figure 4.3c   Evolutionary development of soils at Brean Down from Late Glacial through Neolithic/Bronze Age times. (a) catenary development of alfisol (A, B, and β horizons) formed along the toe slope where *terra fusca* and limestone rubble interfinger with beach/estuarine sand; (b) during the Atlantic period (ca. 5,000 BP) this profile naturally evolved into [Ah, A2(Eb), Bt, and β horizons], which were eventually modified by Neolithic/Beaker period disturbance that mixed horizons (see below) (Macphail, 1990c) (Figure 4.3b).

Box 4.1 (Continued)

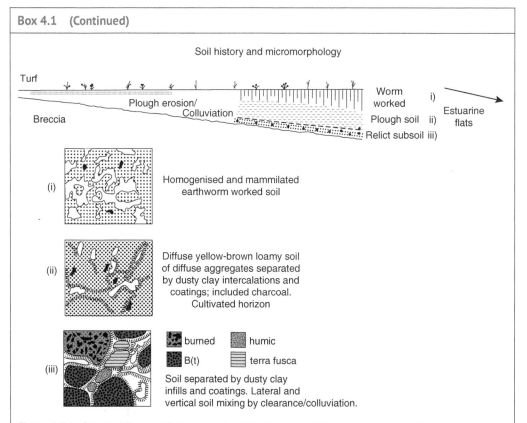

Soil history and micromorphology

(i) Homogenised and mammilated earthworm worked soil

(ii) Diffuse yellow-brown loamy soil of diffuse aggregates separated by dusty clay intercalations and coatings; included charcoal. Cultivated horizon

(iii) burned | humic | B(t) | terra fusca
Soil separated by dusty clay infills and coatings. Lateral and vertical soil mixing by clearance/colluviation.

Figure 4.3d Detail of Figure 4.3, showing Neolithic/Beaker soil development: Upper Figure – three-layer (i, ii, iii) slope soil profile buried by wind-blown sand, and the eventual interdigitation of this slope soil with estuarine sediments. Lower Figure (i) uppermost earthworm-worked and homogenized pasture soil formed in colluvium; (ii) buried ploughsoil (cultivated) colluvium that contains fragmented soil and textural pedofeatures indicative of soil disturbance and mixing; and (iii) lowermost subsoil formed, and then reworked and deposited over the Pleistocene talus substrate; the reworked material is composed of fragments of red terra fusca, clasts of the Holocene Bt horizon, Ah topsoil, and burned soil; the last contains charcoal and probably relict of clearance.

4.3 Differentiating Soils and Sediments

There are clear differences between soils and sediments (Mandel and Bettis, 2001). A sediment has a dynamic history, which encompasses erosion, transport, and deposition over a landscape or area (e.g. glacial till, aeolian loess, beach sand; see Chapter 2, Sediments) (Karkanas and Goldberg, 2018). In contrast, a soil is static in that it has formed *in situ* through various weathering and biological processes that are described below (see Figure 4.1). This soil-forming episode is often termed a period of "stability" ("stasis"). An archaeological deposit is therefore clearly a sediment and not a soil, with a source (e.g. a combustion feature) and a mode of deposition (e.g. dumping, stabling waste accumulation). Like any sediment, an archaeological deposit can undergo post-depositional effects that are geogenic and/or pedogenic in character (Courty et al., 1989: 138–189). When these aspects are understood, the site formation processes involved in any one archaeological context can be identified. As this chapter is focused on soils, post-depositional *soil-forming processes* are discussed in some detail (see section 4.3).

There is of course a hybrid – a soil-sediment, a term coined by Fedoroff and Goldberg (Fedoroff et al., 2018; Fedoroff and Goldberg, 1982). One notable example is colluvium, which is a downslope accumulation of eroded rock, soil and in some cases archaeological construction materials (e.g. mud brick, turf). Although detailed aspects are considered in Chapters 2, 5, and 13, we note here that even in the case of "soil" colluvium or "hillwash", these slope deposits accumulate through sedimentary processes. Thus, colluvium can be described both as a sediment and a soil (Macphail and Goldberg, 2018a: chapter 5). Alluvial soils (soils formed in alluvium) develop in a similar way, with most or all evidence of their sedimentary history also being lost through biological working (Chapter 6). However, before detailing some of the soil-forming factors and processes that contribute to soil profile formation which at one time were linked to geographical regions (i.e. *Zonal Soils*, 4.2.2) – and various human effects on them – some fundamentals are given on how soils are classified and described in the pedological literature (Tables 4.1 and 4.2).

4.4 The Five Soil-forming Factors

The general model of soil formation depicted in Figure 4.1 serves as the basis for conceptualizing the basic processes involved. Nonetheless, it is evident that there are many different types of soils in the world, so there must be an interplay of these processes in order to yield these soil types. Jenny (1941) formulated an expression to account for the various types of soils, the so-called five soil-forming factors:

$$s = f'\left(\text{Cl, O, R, P, and T}\right).$$

Briefly, these are:

> *Cl* – Climate (influences of temperature, precipitation, and seasonality)
> O – Organisms (effects of plants and animals, including humans)
> *R* – Relief (topography, and its control on drainage – see *catena*)
> *P* – Parent material (rocks and archaeological deposits that provide a substrate for pedological activity)
> *T* – Time (length of time over which soil formation has taken place governs degree and maturity of soil formation).

These "factors" have been enumerated and expanded by many authors (Brady and Weil, 2008: 40 *et seq.*) (Box 4.1, Figures 4.3a–4.3d), and they still remain at the core of conceptualizing soil formation (Amundson, 2021). Johnson (2002) stressed that the role of biological activity has been severely underestimated as a soil-forming factor. Nevertheless, the relative weight of each factor in pedogenesis varies from soil to soil, with for example, the oldest (*t*) most weathered soils being *Ultisols* (Table 4.2); while poorly fertile substrates (*p*) and a climate (*c*) that encourages leaching, gives rise to acid unproductive Spodosols (Brady and Weil, 2008: 110–111) (see Figures 4.6a, 4.7 and 4.9 below).

4.4.1 Climate

Very simply, this factor can be first understood from the examples of arctic and arid environments (Figure 4.4), in which little soil is produced. Freezing and thawing generate coarse sediments such as scree (*talus*), and rock fall in caves, but little chemically weathered fine soil because there is a

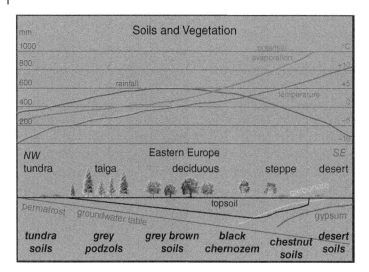

Figure 4.4 Classic soils and vegetation across the globe according to climate (see Bridges, 1970; Duchaufour, 1982).

lack of biological activity and production of organic matter, and chemical processes are inhibited. Equally, cool conditions, including permafrost, permit only modest breakdown of organic matter and peats (Histosols) may form (see Boxes 4.2 and 4.3).

On the other hand, arid conditions lead to little plant growth, and high evapotranspiration can encourage secondary precipitation of salts, such as carbonates, gypsum, and salt (halite) (see Table 7.2). The presence of such neo-formed minerals in ancient deposits has led to interpretations of past desert conditions, as in Aridisols (Fedoroff et al., 2018; Mees and Tursina, 2018), including desert varnish (Cremaschi et al., 2018); a good example of a modern desert soil environment is the Sonoran Desert (Arizona, USA) (McAuliffe, 2000). High temperatures under humid conditions give rise to the deep, highly weathered soils of regions such as the south-east USA, south-east Asia, and parts of the Amazon, especially if formed in very ancient (pre-Pleistocene) landscapes where *time* has been influential and Ultisols dominate (Brady and Weil, 2008: 108–110). Although climate change (interglacials, glacials, stadials. interstadials) is well studied in Quaternary sites, and is a relevant modern topic linked to sea level rise and coastal flooding (Chapter 9), investigations concerning the Holocene are less in fashion compared to previous decades (Ball, 1975). Holocene climate-induced soil changes are important, however, and the amalgamations of European tree ring data, for example, have indicated a marked episode of climatic deterioration from approximately AD 250, possibly linked to soil changes and suspected of triggering the Migration Period in north-west Europe, possibly as early as AD 300 (Büntgen et al., 2011) (see Boxes 4.2 and 18.4). Traditionally, soil types that are governed by climate are called zonal soils.

4.4.2 Zonal and Intrazonal Soils

Some of the earliest classifications of soils employed the concept of zonality. *Zonal* soils, a view stemming from Dochuchayev at the end of the 19th century (Bridges, 1970), are those that simply reflect broad environmental conditions, such as podzols (Spodosols; Figure 4.5) of the boreal climatic/vegetation zone (coniferous forests of the mid-high latitudes). Colder

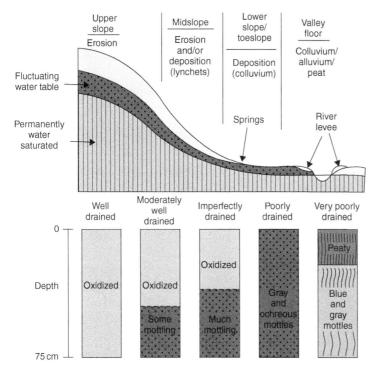

Figure 4.5 The soil catena; soil drainage classes in a moist temperate environment (after Avery, 1990); also showing generalized slope units, areas of predominant soil erosion or deposition and location of springs and river levees (see Table 5.1). In upland and high rainfall areas, gley soils vary according to their position and drainage characteristics on the catena (see Figures 4.7a–4.7c).

environments gave rise to Gelisols (Cryosols), where permafrost and frost turbation predominate. A knowledge of these soils is believed to be fundamental to reconstructing Pleistocene environments and relict soils from the earliest days of geoarchaeology (Zeuner, 1959; see also Bertran and Texier, 1990) (see Box 4.3). Another example is the grassland soil of the dry mid-latitudes, an environment known as prairie, pampas, or steppe, which typically features chernozems (Mollisols) (see Figures 14.2 and 14.3).

Intrazonal soils, on the other hand, are those soil types that are governed by local conditions, such as poor drainage (gleys/hydromorphic soils with oxidation and reduction) or their parent material (e.g. Andisols). Soils developed during one set of soil-forming conditions are termed monogenetic soils, whereas polygenetic soils represent more than one, i.e., superimposed soil-forming events over time. Good examples of monogenetic soils are rendzinas/Rendols (A/C horizons on limestone/chalk; Figure 4.3b) or other Mollisols developed under grassland during the Holocene. Pleistocene and pre-Pleistocene paleosols, on the other hand, are much more likely to be polygenetic in character (Cremaschi et al., 2018; Retallack, 2001). For instance, a Pleistocene soil could have initially developed as an Alfisol during an interglacial, but formed a fragipan because of cold conditions of an ensuing glacial event. A fragipan is a dense, root-impenetrable layer typically within a lower B horizon caused by ice-lensing, freezing and thawing and the infilling of almost all voids due to downwash of fine soil from the elutriated horizons above (FitzPatrick, 1956; Payton, 1983; Van Vliet-Lanoë and Fox, 2018b). Such frost-affected soils are currently gleys because of highly reduced drainage (Fedoroff and Goldberg, 1982; Kemp, 1985a). As described earlier, some European podzols are polygenetic having formed as Alfisols under woodland before Bronze Age clearances led to podzol formation under acidic heath (see Figure 4.9c).

4.4.3 Organisms

The uppermost soil horizon is the most organic because it is the focus of plant growth, litter deposition, and associated biological activity (A1/Ah horizon). Fungi, mosses and lichens invade sediments that are newly exposed to subaerial weathering, and these biota may be soon followed by higher plants such as grasses, shrubs, and trees, which thrive when deeper, rootable soils begin to form. Under poorly sealed conditions, whole wooden ships may be all-but lost to aerobic decay, including fungal attack (see Chapter 15). Close inspection of ash layers produced by burning on archaeological sites often reveals working by soil fauna and growth of lichens and mosses, which initiate weathering of such materials.

Topsoils or A horizons develop through the accumulation and mixing of organic matter by soil fauna. Organic matter is essentially formed of carbon (C) and nitrogen (N). Organic matter breakdown (e.g. humification) is accelerated by the presence of nitrogen, so that plant materials like the grasses, for example, with the greatest amounts of nitrogen (low C:N ratio of <25), break down more rapidly than oak and beech forest litter (C:N of 30–40), and pine and Ericaceous litter (C:N of >60) (Duchaufour, 1982: 44–45). Increasing C:N is reflected in decreasing levels of biological activity present in Mull→Moder→Mor horizons (see Table 4.1). The C:N of the Mull of the Rendzina at Overton Down is ~8–10, whereas at the Wareham podzol the Fermentation and Ah horizons have a C:N of ~23–30, with the *Calluna* Litter recording a C:N of ~30–40; the effect of burial on these is detailed in Chapter 14.2.

An understanding of A horizon formation is also important in archaeology. Artifacts and traces of human activity, such as hearths and charcoal, are jumbled up and scattered by soil mixing. This can occur through root growth, earthworm and insect burrowing, or more destructively by mammals such as moles, gophers, and rabbits. In earthworm worked soils that are also burrowed by small mammals (voles and moles), artifacts become both buried by earthworm casting and mixed throughout the topmost 20–40 cm of the soil, after only a few decades. This was first demonstrated by Darwin, but it has been further elucidated (Armour-Chelu and Andrews, 1994; Atkinson, 1957; Darwin, 1888; Johnson, 2002). The effect of biota on soils and archaeological deposits are homogenization, and porosity formation that can result in a crumb or granular structure of the material. Size, character, and abundance of excrements produced by soil invertebrates, their burrows (sometimes termed biogalleries), and channels formed by roots, reflect types of biota. The analysis of these contributes to the identification of past environments, both local and regional, but may also indicate modern disturbance.

An important example of understanding sedimentation versus bioturbation comes from the Lower Paleolithic site of Boxgrove (UK). In Unit 4b, artifacts left on ephemeral landsurfaces formed on lagoonal mudflats and buried by sedimentation, remain *in situ* (Pope et al., 2020). In contrast, flints deposited later (Unit 4c) on a more mature terrestrial land surface (paleosol) that underwent bioturbation by roots, invertebrates and small mammals became spread both vertically and horizontally (Roberts and Parfitt, 1999). The recognition of bioturbation on a site is therefore essential if dateable materials (e.g. by ^{14}C) and artifacts are to be interpreted correctly, including correlation within and among sites (Sherwood, 2001, 2008; Stiner et al., 2001). Entire stratigraphic columns can be partly or completely modified by bioturbation (Cahen and Moeyersons, 1977; Goldberg and Bar-Yosef, 1998), and if this is not recognized nonsensical analyses and interpretations ensue.

Lastly, an understanding of the factor "organisms" is also important in cultural and environmental archaeology, especially when seeing how a settlement functions in its environs, as in stock, dung, and manuring schemes, for example (Dreslerová et al., 2021; Kenward and Hall, 1995; Linderholm et al., 2019; Macphail et al., 2017a). It is often an essential first step to calculate the natural resources, in terms of vegetation type and fauna that are available to human populations in a given soil landscape, and how they might manipulate it for wet rice cultivation, for instance (Lee et al., 2014; Zhuang et al., 2014; Zong et al., 2007). Similarly, grasslands, for example, provide grazing for both stock and wild herbivores, whereas "managed" woodlands supply fuel, wild fruit, nuts, as well as game (Evans, 1999). Neolithic to modern herders have also used "leaf hay" from trees to provide fodder for their animals (Myhre, 2004; Robinson and Rasmussen, 1989). Equally, coasts can supply seaweed for manuring (Chapter 9).

4.4.4 Relief

Soils are influenced by topography. Steeply sloping ridges are generally eroded and produce thin, droughty (excessively drained) soils, particularly on south-facing slopes in the northern hemisphere. Eroded materials accumulate in valley bottoms producing thick, cumulative soils, which usually reflect wetter conditions (possible gleying) than those along the slope itself (see Box 4.2). Relative relief, i.e. the difference in altitude between the highest ground and the lowest, and steepness of slope, have implications for both the soil and human landscape. In areas of steep slopes, such as 30–40º and marked relative relief, soils of different kinds will be closely juxtaposed, and humans will have access to the varied ecosystems present. In Mediterranean and arid areas, droughty south-facing slopes provide grazing land, whereas north-facing slopes are more moisture retentive and more likely to carry scrub and woodland. The constraints of slope and aspect also govern land suitable for agriculture and pasture in the high latitudes, because they receive more sunlight compared to less sunny slopes, which are best left to forestry.

In hilly and mountainous areas, routeways have to follow the valleys and river systems, and numerous 'mitigation' excavations have found variously dated agricultural and occupation layers, sequentially buried by colluvium, for example (Fechner et al., 2014). Sloping ground may also be managed by terracing to produce agricultural land (Sandor and Homburg, 2015). Low slope areas, such as alluvial plains, may have lower soil variability but supply large areas of flat, easily worked

---

Box 4.2   Some examples of soil degradation along the western maritime and upland regions of England and Wales

The degradation of brown soils (cf. Cambisols) into podzols (cf. Spodosols), peats, and gleys (cf. Histosols, Gleysols and suborder aquents) in Europe, through environmental change and human impact over time, has been under study since the 1960s (Crampton, 1963; Dimbleby, 1962; Duchaufour, 1982: e.g. 53 *et seq.*, 296 *et seq.*). That this transformation could occur without woodland clearance and the development of heath was also recognized early on (Mackney, 1961). Western maritime and upland environments were found to be the most

(Continued)

Box 4.2 (Continued)

susceptible to degradation (Figure 1.12; Avery, 1990; Ball, 1975), and seemingly occurred from the Neolithic onwards in Brittany (e.g. on granite; Gebhardt, 1993). Across the Channel in Cornwall, the brown soil ancestry on the granite of Bodmin Moor was recorded (Maltby and Caseldine, 1982) alongside the Neolithic settlement outer rampart-buried soil at Carn Brea (near Redruth). At the latter, a podzolic A2 and Bs horizon (and associated spodic chemistry) was superimposed on the earthworm-worked acid brown earth formed on granitic head deposits (Figure 4.6a) (Macphail, 1990a; Mercer, 1981). Pleistocene Head is characterized by cool climate-formed "embedded grains", which are sands and gravels originally affected by frost feature link cappings (see Figure 4.6b) (Bullock and Murphy, 1979; Van Vliet-Lanoë and Fox, 2018). The presence of very fine charcoal in a trampled exposed A2 horizon soil indicated that Neolithic landscapes management by fire could have triggered podzolization.

On Exmoor, where the highest point is Dunkery Beacon at 520 m asl, a brown soil heritage can be traced to Mesolithic (argillic brown earths; Alfisols, Luvisols) and Bronze Age (acid brown earth, Cambisols) sites, with blanket peat seemingly developing in some areas from Bronze Age times onwards, sometimes linked to erosion and colluviation (Carey et al., 2020). At Holwell, on Dartmoor, two Bronze Age reaves and a roundhouse were investigated (Carey and Hunnisett, 2019). This site at 375 m asl includes a buried soil below a recumbent "standing stone" and soils forming floor construction make-up for the roundhouse (Figure 4.6c). There is clear evidence of brown soils formed in the Pleistocene regolith; the latter includes "embedded grains" (see Figure 4.6b; Van Vliet-Lanoë and Fox, 2018). Weathered Bw horizons show biological homogenization by earthworms of the brunified fine soil, with later podzolization effects on the buried soil (Figure 4.6c). Moreover, the sequence beneath reave 2 at Holwell, includes a colluvial soil, providing clear evidence of earlier soil erosion and then pedogenesis prior to field division construction, demonstrating a pre-reave phase of landscape use and impacts (Figure 4.5c). Earlier studies of Dartmoor's Bronze Age landscape boundary features, such as reaves and associated buried soils, found that on slopes these reaves could bury a catenary sequence (Figure 4.7; see Figure 4.5) of:

- Blanket peat (bog; Histosols) on the plateau;
- Stagnopodzols ("peaty gleyed podzol"; Histic Placaquept, Humic Gleysol) on the better drained plateaux shoulders and slopes (Figure 4.7a); and
- Peaty gleys (Aquepts, Humic Gleysols) on less well drained lower slopes and peats (Histosols/ valley peats) in valley bottoms, and that these edaphic (soil-vegetation) environments were probably linked to a grazing land use (Figures 4.7b–4.7c) (Balaam et al., 1982).

Past land uses such as grassland for grazing, coupled with climatic change and climatic episodes exacerbated soil degradation tendencies. These hindered woodland regeneration, and decreased evapotranspiration rates, whilst in addition wetter weather increased a potential for surface water holding encouraged by grassland litter accumulation, hence the term for surface soil waterlogging – "stagno" in the resulting soils (Avery, 1990; Ball, 1975; Barrat, 1964). For example, The Short Dykes in upland (450–500 m asl) Mid-Wales are thought to be land boundaries similar to the Offa's Dyke, and are dated to the fourth to sixth centuries and sixth to eighth centuries AD. Buried soils and palynology record acidic, waterlogged surface peat formation (grassland hydromoor), likely linked to environment deterioration during the "Dark Ages" (or European Migration Period) (Blackford and Chambers, 1991; Büntgen et al., 2011; A. Caseldine, pers. comm.).

Box 4.2   (Continued)

Figure 4.6a   Field photo, Carn Brea Neolithic settlement Redruth, Cornwall, UK; sampling the soil buried by the third and outer rampart (Neo Rampart), consisting of the trampled, leached bA2, and bBhs and Bs illuvial podzolic horizons, and underlying bBs/Bw horizons where earlier pre-podzol earthworm working is recorded. The Holocene soil is developed on periglacial "growan" or weathered granite, and includes "embedded grains" relict of periglacial link-capping microstructures (Macphail, 1990a; Mercer, 1981).

Figure 4.6b   Photomicrograph of "embedded grain", a feature relict of periglacial link-capping that originally formed through freezing and thawing (see Box 4.3). Holocene pedogenesis, which predated upland soil podzolization, involved biological fragmentation of these earlier-formed microstructures, that often involved earthworm working. This example occurs in Bronze Age buried soils at Holwell, Dartmoor, Devon, UK (Carey and Hunnisett, 2019). PPL, frame width is ~4.62 mm.

Figure 4.6c   Scan of thin section sample M13 from buried soil 113 below a recumbent standing stone at the Bronze Age roundhouse at Holwell, Dartmoor (Carey and Hunnisett, 2019). The buried soil is composed of the Holocene earthworm-worked Bw acidic brown soil horizon (113Lower), with relict embedded grains of periglacial origin (REG; see Figure 4.5b). Upwards, very fine charcoal may record clearance before round house construction (113Upper). Later Holocene podzolization has also affected 113Upper producing a more pellety fine fabric of Bs horizon character. Moreover, sesquioxides are typically more concentrated in 113Upper (15.1–22.0% Al; 1.76–2.98% Fe) compared to 113Lower (13.4–13.9% Al; 0.94–1.13% Fe); the podzolized upper soil is also characterized by small concentrations of P (0.52%) and S (0.41%). Frame width is ~50 mm.

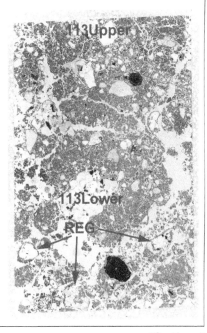

(Continued)

Box 4.2 (Continued)

Figure 4.7   Upland soil drainage catena (see Figure 4.5), with plateau shoulder (4.7a: Stagnopodzol), lower slope (4.7b: Gley) and valley bottom (4.7c: Peat) soils. Soils date to the Bronze Age landscape buried by Saddlesborough Reave, a ditch and bank boundary running across Dartmoor.
Figure 4.7a   Field photo Saddlesborough Reave-buried Ironpan Stagnopodzol with "growan" lower subsoil (~Histic Placaquept) on moderately drained plateau shoulder location; note reddish brown sesquioxide-enriched illuvial Bs subsoil.

Figure 4.7b   Field photo Saddlesborough Reave-buried Cambic Stagnogley, with "growan" lower subsoil (~Aeric Fragiaquept) on poorly drained lower slope location.

Figure 4.7c   Field photo Saddlesborough Reave-buried peat over E horizon (Humic Stagnogley; ~Humic fragiaquept; Humic Gleysol) in very poorly drained valley floor location.

Class 1 land for cultivation, where there are very minor or no physical limitations to use (land capability classification; Bibby and Mackney, 1972; Klingebiel and Montgomery, 1961). Moreover, access to rivers, lakes, and coastlines will permit the gathering of wetland resources, although through time landscape morphology may change due to uplift or sea level rise (Box 18.4).

Relief also strongly governs drainage and depth to groundwater, and their control over soil type. This important concept has been modeled in soil science as the *catena*: the lateral variation in soil profile types across a crest-hollow transect, for a given geological strata (Figure 4.7). Groundwater, which respects the slope of the catena, and which varies seasonally, reaches the surface in the valley bottom where it is expressed as a river, swamps/wetlands (Chapters 5–7). Where permeable strata overly impermeable rock or where the slope profile may coincide with the groundwater profile, a spring line, or boggy ground may occur. Archaeological mapping may detect a coincidence of settlements and specific sandy and sandy loam soil types (Ampe and Langohr, 1996; Fechner et al., 2004), and a spring line, while the existence of boggy ground may promote organic preservation (see Table 4.3).

4.4.5 Parent Material

Rock/sediment type, and even the kind of archaeological deposit present on a site, influences the way pedological processes act. Some parent materials, such as hard and poorly-weatherable acid igneous rocks, or sediments that are derived from deposits that have already been strongly weathered (e.g. glaciofluvial sands) tend to produce acid soils, (e.g. pH 3.5–5.5). On the other hand, soils on soft calcareous rocks like chalk give rise to base-rich alkaline (pH 7.5 to 8.5) soils, as these contain basic cations (Ca^{2+}, Mg^{2+}, K^+ and Na^+). These two extremes produce soils with different humus forms, levels of biological activity, and influence the way cultural and environmental materials may be preserved (Huisman, 2009) (see Table 4.3; Chapter 15). Soft clays are easily eroded, but commonly produce poorly drained ground.

Soil mineralogy is also an important consideration. Much of the coarse (sand and silt) fraction is composed of highly resistant quartz (SiO_2), with small amounts of feldspars and mica that are ultimately derived from igneous rocks. Sedimentary minerals like glauconite are more easily weathered and yield iron into the soil. Also present are the more rare "heavy" minerals (specific gravity >2.9), which can be separated from quartz (s.g. 2.65), and feldspar (s.g. 2.5–2.76) by heavy liquid or magnetic separation, or by centrifuging (see Chapter 17). Heavy minerals include iron oxides, sometimes termed opaque minerals because they are not translucent under the petrological microscope. Some examples are the magnetic mineral magnetite, hematite, and pyrite, although pyrite has often oxidized into limonite. There are over 30 translucent minerals, heavy minerals, such as the micas (biotite and muscovite), amphiboles, and pyroxenes, as well the so-called semi-precious minerals that include garnet and topaz (Chapter 2). Such mineral suites are used to assess the provenance of deposits, and the degree of weathering can be appreciated from the tourmaline:zircon ratio (Krumbein and Pettijohn, 1938). Minerals are now commonly identified on the basis of their chemical composition through SEM/EDS and micro-XRD techniques (Chapter 16). Layers of tephra or volcanic glass have been crucial in relating sediment, including peat sequences, and human events, such as the Landnám occupation of Iceland to specific historically dated volcanic episodes, as well Upper Paleolithic occupation at the Kostenki-Borshchevo sites (Don River Valley, Russia) (Holliday et al., 2007; Sigurgeirsson et al., 2013).

Clay minerals play an important role in governing the structure and fertility of the soil (Brady and Weil, 2008: 310 *et seq.*), and develop through the chemical weathering of rocks, with such rocks termed *saprolites*, and their development into soil materials (Stoops and Schaefer, 2018; Zauyah et al., 2018). They initially form as hydrated silicates of aluminum, iron, and magnesium

Figure 4.8 Field photo of plaggen soil, The Netherlands. Originally a podzol formed on sands, this has become over-thickened through addition of turf and organic manure generally from the dung heap in order to create a horticultural soil; in thin section, soils have a sand and Moder pellety organic microfabric; (Bakels, 1988; Mücher et al., 1990; Pape, 1970) (see Chapter 13); this moderately homogeneous soil profile can be compared to the strong horizonation displayed by podzols (see Figure 4.9c).

by rock weathering, and are termed phyllosilicates. These are layered minerals that are divided into single (1:1) layered (kaolinite-serpentinite group) and double (2:1) layered (e.g. smectite, vermiculite, mica, and chlorite groups); 1:1 clays are less active chemically than 2:1 clays. In addition, 2:1 clay types such as illite and montmorillonite also absorb more water and are known as swelling clays in Vertisols (Kovda and Mermut, 2018). The deep (>1 m) cracking of such soils (Yaalon and Kalmar, 1972) have the potential of markedly affecting vertical distributions of artifacts (Butzer, 1982). Clay types also reflect environment and the weathering/pedological history of landscapes. For example, kaolinite is associated with most weathered tropical soils – Ultisols, while the silicate mineral palygorskite forms under extreme arid conditions (Mees, 2018), but can also occur as an inherited mineral, for example from Cretaceous chalk. Allophane, an amorphous clay mineral, is typically formed in volcanic Andosols (Stoops et al., 2018).

Light sandy soils are easily worked and may have been the first to be cultivated in a region. Over the long term, however, they are often not suitable for sustainable agriculture without manuring, which adds nutrients and increases moisture retention; when over-thickened, these can be termed "plaggen soils" (Figure 4.8; see Chapters 11 and 13). An example of this fertility problem is found across Scandinavia, where Iron Age and later migrating populations needed to employ mixed farming, and in Norway, for example, only some 3% of the land is suitable for arable agriculture. Stock (e.g. cattle) provided the manure necessary to grow barley (Dreslerová et al., 2021; Engelmark, 1992; Myhre, 2004; Øye, 2009; Viklund et al., 2013). Another example is the Neolithic exploitation of the loess belt across north-west Europe (Bakels, 2009). The poorly drained heavy clay soils of Europe could not be cultivated until the Romans introduced ox ploughing using a metal ploughshare (Barker, 1985); the introduction of the efficient "swivel plough" likely spread from the Roman Rhinelands (Germany) to early medieval (10th to 11th century) England (Henning, 2009). In contrast, in the Far East, areas of poorly drained soils were purposely landscaped in order to grow rice using wet cultivation of paddy fields (as opposed to dry rice cultivation; Zhuang et al., 2014) (see Chapters 7 and 11).

4.4.6 Time

The length of time that a sediment or archaeological deposit is exposed to subaerial weathering controls the degree of alteration and the development of soil formation that can be expected. Time in itself is not a soil-forming process, but the passage of time permits the other factors to drive soil-forming processes. Estimating the numerical or chronometric age of a soil is notoriously difficult, but *relative dating* has been identified from some famous *chronosequences* where progressively older soils have been studied from marine and river terrace sequences and where glacial ice progressively retreated, for example in North America, and in central California (Birkeland, 1992; Harden, 1982). The major effect of time can be identified from soil maturity expressed in the formation of soil horizons. Short-lived weathering, termed "soil ripening" on reclaimed polders of Holland (Bal, 1982), produces immature A horizons (Entisols), whereas fully developed soils are recognized by their A, B, C horizons (Figures 4.1 and 4.2). As we have mentioned, soil classification is in fact based upon the identification of a number of well-defined or diagnostic soil horizons (Soil Survey Staff, 2014) (see Table 4.1).

There are several archaeological examples of attempts to "date" cultural phases and specific human activities by the maturity or immaturity of buried soils. A classic example of this is the analysis of an immature buried soil that marked a short-lived standstill phase in the construction of Caesar's Camp (Keston, Kent, UK; Cornwall, 1953:141 *et seq.*). The ephemeral surface became humus-stained in consequence of its being briefly vegetated (during a period of stasis) before the next construction episode. Rates of weathering and soil formation in building debris, modeled in post-Second World War Berlin and London, were utilized in dark earth formation investigations across Europe (Galinié et al., 2007; Macphail, 1994a; Sukopp et al., 1979) (see Chapter 13). Perhaps of wider importance for prehistory is the ^{14}C dating of humus present in the subsoils of podzols in Belgium and France, which effectively increased the age estimate of some podzols from an expected ~3,500 years bp (Bronze Age) to ~6–7,000 years bp (Guillet, 1982). At Hengistbury Head, one Bh horizon has a mean residence date of 1,700±90 BP, consistent with Late Bronze Age/Early Iron Age activity nearby; sandy regoliths with both Upper Paleolithic and Mesolithic finds were luminescence dated (Barton, 1992). Certain soil indices that reflect the age of soils have been suggested. These measurements include surface organic matter accumulation, leaching of $CaCO_3$, Bt (argillic horizon) development, a changing clay mineralogy, and red color (Birkeland, 1992). Soil-sediments (see above) and cumulative soils are nowadays, however, more likely to be dated employing OSL (Optically Stimulated Luminescence) (Fuchs and Lang, 2009; Yi, 2005) (see Figures 5a–5c).

The study of experimental earthworks has shown just how rapidly changes to soils occur after burial, within the first 32 years on both base-rich (chalk rendzinas) and acid (sandy podzols) soils (Crowther et al., 1996; Macphail et al., 2003b) (see Chapter 14). When the fills of ditches, graves, and pits are investigated, soil analyses can be applied to determine the rate of infill. Both slow infilling, e.g. "ditch silting" and rapid dumping to fill pits or graves, can be identified. Likewise, thorough working by biological activity of once heterogeneous deposits may mark a standstill phase or stasis (Box 4.1). When investigating buried soils, one must strive to answer crucial questions: "How long after a certain activity occurred was the site buried? Days? Weeks? Years?"; "Was the site: a) cleared by fire, b) ploughed, c) trampled, or d) razed (burned down)?" (e.g. Forget and Shahack-Gross, 2016; Kruger, 2015). These questions underscore the importance of studying modifications to a site through time after it has ceased to be occupied (Chapter 14).

Soil development as a proxy for age is useful in archaeological surveys. Using soil survey maps, one can avoid looking for "old" sites in an area with Entisols. On the other hand, such sites may be buried in sediments that are capped on the surface by well-developed soils.

4.5 Soil Profiles and Soil Properties

As evident from the above discussions, soil horizons are expressions of soil properties and processes (see Table 4.2) and are encompassed within a particular soil profile. Here we describe several soil properties, associated soil types, and ways that soils are classified.

Essentially, soil profiles have been categorized into soil horizons using the ABCD system of horizon designation (Bridges, 1970, 1990):

i) A or humose topsoil horizon,
ii) B or pedologically formed subsoil horizon (Figure 4.2),
iii) C horizon or weathered parent material and
iv) D, or more commonly the R horizon of consolidated bedrock.

A, B, and C horizons are often given a generic modifier (Table 4.1). For example, humose A1 topsoils can also be termed Ah horizons, while subsoil B horizons enriched with *sesquioxides* (aluminum and iron) are called Bs horizons (Bridges, 1990). In some soils the upper subsoil becomes leached and is commonly termed an A2, E, or Albic horizon. Those soils that have horizons with the subscript "g" (Ag, Bg, Cg), indicates that soils are mottled due to the effect of gleying ("hydromorphism"/oxidation-reduction). They are expressed in the field with ochreous, gray to blue-green colors or a pale-colored (G) horizon, which is completely iron-depleted (see Figure 4.7c).

Different countries have their own classification systems. For instance, in the UK, upper subsoils that have lost clay from downward percolating water enriched with leaf-leachate (translocation) are called Eb horizons, while those chemically depleted in sesquioxides are known as Ea horizons; both are termed A2 horizons in Europe and the USA (Avery, 1990). Fortunately now, there is broad agreement between Europe and North American soil classification systems. Both systems have *diagnostic horizons*, a term first employed in the USA, but now used internationally although with some differences (Avery, 1990; Soil Survey Staff, 2014; USDA, 2014); (see Table 4.1).

Diagnostic horizons, however, are *identified* only after both field and laboratory measurements have been carried out. For example, subsoils enriched in translocated clay are termed Bt ("argillic") horizons and relate to the translocation or "washing" of clay from the A horizons (including the A2 horizon) into the subsoil. Soils with a *diagnostic* argillic horizon are variously called Alfisols (Soil Survey Staff, USA), Luvisols (FAO, Europe), or argillic brown soils (Soil Survey of England and Wales) (Avery, 1980, 1990; Bridges, 1990; WRB, 2006). A further example is the poorly fertile, acid soil commonly called a podzol (see Figure 4.9c; Chapter 14). This has an Ah horizon over a leached upper subsoil (A2/E/Ea) horizon that can also be termed an albic horizon. This is present over a subsoil enriched in humus and/or sesquioxides (Bh/Bs) – termed a spodic diagnostic horizon. Hence these podzols are known as Spodosols in countries employing the American system (Soil Survey Staff, 2014; USDA, 2014). Famously, podzols in The Netherlands were amended (manured etc.) to form cultivatable plaggen soils (Figure 4.8).

Often, soil mapping is carried out using the identification of horizons although such identifications of diagnostic horizons are normally verified later through laboratory analyses (Chapters 16 and 17). For example, soil color, soil structure (peds/aggregates), amounts and type of roots, and approximate grain size (through "finger texturing") can be noted in the field. Later in the laboratory, pH, organic matter status (through organic carbon and loss-on-ignition – LOI analyses), fertility (Cation Exchange Capacity (CEC), N (nitrogen), K (potassium), and P (phosphorus)) and accumulation of sesquioxides (Fe_2O_3 and Al_2O_3), can all be measured. An important characteristic of soils, is their *base status*, as discussed later, and essentially refers to whether a soil is *acid* (pH <6.5) or base rich – alkaline (pH >7.5), which again effects their fertility and what type of humus they develop (*Mull, Moder*, or *Mor* – see 4.3.2; Table 4.1). Of particular interest to archaeologists is the amount and type of phosphate in soils. For example, occupation concentrates phosphate from the release of P from ashes, dung, bones, and burials, for example, and phosphate analysis is one of the oldest geoarchaeological methods (Arrhenius, 1955; Eidt, 1977; Faul and Smith, 1980; Holliday and Gartner, 2007; Sjöberg, 1976) (see Chapters 17 and 18).

Lastly, we note that different types of A and B horizons (see Table 4.2) are formed by *soil-forming processes* (pedogenesis) (see 4.3). Researchers in north-west Europe, for example, should be aware that podzols/Spodosols are developed by *podzolization*, while Alfisols/Luvisols/argillic soils are formed through clay translocation that sometimes is still known by its French name for "washing" – *lessivage*; French speaking soil scientists describe their soils as *sols lessivés* (Duchaufour, 1982; Soil Survey Staff, 2014) (Figure 4.2).

4.6 Important Soil-forming Processes

The recognition of soil types in the field and from laboratory data is essential if correct soil classifications and accurate interpretations of landscape history are to be made. More importantly, the soil-forming processes that govern soil type have to be understood if we are going to be able to comprehend past human environments, plant, animal, and human interactions, site formation processes, and preservation of sites and their artifacts.

4.6.1 Weathering

Soil formation and the development of horizons begin with weathering of the parent material. It is well known that weathering is influenced by temperature, warm conditions favoring chemical weathering (see Figure 4.1).

Physical weathering can take place under conditions of heating and cooling, or salt crystal growth in hot, arid conditions, or freezing and thawing of water in rocks, in humid conditions. In the latter case, temperatures alternate around the freezing point, and rocks broken up by splitting form scree deposits that accumulate along talus slopes (see Box 4.2 Cold Climate Soils). *Cryoturbation* involves physical mixing of the upper regolith (e.g. rotation, horizontal and vertical movement) by differential freezing and thawing, commonly in the seasonally freezing and thawing "active" zone above the frozen substrate or permafrost zone (Gelisols; Table 4.3; Box 4.2). Commonly on slopes, material in this active zone moves down slope as gravity-flows, sometimes as turf-confined solifluction lobes. Where both solifluction and meltwater act, the deposits can be termed as soliflual (Catt, 1986, 1990). At both open-air and cave Paleolithic sites, deposits (including hearths) can be weakly to strongly influenced or disrupted by cryoturbation (Bertran, 1994; Bertran et al., 2010; Bertran et al., 2019; Goldberg, 1979b; Goldberg et al., 2003, 2019; Macphail et al., 1994c; Van Vliet-Lanoë and Fox, 2018; Van Vliet-Lanoë, 1998). Commonly, rock-falls in caves (*éboulis*) reflect major cold episodes in the Quaternary (Courty et al., 1989; Laville et al., 1980; Macphail and Goldberg, 1999).

Table 4.3 Soil characteristics and preservation of artifacts and ecofacts in soils.

Conditions	Parent material	Soil type	Humus form	Soil fauna	Preservation potential	Comments/Other
Acid pH <3.5–6.5	e.g. quartz sand; schist	Ranker; Podzol (e.g. Entisols and Spodosols)	Mor and Moder	e.g. Mites, Enchytraeids, and non-burrowing earthworms	Pollen; macrobotanical material; phytoliths, soil diatoms	Loss of bone; possible damage to pottery; corrosion of iron (loss of magnetic susceptibility signal)
Neutral pH 6.5–7.5	e.g. loams	Brown earth, Luvisols (e.g. Mollisols and Alfisols)	Mull	Collembola, Arionids (slugs) and burrowing earthworms (Lumbricids)	Bone, mollusks, phytoliths, soil diatoms; some macrobotanical material	Pollen is oxidized; large-scale mixing by meso and macrofauna; charcoal can become fragmented; wood remains lost to fungal attack.
Alkaline pH >7.5	e.g. shell sand, chalk, salt lake sediments	Rendzina; Solonetz (e.g. Xerolls and Xeralfs)	Mull	As above but becoming restricted	Mollusks; bone; phytoliths (except for extreme >pH 8)	Pollen is oxidized; salt, carbonate, and gypsum crusts on pottery and other artifacts; large-scale mixing by meso and macrofauna, and fracture by secondary crystal growth.
Waterlogged	e.g. peat, estuarine sediments	Peat (Histosols)	Peat	None	Pollen, mollusks, diatoms, most organic materials, including fragile insects, skin, leather.	Corrosion (loss of magnetic susceptibility signal from iron slags when fluctuating oxidizing/reducing conditions); soft-tissue "pickling" (bog man); secondary mineral formation when water tables fluctuate (gypsum, jarosite, pyrite, siderite, vivianite)

NB: Caveats

1) Fluctuating water tables (i.e., oxidation/reduction cycles) destroy pollen and "rust" iron slags.
2) High biological activity mixes stratigraphy and displaces artifacts
3) Possible serendipitous preservation of strongly residual iron slag, burned soil, fused fly-ash, phosphatized materials
4) Further examples can be found in Huisman (2009). NB: gypsum ($CaSO_4 \cdot H_2O$), jarosite ($KFe_3(OH_6)6(SO_4)_2$), pyrite (FeS_2), etc., see Bullock et al. (1985: table 6.2).

Chemical weathering includes the decay and alteration of minerals, as well as the breakdown of soil organic matter (see below). Typical geochemical processes include hydrolysis, related oxidation and reduction, and solution. As noted by (Duchaufour, 1982: 7), "hydrolysis, or the effect of water containing such entities as Hydrogen ions (H^+) is the most important process of weathering" on minerals. For example, additions of Hydrogen ions and water with feldspars produce clay minerals through hydrolysis (Birkeland, 1999). H^+ can be supplied from atmospheric sources ($CO_2+H_2O \rightarrow H_2CO_3$) and organic acids from plants. In waterlogged/hydromorphic conditions, iron as Fe^{2+} can be lost in solution, and this process is called reduction. The formation of the diagnostic cambic Bw horizon (weathered B horizon of a Cambisol) (USDA, 2014) (see Table 4.2) is expressed by the subsurface neoformation of clays, the replacement of geological bedding by soil structures, removal of carbonates, and development of colors redder than the parent material because of the formation of hydrated iron oxides.

4.6.2   Leaching and Clay Eluviation

Weak carbonic acid forms from the mixing of rainwater and dissolved CO_2: $CaCO_3 + CO_2 + H_2O \leftrightarrow Ca^{2+} + 2HCO_3$. These solutions, along with organic acids derived from plants can weather calcareous materials, as well as human-made lime-based plaster and mortar. They also can leach base cations Ca^{2+}, K^+, Mg^{2+}, and Na^+ by replacing them with H^+. Soil reaction becomes acid (pH <6.5). Soil that was well flocculated because clay was bonded by Ca^{2+} becomes more easily disaggregated and the process of pedogenic clay translocation can be induced. Pedogenic clay translocation is much less likely in calcareous, base-rich regimes (pH 7.5–8.0), and should not be confused with the physical inwash of calcareous fine particles caused by sedimentation and trampling, for example. These non-pedogenic clays are usually coarser, more poorly sorted, and can have fine microcharcoal inclusions.

In addition to measuring pH, base-rich versus acid status of a soil can be deduced from its vegetation, surficial humus type (Figure 4.13), and associated mesofauna. For instance, a Mull topsoil (Barrat, 1964) has only a litter (L) layer (Table 4.1), because biological breakdown of organic matter produced by plants, such as leaf litter, is rapid (C:N of 5–15) (see 4.4.3). Such soils often have a neutral pH (6.5–7.5), and in western temperate countries, earthworms such as *Lumbricus terrestris* make up the main part of the large invertebrate biomass in such soils. Termites and millipedes also thrive in neutral to base-rich soils. Plants requiring Ca, termed calciphyles (e.g. Garrigue in the Mediterranean, and specific chalk and limestone grassland communities in the UK), are typically present (Polunin and Smythies, 1973; Rodwell, 1992). *Mollisols*, the soils of grassland and prairies, have characteristically high levels of biological activity. This is recorded at the micro scale by what is termed a "total excremental or biological fabric", where mm-sized mammilated organo-mineral excrements dominate. This Mull humus (L) topsoil may also carry a population of moles who feed on earthworms, as well as voles and prairie dogs who eat grass roots in Mollisols (Table 4.2). Large grazing animals fertilize these soils with their dung, and this activity may be recorded by the presence of dung beetles and their burrows; hence the importance of insect studies in archaeology to infer pastoralism (Campbell and Robinson, 2007; Robinson, 1991).

It can also be noted that Collembola (Springtails) are more numerous above pH 5.4 (Figure 4.13). If pH decreases, earthworms, which are strong mixers of soil and decomposers of organic matter in and on the soil, die out, and Enchytraeid worms become more common below pH 4.8 (Mücher, 1997). Mineral soil becomes much less mixed upwards from less weathered subsoil horizons, because Enchytraeid worms are concentrated in the topmost few cm of the soil. Organic matter starts to accumulate on the soil surface, and thus more plant acids become available to leach out bases. Overall leaching is thus exacerbated, and bases are lost to illuvial subsoils and/or removed by groundwater ultimately into the fluvial system.

The surface organic matter thickens with the development of a fermentation (F) layer that is composed of mesofauna excrements and comminuted plant remains (Kooistra and Pulleman, 2018). This LF horizon is known as a Moder humus and is associated with Enchytraeids, insects, and their larvae; it is most commonly linked to broadleaved forests with a pH of about 4.5 and a C:N of 15–25 (Babel, 1975). The most commonly associated soils are Alfisols (Figure 4.2; Table 4.2), which are characterized by pedogenic clay eluviation or clay washing (*lessivage*) from the E (USA)/Eb (UK) or A2 horizon, with clay translocation into the underlying clay-enriched subsoil Bt horizon. This then constitutes the "diagnostic" argillic horizon. This horizon exhibits the presence of translocated clay in the form of clay coatings (Kühn et al., 2018) that can be seen in the field but which are more clearly visible in thin section. Past forest soils, either from the Holocene or from interglacial (temperate) episodes within the Pleistocene, have been identified by the presence of such argillic horizons (Cremaschi et al., 2018; Kemp et al., 1994). All horizons that seem to be argillic in character, however, are not necessarily relict of a forest soil development (Brammer, 1971; Goldberg et al., 2001; Kühn et al., 2018). Pre-Quaternary, Pleistocene and Holocene (e.g. archaeologically buried) paleosols have been studied (see 3.1), but it is obvious that long burial leads to diagenetic transformations of original soil features. Experimental archaeology has in fact demonstrated how quickly some changes occur in buried soils (see Chapter 14).

4.6.3 Podzolization

On substrates that contain few weatherable minerals and are freely draining, such as glacial outwash or decalcified dune sands, leaching can rapidly cause the development of acid soils. At increasingly acid pHs, clay is destroyed and Al (aluminum) is released. In such environments (pH 3–3.5), acidophile plants dominate and commonly include *Ericaceae* (heath plants); in boreal conifer forests more rapid podzolization can take place: iron pan formation, for example, can take place in a decade in newly planted areas with high rainfalls. These types of plants produce leachates (e.g. chelates) that chemically combine organic matter and sesquioxides (Al_2O_3 and Fe_2O_3), thus making them water-soluble and mobile. This process of destruction of clay and translocation of sesquioxides and humus, from a leached Albic E horizon to a diagnostic spodic (Bh/Bs) subsoil horizon, is termed podzolization (Figure 4.9). In thin section, the E horizon is purely made up of quartz sand with very small amounts of organic matter. The spodic horizon is composed of two microfabrics, recently formed polymorphic (pellety) material and very much older monomorphic (grain coatings and cement) humic gels of Fe and Al (Van Ranst et al., 2018). The hard, cemented spodic horizons of podzols are therefore easy to recognize both in the field and under the microscope. Sometimes cementation results in localized impedance of drainage and waterlogging (hydromorphism) causes an ironpan (Bf) to develop because Fe^{3+} is converted to mobile Fe^{2+} (see Box 4.2). Soil phosphorus also accumulates in spodic horizons, and phosphate surveys usually sample this horizon to examine ancient patterns of occupation in present-day areas of podzols (i.e., patterns of accumulated P that have been leached from the soil surface for example).

The form of organic matter accumulation that develops on podzols is termed a Mor humus and comprises L, F, and H (humus) layers (see Table 4.1). The H layer is typically composed of amorphous organic matter, and takes a longer time to form compared to the L and F horizons because all recognizable plant material and excrements have been decomposed by bacteria. Under normal conditions of high acidity, e.g. pH 3.5, and high C:N ratio (>25), this form of surface humus accumulation may include organic matter that is over a thousand years old. Extant Holocene spodic horizons yield dates as old as 3,000 years that include both ancient and modern humus, providing what are termed "mean residence dates" (Guillet, 1982). Other ratios can also be of interest. Phosphate fractionation produces a ratio between inorganic and organic phosphate, which can be of note because mineral phosphate such as bone and ashes will have ratios of ~1.0 (see Crowther, 1997, 2007; Macphail et al., 2000); organic matter-rich materials characterized by dung and soil humus produce a Pratio of >1.0. In the latter case, dung-manured agricultural fields have enhanced Pratios (or "PQuota") of ~2.0–4.0, while humic topsoils can achieve Pratios of >20 as measured in examples of buried woodland Ah horizons identified through palynology; organic carbon and Pratios were also found to be statistically correlated (Engelmark and Linderholm, 1996; Macphail, 2011a).

Podzols are typical of heath and coniferous woodland on poor substrates of the mid-latitudes, and across much of the boreal and taiga regions. When buried, for example by Bronze Age barrows, these soils provide useful stratified pollen sequences because of their acidity and slow organic matter breakdown. Dimbleby and others (Dimbleby, 1962; Scaife and Macphail, 1983), showed how heathlands and moorlands formed in once wooded landscapes. Such change is demonstrated by the presence of a relict argillic (Bt) horizons overprinted by spodic horizons. This soil degradation sequence, forest Alfisol → heathland Spodosol, demonstrates an important soil transformation (Gebhardt, 1993; Macphail, 1992b). Equally, because of very low levels of biological mixing, pollen studies of surface horizons in the present-day coniferous woodlands of central Sweden, revealed evidence of cereal pollen from fields that had not been cultivated for 1,000 years (Segerström, 1991).

4.6.4   Calcification, Salinization, and Solodization

In warm and dry regions, or where seasonal precipitation is strongly outweighed by evapotranspiration, soluble salts, such as Ca^{2+}, Mg^{2+}, Na^+, and K^+ that are lost by leaching from more humid soil areas, become concentrated; they occur in a group of soils called *aridisols*. Irrigation and terracing has permitted their agricultural use in the past, however (Sandor et al., 2021). Such soils are typified by the sparse growth of plants that are drought resistant and sometimes salt tolerant, and include grasses and cacti, with mesquite, creosote bush, yucca, and sagebrush in south-west USA and tamarisk in Asia (Phillips and Comus, 2000) (see Chapter 8). Levels of organic matter and biological activity reflect precipitation and can be very low, although a mollic epipedon (dark-colored humic and bioactive topsoil) can be present in prairie areas (Table 4.2). Here, for example, calcic (Bk) and petrocalcic (cemented Km) diagnostic horizons form through the precipitation of calcium carbonate in subsoils; the depth to the $CaCO_3$ enriched horizon increases with increasing precipitation, resulting in the formation of calcretes (e.g. Durand et al., 2018) (Figure 4.9e).

Even in temperate areas, groundwater may supply ions that are drawn up-profile by plant roots, and which are precipitated as root pseudomorphs or rhizoliths, or as nodules within the soil. Past climatic fluctuations, effecting major fluctuations in ground water, for example, have been revealed from the investigation of calcite nodules in semi-arid regions of India (Courty and Fedoroff, 1985). In the Negev and Sinai desert, secondary calcium carbonate precipitation, some associated with archaeological sites, has proven a useful resource for dating and climatic inference (Amit et al., 2007; Amit et al., 2011; Goldberg, 1977a; Itkin et al., 2018; Magaritz et al., 1987; Magaritz et al., 1981; Roskin et al., 2013a; Wieder and Gvirtzman, 1999).

Gypsum (Table 4.1) accumulations (gypsic horizon, By) have proven to be both a useful indicator of past arid conditions and very damaging to archaeological stratigraphy and materials

General process	Chief characteristic and result	*Horizon*
Homogenization and weathering (sometimes with leaching or accumulation)	Biological activity, organic matter accumulation, weathering sometimes with mineral accumulation or loss.	A
Homogenization, weathering, mineral formation (neoformation)(sometimes with leaching or accumulation)	Often breakdown of primary minerals and formation of new ones; biological mixing; vertical additions from other horizons and/or losses to other horizons; vertical or lateral additions or subtractions by groundwater or throughflow.	B
Weathering (see Table 3.b)	Weathered, poorly to unconsolidated parent material (sometimes with minor losses and additions – as B horizon above)	C
	Unweathered hard Bedrock	R

Some general processes	Some characteristics and results	*Horizon*
Ripening (oxidation of organic matter and minor weathering; e.g., decalcification, iron mobilisation) through to homogenization and mineral breakdown	Minor loss of carbonate, iron staining, incorporation of humus through to total biological mixing (biological microfabrics), organic matter incorporation and mineral weathering	A
Mineral breakdown and neoformation	Organic matter incorporation, clay formation, release of iron etc., biological mixing (biological microfabrics)(haploidization)	Bw

(a)

Leaching, clay translocation, and podzolisation

Process	Chief characteristic and result	Horizon
Leaching ▼ ▼ ▼ ▼	↓ ↓ ↓ Leaching and translocation of clay, humus, sesquioxides (Fe_2O_3-Al_2O_3)	A2 (Ea)
Illuviation ⊥⊥⊥⊥	• Clay enriched • Humus enriched • Sesquioxide enriched	• Bt • Bh • Bs

Podzol

(b)

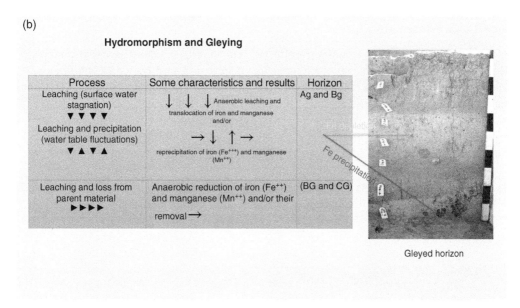

Hydromorphism and Gleying

Process	Some characteristics and results	Horizon
Leaching (surface water stagnation) ▼ ▼ ▼ ▼	↓ ↓ ↓ Anaerobic leaching and translocation of iron and manganese and/or	Ag and Bg
Leaching and precipitation (water table fluctuations) ▼ ▲ ▼ ▲	→ ↓ ↑ → reprecipitation of iron (Fe⁺⁺⁺) and manganese (Mn⁺⁺)	
Leaching and loss from parent material ►►►►	Anaerobic reduction of iron (Fe⁺⁺) and manganese (Mn⁺⁺) and/or their removal →	(BG and CG)

Gleyed horizon

(c)

Calcification and Alkalization

Process	Some characteristics and results	Horizons
Calcium carbonate movement (vertical) ▼ ▼ ▼ ▼ (throughflow) ►►►► and accumulation ⊥⊥⊥⊥	Precipitation of calcium carbonate in voids, impregnating the soil as concretions and root pseudomorphs (rhizoliths) (calcretes)	Calcic Bk and Km
Sulphate (SO₄⁻) movement (vertical) ▼ ▼ ▼ ▼ (throughflow) ►►►► and accumulation ⊥⊥⊥⊥	Precipitation of gypsum (CaSO₄) often as lozenge shaped crystals within voids or within the soil matrix, which then can become fragmented.	Gypsic By
Vertical movement and precipitation of Cl⁻ and Na⁺ ▲ ▲ ▲ ▲ (throughflow) ►►►► and accumulation	Surface salt crystal (halite) efflorescence (salt crust) and subsoil accumulations (halite is not normally preserved in thin sections)	Natric Btn Salic Bz

Calcium carbonate nodules

Figure 4.9a, b, and c   Selected types of soil processes (simplified).

(Goldberg, 1979a; Poch et al., 2018). In the latter case, the growth of gypsum crystals can substantially fragment mudbricks for instance (Goldberg, 1979a). Gypsum has also been shown to be an important constituent of the soils present in the raised fields of Maya sites in Mesoamerica, and the process of gypsum crystal growth has been cited as a pedological mechanism contributing to the present height of these raised fields (Jacob, 1995; Pohl et al., 1990). Gypsum is also a common secondary mineral in many intertidal soil sites, and like pyrite, is linked to the breakdown of organic matter (Mees and Stoops, 2018).

Box 4.3   Cold climate soils

It is important to be able to recognize the effect of cold on soils, sediments, and archaeological deposits and materials. In the first place, cold climate processes, e.g. frost wedge infilling, hummock and hollow, and stone stripe patterned ground formation, can produce natural features that archaeologists should not confuse with anthropogenic features (Bertran et al., 2010). At the micro scale, textural pedofeatures such as dusty and impure void coatings formed by freezing and thawing, must not be mistaken for similar features resulting from cultivation and trampling (Figures 4.10a–4.10c) (Deák et al., 2017; Goldberg, 1979; Rentzel et al., 2017; Van Vliet-Lanoë and Fox, 2018). Finally, fragmentation and mixing of archaeological deposits by cryoturbation may give the impression that a dump is present, whereas the current deposits simply record a reworked occupation floor/combustion zone, for example (Karkanas and Goldberg, 2018) (see Figures 4.11a–4.11d).

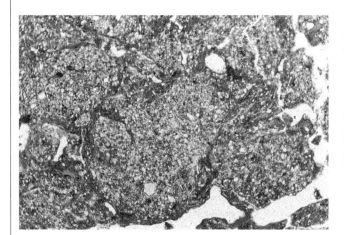

Figure 4.10a   Photomicrograph of Late-glacial hollow infilling, associated with Upper Paleolithic long blade culture at Underdown Lane, Kent, UK (Gardiner et al., 2016). Here, soil-sediments have become fragmented by frost action and have been transported downslope into a hollow. The clayey infillings and coats indicate soliflual conditions. PPL, frame width is ~3.5mm.

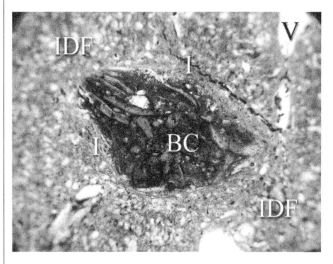

Figure 4.10b   As Figure 10a, photomicrograph of ferruginized bone cluster (BC) from a layer rich in such cemented aggregates; a probable regurgitation pellet from a raptor/or small carnivore scat (Macphail and Goldberg, 2018a: 262–263; cf. Andrews, 1990). Water-saturated (soliflual) sediments include iron depleted microfabrics (IDF) and intercalatory textural features (I) surrounding the "embedded" bone cluster; voids (V) are also present. PPL, width is ~1.75 mm.

Box 4.3   (Continued)

Figure 4.10c   Scan of M1 1995 (Hyena Den Cave, Wookey Hole, Somerset, UK); Laminar/lenticular structured poorly sorted silt to gravel-size deposit, formed by wash and freeze-thaw. Frame width is ~50 mm.

Fortunately, there are numerous classic studies of cold climate geomorphic field and microscopic features, from Quaternary paleosols and currently cold regions of the world (Fedoroff and Goldberg, 1982; FitzPatrick, 1956; Kemp, 1985b; McKeague et al., 1973; Romans and Robertson, 1974; Van Vliet-Lanoë, 1985; Van Vliet-Lanoë, 1998; Van Vliet-Lanoë and Fox, 2018; Van Vliet, 1982), which include features such as "embedded grains" (Bullock and Murphy, 1979) (see Figures 4.6b, 4.12c and 4.12d). The field handbooks by Catt are also important in the investigation of cold climate deposits within the Quaternary (Catt, 1986, 1990). Many papers also relate to the deposition of Pleistocene loess (Kemp et al., 1994; see review in Mücher et al., 2018) and cave sediment formation (Macphail and Goldberg, 1999) (Chapter 10). The disruption of occupation surfaces and reworking of combustion zones (Macphail et al., 1994c), the movement of artifacts by cryoturbation (Bertran and Texier, 1990; Hilton, 2002), and paleosol formation (Cremaschi et al., 2018; Fedoroff et al., 2018) are also topics of interest. As examples, the Mid-Pleistocene early Paleolithic site of Boxgrove, UK (see Chapter 7) and some Late Glacial Upper Paleolithic occupations are cited.

In the UK, an Anglian Period (Isotope stage 12) cold stage soil developed under very cold and dry conditions, as characterized by blown sand infilled ice wedges at Barham, East Anglia (Kemp, 1986). At Boxgrove, West Sussex the cold climate terrestrial slope deposits that bury the early hominid (*Homo heidelbergensis*) coastal land surfaces are formed out of periglacial flint and chalk pellet gravels and brickearth deposits (Roberts and Parfitt, 1999; Stringer et al., 1998). These deposits were thought to broadly date to the Anglian Cold Stage, but this "dating" correlation is hindered by the fact that the "soils" within these sediments show deposition under soliflual – cold and *wet* conditions, not cold and dry ones (Macphail, 1999a). Conditions in the

Box 4.3 (Continued)

Anglian were also supposed to be so harsh that all hominid (*Homo erectus*) populations went south. At Boxgrove, however, refitting flints occur within these deposits, in horizons where ephemeral temperature amelioration (temperate interstadials) led to biological activity – rooting, and soil mixing by earthworms – showing that human populations had not simply disappeared (Roberts and Parfitt, 1999). This site thus records the formation of a cold climate slope deposit where high-energy gravels are interbedded with silty soil-sediments formed under lower energy deposition (Figures 4.11a – 4.11d); earlier sediments were fragmented into granules by cryoturbation, and ensuing meltwater caused the inwash of clay into voids. Mass-movement chalk gravels (Unit 8) also surprisingly contain an *in situ* refitting flint scatter, the surrounding soil matrix recording both the temperate conditions of an Interstadial and renewed cold climate effects. Such microfabrics may well also show seasonal activity within an Interstadial.

Soils and deposits that date to the Late-glacial (see Table 4.4) can be associated with Final Paleolithic artifacts across Europe. Animal bones of lemmings and reindeer indicate cold climate conditions, whereas the presence of red deer has been used to identify a temperate Interstadial. Small mammal bones from the regurgitation pellets of raptors (birds of prey) have also been extensively used to reconstruct paleoclimates (see Figure 4.10b) (Andrews, 1990). In many open-air sites, the nature of the soil-sediments containing artifacts can be enigmatic because of post-depositional Holocene biological working and homogenization. Hence, Final Upper Paleolithic artifacts are found to span enigmatically the Pleistocene-Holocene Boundary (Table 4.4).

In southern UK, the mollusk, insect, and soil-sediment characteristics from a number of lower slope and valley bottom Allerød soils have been studied during deep excavations for the

Figure 4.11a Field photo; Boxgrove, GTP 25; terrestrial slope solifluction deposits (here the "chalk pellet gravel" – Unit 8) burying the early hominid coastal landscape (here the beach); top of the ladder marks the *in situ* flint scatter.

Box 4.3 (Continued)

Figure 4.11b As Figure 4.11a, *in situ* refitting flints within the chalk pellet gravel; image is ~30 cm across.

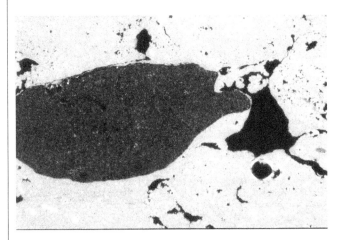

Figure 4.11c As Figures 4.11a–4.11b, photomicrograph of terrestrial slope solifluction deposits; flint artifact within the "chalk pellet gravel" (Unit 8). The *in situ* flint scatter is set in a pedological microfabric composed of chalk and micritic chalk soil. A later phase of freezing and thawing has induced depositional inwash of colloidal chalk, coating some pores and forming a compact soil-sediment. XPL, frame is ~5.5 mm.

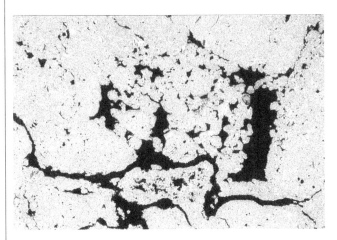

Figure 4.11d As Figures 4.11a and 4.11b, photomicrograph of terrestrial slope solifluction deposits, showing probable mammilated earthworm excrements demonstrating biological activity that was likely contemporary with occupation at this location. This microfabric probably marks an Interstadial during the formation of the generally cold climate Unit 8. XPL, frame is ~ 5.5 mm.

Box 4.3 (Continued)

US 101

US 108

US 120

US 102

US 103

Figure 4.12a Field photo of Azilian site of Rocher de l'Impératrice, Plougastel-Daoulas, Finistère, Brittany, France; US 108 – coarse rocky mass-movement deposit that is believed to pre-date the Neolithic occupation of the rock shelter; US 120: solifluction deposits with imbricated stones with link cappings – Older Dryas/Bølling paleosol (Late Upper Paleolithic Azillian occupation); US 102: more clayey solifluction deposits with banded fabrics – Older Dryas/Bølling paleosol (Late Upper Paleolithic Azillian occupation); US 103: possibly Last Glacial Maximum sediments. (Photo by Marine Laforge, Éveha – Brest Géomer, and Nicolas Naudinot, Université de Nice Sophia-Antipolis and Université de Rennes 1).

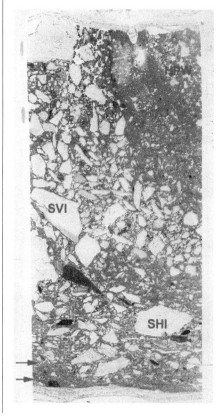

Figure 4.12b Scan of M3/3 Rocher de l'Impératrice, Plougastel-Daoulas, Finistère, US 120–102; diffusely layered basal deposits of lenticular silt loam containing charcoal (arrows) of US 120, overlain by sub-horizontally oriented imbricated stones and gravel (SHI), subvertically oriented imbricated stones and gravel (SVI – talus deposit; US 108), and "hollow" that is pedologically worked (HPW). Frame width is ~65 mm.

Box 4.3   (Continued)

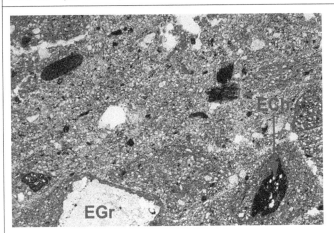

Figure 4.12c   As Figure 4.11b; Photomicrograph of M3/3 (base); compact lenticular structure with embedded grain (EGr) and charcoal (ECh). PPL, frame width is ~4.62 mm.

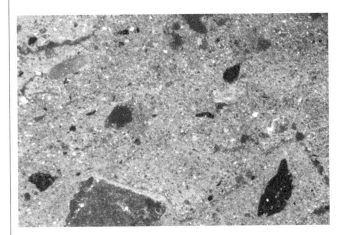

Figure 4.12d   As Figure 4.11d, under oblique incident light (OIL); note iron depleted microfabric and possible burned silt-size mineral material.

Channel Tunnel (Preece and Bridgland, 1998). Such Allerød soils were recognized in the field from their dark humic character, but when workers began studying them at the micro scale they discovered that these "soils" resulted from very "disturbed" conditions that had led to mass-movement (Preece, 1992; Preece et al., 1995; Van Vliet-Lanoë et al., 1992). It can be suggested that these are not Allerød *soils* but Upper Dryas *sediments*. A series of samples from White Horse Stone, Kent, UK (Macphail and Crowther, 2006a) – a site excavated along the Channel Tunnel Link to London – show: a) the bedded character of the Allerød soil', (b) the typical nature of soliflucted (Lower Dryas) deposits, (c) a soliflual chalky mud flow layer *within* the Allerød soil and the character of the Allerød soil, as chaotic mixture of reworked humic (rendzina) soil clasts. The Allerød *soil* here is thus probably an Upper Dryas *sediment*; a stratigraphic reconstruction of the Late-glacial "soils" at White Horse Stone is presented in Table 4.4 (Macphail and Crowther, 2006a). On the other hand, *in situ* Allerød soils have been recorded as

(Continued)

Box 4.3 (Continued)

Table 4.4 White Horse Stone, Kent, UK (Channel Tunnel Rail Link): summary of late-glacial stratigraphic history based upon soil micromorphology (Macphail and Crowther, 2006a).

Period	WHS and Pilgrims Way valley floor	Local hypothetical valley slopes etc
Loch Lomond stadial/"Younger Dryas" Phase 3	Cool climate downslope solifluction and "down-valley" soliflual accretionary deposition of chalk gravel-rich calcareous sediment and *seasonal* formation of very immature calcaric lithosols; *seasonal* post-depositional rooting and burrowing, but little evidence of earthworms.	Inferred cool climate erosion
"Younger Dryas" phase 2	Cool climate solifluction deposition of inferred valley slope (Allerød) soil fragments; *seasonal* post-depositional rooting and burrowing, but little evidence of earthworms; coarse solifluction sedimentation interspersed with likely down-valley soliflual deposition of calcareous mud(s) containing chalk gravel.	Inferred cool climate erosion
"Younger Dryas" phase 1	Cool climate down-valley erosion of any *in situ* Allerød soils. (HIATUS)	Inferred cool climate erosion
Allerød (Windermere Interstadial)	No preserved deposits or soils. (HIATUS)	Inferred temperate humic and brown rendzina formation, earthworms present; possible traces of human activity (fine charcoal)
"Older Dryas"	Cool climate downslope solifluction and "down-valley" soliflual accretionary deposition of chalk gravel-rich calcareous sediment and *seasonal* formation of very immature calcaric lithosols; *seasonal* post-depositional rooting and burrowing, but little evidence of earthworms.	Cool climate erosion

thin Entisols in England, for example as a humic ranker (Westhamptnett, West Sussex) and as a pararendzina on Lower Dryas scree at King Arthur's Cave, Herefordshire; in both cases microscopic bone and fine charcoal are present (ApSimon et al., 1992; Macphail, 1995; Macphail et al., 1999).

King Arthur's Cave recorded an important Final Paleolithic occupation contemporary with the soil (Barton, 1997). Similarly at Underdown Lane, Kent (Macphail, 2004), a Final Upper Paleolithic artifact assemblage is associated with very wet and disturbed Late-glacial soil-sediments formed out of Late Pleistocene silty brickearth, where bone clusters possibly testify to the earlier presence of raptors and where burned flint was moved alongside the fragmentation and re-deposition of probable humic ranker topsoils (see Figures 4.10a and 4.10b). At Hyena Den Cave, near Wookey Hole Cave, Somerset, both the bones and copro-lites of hyenas were found, and here some sediments clearly display typical lenticular and laminar fabrics due to wash and freezing and thawing (Figure 4.10c) (Jacobi, undated; Macphail, 2006).

Box 4.3   (Continued)

At Rocher de l'Impératrice, Plougastel-Daoulas, Finistère, Brittany, France (Figure 4.12) engraved schist tablets date to an early Azilian culture, and deposits show the effects of cold climate on cultural horizons; dating to 14,000 bp, the occupation coincides with the end of the Oldest Dryas and Bølling climatic periods and records various phases of solifluction (Naudinot et al., 2017). These effects also impact combustion features, just as at Arene Candide, Liguria, Italy (Macphail et al., 1994c; Rellini et al., 2013). Charcoal at Rocher de l'Impératrice can be finely fragmented and dispersed along with very fine rubified (burned) mineral material, in what were water-saturated reworked rock shelter sediments; these show microfeatures of elutriation and iron-depletion. Some cultural deposits show only very localized combustion zone dispersion, however, with charcoal being sometimes present as embedded grains (Figures 4.12c and 4.12d). In addition, deposits show macro-features of stone imbrication and banded fabrics, which are alternating elutriated layers, and iron and clay-enriched layers. The site as a whole was also affected by other complicating site formation processes, including rock fall talus deposition and Holocene podzolization, the illuvial effects of which can penetrate deeply along tree roots into the Late-glacial sediments (Naudinot et al., 2017).

Saline soils, *sensu lato*, are formed mainly in arid steppe and semi-desert areas where soluble salts are transported from high ground or in ground water from a source of sodium salts, and which are then concentrated in low-lying ground (Mees and Tursina, 2018). Amounts of salt are measured in terms of the ratio between Na^+ and total exchange capacity (CEC). A natric horizon, for example, contains more than 15% of its CEC (Cation Exchange Capacity) as Na^+ (Brady and Weil, 2008: 418–420). A salic horizon is simply one that is enriched with salts (e.g. halite – NaCl) more soluble than gypsum. Saline soils normally have a pH below 8.5, whereas a natric horizon may develop a pH above 8.5, and such alkaline soils may attain a pH as high as 10. In addition to pH, the presence of salts is measured through conductivity. Saline soils as identified in the UK have a conductivity of the saturation extract of more than 4.0 S m^{-1} at 20º (Avery, 1990). The presence of any salts in groundwater is most important in the preservation or corrosion of artifacts (Goren-Inbar et al., 2018; see Huisman, 2009). Also when sodium ions are introduced in large amounts to stable soils, such as during marine inundation they have an effect of dispersing clay and forming unstable soils (see below and Chapter 9) (French, 2003; Macphail et al., 2010).

In saline soils, salt in solution is drawn to the surface by evaporation and forms a white salt crust or salic horizon, but because the topsoil is still dominated by Ca^{2+}, Mg^{2+} and K^+, the soil remains stable; such solonchak soils were once known as white alkali soils. Where, however, the presence of Na^+ becomes completely dominant, perhaps as high as 70% of CEC, solodization or alkalization takes place, and all clay and humus is dispersed, all aggregate stability is lost, and a non-aerated massive structure forms (Brady and Weil, 2008: 418–420; Duchaufour, 1982: 434). During humid periods, dispersed humus and clay are translocated down-profile and coat the natric horizon in what have been called black alkali or solonetz soils (Bridges, 1970).

Salts such as sodium carbonate ($NaCO_3$) can also form in saline soils and can be found in environments as different as playas and coastal muds. In the latter case, salt-tolerant vegetation can be found, because of the influence of saline tidewater. In estuaries, saltmarsh plants such as glasswort grow on saline *Entisols* (recently formed soils; Table 4.3) formed in marine alluvium (Avery, 1990: 324).

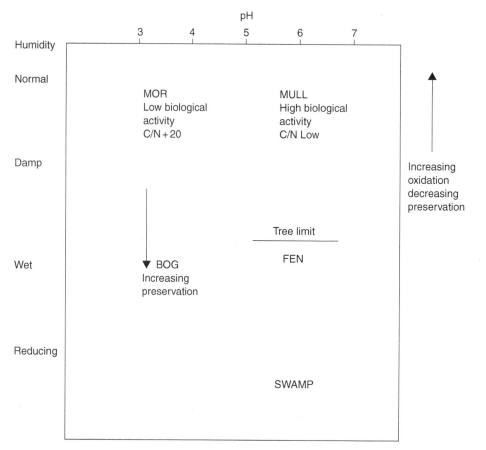

Figure 4.13 Generalized relations between soil pH, humidity, and humus types (after Pearsall, 1952) (from Macphail, 1987a).

4.6.5 Tropical (and Mediterranean) Soils

The two main types of soils formed in warm and humid regions are Oxisols and Ultisols, which develop through what was once known as laterite soil formation, i.e., through the leaching of silica. An understanding of such processes is important when sites associated with *tera preta* have been investigated (Arroyo-Kalin, 2017); a comprehension of the biological circulation of nutrients in subtropical soils occupied by the Maya is also a developing field of research (Graham, 2006; Graham et al., 2016). Other primary pedogenic processes are weathering of primary minerals, rapid biodegradation of the near-surface organic matter, deep leaching and the concentration of free oxides (Fe^{3+} and Al^{3+}) as gibbsite ($Al[OH]_3$), ochreous goethite ($\alpha FeO[OH]$), or reddish hematite ($\alpha Fe2O_3$) (Lelong and Souchier, 1982; Marcelino et al., 2018; Stoops and Marcelino, 2018). Leaching of silica can lead to neoformation of 2:1 lattice clays (e.g. illite) rich in Si. Rubification (reddening) of the subsoil may also occur if there is a marked dry season. A special case of rubification of soils on limestone is the formation of the well-known Mediterranean reddish Ultisol called *terra rosa*. Here, a typical brown Alfisol type known as *terra fusca* has developed from clays inherited from the rock after it has been weathered. In this case, rubification is caused by the presence of some of the iron oxide becoming recrystallized as hematite (Lelong and Souchier, 1982). This process is important because many reddish paleosols have been termed climatic indicators of Pleistocene interglacials (Cremaschi et al., 2018); such red soils have proven to be an ideal raw material for ochre production that was utilized from the Paleolithic onwards (Ferraris, 1997) (see Chapter 12).

The mobility of iron in tropical soils can lead to downslope secondary hydromorphic ferric (Fe^{3+}) iron accumulations; in such cases ferrous iron (Fe^{2+}) in the groundwater precipitates to produce an "iron cuirasse" (Duchaufour, 1982: 405). Another form of iron and clay cemented tropical subsoil is the diagnostic plinthic horizon ("laterite"), that irreversibly hardens after repeated wetting and drying. A typical Ultisol develops under a forest cover when two combined processes occur: strong weathering produces a kaolinite- (1:1 lattice clay) dominated upper profile, while clay is translocated down-profile into the argillic subsoil. Such strongly leached soils have been known as planosols and yellow-brown podzolics. They are differentiated from Oxisols by the presence of clay translocation.

4.7   Conclusions

Pedogenic processes affect most archaeological sites (Courty et al., 1989: chapter 8). These may be short-term or long-term effects (see Chapter 14) and were noted, for instance, at Clovis sites in Arizona and Virginia, and SE USA (Holliday, 2004: figure 9.3; Leigh, 1998, 2001; Macphail and McAvoy, 2008). Basic soil science can also be applied to establishing amount and type of any post-depositional processes that may have affected a site, as well as providing snapshots of past environments through the study of buried soil profiles; a paleosol cover sometimes constituting an anthropogenic feature in itself. Armed with these essentials the geoarchaeologist can apply soil science techniques, as detailed in Chapters 16 and 17, in a focused way to tease out data that can then be applied independently to reconstruct past natural and cultural landscapes.

5

Hydrological Systems I: Slopes and Slope Deposits

5.1   Introduction to Fluvial Systems

Slopes, streams, lakes, and wetlands are the most widespread and among the most important of geoarchaeological environments (see Chapters 6 and 7). They encompass a variety of readily available natural resources, and potentially furnish the right types of conditions for the preservation of archaeological sites and contexts. These environments are also the most written about of all the geoarchaeological situations (e.g. Berger et al., 2008; Brown, 1997; Frederick, 2001; Hajic et al., 2007; Kidder et al., 2012b; May and Holen, 2014; Needham and Macklin, 1992; Sidell, 2003; Waters, 1986; Wetzel and Unverricht, 2020). The focal point of this environment revolves around water, which is not only necessary to sustain life but also serves to attract game, and furnishes water transport and raw materials, including wood, reeds, and sediments. These latter items serve a variety of functions and uses, including fuel, matting, basketry, pottery, and associated construction materials (principally wattle, mudbrick/adobe, or daub). Ancient quays and river-docks can be a foci for waterlogged waste accumulations, including wood working debris (see Box 18.1). Major rivers are important for inland waters such as lakes and managed wetland (e.g. for rice paddy) are also a source for fish, including artificial fish ponds (Lee et al., 2014; Zhuang et al., 2014). In drier climates, management of water can be even more important (Sandor and Homburg, 2015; Sulas, 2018).

As we discuss in Chapter 6, fluvial systems also tend to preserve the context of archaeological sites and remains (Bates and Stafford, 2013; Carey et al., 2018; Ferring, 1986; Ferring, 1992; Ferring, 2001; Howard et al., 2003) (the archaeology of estuarine environments is discussed in Chapter 9). Low-energy inundations of flood plains away from the higher energy channels provide a favorable potential for preserving artifacts and features near to their original depositional contexts. The French sites of La Verberie (Audouze and Enloe, 1997) and Pincevent (Leroi-Gourhan and Brézillon, 1972) are noted for their evident preservation of the physical integrity and paleoanthropological fidelity of the remains; concentrated remains of wildfires have also been found in fine overbank sediments (see Box 17.1). The greatest concentrations of hand axes occur at the probable spring-fed "waterhole" at the early hominid site of Boxgrove, UK (Roberts et al., 1994) (see Chapter 7).

In order to be able to gain an understanding of past human interactions with the fluvial system, and how geoarchaeologists can go about recognizing and interpreting them, it is first necessary to be aware of some of the aspects of how fluvial systems operate. Alluvial environments can be viewed from any one of a number of standpoints, and here, we attempt to link geomorphic aspects with sedimentological ones, emphasizing how they interface with geoarchaeological considerations. Many basic works (e.g. Boggs, 2012; Bridge, 2003; Reineck and Singh, 1986) can be consulted for additional details.

Practical and Theoretical Geoarchaeology, Second Edition. Paul Goldberg and Richard I. Macphail.

5.1.1 Characteristics of the Fluvial System

Fluvial environments are distributed throughout the globe from the subarctic to the tropics, reflecting a wide range of variability as a function of local environmental conditions, such as moisture, temperature, seasonality, and other climatic variables. Arid fluviatile systems are different from those in more humid areas, for example, in their lower frequency but potentially greater intensity of streamflow, and the geometry of the channel patterns (Huckleberry, 2001). In these chapters we examine some of the most important aspects of the fluvial system that would be useful to the geoarchaeologist in understanding why and where sites might be located, eroded, or buried in a given area. Hence here in Chapter 5 we deal with the watershed, fluvial systems on slopes and slope deposits, and associated archaeological features such as terraces, lynchets and human impact such as clearance (see Chapter 11), while in Chapters 6 and 7 the focus is rivers and lakes, respectively, and their deposits associated with wetland environments.

5.2 Fluvial Landscape Studies – Slopes

A prerequisite for investigating any landscape is the ability to identify first, areas of erosion, deposition (e.g. alluvium and colluvium), and locations of zero erosion or deposition. This chapter focuses upon landscapes with climates where flowing water on slopes is the major mechanism for erosion and colluviation. Whereas sloping ground can be affected by water erosion, flat ground can suffer wind erosion (see Chapter 8 Aeolian Environments). This present chapter spotlights slopes and the type of processes prevalent on them, as well as the possible associated archaeology. The susceptibility of certain soil types and their horizons to erosional processes are also outlined. Lastly, ways to recognize colluvium in the archaeological record are given, although types of human impact that initiate or encourage erosion are described later in Chapter 11.

5.2.1 Water Movement on Slopes

Water movement, and associated drainage channels, begin on upland positions of the watersheds in the landscape. On these slopes precipitation falling as rain or snow leads to water flow that follows a number of options (Figure 5.1) (Butzer, 1976). Initially water can infiltrate into the soil or joints or pores within the bedrock. With time – and depending on the environment and local conditions – such water can percolate laterally (*throughflow*) or downward vertically. In the latter case, it eventually reaches the groundwater table where it flows laterally; on the surface, such water flow is expressed in the form of springs or as direct inputs into streams.

A number of factors influence the degree to which water is absorbed into the soil. Some of these factors include (Butzer, 1976):

1) vegetation cover (Figure 5.2) and the way it intercepts and absorbs water: in contrast, bare slopes promote surface water flow that results in various forms of erosion, such as rilling, gullying and sheet erosion (see below);
2) soil texture and structure, and bedrock: water tends to percolate rapidly through sandy and open structured soils for example, and is impeded by dense, clayey soils or subsoils where void space has been infilled by clay, iron, or secondary carbonates. Jointed bedrock also promotes percolation; and
3) soil moisture is also a factor: moist soils, which are limited in their ability to absorb additional water, will quickly become saturated, and any further water will accumulate on the surface or flow overland downslope.

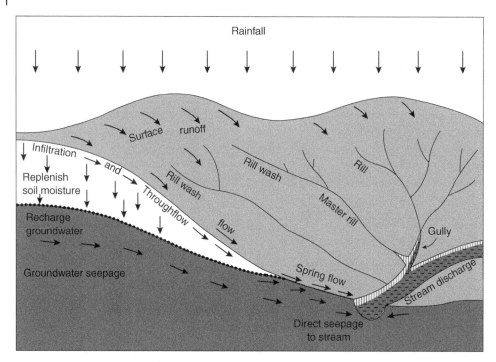

Figure 5.1 Pathways of water movement along slopes, and within soils and bedrock (modified from Butzer, 1976).

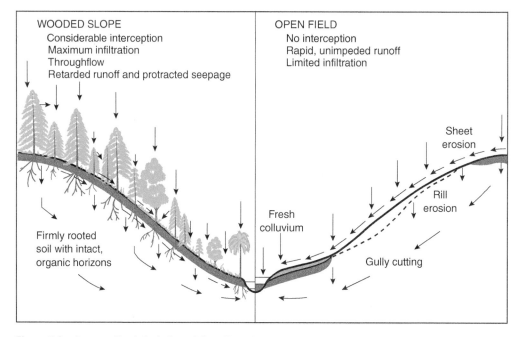

Figure 5.2 A generalized depiction of the effect of vegetation on runoff and infiltration along slopes. Not all slopes have behaved this way in antiquity, as for example, only very minor downslope soil movement has occurred under woodlands in Ardennes, Belgium (Imeson et al., 1980). Commonly, pollen-rich bog sediments are concealed beneath the colluvium (Burrin and Scaife, 1984) (modified from Butzer, 1976).

Water that does not infiltrate into the subsurface may either rest in place, and in cases where there is no relief or there are horizontal surfaces or depressions it can remain in the form of ponds or bogs, for example. The dynamics of water movement in the soil catena – i.e., from plateau to river valley – was investigated by Dalrymple et al. (1968) who developed a nine unit land surface soil water gravity model that helps account for numerous soil features (see Chapter 4). Water standing on the surface can play a major role in soil formation (pedogenesis), resulting in the formation of surface gley soils with associated oxidation/reduction reactions that produce soil mottling overall, as is typical of bogs there is a decrease in biological activity (see Chapter 4 Soils); in calcareous environments, carbonate-rich spring deposits can form (e.g. tufa) (Ford and Pedley, 1996). These can be important archaeological sites (Dabkowski, 2014), as for example, tufa springs providing a probable Upper Paleolithic hunting site at Kostenki, Russia (Hoffecker et al., 2016; Holliday et al., 2007). Alternatively, water can move down slope. As shown in Figure 5.2, this situation is particularly true for bare slopes that are common in arid and semi-arid areas.

Downslope movements take on a number of forms that relate to the concentration of water and whether water flow is confined or not. Near the crest of slopes water that just begins to flow typically takes the form of a mm-thick irregular sheets of water (sheetwash or sheet flow) that can transport dissolved and fine-grained load (Butzer, 1976; Chorley et al., 1984). Movement of particles is also encouraged by the action of rainsplash, which can dislodge and facilitate the transport of silt- and sand-size particles. The effects of rainsplash have been well documented by field studies and by numerous laboratory experiments (numerous experimental studies are summarized in Fedoroff et al., 2018; Mücher et al., 2018) (see below).

Thus, it is when the intensity of rainfall exceeds the infiltration capacity of the soil or substrate that overland flow commences, enabling erosion to take place (Ritter et al., 2011). Irregularities on the surface bring about a transition to conditions where sheet flow no longer operates and water is carried along preferred lines of drainage. The smallest size feature, the rill, typically forms on fine-grained material and is a few cm deep and wide (Ritter et al., 2011). Rills can be rather ephemeral, becoming obliterated by shrinking and swelling or freezing of the soils, or can be destroyed by rainfall itself; since they are in active environments and mobile substrates they can reform and shift positions.

Gullies are larger-scale features that are in the order of centimeters to meters in depth and width (Figure 5.3), and can easily develop into features that are tens of meters long. Both rills and gullies overlap in size and potentially occur in the same area, particularly in places that are undergoing rapid erosion, such as *badlands*. Gullies function ephemerally during rainfall events. Gully erosion advances by channel deepening on the slope and by headward erosion upslope (Butzer, 1976). On the surface, gullies and rills usually develop a dendritic pattern which is partly controlled by the uniform (isotropic) nature of the substrate (e.g. silt, clay, marl). Subsurface channels or pipes serve as conduits to water and sediment and with continued subsurface flow they become enlarged, often to the extent that their roofs collapse (see also Dalrymple et al., 1968). This leads to lateral or headward extension of the gully.

Gullies and pipes, because of their relatively large size, are very important in geoarchaeological studies because deeply buried archaeological sites that are normally undetectable, can be readily revealed in gully exposures or collapsed pipes. One of the few sites that appears to span the typological transition from Middle Paleolithic to Upper Paleolithic, Boker Tachtit (Negev, Israel; Box 6.2) was found by one of the authors while carrying out a foot survey in the Nahal Zin Valley, focusing on exposures in gullies and in eroded banks. In this case, charcoal was found eroding out of a gully that cut through an alluvial deposit containing the occupations (Goldberg, 1983; Goldberg and Brimer, 1983).

Figure 5.3 Gullies formed on soft, easily erodible Pleistocene/Holocene lacustrine sediments from Owens Valley California. Many prehistoric lithic scatters and concentrations were found on these eroded deposits.

An additional noteworthy instance is illustrated from the Lower Paleolithic site of Dmanisi, Republic of Georgia. Here, spectacular faunal and hominin remains (Ferring et al., 2011; Gabunia et al., 2000) owe their survival and remarkable preservation in part to their recovery from gullies and pipes. These features not only formed rapidly, but also were infilled quickly (personal observation and communication with C. Reid Ferring).

Along the upper part of slopes and along slope crests, overland erosion and gullying is relatively minor (Figure 5.1), and gravitational movements remove volumetrically more material than by processes of running water (Boardman and Evans, 1997). On the other hand, gullying and rilling are more concentrated along the midslope positions. Materials derived from along the slopes accumulate at the base of the slope (the footslope; Figure 5.2; see also section on catena and soils along slopes). The generic term for slope deposits is colluvium, which is generally massively bedded, and poorly sorted (Fedoroff et al., 2018; Mücher et al., 2018). In the UK especially, water-worked Pleistocene slope deposits are termed soliflual deposits, and in France, slope deposits are commonly associated with *solifluction* (Bertran et al., 1997; Bertran and Texier, 1999; Bertran and Texier, 1990) (see Box 4.3). In the UK, "colluvium" is normally used to refer to fine and well-sorted hillwash and in many cases it relates to a human impact on the landscape, such as cultivation (Bell and Boardman, 1992; Catt, 1986; Catt, 1990; French, 2016) (Chapter 11). In much of the Mediterranean, colluvium is mainly comprised of both eroded soils and sediments, and consists of heterogeneous, poorly sorted debris, with angular clasts (French and Whitelaw, 1999; Krahtopoulou and Frederick, 2008).

Elsewhere, such as in the USA, colluvium would include any type of unsorted slope mantle. Therefore, the nature of colluvia varies according to geographical region, and local and regional processes of deposition. Since the term is all-encompassing, it is a useful term in the sense of its being specifically vague, especially in the field. On the hand, slope deposits can be studied in detail to reveal the types of processes and activities operating on a slope – periglacial activity, effects of gravity, ratio of water to sediment, land use practices and management (e.g. clearance, cultivation, terracing; see below). Hence, both human and climatic information can be extracted from their study. In any case, since such sediments occur in the lower part of slopes, colluvially derived slope deposits commonly grade into or interfinger with fluvial sediments deposited in the main fluvial system (Figure 5.2).

5.3 Erosion, Movement, and Deposition on Slopes

5.3.1 Slopes

The ability to recognize the relationship between sloping ground and colluvium is important, because such landscapes may provide the best examples of buried land surfaces and stratigraphic records of human activity where erosion has been active. The landscape model of the *catena* (see Chapter 4) is significant here. There are also links between slope unit types and potential past land uses, which can be inferred by applying modern land capability classification. For example, it can be simply recognized that steeply sloping (16–25°/28–47%) land is difficult to plough, so that pasture is a more likely land use practice (without terracing), with the extra constraint that soils are becoming more liable to the risk of erosion (Bibby and Mackney, 1972; Klingebiel and Montgomery, 1961). GIS (see Chapters 16 and 18) is now a common method for integrating this kind of survey work into archaeological models. As a note of caution, it must be remembered that modern slopes are in fact *modern* and may bear little relationship to landscapes of the past. Massive Roman to medieval colluviation may have completely changed soil thicknesses, with once-concave slopes becoming convex as they became infilled; humans have been causing soils to erode since the Mesolithic (see Chapter 11).

Even gentle slopes (2–3°/3–5%) along the plateau edge, as well as moderately (4–7°/6–12%) to strongly inclined (8–11°/13–30%) upper slopes may show evidence of erosion. On the other hand, mid and lower slopes, and the valley bottom itself are the loci of colluvial deposition (see Chapter 2 and Table 5.1). This slope situation produces important archaeological stratigraphy particularly on chalk downlands, which have been studied through molluskan analysis (Allen, 2007; Allen, 2017; Bell, 1983, 1992). Bare, arable ground is highly susceptible to erosion by rain splash and ensuing formation of rills and gullies, and sheetwash is now thought to be of negligible importance in southern England (Boardman, 1992) (see above), but in fact plays a major part on the coastal plain of Djibouti (Sara et al., 2006). Slope bottom colluvium is deposited, at least in the first instance, as a laminated waterlain sediment (Farres et al., 1992). Infrequent, but archaeologically significant storms can generate sufficient energy to erode and transport stone-size material producing gravel fans that can include lithics (Allen, 2005, 2007; Allen, 1992).

The colluvial footslope/valley bottom interface is characterized by interdigitation of colluvial and alluvial deposits, as for example, in the case where poorly sorted silts may interfinger with organic-rich (peaty) clays or bedded overbank deposits (Brown, 1997). Dry valleys are not only typical of karstic terrains in the Dordogne (France), but found commonly in the more northern chalklands of the Chilterns and Yorkshire Wolds (UK) and the Pays de Calais (France); during the Pleistocene frozen ground promoted overland flow (and erosion) on these normally porous rocks. These valleys can contain both periglacial mass-movement deposits as well as overlying Holocene colluvium (Van Vliet-Lanoë et al., 1992). Towards the top of the Pleistocene deposits, a humic Late glacial soil can be found across Europe that marks a temperate Interstadial at ca. 11,000–13,000 bp. This interval is termed the Windermere Interstadial in the UK and in mainland Europe, includes the Bølling, Older Dryas and Allerød climatic episodes (Barton, 1997) (Box 4.3; Table 4.4).

Table 5.1 Generalized characteristics of slope deposits (see Figure 5.4).

Position on slope	Sedimentary/erosional processes	Sedimentary attributes – macro	Soil characteristics	Micromorphology and other environmental indicators	Archaeology
Upper slopes	Rainsplash and rilling; sheetflow (not significant)	Aerobic decomposition Loose Stony	Bare rock and shallow soils Eroded topsoil Exposed subsoil Droughty soils	Possible identification of poorly humic newly formed topsoils and truncated subsoils; negative features and subsoil hollows may contain mollusks.	Eroded features Displaced artifacts Common location of negative features (e.g. pits, ditches, and post holes)
Midslopes	As above, with gullying; tilling effects and creep (gravity movement) Eroded and over-thickened sediment/soil profiles	Cut-and-fill features; coarse (stones and sand) and fine (silts and clay) laminae.	Bedrock exposed in gullies Eroded and over-thickened sediment/soil profiles. Well-drained soils	Possible identification of charcoal and burned soil (clearance), lithorelics, and soil fragments of earlier landscape – e.g. from A horizons and subsoil argillic Bt horizons, *terra rosa.* Pollen and mollusks from feature buried surfaces.	Buried or eroded features and artifacts; common location of lynchets, walls, and terraces.
Lower slopes/ Toeslopes	Minor rilling and gullying with more dominant accretionary soils (colluvium). Possible springhead sapping.	Stone-free and stony layers; downslope-oriented finger-shaped stone (gravel) fans; possible spring associated tufa and deposition of "salts".	Homogeneous colluvial soils with relic/ephemeral stabilized topsoils and stony horizons. Cumulic soils with over-thickened A horizons. Generally well-drained soils, but possible seasonal effects of heightened groundwater (groundwater gleys), possible springs and peats.	Possible identification of soil fragments from eroded "natural" soils, and tillage soils (crust fragments from furrows); slaked soils relating to mass movement; possible relict sedimentary features; features commonly worked or partially worked by soil fauna. Potential of well stratified molluskan sequences.	Increased stratigraphic resolution due to rapid sedimentation; potential location of buried occupations, sites of well and water holes utilizing springs; paddy fields.
Valley floor	Interfingering of colluvium and alluvium/peat; potential combination of local valley and distant – up-river – deposits; likelihood of deep erosion of valley floor sediments by migrating meanders.	Colluvial loams intercalated with silts, clay, and organic sediments; more likely tufa and deposition of "salts", as spring lines/water table are met.	Cumulic soils show evidence of occasional poor drainage; alluvial soils display seasonal high water tables (ground water gleys); intercalated peats result from major changes to base level induced by climate/sea level changes and human activities (clearance).	Possible identification of biologically homogenized colluvium and alluvium, and ground water effects (secondary calcium carbonate (e.g. tufa), iron and manganese mottling/ panning, pyrite formation; salt pans and gypsic horizons). High potential for well stratified sequences of terrestrial and aquatic mollusks, and pollen and macrofossils (plants and insects).	Increased stratigraphic resolution due to rapid sedimentation; possible waterlogging and preservation of organic remains; wells and waterlogged pits; potential for preserved routeways – bridges, causeways, fords; industrial activity – water mills, artificial channels, and ponds; paddy fields.

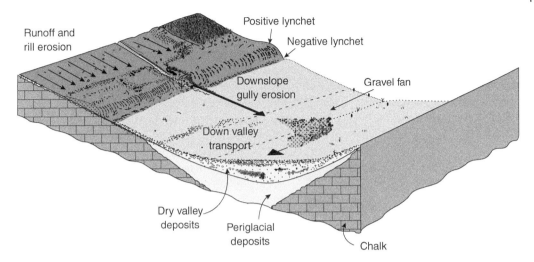

Figure 5.4 Colluvial soil movement down slope and down valley (after Allen 1988); hypothetical soil erosion and deposition on the chalklands of southern England (see Table 5.1; Allen 1992; Bell, 1983; Boardman, 1992).

More recent studies along new pipelines, highways, and high-speed train links across Belgium, Luxembourg, and northern France have enormously increased our knowledge of the relationships between human impact, erosion, and deposition of colluvium and alluvium (Fechner et al., 2014). Mediterranean valley fills, and associated eroded landscapes were studied in detail, as for example in Greece and the Levant (Bintliff, 1992; French and Whitelaw, 1999; Goldberg, 1986; van Andel et al., 1990; Zangger, 1992). Such research has evaluated the roles of humans and climate in slope and alluvial erosion. Climatic change was first identified as a trigger for massive erosion (Vita-Finzi, 1969), but as more archaeological sites and their environments were investigated, land use, expanding populations, local soils and geology were increasingly cited as factors influencing erosion (French, 2015; French et al., 2017; Fuchs, 2007). More commonly now, rare events, such as intensive rainstorms, are seen as the most important cause of major erosional and depositional events (Allen, 2005). On the other hand, landscape instability has been triggered by clearance and development of bare arable ground (see below and Chapter 11).

Lynchets (an earth terrace along the side of a hill) usually occur at slight changes or dips in slopes, through the combined effects of erosion and colluviation (Fowler and Evans, 1967; Macnab, 1965). Erosion leads to a negative lynchet down slope, while colluvial deposition forms a positive lynchet upslope (Figure 5.4). Occasionally, the positive lynchet may itself be formed on eroded ground, and slope soils may be stabilized by wall building along lynchets. At Chysauster (Cornwall, UK), erosion of the negative lynchet reached an estimated depth of 0.50–0.70 m. In many societies where both flat ground is rare and erosion is a constant threat, labor-intensive terracing has been introduced. Examples can be cited from both the Old and New Worlds (French and Whitelaw, 1999; Minnis and Whalen, 2015; Sandor, 1992; Sandor and Homburg, 2015; Wagstaff, 1992; Wilkinson, 1997) (see Chapter 11).

5.3.2 Soil Stability and Erosion

Soil stability and erosion have been studied in great detail throughout the world and especially for Holocene geoarchaeological landscapes (Kwaad and Mücher, 1979; Macphail, 1992a; Mücher et al., 2018; van Andel et al., 1990). The major mechanisms of erosion dealt with here are somewhat simplified but can be ascribed to: (i) erodibility and resistance to erosion by different soil types; (ii) rainsplash impact; (iii) water flow; and (iv) effects of ploughing and cultivation.

Different soil materials erode more easily than others, as for example windblown sand (dunes) vs. silt (loess) (Evans, 1992) (see Chapter 8). In general, silty soils and sands, because of their inherent poor soil structure, are more erodible than soils with a high clay content. Moreover, soil organic matter, land use practices, and general soil type also have an effect on soil erodibility. Although the interaction of these factors is complex, some useful generalizations can be made. Topsoils, for example, have greater aggregate stability than subsoils because of their greater humus content, and in general, decreasing aggregate stability follows the trend: grassland→woodland→arable land (Imeson and Jungerius, 1976).

Extant "natural" topsoils also have good structure and open porosity that allows them to drain well; consequently, they are only slightly influenced by surface water flow and erosion. On the other hand, research from the southern Midwest USA shows that bare arable topsoils may suffer badly: rainsplash may break up soil structures causing surface crusts that lead to overland flow (McIntyre, 1958a, b). The resulting slaked soil is then transported along rills and gullies (West et al., 1990). On flat ground, on the other hand, silty clay and sandy laminated surface crusts are formed as these constituents become redistributed. Such arable soil crusts on arable ground have been extensively studied (Boiffin and Bresson, 1987; Williams et al., 2018a).

Leached upper subsoil horizons of podzols (Albic (A2; Ea in UK)), are comprised of sand with little organic matter and therefore have weak structures and in theory are easily eroded; the presence of fungal hyphae, however, can actually make the sands water-repellent. This phenomenon not only may aid resistance to erosion, but also make such sandy soils droughty because of poor infiltration. Generally, however, when the protective and water absorbent Ah horizons of heath are breached by human and animal activity such as trail formation, massive sand colluviation can take place (e.g. West Heath, Surrey, UK; Drewett, 1989; Scaife and Macphail, 1983).

At one Vestfold E18 routeway site (Hesby, Norway), multi-method analyses of colluvium provide a proxy landscape history of land-use on the plateau and upper slopes, which are eroded areas from modern development (Gollwitzer, 2012). Colluvia from several "geoprofiles", which were mainly dated employing OSL (Fuchs, 2011; Fuchs and Lang, 2009), record a history of grazing followed by cultivation. The latter became intensive and despite use of both organic and inorganic manures, led to soil instability and the formation of thick colluvial deposits downslope (Viklund et al., 2013) (Figures 5.5a–5.5c; Table 5.2).

On argillic brown earths, the upper subsoil horizon (A2; Eb in UK) can become rather compact and poorly structured because of loss of clay and its translocation into the underlying Bt (argillic) horizon. Consequently, this A2 horizon may permit only slow drainage, and is thus very susceptible to surface flow, subsurface "piping" and erosion (Dalrymple et al., 1968). The underlying Bt horizon may also develop slow or impeded drainage because soil voids become increasingly coated or even sealed by translocated clay, forming a surface-gleyed Btg horizon. If soil water stagnates in and on this horizon, there is also the possibility that reducing conditions will lead to the breakdown of this translocated clay here in the Btg and above (A2; cf Ebg horizon). In tropical soils such as Ultisols and Oxisols, topsoils and upper subsoils may become eroded on slopes uncovering an iron and clay-enriched subsoil, which on exposure irreversibly hardens (see Chapter 4) becoming a cemented *ferricrete* or *cuirasse* subsoil horizon that is strongly resistant to erosion.

Figure 5.5a Field photo, Hesby nordre, Vestfold, Norway; toe-slope sediments have migrated across the valley floor – once a marine inlet; valley side colluvium has accumulated over the original toe-slope sediments, changing the morphology of the partially infilled valley. There are round holes from OSL dating and long monolith extraction gaps (21300635) (see Figure 5.5b). At the base are gray marine clays, over which are the humic toe-slope soil-sediments that have been affected by stock trampling. The lighter-colored deposits above, are a series of colluvia relating to different periods of cultivation and upslope erosion – from Roman Iron Age to the Viking and Medieval Periods (Table 5.2).

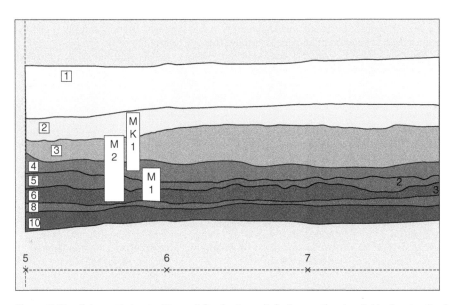

Figure 5.5b Schematic key to Figure 5.2a; 1 – topsoil, 2–3 – medieval soil, Medieval colluvium, 4 – Viking Age colluvium, 5 – Migration Period colluvium, 6 – Roman Iron Age colluvium, 7 – Pre-Roman Iron Age colluvium, 8 – Early Bronze Age(?)/Pre-Roman Iron Age stock trampled surface, 10 – marine clay.

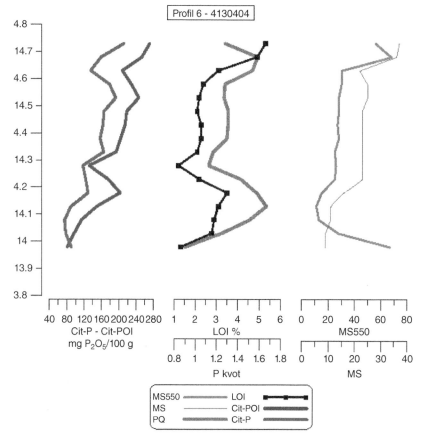

Figure 5.5c Hesby nordre, Vestfold, Norway: Geochemical log of P, LOI, and magnetic susceptibility. Interpretation (see Table 5.2) from the base upwards:

- <14 m asl: waterlogging removes P, but concentrates iron (peak in MS550, proxy iron measurement);
- 14.0–14.2 m asl: marked increases in organic matter (LOI), phosphate and proportion of organic phosphate (Pquota) coincide with dung inputs related to livestock trampling;
- ~14.3 m asl: low phosphate and LOI, but sharp rise in MS, records high-energy layer of stony and sandy cultivation colluvium, which includes burned minerogenic settlement waste (hence high MS);
- 14.3–14.6 m asl: generally high P, LOI, and MS, consistent with manured arable colluvium, which includes burned settlement waste, and dung traces concentrated in silty clay lenses;
- >14.6 m asl: sharp rise in MS results from high-energy manured arable soil colluvium becoming richer in burned settlement waste, which is also employed to amend the soil.

Table 5.2 Hesby, Vestfold, Norway; geoarchaeological data and land use reconstruction from Geoprofiles (GP) 6, 7, and 8, pit fill S1130515, and Viking grave S1130613. For physical/chemical profile at Geoprofile 6, see Figure 5.5c (Fuchs, 2011; Gollwitzer, 2012; Viklund et al., 2013).

Data	Deposit/Process	Period and interpretation
GP 6 and 7: Layered coarse and fine colluvia with high OM, P, and MS, settlement waste, FeP nodules, FeP stained charcoal, and dung traces.	Humic, dung, and phosphate-enriched fine colluvia alternating with coarse colluvia containing settlement waste	Layer 3; Medieval: Renewed manured arable, with both animal dung and settlement waste. Major soil erosion and colluviation recommences.
GP 6: Crumb and blocky structured humic silty loams over structureless sands and gravel (with enhanced MS); Viking grave: inwash of dung-manured humic fine loams into base of grave, infilling fissures in and below wooden chamber framework.	Upper: Fine dung-enriched colluvium, with period of stasis, over Lower: high-energy colluvium, with settlement waste	Layer 4; Viking Age: Manured arable, with both animal dung and settlement waste. Periodic high-energy erosion and colluviation, with (short-lived) period of possible grazed fallow, or pasture development during Viking grave period?; grave chamber left open?

Table 5.2 (Continued)

Data	Deposit/Process	Period and interpretation
GP 6 and 7: As below, with high and low P levels recording different humic fine and coarse colluvia, respectively; settlement waste present and rise in MS; Fe-P nodules in fills.	Arable manured colluvium, with alternating humic fine hillwash alternating with coarse colluvia, reflecting intensive erosion, rainstorms resulting in gullying and colluvial ponding; spread of arable soil onto low ground. Manuring with dung (also pig? husbandry?) and settlement waste. Biologically worked soil layers.	Layer 5; Migration Age: Manured arable, but with erosion problems. Both cattle dung and pig? waste, and increasing amounts of settlement waste employed. Colluvial deposition along break of slope/toe slope.
GP 6: humic silty loam, with sand, coarse sand, and gravel lenses; fine soil lenses (with micro-traces of dung) show marked increase in P, LOI, and Org P; GP 7, as GP 6, with inclusion of coarse constructional and burned waste; GP 8: as below, but partially bio-homogenized.	Mixed dung-enriched arable and trampled colluvium, with arable high-energy coarse colluvial sands, and manured with settlement waste throughout; biologically worked topsoil developing at GP 8.	Layer 6: Roman Iron Age: intensive arable activity becoming dominant land use, with patchy/ephemeral areas of fallow/pasture(?); erosion (and possible droughtiness? at rare times) developing as serious problem(s).
GP 6: humic silty fine sand with marked increase in P, LOI, and Org P, muddy textural intercalations; micro-traces of dung; e.g. of coprolitic bone; GP 7: mesofauna working of muddy soil; GP 8: gravel-rich colluvium; much burned settlement/hearth waste (ash residues).	Muddy colluvial toe-slope animal trample/colluvium; intermittent periods of stasis. Higher upslope (GP 8) high-energy colluvia, from intensive manuring and/or ploughing of settlement activity area.	Layer 7; Pre-Roman Iron Age: differing land uses of intermittent herding and accessing of water by livestock, whereas upslope settlement waste from hearths/industrial activity applied as manure to unstable sand and gravelly drift soils.
GP 6: Silty clay (of local marine clay origin), with humic silty fine sand laminae, including fine charcoal, secondary Fe-stained and Fe-depleted soil; GP 7: muddy textural intercalations and fine charcoal.	Semi-waterlain toe-slope/wetland colluvium/low-energy colluviation; trampling in places, management of open(?) landscape by fire	Layer 8; Young Bronze Age(?)/ Pre-Roman Iron Age: maintenance of already open landscape by fire for herding; access to drinking water by livestock (poorly drained low ground area of marine clay – Layer 10).

5.3.3 Colluvium (Hillwash)

The interpretation of colluvium is not always easy. Mücher et al. (2018) showed that soil micro-morphology could be used to identify soil components within the colluvium. They also elucidated the types of processes that could be identified in thin section, such as the structural instability of medieval arable soils in northern Luxembourg (Kwaad and Mücher, 1979). At Ashcombe Bottom (West Sussex, UK), Beaker colluvium (hillwash) was associated with arable activity that produced surface slaking crusts, and eroded fragments of these became embedded into the colluvium (Figure 5.6; see Figures 11.8a and 11.8b) (Allen, 2005). Similarly, studies of colluvial profiles in north-west Europe revealed a series of paleolandscapes and associated finds (Fechner et al., 2014), Figure 12):

- colluvia cut by Roman cremation graves
- later Neolithic colluvium including Neolithic sherds, and at the base,
- the eroded Mesolithic landscape with tree-throw and worked flints, occurring on Late Glacial soil-sediments

Often, colluvium can display sequences of stone-free and stony horizons. One explanation for this phenomenon was that stones brought to the surface by tillage were periodically concentrated some 20–30 cm below stone-free soil, because of earthworm working of ephemerally stabilized colluvial soils (e.g. during periods of fallow and/or abandonment). The identification of high intensity rainstorm events in the present-day record (Boardman, 1992), however, showed that rill and gully erosion could transport large numbers of stones, which were then deposited downslope as colluvial gravel fans. A study by Allen employed three-dimensional excavation, mollusk analysis, and soil micromorphology at Strawberry Hill (Salisbury Plain, Wiltshire) (Allen, 1994) (Figure 5.7). He found that Iron Age rill erosion on the chalk produced spreading "fingers" of gravel (fan). Mollusks indicated the presence of large amounts of arable ground, whereas soil micromorphology showed that chalk stones had been deposited within a slurry of local chalk soil material (rendzinas); the underlying stone-free horizon was similarly deposited as a slurry. The stone-free horizons here had not resulted from earthworm working. Mass-movement produces specific types of microfabrics, particularly those associated with solifluction (see Box 4.3), a recurrent type of Pleistocene sedimentation in southwest France (Bertran et al., 1994; Bertran et al., 1997; Bertran and Texier, 1999; Courty et al., 1989; Fedoroff et al., 2018; Karkanas, 2019; Mücher et al., 2018). However, such interpretations of Holocene sediments (see Norwegian examples, below) may be more readily taken on board if accompanied by flattened walls with a downslope orientation (*imbricated* fabric), or if they occur in well-known tectonically unstable areas (Joelle Burnouf, pers. comm.).

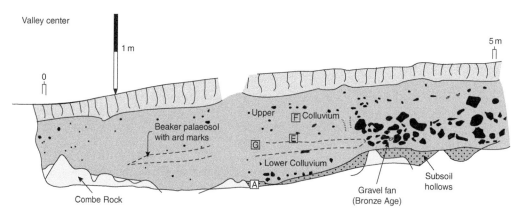

Figure 5.6 Ashcombe Bottom section drawing (courtesy of Mike Allen) showing gravel fan and colluvial deposits, and location of thin section boxes and Beaker paleosol (from Macphail 1992a) (see Figure 9.12a).

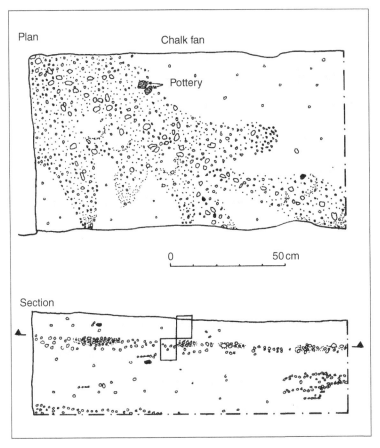

Figure 5.7 Strawberry Hill, Wiltshire, England; plan and section of colluvial fan (from Allen, 1994); rectangular boxes are thin section samples studied by R. I. Macphail.

The over-steepened valley sides of Norway, which are relict of glacial valley erosion, are prone to landslides, especially when becoming water saturated in spring. At Fryasletta, Oppland these both bury alluvial and cultivation soil sequences in the valley floor (Figures 5.8a–5.3c; see Figure 17.5a) (Gundersen, 2016; Macphail et al., 2016). The city of Bergen, an important ancient west coast port, is located on low ground between steeply sloping valley sides. Excavations ahead of railway tunneling for the Fløen-Bergen routeway revealed basal deposits abutting the steep valley side (A1194; 17.737 m asl), which record meandering stream sedimentation (Figure 5.9). Here, interbedded fluvial sands and gravels occur, some likely augmented by mass-movement deposition; on the other hand, inactive river cut-offs produced low-energy silting and peat formation (Figures 5.1–5.5) (Reineck and Singh, 1986: 261, 293). Episodic base level fluctuations led to peat ripening and peaty soil formation (Jongerius, 1962; Jongerius and Jager, 1964); wildfire or human activity produced scorched peat soils dating to the Neolithic. Upwards in the stratigraphic sequence, later prehistoric and early medieval activity on the slopes, including woodland clearance and farmsteads, exacerbated slope instability. Mass-movement slides occurred regularly, induced by slope deposits becoming water-saturated due to locally high precipitation. These commonly buried cultural soils with often a brief period of ponding that produced well-sorted clayey crusts capping the unsorted mudslide material.

Figure 5.8a Scan of M56 (steep-sided Lågen River valley, Fryasletta, Oppland, Norway; see Figure 17.5a), showing steeply sloping layers, composed of mass-movement deposits and slurries, including charcoal-rich cultural soil Layer1146 and muddy 1142b. Frame width is ~50 mm.

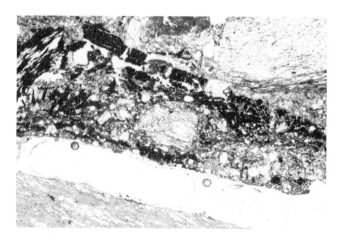

Figure 5.8b Photomicrograph of M56, Layer 1146 (Figure 5.8a); mass-movement, semi-waterlain, layered charcoal, and stones. PPL, frame width is ~4.62 mm.

Figure 5.9a Field photo of excavations ahead Bergen-Fløen rail tunneling, Bergen, Norway. A series of series of fluvial sands and gravels, with likely local inputs from mass-movements (note rock) are interbedded with peats recording low-energy river cut-offs. Basal peats date to the Neolithic and are scorched in places.

Figure 5.9b Field photo, as Figure 5.9a, but further upslope and up in the sequence. Mass-movement poorly sorted sands and gravels and muds have buried several cultural soil horizons. The pre-modern soils date from the Bronze Age (ca. 3000 bp) to the Viking Period (ca. 1100 bp). Note resemblance of these deposits to those found at Fryasletta (see Figure 17.5a) (Søren Diinhoff, Yvonne Dahl, Sofia Cecilia Falkendal and other excavation team members, University Museum of Bergen, are kindly acknowledged).

Figure 5.9c Field photo, as Figure 5.9b, but lateral examples of deposits at Profile C2031 (North Face). Here a number of Bronze Age (2970–3020 bp) layers were investigated (thin section samples 50136A-B and 50137A-B) within this unstable slope deposit sequence. Short-lived soil formation, erosion and colluviation, including waterlaid silt loam deposition along the toe slope, took place under ponded conditions (Layers 6/12), are present. The last deposit could be termed "hillwash". Colluvia also record instances of soil creep, which testify to woodland disturbance by fire/clearance activities(?) (Søren Diinhoff, pers. comm.). Charcoal-wash in the latter may be responsible for date inversion (3020 bp over 2970 bp). These colluvial layers record instances of markedly high magnetic susceptibility values (from burning), low organic matter and low PQuota (max MS = 143 χlf 10^{-8} m^3 kg^{-1}, min PQuota = 1.19, min 2.7% LOI).

Figure 5.9d Scan of thin section 50137A from Profile C2031 (Figure 5.9c; Layers 6/12;); a thick layer of waterlaid silt loam sometimes alternating with silts, and with occasional deposition of clean sands; silt loam formed under probable muddy ponding conditions. Frame width is ~50 mm.

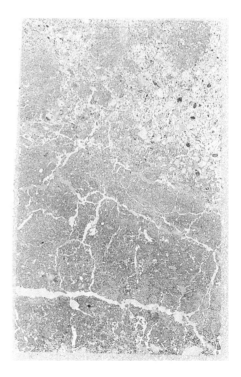

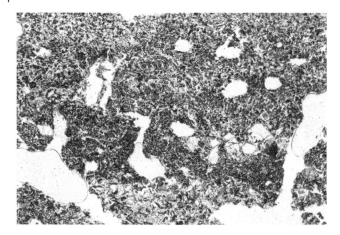

Figure 5.9e Photomicrograph of M50137A (Layers 6/12); silt loam and silt layers, with clayey crusts and fine channels suggesting muddy depositional conditions and occasional plant growth (vegetation), respectively. PPL, frame width ~4.62 mm.

Figure 5.9f As Figure 5.9e, under OIL.

5.4 Low Slope Arid Environments and Sheetwash

The broad nature of desert soils has been presented in Chapter 4, where it was noted that paucity of binding organic matter makes them prone to erosion, and also where high evapotranspiration rates can lead to secondary "salt" formations (McAuliffe, 2000; Mees and Tursina, 2018). When slopes in general were discussed earlier in this chapter, sheetwash sediments were not found to be common. However, in some dry environments in low slope areas they may be much more frequent due to seasonal storms. Low slope areas of coastal plains, for example, may receive sediments from inland uplands due to seasonal rainstorms. The coastal plain of Djibouti, despite having an average precipitation of only 60–70 mm, is subject to flash floods; it is currently under a sparse acacia vegetation (Figure 5.10a). Seasonal high-energy flash flood events have led to the formation of 30–50 cm-thick stratified sandy sediments in places. One 33 cm-thick sediment is variously composed of very poorly sorted well-bedded fine to coarse sands and gravels, and well-sorted fine or medium sands that have dark grayish brown colors when moist. The dark field colors were incorrectly believed to indicate unexpectedly high amounts of organic matter. Large cobbles of basalt could also be present, washed down from the volcanic hills inland. Also occurring are stone tools and pottery sherds as lag deposits, concentrated on a hardpan below the bedded sands. Two AMS radiocarbon dates on these poorly organic (1.39–1.82% LOI) sediments produced historical and modern dates (cal AD 1420 to 1490 (cal 460–540 BP); cal AD 1950 to beyond 1960 (BP 0 to 0)) consistent with recently recorded flash flood events, e.g. as in 1989 (Sara et al., 2006).

Bulk analyses and soil micromorphology revealed the sediments to be very fine to coarse sands and gravels that are poorly to well bedded, with fine ponding pans developing after rain storms (Figure 5.10b). They contain high amounts of volcanic rock fragments and magnetite, the latter contributing to very high magnetic susceptibility values (Gale and Hoare, 2011). Of particular interest is that sometimes very fine sand-size magnetite is concentrated (gravity separated) into thin laminae (cf. placer deposits), probably due to its relatively high specific gravity (magnetite = 5.18) compared to quartz (2.61) and calcareous material (~2.71), producing layers with very high natural magnetic susceptibility enhancement due to high magnetite content (max. 666×10^{-8} m^3 kg^{-1}) (Figures 5.10c and 5.10d).

Figure 5.10a Field photo; Djibouti, sheetwash sand and gravel-covered low slope coastal plain, with Acacia shrubs in the background. Test pit profiles dug into these deposits exposed dark, but very poorly humic sands with historic and modern ^{14}C dates (see text).

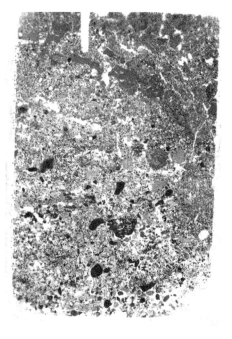

Figure 5.10b Scan of thin section M1B; note junction between calcareous fine loamy sands (LA2) and sands (LA3); upper layer of muddy (ponded) sheetwash sediments, which are calcareous due to erosion of exposed Pliocene-Pleistocene carbonate bedrock, with below poorly sorted sands characterized by sand-size magnetite, and volcanic gravels – together producing a naturally very high magnetic susceptibility (max. 666 × 10^{-8} m^3 kg^{-1}) at the studied profile. Frame width is ~50 mm.

Figure 5.10c Photomicrograph of M1B; quartz sand, sand-size calcareous sediment clasts (eroded from Pliocene-Pleistocene substrate) and opaque minerals. XPL, frame width is ~4.5 mm.

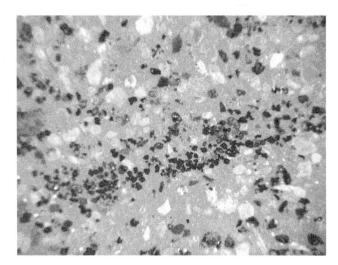

Figure 5.10d As Figure 5.10c, under OIL; opaque minerals are magnetite and form a gravity separated pan (cf. placer deposit).

5.5 Conclusions

In this introduction to slopes, slope processes, and colluvium, the importance of recognizing all these at both the large and small scale has been noted. These findings contribute to arguments for past land use patterns of perhaps arable vs. pasture, or the location of wells, as well as more obvious considerations of patterns of field/lynchet boundaries or the presence of terraces. They also contribute to the desktop assessment of where trenches may be best cut to find environmental information (Bell, 1983) (see Table 5.1). Of further note are the less frequently encountered mass-movement slumps and sheetwash sediments associated with arid environments and low slope sedimentation; both can be associated with seasonal high intensity precipitation events. Ensuing Chapters 6 and 7 examine the hydrological system as it manifests itself as rivers, lakes, and wetland.

6

Rivers

6.1 Introduction

We have already seen how a fluvial system can be traced from the watershed, via the slopes to the valley bottom (Figure 5.1). On the slopes there is both subsurface groundwater flow, which can affect both soils and the underlying geology, and overland flow, which can produce rill and gully features and slope deposits. In the case of slopes, different areas of the slope are more prone to erosion than others, while other portions of the slope may be wetter (e.g. springs) and/or more likely to accumulate colluvium (Table 5.1). The lowermost valley slope environment, however, is commonly the wettest and here colluvium may interfinger with fluvial sediments, producing interdigitated colluvium and alluvium that may well be influenced by groundwater. In this chapter we focus on the valley bottom channel and floodplain environment. We discuss a series of geoarchaeological studies to demonstrate the multi-scalar dynamics between fluctuating fluvial settings and ancient cultures. Since many streams flow into closed depressions forming wetland and lakes, we further discuss lacustrine sediments and their implications to understand human adaptations in Chapter 7.

6.2 Stream Erosion, Transport, and Deposition

Water and sediment from the slopes eventually arrive at valley floor where they enter the channel system. There, water and sediment inputs reflect physical characteristics of drainage basin (Ritter 1986: 204). Furthermore, clastic materials in fluvial systems, as with slopes, can be further subjected to erosion, transportation, and deposition.

Rivers flow at various volumes during the course of the year. During periods of low flow (*base flow*), in which most of the water entering the stream is from springs or groundwater, volumes of water and *discharge* (the volume of water carried per unit time, e.g. $m^3 s^{-1}$) are relatively low; little change in the stream morphology occurs during these periods of base flow (Butzer, 1976). In northern temperate areas, base flow is particularly low in summer months; in drier areas, flow ceases for much of the year except for the rainy season, when it essentially results from contributions by runoff (*ephemeral* flow). However, in periods of flood, such as in spring with melting of snow or during major storms (e.g. hurricanes) associated with major precipitation events, discharges can increase dramatically resulting in the condition when the stream channel is full (*bankfull* stage), even overtopping its banks and resulting in *overbank* flow. Fluctuation of river water table is particularly high in monsoon regions of East and Southeast Asia where concentrated

Practical and Theoretical Geoarchaeology, Second Edition. Paul Goldberg and Richard I. Macphail.
© 2022 John Wiley & Sons Ltd. Published 2022 by John Wiley & Sons Ltd.

seasonal rainfall causes excessive overbank flow and inundates floodplains and even terraces for a prolonged period of time (Yi, 2005) (also see Figures 17.4a–17.4d). Most fluvial changes, however, take place with frequent and continual flow when at the level of ½ to ¾ of the bankfull level (Butzer, 1976).

Within the stream channel, water flows at different velocities, depending on location. The maximum velocity is near the surface of the water, above the deepest part of the channel. However, next to the channel walls and along the channel bed, where friction is greatest, flow velocities are much lower. Thus, all parameters being equal, greater velocities occur in deeper narrower channels than in shallow, broad ones, which have a proportionately larger wetted perimeter and a greater amount of external friction along the channel.

A stream transports different material in different ways, and these substances are known as *load*. *Dissolved load* refers to various materials such as salts that are transported in solution, such as carbonates, sulfates, nitrates, and oxides. These are mixed throughout the water column. The *suspended load* is composed of finer particles (generally silt, clay, colloids, and organic matter) that are kept in suspension by turbulence. Coarser materials are carried as part of the *bed load* along or close to the bottom of the channel. These sand-size and larger particles move by bouncing (*saltation*), rolling, or sliding. Associated with this movement is the overall organization of the material being transported, and we see that the movement of sand-size grains is associated with different types of *bedforms* (see Chapter 2) that change according to flow velocity and mean sediment size (Figure 6.1).

The load that a stream can actually transport is expressed in two ways. *Capacity* is the total amount of material that a stream can transport and depends on the current velocity and the

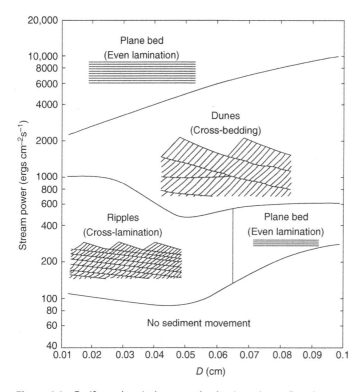

Figure 6.1 Bedforms in relation to grain size (mostly sand) and stream power (modified from Allen, 1970).

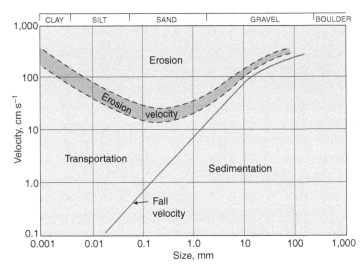

Figure 6.2 Hjulström's diagram showing velocity associated with the erosion, transport, and deposition of different sizes of particles.

discharge. *Competence* refers to the size of the largest particle that a stream can carry under existing conditions. It depends mostly on velocity.

The proportion of the types of load that a stream carries varies from year to year, and within the year; it is also a function of climate. For example, streams in arid climates transport little dissolved load, but much bed load, especially during flooding; dissolved load is obviously more important in humid climates. Thus, the ability of a stream to erode and transport sediment is related to velocity and discharge, it is evident that most erosion and transport is carried out during major flow events, such as spring flooding or during exceptional events such as hurricanes and other storms.

In a classic study, Hjulström (1939) related erosion, transportation and deposition to particle size and mean velocity (Figure 6.2). This figure shows that coarse sand size (ca. 0.5 to 1 mm) is the most readily eroded sediment by water, the river requiring more energy for both coarser and finer sediments; the latter is due to the cohesion of finer particles requiring higher energies to entrain them and the lower drag force. The diagram also shows the relatively low velocities to keep finer material entrained and transported.

6.3 Stream Deposits and Channel Patterns

Deposition takes place predominantly along the channel beds where discharges fluctuate in response to inputs of water or sediment. In the case of point bars, for example (see below), which are built vertically during bank full and high stage events, deposition takes place along channel sides. In addition, different types of deposits occur in response to certain related conditions, such as energy of the stream (e.g. low flow vs. flooding conditions), whether flow is confined to the channel or outside the channel, and extraneous inputs from the valley sides (Table 6.1).

The type of sediment being transported is generally linked to the morphology of the stream, which in turn is related to the flow conditions; both can ultimately be conditioned by climate (e.g. precipitation regime, vegetation). Geomorphologists and sedimentologists recognize several types

Table 6.1 Characteristics of valley sediments (modified from Summerfield, 1991).

Locus of deposition	Type	Characteristics
Channel	Transitory channel deposits	Mostly bed load temporarily at rest; may partially be preserved in more long-lasting channel fills or lateral accretions
	Lag deposits	Segregations of larger or heavier particles, more persistent than transitory channel deposits
	Channel fills	Accumulation in abandoned or aggrading channel segments; range from coarse bed load to fine-grained oxbow lake sediments
Channel Margin	Lateral accretion deposits	Point and marginal bars resulting from channel migration
Overbank floodplain	Vertical accretion deposits	Fine-grained suspended load of overbank flood water; includes natural levee and backswamp deposits
	Splays	Localized bedload deposits spread from channel onto adjacent floodplain
Valley margin	Colluvium	Slope deposits, poorly to moderately sorted, stony to fine-grained material, accumulated by slopewash and gravity (e.g. creep); commonly interfingers with channel margin and floodplain deposits
	Mass-movement deposits	Debris avalanche and landslide deposits intermixed with colluvium; mudflows generally in channels but spill over banks

of stream patterns (Figure 6.3). Familiarity with these types of patterns and associated deposits is important; it can aid in the geoarchaeological interpretation of specific fluvial deposits and facies (see Chapter 2) that can then be valuable in looking for (or avoiding) certain loci for past human settlement. Low-energy deposits occurring outside the channel, for example, are more likely to preserve *in situ* archaeological material compared to those from high-energy gravel bars that are found within active channels (Figures 6.4, 6.5, 6.6; see below). Moreover, an understanding of stream morphology and processes, including associated rates of change, may help explain an absence of surface sites in some areas (Guccione et al., 1998). A striking example on the alluvial geomorphology of major rivers and its impact on the preservation of archaeological sites comes from some recent surveys and excavations on the floodplains of the Yellow River, known for its high sediment loads and frequent avulsion and channel shifting (Qin et al., 2019). Prehistoric and historic sites which were once situated on vast floodplains of the Yellow River are often covered by deep flood deposits up to 5–10m thick (see Box 6.1).

Channels can be characterized as single or divided, with shapes that vary from straight, sinuous, meandering, braided, and anastomosing (anabranching) (Figure 6.3; Table 6.2). These different types represent an average condition as channel morphology can change from season to season and year to year, and often grade into each other (Bridge, 2003; Ritter et al., 2011). Such short-term changes are not normally visible on the geoarchaeological scale, especially with older, Lower Paleolithic sites, for example, in East Africa (Rogers et al., 1994; Stern et al., 2002). However, as shown in the following sections, Holocene channel changes and migration profoundly shaped alluvial landscapes and human occupation patterns in especially arid and semi-arid regions.

Straight, single channels tend to be rare and carry a mixture of suspended load and bedload. The latter commonly accumulates on opposite sides of the channel (Ritter et al., 2011), called *alternate bars*; shallow zones are called *riffles*, whereas deeper areas are called *pools*. The path that connects the deepest parts of the channel is called the *talweg*. Erosion is relatively minor, with slight lateral widening or vertical incision. Deposition occurs along bars during flood stage.

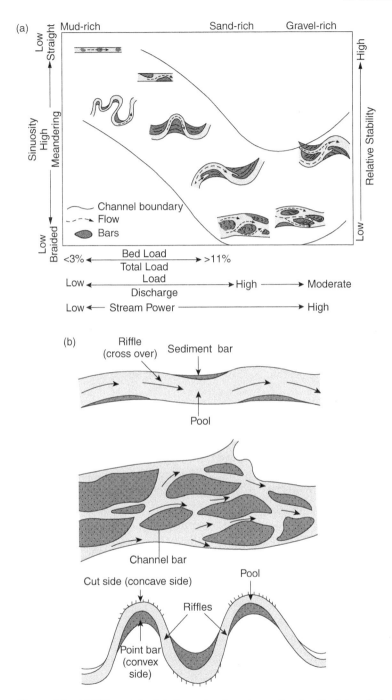

Figure 6.3 (a) Different fluvial channel pattern types showing the interrelation between channel form, sediment load, stream power, and stability (modified from Boggs, 2001). (b) Detailed view of straight, braided and meandering patterns (modified from Reineck and Singh, 1986: figure 372).

In *braided systems* rivers tend to be straight and many small channels are split off from the main channel; separating the channels are raised portions (bars), which are covered during periods of high water. Braided channels are characteristically found in arid and semi-arid areas (Figures 6.4a, b), especially in alluvial fans (see below), and in areas of glacial outwash. Braiding occurs in such

---

Box 6.1   Geoarchaeology and alluvial landscape at the Han Dynasty Sanyangzhuang site, Henan Province, China

---

The Middle and Lower Yellow River channel, for instance, is long known for its propensity to become infilled, migrate, and break out due to its silt-laden water and relatively flat basal geomorphology. Such characteristics rendered China's mother river also being considered as China's especial "sorrow". In the past 4000 years, the river experienced at least eight events of major channel migration and avulsion, with more than 1000 episodes of floods. Some of these devastating events of channel avulsion and migration coincided with the most populous early imperial and medieval periods and caused dramatic impact on the society when large populations began to settle along the river. The geological record of these floods is sometimes captured by deep alluvial sequences, some of which were recently exposed through quarrying and systematic archaeological excavations. Between 14–17 AD during the transitional Xin Mang regime sandwiched chronologically between the Western (202 BC–8 AD) and Eastern Han (25–220 AD), a massive flood broke out after channel avulsion. Topographic mapping shows that the crevasse splay covered a massive area of around 100 km^2. The Sanyangzhuang site and many other contemporary Han Dynasty sites (ca. 140 BC–23 AD) in Neihuang County of Henan Province, were buried by thick Yellow River flood deposits; Warring States and Late Neolithic layers were covered in more flood deposits, reaching up to 12 m thick in the excavated sediments (Kidder et al., 2012a, 2012b) (Figure 6.8). Excavations revealed 14 residential compounds, large areas of farming fields, roads (with cart tracks), and many other activity areas and artifacts (Qin et al., 2019). The roof tiles that were preserved almost intact, stone mills and mortars, cooking vessels, fossils casts of mulberry leaves – and most exceptionally, the well-preserved ridge and furrow within fields surrounding the houses – all pointed to a flourishing local community that was engaged with a wide range of economic activities including farming, and mulberry tree cultivation, and silk production. The establishment of the Sanyangzhuang and many other farming villages on the floodplains of the Yellow River represented an unprecedented development during the Han period when population grew to a colossal scale, forcing the government to allow population migration to the floodplains, which were previously reserved as the flood buffering zone (Zhuang and Kidder, 2014). However, Kidder et al.'s (2012a, 2012b) survey showed that people already began modifying the alluvial landscape since the Warring States era (475–221 BC) with the construction of possible levees. Indeed, at the Sanyangzhuang site too, Warring States farming fields were constructed in deeper positions of the excavated stratigraphy and were also overlain by flood deposits before the Han people occupied the area. The burgeoning Han agricultural community was put to an abrupt end by the floods. It appeared that the floods occurred several times but importantly, the flood water did not rise to a very high level overnight. Rather, it took days, if not longer, for the flood water to inundate the entire area. This meant that the Han community must have had sufficient time to retreat and take valuables with them. Even though the Han population had never made a return to the area, the Tang (618–907 AD) and later period occupation layers vividly denote social resilience in the long and diachronic relationship between human occupation history and alluvial evolution along the Middle and Lower Yellow River.

Table 6.2 Types of river patterns and characteristics (modified from (Morisawa, 1985)).

Type	Morphology	Load type	Width/ depth ratio	Erosion	Deposition
Straight	Single channel with pools and riffles; meandering talweg	Suspension- mixed load or bedload	<40	Minor channel widening and incision	Shoals
Sinuous	Single channel with pools and riffles; meandering talweg	Mixed	<40	Increased channel widening and incision	Shoals
Meandering	Single channels	Suspension or mixed load	<40	Channel incision, meander widening	Point bar formation
Braided	Two or more channels with bars and small islands	Bedload	>40	Channel widening	Channel aggradation, mid-channel bar formation
Anastomosing	Two or more channels with large, stable islands	Suspension load	<10	Slow meander widening	Slow bank accretion

Figure 6.4a Braided channel during a flood in the Gebel Katarina area, south-central Sinai, 1979. Note the gravel bar between two branches of the channel.

Figure 6.4b Braided channel from Qadesh Barnea area, Sinai. The braided channels are shown in the central part of the photograph with the main channel carrying water, and abandoned channels adjacent to it. The entire area between the vegetated areas on the right and left is covered during occasional flooding events that occur in winter; this area currently receives <100 mm of rainfall per year. On the left is a terrace riser ("MP") composed of gravel and associated with Middle Paleolithic artifacts. The height of the terrace is ca. 19 m above the present valley floor and indicates that during the Middle Paleolithic the valley was filled with gravels to this height (Goldberg, 1984).

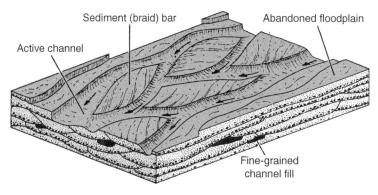

Figure 6.5 Block diagram showing surface morphology of a braided stream with active channels and bars. The depth dimension shows deposition of sediments within lenticular bodies, beginning with gravels that grade upward to finer sand and gravel. An abandoned floodplain on the right exists above the braided channel area (from Allen, 1970: figure 4.7). The preservation potential of intact sites within such deposits is low.

locales where there are rapid shifts in water discharge, in the supply of sediment (generally coarse), and where the stream banks are easily erodible (Boggs, 2001); the ratio of bedload to suspended load is high. Deposition of coarse material results in the formation of mid-channel bars, which direct the flow around them during high water discharges. Through time, valley floors can accumulate several meters of braided stream deposits, which appear as nested, lens-like masses of gravels (Figure 6.5). It is not surprising that in light of the periodically high discharges, the likelihood of finding intact archaeological sites within such deposits is low. Nevertheless, archaeological sites can be found on *abandoned* braided stream deposits, such as those on old allu-vial fan surfaces. In Sinai, for example, deflated remains of coarse bouldery deposits (*lags*) are commonly exploited as large building stones; these can be difficult to distinguish from the natural bouldery surface cover (Figure 6.6). Furthermore, on the floodplains of the Lower Nile, with the limited sinuosity, the braiding river is confined by "one of two possible channels" separated by an island which formed through continuous accretion of sandbars. In their recent surveys, Bunbury and colleagues found that some of these islands were attractive habitats for ancient Egyptians. The relatively brief timespan (< 100 years) of island formation and abandonment and consequent short human occupation, nonetheless witnessed some of the earliest attempts to adapt to and engi-neer the river geomorphology for human inhabitation (Bunbury, 2019: 106–110). These ambitious anthropogenic alterations of alluvial landscape included segmenting the "waning channel" in New Kingdom Memphis to reclaim more land, predating similar actions only prevalent in much later periods elsewhere (e.g. the Dujiangyan irrigation project during the Qin Dynasty in China) (Zhuang, 2017).

In *anastomosing channels*, in contrast to braided channels, flow is around bars that are relatively stable and not readily eroded. In the Welland Valley area of eastern Britain, anastomosing chan-nels lean toward representing cold, periglacial conditions between 10,900 and 10,000 years ago (French, 2003). Overall, however, they are relatively infrequent in the archaeological record.

Meandering channels and systems are widespread in geoarchaeology, and many sites are associ-ated with them. Unlike braided streams, in meandering ones, flow is within a single channel. They also differ in having lower sinuosity, shallower gradients, and finer sediment loads (Boggs, 2006). The principal aspects of a meandering river system are illustrated in Figure 6.7.

In a meandering system, coarse, gravelly material is transported within the channel, generally during times of flood; during average conditions, sandy bedload is transported (Walker and Cant, 1984) (Table 6.1). Erosion takes place along the outer reaches of meander beds where veloci-ties are higher. In contrast, deposition occurs in the inner part of the meander loop, leading to the formation of the point bar (Figures 6.3 and 6.7). Thus, through time the point bar can be seen to shift laterally across and downstream along the valley bottom and floodplain. This process of *lat-eral accretion* leads to the formation of a package of sediment that fines upward from gravel, sand, and finer silts and clay.

Increased erosion along the outer cut banks can lead to meanders being cut off and their aban-donment, ultimately leading to the formation of an oxbow lake (Figure 6.7). The sediment in these

Figure 6.6 Coarse alluvium exposed at surface of an alluvial fan in south-western Sinai. This coarse lag of boulders and gravels were original deposited by braided channels of the fan when it was active. The two figures (O. Bar-Yosef and N. Goren-Inbar) examine the remains of a Bronze Age building whose inhabitants exploited these boulders for construction.

(a)

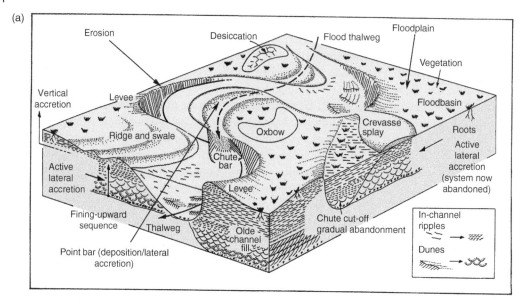

(b)

Figure 6.7 (a) Block diagram illustrating the major features of a meandering river system (from Walker and Cant, 1984: fig. 1). (b) Aerial view of a meandering river system showing flood basin (FB), present channel (Ch), and abandoned meander (MS) and associated ridge and swale topography along the point bar.

depressions consists of fine silt and clay, and are typically organic-rich; they contain diatoms, mollusks, and ostracods. Since they are away from the channel and contain water before being silted up, they are attractive to vegetation, game, and human occupation. A similar type of depression and infilling can be formed through avulsion, in which a channel breaks through its levee and abandons its channel. Such abandoned and isolated basins are not uncommon in the rivers that

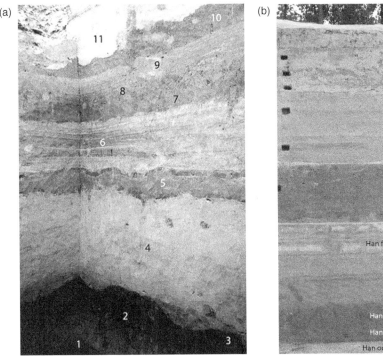
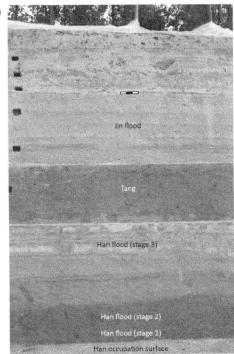

Figure 6.8 (a) Stratigraphy at Sanyangzhuang showing anthropogenic paleosols alternating with Yellow River flood deposits. 2(3), 5 and 7: Late Neolithic paleosol, Warring States Period field, and Han Dynasty occupation layer, respectively; 1, 4, 6, 8, 9, and 10: Yellow River flood deposits, 8–10 are dated to Western Han Dynasty (Kidder et al. 2012b/with permission of John Wiley & Sons). (b) Another stratigraphic section near Sanyangzhuang showing Tang Dynasty occupation deposits sandwiched by Han Dynasty and Jin Dynasty flood deposits (Kidder et al. 2012a / with permission of Cambridge University Press).

drain eastern Texas to Gulf of Mexico. A considerable part of the Late Paleoindian occupation at the Wilson-Leonard site (near Austin, Texas, USA), for example, took place next to an *avulsed channel* (Figure 3.6). A banana-shaped accumulation of organic silty clays within the avulsed channel was accompanied by bison kills (Bousman et al., 2002; Goldberg and Holliday, 1998). At the Aubrey Clovis site on the Trinity River in north-central Texas, a cut-off channel was the scene of a groundwater spring-fed pond which slightly pre-dates and is partly contemporaneous with the Clovis occupation (Humphrey and Ferring, 1994). The vast Amazonian alluvial plains are characterized by active lateral channel migration (Rozo et al., 2012) which has also fundamentally influenced prehistoric human occupation in these tropical and subtropical environments. Building on the pioneering work by Lathrap (1968), Parssinen et al. (1996) further investigated larger-scale interaction between evolution of meandering rivers and occupation dynamics in the Middle Ucayali River. The channel avulsion, which was well documented in early missionary records and oral history during late 1700s, was caused by long-term tectonic movements and occurred during a flood period. The avulsion left abandoned floodplain, oxbow lake, meander ridges, and swamps in the local landscape. The traditional communication route was cut off, and local hydrological and vegetation conditions were significantly altered, all contributing to adaptation and expansion patterns of indigenous cultures in the region. In addition to the brief cases discussed above, there are also ample examples of oxbow lake formation and channel avulsion resulted from evolution of meandering channels in the Old World, some of which are elaborated in Box 6.1.

6.4 Floodplains

Floodplains, flat to gently sloping areas adjacent to stream channels, are notable in geoarchaeology, for not only are they one of the most widespread of fluvial landscapes – past and present – but they served as significant loci for past activities, which can be well preserved. They are dynamic landscapes that exhibit a variety of local sedimentary environments and processes (Brown, 1997) that change over time as the floodplain evolves (Figure 6.9). Close to the channel, for example (Figure 6.10), sand and silt are found in raised, natural levees. These deposits grade laterally into finer silts and clays that accumulate in lower, backswamp areas where drainage is poor and only finer material accumulates during a flooding event. Consequently, although the backswamp might appear as an attractive locale for game, actual habitation is more likely to occur closer toward the levee, where the deposits are coarser and drainage is better. On the other hand, the position of these environments continually changes as the meanders sweep across the floodplain, covering or erasing previous deposits. Along the meanders of the Red River, Arkansas, for example, Guccione et al. (1998) found sites of different ages associated with different positions of the meander. The most recent meander belt (limits of meanders within the valley bottom) is about 200 to 300 years old and has covered any previous artifacts or sites with 1 to 2 m of alluvium. An older prehistoric site, on the other hand, is located on the surface associated with a 500- to 1,000-year-old abandoned meander belt. Finally, they indicate:

> At locations proximal to the river, the site may be buried by overbank sediment 0.4 m thick, but at more distant locations the site is at the surface or only buried by thin overbank sediment because of low sedimentation rates (0.04 cm yr^{-1}) over the span of a millennium. Sites, such as 3MI3/30, [Figure 6.10] that are occupied contemporaneous with overbank sedimentation may be stratified; however, localized erosion and removal of some archaeological material may occur where channelized flow crosses the natural levee.

Guccione et al., 1998: 475

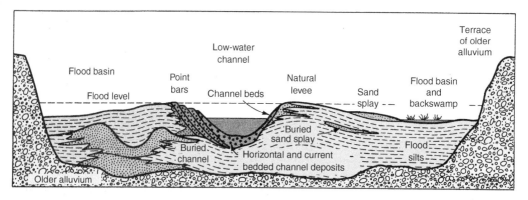

Figure 6.9 Lateral variations in deposits and facies are clearly shown in this schematic view of a floodplain. Buried and active channel gravels interfinger with finer grained flood silts, some associated with point bar deposits which here are migrating from left to right. In this example, the entire floodplain sequence is set into an eroded valley consisting of older, gravelly alluvium, which forms a terrace on the valley flanks (modified from Butzer, 1976: Figure 8-3).

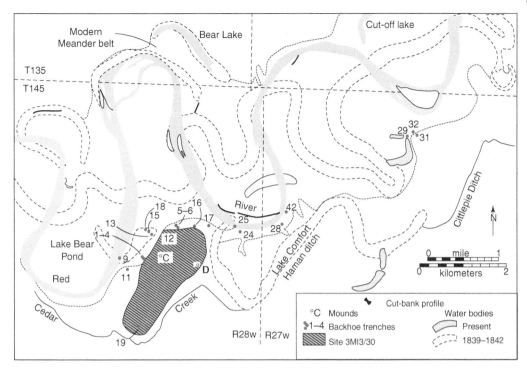

Figure 6.10 Changes of positions of meanders within the Red River during the past 160 years as determined by trenching and excavation. It is clear that any sites situated next to the older channel (e.g. 3MI3/30, a mound site) could be buried under point bar deposits of the present meander (from Guccione et al., 1998: Figure 3/with permission of John Wiley & Sons).

Their study elegantly demonstrates the active nature of meander belts, even within the past few centuries. Furthermore, it establishes that one must take into account issues, such as the vigor of the meander belt when working in such geomorphologically dynamic areas. Many of the techniques discussed in Chapter 16 (e.g. aerial and satellite photos, trenching, soil surveys) are helpful in evaluating the age of a fluvial landscape and rates of change, and whether sites of a given age will be likely be preserved if at all, either on the surface or beneath it.

6.4.1 Soils

Soils play an important part in fluvial geoarchaeology. They provide windows into understanding the relationships between sedimentation, pedogenesis, and archaeological site formation processes (Ferring, 1992). Furthermore, their formation can reflect regional landscape and climatic conditions, which in turn can be used as time parallel (roughly synchronic) stratigraphic marker horizons (see Chapter 3).

Soil visibility on floodplains involves a balance between rate of pedogenesis and geogenic sedimentation. At one extreme, in cases with continuous deposition in the same location, there is little chance for a mature profile to form: A horizons (perhaps as defined in the strictest sense) are continually forming with each subsequent flooding event, whether this interval is weekly, monthly, or annually. Such types of alluvial soils or *cumulic soils* straddle the area between sediments and soils (Figure 6.11). In such cases of relatively rapid rates of deposition, any archaeological occupation

Figure 6.11 Photograph of late Holocene floodplain Ford alluvium from Cowhouse Creek, central Texas, USA. Shown here is sequence of buried A and A-Bw horizons (Entisols) within vertical accretion flood plain deposits consisting of silts and clays (Nordt, 2004).

has a reasonably good chance of being buried, with the net result that several discrete occupations can be observed throughout out this over-thickened A horizon.

At the other extreme, where there is little or no sedimentation, repeated occupations occur on top of each other, resulting in the formation of *palimpsests* (Bailey, 2007; Ferring, 1986). As a result, it is virtually impossible to isolate individual occupations. Furthermore, inasmuch as the archaeological material lies about the surface, it can be subjected to modification and disturbance by a host of post-depositional processes and pedogenesis, such as displacement by water, burrowing and gnawing by animals (*bioturbation*), chemical weathering, and pedoturbation (e.g. shrink-swell) (Leigh, 2001; Limbrey, 1992). If the surface is exposed for a long enough period of time (depending on factors such as climate, parent material; see Chapter 4) soil horizon development will take place. If such a soil (now paleosol) profile ultimately becomes preserved and protected by renewed deposition, future excavation will reveal archaeological material occurring within the soil profile. Finally, since soils can be faithful recorders of local or regional (climatic) conditions (see Chapter 4 on soils) the type of soils can therefore provide paleoenvironmental information during or close to the time of one or more of the occupations. Such is the case for a calcic horizon, which is developed on red clayey alluvium in the Jordan Valley and which is universally associated with Epi-Paleolithic (Geometric Kebaran; ca. 14.5 to 13.7 ky uncal.) sites (Bar-Yosef, 1974; Maher, 2004; Schuldenrein and Goldberg, 1981).

Another indicator of landscape stability on fine-grained floodplain deposits is the formation of surface gleys. These result from poor drainage and associated oxidation/reduction conditions with the accumulation of organic matter. In addition, under wet conditions such formerly humic topsoils and turbated topsoil material become preferentially impregnated with iron and manganese compounds (mottling), and commonly show the occurrence of pyrite or pyrite pseudomorphs (g horizon) (Mees and Stoops, 2018; Miedema et al., 1974; Wiltshire et al., 1994).

Paleosols such as those described above also serve as temporal markers because they may have formed at roughly the same time interval in many places. In the Southern Plains, for example, soil formation took place during a period of flood plain stability between 2000 and 1000 BP

(Ferring, 1992). Although the paleosol goes by different names in different places (e.g. Caddo in Oklahoma; Quitaque in Texas) it serves as a useful stratigraphic marker for placing late Archaic, Plains Woodland, and Late Prehistoric sites in a firm stratigraphic and temporal framework (Ferring, 1992).

Alluvial archaeology is equally well established in Europe, particularly since the 1980s when sand and gravel extraction increased exponentially the number of CRM/"Rescue" sites (Brown, 1997; Howard et al., 2003; Needham and Macklin, 1992). These studies provided insights into landscape and climatic changes. For instance, in the Fens (East Anglia), and along major rivers such as the Nene (Northamptonshire), Ouse (Bedfordshire), and Upper Thames (Oxfordshire), whole prehistoric landscapes have been buried by fine alluvium dating to Iron Age through to Medieval times (Healy and Harding, 2007; Robinson, 1992) (Figure 6.12). The original landscape consisted of coarse loamy Pleistocene/early Holocene river terrace deposits on which the soils had developed by mid-Holocene times. Ritual and burial monuments and landscape divisions (e.g. avenues, barrows, cursuses) exhibited tree-throw features that could date back as early as the Late Mesolithic-Early Neolithic transition, some 6,000 years ago (French, 2003; Lambrick, 1992; Macphail, 2011a; Macphail and Goldberg, 1990).

Such sites thus provide the opportunity to investigate the archaeology and environment over long periods. Tree throws provide windows into the formation and state of prehistoric landscapes, such as Early Neolithic and Early Bronze Age monuments and Roman and Saxon settlements at Raunds (Northamptonshire) (Harding and Healy, 2011), and at Drayton Cursus (Upper Thames). In the latter, the Neolithic Cursus seals tree-throw hollows, and subsequent prehistoric soils are buried by Roman and Saxon alluvium (Barclay et al., 2003). At both sites the alluvium buried early to mid-Holocene Alfisols, which provide "control soils" that can be used to compare soil development in later prehistory and historic times (Macphail and Goldberg, 1990). At Raunds, the consensus interpretation of the environmental data is for a Neolithic and Bronze Age grazed floodplain

Figure 6.12 Section through charcoal and burned soil-rich remains of prehistoric tree-throw hole fill buried by a meter of Saxon and Medieval fine alluvium at Raunds, Northamptonshire, UK. 4 – subsoils and gravels of Holocene alfisols formed in Nene river terrace sandy loams, sands and gravels; 3 – truncated burned tree hollow with burned soils with high magnetic susceptibility values (e.g. χ = 894 × 10^{-8} m^3 kg^{-1}); 2 – Saxon-medieval fine alluvium (e.g. clay loam); 1 – modern pasture soil with grass rooting and earthworm burrowing into top of fine alluvium.

landscape on which stock concentrations produced trampled and phosphate-enriched soils that were sealed by numerous earth monuments (Macphail and Linderholm, 2004a). Textural pedofeatures associated with this soil disturbance were readily differentiated from "Saxon and medieval" clay inwash induced by alluviation (Brammer, 1971; Kühn et al., 2018; Macphail, 2011a) (see Table 17.1b).

A number of other riverine sites suffered the effects of increased wetness during the Holocene. For example, at Three Ways Wharf (Uxbridge, UK) Upper Paleolithic artifacts (*c.* 10,000 BP) were biologically worked throughout the earliest Holocene soil whereas flints from the Early Mesolithic (*c.* 9–8,000 BP) occupation (ca. 1000 years later) remained near or on the surface because they were sealed by a fine charcoal-rich humic clay (and peat) associated with wetter conditions. Similar peaty deposits formed throughout the Colne River valley during the early Boreal period (Lewis et al., 1992; Lewis and Rackham, 2011). This wetness essentially stopped bioturbation of the Mesolithic artifacts at this site. The Upper Paleolithic site was located on a low bar of a braided stream and functioned as a likely ambush camp at a river crossing. In contrast, by early Mesolithic times the site was swampy and humans were seemingly affecting the coniferous landscape by the use of fire (for comparable wildfire site, see Box 17.1).

The Upper Paleolithic site of Pincevent in France, famous for the preservation of its living floors and activity areas (Leroi-Gourhan, 1984) exhibits exceptional preservation, largely due to its location on the floodplain of the Seine. Here, low-energy overbank flooding led to internment of the occupational remains, without movement of the artifacts. Lastly, it can be recorded that alluviation of the Lower Thames in London led to the deposition of highly polluted (P, Pb, Cu, and Zn) moat deposits at the Tower of London (Macphail and Crowther, 2004).

Recent studies of the 0.7–1.4 million-year-old Early Stone Age Olduvai Middle Bed II site EF-HR uncovered a hundred hand axes and thousands of fossil remains (de la Torre et al., 2018; Macphail and Crowther, 2015a). Alluvial sediments within the deepest parts of a river-cut valley, which retain small amounts of organic matter (1.12–1.65% LOI), indicate an originally humic deposit (Macphail and Crowther, 2015a). The sediments also seem anomalously phosphate-rich (3.09–4.54 mg g^{-1} P), until a sediment micromorphology study revealed concentrations of poorly preserved fine bone, including likely fish bone in a muddy matrix (Figure 6.13). These results indicate drying out of the river and a death assemblage of fine bone. Also in the valley (T2-Main Trench), a claystone occupation with surface concentrations of both poorly preserved small and large fossils, and artifacts was sealed by a post-depositional carbonate crust (archaeological unit T9L10).

Holocene fluvial dynamics played a decisive role in the long flourishment and eventual decline of the Harappan Civilization as demonstrated by some recent geoarchaeological investigations. Unlike many other fluvial systems in semi-arid and arid environments that experience frequent alluvial incision and aggradation, the fluvial landscape of the western Indo-Gangetic plain remained remarkably stable for a relatively long period of time (between c.7700–3900 BP). Whilst the late Holocene climate was becoming drier, the stable fluvial landscape would have provided reliable water source for the floodwater agriculture of the Harappan Civilization (Petrie et al., 2017). The Harappan settlements were located on a diverse range of micro-environments on such broad fluvial landscape. Neogi et al.'s (2020: 137) geoarchaeological investigation confirms that Harappan settlements situated on "sandy levees and/or riverbank" that were linked with former channels were sustained by annually replenishing river water with rich nutrients for agriculture (Figure 6.14). Such sedimentation represents "an important aspect of Indus cultural adaptations to diverse, variable, and changing environments". According to Giosan et al. (2012) as monsoon weakened and rivers became seasonal or dried, the Harappan urbanism declined, punctuated by a pronounced trend of eastward migration.

Figure 6.13 (a) SEM X-ray backscatter image of HWK M26A (Layer 2 Lower); bone fragment(s) within sands and small amounts of matrix sediment. Scale bar = 1mm. (b) Figure 6.13a, under BL, showing small amounts of residual bone autofluorescence. Frame width is ~1.5mm.

(a)

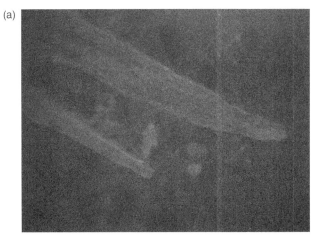

(b)

1 mm   Electron Image 1

Figure 6.14 Digital Elevation Model (DEM) around the Sutlej-Yamuna interfluve showing alluvial geomorphology with the form and relative elevation of the levees (yellow raised lines), lower terrains (light blue), and paleochannel courses (dark blue). Frame width ca. 100 km (modified after Neogi et al., 2020 / with permission of Cambridge University Press).

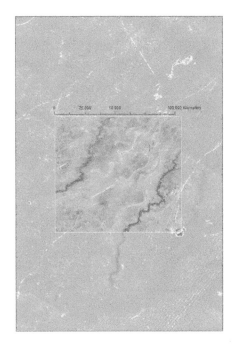

6.5   Stream Terraces

As discussed above, channels constantly shift their position, either on a seasonal or yearly basis. In meandering channels, the stream sweeps across the valley floor resulting in the lateral accumulation of point bar deposits that over time can lead to vertical net accumulations. Similar net accumulation of deposits can occur in braided systems.

6.5.1   Characteristics of Terraces

In many environments and locales, this net sediment accumulation is interrupted by a marked change in hydrological regime, resulting in the vertical incision (*entrenchment*) by the stream into previously deposited alluvium, or even bedrock. Such incision often leaves behind a geomorphic feature, a *terrace*, which consists of a flat portion (the "tread") and a steeper sloping surface (the "riser") that connects the tread to the level of the new floodplain or a lower terrace surface (Figure 6.15). Thus, in many cases, terraces represent abandoned floodplains that existed when the river flowed at this higher elevation. However, it should be kept in mind that terraces are essentially a geomorphic/topographic feature that can form on bedrock, or previously deposited alluvium or other sediment(s) (Figure 6.16).

Two types of terraces are commonly recognized, erosional and depositional. Erosional terraces are produced by downcutting associated with lateral erosion, producing stepped surfaces as shown in Figure 6.15. If the underlying material is bedrock, they are commonly called *strath terraces*. With depositional terraces the tread and riser are composed of valley alluvium (Figure 6.16).

Although processes involved in the formation of terraces is complex, varied, and often a product of a combined factors, a phase of valley accumulation (*aggradation*) must be followed by a

Figure 6.15   Photo of terraces from Nahal Ze'elim, western Jordan Valley. This ephemeral stream flows into the Dead Sea from the Judean Mountains during major rainfall events. During Pleistocene times a large lake, Lake Lisan, over 200 m higher than the present-day Dead Sea (ca. 405 m below sea level), existed in the Jordan Valley. When this lake dried up at the end of the Pleistocene (Bartov et al., 2002), base level lowered rapidly, and streams such as this one incised into their channels relatively rapidly as they moved across the valley floor. The result is this flight of terraces developed on bedrock, fluvial, and lacustrine deposits.

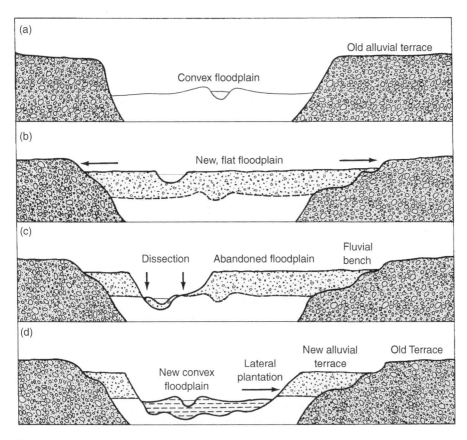

Figure 6.16 Idealized view of alluvial terrace formation developed in alluvium. (a) As shown here, this first stage depicts build-up (aggradation) of fine-grained alluvium within an eroded valley that formed by the downcutting through previously deposited gravelly alluvium; (b = OLD c). Coarser alluvium (braided stream, perhaps dryer conditions) covers these initial deposits, enlarging the floodplain by lateral planation; (c = OLD d) initial incision and downcutting (degradation) take place due to such factors as climatic change or base level lowering, leaving behind an abandoned floodplain surface; (d = OLD e) downcutting ceases and renewed alluviation takes place that is characterized by fine-grained sediments (as in (a)), with lateral planation resulting in the widening of the floodplain and removal of some of the previous alluvium from stages (a) and (b = OLD c). As a consequence, a second alluvial terrace is formed at a lower elevation than the first, but it composed of two different types of deposits (modified from Butzer, 1976: figure 8-13, p. 170).

phase of downcutting (entrenchment). Similarly, aggradation occurs when the supply of material is greater than the ability of the system to remove it. This in turn can be brought about by (a) inputs of glacial outwash, (b) changes in local or regional base level caused by changes in sea level (brought about by eustatic or tectonic causes), (c) inputs of coarse sediment related to tectonic uplift, or (d) climate change (Ritter et al., 2011) (see Box 6.2). Entrenchment arises from similar causes, principally tectonism or climate change. In valleys adjacent the Dead Sea, for example, terraces were formed as the streams incised their channels in response to a lowered base level resulting from the shrinking of the Pleistocene Lake Lisan, the precursor to the modern Dead Sea (Figure 6.15) (Bartov et al., 2002; Bowman, 1997; Niemi, 1997; Schuldenrein and Goldberg, 1981).

6.5.2 Archaeological Sites in Terrace Contexts

Depositional terraces provide a wealth of stratigraphic, paleoenvironmental, and geoarchaeological information as they represent fossilized loci of human activities focused on alluvial environments. Furthermore prehistoric sites associated with terraces provide materials (organic matter/charcoal/hearths) that are datable by ^{14}C or cultural remains that can be temporally bracketed. Within recent years the success and refinement of OSL/TL techniques shows that they can be used to date the sediments directly (Fuchs and Lang, 2001), thus avoiding the problems of the charcoal being reworked or derived from old wood.

6.5.2.1 New World Sites

Sites found in terrace contexts exist over the entire globe and many have been extensively studied. In Central North America, terrace sequences are common along major and minor drainages, and many contain prehistoric sites with among the earliest Paleoindian sites there (Blum et al., 1992; Blum and Valastro, 1992; Ferring, 1986, 1992; Mandel, 1995, 2008; Nordt, 1995). The common practice is to label the lowest and youngest terrace, i.e., the one closest to the level of the modern channel, T_0 and successively older and higher terraces, T_1, T_2, and so forth.

Geoarchaeological research in the American Southwest has been intensive and extensive for over 65 years with the early work of Kirk Bryan and E. Antevs (Waters, 2000), followed by the dedicated work of C. V. Haynes (e.g. Haynes, 1991, 1995) and his student, M. Waters (Waters, 1992; Waters and Ravesloot, 2000). Waters' lifetime work in the area has sought to address a number of issues relating to alluvial stratigraphic sequences, how they reflect upon paleoenvironments and the completeness of the archaeological record (Waters, 2000). For example, the occurrence of Clovis sites in the San Pedro Valley of southern Arizona is not matched in neighboring valleys where Clovis remains are absent. The question is of course, does this distribution represent a cultural preference or another, perhaps geological control of site distribution?

Detailed geoarchaeological work in these valleys is shown in Figure 6.17, which reveals some interesting patterns. It is clear that the absence of sites in the Santa Cruz River and Tonto Basin, for example, is related to the absence of sedimentary traps to store these sites; even if they were there, they are long been eroded. Secondly, even where deposits do exist (e.g. Whitewater Draw, Gila River) the braided stream hydrology would result in cultural material being eroded and reworked throughout the gravelly deposits. Finally, we see that valley deposition and erosion in the Late Pleistocene and early Holocene were not uniform throughout the area. This pattern would be expected if each valley had specific and different geomorphic variables (e.g. rock type, slopes), or if there were climate differences among valleys, or the climate was the same but each valley reacted differently to climate changes (Waters, 2000). In any case, this research shows that a geological filter (viz., erosion) is responsible for the apparent absence of sites beyond the San Pedro River.

In the same area, Waters evaluated the effects fluvial landscape change on Hohokam prehistoric agriculturalists who occupied the Gila and Salt River valleys from AD 100 to 1450 (Waters, 2000). The major questions involved the change in pattern in public areas around AD 1250 (e.g. ballcourts being replaced by platform mounds; the reduction in number of settlements but increase in size), and the collapse of the Hohokam at AD 1450 and the abandonment of canals. A summary of thorough geoarchaeological examination of natural exposures and those from backhoe trenches is given in Figure 6.18. As can be seen several aggradational phases could be recognized, accompanied by downcutting and infilling. Waters (2000) noted that the during the period AD 1050 to 1150 the Gila River channel incised and widened (Unit IV), which appears to have coincided with a stage of high magnitude flooding in this area. The braided channel cut into the floodplain that had been previously stable for close to 1000 years; such conditions made it difficult to organize the

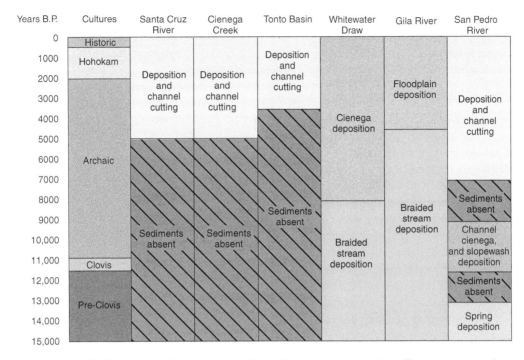

Figure 6.17 Alluvial sequences from several drainages in south-western Arizona. These sequences show different events (e.g. erosion, deposition in streams, and cienegas) taking place in different valleys at different times. Erosion and lack of sediments in some valleys explains the differential distribution of Clovis sites in the area (from Waters, 2000: Figure 3/with permission of Cambridge University Press).

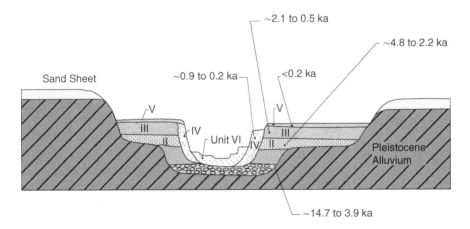

Figure 6.18 Composite cross-section from Middle Gila River, AZ, illustrating terraces, stratigraphic units and associated ^{14}C dates (modified from Waters, 2000: Figure 4).

transport of water over the highly porous stream gravels as channel location shifted continuously: "Any temporary diversion dams constructed to get water into headgates [of the canals] would have been vulnerable to being washed out with such channel environment" (Waters, 2000: 552–553). Waters links these hydraulic issues to changes in labor organization, canal engineering, and increased and more diverse food production. Furthermore, he notes that after this time, the channel appears to have stabilized with the deposition of silty clay and the formation of a paleosol, which was accompanied by probable farming and the establishment of villages on the stable sur-face. Little or no geomorphic change appears to have taken place at the end of Hohokam occupa-tion, and Waters concludes that ". . . the reorganization of the Hohokam at the end of the Classic must have been the result of other factors" (Waters, 2000: 554).

In the Central Plains (Kansas, Nebraska, and Texas USA), extensive work by Mandel has been instrumental in reconstructing the geomorphic history of this large region and how this knowl-edge can be used to understand the presence and temporal distribution of Paleoindian and Archaic sites in the area (e.g. Holliday and Miller, 2013; Mandel, 1992, 1995, 2000b, 2008; Mandel et al., 2018). He used data from his own field work and those from the literature (e.g. unpublished dissertations, CRM reports) to document Holocene alluvial stratigraphy within parts of the Platte, Missouri, Kansas, and Arkansas River drainages. As in the southwest, perceptible gaps in the archaeological record can result from burial of the archaeological material, and erosion and removal of deposits containing cultural materials.

His results showed significant differences in the presence of deposits according to stream size, in that large streams and small streams exhibited different age distributions of the sediments (Figure 6.19). He noted, for example, that large streams contained early, middle, and late Holocene alluvium, whereas small streams exhibited only late Holocene deposits; early and middle Holocene deposits can be found within alluvial fans where small streams enter major valleys. Furthermore, the overall pau-city of radiocarbon dates (usually determined on soil organic matter) between 7 and 6 ka (roughly the time of the Altithermal in this area) is ascribed to the dynamic nature of valley sedimentation, which prevented the formation of soils. It appears that erosion of the uplands and much of the early Holocene alluvium in small streams was removed and redeposited in the alluvial fans through the middle Holocene (ca. 8 to 4 ka). These patterns can be documented elsewhere in the Midwest, and when combined with paleoclimatic data (pollen, paleobotanical evidence, computer models of climate), the inference is that they result from climatic change over the entire region. Specifically, they point to warmer and drier conditions associated with zonal (i.e., west to east) atmospheric circulation. In the later Holocene, in contrast, flow became more mixed with a shift toward more meridional (south to north) flow, and greater influence of moisture coming from the Gulf of Mexico. This shift corresponds to increased precipitation, and a concomitant change in amount and type of vegetation, such as short-grass prairie to mixed and tall-grass prairies, and increased forest cover. In turn, this led to ". . . reduced erosion rates on hillslopes, which in turn would have reduced mean annual sediment concentrations in small streams" (Mandel, 1995: 60) and greater storage of alluvium.

The importance of such detailed work over a large region is to demonstrate that the lack of Archaic sites in the interval 8 to 4 ka is a function of geomorphic controls rather than prehistoric lifeways. Such geomorphic processes either removed sites or buried them. Mandel points out that prior work relied on shallow testing of the landscape (e.g. shovel testing or surface survey) to locate sites. It is clear that such strategies will lead to a disproportional amount of younger sites that occur at the surface, and will ignore the potential for locating earlier archaeological material. Furthermore, he concludes, that a basin-wide approach is needed to reveal the geomorphic and archaeological history, including the major gaps.

In the Southern Plains, alluvial sequences and terrace formation have been documented along many rivers in Texas and Oklahoma (e.g. Ferring, 1990, 1992, 2001; Nordt, 1995, 2004; Waters and Nordt, 1995), revealing deposits that span what appear to be the bulk of human settlement in this part of North America. The Aubrey site, North Texas, for example, contains among the earliest dated Clovis remains in North America (Ferring, 1995, 2001; Humphrey and Ferring, 1994). It consists of a variety of sediments that include a basal channel lag/bar, overlain by lacustrine/spring deposits and flood plain silts, capped by soil formation.

A very similar setting is found at the Wilson-Leonard site, north of Austin (see Box 3.1). Here, along a lower terrace of Brushy Creek, is a ca. 5m thick sequence spans Paleoindian through

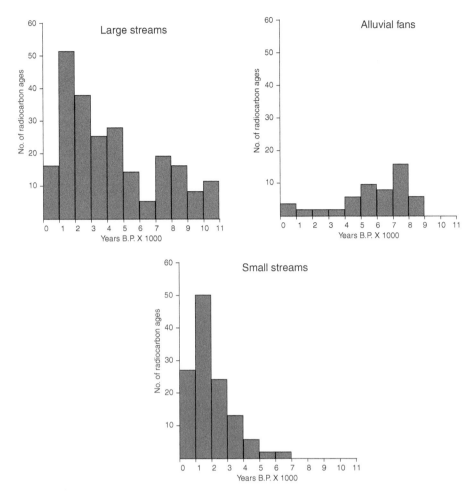

Figure 6.19 Histograms showing number of radiocarbon dates for alluvial fills from large and small streams, and alluvial fans in the Central Plains (modified from Mandel, 1995: figure 8).

Late Prehistoric archaeological remains, roughly 11 ka through 4 ka (uncalibrated) (Bousman et al., 2002; Goldberg and Holliday, 1998). The deposits consist of alluvial gravel that grades upward and laterally into organic-rich cienega deposits and flood plain silts, capped by the formation of a weak soil. These in turn, are covered by colluvial silt and fine gravel that become increasingly anthropogenic toward the top of the sequence, where numerous burned rock ovens occur in association with Early Archaic hunters and gatherers. The sequence points to an overall decrease of alluvial activity during the late Quaternary with a drier spell occurring at the beginning of the Holocene. This is also confirmed from paleobotanical and molluskan data (Bousman et al., 2002). In any case, the site was occupied on and off for 7,000 years of fluvial and colluvial deposition.

Elsewhere in Texas, two sites have strongly figured into the debate about the peopling of the Americas and pre-Clovis occupation: the Gault site (Williams et al., 2018b) and the neighboring Debra L. Friedkin site (Waters et al., 2011) within the Buttermilk Creek Valley. At both sites, alluvium underlying Clovis-rich layers date to ~16–20 thousand years ago (OSL) at Gault and 13.2 to 15.5 at Friedkin. The sealed nature of the associated lithic complexes strongly supports the picture of Pre-Clovis occupation in the US.

6.5.2.2 Sites in the Old World

In the Old World, terrace sequences are abundant in the southern and eastern half of the Mediterranean Basin, where numerous prehistoric and historical archaeological sites have been well documented in fluvial and fluvio-lacustrine deposits (e.g. Goldberg, 1986, 1987, 1994; Hassan, 1995; Macumber et al., 1991; Rosen, 1986; Schuldenrein, 2007; Schuldenrein and Clark, 1994, 2001; Zilberman, 1993) (see Box 6.2). The earliest and best preserved ones found with prehistoric materials in clear context date to late middle Pleistocene or Late Pleistocene times and are associated with Middle Paleolithic sites and artifact scatters (Figure 6.7b). The earliest part of the sequence begins with the accumulation of well-rounded gravels linked to Middle Paleolithic artifacts that locally occur within or on top of them. These gravels occur within well-defined terrace morphologies (e.g. Figure 6.7b) or as remnants within or on the side of valleys. Unfortunately, chronometric dating of these early terraces has been elusive, and in need of clarification. The presence of fossil springs and relatively high amounts of arboreal pollen with some of these sites (D-35 in the Central Negev) suggested markedly wetter climates at this time (Goldberg, 1977b; Horowitz, 1979).

Most of these gravels were eroded with up to 20 m of vertical downcutting during the later part of the Middle Paleolithic, roughly between 70 ka and 45 ka, although the date of the onset of incision is approximate. In turn, incision was followed by renewed and widespread aggradation of massive fine-grained silts and clays that began at the very final stages of the Middle Paleolithic and beginning of the Upper Paleolithic, roughly 45 ka. These deposits can be found over the entire area, from southern Sinai (Gladfelter, 1992; Phillips, 1988), the Negev, and into Jordan. Their lithologies and bedding characteristics indicate that they represent massive inputs of originally aeolian derived sediments that were washed out of the atmosphere and concomitantly eroded off the slopes and introduced into the fluvial system. Damp and swampy (*paludal*) conditions existed in a climate that was overall wetter than today's; the reasonable possibility of summer rainfall (at present only 50–200 m of precipitation falls in the winter) have been suggested (Sneh, 1983).

Upper Paleolithic aggradation ceased about 22 ka and in most areas was followed by incision until about 15–14 ka, although of a more modest scale than the previous downcutting cycle. The end of the Late Pleistocene (ca. 15–14 to 10 ka) is overall marked by renewed sedimentation but on a smaller scale. In many places, the deposits are locally reworked alluvium and colluvium from the previous depositional cycle but locally, high water tables attest to overall wetter climates for the Epi-Paleolithic. In fact, Epi-Paleolithic sites are widespread over the entire zone covering much of Sinai and the Negev (e.g. Goring-Morris, 1987). The early Holocene is represented by marked incision: in Nahal Besor ca. 20 m of incision took place within <8000 years, and was interrupted by a silty terrace deposits that interfinger with Chalcolithic sites and cultural deposits (Shiqmim; Goldberg, 1987; Goldberg and Rosen, 1987) dated to about 7 to 5 ka. It is not a coincidence that the stream was aggrading at this time of occupation of this relatively large village in the desert (Levy, 1995), and it is likely that climate was wetter than at present. Much of the later part of the Holocene is characterized by downcutting and erosion except for a period of extensive alluviation of ca. 1 m thick silts with *in situ* Byzantine remains throughout the Negev and E Sinai. Many of these deposits appear to have been cultivated by the Byzantines.

Causes of aggradation and incision in this region - and in many places - appear to be clearly driven by climate. This is shown by the fact that similar depositional and erosional features occur over a large area (Sinai, the Negev, and W Jordan) in which streams drain both to the Dead Sea (then about 175 to 200 m below sea level, vs. ca. −405 m today) and to the Mediterranean. This latter aspect rules out base level changes as an influence. Similarly, a marked period of incision along Nahal Besor, a major stream that drains much of the Western Negev, took place at the very beginning of the Holocene, just when worldwide sea levels were rising. Furthermore, it is interesting to note that excavation and survey data suggest that for many locations, particularly for the Upper Paleolithic and Epi-Paleolithic there is a general trend toward increased number of sites including surface sites, not associated with fluvial deposits during periods of sedimentation. The suggestion again, is that in most general terms, these aggradational phases point to wetter climates in what is now a hyper-arid to arid terrain.

Rapidly accumulating geoarchaeological data in East and South Asia have provided fresh insights to understand diverse, long-term relationships between terrace evolution and human activities and assess how fluvial landscape change profoundly impacted ancient lifeways. On the loess plateau, for instance, terraces on loess-mantled landscapes mainly form through the process of alluvial incision during relatively cooler and drier conditions, although tectonic activities might have also contributed to the formation of alluvial terraces (Pan et al., 2007; Wang et al., 2010). Such stable alluvial geomorphology during terrace formation provided desirable condition for settlement location. A good case in point is the Late Paleolithic site at Longwangchan in the Hukou Area of Shaanxi Province which was excavated recently. Two strath terraces were recognized in this loess-mantled landscape. The bedrock surface on the higher one is ca. 25 m above present water table. The exceptionally thick (16 m) loess deposits covering on top of the bedrock surface point to continuous loess accumulation on the stable terrace surface, only punctuated by a sandy layer of "hyperconcentrated flow deposits" (Zhang et al., 2010, 2011) (Figure 6.20). The onset of human occupation coincided with loess deposition after an episode of river incision at around 29 ka. The revelation of human occupation on alluvial terraces during the late MIS3 and early MIS2 with strong winter winds at Longwangchan provides an important case to understand cultural adaptation to adverse environmental condition in the transition to agriculture.

Alluvial terraces continued to be attractive places that supported growth and expansion of prehistoric settlements across diverse environments of North China during the Holocene. In their influential studies, Xia and colleagues demonstrated the vertical distribution of prehistoric settlements in relation to evolution of alluvial terraces in the Xar Moron River of temperate Eastern Inner Mongolia. The two-order terraces (T_1 and T_2) formed after river channel incised loess tablelands around 6500 and 4000 BP that are about 180–150 m higher, respectively than the present river. Together with the original tablelands, the terrace surfaces were occupied by the Xinglongwa-Zhaobaogou (8000–6000 BP), Hongshan-Xiaoheyan (6000–4000 BP), and Xiajiadian cultural periods (4000–2800 BP) respectively, while the most recent terrace and floodplains were lived by late historical people (Xia et al., 2000) (Figure 6.21). The migration of prehistoric and historic populations from higher to lower altitudes was not only intrinsically related to Holocene climate – especially hydrological fluctuations – but also closely linked to alluvial evolution and changing subsistence strategies.

Contrary to the prolonged stability of a terrace surface at Longwangchan and vertical movements of prehistoric settlements due to alluvial incision in the Xar Moron River, alluvial landscapes on the Central Plains of China experienced more frequent cycles between alluvial incision and aggradation and dynamic temporal-spatial variation in alluvial landscapes on different catchments of the rivers. Such cyclic geomorphic processes were determined by combined factors of climate change, tectonism, and base-level change. For instance, in their recent surveys, Lu et al. (2020) have reconstructed at least three major alluvial aggradation-incision cycles in the southeastern Songshan region, the heartland for the cradle of Chinese civilization. The first major alluvial incision occurred ca. 9000 BP on the loess tablelands, with the formation of typical loess terraces. In some places (e.g. the Shumuyuan location in Lu et al., 2020), the river valley incision was up to 10 m deep. The relatively low-altitude plains area of the region experienced continuous alluvial-lacustrine aggradation which sustained the formation of large lakes. In the loess area, the early-Holocene river incision gave way to a prolonged episode of alluvial aggradation, whilst rivers on the plains area continued to silt up. Alluvial downcutting resumed after 4000 BP in the loess area. The remarkably deeply incised valley shapes the basic characteristics of modern geomorphology. Throughout the Holocene, terraces on the loess area were the optimal places for the Peiligang, Yangshao, and Longshan culture (ca. 8500–4000 BP) occupation, whilst the plains area only became suitable for habitation after the waterbodies began to shrink under a relatively cooler and drier late Holocene climate regime (Figure 6.22) (see also Rosen, 2007, 2008; Rosen et al., 2017).

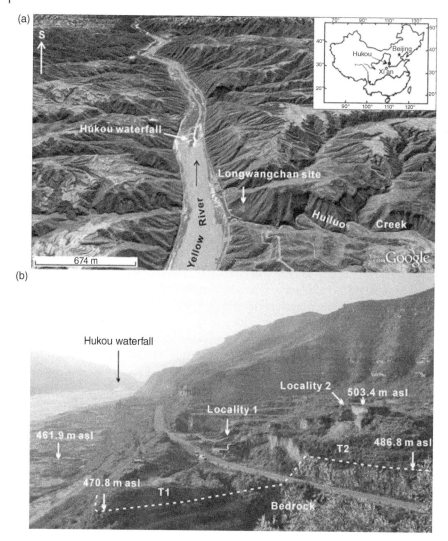

Figure 6.20 (a) Google Earth image showing the location of the Longwangchan site along the Yellow River on a typical loess landscape. (b) Photograph showing positions of localities 1 and 2 at the site on the Yellow River terraces. Captions and figures from Zhang et al. 2011/with permission of Elsevier.

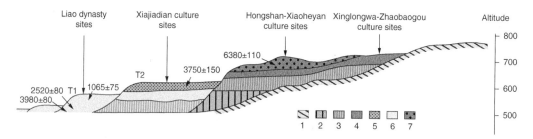

Figure 6.21 Schematic illustration showing locations of Neolithic, Bronze Age to historical sites on different alluvial terraces situated on different altitudes of the Xar Moron River. 1: geological bedrocks; 2: red clay; 3: Lishi loess; 4: Malan loess; 5: Holocene loess; 6: alluvial sediments; 7: aeolian sediments. Dates are in BP. (after Xia et al., 2000).

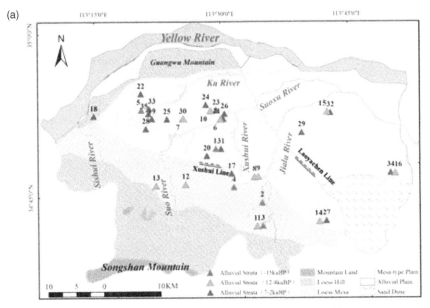

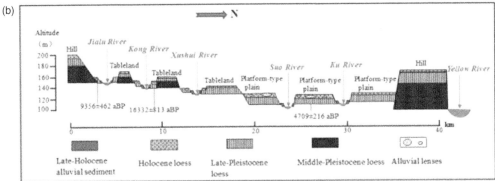

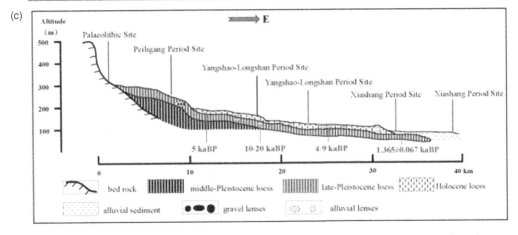

Figure 6.22 (a) Spatial boundaries of different geomorphological types in the Northeastern Songshan Region of China's Central Plains with numerical numbers showing locations of examined stratigraphies. (b) The south-north cross-section showing different geomorphological types and their representative sedimentary sequences and rivers. (c) The east–west cross-section showing typical sedimentary sequences and archaeological sites on different geomorphological types, especially the alluvial terraces. Captions and figures from Lu et al., 2020/with permission of Elsevier.

A final example comes from the Upper Indus region of Pakistan. Geoarchaeological work by Schuldenrein et al. (2004) has shown that alluviation slowed and pedogenesis ensued during the early Holocene; it appears that at this time lateral accretion deposition gave way to entrenchment in these larger floodplains. Such changes appear to be climatically induced and related to changes in intensity of the SW monsoon. Indeed, as we have discussed in section 6.4, such a stable alluvial landscape was conducive to the flourishment of the Harappan Civilization when the late Holocene climate condition as a whole was deteriorating.

6.6 Conclusions

This chapter touched on only some of the major aspects of fluvial environments, and the geoarchaeological issues that might be associated with them. Some of these features are summarized in Figure 6.23 and Table 6.3, which portrays some of the archaeological contexts that can be found in a meandering river. While specifically aimed at the meandering river context, it illustrates some of major questions concerning fluvial geoarchaeology. These include: (a) site visibility – whether archaeological material is not present on the surface because of erosion or burial (the value of examining a large region both on the surface and with depth was demonstrated by Mandel's work); (b) dynamics of the landscape – how active, for example, is lateral planation and accretion, and will sites only a few hundred years old be visible on the surface due to this "etch-a-sketch" erasure of flood plain features? Clearly, older landscapes on higher terraces are liable to have thick, polygenetic soil formation, which may be associated with multiple, likely mixed, occupations from different time periods. Thus, when working within the geoarchaeological contexts of alluvial deposits, one must observe closely the deposits and facies on the surface and at depth and obtain

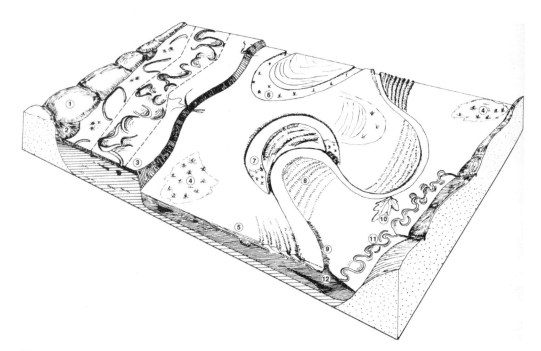

Figure 6.23 Schematic summary of the types of archaeological context that can be found in a meandering river system (modified from Gladfelter, 2001: figure 4.2).

proper chronometric control. The latter can be done preferably by the indirect dating of their contents (^{14}C dating of organic matter or other archaeological materials); advances in OSL dating offer promise to date the deposits directly (Chapter 16). In addition, the growing application of deep drilling techniques in geoarchaeological campaigns, especially in regions which experience rapid urbanization and increasing infrastructural construction, has enabled geoarchaeologists to understand the diachronic relationship between thick Holocene sediment facies and occupation histories in diverse fluvial settings.

Table 6.3 Associations of archaeological material with alluvial contexts (numbers are keyed to Figure 6.23) (modified from Gladfelter, 2001: table 4.1).

Geomorphic aspect	Possible archaeological associations
1. Dissected upland expressed as a bluff above the valley.	Sites of any age possible. Soils may be polygenetic and palimpsests of several occupations are possible making it difficult to distinguish individual occupations.
2. Alluvial fan emanating from dissected upland	Archaeological material on and in the fan can have been reworked from the upland. They are also younger than #3.
3. Alluvial terrace with older, inactive meanders. The meander belt is shown by dashed lines. A paleosol is developed on the surface of the terrace.	Sites within the meander belt cannot be older than the meandering activity, and sites inside of a particular meander loop cannot be older than the meander loop. Buried sites that survived past meandering activity will be old.
4. Backswamp along lower, interior setting on the floodplain and flood basin.	Surface and near-surface sites may be preserved and the oldest sites on the floodplain are likely to be found in these settings.
5. Contemporary floodplain, with convex shape and meander scars of a former river channel. Some features outlined in greater detail below (e.g. 8).	Location and age of sites is a function of the degree of geomorphic activity. For active meanders, older surface sites may be rare; age of buried sites conditioned by rate of vertical accretion. Within a meander loop (e.g. 8) surface sites become younger toward the outside of the loop.
6. Infilled scar of meander cut-off and degraded point bar deposits.	Sites pre-date location of present channel. Deposits within scar are likely to be organic-rich.
7. Oxbow lake in meander cut-off plugged at both ends with clay.	Sites associated with lake can be any age that postdate its formation. Sites can be buried in levee deposits or on surface of point bars.
8. Point bars	Sites on point bars or in swales between them. No site can be younger than the expansion of the meander loop.
9. Natural levee on outside of meander loop	Sites within levee deposits of active channel are penecontemporaneous with active channel; burial at great depth is not likely.
10. Crevasse splay formed where natural levee is breached	Occupation likely brief while splay is active, but more attractive when it becomes inactive.
11. Yazoo-type river that flows in lower, interior margin of the floodplain near the base of the bluff. Source of water comes from bluff with much lower discharges than in major river.	Sites older than meandering of Yazoo stream may survive beyond its meander belt in vertical accretion deposits of major stream.

Box 6.2 Upper and Middle Paleolithic sites of Nahal Zin, Central Negev, Israel

Fluvial environments contain rich records of archaeological sites, and arid environments are particularly effective in preserving and displaying sites because the stratigraphic profiles are exposed, and with little concealment by vegetation they are more readily observable. Such conditions enable us to recognize buried sites that are commonly concealed in more humid environments where vegetation cover is more extensive. An area that has yielded a wealth of prehistoric sites is the Nahal Zin area of the Central Negev in Israel (Figure 6.24) (Marks, 1983). Sites ranging from Lower Paleolithic through Chalcolithic Periods were uncovered during the extensive surveys and excavations made in the 1960s and 1970s. Most of all, numerous Middle and Upper Paleolithic sites were found within discreet fluvial deposits representing succinct aggradational (depositional) events, separated by erosion and vertical downcutting. Younger, Epi-Paleolithic, and Neolithic sites tend to occur on the surfaces of these older deposits.

The Nahal Zin[i] drainage originates at about 1000 m in elevation, in the upper part of the Central Negev Highlands, a plateau composed of Eocene chalk and limestone with abundant layers of chert. Along the northern edge of the plateau, in the area of Sede Boker (Figure 6.24), the stream has incised into the channel forming a canyon and a series of *knickpoints* that are associated with springs (e.g. Ein Mor, Ein Avdat) along the ca. 100 m of vertical drop from the plateau down to its lower course. At this lower elevation, the stream flows on softer marl and chalk, and continues to flow NE for about 1.75 km, after which it turns E and flows eventually into the Dead Sea. Along this NE stretch two important Middle and Upper Paleolithic sites were found, Boker Tachtit (**BT**) and Boker (**B**) (Figures 6.24 and 6.25).

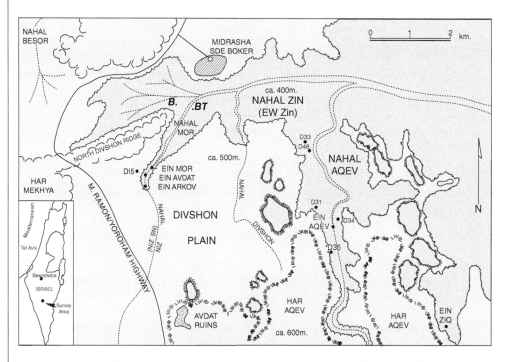

Figure 6.24 Nahal Zin area in the Central Negev, Israel. The Upper Paleolithic site of Boker (B) and Middle/Upper Paleolithic site of Boker Tachtit (BT) are shown (modified from Goldberg, 1976).

Box 6.2 (Continued)

Figure 6.25 Nahal Zin valley looking SW toward the Central Negev Plateau from the area of Sede Boker (cf., Figure 6.24). The site of Boker Tachtit (BT) is situated within terrace deposits adjacent to the road, about 7 m above the floor of the channel. The flat surface of the terrace continues along to the southwest and northeast. The site of Boker (B) sits across the channel from it. The hill immediately above and to the right (west) of Boker is a remnant of a landscape surface and gravelly deposits that contain the remains of an undated Middle Paleolithic site (D6). During the Middle Paleolithic occupation of the site, Nahal Zin was flowing about 60 m above the present-day channel and was more continuous with Nahal Havarim, which flows in from the right (Goldberg and Brimer, 1983). Compare with cross-section in Figure 6.28.

Both sites are contained within a sequence of fluvial deposits that include soft, marly and chalky silts, gravels, calcareous silt, and colluvial clay. The site of Boker Tachtit is the older and is situated on the east side of the valley (Figures 6.24 and 6.25) within the lower part of a terrace remnant. The archaeological material, which spans the Middle Paleolithic/Upper Paleolithic boundary, is composed of predominantly fine marly silts, with lenses and bands of gravel that accumulated as overbank deposits not far from the channel (Figures 6.26). These deposits became somewhat consolidated before they were eroded by gravels that truncate (erode) them with a sharp contact. At this time, the channel – and associated higher energy gravels – had shifted to a new position directly over the former finer-grained flood bank. Occupation at this location after the shift of channel was not very likely as it would have corresponded to the center of the gravel-laden stream channel.

Boker consists of a number of sites or individual occupations that are found at different elevations and locales within a similar terrace vestige of similar height on the west side of the valley (Figures 6.24, 6.25, and 6.28a). The deposits that constitute the terrace are exposed over a length of about 85 m and vary from 3 to 9 m in thickness. The sites are enclosed within interbedded sands, silts, and clays (Figures 6.28b) and locally some gravel lenses. The gravels and sands are predominantly of alluvial origin from Nahal Zin. On the other hand, the silts are

(Continued)

Box 6.2   (Continued)

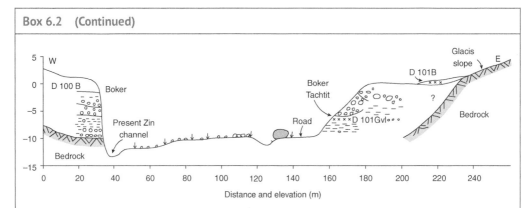

Figure 6.26   Schematic W-E cross-section across Nahal Zin showing the topographic and stratigraphic relationships between the sites of Boker and Boker Tachtit (modified from Goldberg, 1983). Note the well-preserved form of the terraces, particularly Boker Tachtit, which shows the classic tread and riser morphology. The Epi-Paleolithic site of D101B rests within a colluvial and aeolian loess mantle on top of the Boker Tachtit tread surface and is shown in Figure 6.27a (modified from Goldberg, 1983).

Figure 6.27a   The site of Boker Tachtit as seen from the site of Boker. The two major sedimentary units exposed during the excavation can be seen within the terrace. The upper surface of the terrace (the tread) is actually covered with a thin veneer of colluvial and aeolian tan silts (arrow) that contain Epi-Paleolithic artifacts that date to about 13 ka.

both alluvial and aeolian in origin. Windblown silts are common in the Negev and recognizable by their good sorting and massive character. Hard, gray clay lenses are colluvially derived from the surrounding gray Paleogene marls into which the terrace sediments are set (Figures 6.28a, 6.28b, and 6.28c).

Box 6.2 (Continued)

Figure 6.27b The terrace sediments of Boker Tachtit are composed of two major fluvial units as exposed in a gully (at left) and excavations: (a) a basal chalky silt with lenses (stringers) of gravel, which are marked by a sharp erosional contact of (b) the overlying silty gravel. The sharp and undulating contact clearly indicates that the lower deposits were somewhat consolidated before the gravels truncated them. The transition between the Middle and Upper Paleolithic, which occurred about 40 to 45,000 years ago, takes place within the lower, chalky unit. Scale is 50 cm.

The amount of geomorphic change in this area can be appreciated from the cross-section through the Nahal Zin. Figure 6.29 shows the elevations of different surface remnants that can be dated by the presence of *in situ* archaeological sites, including those of Boker, Boker Tachtit, and D6, mentioned above. We see a progressive downcutting of the Nahal Zin from Mousterian times, here estimated to be about 200,000 years ago (A. E. Marks, personal communication; this time estimate is more up to date than that in Goldberg and Brimer, 1983). Moreover, we see that from Mousterian time to that of the Middle/Upper Paleolithic transition about 45,000 years ago, the stream cut down to roughly its present elevation, and then began to fill up the valley resulting in the alluvial accumulation associated with the sites of Boker and Boker Tachtit. The latest occupation phases at Boker some 25,000 years ago marks the termination of accumulation of Nahal Zin sediment, which was followed by another phase of downcutting that continues to the present.

The causes of such periods or cycles of downcutting (incision) and infilling (aggradation) for areas such as these has been much discussed over several decades. Changes in base level, climate change, tectonic movements, and intrinsic changes in the fluvial system (see Chapter 5) are commonly cited to produce such changes in fluvial landscapes. In the Nahal Zin area, fluctuations in Pleistocene climate – mostly precipitation – appear to be the most reasonable cause, although in light of the complexity of alluvial environments it is difficult to reconstruct specific processes and responses. Other data from the southern Levant (e.g. similar

(Continued)

Box 6.2 (Continued)

Figure 6.28a The site complex of Boker, situated on the west bank of Nahal Zin across from Boker Tachtit, looking southwest. The terrace sediments are set into eroded soft gray chalks and marls which can be seen in the foreground and immediately to the right of the site. Eroded and colluvially reworked portions of these marls make up the grayish layers in the terrace sediments shown in the excavated areas of the site (Area BE of the original excavation; Figures 6.28b and 6.28c). The terrace form here is less pronounced and somewhat more eroded than at Boker Tachtit. Its continuation to the south is visible in the upper part of the photograph where a well-defined remnant can be sloping to the east.

Figure 6.28b Area BE of the original excavation. Layers of pale, brown-colored sand and silt alternate with wedges of grayish clay. The sand is alluvial origin with some aeolian loess layers. The grayish clay layers are colluvially derived marls from the surrounding hills (cf. Figure 6.28a). In this regard note that the clays thicken in the colluvial source to the right, whereas the pale, brown-colored alluvium thickens to the left, in the direction of the Nahal Zin. The black streaks are thin lenses of charcoal, representing dispersed hearth components. The height of the profile in the center of the photograph is about 110 cm.

Box 6.2 (Continued)

Figure 6.28c Detail of profile shown in Figure 6.28b showing the alternation of the sediments and lateral thickening and thinning of the tan loess and silt layers (Ls) and the gray clay (Gc). Note how the gray clays coalesce to the right but become separated by the sand silts and sands to the left, both reflecting the different lithological source areas.

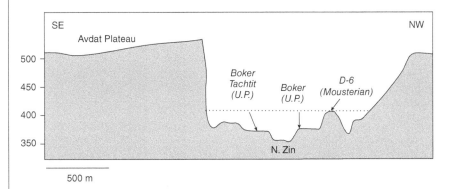

Figure 6.29 Generalized cross-section through Nahal Zin showing the changes in elevations of landscape surfaces through time. The Mousterian surface is coeval with site D6 discussed above and can be traced throughout much of the Nahal Zin area. Its age is not definitively known, but recent estimates put it at about 200,000 years old, indicating that at this time, the valley was about 60 m higher than it is today. The stream subsequently incised its channel eroding into the underlying limestone, chalks, and marls of the local bedrock. This downcutting ceased with the onset of alluvial deposition associated with the Boker/Boker Tachtit terraces, which began about 45,000 years ago as based on the radiocarbon dates from Boker Tachtit (Marks, 1983). Alluvial build-up continued through the middle part of the Upper Paleolithic about 25,000 years ago, as marked by the presence of the upper levels of Boker. Shortly after this time, the stream began to incise its channel, a phenomenon which continues until today.

(Continued)

Box 6.2 **(Continued)**

geomorphological and stratigraphic sequences over large regions, distribution of archaeological sites) lend support to the influence of climatic fluctuations (see Goldberg, 1986; Goldberg and Bar-Yosef, 1995).

[i] *Nahal* is Hebrew for ephemeral stream

7

Lakes

7.1 Introduction

After discussing fluvial systems and a special focus on rivers, we now turn to lakes. Lacustrine environments and associated wetlands are linked to fluvial terrains by the fact that many are fed by streams. The chain of lakes that dotted the East African and Levantine Rift systems served as attractive locations to early hominins coming from Africa, through to the Middle East (de la Torre et al., 2018). Lakes continued to be a focus well into the Holocene, when for example, numerous habitations built out into lakes in Switzerland, and along the margins of the Great Lakes in the USA (Ismail-Meyer et al., 2013). In this chapter, we briefly examine some of the principal aspects of lacustrine systems, categorize several types of lakes and their geneses, and characterize their sedimentation, hydrologies, and ecosystems. In line with the structure adopted in other similar chapters, we discuss relevant examples of the geoarchaeology of lakes in different regions of the world. Because of the conceptual ambiguity between lakes and wetlands under certain circumstances (e.g. Loffler, 2004: 8), especially when dealing with shallow lakes or lakes with dramatic seasonal variations in water levels, we also briefly discuss hydrology and soils of wetlands and provide several geoarchaeological examples.

7.2 Origins and Types of Lakes

Lakes form in depressions with various geomorphological and geological settings. They are closed freshwater bodies of standing water that vary considerably in size, both during past decades and centuries but also over the course of the Quaternary. A lake is a sophisticated ecosystem, cutting across multiple interfaces between sediment, water, and light. A typical lake system will include littoral zone, sublittoral zone, and shoreline or lake edge, of which the littoral zone is typically subdivided into several subzones according to water depth and light penetration (Tundisi and Matsumura Tundisi, 2012) (Figure 7.1).

Lakes can be classified in diverse ways. The basins in which they are formed have numerous origins, including rift valleys, volcanic and meteorite craters, glacial depressions left by decaying ice (kettles) or retreating ice (moraines), alluvial floodplains (oxbows and avulsed channels), or karstic depressions (sinkholes). Nevertheless, lakes can be short-lived features on the earth's surface, and are subject to desiccation or infilling.

Lakes are fed by melting glacier and snow, direct surface runoff through precipitation and stream flows, and groundwater. The water budget of a lake is determined by its inflow and outflow.

Practical and Theoretical Geoarchaeology, Second Edition. Paul Goldberg and Richard I. Macphail.
© 2022 John Wiley & Sons Ltd. Published 2022 by John Wiley & Sons Ltd.

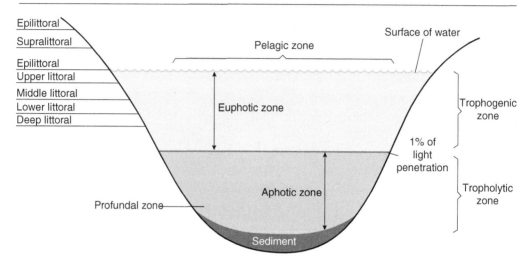

Figure 7.1   Zonation and terminologies used to designate different regions in a lake system (from Tundisi and Matsumura Tundisi, 2012/with permission of Taylor & Francis).

The former refers to sources of water just described, whilst outflow mainly include water discharge (e.g. stream flows and groundwater sinks) and evaporation. In terms of hydrological cycles, lakes are categorized as either open lakes or closed lakes. Open lakes (*exorheic*) are those that have an outlet, and consequently remain fresh, without concentrations of salts. They also tend to have stable shorelines (Boggs, 2012) with only some slight, short-range fluctuations in lake level taking place. Closed lakes (*endorheic*), on the other hand, have no outflow and dissolved solutes are concentrated; in arid and semi-arid areas, evapotranspiration generally exceeds any inputs from rivers or springs. In these same areas, lakes can be only temporary features on the landscape (ephemeral lakes), where their basins are filled for only short periods of time, much of the time being dry. Such a closed water regime, if prolonged can lead to the formation of salt lakes. In more extreme areas, lake bottoms can be seen but are essentially dry for most of the time (playa lakes), although in more humid episodes during the Quaternary, they were perennial, or at least seasonal features of the landscape. This is a common hydrological situation for lakes in arid inlands (Winter, 2004: 73–74) or deserts where lakes experience extreme seasonal fluctuations.

Closed lakes are unstable and subject to large inter- and intra-annual fluctuations in volume and position of the shoreline. Likewise, in permanent lakes, water levels also differ from year to year. Due to this sensitivity of external inputs of water and sediment, and associated changes in water chemistry that influences the biota (e.g. pollen, diatoms), lakes tend to provide valuable paleoclimatic records, either preserved within the lake (lacustrine) deposits themselves (Begin et al., 1980) or in the record of shoreline changes (e.g. Enzel et al., 2003; Klein, 1986). Indeed, limnologists and geoarchaeologists pay great attention to lake water level fluctuations as a crucial proxy for the reconstruction of hydrology and evolution of lakes and understanding how they interacted with ancient populations living by lakes.

According to their geological origins, lakes commonly include *glacial, tectonic, fluvial,* and *volcanic* types; other kinds, such as aeolian lakes and meteorite lakes, are less common (Hakanson and Jansson, 1983: 5–31). Relatively speaking, glacial and tectonic lakes are large and deep (but exceptions exist too, such as the shallow tectonic lake of Chad), in comparison to fluvial lakes that are generally shallow and vary greatly in size. Trophic level refers to water quality of a lake and it is another way of categorizing lakes. Based on the nutrient levels, lakes are divided into oligotrophic, eutrophic, mesotrophic, and dystrophic (Herschy, 2012b). These terminologies, perhaps unfamiliar to many readers, are useful indicators to monitor the nutrient balance and dynamics in a lake system. Another parameter that greatly affects lacustrine chemistry is temperature. The thermal regime oscillates from the surface to the bottom of the lake and between different seasons, forming a so-called thermal stratification (Bengtsson, 2012), which significantly affects chemical properties and biological status of limnetic ecosystems.

The majority of the world's largest lakes are of tectonic or glacial origins (Herdendor, 1990). Most large glacial lakes are located north of 40°N where glacial activities were at their maximum, whilst large tectonic lakes occur where major continental rifts take place. The former includes the Great Lakes in North America, whereas good examples of the latter are Lake Baikal and the great lakes in the East Africa's Rift Valley (Johnson, 1984). In terms of age, compared to tectonic and volcanic lakes, which sometimes are millions of years old, glacial lakes are generally younger than 15,000 years. These lakes are great reservoirs for the world's surface fresh water. The Great Lakes, for example, have a total surface area of ca. 244,000 km^2 and a volume of ca. 22,670 km^3, only exceeded by Lake Baikal. Comparatively, smaller lakes are more numerous, scattered on large-sized continents such as China, Southeast Asia, and Europe. Unlike its enormous rivers, lakes play a less important role in storing water resources in China. Nonetheless, there are at least 2,300 lakes larger than 1 km^2 in area. Although they amount to only 0.8% of China's territory, they hold ca. 708 km^3 of water and receive 11% of annual rainfall in the country (Zhang and Yang, 2012). Poyang, Dongting, and Taihu are amongst some of the important lakes located in the middle and lower Yangtze River valleys. Historically, these regions were said to have been concentrated with many more lakes of various sizes. Excluding the Black Sea and the Caspian that are sometimes considered as "lakes", Europe has many glacial lakes in its Alpine areas. There are many other categories of lakes along the coast and other geological areas of Europe (Fairbridge and Bengtsson, 2012).

7.3 Characteristics of Lakes

7.3.1 Sedimentation

7.3.1.1 Clastic

Sediment deposition in lakes occurs as both solutes and particles (or particulate matter) through the so-called *allochthonous* and *autochthonous* sedimentation processes. The main sources of lake sedimentation come from input of clastic particles from tributaries, atmospheric particles, eroded materials, and biochemical sediments derived from biological, physical and chemical weathering in lakes (Stumm, 2004). In addition to suspension and saltation processes commonly seen in riverine sedimentation environments (see sections 6.2 and 6.3), limnic sedimentation is also influenced by often deep and vertically fluctuating water velocities (e.g. from surface water to water near the lake bottom) and changing from the lake bottoms to lakeshores. These cause resuspension and mixing of sediments near lake bottoms or shallow-water areas, and consequent redeposition in various positions of the lake. Thus, clastic sediments, once they enter the lake system are subject to vertical and horizontal transportation and reworking.

Lacustrine sediments are varied, and several facies can be recognized. Knowledge of these variations can aid in understanding the location and function of sites in now exhumed fossil lacustrine deposits. As shown by the schematic illustration in Figure 7.2, much of the coarser load of sand

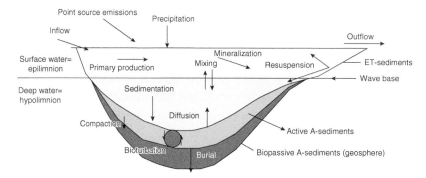

Figure 7.2 Illustration demonstrating transportation processes to, within, and from lakes (from Hakanson, 2012/with permission of Springer Nature).

and silt-sized particles transported into lakes through river streams and canalization, is dropped directly on lake floors along the margins. It is shown that current velocities exceeding 10 or 20 cm s^{-1}, which occur only in deep water during rainstorms, are required to further transport such heavy sediments (Rea et al., 1981) onto lake floors. Away from the lake margins, the finer material (e.g. clay and carbonate minerals) and light organic remains (e.g. diatoms) are carried in suspension by currents, with coarser, silt-sized material being transported closer to the shore, and finer material farther lakeward, where it eventually settles to the bottom. Winds promote surface turbulence and currents, which help keep finer material suspended. At the same time, wind-induced waves and currents may also redistribute coarser materials around the coastal margins (Nichols, 1999). As pointed out by Hakanson (2012), horizontal velocity is 10 times or "up to 10,000 times larger" than the vertical force needed to dislocate finer materials in lakes. Finer materials are therefore much more easily subject to further mixing and redeposition, depending on current velocities and lake floor terrains.

The deep-water facies consists of silts and clays, and organic matter, which are subject to alteration by further sedimentary processes. Lake floors are divided into three zones: the erosional zone, transportation zone, and accumulating zone (Hakanson, 1977). Mixing and redeposition activities on lake floors include focusing, resuspension, and so forth. Sediment focusing refers to transportation of materials by water turbulence from shallower to deeper positions on the erosional and transport zones of the lake floor (Blais and Kalff, 1995). It is a complicated process to model as it involves sliding, slumping and other physical redistribution of sediments and considerable debate exists regarding whether it is deep or shallow lakes that exhibit active sediment focusing (Blais and Kalff, 1995; Hilton, 1985). It is nonetheless an important aspect to consider when estimating sediment balance and lake contamination (e.g. van Metre and Fuller, 2009). Sediment resuspension, sometimes referred to as secondary flux (Bloesch, 2004: 217), is a similarly complex process. Shear stress exceeding settling velocity is required to lift sediments from lake bottoms. Sediment resuspension takes place in both shallow littoral and profundal zones and is determined by wind, currents and lake morphometry (Bloesch, 1995; Zhang et al., 2015) (Figure 7.3). Bloesch (2004) shows that current speeds higher than 7 cm s^{-1} are able to "resuspend organic particles up to 100 μm in diameter" but too weak to stir up clay minerals a few micrometers in diameter. According to Hawley et al. (1996), current speeds exceeding 20 cm s^{-1} are strong enough to resuspend material trapped in lake bottoms. Resuspension of organic and inorganic non-cohesive particles has great ecological implications due to the cycling of particles and nutrients as well as contaminants which are critical for algal production and other biological activities in lakes (Bloesch, 2004: 219).

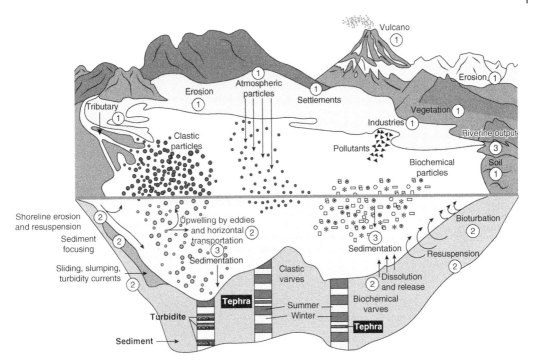

Figure 7.3 External and internal factors on sedimentation in a lake system. ①: sources of suspended particles; ②: transportation and transformation of suspended particles; ③: removal of suspended particles (from Bloesch, 2004/with permission of John Wiley & Sons).

Sedimentation rates vary greatly within and between lakes and over a longer term. In lakes rich in allochthonous input, for instance, sedimentation in lake bottoms is increased (Hilton et al., 1986). Estimation of sedimentation rate is very important in paleolimnologic studies as modeling sedimentation budget is fundamental to understand and quantify mass balance of long-term, and external and internal transportation (Hakanson, 2012). Annual sedimentation in lakes typically falls between 1–5 mm, which is much greater than that of oceans. Lakes rich in particles have yearly sedimentation rates up to 1 cm, whilst many volcanic oligotrophic lakes and crater lakes (e.g. Lake Keilambete and Lake Buchanan from Australia) have annual sedimentation rates as little as 0.008 mm yr^{-1} (Loffler, 2004: 52). It is well established that intensified agricultural activities around lake basins will result in significant increases in lake sedimentation (e.g. Heathcote et al., 2013). After compaction, these yearly sediments become laminated *varves*. In lakes in colder climates such as Norway and Sweden, alternations of mineral and organic layers form on a seasonal basis, with coarse material accumulating during periods of snow melt in spring and summer, whereas the finer, organic fraction builds up in winter when the lakes are frozen. The production of these varves is widespread in Late Weichselian lake sediments (Ringberg and Erlstrom, 1999) and is instrumental in dating glacio-lacustrine deposits there (Ringberg, 1994). In addition, the organic fraction can be dated by radiocarbon. Such a distinctive characteristic of lake sediments makes them valuable archives for high-resolution dating and paleoenvironmental reconstruction (Zolitschka et al., 2015). Lake sediments are also generally characterized by their fine texture and light color compared to alluvial sediments, which results from relatively lower hydrodynamics in lakes and prolonged inundation that promotes chemical reduction.

7.3.1.2 Chemical

Solutes that flow into lakes contain diverse kinds of chemical elements and compounds. Some of these inorganic materials dissolve or precipitate in various locations of the lake during and after transportation. Together with active biochemical weathering of autochthonous materials, large amounts of chemical compounds are concentrated in lakes. This section provides a brief summary of main chemical processes and how they affect the deposition of chemicals in lakes.

Chemical processes originate at the atmospheric level, which involves the reaction of "atmospheric pollutants and natural components in the atmosphere" with rain through the so-called acid rain process, although understandably such a process prevails mostly in lake systems functioning during industrial and post-industrial eras. The process leads to formation of carbon, sulfur, and nitrogen oxides and to the absorption of gas and other materials in water (Stumm, 2004: 80). Redox reactions and photosynthesis are among the most common chemical processes in lake water, sediments, and plants. Photosynthesis converts CO_2 to organic matter and oxygen by plants and algae, which consume phosphate and nitrate. Redox processes take place when one chemical species donates electrons whilst another one receives these electrons. Carbon, nitrogen, oxygen, sulfur, hydrogen, iron, and manganese are the major chemical elements in aquatic redox processes (Stumm, 2004: 94). These chemical processes are responsible for cycling of different elements, between different interfaces (e.g. sediment and water interface) and on various scales in lakes (Hamilton-Taylor and Davison, 1995; Santschi et al., 1990).

A general classification of chemical composition in Nordic lakes provides a useful reference for temperate lakes (Table 7.1). Apart from major elements, silicates and carbonates make up a considerable proportion and come from local rocks and allochthonous inputs. In calcareous regions, calcium and carbonates are particularly rich in lake sediments; *calcite* is the most dominant form of carbonates. Calcite precipitates in warm surface water in lakes with active photosynthetic activities, whilst biogenic calcite precipitates in lakes that are supersaturated with bicarbonate (HCO_3^-) (Stabel, 1985). Significant variables related to calcium carbonate precipitation include temperature, photosynthetic rate, depth of lake water, and pH values (Karami et al., 2019; Kelts and Hsu, 1978). A variety of other forms of calcite also precipitate in lakes, such as magnesian calcite, aragonite, and dolomite. In many cases, Mg/Ca ratio is considered a parameter for water salinity; the formation of each type of carbonate, if examined quantitively, might reflect sensitive temperature, water quality and other parameter changes in lakes (Kelts and Hsu, 1978). *Silicates* tend to remain stable in lakes, although the dissolution and absorption of some silicates is not uncommon (e.g. through diatoms) (Bloesch, 2004). Silicate fractions therefore reflect "bedrock composition of the catchment area" (Hakanson and Jansson, 1983: 104). Some of the carbonates and silicates are also clay sized and to avoid overlap, clay minerals here refer to less-abundant minerals such as chlorite, illite, and smectite. In high-latitude lakes such as the Great Lakes in North America,

Table 7.1 Chemical classification of elements in lake sediments (after Hakanson and Jansson, 1983).

Major elements (Si, Al, K, Na, and Mg) make up the largest group of the sediment matrix.

Carbonate elements (Ca, Mg, and CO_3-C) constitute the largest group in sediments, about 15% of the materials by weight.

Nutrient elements (org.-C, N, and P) account for approximately 10% in recent lake deposits.

Mobile elements (Mn, Fe, and S) make up about 5% of the total sediment weight.

Trace elements (Hg, Cd, Pb, Cu, Zn, Ni, Cr, Ag, V, etc.), the smallest group, account for < 0.1% of the sediments.

Table 7.2 Major evaporate minerals (after Reineck and Singh, 1986: table 19).

Mineral	Composition	Mineral	Composition
Carbonates		*Chlorides*	
Calcite	$CaCO_3$	Halite	NaCl
Aragonite	$CaCO_3$	Sylvite	KCl
Dolomite	$CaMg(CO_3)_2$	*Borates*	
Magnesite	$MgCO_3$	Borax	$Na_2B_4O_7 \cdot 10H_2O$
Natron	$Na_2CO_3 \cdot 10H_2O$	*Nitrates*	
Trona	$Na_2CO_3 \cdot NaHCO_3 \cdot 2H_2O$	Soda Niter	$NaNO_3$
Sulfates		Niter	KNO_3
Anhydrite	$CaSO_4$		
Gypsum	$CaSO_4 \cdot 2H_2O$		
Glauberite	$CaSO_4 \cdot Na_2SO_4$		
Epsomite	$MgSO_4 \cdot 7H_2O$		
Thenardite	Na_2SO_4		

species and abundance of clay minerals are determined by source lithologies (Jones and Bowser, 1978) but in many cases (e.g. Lake Baikal and tropical lakes) they are useful proxies for the reconstruction of climate change (Yuretich et al., 1999).

Nitrogen, sulfur, and phosphorus are among the most abundant and mobile elements in lake systems. Nitrogen and phosphorus, for instance, are key participants in lake eutrophication, leading to element enrichment and undesirable impact on plant growth. Similarly, heavy metal elements are also concentrated in lakes which receive abundant contaminants and industrial pollutants. The sedimentation of these elements as indicators of industrial pollution and lake management has become a focal point in recent limnological studies (Panda et al., 1995). Such studies have also revealed that basin-scale environmental pollution caused by mining, metallurgy and other land use might have predated the industrial revolution (Bindler et al., 2012; Lee et al., 2008). In arid and semi-arid areas, evapotranspiration is strong, and these lakes and playas contain abundant *precipitated salts*, with calcium, sodium, potassium, and magnesium serving as the principal cations (Table 7.2); carbonates, sulfates, chlorides, borates, and nitrates represent the major anions. Different minerals have different solubilities, so calcite, for example, is less soluble than gypsum or halite and would precipitate first; gypsum is next, whereas halite would precipitate last. In arid soils, spatial partitioning can be seen of these minerals, with calcite occurring closest to the surface and halite at increased depth.

7.3.1.3 Organic Matter

Lakes not only receive large amounts of transported organic matter but they are also habitats for the growth of abundant aquatic plants, plankton, and algae, as well as numerous microorganisms. Readers can find more comprehensive reviews from relevant studies (e.g. Tundisi and Matsumura Tundisi, 2012). In addition, a wide variety of aquatic animals also live in lakes and play an equally important role, as do aquatic plants, in maintaining ecological equilibrium. After they cease to grow, most of their body mass becomes the main source of organic matter. It is suggested that small and shallow lakes tend to obtain most organic matter from allochthonous sources whilst large lakes, especially those with high productivity produce most organic matter (Barnes and Barnes, 1978). Most organic remains, either *allochthonous* or *autochthonous* in origin, are subject to further alteration by *chemical decomposition*, *bioturbation*, *physical compaction*, and other processes. They are converted into organic compounds, including fatty acids, isoprenoid compounds, hydrocarbons, carbohydrates, and so on, some of which further interact with inorganic components of lake sediments. For instance, lakes are becoming global hotspots to understand carbon cycling through organic matter accumulation and decomposition (Kellerman et al., 2014). Detailed analysis of organic compounds might also reveal the biological history of lakes through identification of markers diagnostic of specific taxa (Barnes and Barnes, 1978). We also note the abundance of climatic reconstructions made from pollen, diatoms, and other microbiological records that have been extracted from cores taken from lake bottoms. These are valuable archives of information that indirectly place archaeology in an accurate chronological and paleoecological setting. Organic lake fills have often been the focus of pollen studies in order to recreate the vegetation of local landscapes. A case in point is the analyses of small montane lakes in the Italian Apennines (Liguria), which revealed, for example, the change from predominant white fir to beechwoods during the Holocene; Chalcolithic clearance was followed by major exploitation for grazing during Roman to medieval times (Cruise, 1990; Cruise et al., 2009). Changes to base level, peat weathering, animal trampling, and inwash of high-energy colluvium associated with mining were all recorded (see Figures 18.23a–18.23c).

7.3.1.4 Hydrology

Section 7.2 briefly outlines the three main sources of water to lakes: *precipitation, stream inflow,* and *groundwater* (Figure 7.4). Whilst it is evident that water supply to lakes through precipitation and stream inflow is offset by loss through evaporation and stream outflow, it is important to note that groundwater also suffers water loss. The concept of water residence time has been developed to estimate the time needed for water in a lake to be recycled. Three parameters related to water gains and losses are classified in a triangular diagram to quantify water budgets in some large lakes (Winter, 2004: 70) (Figure 7.5). The importance of each of these parameters to lake water losses (and gains with a similar triangular diagram) varies greatly, appearing on distinct positions in the diagram (e.g. Lake Baikal and Lake Mirror are very different). Seasonal and annual hydrological fluctuations are particularly pronounced in terminal lakes. Lake Devils in North Dakota has a

Figure 7.4 Simple schematic diagram illustrating hydrological processes associated with the water budget of a lake. P: precipitation; E: evaporation; SWI: surface water in; GWI: groundwater in; SWO: surface water out; GWO: groundwater out (from Winter, 2004 / with permission of John Wiley & Sons).

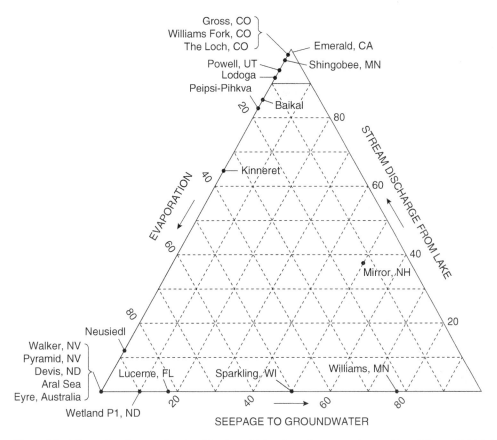

Figure 7.5 Triangular diagram for classification of lakes based on water losses. (from Winter, 2004/with permission of John Wiley & Sons).

water residence time of only 5.6 years whilst that of Lake Eyre in Australia is ca. 1639 years (Winter, 2004: 74). For lakes whose water gains are obtained through stream inflows, it is apparent that their hydrological situations are closely related to or controlled by alluvial hydrological fluctuations, and consequently increasing human exploitation of river water resources also have a strong impact on lake hydrology. Similarly, intensifying human utilization of groundwater in heavily industrial regions also expose a severe threat to hydrological situations in lakes that are reliant on groundwater recharge. On top of these, climate change and its impact on precipitation will also affect water residence time and thus, both the recycling time and water quality in lakes (Blenckner, 2012). In short, hydrological situations in lakes are dynamic and extremely difficult to model and predict. Readers can find more detailed discussion from several sources (e.g. O'Sullivan and Reynolds, 2004).

In addition to the basin-wide scale water budget and hydrological fluctuations of lakes described above, water motion within lakes contributes more complexity to the understanding of internal hydrological fluctuations. Water motion is mostly controlled by currents that occur at different depths and topographies in the lake and are related to wind shear and heat flux, as well as other factors (Imboden, 2004: 119). For the former, studies on sloping topography of lakes and water circulation further show that downslope flow creates thick bottom mixed layers whilst upslope flow is often thin and stable (Rhines, 1998). Several numerical models to compute water flow (e.g. breaks and waves) are used to monitor those alpine lakes (Imberger, 1998) and include changes of water intensity due to addition of dissolved substances or mixing of different water masses.

Horizontal motion of lake water is generally stronger than vertical motion since lake widths are generally larger than their depths. Additionally, density changes of water from top to bottom will also offset vertical vector motion (Imboden, 2004: 116–120).

7.3.2 Shorelines

A shoreline is the location where lake water recedes and sand, gravels, and/or coarser particles accumulate. Lagoons and wetlands often develop along shorelines, on the so-called backshores or backlands (Provencher and Dubois, 2012) (Figure 7.6), or further towards the inland. They are places where rich biomass is produced and sustained. The high productivity attracts not only mammals, birds, and other animals, but also humans (de la Torre et al., 2018; Ismail-Meyer et al., 2013; Karkanas et al., 2011). The extensive shoreline of Lake Baikal, for example, is an attractive place to animals, but anthropogenic pressure has started to exert significant impact on it (Touchart, 2012: 87). Similar problems also face management of shorelines of the Great Lakes (Herschy, 2012a: 313). Shoreline degradation includes erosion, water and soil pollution, and many other physical and chemical changes that directly affect ecological sustainability. Reconstructing and modeling shoreline evolution are therefore vital to understand how a shoreline has changed in the past, how it will be changing in the future, and ultimately, how we can better manage lakes and shorelines.

Shorelines evolve in response to lake level fluctuations, which are controlled by a series of mechanisms such as climate change, tectonism, and geomorphological changes, very similar to those that are responsible for the formation of river terraces (see section 6.5 in Chapter 6). Pleistocene glacial and interglacial cycles profoundly altered shoreline morphologies of glacial lakes such as the Great Lakes and the Baltic Ice Lakes (Andren, 2012: 95–96; Herschy, 2012a: 310). Shorelines of small lakes, closed lakes, or karstic lakes are particularly sensitive to limnetic hydrological changes.

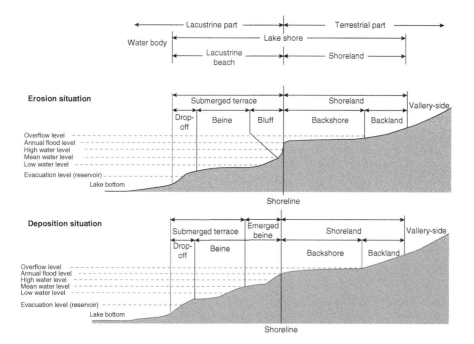

Figure 7.6 Terminologies and classification of erosion and deposition processes at different parts of lakes, from Provencher and Dubois (2012)/with permission of Springer Nature.

A small reduction in lake level will result in large-scale change in shorelines. Examples include the Salt Lake and many paleo-lakes that have now vanished, some of which are discussed below (e.g. the Nihewan Basin in section 7.4). Geomorphological processes affecting shoreline evolution include erosion and deposition. Provencher and Dubois further define different geomorphological positions and hydrological fluctuations on lakeshores, corresponding to erosion and deposition situations, respectively. The main difference, according to them, is that a bluff zone occurs with the erosion, whilst this is replaced by an emerged "beine zone" during deposition; beine here refers to the location "between the mean water level and the inferior limit of wave action" (Provencher and Dubois, 2012: 465) (Figure 7.6). Lastly, winds and currents directly interact with these positions, creating diverse morphologies of limnetic shorelines.

Of more direct interest to archaeologists, however, is the fact that former human occupation took place along or close to lake margins (Feibel, 2001). This setting can include not only the lake-shore itself but also the streams and wetland areas that adjoin them. Thus, stratigraphic sequences in these fluvio-lacustrine settings may include a variety of facies and lithologies that result from contributions from both neighboring streams and the lake. The latter can be triggered by rising or lowering of the lake level, which in turn can be caused by tectonic movements within the basin or drainages, or by climate changes. Consequently, deposits and archaeological sites in lake-margin settings may be influenced by a variety of depositional and post-depositional factors. We now outline some of these.

Changes in lake level may be expressed in different ways. A lowering of lake level exposes any sediment that was previously deposited at or below water level. This fosters pedogenesis and allows for any deposits or archaeological materials previously emplaced to be covered by alluvium or be eroded if channels migrate, either by meandering, avulsion, or braiding, over the occupation area. In such situations, we may see a stratigraphic shift from finer beach sands, for example, to well-rounded, imbricated channel gravels. Sand dunes, if present, may also encroach upon sites.

On the other hand, with rises in lake level, shorelines may transgress over occupation surfaces and deposits. Depending upon the rate of lake level rise, sites can be quickly inundated and preserved with fine-grained deeper water lacustrine deposits, perhaps richer in organic matter. In contrast, with slow rise in lake level and high-energy beach settings, previous deposits and artifacts can be subject to erosion, transport, or destruction by wave activity. The latter is particularly true in the case of major storm events.

Rises in lake level also are associated with higher groundwater tables, possibly transforming previously dry areas into swampy, less habitable ones. At the same time, diagenesis becomes more prominent, and depending on the specific environment, it can be expressed by increased swampy (reedy) vegetation and rooting, cementation by carbonates, or the formation of iron pans (see Fig 4.9, and Boxes 9.1 and 9.2).

7.4 A Short Summary of Wetland Hydrology and Soils

Ecology and geomorphology of wetlands are similarly tied closely to fluctuating water regimes. Based on hydrogeomorphic regimes and plant ecology, wetlands are divided into several types, including swamp, marsh, bog, fen, wet meadow, shallow water, mires, peatland; some of these therefore overlap with the terminology used for lakes (e.g. temporary lakes; Keddy, 2010: 13) and floodplains (e.g. fen, see section 6.4). To avoid repetition, here we note and briefly explain characteristics of wetlands soils as they differ greatly from floodplain sediments and soils. Coastal wetlands and lagoons are discussed in Chapter 9 (section 9.4). This will be followed by a discussion in section 7.6 on geoarchaeology of wetlands to investigate how ancient populations interacted and transformed wetlands (one of most productive ecosystems on earth), with a special focus on wetlands in low-lying areas such as Mesoamerica and Lower Yangtze.

The geomorphic settings for the formation of wetlands primarily include geological depression basins, low-position or discharge areas of riverine landscapes, and estuarine fringes (Noble and Berkowitz, 2000). Influenced by the depth and seasonality of surface and groundwater, soils in these wetlands develop characteristic features that qualify as so-called "hydric soils" (Kolka and Thompson, 2006; Vepraskas et al., 2018). Chapter 4 provides more general discussion of how poor drainage on low-lying positions promotes the gleying process and the formation of gleyed soil types as well as peat (see Chapter 4, especially Table 4.1). Wetland soils are generally saturated for a protracted period of time under different hydrological regimes (periodic and continuous satura-tion, flooding, or ponding), which lead to intensive chemical reduction in the soils and the forma-tion of distinctive redoximorphic features. These features form along plant roots and soil voids, involving mainly the depletion, movement and/or reprecipitation of Fe, Mn, and several other elements. The color, dimension, and morphology of these redoximorphic features, when examined microscopically, might be used as an indicator for the duration of water saturation of the soil (Vepraskas et al., 2018). Two types of redoximorphic features are commonly present in wetland soils: iron and manganese concentration features such as nodules and iron-oxide rich coatings, often of concentric structure, and associated iron and manganese depleted areas in the soil matrix (Vepraskas et al., 2018). Another characteristic of wetland soils is that they are often rich in organic matter (Fox, 1985; Ismail-Meyer et al., 2018). Owing to the high productivity and relatively slow decomposition, mainly due to oxygen scarcity for biological respiration, organic matter accumu-lates faster than in other systems and sometimes form dark-colored peat or peat-like layers, which are often exploited by humans for all kinds of economic purposes (Carter, 1998).

7.5 Geoarchaeological Examples of Lakes

Archaeological sites in lacustrine and fluvio-lacustrine settings can be found throughout the world, particularly in the mid-latitudes and tropics where some of the earliest human sites can be found. The East African Rift (Kenya and Tanzania) and Jordan Valley, for example are associated with prehistoric sites for over the past 1–2 million years (Bar-Yosef and Goren-Inbar, 1993; Bar-Yosef and Tchernov, 1972; Feibel, 2001; Goren-Inbar et al., 2018; Mallol, 2004; Potts et al., 1999).

Scattered along the East African Rift are many large to medium-scale lakes. Surging geoarchaeo-logical studies have provided ample examples of how these lakes evolved during the Quaternary era and how such diverse lacustrine environments interacted with human occupation. Lake Turkana is one such lake found in present-day Kenya and Ethiopia. Early hominin sites have been found at several locations such as Koobi Fora with the presence of some important cranial fossils (Spoor et al., 2007). Survey by systematic drilling has also been carried out to reconstruct paleo-lacustrine landscape of Lake Turkana and other lakes in the region (Cohen et al., 2016) and eluci-date its relationship with early hominin evolution. This line of inquiry has been extended into the late Pleistocene and Holocene era in some recent studies. Bloszies et al. (2015) traced relict beaches along the lake and systematically reconstructed lake level fluctuations of the lake since the late Pleistocene (also see Beck et al., 2019; van der Lubbe et al., 2017). After the dramatic lake level oscillations (30 m) during the late Pleistocene, several episodes of high lake level were sustained after the Younger Dryas event and during the middle Holocene humid period, which were linked to African monsoons. Some of these high lake-level events coincided with human occupation along the lake. Recording of sedimentary history along the wave-dominated clastic shoreline provides an important baseline to understand how prehistoric populations adapted to these diverse, heterogenous lacustrine landscapes that include dunes, backshore ponds, marshes, and lagoons (Prendergast and Beyin, 2018; Schuster and Nutz, 2017).

The significant sites of Kokito 1 and 2 were situated chronologically between the Pleistocene and Holocene when the lake level experienced some fluctuations. The presence of geometric backed stone tools and barbed bone points indicates a flourishing hunting and fishing economy along the lakes, although this was possibly interrupted by the Younger Dryas (Beyin et al., 2017). As the late Holocene aridification began from 5000 BP onwards, the lake became an even more important source of food for prehistoric pastoralists, and hunting-gathering and fishing groups who turned to more flexible subsistence activities and exploited relatively rich resources along the lake (Ashley et al., 2017; Wright et al., 2015). Many recent studies of other large lakes of northeast Africa and elsewhere (e.g. Lake Malawi) show that they were also optimal places that sustained social and economic growth of prehistoric populations (MacEachern, 2012; Wright et al., 2017).

In Israel, the Lower Paleolithic site of 'Ubeidiya is situated on the west bank of the Jordan River in the Central Jordan Rift Valley of Israel, ca. 3.5 km south of the Sea of Galilee (Lake Kinneret). It dates to about 1.4 my and is the oldest early Acheulean site outside of Africa (Bar-Yosef, 1994). The stratigraphy of the site is rather complex (Picard and Baida, 1966; Tchernov, 1987). In brief, the lithic materials are found within the 'Ubeidiya Formation, fluvio-lacustrine deposits composed of shoreline and fluvial conglomerates interfingering with freshwater lacustrine clays and marls; incipient soil formation punctuate these deposits (Mallol, 2004), which were folded and faulted during the Middle Pleistocene. Multiple occupations took place along this fluctuating shoreline/fluvial environment (Figure 7.7).

Figure 7.7 Field photograph of tilted sediments from 'Ubeidiya Formation, Jordan Valley, Israel. These chalky lacustrine marls are stained with goethite and overlain on the right by gravelly clays of alluvial origin (see also Figure 7.8). Scale is 20 cm.

Although the general scenario outlined above has been known for close to 40 years, the detailed aspects of the location of individual occupations have been made clear relatively recently by the microstratigraphic study of the deposits using micromorphology (Mallol, 2004) (Figure 7.8). Results for the Layer I26 complex, for example, one of the most striking units in the field, show that it is essentially mudflats and beaches that were subject to periodic desiccation, allowing for the formation of gypsum and other evaporites, as well as prehistoric occupation. These margin deposits were also subject to wave action that reworked some of the geological and archaeological materials. These findings enabled Mallol (2004) to follow lateral facies changes within lithostratigraphic units and make detailed correlations not previously apparent.

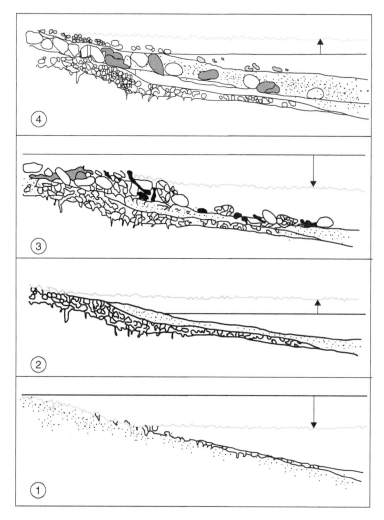

Figure 7.8 Reconstruction based on micromorphology of the sedimentary environment associated with layer I 26 at 'Ubeidiya. (1) low lake level exposing mudflat which is subject to desiccation, cementation by carbonate, and break up by wave action; (2) rise in level of the lake resulting in reworking of carbonate gravel and other surface clasts and covering by lacustrine silts; (3) drop in lake level with weathering and slight reworking of shoreline deposits, including remains of abandoned occupation that were situated on a pebbly beach; (4) rise in lake level which reworked and redeposited small clasts (as well as lithics) on top of the original occupation; this wave reworked material grades laterally lakeward into lacustrine deposits (after Mallol, 2004)

The Jordan Valley also contains lacustrine relics of late Pleistocene and Holocene age, many of which articulate with archaeological sites. The Epi-Paleolithic site of Ohalo II (Israel) (ca. 19.5 ka BP), for example, is remarkable for its preservation of architectural and botanic remains, in large part due to submersion by the late Pleistocene Lisan Lake (cf. Figure 2.10d), and one of its present relics, the Sea of Galilee (Nadel and Nadler, 2006; Nadel et al., 2004; Tsatskin and Nadel, 2003). Archaeological material rests upon the Lisan lacustrine deposits; through time, the lake became increasingly shallow as shown by the inputs of basalt-derived near-shore sandy deposits. Water level fluctuated, resulting in periodic submergence of the sediments, accompanied by truncation; a greater overall exposure is found in higher, more shoreward parts of the site, which shows bioturbation and incipient soil formation. Further south in the Dead Sea Basin archaeological structures of known age show impacts of shorelines much above those of present levels (Enzel et al., 2003; Klein, 1986).

In Europe as elsewhere, ponds and natural waterholes served as foci for animals and people to access drinking water, plant materials, and other food resources. In the case of both large predators and humans, waterholes make an ideal hunting site. In the geological record, marl and tufa formations may embed bones and artifacts. One example is Lower Paleolithic Boxgrove in the UK, where just inland of the beach and mudflats, the only identified source of freshwater is a small pond. Possible springhead sapping had eroded the underlying marine beach sediments (Unit 3), and the depression had become infilled with alkaline carbonate-rich (chalky) sediments (Roberts and Parfitt, 1999). Of taphonomic interest is that both terrestrial (earthworm granules and slug plates) and marine (foraminifera) microfauna are present, the latter eroded into the pond from earlier-deposited mudflat sediments (Macphail and Goldberg, 2018a: figs Ib, IIIc–IIIf). The pond sediments have the highest concentration of Acheulian hand axes and a concentration of large faunal remains at Boxgrove, such as elephant, horse (*Equus ferus*), extinct rhinoceros (*Stephanorhinus hundsheimensis*); a tibia and two teeth from *Homo heidelbergensis* were also found (Stringer et al., 1998). Intact and bulk samples were used for geochemical, grain size, and magnetic susceptibility analyses, along with soil micromorphology, microprobe and image analysis studies (Macphail et al., 2001) (see Figures 17.7a–17.7d). In brief, the truncated marine sands (Unit 3) were only moderately calcareous (18.1–20.2% $CaCO_3$) compared with overlying fills (23.4–31.6% $CaCO_3$ in 4*; 55.4–56.5% $CaCO_3$ in Unit 4d1); in fact, Unit 4d1 was alkaline (pH 8.6–9.1), probably due to the presence of both relict Cretaceous materials and inwashed marine sediments. The uppermost pond infill was through mass-movement (soliflual) deposition which brought in the hominin tibia in Unit 8ac (64.4–67.3% $CaCO_3$).

Recently, some synthetic overviews on the archaeology of Lake Baikal have provided fresh insight into the Holocene occupation history along this world's largest lake. The Cis-Baikal region of Eastern Siberia has been home to prehistoric hunting-gathering populations since at least ca. 9000 BP (Weber and McKenzie, 2003), represented by the Kitoi, Serovo-Glazkovo, and other cultural groups. The Baikal Archaeology Project (BAP) has, in the past decades, been dedicated to reconstructing spatiotemporal variations of prehistoric groups and their cultures and diets in diverse macro-and-micro lacustrine ecosystems. Amongst the BAP's many recent works, several results are worth highlighting. First, there is an "intriguing" chronological hiatus (ca. 6600–6000 BP) in the archaeological record. It has been suggested that this hiatus might have been related to abrupt change in atmospheric precipitation that caused prolonged snow cover (Kobe et al., 2020; Weber et al., 2002). Second, the middle Holocene populations in the Cis-Baikal region were engaged in various kinds of social reorganization, in terms of mobility, migration and marriage (Moussa et al., 2021; Scharlotta, 2018). Lastly, the diets of these prehistoric groups also underwent active change; isotopic records of human skeletons from several cemeteries have revealed some interesting intra-regional dietary differences in this lacustrine environment. For instance, cemeteries (ca.5300–3550 BP) in the so-called Little Sea microregion displayed a dietary difference between the islanders and mainlanders, with the former focusing more on seals and the latter eating more shallow-water fish. After 4100 BP, the mainlanders were consuming less fish and more terrestrial herbivores (Waters Rist et al., 2021; White et al., 2020). Despite just a small research budget, the archaeology of Baikal Lake continues to contribute some of the finest archaeological and paleoecological data to understand patterns of prehistoric lacustrine occupation.

Moving further east on the Eurasian continent, the Nihewan Basin on the present Loess Plateau of China is considered one of the most important cradles of early hominin evolution in East Asia (Dennell, 2013). The so-called Nihewan Stratum represents the dynamic lacustrine history in the basin. Recent advances in OSL and other Quaternary dating methods have significantly enhanced our understanding of the exceptionally long history of human occupation around lakeshore environments in the basin. The Nihewan paleo-lake first appeared at the end of the Pliocene and underwent several stages of shrinkage and expansion during the Pleistocene as a result of tectonism and climate fluctuation; its size reached ca. 10,000 km^2 at its maximum (Nian et al., 2013; Xia, 1992). Interestingly, some of the lakeshore sites were occupied during relatively cool and dry intervals of the Quaternary. For instance, the excavation and paleoenvironmental reconstruction at and around the Donggutuo site confirmed that early hominins (ca. 1.1 Ma) lived by the lake when the water level was low during cool and dry climates (Pei et al., 2009). Similarly, the occupation at the Houjiayao site (220 to 160 ka, MIS6) also corresponded to cold climate conditions (Li et al., 2014). Further unpacking the correlation between lacustrine history and human occupation in the Nihewan Basin therefore provides a key to understanding adaptations of early hominins in East Asia.

Amongst the rapidly developing Quaternary science and geoarchaeology of lacustrine evolution and human adaptation in China, recent surveys on the Tibetan Plateau have supplied some of the most interesting examples. The plateau is blanketed with numerous glacial and tectonic lakes that are of various sizes and situated at different altitudes (Zhang et al., 2020). Broadly speaking, Holocene lake level fluctuations of these lakes responded to changes of Holocene precipitation, which, according to Chen et al.'s (2020) recent review in south-western Tibet was related to Indian Summer Monsoon, whilst in northeastern Tibet it was under the combined effect of westerlies and the East Asian Summer Monsoon. Systematic surveys and excavations by colleagues from the Institute of Tibetan Plateau Research of Chinese Academy of Sciences have found that some of the high-altitude lakes were occupied by prehistoric populations who subsisted on diverse food webs (Professor Xiaoyan Yang, pers. comm.). These lacustrine environments arguably played a vital role in the movement of crops and populations on prehistoric Tibetan Plateau and represented some of the most dramatic events of adaptation to high-altitude conditions.

In Australia, the earliest sites on the continent are found in *lunettes*, formed on the lee side of lake basins, particularly in Lake Mungo in the Willandra Lake system (see Figure 1.3; Chapter 6) (Bowler, 1998; Fitzsimmons et al., 2014; Jankowski et al., 2020). At Lake Mungo, for example, a human burial occurs within a complex sequence of lake and aeolian deposits separated by distinct paleosols (Bowler et al., 2003). Fluctuations in lake levels occurred between 50 to 40 ka, which coincided with human occupation. Bowler is largely responsible for establishing the Lake Mungo sequence in regional detail. However, recent studies at microstratigraphic scales using micromorphology and OSL dating have shown more subtle depositional and pedological histories (Jankowski et al., 2020)

In a somewhat comparable setting and latitude in the New World, many Paleoindian sites occur in association with playa, lacustrine, and palustrine deposits (e.g. diatomites) on the Great Plains (Holliday, 1997). Here, intact and bulk samples were used for geochemical, grain size, and magnetic susceptibility analyses, along with soil micromorphology, microprobe, and image analysis studies. The two principal types of deposits exposed at the surface are dark gray, slightly calcareous mud, and a light gray, loamy, highly calcareous deposit. These types of deposits appear to represent wetter ("pluvial") conditions during which the basins were perennially filled or nearly so.

7.6 Geoarchaeological Examples of Wetlands

Some typical Pleistocene wetland sites in the UK include the Clactonian Elephant butchery site at Ebbsfleet (Kent, UK). Here, an elephant may have been hunted in a marsh. Associated but anomalous concentrations of iron-phosphate staining (e.g. 55.3% Fe, 1.07% P) possibly record this butchery event, because the sediments here are generally strongly iron depleted, as shown by SEM/EDS analyses on specific microfeatures (Wenban-Smith, 2013; Wenban-Smith et al., 2006). A typical glacially-formed pond, where a melted body of ice has produced a water-filled depression is termed a pingo. One formed in a chalk substrate at Boxmoor (Hertfordshire, UK) has a biochemically precipitated fill, possibly produced by charophytes, a fresh water green alga (PEJ Wiltshire, Southampton University, pers. comm.), with only very small amounts of fine clastic material and detrital plant traces (Figure 7.9).

In Italy, the large Bronze Age settlement of Fondo Paviani (14th century – 12/11th century BC) is located in the Valli Grandi Veronesi lowlands. Recently, Longa et al. (2019) have reconstructed the paleo-hydrology and paleoecology at and around the site, providing important data to assess

Figure 7.9 Photomicrograph of Boxmoor pingo pond fill, Hertfordshire, UK; PEJ Wiltshire, Southampton University suggested that this is a charophyte, biochemically formed carbonate. Note lack of clastic material apart from a sand-size flint. Small amounts of "browned" partially humified fine detrital plant material are present alongside opaque pyrite framboids, which indicate anaerobic conditions: Frame width is ~6.4 mm; XPL; note high interference colors consistent with calcite.

occupation dynamics in response to wetland fluctuations. The sites were mostly distributed along spring-fed rivers and low-plain areas, which supported the flourishing of the settlement that displayed increasing social complexity. Pollen records indicate extensive deforestation, farming, and animal-raising activities in the local environment. Interestingly, whilst a well-attested dry phase had caused settlement contraction, continuous settlement shift towards places with optimal hydrological regimes might have played a key role in local adaptation to climate-induced hydrological fluctuation (Longa et al., 2019).

Comparatively, there are several archaeological examples of prehistoric and historic wetland occupations from China and the Maya region. History of wetland occupation in the Lower Yangtze River can be traced back to at least 8000 years ago. Succeeding the Shangshan culture (10,000–8500 BP) in the mountainous regions of Zhejiang Province to the south of the Yangtze Delta, the Kuahuqiao (8000–7000 BP) people began to live on the estuarine wetlands. This occupation coincided with the beginning of a reduced-rate but continuous rise of Holocene sea level, after it experienced a rapid early-Holocene rise from ca. 25 m lower than the present sea level to ca. −5 m by around 8000 BP. The basal late Pleistocene alluvial landscape on the delta gave way to rapid marine inundation with increased sedimentation rate. The low-lying (average 2–5 m above sea level) landscape surrounding the delta was dominated by clayey silt and muds formed in tidal, lagoonal, and estuarine environments (Wang et al., 2012) (Figure 7.10). In their paleoecological study analyzing pollen, algal, fungal spore, and micro-charcoal remains at the site, Zong et al. (2007) proposed that the Kuahuqiao people used fire to clear land and "built bunds to control receding-flood" on the "brackish coastal reed-swamps" for early rice farming (Zong et al., 2007: 461). The wetland ecosystem also provided abundant other resources for subsistence (ZPICRA and Museum, 2004). Although this "slash and burn" farming regime has been questioned in a recent study (Hu et al., 2020), there is little doubt Kuahuqiao represented one of the earliest and most pioneering exploitations of wetland environments in prehistoric China.

Such wetland occupation was further intensified during the Hemudu culture period (7000–5300 BP) with the invention of a series of architectural technologies and subsistence strategies. Wooden architecture with raised floors was erected on wetland edges, while bridges and other wooden structures were also built (ZPICRA, 2011) (Figure 7.11); these facilities were possibly built for boiling nuts to leach out tannins before they can be consumed. These wetland occupants were exploiting a wide range of aquatic resources and also started to systematically cultivate rice. According to Fuller and colleagues' research at the Hemudu site of Tianluoshan, rice domestication was a protracted process, in which wetlands must have played a key role in sustaining the survival and development of these early rice farmers (Fuller et al., 2009).

Holocene sea level reached its peak at around 7000–6500 BP and gradually stabilized afterwards (Zong, 2004). The subsequent emergence of *chenier* ridges blocked marine water and resulted in the formation of freshwater wetlands on the delta around 6000 BP. The eastern part of the massive, previously brackish Taihu Lake also became freshwater during this time. The emergence of land and the radical shift of hydrological regime from brackish water to freshwater created an opportunity for the rise of the Songze (6000–5300 BP) and Liangzhu Cultures (5300–4300 BP). The Songze people stacked earthen layers on wetlands to elevate the living surface. As revealed by several recent excavations, families living in such wetland environments were organized according to separate artificially stacked mounds and engaged in rice farming as well as other economic activities (ZPICRA and Museum, 2015). By the Liangzhu period, mound construction was significantly expanded, including the construction of large palatial buildings, altars, and hydraulic infrastructure at the Liangzhu City, through the development of new architectural technologies such as

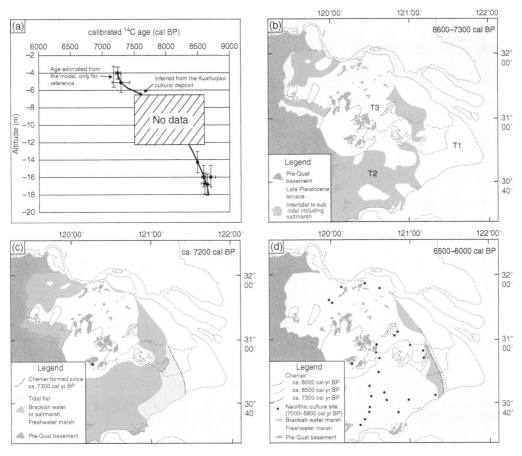

Figure 7.10 (a) Sea level curve for the southern Yangtze delta from 8700 to 7200 cal BP. (b) Distribution of subtidal to intertidal zone including saltmarsh during ca 8600–7600 cal BP. The late Pleistocene terraces at ca 5 m depth are also shown. (c) Freshwater, brackish water or saltmarsh, tidal flat, and chenier ridge at ca 7300 cal BP. (d) Freshwater/brackish water marsh and chenier ridges at ca 6500–6000 cal BP. From Wang et al. (2012)/with permission of Elsevier.

Figure 7.11 Wooden structure showing a possibly wooden wall outside raised floor building and a bridge over a channel (courtesy of Professor Guoping Sun).

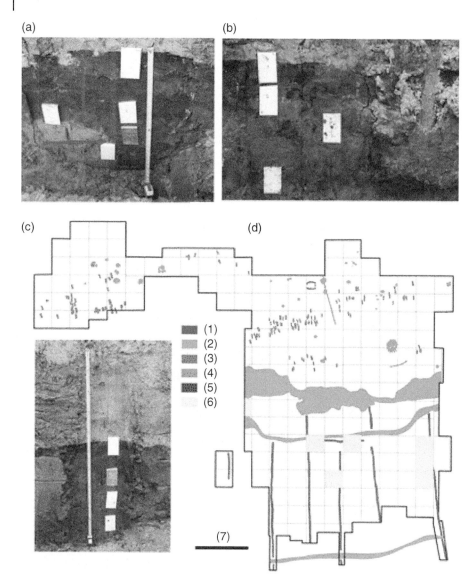

Figure 7.12 (a) Profile showing the grayish–darkish later period paddy field deposits (two-headed arrow) and earlier period paddy field deposits (red arrow); metal boxes show sampling locations for micromorphology, same below; (b) profile examined at the park east of the Maoshan site in roughly the same stratigraphic order as examined in the field: the grayish–darkish alluvial materials overlie pale-yellowish alluvial materials and are covered by light-yellowish flood deposits; (c) profile 2 showing the peat-like deposits and the underlying grayish, later period paddy field deposits; and (d) excavation outline and sampling plans. (1) Small pathways across the paddy field, (2) later period ditches, (3) burials, (4) pits, (5) houses, (6) sampling locations (those for off-site samples and modern rice paddy soils are not shown here) and (7) bar length = 30 m. From Zhuang et al. (2014) / with permission of Sage Publications.

stone beddings at the base of earthen walls and the so-called "prehistoric sandbags" (silty clay wrapped with grass) technology (Liu et al., 2017). The number of Liangzhu period settlements surrounding the city increased rapidly to several hundred across the wetland environments around the Yangtze Delta region. These settlements were sustained by intensifying rice farming, the development of which is best illustrated by the excavation and geoarchaeological research at the Maoshan paddy field site. Two phases of rice paddies were built directly on top of alluvium with bunds, irrigation channels, water outlets, and other field facilities. The transition from the early to late phase saw a dramatic increase in the size of paddy fields (up to 2000 m^2 per field plot by the late phase) and intensified farming practices. The latter are confirmed by the presence of abundant Fe oxide features and concentric dusty clay features derived from frequent oxidation and reduction, micro-charcoal, and other features in the cultivated soils (Zhuang et al., 2014) (Figure 7.12), some of which resemble micromorphological features of modern paddy soils. The Liangzhu Civilization (and its predecessors) might be therefore considered a successful example of wetland civilization whose rise relied on advanced architectural and farming technologies in transforming wetland conditions.

The Southern Lowland Maya (1000 BC–1500 AD) cities were located in the lowland areas of present Mexico, Belize, and Honduras, including the slightly elevated Yucatán Plateau and the low-lying coastal plains. The functioning of most Lowland Maya sites was closely related to their wetland ecosystems, characterized by mosaic habitats with diverse groundwater systems and soil conditions on karst landforms (e.g. depressions) (Dunning et al., 2012). Most Maya monumental sites were situated on the "escarpment edges" adjacent to the wetlands, whilst smaller and medium sites were scattered around the local environment with surrounding wetland fields on relatively "high points" (Luzzadder-Beach et al., 2012). Precipitation in the Maya region is defined by strong seasonality, punctuated by severe, recurrent droughts (Kennett and Hodell, 2017; Luzzadder-Beach et al., 2016). In contrast to the water management system adopted by urban centers in Mexico highlands that targeted the harnessing of groundwater (drainages and springs), Lowland Maya cities and hinterlands focused more on collecting rainfall and surface water through building reservoirs and related infrastructures (Scarborough and Gallopin, 1991). At the Blue Creek site, as well as many other medium to small sites, wetlands provided sufficient access to shallow or near-surface water during the prolonged dry season (December to April) and recurrent droughts. Moreover, they sustained cultivation of a diverse group of crops in wetland fields (Beach et al., 2015; Dunning et al., 2012) supported by long canals, which reached a large scale (ca. 7 km^2) during occupation (Luzzadder-Beach et al., 2012). Although the wetland field systems also experienced a "collapse" during the Terminal Classic period, coinciding with population decline and radical change in social structure, wetlands arguably sustained Lowland Maya society's resilience to external change for a long time.

7.7 Conclusions

As with rivers, the formation and evolution of lakes are also inherently connected to climate change, tectonism, and geomorphological processes. Together with rivers, lakes, and associated wetlands, they form a vital part of terrestrial ecosystems and drainage systems. Despite often comparatively limited in size and catchment and with standing water, lakes and wetlands provide a critical lifeline in the forms of water, ecological resources, and transportation routes, supporting the survival and flourishment of our ancestors. From the environmental and geoarchaeological research point of view, lacustrine and wetlands sediments hold great academic significance. Their (relatively speaking) uninterrupted sedimentation record, provides reliable environmental proxies for dating and reconstructing regional climate history and for understanding how our ancestors interacted with lacustrine and wetland environments. This is especially the case in arid regions where

Quaternary sediments tend not to be preserved due to continuous erosion; in such regions lakes and wetlands are also particularly sensitive to human- and climate-induced modification and intervention. This chapter shows only some of the recent environmental science and geoarchaeological research advancements on lakes and wetlands (also see Menotti and O'Sullivan, 2012), and we expect to see more interdisciplinary methods being applied to study them.

8

Aeolian Environments

8.1 Introduction

When one thinks of the aeolian setting, notions of sand dunes and vast sand seas come to mind. Indeed, windblown sand covers about 25% of the earth's surface (Cooke et al., 1993) (Figure 8.1). Yet aeolian activity is not just confined to blowing sand around: it is responsible for eroding and depositing vast quantities of silt and clay-sized dust; the Sahara alone has been estimated to produce between 60 and 200 million tons of dust per year (Cooke et al., 1993). Similarly, aeolian processes are responsible for producing markedly eroded landscapes composed not only of softer aeolian, fluvial, and lacustrine sediments (Brookes, 2001), but also of harder bedrock.

The aeolian environment and associated deposits constitute a significant geoarchaeological resource. Aeolian deposits contain and bury many sites that played an important part in human history and evolution, especially in lower latitude arid areas, where there are drifting sand dunes and desert expanses. Likewise, in the higher latitudes of Western, Central, and Eastern Europe, and Central and NE Asia, finer grained aeolian dust accumulations (loess) derived from or associated with glacial activity, blanketed many of these areas when glaciation was more active and extensive, and winds were stronger. Set into these loess deposits are numerous Paleolithic sites, and because of the lower energies associated with dust accumulation the degree of preservation of the archaeological record and its contextual framework is generally high, especially in comparison to higher energy fluvial deposits.

Furthermore, aeolian sediments and the preservation of archaeological sites are not only associated with Pleistocene sequences; aeolian sediments have buried numerous sites during the Holocene, again providing a mechanism for the preservation of a huge range of geoarchaeological resources. One only has to think of the deserts of North Africa and the remains of ancient Kingdoms in Sudan and Egypt, to see both the burial and erosion of Holocene archaeological resources by aeolian processes. In this chapter, we discuss some of the basic elements of aeolian settings, their landscapes and deposits, and relate these to their geoarchaeological and paleoenvironmental records.

8.2 Aeolian Processes and Sediments

For the geoarchaeologist, awareness of aeolian sediments, their diagnostic characteristics, and their potential to contain geoarchaeological resources is of key concern, due to their widespread distribution. The pioneering work of Bagnold (1941) demonstrated processes of erosion, transportation, and deposition of sediments by wind and this subsequently became a more active field of sediment

Practical and Theoretical Geoarchaeology, Second Edition. Paul Goldberg and Richard I. Macphail.
© 2022 John Wiley & Sons Ltd. Published 2022 by John Wiley & Sons Ltd.

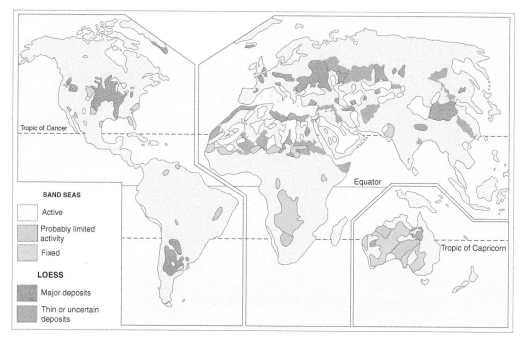

Figure 8.1 Distribution of sandy and silty aeolian deposits over the globe (modified from Thomas, 1989: Figure 11.1).

transport research during the second half of the 20th century. Sandy terrains are generally located in the arid parts of the globe (Figure 8.1), where the aridity has a number of causes. These factors include (a) subsidence of air associated with subtropical high pressure belts (most of the world's deserts are found here, such as the Sahara); (b) the positioning in the center of large continental land mass (e.g. Chinese deserts); and (c) orographic factors related to uplifted terrains (e.g. Californian deserts, Dead Sea Rift) (Cooke et al., 1993). On the other hand, geoarchaeological (e.g. Breeze et al., 2017; Scerri et al., 2015) and pollen evidence (Dinies et al., 2015) show that environmental conditions have changed dramatically in the past, permitting earlier habitation in what are now among the driest parts on earth. Furthermore, aeolian sand sediments are also found within non-arid areas; the extensive temperate dune systems of the Atlantic coast being a prime example (also see section 9.2.2).

The movement of sediment by aeolian processes has inherent similarities to the movement of sediment by water (see Chapter 6). Sand moves when the critical threshold friction velocity of the wind is exceeded, and the entrainment of grains is generally related to particle size and wind velocity, although other factors such as surface roughness, soil moisture, salt precipitation, and surface crusting also play a role (Ritter et al., 2011). The work of Bagnold (1941) showed graphically the relationship between grain diameter and critical threshold. As such, the most easily erodible size is about 0.1 mm (Figure 8.2). Below this size, greater wind velocity is required to entrain particles because of cohesion and low surface roughness. On the other hand, once grains are set in motion, they impact stationary grains whose entrainment velocities ("Impact threshold" in Figure 8.2) are lower than the fluid threshold for equivalent sized grains.

Once grains begin to move, they are transported by different processes, which are primarily related to grain size (Figure 8.3; Holden 2017). *Creep* generally involves coarser sand (ca. > 150 µm) and movement by rolling at the near surface. *Saltation* is the ejection of medium size grains (finer sands) into the air at a relatively steep angle. These grains pick up additional momentum from the wind and

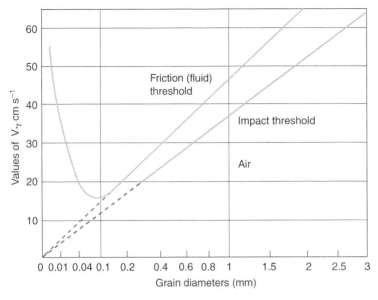

Figure 8.2 Diagram showing the relationship of particle size in mm to threshold velocity as determined by Bagnold, 1941 (modified from Ritter et al., 2011: Figure 8.7 after Bagnold, 1941).

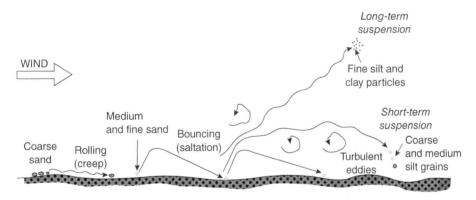

Figure 8.3 Schematic view of aeolian transport of sediment grains (from Pye, 1995 / with permission of Elsevier).

land at a shallower angle, causing a hopping motion, ca. <1.5m above the ground. The impacts of sand grains caused by the hopping motions of saltation, lead to *reptation*, where finer sands have a low hopping motion, caused by the descent and impact of high-energy particles. Finally, there is *suspension*. Suspension acts on the smallest grain sizes (silts-clays; < 62 μm), carrying the grains in suspension with turbulent motion (Cooke et al., 1993). Sediments transported by suspension are capable of being transported over large distances and can end up being embedded in soils or end up as large accumulations of *loess* (see section 8.5) (Amit et al., 2011; Jewell et al., 2021; Pye, 1987).

8.2.1 Aeolian Erosion

Erosion by wind is visible at both the smaller, site scale and the larger, regional scale, with both scales affecting the geoarchaeological record. At a small scale, wind erosion by abrasion can result in the grinding away of rock fragments, including lithic implements. Abrasion, usually by

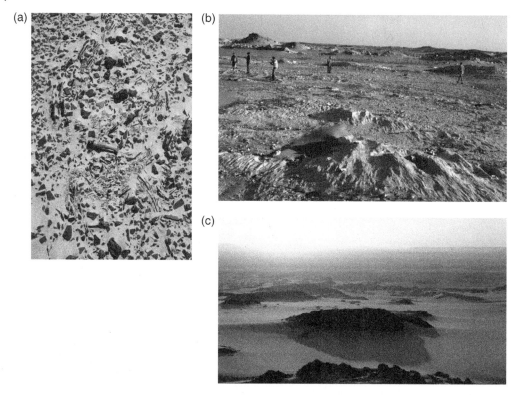

Figure 8.4 Erosional effects of the wind: (a) ventifacts from the Western Desert of Egypt, exposed on the surface and covered with a thin veneer of sand; knife handle is 12 cm long; (b) wind-polished limestone showing elongated grooves that parallel the strongest or prevalent wind directions; in the background on the left is a small yardang; (c) large yardang in Western Desert of Egypt, north of the Kharga Oasis and not far from the location in (b).

sand-sized particles, can result in the formation of *ventifacts*, typified by facetted and pitted rocks and boulders, or smoothed and polished surfaces with elongated grooves (Figure 8.4a, 8.4b). Measurement of the alignment of the latter can be used to infer the direction of the strongest or most prevailing wind direction (Ritter et al., 2011). At a larger scale, erosion may sculpt entire exposures of bedrock resulting in the formation of *yardangs*. These are elongated hills, which parallel the prevailing wind direction (Figure 8.4c). Typically, the windward side is blunter than the leeward end, which tapers downwind.

From a more geoarchaeological perspective, wind erosion may play an important role in the integrity of archaeological assemblages. The principal risk is the effect of *deflation*, whereby fine-grained material is blown away leaving a lag deposit of heavier, stony objects, including artifacts (Figure 8.4a) (Goudie, 2008) (see Box 8.1). Consequently, artifacts originally deposited in different sediments from successive occupations can be found together within the same "assemblage", after the finer interstitial material has been eroded. Deflation surfaces are common (e.g. Coronato et al., 2011; Rick, 2002). This is especially important in Old World settings, where time depth provides the opportunity for repeated and longtime deflation. An abundance of sites in Egypt (and Sinai), Israel, and Jordan have been surveyed and excavated, with many showing signs of deflation (Goring-Morris and Goldberg, 1990; Wendorf and Schild, 1980).

Evaluating the effects of deflation is essential for the correct interpretation of the archaeological record within these environments. In cases where distinct lithic tool types are found together (e.g. Middle Paleolithic Levallois core mixed with Epi-Paleolithic bladelets), the effects of deflation are easily recognized. In some cases, however, where non-diagnostic lithics occur, it is more difficult to detect mixing of occupations and assemblages. Fortunately, in most desert terrains, rocks exposed on the desert surface develop *desert varnish*, which is a shiny, dark reddish or blackish coating about 100 μm thick, and forms on exposed surfaces. It consists of varying proportions of iron and manganese (which influence the color) and various clay minerals. Its origin is still debated but it is thought to form from physico-chemical and biogeochemical processes, with the main source of varnish components coming from aeolian dust (Dorn, 1991; Dorn and Oberlander, 1982; Goldsmith et al., 2014; Lang-Yona et al., 2018; Watson, 1989a). In any case, the occurrence of desert varnish on artifacts can provide an indication of past landscape stability or deflated material.

The chronometric dating of desert crusts has proven to be problematic (Bierman and Gillespie, 1994; Dorn, 1983), but some measures can be made by radiocarbon dating of organic carbon in the varnish (Martínez-Pabello et al., 2021). In the latter case, radiocarbon dates were much younger than the archaeological rock art on which the varnish developed. Relative chronological frameworks have been attempted, comparing cation-ratios (Ca + K)/Ti with the degree of stability of different geomorphic surfaces (e.g. pediments, alluvial fans, flood plains) in Iran (Sarmast et al., 2017). Similarly, qualitative evaluation of mixing of archaeological surface finds in deflated terrains may be undertaken by examining different degrees of development on lithics (cf. Helms et al., 2003). In addition, a lack of small lithic debris or a high proportion of cores and other heavy, large elements would suggest some removal of fine material by deflation or runoff; context is the best means to evaluate which of these alternatives is most reasonable.

Despite the problems of mixing assemblages, deflation is also a significant process in the past, as well as in the present. The present-day formation of deflation basins, or *blowouts*, is helpful in exposing stratigraphic sequences of deposits that would normally be concealed, thus providing windows into former landscapes, deposits, and sites. In antiquity, the formation of blowouts results in depressions in which water collects, even if seasonally or during periods of wetter climate. Therefore, they acted as loci for floral, faunal, and human activity, and ultimately as traps for sediments and archaeological remains. At the Casper site in Wyoming, for example, an elongated depression (ca. 93 × 22 × 2 m) that formed about 10,000 radiocarbon years ago, was filled in with lenticular accumulation (Unit B) of well-rounded sand with frosted and pitted surfaces that exhibited some low angle cross-bedding (see Chapter 2; and below). Within the deposits were the remains of 74 *Bison antiquus* representing a "communally-operated bison trapping, killing, and butchering station which can be associated with the Hell Gap cultural complex" (Frison, 1974). These deposits and setting were interpreted as a deflation hollow, similar to ones currently forming in the same area (Albanese, 1974), and were protected from subsequent deflation by the overlying lacustrine deposits of Unit C which filled in the original blowout depression.

Similar aeolian situations are also documented in northeastern Sinai (Egypt) and the Negev (Goring-Morris, 1987; Goring-Morris and Goldberg, 1990; Hill, 2009; Wendorf et al., 1993; Wendorf et al., 1994). In the Nabta Playa area of Southern Egypt, Holocene deflation of over 6 m has exposed lacustrine deposits, which overlie Terminal Paleolithic artifacts that rest on a cemented, fossilized dune. Not far away, in the Bir Sahara/Bir Tarfawi area, deflation is also responsible for excavating more than 8 m of sediment below the plateau level; this was later filled with aeolian sand with "numerous but heavily sandblasted Mousterian artifacts. . .." (Wendorf and Schild, 1980: 25).

Box 8.1   Aeolian features in desert environments of the Southern Levant[i]

The Eastern Saharan and Arabian Deserts run from Sinai, Egypt, across the Negev in southern Israel into Jordan and the Arabian Peninsula of SW Asia (Figure 8.5). This terrain is among the most arid parts of the world, with precipitation being << 100 mm per year and averaging < 5 mm per year in the Arabian Peninsula. Although prehistoric sites have been recorded from the entire region (e.g. Breeze et al., 2017; Phillips, 1988; Wendorf et al., 1994; Wilkinson, 2003),

(a)
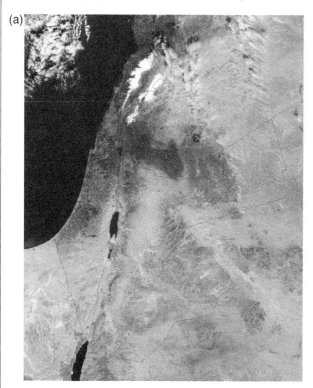

Figure 8.5   Sites and aeolian deposits in the Sinai and Negev Deserts. (a) Satellite photograph of SE Mediterranean area showing the largely unvegetated nature of the region except for a stretch along the northern Levantine coastline of Israel and Lebanon with a typical Mediterranean climate; (b) map of localities mentioned in the text.

(b)
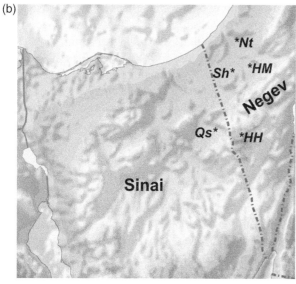

Box 8.1 (Continued)

eastern Sinai in Egypt and the western Negev Desert of Israel, have received intense attention revealing numerous archaeological and prehistoric sites in aeolian terrains (Goring-Morris and Goldberg, 1990; Goring-Morris, 1993; Goring-Morris and Belfer-Cohen, 1997; Roskin et al., 2011b, 2013b, 2017). These include both sandy dune and loess deposits, as well as fluvially and colluvially reworked sands and silts.

During the Late Pleistocene (within the past 50,000 years), a good part of the area received large inputs of aeolian sand and dust (Crouvi et al., 2009; Roskin et al., 2011a). This accumulated as primary dust fall on upland surfaces (Bruins and Yaalon, 1979; Crouvi et al., 2009; Rognon et al., 1987; Wieder and Gvirtzman, 1999) (Figure 8.6). In addition, loess was pene-contemporaneously reworked along slopes and into most drainages from Wadi el-Arish in Sinai to Nahal Besor in Israel, and all drainages in between (Goldberg, 1986; Goring-Morris and Goldberg, 1990; Zilberman, 1993) (Figure 8.7). These alluvial deposits are actually paludal (swamp-like) and have been interpreted to indicate during the past ca. 40,000 years or so, more gentle and sustained rainfall patterns, possibly throughout the year (Horowitz, 1979; Sneh, 1983). Such conditions were obviously attractive to prehistoric settlers, and we see many

(a)

(b)

Figure 8.6 Loess deposits. (a) Section of loess and paleosols in an upland position of the Netivot area of the western Negev, Israel (Bruins and Yaalon, 1979) (Figure 8.5b). The bulk of the profile is composed of calcareous clayey silty loess, which has a relatively high proportion of finer silt and clayey washout dust (see text). It is interspersed with a number of calcic paleosols (nodules and calcic horizon) that point to intermittent periods of leaching and wetter climates. Only a few scatters of non-distinctive artifacts, however, were found within these deposits. Meter and hammer are for scale. (b) Holocene loess deposits also occur in the region, but they are much rarer. Shown here is a thin (ca. 10 cm) accumulation of loess on the high plateau (ca. 1000 m elevation) of Har Harif (Figure 8.5b). The loess is below the feet of prehistorian Nigel Goring-Morris and overlies a concentration of Chalcolithic (ca. 6–7,000 BP uncalibrated) flint artifacts that can be seen here in the foreground eroding out from underneath the loess cover. Unlike in glacial terrains, loess in this part of the world is commonly associated with wetter conditions, in which aeolian dust is washed out of the atmosphere during rainfall events; these conditions also promote a macro- or micro-vegetative cover (Danin and Ganor, 1991), which aids in trapping the loess.

(Continued)

Box 8.1 (Continued)

(a)

(b)

Figure 8.7 Two views of sediments in the Hamifgash area, the confluence of Nahals Beer Sheva and Besor, both major drainages in the Western Negev of Israel (cf. Figure 8.5). (a) Light-colored silty and sandy alluvium are visible. Persons in the center are standing next to the excavation of a late Middle Paleolithic site, Farah II. This is found at the base of ca. 18 m of fluvially reworked loessial and sandy deposits, consisting of cm-thick sandy and silty beds and dated to 41–50 kya (Goder-Goldberger et al., 2020). (b) These sediments represent paludal/fluvial deposition. Within these overlying deposits, many Upper Paleolithic and Epi-Paleolithic sites can be found here and elsewhere along this broad coastal plain, from the western Negev, SW into the northern part of Sinai. They testify to the attraction of these water-rich environments during the wetter conditions of the late Pleistocene (Goring-Morris, 1987).

Box 8.1 (Continued)

Upper Paleolithic and Epi-Paleolithic sites intercalated within these deposits from this region at this time.

In many places in Northern Sinai and the Western Negev, the loesses have been fluvially reworked with silts filling major drainages (Goldberg, 1994; Goring-Morris, 1987; Zilberman, 1993). In the latest part of the Pleistocene (starting at about 25,000 BP) we see the relative reduction in dust accumulation and its fluvially reworked counterpart, and a concomitant increase in sand. The latter, early on is expressed as sandy additions to the alluvium. Incision into alluvium becomes widespread and is also somewhat coeval with the deposition of well-defined aeolian sand deposits, either as localized sand accumulations (Figure 8.8) or as more massive linear dune features. The latter are strikingly visible in the field and on aerial photographs today (Figure 8.9) from where there it is obvious that these sand dune bodies and morphologies are plastered onto the existing Upper Paleolithic fluvial terrain, which existed during wetter times.

Moreover, we see that these lithological and geomorphological changes took place across the landscape in a time-transgressive fashion: in northern Sinai (e.g. Qseima area, Qadesh Barnea; Figures 8.5b and 8.8), the first sand incursions occur about 20 ka, whilst in the Hamifgash region in the western Negev, increased sand is noted above an early Epi-Paleolithic site (Hamifgash IV; ca. 16,000 BP); here, a veneer of sand covers Neolithic remains. In the nearby Shunera Dunes area just to the south (Figure 8.9), Terminal Upper Paleolithic material dated to just before 16,000 BP is well contextualized within dune sand (Goring-Morris and Goldberg, 1990). Thus we see an overall aridification taking place at this interval (roughly 23 ka – Roskin et al., 2011a), culminating just after the Late Glacial maximum.

Figure 8.8 The hammer here is resting on the remains of cross-bedded sandy dune deposits (center) in the area of Qseima near Qadesh Barnea in Eastern Sinai. In this area, these sands pre-date 19 ka and represent an incursion of sand that transgressed across northern Sinai and into the Western Negev at this time. Glistening against the reflection of the sun and resting on these sands are deflated scatters of Epi-Paleolithic (ca. 13–14,000 BP) lithics of a large, partially eroded site. Note that in eroded terrains such as these where deflation is prominent, many of the lithic assemblages are likely to be mixed and it is not possible to discern individual occupation events.

(Continued)

Box 8.1 (Continued)

(a)

(b)

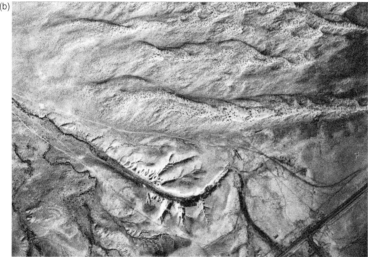

Figure 8.9 (a) Google Earth image detail of Negev and Sinai showing linear dunes that cross the border. To the east (areas of Tze'elim, Gvulot, Ein HaBesor) are extensive fluvially reworked sand and loess deposits. (b) Aerial photo of part of Shunera area (Figure 8.5b) above, showing linear dune in upper half and incised channel in lower half. Exposed next to the incised channel are late Quaternary fluvial silts (7–8 m thick) marked by an upward increase in sand and containing Upper Paleolithic artifacts. These deposits were eroded and followed by the formation of a calcic paleosol that is associated with the an eroded late Upper Paleolithic site (Shunera XV; ca. 18,000 BP) and a Kebaran site (Shunera XVII; ca. 17,000 to 15,000 BP) that are found within thin sandy veneers that cover the terrace developed on the Upper Paleolithic silty alluvium (Goring-Morris and Goldberg, 1990). Dune formation had already begun by ca. 16,000 BP as demonstrated by the presence of the Terminal Upper Paleolithic site of Shunera XVI, which is directly associated with the accumulation of these sands (photo courtesy of N. Goring-Morris). (c) Field view of excavation of Shunera VII, an Early Natufian site eroding out of sands in the foreground. In the background, the southernmost dune shown in Figure 8.6b is associated with numerous prehistoric sites that range from Geometric Kebaran (ca. 14,500 BP) through Harifian (ca. 10,700 to 10,000 BP); a deflational surface that post-dates the Harifian contains Chalcolithic or Early Bronze Age pottery (ca. 6000 to 4500 BP) (Goring-Morris and Goldberg, 1990) (photo courtesy of N. Goring-Morris).

Box 8.1 (Continued)

(c)

Figure 8.9 (Continued)

ⁱ All radiocarbon dates are uncalibrated.

8.2.2 Sand-size Aeolian Deposits

Coarse-grained aeolian deposits are usually sand-size and larger (>62 μm), and overall, consist of quartz, which is most resistant to weathering. Depending on locale and age, other components are feldspar and *heavy minerals* (e.g. hornblende, zircon, garnet, magnetite, hematite; see Chapter 17), although these are subject to weathering especially if the material has been recycled during a long geological history. In the case of dunes (see below), wind is selective in the size of the material it transports and either finer sizes are removed (*winnowed*), or coarser sizes remain as lag deposits. Consequently, aeolian sands tend to be well sorted (see Chapter 2) and increasingly negatively skewed (proportionately more finer material and less coarser material) (Watson, 1989b); however, much variation can be seen within the same deposit (Ahlbrandt and Fryberger, 1982). In addition, quartz sand grains tend to be subrounded and larger sizes tend to be increasingly rounded, although shape is commonly a function of the nature of the material at the source; finer sand and silt sizes (<100 μm) tend to fracture during production and transport and consequently are angular to subangular (Watson, 1989b).

8.2.2.1 Bedforms

Sandy deposits accumulate in a variety of forms that are function of source and supply of sand, wind strength and direction, nature of substrate, and seasonal changes (Reineck and Singh, 1986). These bedforms occur at various scales ranging from ripples on the cm/dm scale up to large sand seas (draas), which cover large areas of the landscape, in the order of kilometers.

Wind ripples range in size from 0.5 to 1.0 cm in height with wavelengths from 1–25 cm, with the size of these features controlled by wind velocity and particle size. They are composed of fine to coarse sand (ca. 62 to 700 μm) (Boggs, 2012). In section, they are asymmetrical, with the windward side exhibiting slopes of about 10°, whereas the lee side has slopes of about 30° (Ritter et al., 2011). Grains move by saltation, creep, and reptation, and climb up the windward slope of the ripple; finer grains are removed, leaving behind relatively coarser sand on the surface of the ripple. The coarse sand moves along to the crest, where it ultimately rolls down the windward side of the ripple.

Through time, the ripple moves downwind, similar to dunes (see below) and as a result, the internal organization of a ripple resembles those of dunes with foreset bedding (Livingstone and Warren, 1996). Fine archaeological lithic debris can be moved along ripple surfaces, whereas coarser artifacts (e.g. cores) can be left immobilized with drifting material moving past it as the ripple migrates downwind.

8.2.2.2 Dune Forms

Sand dunes and associated dune forms are more significant to geoarchaeology from a landscape and evolutionary perspective. On this scale, their presence/absence and form may indicate long-term environmental changes, such as climatic amelioration or desiccation (Roskin, 2012a, b; Roskin et al., 2011b). Furthermore, the movement of sand may result in the erosion or burial of archaeological sites, thus removing them from the stratigraphic record or preserving them for later excavation.

Dunes are the products of deflation and deposition of sand-size material that is sculpted into a variety of forms that are controlled by factors such as wind direction and speed, sand supply, vegetation and grain size (Ritter et al., 2011; Thomas, 1989). They differ from ripples, which are small-scale bedforms that form at lower flow velocities (Boggs, 2012). Numerous classifications of dunes exist that reflect shape, origin, wind characteristics, surface conditions, and whether they are mobile ("free") or fixed. One such classification by Livingstone and Warren (1996) is given in Table 8.1. A similar classification makes use of form and also characteristics such as wind regime and number of slip faces (the leeward face of the dune which is typically between 30–34°, the angle of repose for sand) (Ritter et al., 2011; modified from McKee, 1979) (Table 8.2 and Figure 8.7a). The section below gives a broad overview of the diversity of these dune types, before their geoarchaeological significance is discussed (see section 8.3).

Free Dunes

Transverse dunes have slip faces oriented in the same direction due to a unimodal wind regime. Sand is transported in a direction perpendicular to the orientation of the dune crest (Figure 8.10a, b). Forms range from elongate, continuous ridges to discontinuous, individual dunes (*barchans*) (Figure 8.11a, b). These latter types, with crescentic shape and horns pointing downwind, form in areas of restricted sand supply. They typically have tabular to planar cross-beds in the center of the

Table 8.1 Classification of aeolian dunes by Livingstone and Warren (1996), which is essentially based on morphology.

Dunes					
Free				Anchored	
Transverse	Linear	Star	(Sheets)	Vegetation	Topography
Transverse	Sandridge	Star	Zibar	Nebkha	Echo
Barchan	Seif	Network	Streaks	Parabolic	Climbing
Dome				Coastal	Cliff-top
Reversing				Blowout	Falling
					Lee
					Lunette

Table 8.2 Classification and illustration of major aeolian dunes types (modified from Reineck and Singh, 1986; Ritter et al., 2011; Thomas, 1989), based on morphology, number of slip faces, and wind regime.

Form/Dune	Number of slip faces	Major control on form	Formative wind regime
Transverse dunes			
Barchan dune	1	Wind regime and sand supply	Transverse, unidirectional
Barchanoid ridge	1	Wind regime and sand supply	Transverse, unidirectional
Transverse ridge	1	Wind regime and sand supply	More directional variability than for barchans
Linear dunes			
Linear ridge/seif	1–2	Wind regime and sand supply	Bidirectional/wide unidirectional
Reversing dune	2	Wind regime and sand supply	Opposing bimodal
Star dune	3+	Wind regime and sand supply	Complex, multidirectional
Zibar	0	Coarse sand	Various
Dome dune	0	Coarse sand	Various
Blowout	0	Disrupted vegetation cover	Various
Parabolic dune	1	Disrupted vegetation cover	Transverse, unidirectional

(a)

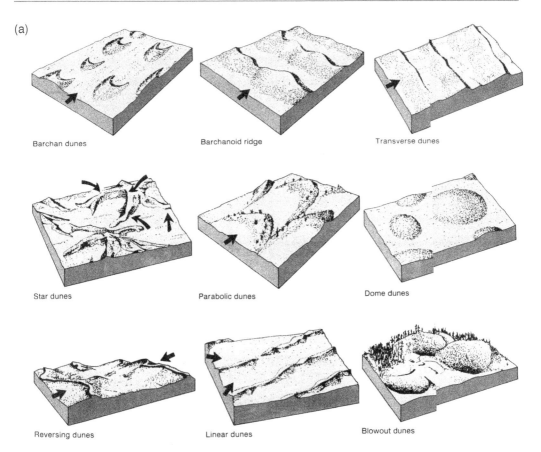

Barchan dunes

Barchanoid ridge

Transverse dunes

Star dunes

Parabolic dunes

Dome dunes

Reversing dunes

Linear dunes

Blowout dunes

Figure 8.10 Dunes and stratification. (a) Major dune types with arrows representing the principal wind directions (modified from Ritter et al., 2011 after McKee, 1979).

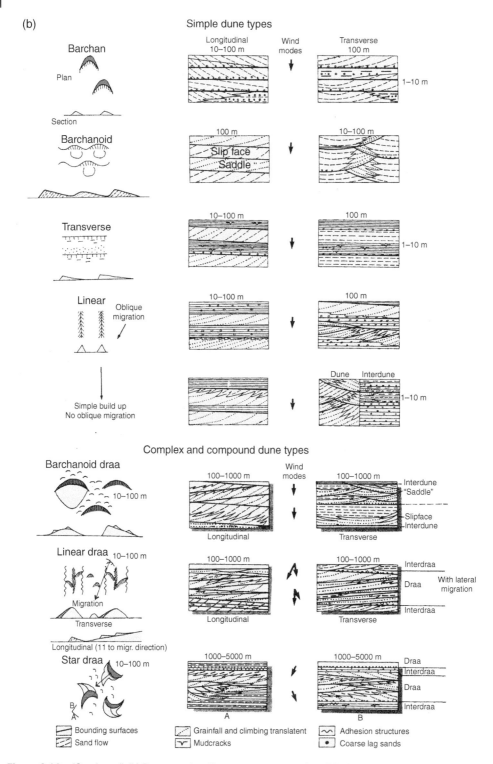

Figure 8.10 (Continued) (b) Dunes and sedimentary structures (modified from Brookfield, 1984: figure 7).

(a)

(b)

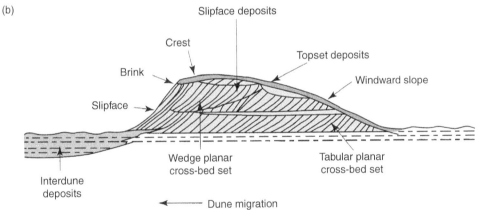

Figure 8.11 Barchan dune. (a) Photograph of barchan dune in Sinai, Egypt. Note the smaller scale ripples on the surface of the dune. The double arrows in the middle ground point to fluvial terrace consisting of reworked sandy deposits that are capped with a rubified calcic horizon overlain by Epi-Paleolithic (Kebaran; ca. 13 ka) implements (sites Lagama 1C and 1F; Goldberg, 1977); slightly younger Harifian (ca. 10.5 ka) material is scattered on the terrace riser between the terrace tread and the barchan; (b) schematic view through a barchanoid dune ridge showing types of bedding and relation to lateral, interdune deposits which are finer grained (modified from Boggs, 2001: Figure 9.21).

dune at about 34° (Figure 8.10b). Coalesced barchans result in barchanoid dunes and form with greater sand supply. Heights of transverse dunes range from as little to 0.5 m, up to >100 m (Livingstone and Warren, 1996).

Linear ridge dunes entail sand transport along the length of the crest of the dune, with slip faces found on both sides (Figure 8.10a). They are relatively long and narrow (see Box 8.1) and may reach up to tens of km in length (Cooke et al., 1993; Livingstone and Warren, 1996). A variant of the linear form is the *seif* dune, which has sharp, sinuous crests and can have lengths up to hundreds of km (Ritter et al., 2011). They are widespread in the Namib Desert, the Eastern Sahara and in Arabia, and the Simpson Desert of Australia (Thomas, 1989). Their origin is complex but it is generally accepted that they form in a bidirectional wind regime.

Reversing dunes are a form of dune networks in which the wind regimes are completely opposite and the ridges are asymmetrical (Boggs, 2012) (Figure 8.10a). As described by Cooke et al. (1993) for Oman, larger dunes are formed during the summer when the southwest monsoon is active. In winter, on the hand, the wind direction is reversed, with smaller dunes being formed.

Star dunes appear to have a chaotic set of slip faces, with many pointing in several directions; they tend to have a somewhat radiating pattern with ridges emanating from a central higher area. Star dunes occur in sand seas, where the source of sand is abundant, such as the Grand Erg Oriental in North Africa (Livingstone and Warren, 1996). They appear to form in areas with multidirectional wind regimes.

Zibar are extensive dunes that have no slip faces and are characterized by their composition of coarse sand and hard surfaces. They tend to have low relief and are straight to parabolic in form (Cooke et al., 1993).

Sandsheets are areas that have very little relief and subtle dune morphologies. The largest sandsheet occurs along the border between Egypt and the Sudan, which has been the focus of prehistoric research for several decades, formerly by F. Wendorf and more recently by C. Hill (see above). In this area, the sand sheet is capped by a soil developed in Neolithic times, clearly attesting to wetter conditions at this time. Ripples, consisting of coarse sand, are the predominant bedform.

Anchored Dunes

Anchored dunes are formed by topographic barriers or by fixed vegetation, with these obstacles affecting the flow conditions. In the case of *climbing dunes*, for example, sand is trapped on the upward slope, usually 30° to 50° (Livingstone and Warren, 1996). This is the context for the coastal, Middle Stone Age site of Blombos Cave (South Africa) where climbing dunes were dated by OSL to about 68 ka (Haaland et al., 2020; Jacobs, 2008; Jacobs et al., 2003); aeolian sands of similar date can be found along the southern coast of South Africa (Karkanas and Goldberg, 2010) and are related to sea level regression. Likewise, on the lee side of wide obstacles, calm air leads to the accumulation of sand along the lee slope of the obstacle (*falling dunes*) (Livingstone and Warren, 1996).

Plants serve as anchors to dunes and are commonly used to stabilize moving sand in cases of coastal erosion. Trapping vegetation can range from isolated plants to more extensive areas, such as along coasts. Small clumps of vegetation will tend to produce individual mounds, whereas plants that spread laterally, will create more undulating topography (Livingstone and Warren, 1996).

Parabolic dunes are U- or V-shaped in plan view, and the open side points into the wind. They are associated with vegetated areas at the margins of deserts, such as the Thar Desert in India. They appear to develop from *blowouts* whereby material is eroded from the underlying sediment, while at the same time the flanks are stabilized by vegetation (Cooke et al., 1993).

Coastal dunes (see also below; Chapter 9) are found not only along vegetated coasts, such as those of South Africa cited above, Cape Cod, USA, but also in areas flanking desertic terrains, such as in the Negev/Sinai (Figure 8.9a). In this case, coastal sand is transported northward along the coast by longshore drift from the Nile. This littoral transport, coupled with low relief and relatively shallow shelf and constant winds, results in the sand being blown up off the beach and accumulating on coastal cliffs or behind them. Many Holocene archaeological sites occur just beneath these coastal sands, while resting on Pleistocene aeolianites that also contain archaeological sites (e.g. Frechen et al., 2001, 2004; Neev et al., 1987; Taxel et al., 2018; Tsoar and Goodfriend, 1994).

Interdunal areas – these are areas between dunes and are confined by dune bodies (Boggs, 2012) (Figure 8.11b). They may be deflationary as in the case of *blowouts* (see above, Figure 8.10a) in which case little sediment accumulates, with both wet and dry interdune areas occurring. In dry interdune areas, the deposits tend to be coarser, poorly sorted and poorly laminated. In wetter interdune areas, which are commonly moist throughout the year, the sediments are finer grained silts and clays, with freshwater mollusks, ostracodes, and diatoms (Boggs, 2012); they commonly serve as sources for aeolian dust (Enzel et al., 2010). Where strong evaporation occurs or in areas undergoing desiccation, evaporates (carbonates, halite, and gypsum) precipitate, forming sebkhas. As discussed above, wet interdune areas are attractive to game and humans in the past (Roskin et al., 2017).

8.3 Examples of Sites in Aeolian Sand Contexts

Many interesting and significant archaeological sites are found in dune and associated arid-land contexts. These geoarchaeological examples present stratigraphic sequences that reflect both local and regional environmental changes, which ultimately are linked with former climates.

In Wyoming, USA, Quaternary dune fields are characterized by thin, localized sandy deposits such as the Ferris Dune field where stabilized and parabolic dunes occur (see also comments above about the Casper site) (Albanese and Frison, 1995). During the interval of ca.7,660 to 6,460 BP, 9 m of sand accumulated during a marked drought. Overlying this are 8 m of aeolian sand containing lenses of interdunal pond deposits (ca. 2155 BP), attesting to more moderate conditions. Nearby, a 4 m sequence of sands associated with archaeological remains shows at least two buried, truncated calcic horizons, attesting to stabilization and climatic change. In the Killpecker Dune Field, dome, transverse, barchan, and parabolic dunes occur. An iron-stained basal sand resting on Pleistocene deposits is overlain by sand with interdunal pond deposits at the top containing Cody Complex cultural material at the Finley site (ca. 9,000 BP); the upper unit is 3,000 years younger at its base, and much younger at the toe (ca. 755 BP) (Albanese and Frison, 1995).

Older and more sweeping geoarchaeological sequences occur in North Africa and south-west Asia, in areas of Egypt, Sinai, and Israel. In the Bir Tarfawi and Bir Sahara areas in Southern Egypt, lacustrine and aeolian deposits associated with Lower and Middle Paleolithic artifacts testify to human occupation alternating and marked differences in geological and regional environments and climates (Hill, 2009; Wendorf et al., 1993). At site BT-14, for example, the stratigraphic sequence begins with topset and foreset beds of a large dune whose eroded surface displays wind-blasted Aterian artifacts ($\approx$ Middle Paleolithic). In turn, these sands were deflated and the resulting depression filled with carbonate- and organic-rich lacustrine silts (ca 44,190 BP; Wendorf and Schild, 1980) that grade laterally into sandy beach deposits with numerous Aterian artifacts; a wetter climate is clearly indicated. These deposits were subjected to late Pleistocene-early Holocene deflation, and eroded surfaces are dotted by recent dunes that tend to fill depressions (Wendorf and Schild, 1980: 38).

Similar types of events are well documented to the northeast, in Sinai, Egypt, and the Negev, Israel, although the record of aeolian and fluvial activity is more pronounced (see Box 8.1). Gebel Mughara in northern Sinai is a ca. 40 km long breached fold structure that contains aeolian (Figure 8.11a) and fluvially reworked sand and lacustrine deposits, with rubified and calcified paleosols and associated Upper Paleolithic and Epi-Paleolithic sites (Bar-Yosef and Phillips, 1977; Goldberg, 1977). In the Wadi Mushabi valley, the base of the Later Quaternary sequence is represented by massive, hard/compacted sand, which cover the center and SW part of the basin. The surface is locally rubified (reddened) and is covered with ventifacted stones; carbonate nodules can be seen eroding out of these sands. Due to poor exposures, no artifacts were observed in the sands, yet they are dotted with Epi-Paleolithic (Geometric Kebaran and Mushabian, ca. 13–14,000 BP) sites.

Elsewhere in the area, similar types of massive sands contain Upper Paleolithic sites dated to about 31–34,000 years ago (Bar-Yosef and Phillips, 1977). Overlying the basal deposits are thin, patchy veneers or more massive sands, some of which are linear dunes. These sands are of various ages: the Epi-Paleolithic site of Mushabi V, for example, occurs within in loose, massive, coarse sand that rests upon the soil of the basal sand; the material is *in situ* as demonstrated by the presence of three intact hearths. On the other hand, the slightly younger, Late Epi-Paleolithic site of Mushabi IV is found in a more compact and poorly sorted sand. Both sites indicate a minor movement of sand that rests upon on a stabilized, pedogenic surface that pre-dates these later dunes by several thousand years. Moreover, the soil developed on the basal sand points to a wetter regime in which clay translocation and carbonate nodule formation could develop (Goldberg, 1977). Correlative wet intervals can be seen over much of the Sinai and Negev at this time (Belfer-Cohen and Goldberg, 1982; Goldberg, 1986; Goring-Morris and Goldberg, 1990; Roskin et al., 2011a).

Eastward into the Qadesh Barnea area of easternmost Sinai, and across into the Negev, aeolian deposits play an important part in documenting the geological and environmental history of this area during the late Pleistocene (Upper Paleolithic, Epi-Paleolithic), and early Holocene (Neolithic) (Figure 8.12; Box 8.1). At present, this area is covered with a system of SW-NE or W-E trending linear dunes that cover an area of about 10,000 km² (Goring-Morris and Goldberg, 1990). Intensive and extensive prehistoric research in this area has provided high-resolution stratigraphic sequences, which owing to the associated presence of datable materials (e.g. charcoal, shell, carbonate nodules) has resulted in a clear picture of the landscape and the people that occupied it (Goring-Morris, 1987). We note for example, that dune encroachment began prior to the Late Glacial Maximum, reaching north-central and northeastern Sinai about 25,000 years ago from the west. It steadily progressed eastward into the Negev, arriving in the Nahal Sekher (see Box 8.1) region about 5000 years later.

Such progressive landscape and climatic aridification led to the blockage of many wadis (ephemeral streams) and the formation of localized playa/sebkha-type environments (Roskin et al., 2017). This situation is in contrast to the preceding interval encompassing late Middle Paleolithic and early Upper Paleolithic sites. During this interval of ca. 50–40,000 to ca. 25,000 BP wetter climates prevailed as expressed by massive fluvial accumulations of silts and reworked sands within most of the drainages of the area; the dunes can be clearly seen in the field and in aerial photographs to rest unconformably on the previously formed fluvial terrain (Roskin et al., 2011b). Furthermore, *in situ* sites demonstrate that dunes were also locally mobilized in Holocene times (Neolithic, Chalcolithic, and Byzantine Periods) (Goring-Morris and Goldberg, 1990; Roskin et al., 2017). Indeed, human landscape disturbance through the removal of vegetation and disturbance of the biogenic crusts, has been suggested as a mechanism for further mobilization of the dune sediments during the Holocene, with dates in the northwestern Negev dune fields clustering around 900–600 BCE, for dune mobilization, associated with increased human activity (Roskin et al., 2013b).

In Sudan, northern Africa, excavations across the Nubian Desert demonstrate the ongoing burial of archaeological sites via aeolian sandsheets and dunes, such as at the Kushite town of Kawa and the site of Dongola. It has been suggested that the location of important sites in the past were deliberately positioned to avoid inundation from aeolian sand deposition, such as Jebel Barkal on an upstanding massif and the Meroe pyramids located on a ridge above the dunes. The burial of sites has also been recorded in the recent past, with Meinarti being abandoned and buried during the 15th century (Munro et al., 2012).

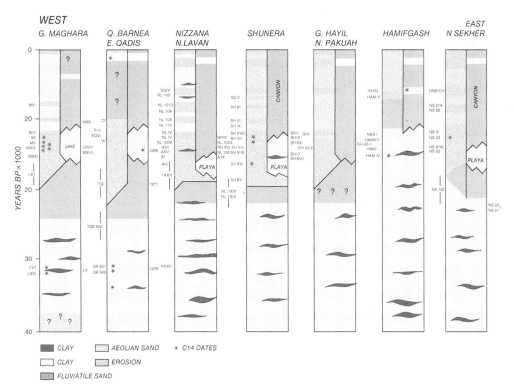

Figure 8.12 Geomorphic evolution of Northern Sinai and the Western Negev, showing the time-transgressive nature of Late Pleistocene dunes from West to East. Indicated on the sides of the panels are the names of prehistoric sites; radiocarbon dates are also indicated. For details, see Goldberg, 1984, 1986; Goring-Morris, 1987; Zilberman, 1993 (modified from Goring-Morris and Goldberg, 1990: Figure 4).

Across northwestern Europe, burial of sites by coastal dune systems is well attested. The burial and preservation of a diverse range of archaeological sites during the Holocene has provided a disproportionately rich archaeological resource within the coastal fringe (also see section 9.2.2 coastal dunes). Examples include the famous Neolithic houses at Skara Brae (Scotland, UK), initially buried by coastal sand dunes, before subsequent erosion of the sand dunes revealed around 10 houses (Darvill 2010) and the site at Brean Down (Somerset, UK) associated with Beaker activity (Bell 1990; see section 9.2.2). Likewise, at Gwithian (Cornwall, UK), a complex multi-period site, with evidence from the Mesolithic through to Medieval was periodically buried and eventually preserved within and under the coastal dune system. Here, significant phases of Bronze Age occupation are present between periods of coastal sand deposition. The preservation of artifacts and materials was exceptional, due to the alkaline conditions of the sand dunes, compared to the acidic nature of the surrounding bedrocks, providing a rich paleoenvironmental contrast. The excavations revealed well-preserved houses dating to the second millennium BC, complemented by evidence of agriculture through both spade and ard marks, (Nowakowski, 2007; see Box 8.2).

Across northwestern Europe, a further type of deposit known as coversands are reasonably widespread, occurring in South Denmark, Poland, North Germany, Eastern Britain, Belgium, and the Netherlands. These deposits are essentially aeolian peri-glacial sediments, dominated by sand fractions (150–200 µm) with a low silt component and form a relatively flat and thin mantle (c <5m) over earlier deposits. Some of these can be associated with activity from the Late Upper

Box 8.2   Aeolian sand dunes and the burial of the Gwithian Bronze Age settlement, Cornwall, UK

For over 20 years, from the late 1940s to the early 1960s, a landscape archaeological survey was conducted across some 15 km² at Gwithian on the north coast in the far west of Cornwall, UK. The deposition of alkaline aeolian sands preserved a remarkable archaeological record, with successive occupation layers of prehistoric and later settlements having been sealed and preserved by windblown sand from the coastal dunes. The full story of these excavations has recently been revisited and the remarkably rich data sets from this archive has come to light (Nowakowski, 2007; Nowakowski et al. 2007). Small-scale fieldwork re-examining some related areas of the Bronze Age deposits has also taken place in the last 15 years (Nowakowski et al., 2007; Walker et al., 2018), using dating techniques (AMS radiocarbon and OSL) and paleoenvironmental (mollusks) and geoarchaeological analysis of these sequences.

This Box focuses on the evidence for settlement dating from the Early Bronze Age through to the 1st millennium BC (Nowakowski, 2009; Nowakowski et al., 2007). Structural evidence for wooden and stone buildings, field walls and middens were found alongside fields, in which criss-cross patterns of ard marks were preserved. A deposit sequence was recorded in occupation bands or horizons, which were recorded as distinctive "Layers" and have subsequently been reanalyzed by a programmed of radiocarbon dating and OSL dating using Bayesian modeling (see Sturgess in Nowakowski et al., 2007 and Hamilton et al., 2007d) (Figure 8.13a):

Figure 8.13a   The original cutting of GMX at Gwithian showing the key Phases (1–7) and associated Layers (8-2). The interleaving of the aeolian sand deposits stands out clearly against the deposits associated with Bronze Age occupation and farming. The preserving, stratifying effect of the intermittent aeolian sand deposits is clear; without these, the archaeological deposits would be mixed up, with successive phases of ard ploughing destroying the earlier evidence (image reproduced with the kind permission of the Royal Institution of Cornwall, Truro, Jacqueline Nowakowski and Jo Sturgess; original photograph by JVS Megaw, the Gwithian Archive).

Box 8.2 (Continued)

- **Phase 1:** Early Bronze Age ca. 1800 cal BC. Small farmstead with terraces and fields ("Layers 7 and 8").
- **Phase 2:** Aeolian sand deposit ("Layer 6").
- **Phase 3:** Middle Bronze Age ca. 1500–1200 cal BC. Settlement, terraces, fields, human burials ("Layer 5").
- **Phase 4:** Aeolian sand deposit ("Layer 4").
- **Phase 5:** Middle Bronze Age/Late Bronze Age ca. 1300–900 cal BC. Settlement and fields, human burial, middens ("Layer 3").
- **Phase 6:** Aeolian sand layer ("Layer 2").
- **Phase 7:** Modern turf and topsoil.

Phase 1: This phase of settlement contained the remains of a circular wooden building set within a wooden fence enclosure located amid terraces and accompanied by fields. Some of these fields were under arable cultivation, with ard marks recorded. The circular wooden building had two hearths and contained a range of structured deposits, including broken quern fragments, a perforated dog whelk, a bronze awl, and worked bone. An archaeomagnetic date from the hearth at Gwithian provided a date of ca. 1700 BC, with more recent radiocarbon dates supporting an Early Bronze Age date (see Hamilton et al., 2007d). Ard marks from an early wooden plough were recorded in Layer 7 and *Hordeum spp.* (barley) seeds were recovered. Analysis of terrestrial mollusk assemblages suggested a somewhat open environment, consistent with a managed farming regime. Fragments of marine species hinted at the possibility of local seaweed collection for manuring the nutrient-poor soil. Wood charcoal analyzed from Layer 7 indicated the presence of woodland and scrub close by with oak, hazel, hawthorn, gorse, and broom identified. Limited pollen analysis identified plantain (*Plantago*), a key ground disturbance indicator species.

Phase 2: Layer 6 is an aeolian sand layer that was discontinuous across the original area of excavation, with low density mollusk survival defining open country species. An OSL sample from this layer produced a date of 1410 BC +/− 160 years (see Roberts in Hamilton et al., 2007d).

Phase 3: Layer 5 is characterized by fields and a prolonged period of arable cultivation. Two major north-south boundaries were identified that became fixed features within the landscape. On the eastern side of these boundaries, ploughing seemed almost continuous and intense. The area around and between the major field boundaries was marked by surviving blocks of criss-crossing patterns scored into the soils by a stone ard (Figure 8.13b) (cf. Deák et al., 2017; Lewis, 2012). Geoarchaeological analysis of samples from Layer 5 confirmed a heavily worked horizon, with seeds of emmer wheat (*Triticum* cf. *diococium)* and field madder (*Sherardia arvensis*), indicating arable farming. Analysis of terrestrial mollusk species from this layer defines an open landscape of fields. However, this phase was not only a domestic and agricultural landscape. A line of four pits, containing cremated human remains, with other small quantities of animal bone and marine shell were placed alongside one edge of a field division. This phase dated to between 1500–1200 cal BC, showing some longevity and provides some of the richest archaeological evidence for Middle Bronze Age farming in the UK.

Phase 4: Layer 4 was a further aeolian sand deposit, albeit one that is discontinuous over the landscape, which again preserved and sealed earlier phases of landscape use. The analysis of terrestrial mollusk species layer defined a transition from open fields into woody scrub. Interestingly, sand filled ard marks were visible at the base of this unit and more recent analysis revealed that there had be some cultivation of this layer (Sturgess in Nowakowski et al., 2007: 29–30), demonstrating that people were attempting to maintain their environment and continue to use this locale, in the face of increased sand deposition.

(Continued)

Box 8.2 (Continued)

Figure 8.13b The criss-crossing ard marks preserved at the base of layer 5, phase 3 attest to the arable cultivation during the Middle Bronze Age, something that is rarely preserved on most Bronze Age sites in southern England. The preservation of these remarkable features was through burial (and hence preservation) by aeolian sand deposition (image reproduced with the kind permission of the Royal Institution of Cornwall, Truro, Jacqueline Nowakowski and Jo Sturgess; original photograph by JVS Megaw, the Gwithian Archive).

Phase 5: Layer 3 is a major phase of settlement, with houses and fields, dated through large quantities of Trevisker pottery and subsequent radiocarbon AMS dates, to ca. 1200–900 cal BC. It is a more substantial settlement, where wooden and stone buildings replace one another, with some residential and some workshops, alongside evidence for crafting (pottery making, shale, metal, bone, wood and leather working) (Nowakowski, 2009). At least six buildings were identified, centered around a probable communal area or "town-place". During this phase more ard marks were found alongside the imprints of wooden spades within the fields, defining a cleared and arded soil (Figure 8.13c). The mollusk species from this layer indicated a diverse environment, with both shade and open loving species present. Charred remains of emmer wheat (*Triticum* cf *dicoccum*) were found, alongside an extensive animal bone assemblage that included both domestic (cattle, pig, and sheep/goat) and wild (roe deer) species. Oyster, marine mussels and whale bone revealed some exploitation of marine resources within this farming community.

Phase 2: This is a further aeolian sand deposit. After phase 2 there is a subsequent phase of post-Roman activity (not discussed here) at several sites within this landscape.

The excavations at Gwithian demonstrate the richness of the archaeological record that can be preserved in these coastal aeolian environments. The interleaving of aeolian sand deposits between periods of occupation and land use, has provided a successive narrative for Early to Middle Bronze Age life and death in Cornwall, and southern England more generally (see Bell, 1990). The preserving and stratifying nature of the aeolian sands has provided evidence of continuity from the Early and into the Later Bronze Age, evidence that has been lacking across nearly all of southern England. The boundary between the Early and Middle Bronze Age has been argued to represent a marked shift in society (Bradley 2007); however, the excavations at Gwithian suggest the fields and roundhouses, that are so diagnostic of the Middle Bronze Age, have their roots in the Early Bronze Age.

Box 8.2   (Continued)

Figure 8.13c   The preservation of spade marks in phase 5 again attest to the remarkable preservation at Gwithian. The spade marks are interpreted as associated with ground clearance and agriculture (image reproduced with the kind permission of the Royal Institution of Cornwall, Truro, Jacqueline Nowakowski and Jo Sturgess; original photograph by JVS Megaw, the Gwithian Archive).

Furthermore, the preservation of animal bone, antler, terrestrial mollusks, due to the alkaline sand conditions, alongside seeds and charcoal, has provided insights into the ritual and domestic lives of the people who lived at Gwithian for over 1,000 years. The preservation of Middle Bronze Age cultivation evidence, through both ard and spade marks, stands in stark contrast to the majority of Bronze Age sites in southern Britain, especially the upland zones, such as Dartmoor and Bodmin Moor. Gwithian clearly demonstrate the presence of relatively extensive of arable faming during the Bronze Age, something that is often suggested for Middle Bronze Age settlements, but is rarely demonstrable due to a lack of evidence. Whilst the original excavations at Gwithian are of "their time", the evidence from this project stands as a testament to the unique archaeological records that can be preserved due to aeolian coastal sand deposition.

(Jacqueline Nowakowski and Jo Sturgess are thanked and acknowledged for their contributions to this section.)

Paleolithic, such as at Farndon (UK) (Tapete et al., 2017). Critically, many of these sand deposits have been remobilized through aeolian weathering during the Holocene, sometimes potentially because of earlier phases of human activity, such as vegetation removal, and as such, they have a high potential to bury and preserve sites of Holocene age (Bateman 1995; Bateman and Godby 2004). Under dry weather conditions, when humic topsoils on sands are breached, they are prone to deflation; the Pleistocene sands of Eastern England are well known in this respect, with both deflation and sand deposition evident in the Holocene (Corbett, 1973; Murphy, 1984; Radley and Simms, 1967). A number of excavations of blown sand-buried sites and landscapes have also taken place in Jutland, Denmark, including both coastal dune and inland locations where Bronze Age management of manured cultivated and grassland pasture soils delayed the onset of wind erosion until less effective land use practices were carried out (Courty and Nørnberg, 1985; Dalsgaard and Odgaard, 2001; Mikkelsen et al., 2007).

Coversands have been extensively studied in the UK with the sequences in East Lincolnshire containing sites dating throughout the Holocene (McDonnell 2012). Likewise, the landscape scale surveys at the Vale of Pickering have demonstrated a diverse multi-period archaeological landscape, with sites both cutting into and located on top of the coversands, as well as some sites being buried during the Holocene. This project has used multiple prospection methods over large areas including multispectral remote sensing, aerial photography, and geophysical surveys (see Chapter 16). Models of the overall coversand depth have been created through coring to help explain the poor visibility of archaeological remains to prospection methods in some areas of the landscape, again demonstrating remobilization of these sediments during the Holocene (Powlesland et al., 2006).

Finally, it should be noted, that although luminescence dating of such deposits has progressed significantly in the last few decades, the presence of datable materials from archaeological sites serves as an independent, often high-resolution dating source. Moreover, their excavation has the potential to provide higher resolution stratigraphic information than can be obtained from examining the aeolian deposits in isolation.

8.3.1 Coastal Sands and Caves

Coastal cave sites are remarkable in their ability to trap aeolian sediments from the outside (see Chapter 9). Many of cave sites along the Mediterranean and shores of South Africa (e.g. Blombos, Die Kelders, Pinnacle Point) are noted for their proximity to the sea, causing deposition of aeolian sediments (Goldberg, 2000b; Haaland et al., 2020; Karkanas et al., 2015; Karkanas and Goldberg, 2010; Marean et al., 2000). Tabun Cave in Israel, for example, is one of the major Paleolithic sites in southwest Asia, having a long archaeological and geological sequence, as well Neanderthal remains. It contains over 25 m of sandy and silty aeolian sediments associated with Lower and Middle Paleolithic artifacts, overlain by burned presumably anthropogenic layers and stony terra rossa-rich chimney fill (Garrod and Bate, 1937; Goldberg, 1973; Shimelmitz et al., 2014). In early investigations the infilling of the cave sequence was framed within a "short" chronology (Farrand, 1979) and tied to sea level fluctuations that date to the last interglacial and younger. More recently ages furnished by TL and ESR techniques (Mercier et al., 2000; Mercier et al., 1995; Rink et al., 2003, 2004), point to much greater ages that span over ca. 200,000 years of accumulation, beginning more than 415 ka ago.

On the other side of the Mediterranean, sandy deposits are also a major constituent of the infillings in Gorham's and Vanguard Caves in Gibraltar. Together, these sites have over tens of meters thick deposits, associated with Middle and Upper Paleolithic occupations and provide evidence for a late Neanderthal presence in the area, including rock engravings, thus contributing to the understanding of the behavior of Neanderthals and their extinction in Iberia (Barton et al., 1999; Finlayson et al., 2008; Goldberg and Macphail, 2000; Macphail and Goldberg, 2000; Rodríguez-Vidal et al., 2014; Stringer et al., 1999). However, the dating of the sediments initially proved problematic, both from the technical viewpoint of securing accurate TL dates, but also from the standpoint of site formation. The key strata that span the Middle Paleolithic and Upper Paleolithic units were bioturbated and slumped, rendering any radiocarbon dates suspect from these deposits (cf. Pettitt, 1997).

More recent investigations within the cave interior have demonstrated that the areas excavated affected by bioturbation were located in parts of the cave more exposed to external conditions. In these areas, the cave infill was formed through aeolian coastal dune sands that penetrated upwards towards the interior of the cave. In effect, this created a sloping entrance, with a strong gradient (15–20%), with its sedimentary stratigraphy showing evidence of disturbance by animals and hominins. The more recent excavations in the deeper part of the cave, which were less exposed to external conditions, defined a depositional surface, with sub-horizontal stratigraphy and lower aeolian sand content. The main post-depositional effects were compaction through trampling, rather than bioturbation. These excavations produced a radiocarbon chronology that demonstrated level IV exclusively records Neanderthal occupation and is characterized by Mousterian technology dating from between 33–24 kpa BP. The overlying Level III is Upper Paleolithic, and the earliest diagnostic culture is Solutrean at ca. 18.5 ka BP. Critically, a dating interval of over four thousand years separates these two horizons (Finlayson et al., 2008). However, it should be noted that the persistence of the Neanderthal population here seems very late (Galway-Witham et al., 2019) and only serves to highlight the need for future research.

Similar types of sandy aeolian deposits are found in many coastal sites from South Africa. These range from sandy intercalations with cultural deposits during the Middle Stone Age (MSA) to more massive, typically sterile dune sands between the MSA and Late Stone Age (LSA). An excellent example is Die Kelders Cave, situated about 100 km east of Cape Town. Here detailed granulometric and micromorphological analysis was key to revealing the nature of site formation processes at the site, including its aeolian history (Avery et al., 1997; Goldberg, 2000b; Marean et al., 2000; Tankard, 1976; Tankard and Schweitzer, 1974, 1976). Research in the Mossel Bay area, has refined the chronology and occupation of some of the cave sites, whilst documenting the type and age of sandy incursions in coastal prehistoric caves and settings during the Quaternary (Karkanas et al., 2020).

In particular, Cave 13B has recently had is depositional history investigated, with micromorphological analysis demonstrating a range of natural sedimentation processes, ranging from water action to aeolian activity and from speleothem formations to plant colonization and root encrustation. In addition, anthropogenic sediments, mainly in the form of burned remains (e.g. wood ash, charcoal, and burned bone) were deposited. Several erosional episodes have resulted in a complicated stratigraphy, as discerned from different depositional and post-depositional features. The sediments demonstrate a fluctuating coastal environment, changes in sea level, and climate-controlled patterns of sedimentation, and the presence (and absence) of humans (Karkanas and Goldberg 2010). Significantly, the dating of the occupation deposits within the cave demonstrates that Pinnacle Point Cave 13B (PP13B) has the earliest evidence for the exploitation of marine shellfish, along with early evidence for use and modification of pigments and the production of bladelets, all dated to approximately 164 ka (Marean et al., 2007). This makes PP13B a key site, showing evidence of human occupation in Africa during MIS 6 (Marean et al., 2010).

In another site within the Pinnacle Point Complex, site PP5–6 (Karkanas et al., 2015), a careful and detailed microstratigraphic study of the geogenic and anthropogenic deposits involving the documentation of microfacies in the field coupled with micromorphology was undertaken. Several facies were recognized, including aeolian sand, roofspall, combustion, and black sand. The formation of this deposit sequence was used to interpret a change from short occupations by small groups to more intense occupations. In the upper part of the sequence, aeolian sands are more common and demonstrated relatively rapid accumulation rates, which helped to preserve the anthropogenic deposits.

8.4 Bioturbation in Sandy Terrains

It is important to mention a recurrent theme that arises with archaeological sites in sandy sediments: does a buried artifact assemblage result from aeolian sedimentation, colluviation, or bioturbation (Frederick et al., 2002; Leigh, 1998, 2001). Deciphering the effects of these processes is crucial if we are to interpret the archaeological record correctly.

In spite of difficulties, appraising such problems can be accomplished in the field through detailed documentation of the site's geomorphic setting and microstratigraphy, and in the laboratory by investigating bulk properties such as granulometry, heavy mineral content, magnetic susceptibility, and phosphate content. Micromorphology and OSL measurements (Bateman et al., 2007) are quite effective in evaluating the effects of bioturbation, as microscopic signatures are often quite distinctive. A lack of correspondence of different measurements (e.g. radiocarbon dates, artifact abundance, phosphate content) with depth, for example, suggests that bioturbation may indeed be responsible for displacing artifacts and burying them; similarly the co-occurrence of lithic tools not normally associated (e.g. Middle Paleolithic and Upper Paleolithic artifacts) might point to bioturbation (Stiner et al., 2001).

8.5 Fine-grained Aeolian Deposits

Fine-grained aeolian deposits, including primary windblown loess and its derivatives (e.g. colluvial loess), have been estimated to cover up to 10% of the earth's land surface (Muhs and Bettis 2003). The thickest and most widespread deposits are in China, Central Asia, Ukraine, Central and Eastern Europe, the Great Plains of North America, and Argentina (Muhs et al., 2014) (Figure 8.14). Most geoarchaeology-related contexts are found in the Old World, due to the relatively late arrival of humans in North and South America, although some sites are associated with Holocene loess (Thorson and Hamilton, 1977). During the Holocene, loess may comprise a sediment source that can be part of the sedimentary record of archaeological sites (e.g. Dust Cave, Alabama; Goldberg and Sherwood, 1994; Sherwood, 2001), or the substrate upon which occupation takes place. In addition, some soils from the southern Plains, in the Edwards Plateau, USA, for example, contain embedded aeolian silt (Rabenhorst et al., 1984) derived from the High Plains; similar processes of aeolian embedding of silts is widespread in North Africa and the Levant (Coudé-Gaussen, 1988; Fedoroff et al., 2018; Yaalon and Ganor, 1975).

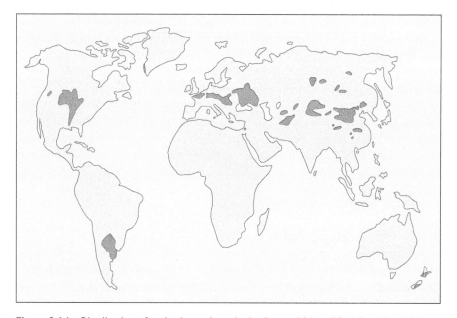

Figure 8.14   Distribution of major loess deposits in the world (modified from Pye, 1987: Figure 9.1).

The geoarchaeological importance of loessic deposits is that they are widespread, have accumulated over long time periods, and contain abundant soils of varying types that have paleoenvironmental significance which can be retrieved relatively easily. This integration of aeolian loessic sediments with paleosols is often referred to as loessic paleosol sequences (LPS). Important prehistoric sites, such as Willendorf in Austria (Haesaerts and Teyssandier, 2003; Nigst et al., 2008), are found within loess deposits and associated soils. Thus not only do these loesses and soils constitute one of the best and most continuous records of continental deposition, but also they supply high-quality, detailed paleoenvironmental information, before, during, and after prehistoric occupation (Zeeden and Hambach, 2021). Such information is crucial in reconstructing prehistoric lifeways and resolving thorny issues, such as the arrival of earliest Modern Humans in Europe (Conard and Bolus, 2003) and landscape reconstruction and subsistence strategies in the Upper Paleolithic (Leonova et al., 2015).

Loess can be characterized as "a terrestrial (i.e., subaerial) windblown silt deposit consisting chiefly of quartz, feldspar, mica, clay minerals and carbonate grains in varying proportions" (Pye, 1987: 1999). It can also contain phytoliths, heavy minerals, volcanic ash, and salts. Loess is unstratified and typically consists of silt-size grains (ca. 20 to 40 μm) and is positively skewed. It may also contain clay (<2 μm) and sand (>63 μm) size material of varying proportions, and is commonly, but not always, calcareous (calcite and dolomite). It is porous, tends to keep vertical faces when exposed, and is easily erodible (Pye, 1987).

The origin of loess particles can be varied and there are several mechanisms that can produce silt-sized grains, which can be transported by wind (see Pye, 1987 for details). The formation of loess has been previously reduced to a glacial loess versus desert loess polarization, but it is clear this is an oversimplification for a range of processes (Muhs and Bettis, 2003). The principle processes include (a) glacial grinding and frost-related comminution; (b) liberation of silt-sized material from original rocks; (c) aeolian abrasion (Crouvi et al., 2010); (d) salt weathering; (e) chemical weathering; (f) clay pellet aggregation; and (g) biological processes (Pye, 1995). Glacial grinding, for example, is clearly pertinent to silt and loesses produced proximal to glaciated terrains, whereas aeolian abrasion and salt weathering are more closely linked to the formation of silt-size particles in desertic areas. As in the case of western and central Asia, weathering of rocks takes place in the high mountains, and silt is transported and deposited on to the lowlands as alluvial deposits. Here, it is eroded and transported in the atmosphere eventually being deposited and accumulating as loess. Typical areas that can supply dust for transport and eventually end up as loess deposits include: (a) areas proximal to glaciers: glacial outwash and braided fluvio-glacial channels; (b) dry lake and stream beds, sebkhas, lunettes, alluvial fans, and existing dune and loess deposits that are being deflated due to lack of vegetative stabilization (Crouvi et al., 2008); (c) alluvial floodplains.

Aeolian transport of dust occurs when the wind exceeds fluid threshold (Bagnold 1941; Pye, 1987). As pointed out above, greater velocities are needed to move both coarser and finer grains, as the latter have internal cohesive forces that require greater velocities to overcome this cohesion; finer material also tends to have higher moisture contents. Similarly, the binding effects of salts, clays, and microbiological crusts tend to inhibit deflation of finer particles (Pye, 1987). In fact, algae and mosses serve as dust traps in arid areas. Surface crusting, whether biological (Danin and Ganor, 1991) or physical, produced by slaking, inhibits dust entrainment. Yet, once the crust is broken (e.g. by fluvial activity, ploughing, construction, trampling, vehicle movement), deflation is greatly facilitated.

For a typical windstorm event, the smallest particles (< 10 μm), with low settling velocities, are carried into the atmosphere. The coarser silt and fine sand fractions have higher settling velocities and are transported for shorter distances and periods of time; coarse and medium silt grains can be

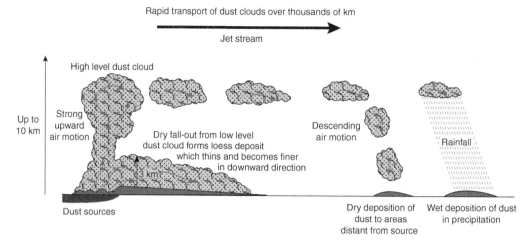

Figure 8.15   Generalized model showing methods of transport and deposition of aeolian dust (modified from Pye, 1995: figure 10).

transported on the order of tens to hundreds of kilometers (Pye, 1995) (Figure 8.15 – from Pye 1995, figure 10).

Dust can be transported in both the lower and upper levels of the atmosphere, and is a function of local vs. regional wind systems. In general, trade winds, which do not extend high into the atmosphere, transport coarser material over shorter distances where it is deposited as dry fallout (Figure 8.17). On the other hand, thunderstorms and strong advection transport large amounts of finer dust into the atmosphere where it can be transported for long distances, up to thousands of kilometers (Goudie, 2008; Goudie and Middleton, 2001).

Deposition of dust occurs through a number of processes, dependent on local and atmospheric conditions. These comprise (a) gravitational settling of individual grains, or aggregates of grains (bound by electrostatic charges or moisture; (b) downward turbulent diffusion; (c) downward movement of air containing dust; and (d) particles washed out of the atmosphere by precipitation. Under dry accumulation, deposition takes place with reduced wind velocity or an increase in the surface roughness. However, if such material falls on a bare, unvegetated, or smooth surface, it is unlikely to be permanently trapped and can be subjected to further erosion and redeposition. On the other hand, dust is likely to remain if it falls on a moist surface (e.g. lake, playa, sebkha, interdunal areas), a surface with vegetative cover (including both larger plants such as trees and cryptogams such as algae and mosses; Danin and Ganor, 1991); on rough surfaces, such as alluvial fans, dust can settle between boulders (Pye, 1995).

Such dry loess accumulations can be relatively coarse, with grain size and thickness decreasing away from the source. This type of loess (Figure 8.15) is typical of dust derived from outwash plains or from large streams, such as the Mississippi or Missouri Rivers (Muhs and Bettis, 2003). In desert areas, dust may be laterally contiguous with sand dunes blown from the sedimentary (fluvial source). In this case, sand accumulates adjacent to the source and the amount of finer material increases downwind and away from the source (Crouvi et al., 2008) (Figure 8.16). This situation is evident in southern Israel and Sinai (Egypt) where sandy deposits grade into finer, siltier and clayier loess and reworked loess as one proceeds away from larger wadis (ephemeral streams) (Yaalon and Dan, 1974). In this latter case, fluctuations in climate can shift the lithological and depositional boundaries and thus the stratigraphic loess records in these localities can be

(a) Proximal loess accumulation

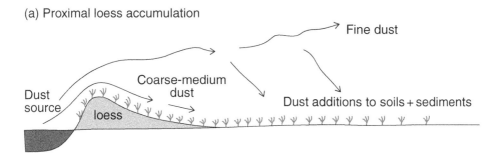

(b) Loess accumulation contiguous with sand sheet–sandy loess transition

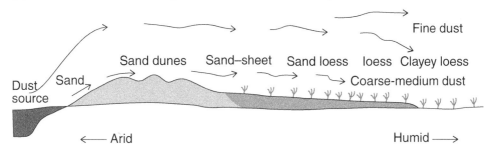

(c) Distal loess accumulation against topographic obstacle

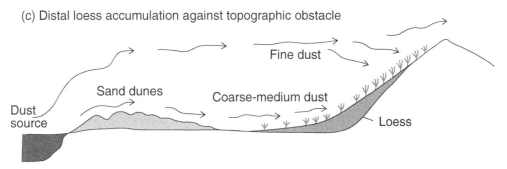

Figure 8.16 Mechanisms of dust entrapment and accumulation (modified from Pye, 1995: figure 11).

sensitive paleoclimatic indicators as registered by the formation of soils and in isotope studies (Bruins and Yaalon, 1979; Goodfriend, 1990; Magaritz et al., 1987; Magaritz et al., 1981; Muhs et al., 2014).

In both cases above, finer-sized dust remains in suspension and is only deposited when it is washed out of the atmosphere through precipitation (Figure 8.16). This type of accumulation is widespread, and is well documented in Europe, where dust of both Sahelian and Saharan origin can be found accumulating on surfaces (Littmann, 1991). In the Mediterranean region, terra rossa soils have high quartz contents, which are clearly not derived from the underlying limestones that have little insoluble residues (Rapp and Nihlén, 1991; Yaalon and Ganor, 1975). In addition, many Vertisols in Israel owe their properties to the imbedding of smectitic clays of aeolian origin (Yaalon and Dan, 1974; Yaalon and Kalmar, 1978), whose shrink-swell properties and mixing can have disastrous effects on the integrity of archaeological assemblages. In Hawaii, studies by Jackson et al. (1971) revealed the presence of quartz in soils developed on basaltic substrates.

Finally, topographical factors may lead to dust accumulation, such as that on the windward side of obstacles where a reduction of wind velocity may take place as well as an increase in trapping vegetation (Figure 8.16) (Rognon et al., 1987). In the Negev Desert, Gerson and Amit (1987) found significant accumulations (0.5 to 0.8 m) of aeolian silt within roofless Middle Bronze Age I buildings, which took between 1,000 to 2,000 years to accumulate. These values are much higher than present-day rates of accumulation in the Negev, which are influenced by human activities (Yaalon and Ganor, 1975).

In many parts of Europe during the Pleistocene, dust was deflated from outwash belts and river bars, and accumulated in a large band situated between the Alpine glaciers and those from the Scandinavian ice sheets. Dust accumulation took place during principally cold phases and was commonly syn-depositionally modified by cold climate processes such as solifluction, cryoturbation, and ice wedging. During somewhat milder Interstadial events, modification of the accumulated deposits took place, locally represented by various types of pedogenesis (e.g. Nigst et al., 2014), depending on climatic conditions.

Recently, there has been continental scale synthesis of Late Pleistocene loess–paleosol sequences across Europe (Lehmkuhl et al., 2020). This synthesis and mapping of loess–paleosol sequences recognizes 6 main domains and 17 subdomains of loess–paleosol sequences across Europe. Three main depositional regimes are proposed for this loess formation, being (I) periglacial and tundra loess formation with periglacial processes and permafrost in the high-latitude and mountainous regions; (II) steppe and desert-margin loess formation in the (semi-) arid regions; and (III) loess and soil formation in temperate and subtropical regions.

Although loess is strictly speaking an aeolian deposit, other silty deposits are either derived from, or simply associated with, loess accumulations. These are secondarily reworked loess deposits, and naturally have similar overall textural characteristics, although they tend to be more poorly sorted and occur in the form of aggregates, with some being gravelly. These "secondary" deposits have a variety of names, depending on the language and culture. In English, "alluvial loess", "colluvial loess" are used, some of which have demonstrably Holocene dates, associated with reworking (Vincent et al., 2011). In Germany, where loess deposits are well expressed, the following distinctions are made: (a) "Fließlöß" (literally "running loess") in which loess is redeposited by sheetflow, resulting in bedded silty deposits; (b) "Schwemmlöß" is equivalent to alluvial loess, occurs in fluvial channels, and may contain boulder beds, plant remnants, and snail shells (Schreiner, 1997).

Loess deposition was not continuous, and during frequent breaks of deposition different types of paleosols formed on these stabilized surfaces (e.g. Fedoroff and Goldberg, 1982; Haesaerts et al., 1999, 2003). The differences in these soils are quite evident, and testify to evolving paleoclimatic conditions over large areas and within a specific locality. In Belgium, for example, (Haesaerts et al., 1999) used macro-, meso-, and micromorphological observations to document 14 phases of pedosedimentary development. Early pedogenesis is characterized by alfisols with textural B horizon and clay translocation. Subsequent soils were developed on loess, colluvial loess, and soliflucted materials. These soil types include among others, (a) a Greyzem (US: Mollisol) deep, dark, organic-rich soils, with deep leaching of carbonates formed "in a continental climate with cold snowy winters and warm summers" (Haesaerts et al., 1999: 17); (b) a humiferous, chernozem-type soil formed in humid conditions "with a steppe or forest-steppe vegetation"; (c) a subarctic humiferous soil, in which oxidation-reduction and polygonal patterning and platy soil structure attest to seasonally humid conditions and cold conditions.

In Central and Eastern Europe, loess deposits are up to tens of meters thick and are associated with paleosol stabilization horizons, with aeolian dust derived from the outwash plains of the

Alpine foreland, as well as the Danube Basin. There have been numerous studies of these sequences and attempts to tie loess–paleosol sequences across wider areas into more refined chronostratigraphic sequences. Magnetic susceptibility has been used as a basis for differentiating widespread loess and paleosol units and correlating them regionally, whilst these sequences have also been related to deep-sea isotope stratigraphy (Marković et al., 2011), as well as using absolute dating such as infrared stimulated luminescence (Wacha and Frechen 2011). For example, loess–paleosol sections were analyzed and compared at Batajnica/Stari Slankamen (Serbia), Mircea Voda (Romania) and Stary Kaydaky (Ukraine). Analysis of pedostratigraphy and correlation of magnetic susceptibility records with susceptibility data of other sections in the area (Koriten, Mostistea, Vyazivok), in the Chinese Loess Plateau, and with the benthic $\delta^{18}O$ record of ODP 677, demonstrated loess–paleosol sequences dating to Marine Isotope Stage (MIS) 17 at Batajnica/Stari Slankamen and Mircea Voda, and MIS 13–15 at Stary Kaydaky (Buggle et al., 2009).

At Drmno, northeastern Serbia, a 25 m-thick loess–paleosol sequence was deposited during the last two glacial–interglacial cycles, associated with faunal remains (Marković et al., 2014). A layer containing a rich paleofaunal assemblage, including mammoth bones, was associated with the final part of MIS 7a and was interpreted as a dry grassland environment characterized by a decrease of pedogenic intensity and increased dust deposition. The gastropod malacofauna indicated open vegetation and mostly dry steppe-like grassland conditions that were contemporary with the deposition of the mammoth remains.

Extensive and thick loess deposits and paleosols are found along the Danube in Hungary (Figure 8.17a) and in the arid and semi-arid Central Asian areas of Tajikistan and Uzbekistan. In the latter, loess thicknesses total up to 200 m and range from Upper Pliocene to Upper Pleistocene, dating back to over 2 million years (Dodonov, 1995; Dodonov and Baiguzina, 1995) (Figure 8.17b). They have been studied in detail, including their paleomagnetism, macro- and snail fauna, palynology, and pedology. Sections from the area all contain paleosols, grouped into pedocomplexes that can correlated; the Brunhes/Matuyama boundary is systematically located between paleosols 9 and 10, and the Laschamp excursion occurs between paleosols 1 and 2 (Ranov, 1995). The soils in the lower part of the sequence consist of red-brown soils, whereas those in the upper part are brown soils and are thicker, due to penecontemporaneous loess accumulation during pedogenesis (Dodonov, 1995).

Pollen analysis at the Lakhuti section, Tajikistan, has shown a systematic paleoclimatic change from hot and occasionally moist conditions in the Early Pleistocene to systematically cooler and more arid climates in the Late Pleistocene. At this same profile, several important Paleolithic localities have been found within the loesses in association with paleosols that are up to 50 to 60 m below the surface, and which pre-date the Brunhes-Matuyama boundary. At the Karatau site (ca. 200,000 BP; Dodonov, 1995) choppers, "Clactonian" flakes, and unifaces occur within the 6th pedocomplex. At site of Lakhuti (ca. 130,000 BP) Mousterian elements with rare Levallois technology are found within the 6th and 5th pedocomplexes (Dodonov 1995: fig. 5). Upper Paleolithic material is normally found within the 3rd to 1st pedocomplexes.

Younger loesses of Western and Central Europe, datable by ^{14}C of charcoal and bones, mostly found in association with archaeological sites, have been documented with exceptional clarity by P. Haesaerts and co-workers. Their stratigraphic drawings (Figure 8.18) are outstanding, which document the sedimentological and stratigraphic information concisely, conveying the geological (depositional and post-depositional) story at a glance (Haesaerts et al., 1999, 2003; Nigst et al., 2014). In the East Carpathian region for example, they reconstructed a precise pedosedimentary and paleoclimatic sequence (spanning ca. 33,000 to 10,000 BP) that incorporates Middle Paleolithic (Moldova V, Ukraine) through Late Paleolithic (Epi-Gravettian; Cosautsi, Moldova)

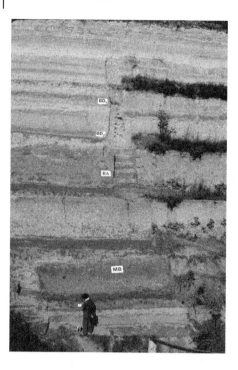

Figure 8.17a Thick sequence of loess deposits interspersed with simple and complex paleosols along the Danube at the Paks locality spanning the last 800 ka (Újvári et al., 2014). The lighter layers are aeolian, whereas the darker bands are paleosols.

Figure 8.17b Thick accumulation of loess blanketing the landscape in Uzbekistan. These deposits can attain thickness of up to 200 m and contain numerous paleosols.

sites (Haesaerts et al., 2003). Such sequences not only depict the nuances of rapid climate change during this time interval but also indicate that the area was more or less continually occupied. The types of sites changed as did the environments: Aurignacian occupations, for example, consist of workshops, whereas Gravettian occupations (ca. 26,000 to 23,000 BP) are distributed over larger areas and appear to be associated with long-term seasonal encampments. During a colder and then drier period between 23,000 and 20,000 BP, occupation is scarcer and represented by small concentrations that seem to represent short-term occupation. Epi-Gravettian (20,000–17,200 BP) settlement represents a return to long-term seasonal occupation in a "contrasted and relatively moist climatic context" whereas few remains are visible for the very cold conditions between 17,200 and 15,000 BP.

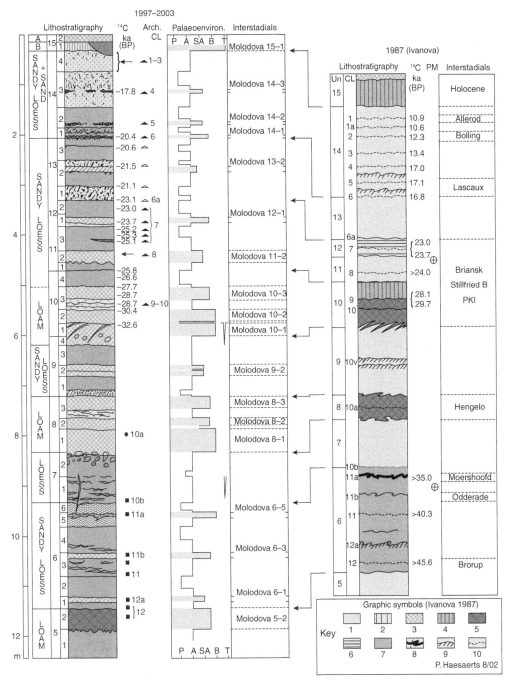

Figure 8.18 Stratigraphic drawing of loess section from Moldova V (Ukraine) showing various lithostratigraphic units, soil horizons, and soil features (e.g. gleying, banded B horizon, biogalleries). The illustration provides the essential depositional and post-depositional elements of the profile, along with environmental data and environmental correlations. These are all readily understandable just through observation. Such drawings – a combination of science and art – should be aspired to in geoarchaeology (from Haesaerts et al., 2003).

Loess deposits that cover parts of Brittany, Normandy, and Belgium originate from the English Channel, which was repeatedly exposed as sea level fluctuated in response to glacial growth. Loess deposits in northwest Europe are generally thin and less expressed on the landscape, lacking the vast loess covers seen further eastward. Yet, notable loess exposures do exist associated with pre-historic remains. The legendary prehistorian François Bordes was a geologist by training, and his doctoral dissertation was concerned with the lithic industries found in the Paris Basin north-west of Paris (Bordes, 1954). Early on, he defined a *loess cycle* that started with a band of pebbles at the base accompanied by solifluction; both represent the onset of a cold deterioration of the climate. As the climate continues to deteriorate, loess is deposited under cold and dry conditions. At the end of the loess cycle, climatic amelioration associated with an interstadial or interglacial climatic warming, results in a cessation of loess deposition and the formation of various types of paleosols described above. Although he clearly documented the association of lithic artifacts with the different loesses and paleosols (e.g. Acheulean industries were found in the "Older Loess"), we now know that the dating and stratigraphic and paleoclimatic sequencing of these deposits was not correct. Nevertheless, his early work showed the presence of lithic industries within the loess sequences.

In Britain, loess was originally deposited as a thin blanket in the south, with subsequent erosion having created a now patchy distribution. There is some variation in these loessic deposits between the southeast and southwest England, with a general decrease in modal particle size, from of 45 μm in the southeast to 25 μm in the southwest. To the southwest, the amount of chlorite and muscovite increases compared to the southeast loess, suggesting two different sources of loessic sediments (Catt 1978; Antoine et al., 2003). Originally, loessic sediments in the UK were termed "brickearths" and were interpreted as being associated with the last glaciation (Weichselian). However, brickearth is not a useful name for these loessic deposits, and it is clear that many brickearth deposits are polygenetic in origin (Bell and Brown 1997). In addition, analysis of sites such as Northfleet and Boxgrove has provided dating evidence of phases of loessic sediment deposition associated with pre-Devensian cold stages (Parks and Rendell 1992) and as such, they provide important pre-serving blankets for earlier periods of hominin activity. At Boxgrove, along the south coast of England, the shoreline sandy deposits associated with the hominin occupations are covered by colluvially reworked loess (Macphail, 1999a), which have associated paleosols that have been ana-lyzed to understand the paleoclimate (Kemp, 1985, 1986; Kemp et al., 1994).

In most of North America, extensive loess deposits occur over much of the mid-continent region in the area of the Mississippi Basin, with patches of loess being found scattered locally over the eastern third of the US. Most of the loess is essentially Pleistocene in age, although Holocene loess is recognized (Mason and Kuzila, 2000), and thus pre-dates the arrival of most humans into North America and consequently it has only minor geoarchaeological relevance. In Alaska, however, close to a major source of human migration from Asia Thorson and Hamilton (1977), excavated the stratified Dry Creek site in central Alaska. This site occurs within a sequence of loesses and interstratified sand units and paleosols that overly Pleistocene glacial outwash. The three oldest soils are immature tundra soils (Cryepts) with organic A horizons on mineralogenic B horizons; these are not forming in the region today. Archaeological compo-nents consisting of artifacts, vertebrate faunal remains, and pebbles displaced by humans are found in four distinct layers within the loesses. The lower three are similar to material in upper Paleolithic sites in NE Siberia, but here date to about 11,000–8500 BP (uncalibrated) and are associated with conditions of the Late Glacial. The youngest archaeological occupation, which was less intense than the earlier ones, is associated with a forest soil and was occupied between ca. 4700 and 3400 BP (uncal).

The deposition of loessic sediments and its presence during the Holocene also has clear importance for the distribution of archaeological sites. This issue is well documented in central Europe, in what is described as the loessic belt, with Neolithic LBK sites occurring on soils derived from loessic parent material (Piggot 1965; Langohr 2019). Likewise, in southern England, sites from the Early Neolithic – Bronze Age are commonly found on sediments of loessic origin or soils derived from loessic sediments. Classic examples include paleosols associated with Neolithic – Bronze Age monuments in the southwest uplands of Britain (Macphail 1987b; Macphail 1990; Macphail 2019a) and loessic contributions to dry valley colluvial deposits on the southern chalklands (Bell 1983).

8.6 Concluding Comments

Aeolian deposits, whether in sandy or silty terrains, constitute a remarkable geoarchaeological resource, and do not always occur in arid environments *sensu stricto*. On one hand, loess sites in Central Europe and Western Asia, for example, contain among the most significant sites for documenting important archaeological questions, such as the replacement of Neanderthals by anatomically modern humans. Because of fine-grained sedimentation, coupled with various types of soil formation, they represent faithful indicators of past environments and archives of human activities. On the other hand, sandy deposits, particularly those in coastal settings, such as those in the Mediterranean and South Africa, are notable for their connection to eustatic (worldwide) sea level changes. In these sites (e.g. Gorham's Cave, Die Kelders, Blombos) the sandy intercalations form the substrate for human occupations and they also seal the deposits, thus permitting high-resolution temporal images of living surfaces, and windows overlooking specific human activities. Likewise, ongoing sand deposition across arid and coastal areas has preserved a remarkable range of archaeological sites and resources throughout the Holocene.

9

Marine Coasts

9.1 Introduction

Marine coasts are important in archaeology because they were often preferentially occupied by humans, for their resources and for trade. It is therefore important to be able to identify the different types of settings and deposits formed in coastal areas in order to understand past coastal morphologies. For example, the 500,000-year-old early hominid and coastal site of Boxgrove (West Sussex, UK), is now located along a former coastline (paleocoastline) that is 46 m above present sea level, and some 11 km inland from it (Roberts and Parfitt, 1999; Roberts and Pope, 2018) (Box 9.1). The Roman coastal port and fort of Pevensey is now nearly 2 km from the sea, but was at the head of a peninsula adjacent to a 10 km deep inlet – now infilled with (pelo-calcareous alluvial gley) soils formed in marine alluvium (Jarvis et al., 1983, 1984).

On the other hand, as sea levels rose during the Holocene many coastal sites were drowned. Such rising sea levels not only preserved some unique and important sites (such as the submerged Neolithic village of Atlit-Yam off the coast of Israel; Galili, and Nir, 1993; Galili et al., 2017) and organic remains, such as boats, but they have also buried evidence for former shoreline occupations associated with the colonization of the New World (Fladmark, 1982). Equally, coastal resources, which include foodstuffs, fodder plants (reeds and other plants for matting and baskets), seaweed (for manure; Conry, 1971), and salt (briquetage), are often recorded in near-coast terrestrial sites (Biddulph et al., 2012; Graham et al., 2017; Macphail et al., 2017b). Due to glacio-isostatic rebound of the FenoScandinavian region over the last 13,000 years, coastline settlements are presently found up to 100 km inland (Linderholm, 2003); a process also recognized in New England for example (Kelley et al., 2010). As marine and glacial sediments emerged, and as soon as they became stabilized, prey animals such as reindeer occupied these landscapes to be soon exploited by Mesolithic peoples and later cultures, for example (Linderholm pers. comm.) (Kristiansen et al., 2020). In some regions, coasts and their sediments and populations have also been subjected to the threat of Tsunami ("tidal" waves induced by shifts in the sea bed by seismic/earthquake activity) and major environmental alternations such as El Niño (Nunn, 2000). Underwater archaeology, emerging features, and paleofeatures cannot, however, be simply termed "time capsules" because they were inundated by the sea, since a wide variety of processes can affect underwater sites (Stewart, 1999). Nevertheless undisturbed drowned stratified sites have been found in some quiet "backwaters" of the Baltic, for example, where they were inundated during the Holocene (Grøn et al., 2021; Skaarup and Grøn, 2004).

In this chapter, we discuss marine coasts in terms of their cultural importance. In addition, we examine coastal morphology, sea level dynamics, and sediment types before providing examples of past utilization of coastal resources by humans.

Practical and Theoretical Geoarchaeology, Second Edition. Paul Goldberg and Richard I. Macphail.
© 2022 John Wiley & Sons Ltd. Published 2022 by John Wiley & Sons Ltd.

Box 9.1   Boxgrove (UK) – the marine and salt marsh sequence

The 500,000-year-old early hominid and coastal site of Boxgrove, West Sussex, UK is now located along a former coastline (paleocoastline; Figure 9.1a) as traced by a borehole survey, and which is 46 m above present sea level, and some 11 km inland (Roberts and Parfitt, 1999; Roberts and Pope, 2018). Here, flint freshly weathered from the chalk cliffs and beach cobbles provided both cortical nodules for hammerstones and raw material for Lower Paleolithic Acheulian hand axe production (Figures 9.1b and 9.1c). Three interglacial marine cycles were identified at Boxgrove based upon beach sand, beach gravel, and cliff fall deposits, the last cycle being contemporary with early hominid occupation (Colcutt, 1999). In the fine sand-size (63–125 µm) beach sands (Unit 3), iron stained root traces of terrestrial plants, reflected periods of marine regression. The overlying lagoonal sediments (Slindon Silt; Unit 4b) are composed of both thick (10 mm) and thin (0.5 mm) sandy laminae that are interbedded with very thinly laminated (0.15–0.30 mm) mainly coarse silt (31–63 µm) and calcareous clay (Figure 9.1d). This bedding shows very little bioturbation, the reasons for which are still being studied. More important, however, is the evidence of *in situ* butchered large animal remains (e.g. horse) and hand axe manufacturing debris within these mudflat sediments (Figures 9.1e and 9.1f) (Pope et al., 2020). The undisturbed nature of these artifacts within intact laminated sediments, implies that butchery and artifact manufacture could have taken place over a very short space of time, possibly as little as a 4–12 hour period of low tide when the mudflat was exposed. Fine sediments deposited during the ensuing high tide then sealed the scatter. In such ancient sediments as at Boxgrove, minerals such as black pyrite framboids become oxidized into reddish iron oxide pseudomorphs of pyrite. Organic laminae have similarly been transformed into "ironpans".

The mudflats deposits marked the final marine regression at Boxgrove and a developing freshwater and terrestrial environment. Subaerial weathering, that included sediment ripening, the homogenization of sedimentary bedding, the removal of carbonates, structural formation, and activity by small mammals and soil invertebrates produced an Entisol, which was very

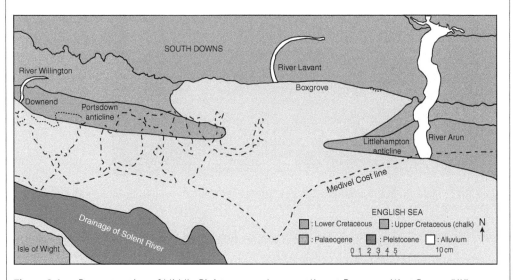

Figure 9.1a   Reconstruction of Middle Pleistocene paleo-coastline at Boxgrove, West Sussex (UK) (modified from Roberts and Pope, 2018).

(Continued)

Box 9.1 (Continued)

Figure 9.1b Field photo of Boxgrove paleobeach at GTP 25, looking "seawards"; bedded beach sands (Unit 3) occur over beach cobbles, with cool climate (Anglian) chalk pellet gravels sealing the temperate climate beach. Note iron staining of beach sediments. (Roberts and Parfitt, 1999).

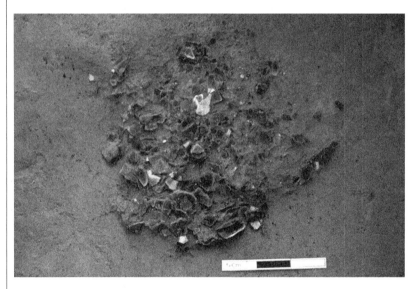

Figure 9.1c Field photo, Boxgrove *in situ* flint scatter Unit 4b, mudflat and lagoonal deposits (cf. Pope et al., 2020).

difficult to demonstrate both in the field and in thin section because of post-depositional geogenic processes (Macphail, 1999b). On the other hand, the important presence of shrews, moles and voles (*Microtus*), the last aiding the biostratigraphic dating of the site, demonstrated that a terrestrial soil (Unit 4c) had formed. This soil was succeeded by a rise in ground water,

Box 9.1   (Continued)

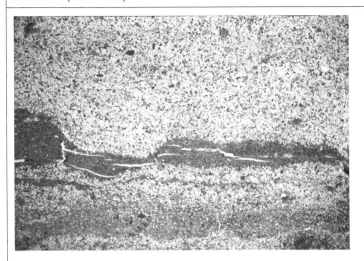

Figure 9.1d   Photomicrograph of Unit 3 – fine beach sands with intercalated silty clay layers resulting from marine regression (plane polarized light (PPL), frame ~9 mm).

(e)

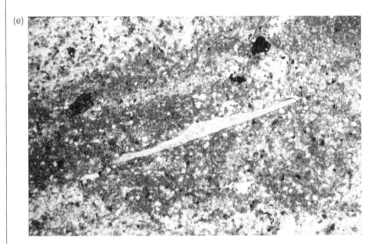

Figure 9.1e   As Figure 9.1b, photomicrograph of flint flake within the scatter; flint is embedded in Unit 4b intercalated silts and chalky muds indicative of tidal sedimentation sealing the flint scatter (PPL, frame length ~5.5 mm).

peat formation and accompanying terrestrial drainage, an environment where both amphibians and fish are recorded. This environmental change caused the Boxgrove paleosol to become totally transformed, through Na^+-induced slaking and loss of iron through reduction (see Box 9.2). Iron is now only preserved as infrequent root mark pseudomorphs reflecting the presence of salt marsh plants. Most of the iron, in fact, became concentrated into a form of bog iron, the multi-layered salt marsh peat being transformed into a 20 mm thick iron and manganese pan (Unit 5a). Most of the hominid activity was focused around a freshwater pond, where a tibia and two teeth were recovered (Roberts et al., 1994).

(Continued)

Box 9.1   (Continued)

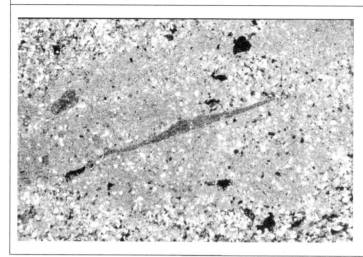

Figure 9.1f   As Figure 9.1d, but under crossed polarized light (XPL) showing high interference colors of quartz silt and medium interference colors of chalky (calcitic) mud. (frame length ~5.5 mm).

9.1.1   Paleo-sea Shores and Paleo-coastal Deposits

Coastal sediments can best be described according to depositional energy and degree of salinity. For instance, the highest energy is recorded at the beach through wave action, whereas lower energies can be found with increasing depths seaward. *Deltas* and *tidal estuaries* are less saline than totally marine environments. Depositional energy and character are investigated, for example, by their textural characteristics and bedding types (see Chapter 2). *Salinity*, on the other hand, is mainly studied through analysis of faunas because sodium (Na^+) salts (halite) are extremely labile (mobile) materials, and rapidly leached from ancient sediments; salt concentrations have, however, been identified in salt processing deposits through specific conductance measurements (see below 9; Chapter 17). Fauna are very sensitive to both salinity and sedimentary environment, and can include fish and mollusks, with microfossils such as diatoms, Foraminifera, and Ostracoda that produce very sensitive datasets of environmental change. Foraminifera, for instance, provided key data at coastal Boxgrove (Box 9.1) and at experimental intertidal sites associated with salt working (see below). Further, the remains of large marine animals that range from turtles and seals to whales, can be found in coastal occupation deposits; deposits in Norse Orkney, for example, link coastal farming with the fishing industry (Linderholm, 2003; Simpson et al., 2000, 2005). These have the potential of providing important cultural and environmental information.

Former shorelines are commonly preserved in Quaternary settings. They can be documented in a number of ways that include raised beaches, erosional notches (knickpoints) and platforms, as well as beach sands, gravels, and remains of marine fossils. Studies of raised beaches and associated deposits such as aeolianites (Ramsay and Cooper, 2002) – especially those associated with older prehistoric remains – have been the main method of investigating *eustatic* (worldwide) paleo-sea levels, especially before developments in bathymetry (Zeuner, 1959); see review in (Jedoui et al., 1998). The rise of chronometric methods increased the spatiotemporal resolution of these changes (e.g. Amorosia et al., 2017). Investigations of coastal shell middens (Marean, 2014; Villagran et al., 2009) and deposits rich in fish bones have shown that humans exploited various coastal niches (see middens, 9.5).

When such studies are undertaken, the changing morphology of coasts through time has to be considered. Rias (drowned valleys) occur in many locations worldwide (e.g. East coast N. America – Sanger et al., 2003; Australia – Jankowski et al., 2015). In south-west England and Brittany, they formed as sea levels rose during the early Holocene, with local affects sometimes being exacerbated by local tectonic activity. In Scandinavia, land upheaval and exposure were not always a continuum (Kristiansen et al., 2020) where at Norum, County of Bohuslän, Sweden, for example, uplift was halted by a marine transgression during the Mesolithic (ca. 9.000 BC), when coastal wetland was occupied and exploited (Linderholm et al., 2020). The former coastal plain along the Italian Riviera exploited by Upper Paleolithic and Neolithic populations, now no longer exists: the cave of Arene Candide that had been previously occupied by these populations, now lies 300 m directly above the Mediterranean sea (Maggi, 1997). The Gokstad Viking ship (Sandefjord, Norway), which both rests on, and is sealed by a mound employing marine mudflat sediments coastal wetland turf, now lies a kilometer or so from the nearest inlet (Cannell et al., 2020; Macphail et al., 2013b; Nicolaysen, 1882). Similarly, land bridge avenues of migration of south-east Asia and the Alaskan arctic became fragmented into islands as sea levels rose.

9.2   Coastal Environments

Coasts can be divided into a series of environments that are important to the human use of shorelines and their immediate hinterland (Figure 9.2). These are:

- the high-energy cliff/beach zone;
- the coastal dune area;
- low-energy estuarine mudflat and lagoonal environments; and
- saltmarsh, mangrove, and other swampland.

9.2.1   High-energy Coastal Environments

9.2.1.1   The High-energy Cliff/Beach Zone: Sediments and Sedimentary Structures

An idealized coastal sequence could comprise a gentle (ca. 10°) seaward sloping wave-cut platform formed out of the local bedrock (Figure 9.3). The high water mark is indicated by a sea cliff formed out of the local geology, the presence of mollusks such as limpets and marine deposits. These include:

- meter-size beach boulders at and near the base of the cliff, derived from the cliff;
- beach "gravels" composed of cobbles (>60 mm), pebbles (4–60 mm), and gravel *sensu stricto* (2–4 mm), stretching some 50 m from the base of the cliff; and
- beach sands (50–2,000 µm) that can extend 100s of meters seawards.

Coarse beach deposits are derived from the reworking of fluvial gravels, conglomerates, tills, etc., or from bedrock and materials eroded from coastal cliffs. The composition of ancient beach gravels at Boxgrove (UK), for example, has been studied, in order to trace their origins (provenance) including autochthonous nearby cliff fall, versus local longshore movement, or distant deposition by glacial ice (Roberts and Parfitt, 1999). In addition, the shape characteristics (angularity/roundness;

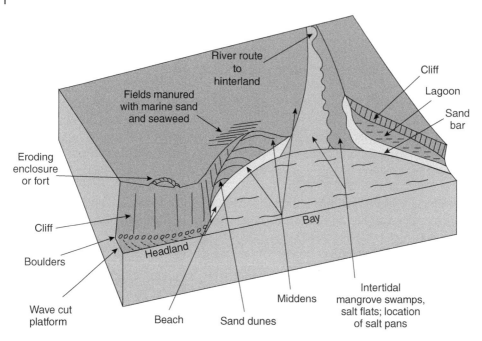

Figure 9.2 Diagram showing possible coastal zones – high-energy cliff/beach zone, coastal dune area, low-energy estuarine mudflat/salt flat, mangrove, lagoonal, and saltmarsh areas, and the wide variety of possible archaeological features.

see Chapter 2) of beach gravels (e.g. subangular modal class) are generally different from those of coarse fluvial deposits that often involve larger proportions of well-rounded clasts. As a caveat, however, it must be mentioned that some ancient beaches can contain large numbers of extremely well-rounded material, testifying to considerable high-energy working (Bridgland, 1999). In any case, it should be remembered that although it has been claimed that beach pebbles may tend to become flattened in the swash zone (zone of back and forth movement of breaking waves), particle shape is strongly influenced by the lithological characteristics of the original material (Boggs, 2006; Pettijohn, 1975).

Beach sands that vary in grain size from 100–500 µm can display excellent sorting and are generally negatively skewed (skewness of >1); in the offshore direction where water deepens, sediments become finer grained and less well sorted (Reineck and Singh, 1986: 137, table 9). Individual sand grains also have distinctive surface features when observed under the SEM (see Chapter 2, Figure 2.4). A variety of bedding types may be observed within beach sands. These could include planar and cross-bedding, which may help to identify upper and lower beach facies, and direction of wave energy (Reineck and Singh, 1986). Such beach sands, when viewed in thin section, have a massive structure and a porosity consisting of packing voids (Figure 9.1c). Beach sands in some cases can be thrown up by storm events occurring anomalously within fine mudflat deposits for example, and beach boulders may form uncharacteristic inclusions (e.g. benthic foraminifera – Vött et al., 2011) through the impact of tsunamis in tectonically active areas, such as Greece, the Caribbean, and Japan (Okinawa) (Pepe Orthiz pers. comm. and Akira Matsui, pers. comm.). As a caveat, however, tropical storms (hurricanes and typhoons) can also readily shift and deposit coarse materials.

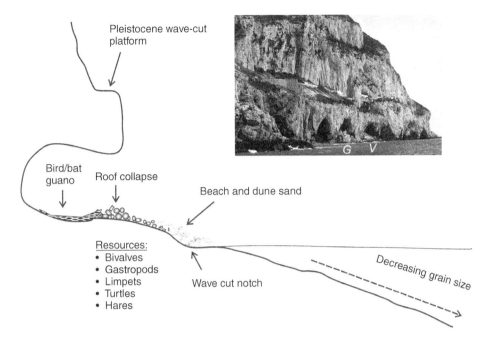

Figure 9.3 Section across cliff/beach zone showing sediment lithology, features, and resources; Gibraltar Caves – Middle and Upper Paleolithic Gorham's Cave (G) and Middle Paleolithic Vanguard Cave (V) (see Figure 10.2).

9.2.1.2 Fauna and Resources

Some beach sands typically contain large numbers of faunal remains, termed *bioclasts*. When formed of mollusks (bivalve and gastropoda), foraminifera tests, and ostracod valves all composed of calcite and aragonite, these sands are calcareous, and called bioclastic sand. Such fossils provide information on environments (e.g. sea temperature, energy, salinity) and biostratigraphy (e.g. microscopic nannofossils such as coccoliths) (see Chapter 3), with the latter providing proxies for Quaternary dating. The analysis of coastal middens (see below) shows that, although foodstuffs like limpets (*Patella* sp.) are less often eaten nowadays (Western Isle of Scotland, southern Italy), they were commonly consumed in earlier times, as for example at Arene Candide on the Neolithic Mediterranean coast, as well as in S. Africa (Jerardino and Marean, 2010; Marean et al., 2007). Other edible shellfish of this environment include gastropods, whelks (e.g. *Nucella* sp.), winkles (*Littorina* sp.), and the bivalve mussel (*Mytilus* sp.). Beaches are also the habitat of many types of edible bivalves, such as cockles (e.g. *Cerastoderma* sp.), razor shells (*Ensis* sp.), and clams. Shellfish were also used for fishing bait. Marine Argopecten and Chione shells at St. Elijo Lagoon, California provide much of the dating evidence for the ca. 4- to 7,000-year-old middens there (Byrd, n.d.).

It should be noted that concentrations of mollusk shells are not always middens. Channels sometimes contain abundant shells, but these are death assemblages concentrated by flowing water, with shells sometimes displaying a preferred orientation; equally, sand worms, and sea birds are known to selectively concentrate shells, the last termed ornithogenetic shell concentrations (Balaam et al., 1987; Reineck and Singh, 1986: 155;). Rocky shores and beaches are also foci of sea bird activity, which may provide eggs, meat, and oil. The last is also obtainable from cetaceans (seals), and both prehistoric (Norway) and recent (Antarctica) examples have been studied

(Villagran et al., 2013). The exploitation of turtles would also yield meat and eggs. Cliff-dwelling birds would include raptors and scavengers, and bird bones concentrated in their pellets should not be confused with that of birds consumed by humans. Zinc (Zn), an element that is relatively high in the marine environment, may be preferentially concentrated in residues from seal exploitation and conceivably in marine bird guano (iron pan layer with 45.5–58.9% Fe, 1.11–2.97% P and 7.20–9.00% Zn at the Mousterian stie of La Cotte de St Brelade, Jersey) (Bates et al., 2013; Brunborga et al., 2006; Drescher et al., 1977).

9.2.1.3 Post-depositional Changes

Relict beach deposits may be cemented to varying degrees. Such cementation – either of marine or freshwater origin (Flügel, 2009) – can range from small amounts of microcrystalline secondary calcium carbonate (both calcite and aragonite) that coats packing voids or forms micrite bridges between particles, to completely micrite- to sparite-cemented sandstones (beachrock), as found at Brekstad (Ørland, Norway) (Figure 9.4). Beach ridges record reworking of cool climate, Late Glacial-earliest Holocene dissolution of carbonate from shell sand and its secondary deposition forming "beach rock", along with non-calcareous sands and gravels and shell. These relict calcareous beach ridges were often used for Iron Age house platforms, where they led to poor macrofossil preservation with a strong buffering effect on acid extraction of phosphate (see Figures 18.19–18.21). The latter may form rather quickly, being penecontemporaneous with the accumulation of the sand. Sediments stained by secondary iron and manganese may highlight both bedding and bioturbation features produced by burrowing mollusks or root traces developed during ephemeral exposure to subaerial weathering (Colcutt, 1999).

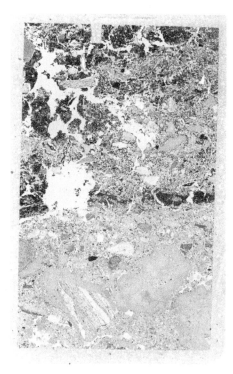

Figure 9.4 Scan of M616333 (Brekstad, Ørland, near Trondheim, Norway; Site D, Water hole 606502, 614956); coarse shell sands and gravel including shell fragments, with coarse fragments of very pale orange "beachrock" where sands and gravel had been cemented with concentric infills of secondary micrite. Overlying, are organic sands and gravels with pellety and aggregated pellety amorphous organic matter and plant fragments, resulting from animal trampling of the waterhole at this location. Frame width is ~50 mm.

9.2.2 Coastal Dunes

On beaches, sand located above sea level is exposed to wind action, and on stable shores, wide belts of sand dune can form; these are 4.5–9 km wide along Pacific coast from Oregon to California, USA. Coastal sand dunes also characterize Les Landes, the coastal area north-west of Bordeaux in France where large tracts of coniferous woodland have been inundated (FAO map soil type Calcaric Regosols on recent dune sand) (Reineck and Singh, 1986: 350). Coastal sand dunes can also accumulate against cliffs, (climbing dunes), and many cave sites along the south coast of South Africa were sealed by such dunes during the Late Pleistocene (Karkanas and Goldberg, 2010; Karkanas et al., 2015, 2020). One notable example is at Arene Candide, Liguria, Italy, where the cave was hidden by a sand dune (now quarried away) that was nearly 300 m high (Maggi, 1997) (Figure 9.5). Caves at Gibraltar (Gorham's and Vanguard) are similarly partially infilled with dune sand (Macphail et al., 2012b).

Coastal sand dunes normally show cross-bedding and are characterized by exceedingly well-rounded sand grains compared to beach sands and offshore beach deposits, and are only a little less rounded than terrestrial sand dunes (see Chapter 2). At Vanguard Cave, sands dip (slope) from the shore into the cave and are indicative of a coast characterized by sand dunes when sea level was lower between 25,000 and 100,000 years ago (Figure 9.6). Here sands are interdigitated with silts originating from surface water flow. Mousterian combustion zones are found on both silty and sandy substrates, and at one combustion zone, sand dune burial was so rapid that the ash layer was perfectly preserved, despite wind activity commonly reworking and deflating ash crystals at such cave sites (Macphail and Goldberg, 2000).

At the coastal site of Brean Down, Somerset, UK, both Quaternary and Holocene dune sands that had accumulated against a limestone promontory were studied by Ian Cornwall and Richard

Figure 9.5 Field photo of Arene Candide, Liguria, Italy; 19th century photo of cliff and white coastal sand dune (now quarried away); Arene Candide, roughly translates as "hidden by sand". Note fort built against the *Saraceni* at the top of the cliff.

Figure 9.6 Field Photo of Vanguard Cave, Gibraltar, Upper Area B; blown dune sands (from outside) alternate with silt loam wash deposits (from inside). Within these is a Middle Paleolithic (Mousterian) combustion zone.

Macphail (Macphail, 1990) (see Figures 4.2–4.4). Sand dune formation seems to have restarted during the Beaker period (ca. 4,000 BP), burying a Beaker ploughsoil and infilling some possible ard marks in one location and inundating a probable grassland surface in another. Ensuing Bronze Age settlements occur between sand blow phases which seem to mark marine regression affecting the local estuary of the River Severn (Bell, 1990). *In situ* combustion zones are characterized by stratified sequence of wood charcoal and wood ash crystals, while locally reworked combustion zones are composed of blown sand, ash crystals, and charcoal fragments.

Thus, dune sand deposits, although primarily composed of resistant minerals such as quartz, can include local material such as ash, shell fragments, and sand-size bone and mineralized coprolites – as at the Gibraltar caves. It is also noticeable that when charcoal is present it is also rounded, but often with a larger grain size compared to the quartz sand matrix, as wind is able to transport larger pieces of charcoal compared to sand because of their lower specific gravity. Commonly dune sands include fine shell (e.g. mollusks and foraminifera) and other marine carbonate, typically giving them an alkaline pH in an unweathered state (e.g. max. pH 8.7 at Vanguard Cave).

9.2.2.1 Post-depositional Changes and Soil Formation in Coastal Dune Sands

Like beach sands, dune sands can become cemented by iron or calcium carbonate according to geogenic conditions, such as groundwater levels. Sand dunes may be stabilized by vegetation, which is often nowadays purposely carried out using marram grass (*Ammophila arenaria*), blown sand losing velocity in grass and becoming deposited. Fungal hyphae and some lichens are also able to begin the stabilization process of sand dunes. Soil invertebrates living on the surface begin to inhabit the litter layer and the 1.5–3.5 cm-thick humic AC horizon that can develop over a period of 13 years. This has also been shown by analogous studies of truncated acid sands (Mücher, 1997). Calcareous sands are more fertile and likely undergo a more rapid soil formation, but this is accompanied by decalcification. Various workers agree to a rate of 0.03–0.04% removal of calcium carbonate per year. However, decalcification has been shown to be favored by low temperature and high percolation rates, as for example, during the early Post Glacial period in present-day temperate regions (see Figure 9.4). In later Holocene times in contrast, warmer temperatures caused this rate of dissolution to decrease (Catt, 1986; van der Meer, 1982), hence the occurrence of early Holocene tufas, for example at Westhampnett (Hampshire, UK) and Gough's Cave (Cheddar, Somerset, UK) (Allen, 2008; Macphail and Goldberg, 2003). In areas such as Romney Marsh (Kent, UK) there is a prograding (expanding) shoreline, and typical sand-pararendzinas (e.g. Typic Udipsamments, Calcaric Arenosols) occur on recent sand dunes with a sparse cover of Marram grass. On the older, inland dunes, weathering has induced the formation of sand-rankers (Quartzipsamment; Arenosol); these have an acid flora that includes gorse (*Ulex europaeus*) (Avery, 1990; Jarvis et al., 1984). In these latter soils, the calcium carbonate content was essentially lost, and the original alkaline pH's were lowered to acidic pH levels, possibly since medieval times.

9.3 Low-energy Estuarine Mudflat and Lagoonal Environments

Fine sediments composed of silt and clay are deposited in protected coastal environments where there are offshore bars (e.g. lagoons), and in and near estuaries (Figures 9.7 and 9.8). Such environments have proven to be rich in archaeological remains, first because of incremental Holocene sea level rise buried sites under marine and intertidal muds, and secondly, because localized marine regression (retreat) and transgression (advance) led to changes in morphology of coastlines that were often utilized by humans. Modern examples of such coastlines (Boorman et al., 2002; Mees and Tursina, 2018; Reineck and Singh, 1986) show a series of sandflat and mudflat deposits; at the top of the shoreward sequence, salt marsh deposits (see below), were variously utilized as a resource base from prehistory to the modern period (Bell, 2007; Bell et al., 2000; Wilkinson and Murphy, 1986; Wilkinson et al., 2012). The Gwent coastline to the Severn estuary, South Wales, has been a focus for investigating Mesolithic footprints for example (Bell, 2007).

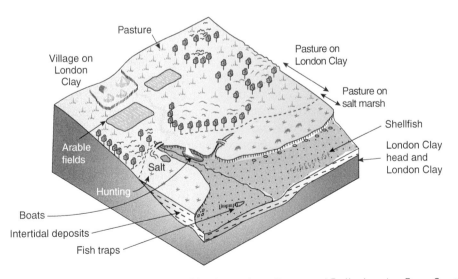

Figure 9.7 Estuarine environment and land-use – Later Bronze and Earlier Iron Age Essex Coast; a boat, and human and animal footprints were found in the intertidal deposits at Goldcliff, Gwent, Wales, UK.

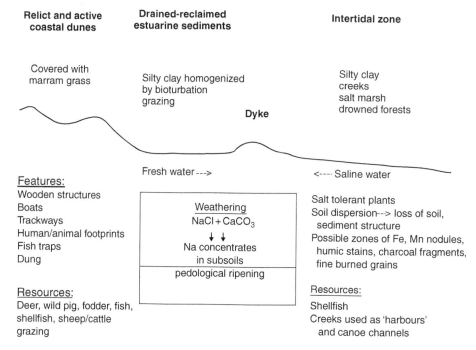

Relict and active coastal dunes	Drained-reclaimed estuarine sediments	Intertidal zone

Covered with marram grass

Silty clay homogenized by bioturbation grazing

Dyke

Silty clay creeks salt marsh drowned forests

Fresh water --->

<---- Saline water

Features:
Wooden structures
Boats
Trackways
Human/animal footprints
Fish traps
Dung

Weathering
$NaCl + CaCO_3$
↓ ↓
Na concentrates
in subsoils
pedological ripening

Salt tolerant plants
Soil dispersion --> loss of soil,
sediment structure
Possible zones of Fe, Mn nodules,
humic stains, charcoal fragments,
fine burned grains

Resources:
Deer, wild pig, fodder, fish,
shellfish, sheep/cattle
grazing

Resources:
Shellfish
Creeks used as 'harbours'
and canoe channels

Figure 9.8 Low-energy estuarine mudflat and lagoonal environments: sediment lithology, features, and resources.

9.3.1 Intertidal Sediments and Sedimentary Structures

Horizontal, thinly laminated fine tidal deposits contrast with the massive and sometimes cross-bedded sands of beach environments. Lagoonal and mudflat deposits can include coarse (fine sand) as well as fine (silt and clay) laminae, with some laminae rich in detrital organic matter. Under some circumstances lagoons have a higher salt content than fully marine environments, and biological activity can be greatly decreased; nevertheless, stromatolites can form under such conditions (see Figure 9.10). An example of a marine beach – intertidal mudflat/lagoonal succession is given from the Lower Paleolithic site of Boxgrove (UK; Box 9.1). Such lagoonal/estuarine sediments have been found sealing old terrestrial landscapes, where they reflect sea level rise during the early Holocene (Macphail et al., 2010). Coastal peats, the tree-stump remains of forests, Mesolithic middens, hearths, and Neolithic through to Iron Age occupations have all been found along the present intertidal zone of southern England. Extremely well preserved Mesolithic forests are reported from Danish Baltic (Skaarup and Grøn, 2004). Holocene inundation and sea level rise is similarly recorded by drill cores from the Yangtze delta, in China, where it is argued that such environmental stress around 4,000 years ago caused a major cultural discontinuity (Stanley et al., 1999).

One of the first geoarchaeological and environmental studies of a coastal midden was carried out at Westward Ho! (Devon, UK), where the late Mesolithic site (ca. 6,000 BP) at −2.20 m OD became exposed only when storm scouring of the modern sandy beach coincided with spring tides, roughly once every one or two years (Balaam et al., 1987) (Figure 9.9). The site is located near the confluences of the Rivers Taw and Torridge. Here, marine "blue clay" (Munsell color 5Y5/1) up to 0.9 m thick occurs over an eroded terrestrial substrate composed of fine clay loamy Pleistocene (Ipswichian?) alluvium (29% clay, 34% silt, 38% sand) that is characterized by polygonal patterning

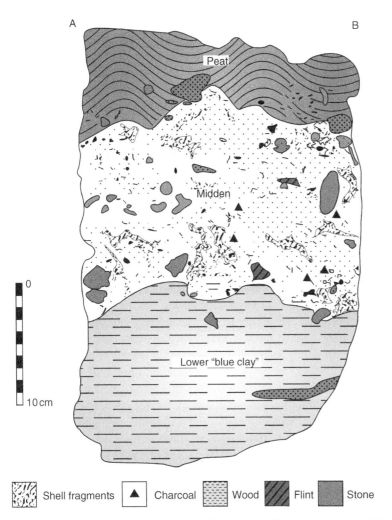

A   B

Peat

Midden

0

10 cm

Lower "blue clay"

Shell fragments   ▲ Charcoal   Wood   Flint   Stone

Figure 9.9   Westward Ho!, Devon, UK; section drawing through estuarine sediments and Mesolithic midden (Modifed from Balaam et al., 1987).

(periglacial patterned ground). In contrast, the overlying "blue clay", a fine silty clay loam (31–34% clay, 50–53% silt, 16–17% sand), is massive and strongly influenced by pyrite (3.2%; point counting), which produces pH's as low as pH 2.6 (cf. acid sulfate soils). Archaeomagnetic dating of this "blue clay" yielded a chronological range of 8400–7800 cal BP. Mesolithic occupation took place on the "blue clay" during an assumed marine regression. Ensuing freshwater/terrestrial conditions led to the formation of an "Atlantic" woodland peat as evidenced by pollen of lime, oak, elm, ash, hazel, and willow trees, as well as an associated woodland insect and land snail fauna (Balaam et al., 1987).

   Such intertidal "blue clays" have also been encountered in Vestfold (Norway). At the probably seasonally occupied Heimdaljordet Viking Age coastal settlement, parcel ditches cut into this sediment and accompanied by marine flooding, produced infills characterized by diatomaceous death assemblages; 0.5 km away the mound-buried Gokstad Ship was sealed by a layer of "blue clay" (Cannell et al., 2020; Macphail et al., 2013a). Here sediments were clay loams or silty clay loams (25–26% clay, 45–63% silt, and 10–33% fine sands). In contrast, the shallow waters of Oslo harbor are characterized by mainly finer silty clay loam sediments (late medieval B13 wreck site: 18–34% clay, 65–80% silt, and

2–7% fine sands). In contrast, at the Sørenga site, the influence of the Alna River delta was revealed by deep coring, which showed a complicated set of silty clay loam layers interbedded with medium and coarse sands; these lithologies suggested beach and intertidal sediments supposedly from deep-water environments (see Box 18.1). Dating, and studies of other fjord environments suggest such anomalous findings are due to typical slumping of fjord sediments (Holtedahl, 1975).

9.3.2 Post-depositional Changes to Intertidal Sediments

Massive silty clay deposits and laminated silts are susceptible to bioturbation by plants and animals. They can also be modified by the formation of mud cracks and the development of crystals of salt (halite) and gypsum along with other secondary minerals such as trona (sodium bicarbonate $Na_3H(CO_3)_2 \cdot H_2O$), siderite ($FeCO_3$), and pyrite (FeS_2) (Kooistra, 1978; Mees and Tursina, 2018; Miedema et al., 1974; Reineck and Singh, 1986). These minerals are typical of such environments. Much archaeological work has been carried out on the Thames foreshore in London (Sidell, 2003), and it is worth noting that the mortar floor and its "foundations" of slag and charcoal at the Shakespearean Rose Theatre, rest on fine intertidal deposits that contain organic fragments and secondary pyrite and gypsum (Macphail and Goldberg, 1995). When the Tower of London Moat deposits were investigated and interpreted, it had to be remembered that there were inputs both from the London Clay substrate (excavated for Henry III's constructions), and the River Thames itself (Keevill, 2004). The London Clay supplied Tertiary Period relics of pyrite whereas the Thames estuarine sediments contained contemporary sulfate that also produced secondary gypsum crystals and pyrite (Macphail and Crowther, 2004). The heavy metal content, however, was mainly associated with medieval and post-medieval contamination associated with cess and industrial activity. At Westward Ho! and on the Essex coast, monocotyledonous plants are also recorded. The many creeks and first order channels that often drain such environments produce gullies, leading to the development of cut-and-fill features within the sedimentary sequence. These features may contain death assemblages and also be the locations where organic artifacts such as fish traps and even boats are preserved; the footprints of Mesolithic people have also been found (Bell, 2007; Bell et al., 2000).

When terrestrial soils are affected by saline water and estuarine sediments, some marked changes occur. The primary transformation is the slaking of fine soil, which is produced by Na^+ ions that produce the collapse of soil structure and the dispersal of fine soil down-profile (Macphail, 1994a); modern-day soils are given a dispersion ratio (see below; Box 9.2). Arte- and ecofact distributions are unaffected by this pedological process, however, and remain as well-sealed *in situ* material. A model of the effects of drainage on lagoonal deposits have been monitored from Dutch polders, where the tree growth, invertebrate and small mammal activity produced "zoologically ripened" Entisols (6–12 cm thick) over a 10 year period (Bal, 1982).

9.4 Salt Marsh, Mangrove, and Other Swamplands

Along low-lying coastlines, saltmarsh occurs above the tidal mudflats of western Europe. This setting is the equivalent to areas of mangrove swamp in the tropics and sub-tropics. Swamplands with mangrove or the swamp/bald Cypress (*Taxodium distichum*), are found along the margins of the coastal plain in the Southeast United States and were occupied by hunter-gatherers, including more recently the Seminole, for example. Such environments are also found in Mesoamerica and are the location of Mayan "raised fields" (Jacob, 1995). Salt marshes have also provided important grazing land. The low-lying island of Marco Gonzalez, Ambergris Caye, Belize where Late Classic

Period Maya (ca. AD 550/600 to 700/760) now occurs within mangrove swamp and mangrove wood was a fuel component in salt processing (Graham et al., 2017; Macphail et al., 2017b). Mangrove provides essential coastal habitats and aids in flood defense, but is on the decline worldwide.

9.4.1 Nature of Deposits

The deposits of the salt marsh and other swamplands are very similar to those of mudflat and lagoonal environments, but have been variously affected by subaerial weathering, surface and channel water flow, and biological activity. Different marine animals exploit sand and clay (mud flat) niches, and most salt marshes display a succession of less salt-tolerant plants landward. This is because the landscape receives freshwater from the landward side, but is also influenced daily by saline water, which at high maximum tide may well flood the whole area; hence, one can see the parallel practice of building sea defenses (e.g. dikes) along the coast of the North Sea in Europe at least from the medieval period onwards. Traces of mangrove colonization were encountered at coastal Djibouti, and small areas are being conserved on the west coast of the Ningaloo Reef National Park (Western Australia), where they are protected by the reef itself. Such environments are useful for trapping muddy sediments in otherwise sandy shore areas, and these more clayey sediments can become intercalated with these sands.

Coastal lagoons may also develop saltier conditions, and as such, a restricted fauna may exist, allowing cyano-bacterial activity to form stromatolites. These conditions occur at Hamelin Bay and Thetis Lake (lagoon) (Western Australia), for example, where living – not fossil – stromatolites occur (Figure 9.10). At both these examples water is ~1.5 times saltier. At Thetis Lake (Cervantes), sea levels were higher at 5,000 BP, and when they dropped, saline lagoons were left behind, cut off from the sea by coastal dune systems. The cyanobacteria convert CO_2 into organic carbon, freeing oxygen, while the stromatolites formed through the secretion of calcium carbonate and trapping of fine sediment by the microbial mat.

Commonly relict intertidal deposits retain some sedimentary elements, such as laminae, but because of biological activity, some or total homogenization takes place. Subaerial weathering often removes some of the salt (NaCl) and calcium carbonate ($CaCO_3$) from the upper ripened horizon of the soil/sediment, resulting in calcareous and non-calcareous pelo-alluvial gley soils (Vertic Fluvaquent; Eutric Fluvisol; Avery, 1990). These soils have a saturation extract conductivity ($S\ m^{-1}$) increasing from 0.16 (0–5 cm) to 0.47 (98–120 cm), and show an exchangeable Na increasing from 7.4–10.6% (0–15 cm) to 22.7–30.2% (70–85 cm) (Hazelden et al., 1987; Jarvis et al., 1984). These workers also found that microrelief has an effect on Na content of topsoils, with pools and creek sites containing more Na (12–13 % exchangeable Na) compared to neutral ground and levees (9% exchangeable Na). They also calculated a dispersion ratio when investigating structural stability of the soils, based upon % exchangeable Na, organic C, and $CaCO_3$, with organic matter and calcium carbonate promoting better-structured, well-flocculated soils and Na-promoting dispersion (see Chapter 4; Figures 9.11b–9.11c). As noted above, when early Holocene soils became inundated they became de-flocculated ("dispersed"). Near-shore terrestrial soils may also be influenced by sea spray-transported Na, leading to accelerated down-profile clay translocation (V. Holiday, pers. comm.).

Bronze and Iron Age salt marsh deposits have been studied from the Severn Estuary (between South Wales and England, near Bristol). It was found that magnetic susceptibility measurements only reflect amounts of burned soil present because the iron minerals that contribute to "topsoil"

Figure 9.10 Field photo of living stromatolites at Hamelin Bay, Western Australia; very slow-growing stromatolites develop as cyanobacteria convert CO_2 into organic carbon, freeing oxygen, while the stromatolites form through the secretion of calcium carbonate and trapping of fine sediment by the microbial mat.

magnetic susceptibility enhancement have been transformed, or the iron forming them removed by gleying (Crowther, 2000, 2003). A common soil micromorphological phenomenon is the presence of intercalatory and other textural features that infill relict void space between now-poorly preserved peds. This is again the result of de-flocculation of ripened "soil" surfaces. Voids tend to be either vesicles (complete water-saturated collapse) or closed vughs (massive inwash of fines). Additionally, channels and burrows can become infilled with silty estuarine material. The remains of organic matter from the ripening "humic" Ah horizon, because of post-depositional oxidation-reduction, is often only represented by iron and manganese staining, although this sometimes may be as pseudomorphs.

The effect of inundation and renewed sedimentation is less clear when salt marsh deposits are flooded as compared to terrestrial soils, because the ripened sediment is very similar to the new deposit, and de-flocculation masks the boundary. Often, an old land surface may only be vaguely identified in the field through a concentrated horizontal zone of iron and manganese nodules (Allen et al., 2002). It then takes soil micromorphology to reveal a buried layer more precisely. It can also be noted from the Wallasea Island experimental inundation (River Crouch estuary, Essex), that reclaimed salt marsh employed for arable agriculture (behind sea walls) has a saturation extract conductivity (S m^{-1}) as low as 0.046 (Ap horizon). Moreover, it contains few relict salt marsh foraminifera compared to new salt marsh inundation sediments (6.93 S m^{-1}) in which salt marsh foraminifera are numerous and well preserved (Macphail, 2009a; Macphail et al., 2010) (see Chapter 14). Pollen within both these new and ancient marine flooding sediments can be rich in tree pollen derived from the whole catchment area of the North Sea (Scaife, 1995; Wilkinson et al., 2012).

Box 9.2   Drowned coasts of Essex and the River Severn, UK

Recent coastal erosion and industrial development projects in the UK have been surprisingly effective in uncovering artifacts, features and land surfaces that became drowned as sea levels rose during the Holocene, a well publicized example being "Sea Henge", a circle of well-preserved large wooden posts at Holme-Next-The-Sea on the Norfolk coast. In fact, submerged wetland landscapes have been surveyed in much detail in the UK, and include the Humber Wetlands, the Lincolnshire and Cambridgeshire Fenlands, the drowned coastlines of Essex and Thames estuary, the Somerset Levels, the Severn estuary (e.g. Goldcliff on the Welsh side), and the North-West Wetlands (French, 2003; Wilkinson and Murphy, 1986, 1995). The first evidence of this phenomenon, in the form of submerged forests, was noted in the 18th century, although historically recorded from Wales in AD 1170 by Giraldus Cambrensis (Bell et al., 2000: 3). Major coastal survey in Essex (the Hullbridge Survey) began in the 1980s (Wilkinson and Murphy, 1986; Wilkinson et al., 2012). Much of the Baltic and the North Sea between England and Europe was dry land during the early Holocene, but as the ice sheets melted a sea level that was some 45 m below present mean sea level (9500–9000 BP) began to rise rapidly at around 2 m a century (Devoy, 1982; Wilkinson and Murphy, 1995) (Ole Grøn, pers comm.). Estimates of the rising sea level at the River Severn site of Goldcliff, Gwent, have been calculated at this time at 16 mm per year (i.e., 1.6 m/century), diminished to 8.5 mm per year by c 6600 BP (Bell et al., 2000). The earliest excavated occupations date to the later Mesolithic, and include a hearth and footprints (Bell, 2007). An important Mesolithic midden was also found at Westward Ho!, Devon, approximately 100 km to the southwest of Goldcliff (Balaam et al., 1987). While the Neolithic (The "Stumble" and other River Blackwater sites) and Bronze Age are well represented on the Essex Coast, the next major period of occupation at Goldcliff is the Iron Age. Typically, at sites in both Essex and Gwent, occupations mainly occur in loamy "soil" substrates formed in sand and clay-rich Pleistocene Head deposits, and are sealed by the lowermost silt-rich clay deposits that record estuarine inundation of the sites.

At Goldcliff, the Mesolithic population occupied the site during a period of marine regression (from around 5600 Cal BC), and probably hunted wild pig, deer, waterfowl and trapped fish (Bell et al., 2000). The geoarchaeological study of the bone and charcoal-rich Mesolithic hearth site included a loss-on-ignition (LOI), phosphate and magnetic susceptibility (χ) survey (Crowther, 2000) and soil micromorphology (Macphail and Cruise, 2000). Although most phosphate concentrations were low (P, <0.400 mg g^{-1}), around background levels, there were localized patches of high concentrations (P, >2.00 mg g^{-1}; max. 3.46 mg g^{-1}); and there is a highly significant statistical correlation between P and LOI for the occupation as a whole. On the other hand, the magnetic susceptibility values (χ and χ_{max}) (see Chapter 17) are low, and there is no correspondence with the concentrations of charcoal that were believed, at the time, to represent *in situ* burning. The main cause of these results from magnetic susceptibility could either be the poor correlation of the χ survey and the mapped charcoal scatter, or simply the marked post-depositional effect of gleying and associated low concentrations of iron (Crowther, 2000). The soil micromorphology investigations here and in Essex, may also help understand these findings. The Mesolithic hearth at Goldcliff is represented by compact dark reddish brown soil that includes charcoal and burned bone, and although biological working

(Continued)

Box 9.2 (Continued)

fragmented this hearth/old land surface, its burned state allowed it to resist many of the effects induced by marine inundation. On the other hand, charcoal can also be found both embedded across the boundary between the old ground surface and the estuarine clay, and at the base of this layer. This, phenomenon, which was also recorded at The Stumble in Essex, demonstrates the lateral mobility of charcoal when sites become inundated. The present distribution of charcoal at such sites does not therefore necessarily correspond to its original pattern.

Once-terrestrial soils are normally completely transformed by increasing site wetness and marine inundation. This phenomenon is recorded both in the geological record (Retallack, 2001), and locally in the Fenlands and Essex coastline of the UK (French, 2003). At a series of Neolithic coastal sites in Essex, both inundation and regression are recorded. For example at Purfleet on the River Thames (London), an estuarine clay loam (21% clay, 55% silt, 24% sand) sediment was deposited that related to the Thames II transgression of ca. 6500–5400 BP; a Neolithic woodland soil developed in this during the Tilbury III regression (ca. 4930–3850 BP) (Wilkinson and Murphy, 1995; Wilkinson et al., 2012). The latter was evidenced by a small number of finds, including polished axes, a woodland land snail fauna that had developed over some centuries, and a microfabric relict of a possible woodland topsoil. Biological activity removed any evidence of original sedimentation. Subsequently (after 3910 BP) the site became wetter and was sealed by a 0.77 m thick freshwater wood peat containing ash, alder, and yew trees. The buried soil microfabric of organo-mineral excrements, thin (<1 mm) and very broad (3–5 mm) burrows, and woody roots, is presently partially preserved by iron stained sodium carbonate and micritic calcium carbonate (14.0% carbonate) and typical minerals of estuarine deposits, such as pyrite are present (Macphail, 1994a).

On the "estuaries" of the Rivers Blackwater and Crouch presumed argillic brown earth soils formed (in Head of London Clay origin) under an oak and hazel dominated woodland (pollen evidence) (Wilkinson and Murphy, 1995). This woodland was probably influenced by local clearance around occupations such as at Neolithic site of The Stumble, where flints, pottery, burned flint and charcoal of cereals and wild plant food were found (Wilkinson et al., 2012). The overlying estuarine clay loam deposits that were deposited after 3850 BP, are massive with coarse vertical plant root channels, and contain few horizontally oriented planar voids relic of decayed detrital plant matter (Figure 9.11a). At the five sites investigated, the underlying 100 mm thick "topsoil" is composed of very pale massive silt loam (10–12% clay, 76–77% silt, 12–13% sand) with a very low porosity (5–10% voids). In marked contrast, the underlying "subsoil" is a silty clay loam and clay loam (e.g. 21–34% clay, 27–69% silt, 10–39% sand), with very abundant finely dusty, and moderately to poorly oriented clay void coatings and infills (Macphail, 1994a; Macphail et al., 2010). It is quite clear that that these textural pedofeatures owe nothing to an earlier and assumed terrestrial woodland soil here, but rather totally reflect a transformation induced by saline water inundation (Figures 9.11b–9.11c). Here, Na^+ induced complete slaking (soil dispersion) of the Neolithic topsoil, as in saline soils generally (see Chapter 4). The current presence of pyrite framboids and gypsum also mirror this transformation into

Box 9.2 (Continued)

estuarine soils (saline and acid-sulfate alluvial gley soils; Wallasea 1 soil association) (Avery, 1990; Jarvis et al., 1984). This has meant that most of the clay has been dispersed down-profile, while archaeological materials such as pottery and coarse charcoal were unaffected and remain at or near the surface of the once-biologically worked Neolithic soil. Finally, at one location, now sealed by peat, inundation led to "waves" of humic matter being washed down-profile into the newly formed massive "topsoil". These formed pseudo-laminae that could be mistaken for relict sedimentary bedding (Lewis et al., 1992). Experimental multiproxy geoarchaeological investigations of newly inundated soils at Wallasea Island, Essex, which complement the archaeological finding given above, are presented in Chapter 14 (e.g. Figures 14.22–14.24).

Figure 9.11a Estuarine inundation silts burying prehistoric old ground surface (Kubiëna box sample) at the Blackwater estuary, Essex, UK.

(Continued)

Box 9.2 (Continued)

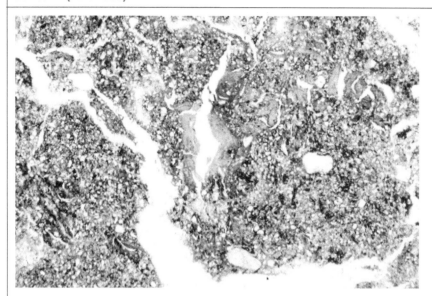

Figure 9.11b Photomicrograph of Blackwater: "subsoil" (300–370 mm depth) at The Stumble Neolithic site. Original soil porosity was infilled with translocated clay and very fine silt, resulting in a 29–34% clay content. Dark coloration of the matrix is related to iron-staining. This clay deposition implies down-profile drainage, and thus that the site was flooded from the surface. It also suggests the soil was sufficiently affected by saline salts to allow soil dispersion (see text). PPL, frame width is ~3.5 mm.

Figure 9.11c As Figure 9.11b, under XPL; note moderately well-oriented void clay infills.

9.4.2   Intertidal, Salt Marsh, and Swampland Resources

Many kinds of fish, shellfish, Amphibia (turtles) and game were exploited from the intertidal zone. Late Mesolithic occupations in the UK show that deer and wild pig were hunted among others. In later prehistory to the modern period, an additional important land use was grazing (Wilkinson and Murphy, 1995, 220). It is likely that both sheep and cattle were grazed on salt marshes, with the exploitation of cattle that stabled on the salt marsh itself, being indicated by cattle bones, cattle lice and cattle foot prints at Goldcliff (Bell et al., 2000); coastal wetland turf employed to construct the Gokstad Ship Mound was probably grazed by sheep. The intertidal deposits studied at Goldcliff showed that creeks (paleochannels; see Chapter 6) were used as tracks, as further suggested by the presence of fragments of dung. Hoof prints show slow inwash of muddy sediments, and paleo-channel infills demonstrate how it is likely that cattle were involved in eroding the contemporary peat surface. The salt marsh became dry enough at times for heather (*Calluna*) to grow on the ancient sphagnum peat surface; the presence of *in situ* charred heather roots suggest that the pasture was managed by fire. Such locations are also essential for communications by coastwise traffic, and boats were used to move cattle and sheep.

Salt production, for example, using salt pans and sometimes producing deposits rich in bri-quetage (Bell, 1990) ("red hills"), have commonly been reported in the literature (Biddulph et al., 2012; Wilkinson and Murphy, 1995). Paleoenvironmental studies of the geographically con-trasting locations of Stanford Wharf (River Thames estuary, Essex, UK) and Marco Gonzalez (Ambergris Caye, Belize) suggest possibly similar techniques to produce brine from marine sedi-ments ("sleaching") rather than simply using sea water (Graham et al., 2017; Macphail et al., 2012a). This method produced thick red waste deposits composed of heated salt marsh sediments and fuel ash (see Chapter 12). Mangrove woodland provided fuel – especially buttonwood (*Conocarpus erectus*) and white mangrove (*Laguncularia racemosa*) – and the waters within reefs in general provide opportunities for fishing and collection of shellfish (e.g. conch, especially at Marco Gonzalez, where large middens occur; Graham et al., 2016). Conch seems to have been employed for ground raising and as a source of lime – lime mortar floors/hearths.

9.5   Middens

Anthropogenic deposits are found in coastal sites throughout the world, reflecting the many activities associated with the sea for tens of thousands of years. The stratigraphy of these types of sites is com-plex and they are rather difficult to decode and interpret. Many earlier studies have taken more tradi-tional (geo)archaeological approaches that concentrate on the objects and stratigraphy. Until recently however, only a few studies dive into the microstratigraphic aspects of midden formation (Aldeias and Bicho, 2016; Aldeias et al., 2016; Balbo et al., 2010; Macphail, 1999a; Villagran et al., 2011a, 2011b, 2013).

Among the earlier occurrences of fish and shellfish exploitation are Paleolithic cave sites in the Mediterranean and Middle Stone Age coastal sites in South Africa (Chapter 10). Gorham's and Vanguard Caves, Gibraltar contain numerous shellfish remains in spite of the fact that the sea was over four km away (Barton et al., 2012). A similar setting and type of exploitation was noted for Die Kelders Cave (South Africa) where numerous shell fish remains survived extensive diagenesis after having been taken from the sea which was 10s of km distant at that time (Goldberg, 2000b). Further, the remains of large marine animals, that range from turtles and seals to whales, can be found in coastal occupation deposits; deposits in Norse Orkney, for example, linking coastal farming with the fishing industry (Linderholm, 2007; Simpson et al., 2005; Simpson et al., 2000). These have the potential of providing important cultural and environmental information. Although, for example Norwegian west coast Mesolithic and Neolithic structures which may have been occupied seasonally may include concentrations of burned fish bone, as well as medieval urban centers such as Trondheim, and east coast Oslo, and Viking Heimdaljordet (Figures 9.12a–9.12d) (for fish bones in another dark earth site, see Box 13.1). Less obvious inferences can be made to indicate use of marine resources (e.g. use of driftwood as fuel) from the enigmatic occurrence of elements such as Ba, Br, Cl, I, and Zn, found in absorbent anthropogenic inclusions such as char and charcoal.

Middens form an archetypal coastal occupation deposit that is found worldwide. For example, gastropods (e.g. winkles and limpets) were collected along the high tide limit of rocky shores, while bivalves (e.g. clams) were gathered from lower energy environments of sandy beaches and mud

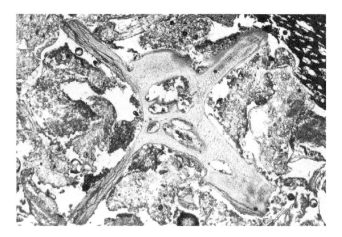

Figure 9.12a Photomicrograph of M11971 (Medieval Oslo, Norway; Profile 11498, North; Pit layer 11780); general urban discard includes distinctive fish bone vertebrae; humic staining probably derives from dumping background of humified plant residues. PPL, frame width is ~4.62 mm (Macphail, 2016a; Follobaneprosjektet: Follobanen FO3 by NiKU).

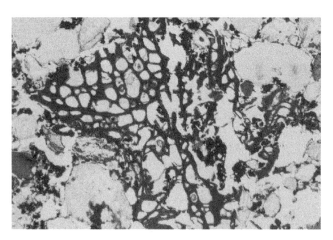

Figure 9.12b Photomicrograph of M2267 (Middle Neolithic Korsmyra in Bud, Møre og Romsdal, Norway; Lower Cultural layer, Profile 2275); occupation floor deposits, with burned (calcined) fish (bone?) remains. EDS analysis of burned bone in same thin section measured ~18% P and ~38% Ca, while charcoal examples recorded ~1–5% Cl, hinting that driftwood had been employed as a fuel. PPL, frame width is ~4.62 mm. (Linderholm et al., 2019; Macphail et al., 2017c; NTNU Projects).

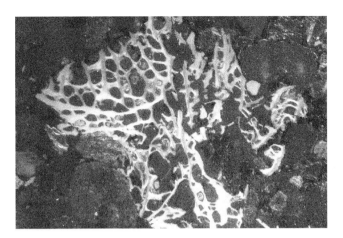

Figure 9.12c As Figure 9.12c, under OIL.

Figure 9.12d Photomicrograph of M11957-2 (Sondre gate, Trondheim, Norway; Context 10383); medieval dark earth-like deposits post-dating a whole series of wooden churches, and recording fish processing on an industrial scale. Fine bone is embedded in phosphate, concentrating and cementing fish processing residues (e.g. matrix 4.37 P_2O_5, 12.0% CaO, 5.84% FeO; Macphail, 2018; NiKU Project). PPL, frame width is ~0.90 mm.

flats, examples of which can be found in South Africa, California, Portugal (Aldeias and Bicho, 2016), and Gibraltar (Fa and Sheader, 2000). Uzzo Cave (Sicily), also records the seasonal gathering of mollusks during the Mesolithic, when cave deposits are very dominantly natural in character compared to the ensuing Neolithic occupation (Mannino and Thomas, 2004–6). The NW coast of North America contains numerous shell middens that have been painstakingly investigated by Stein and her team (Stein, 1992). Their work at the time set the geoarchaeological standard for shell midden study. More recently, soil micromorphological studies of Californian shell middens have shown the importance of identifying the effects of bioturbation and weathering on these deposits (Byrd, n.d.). Food waste such as shell and bone, hearth debris (charcoal and ash), and contemporary environmental conditions (e.g. recognized from pedofeatures analysis) can all be differentiated from "simple" post-depositional events. For example, scavengers comminute bone and leave coprolitic material.

Similarly in the 1980s at Westward Ho!, Balaam et al. (1987) carried out detailed analysis of a midden. Some characteristics can be noted (see Figure 9.9). The midden is both more poorly sorted and coarse (14% clay, 37% silt, 48% sand) compared to the underlying more clay and silt-rich "blue clay" (see above), although this mineral material is of alluvial/fen carr origin. In addition, it is more organic (9.2% organic C) and contains, for example, 2.7% rock fragments and 18.7% shell inclusions (point counting). Despite being located in a wetland (land snails present), this midden was sufficiently bioturbated not to reveal any archaeological stratigraphy. On the other hand, fine mineral and organic laminae recognized under the microscope show how the midden built up alongside alluvial sedimentation and peat formation.

In Orkney, other Northern Isles, and in Scandinavia, midden deposits, especially within urban areas (e.g. Oslo and Trondheim) contain fish bone and phosphatic fish bone residues that reflect the industrial scale of fishing in medieval times (Simpson et al., 2000) (Figures 9.12a–9.12d). Seaweed has been a traditional ingredient for manuring to aid the formation of "cultosols" such as *plaggen* soils, for example in Ireland (Conry, 1971). The possibility that seaweed was utilized as a manure in inland Sussex (UK) during the Bronze Age was suggested by Bell (1981) from the presence of small marine shells. Similarly, links have been made between humans and the sea from the identification of chemical elements and isotopes ($^{12}C/^{13}C$) in human bones that indicate a diet rich in marine resources (Reitz and Wing, 1999: 247 *et seq.*).

9.6 Conclusions

Many sites from coastal areas have been preserved, some as deeply buried Quaternary sites such as Boxgrove, while others, such as intertidal sites, can only be explored at low tide. It has been important to understand the processes that form coastal sediments, and how they can be transformed by changing sea levels, storm inundation and post-glacial uplift in order to investigate associated human activities accurately. In addition, some Scandinavian coastal sites can now be found inland due to uplift, a phenomenon that also affected the northeast coast of the USA. Experimental investigations have also allowed visions of the effects of marine inundation and sedimentary processes that aid the interpretation of coastal sites in prehistory, dating back to the Middle Pleistocene. As research has progressed, we are achieving greater insights into the resources of the coast and how these were utilized and exploited, with fish remains being found in some coastal cities, where they can contribute to dark earth deposits. Use of seaweed as a manure has long been identified, but now the study of middens, occupation, and industrial deposits is allowing more fine-tuned interpretations of seasonality (mollusks), technology (salt processing), and wholesale exploitation of fish resources. Moreover, the influence of coasts on humans can be detected down to the elemental level, as for example the presence of Zn in fuel waste possibly associated with marine mammal exploitation and concentrated in sea bird guano; relict Cl is also apparently present in driftwood charcoal. Finally, working along shorelines has its own peculiar dangers, e.g. from rock fall to rapidly rising tidewater risks (Bell et al., 2000).

10

Caves and Rockshelters

10.1   Introduction

One of the most captivating of geoarchaeological environments and one that has attracted the attention of professional and amateur archaeologists and the average person in the street is that of caves and rockshelters. Much of prehistory – particularly in Western Europe – began in caves and rockshelters, whose form and contents can be readily seen in the landscape. These cave sites not only have served as the lynchpins of culture history for this area but all furnished a wealth of Neanderthal remains, such as at La Ferrassie, in the Dordogne region of France (Bertran et al., 2008; Delporte et al., 1984; Guérin et al., 2015; Laville and Tuffreau, 1984; Peyrony, 1934; Talamo et al., 2020; Texier, 2006). Since caves provide a "protected" environment in landscapes, which through time have often become lost through erosion, we see that an apparent disproportionate amount of information about human evolution, art, lithic traditions, and faunal exploitation are evidenced from famous cave sites. Some significant examples in the Old World with geoarchaeological investigations include Zhoukoudian (China), and numerous sites in France (way too numerous to list, but include Combe Grenal, Pech de l'Azé, le Moustier, La Quina). Key Middle Eastern caves noted for early fire, key lithic sequences, fauna, and again human remains encompass Tabun, Amud, Kebara, Qesem, Qafzeh, Hayonim, Manot in Israel, Shanidar (Kurdistan), Yabrud (Syria), Liang Bua (Indonesia), Niah Cave (Malaysia), and Denisova in Siberia (Russia) (Table 10.1), to name just a few. Some African sites are equally important, for instance Wonderwerk and Sibudu (South Africa) and the classic site of Taforalt (Morocco). Equally signigicant sites for the New World, which are much younger, include Meadowcroft Rockshelter (Pennsylvania) and Paisley Cave (Oregon) in the USA, Lapa do Santo (Brazil), and Cuncaicha Rock Shelter (Peru). Caves and rockshelters have also proven to be some of the best sites for recording the often seasonal activity of pastoralists from the Neolithic onwards, especially in the Mediterranean region, where stratified Neolithic open-air sites are rare (Angelucci et al., 2009; Macphail et al., 1997; Polo-Díaz et al., 2016); Near Eastern and North African site studies have also contributed information on how stock management varied (Cremaschi et al., 1996; Rosen et al., 2005) regionally.

Other than being protected and thus attractive places to use and live in, caves and rockshelters constitute intriguing depositional environments, especially from the geoarchaeological point of view. Foremost, cave sites provide distinct advantages over open-air sites, which are exposed and are more prone to erosion. In addition, they are subjected to weathering and soil formation that directly affect organic artifact preservation (bone, shell, etc.). Caves tend to act as sedimentary and behavioral data traps, and exhibit reduced plant growth, soil formation, and exposure to precipitation and ultraviolet rays, for example. On the other hand, caves are prone to be loci of diagenesis, and in

Practical and Theoretical Geoarchaeology, Second Edition. Paul Goldberg and Richard I. Macphail.
© 2022 John Wiley & Sons Ltd. Published 2022 by John Wiley & Sons Ltd.

Table 10.1 Some recent references to important caves and rockshelters discussed in the book: their location, ages, and significance. LP = Lower Palaeolithic; MP = Middle Palaeolithic; UP = Upper Palaeolithic; EP = Epi-Palaeolithic; NA = Natufian; N= Neolithic; MSA = Middle Stone Age; LSA = Later Stone Age.

Site	Location	Time period	Significance	Selected references
Zhoukoudian, Locality 1	China	LP	Once thought to have among the earliest uses of controlled fire	Gao et al., 2017; Jia, 1999; Jia and Huang, 1990; Weiner et al., 2000; Xu et al., 1996
Combe Grenal	France	MP	Thick sequence of deposits; the virtual type section of the Middle Paleolithic in SW France	Bordes, 1972; Discamps et al., 2016; Faivre et al., 2014
Pech de l'Azé	France	MP	Three collapsed caves/ rockshelters (I, II, IV) in the same area of the Dordogne with abundant Middle Paleolithic lithics, fauna, and geological studies	Dibble et al., 2018b; Goldberg, 1979a; Goldberg et al., 2018; Laville et al., 1980; Soressi et al., 2007; Texier, 2009; Texier et al., 2006
La Micoque	France	MP	The type site of the Micoquian industry	Texier, 2009; Texier et al., 2006
Gorham's Cave	Gibraltar	MP; UP	A thick sandy sequence with sandy, anthropogenic, and biogenic layers, with localized diagenesis	Barton et al., 1999; Finlayson et al., 2006; Goldberg, 2001a; Goldberg and Macphail, 2000; Macphail and Goldberg, 2000
Vanguard Cave	Gibraltar	MP	A large heap of sand with localized finer silty sands containing small hearths and occupations	Barton et al., 1999; Carrión et al., 2018; Doerschner et al., 2019; Goldberg, 2001a; Goldberg and Macphail, 2000; Macphail and Goldberg, 2000; Macphail et al., 2012b
Tabun Cave	Israel	LP; MP	A thick sequence of sandy, silty sand, and anthropogenic deposits that have been heavily modified by diagenesis. Neanderthal remains	Albert et al., 1999; Bull and Goldberg, 1985; Goldberg, 1973; Goldberg and Bull, 1982; Goldberg and Nathan, 1975; Jelinek, 1982; Jelinek et al., 1973; Mercier et al., 1995; Rink et al., 2003; Shimelmitz et al., 2014
Kebara Cave	Israel	MP; UP; EP; NAT	Recently excavated site with great detail concerning Neanderthal burial, anthropogenic deposits (many combustion features), mineralogy, and diagenesis and lithics	Albert et al., 2000; Bar-Yosef et al., 1992; Bar-Yosef, 1996; Bar-Yosef and Meignen, 2007; Goldberg and Bar-Yosef, 1998; Laville and Goldberg, 1989; Schiegl et al., 1996; Weiner et al., 2007
Cave of Pigeons	Morocco	MP; EP	Long history, human occupation	Barton et al., 2016; Macphail, 2019b
Qafzeh Cave	Israel	MP; UP	Anatomically modern humans that pre-date Neanderthals from Kebara in stony and anthropogenic diagenetically altered deposits	Hovers, 2009; Rabinovich and Tchernov, 1995; Valladas et al., 1988; Vandermeersch and Bar-Yosef, 2019
Amud Cave	Israel	MP	Neanderthal finds and phytoliths in an ashy matrix	Hartman et al., 2015; Hovers et al., 1995; Madella et al., 2002; Shahack-Gross et al., 2008a

Table 10.1 (Continued)

Site	Location	Time period	Significance	Selected references
Hayonim Cave	Israel	LP; MP; UP; EP; NA	A cave with a deep cultural sequence, anthropogenic deposits, and marked diagenesis	Bar-Yosef et al., 2005; Goldberg, 1979b; Goldberg and Bar-Yosef, 1998; Meignen et al., 2009; Meignen et al., 2017; Stiner et al., 2005; Weiner et al., 2002; Weiner et al., 1995a
Arene Candide Cave	Italy	UP; N	A site showing clear stabling activities during the Neolithic	Courty et al., 1991; Macphail et al., 1997; Macphail et al., 1994a
Die Kelders Cave	South Africa	MSA; LSA	Interbedded sand and anthropogenic units with many ashy combustion features	Goldberg, 2000; Marean et al., 2000; Tankard and Schweitzer, 1974, 1976
Klasies River Mouth	South Africa	MSA	Site noted for its succession of hearths and last interglacial date	Larbey et al., 2019; Singer and Wymer, 1982; Wurz et al., 2018
Blombos Cave	South Africa	MSA; LSA	A remarkable cave noted for its early use of ochre, bone points, and lithic industry	Haaland et al., 2017; Haaland et al., 2020; Henshilwood et al., 2001; Jacobs et al., 2006
Hohle Fels Cave	Germany	MP; UP	Site with deposits that span the MP/UP boundary and anthropogenic burned deposits	Goldberg et al., 2003; Miller, 2015; Schiegl et al., 2003
Geißenklösterle Cave	Germany	MP; UP	Site with deposits that span the MP/UP; early ivory figures	Goldberg et al., 2019; Miller, 2015
Gough's Cave	UK	UP	Controversial site relating to cannibalism in the Upper Paleolithic	Bello et al., 2015; Currant et al., 1991; Macphail and Goldberg, 2003; Stringer, 2000
Meadowcroft Rockshelter	USA	Paleoindian-Archaic	Noted for its controversy concerning early radiocarbon dates and the arrival of humans in North America	Adovasio et al., 1984; Adovasio et al., 1999; Beynon and Donahue, 1982; Donahue and Adovasio, 1990; Goldberg and Arpin, 1999
Dust Cave	USA	Paleoindian-Archaic	A remarkable sequence of geogenic and anthropogenic deposits, including Paleoindian prepared surfaces	Driskell, 1994, 1996; Goldberg and Sherwood, 1994; Goldman-Finn, 1994; Homsey and Capo, 2006; Sherwood and Goldberg, 2001
Denisova Cave	Siberia, Russia	MP, UP	Hominin remains (Denisovans) who interbred with Neanderthals according to DNA evidence; long sequence of deposits	Derevianko et al., 2003; Jacobs et al., 2019; Leonov et al., 2014; Morley et al., 2019; Shunkov et al., 2018; Slon et al., 2017; Slon et al., 2018
Liang Bua	Flores, Indonesia	120 ka to present	Type locality of *Homo floresiensis*	Morley et al., 2017; Sutikna et al., 2018; Sutikna et al., 2016; Westaway et al., 2009a; Westaway et al., 2009b
Lapa do Santo rockshelter	Brazil	~12.7 cal ka to present,	Burials within mostly anthropogenic ashy and red clayey layers	Villagran et al., 2017a
Abrigo do Lagar Velho	Portugal	UP	Skeleton thought to be a hybrid between Neanderthal and anatomically modern human	Angelucci, 2002

certain instances, the mineralogy and archaeological remains (particularly fauna) can be extensively transformed (Goldberg and Nathan, 1975; Karkanas et al., 2000; Stiner, 2005; Weiner et al., 2007).

Typically, materials that accumulate in them, both geogenic and anthropogenic, are generally inclined to stay there. Consequently, caves commonly offer the opportunity to preserve thicker, and more extensive and richer archaeological sediments than open-air hunter-gatherer sites (as opposed to ancient cities and tells, for example). Cave sediments are commonly very rich in anthropogenic and biogenic components (e.g. ashes, hearths, organic-rich remains, guano; cf. Chapters 12 and 13), and therefore they can provide detailed insights into past human activities (controlled use of fire – or not; Dibble et al., 2018b; Sandgathe et al., 2011), treatment of the dead (the debate about Neanderthal burials, e.g. Dibble et al., 2015; Rendu et al., 2014), and stock management practices (Angelucci et al., 2009). This chapter examines cave and rockshelter formation, their sediments, human activities recorded in them, and the ways their sediments have been used to reconstruct past environments.

10.2 Formation of Caves and Rock Shelters

Most caves and rockshelters occupied by prehistoric people are developed typically within carbonate rocks (dolomite and limestone) or sandstones/quartzites (e.g. South Africa – Goldberg, 2000; Karkanas and Goldberg, 2010; Marean et al., 2010); less commonly, they form in other rock types, such as lava tubes (Frumkin et al., 2008), salt domes (Frumkin, 2013; Frumkin et al., 1991) or rarely, in soft marls or other unconsolidated sediments (e.g. Qumran, Israel with the Dead Sea Scrolls – Frumkin, 2001). The vast majority of archaeological caves, however, are in carbonate environments.

We distinguish here rockshelters from caves, as their differences in morphology also influence the types of depositional and post-depositional processes that are recorded. *Rockshelters* tend to be wider than they are internally deep and have a bedrock overhang that serves as protection from sun, wind, and rain (Mentzer, 2017a). They can develop in limestone, as in south-western France (Laville et al., 1980) and desert terrains (Ramon Crater, Israel – Macphail and Goldberg, 2018a), or in clastic rocks as in much of the south-eastern US (e.g. Meadowcroft; Donahue and Adovasio, 1990) and in sandstones or quartzites (Goldberg et al., 2009b; Karkanas and Goldberg, 2010; Miller et al., 2013). In many places such as SW France (Figure 10.1), they are produced by differential weathering of the bedrock, in which weaker strata (generally fine-grained sediments, e.g. shales) are weathered more intensely than harder strata. In the fluvial environment, they can be produced by streams undercutting cliff edges, and in coastal areas, wave-cut niches are also well known. In either case, unlike caves (see below) rockshelters are more subaerially exposed and subjected to atmospheric weathering processes, such as freeze-thaw, solifluction, leaching, and cementation; these processes are more typical to pedological post-depositional modifications found in open-air sites (Mentzer, 2017a). Hearths for example may become fragmented and reworked (Macphail et al., 1994b) (see Box 4.3). They resemble the passive karst settings discussed in Woodward and Goldberg (2001) in which original karst processes, such as solution/dissolution, dripping vadose water, etc. are reduced.

Caves, on the other hand (Figure 10.2), tend to be much larger scale features. Overlooking caves formed along structural weaknesses (e.g. fractures) in the bedrock (Karkanas and Goldberg, 2010), most caves are found in karst terrains where, owing to dissolution, their networks can extend for tens or hundreds of kilometers and can be more than a km below the earth's surface (Ford and Williams, 2007; Gillieson, 2009; White et al., 2019). Although most human activity is concentrated at and towards the entrances, it is quite clear that rituals or other incomprehensible

Figure 10.1 Rockshelters. (a) Laugerie Haute (Dordogne France) with smooth rockshelter cliff face at left and gigantic detached rock at right, which rests on the profile in (b). (b) Laugerie Haute – full frontal profile underneath large rock shown in (a), with Harold Dibble (with hat) looking at the flat-lying sediment composed of stony layers rich in limestone clasts (éboulis) and darker, more anthropogenic ones. (c) Pech de l'Azé I – view of bedrock wall at left and infilling of wall and roof collapse (Soressi et al., 2013; Soressi et al., 2008). (d) Diepkloof Rockshelter (South Africa) formed by the collapse from the roof of a large block of bedded, quartzitic sandstone, as can be seen in the middle of the photograph. (e) The excavations of Diepkloof in 2010, with prehistorian Guillaume Porraz wearing the green shirt. The deposits in the long trench in the center consist of diagenetically altered combustion features (Miller et al., 2013). (f) Caves along Pinnacle Point (South Africa). These caves are formed within enlarged shear zones of recrystallized quartzite (Table Mountain Sandstone). See Karkanas et al. (2020) for details.

activities were carried out very deep within some caves (Jaubert et al., 2016; Sherwood and Goldberg, 2001). Their atmospheres tend to be more humid than found in rockshelters, a factor which favors biological activity and various forms of chemical and biological weathering (Woodward and Goldberg, 2001).

(a)

(b)

Figure 10.2 Caves. (a) Hayonim Cave (Israel) in center with remains of collapsed cave at the left, now seen as an indented cliff face. (b) Coastal karstic caves on Gibraltar viewed from the sea. The three prominent ones from left to right are Bennett, Gorham's, and Vanguard Caves.

10.3   Study of Caves and Rockshelters

Human (anthropogenic) deposits represent the effects of former human activities, whether they be lighting a fire, preparing a floor for sleeping, or simply throwing out the trash and remains of dinner. Such anthropogenic deposition can occur independently of – or pene-contemporaneously with – geogenic deposition, or it can be interspersed with geological depositional events over time. The latter often represent periods of human abandonment in caves, when geogenic depositional processes dominate (Goldberg et al., 2012). Furthermore, human activities such as trampling, sweeping, or digging can modify previous deposits that create new and complex sedimentological signatures.

The study of the human use of karstic caves and rockshelters has a long history that spans continents and time intervals. A geoarchaeological perspective began with excavators in Europe in the first half of the 20th century as interdisciplinary researchers documented the stratigraphy of human, lithic, and faunal remains at varying degrees of detail and success (see discussion and references in Texier (2009). The debut of more systematic geoarchaeological research started in the 1950s, especially in the Old World – particularly France, Germany, and Switzerland – when sedimentological studies of cave deposits were directed at extracting paleoclimatic data, which in turn, formed part of a climato-chronostratigraphic system that could be used in correlating the many cave sequences that existed (e.g. Bonifay, 1956, 1975; Bordes, 1972; Farrand, 1975; Goldberg, 1969; Laville et al., 1980; Le Tensorer, 1972; Miskovsky, 1969; Schmid, 1958, 1963).

In the New World, where archaeologists are traditionally trained as anthropologists, geoscience approaches to prehistoric caves in the mid-20th century were more eclectic, focused primarily on reconstructing local paleoclimate and discovering Late Pleistocene Paleoindian sites. Paleoindian studies in particular, in their search to discover the earliest inhabitants in the Americas, recognized the need for a geoscience approach in order to establish chronology (e.g. Haury, (1950) at Ventana Cave). As Paleoindian research in the Americas evolves it has continued to maintain close ties with the geosciences (e.g. Huckleberry and Fadem, 2007). Other geoarchaeological research in caves concentrated on reconstructing local paleoclimate, especially in places like the Great Basin where arid conditions preserved a long sequence of datable environmental materials (e.g. Rogers Shelter – Ahler, 1973, 1976); Danger Cave (Jennings, 1957)).

After these pioneering efforts, researchers have become increasingly aware that in addition to the geogenic/paleoclimatic aspects of prehistoric caves, a significant amount of cave infilling is anthropogenic, and it has become clear over the past few decades that these types of sediments represent artifacts – much like stone tools or faunal remains – that could be interpreted in terms of human activities and behaviors (Berna and Goldberg, 2008; Courty et al., 2012; Goldberg and Berna, 2010; Goldberg et al., 2009b; Karkanas et al., 2004; Miller et al., 2013; Vallverdú et al., 2001, 2005).

10.4 Cave Deposits and Processes

Deposits from both inhabited and uninhabited caves from throughout the world have been investigated for decades. In what follows, we emphasize geogenic and biogenic deposits, since anthropogenic accumulations are treated in Chapter 13. We distinguish natural deposits that originated inside the cave (autochthonous or endogenous) from those that were transported into the cave from the outside (allochthonous or exogenous) (Tables 10.2 and 10.3) (Figure 10.3).

10.4.1 Autochthonous Deposits

A major depositional component of cave deposits are clasts of bedrock derived from the walls and ceiling (roof fall), and deposited mostly by gravity; in the French literature this rock rubble is called *éboulis* (Figures 10.1, 10.4). Such material accumulates on the surface and includes not only larger cm-sized and larger clasts, but also individual sand-sized grains (generally quartz and limestone) that have been liberated by the disintegration of the bedrock by dissolution (Donahue and Adovasio, 1990; Goldberg and Arpin, 1999). The mechanism of detachment is commonly related to freezing and thawing of water that has percolated into cracks and fissures in the bedrock (*cryoclastic deposition*; see "mechanical weathering" in soils Chapter 4.2.1) (Bonifay, 1956; Laville et al., 1980; Schmid, 1963); this process is particularly prominent in Pleistocene western and central Europe. Loosening of bedrock blocks by dissolution also occurs, particularly during warmer periods and in warmer climates. In other instances, accumulation of sand from human abrasion of

Table 10.2 Types of cave sediments (modified from Sherwood and Goldberg 2001).

Sediment type	Autochthonous (endogenous)	Allochthonous (exogenous)
Clastic sediments (geogenic)		
● Weathering detritus	●	
● Block breakdown (éboulis)	●	
● Grain breakdown by dissolution, abrasion of rock walls by humans[a]	●	
● Entrance talus	●	●
● Infiltrates (drip)	●	●
● Fluvial deposits	●	●
● Glacial deposits		●
● Aeolian deposits		●
Biogenic inputs		
● bird and bat guano	●	○
● gastroliths	●	○
● carnivore coprolites and bone	●	●
● wood, grass	○	●
● humus	○	●
Anthropogenic deposits	○	●
● microartifacts (bone, shell, lithics)	●	
● bedding	●	
● transported soil/sediment	●	○
● grass/	●	●
● wood/charcoal		●
● ash	●	
Chemical sediments		
● Travertines, flowstones, "cave breccia"	●	
● Phosphate and nitrate minerals	●	
● Evaporites	●	
● Resistates	●	
● Ice	●	

Locale: Likelihood ● = high; ○ = low
[a] (Hughes and Lampert, 1977)

the walls of sandstone shelters has been documented in Australia (Hughes and Lampert, 1977). In South Africa, tectonically induced fracturing of the quartzitic bedrock promotes the splintering and disaggregation of the bedrock (Karkanas and Goldberg, 2010).

As mentioned above, limestone clasts – common in many of the deposits in caves and rockshelters – have been studied in considerable detail (Figure 10.1, 10.4). Following a long tradition of sedimentological studies of French cave and rockshelter deposits (see Texier, 2009), Laville et al. (1980) used a variety of techniques to evaluate past climatic fluctuations. These included

Table 10.3 A sampling of the more important syn- and post-depositional processes and agents acting in prehistoric caves.

Physical

Slumping and faulting

Import and accumulation of bones and organic matter by animals

Vertical translocation of particles by percolating water

Penecontemporaneous reworking by wind, water, frost, and humans

Trampling by humans and other animals

Hearth cleaning and rake-out

Burrowing and digging by humans and other animals

Dumping of material, midden formation

Chemical

Calcification (speleothems, tabular travertine, tufa, cemented "cave breccia")

Decalcification

Phosphate mineralization and transformation, and associated bone dissolution

Formation of other authigenic minerals: gypsum, quartz, opal, hematite

---

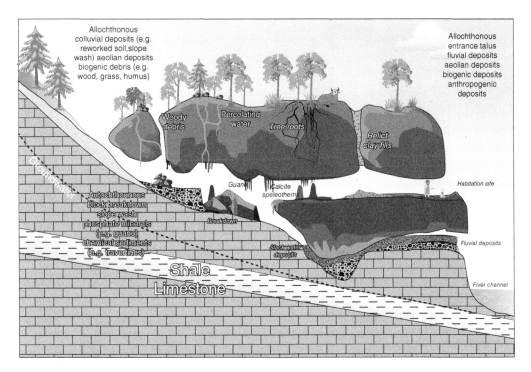

Figure 10.3 Schematic representation of the types of processes operating in a karstic cave environment. Indicated are physical and chemical inputs and outputs, as well as transformations, such as precipitation and dissolution (modified from Gillieson, 2009: Figure 1.1).

Figure 10.4 *Eboulis* from the Middle Paleolithic site of Pech de l'Azé IV, France, excavated by H. Dibble and S. McPherron. The smaller blocks of *éboulis* represent a relatively steady rain of roof fall, while the large blocks at the very bottom and in the upper part represent large-scale roof collapse that took place at more punctuated intervals. The distance between the two vertical lines is ca. 1 m.

analysis of grain size and shape (angularity and roundness; Chapter 2) as indicators of weathering and frost-shattering, measurements of acidity (pH) and calcium carbonate content as signs of weathering/chemical alteration and "soil formation" as exemplified by Pech de l'Azé II for example (Laville et al., 1980). The results of these studies were originally thought to provide paleoclimatic sequences that could be used to correlate cave/rockshelter sedimentary sequences – and their lithic industries – throughout the Dordogne region of south-western France. Thus, layers with abundant coarse éboulis, for example, were considered to represent relatively cold climates, whereas red layers were thought to be paleosols that developed during more temperate episodes (cf. β-clay formation in weathered limestone and terra rossa paleosols; Chapter 4). This strategy not only supplied background paleoenvironmental information to the archaeological context, but also served as a means of indirectly dating the caves and their artifact sequences, as dating other than radiocarbon and ESR (electron spin resonance) were not widespread at the time.

We now know that these pioneering efforts were a bit overzealous, as climatic factors responsible for the production/breakdown and alteration of limestone clasts in caves are more much complex than originally envisioned (Bertran, 1994; Lautridou and Ozouf, 1982). In addition, TL dating of burned flints demonstrated that the inferred climato-correlations were largely inaccurate by several millennia (Valladas et al., 1986). Similarly, micromorphological work showed that reddish stratigraphic units are not soils but deposits of reworked Tertiary weathering products derived from the plateaux above the caves and rockshelters (Goldberg, 1979a); 'soils' in the strict sense, as opposed to exposed occupation or weathering surfaces, do not form in caves (see below). Lastly, Campy and Chaline (1993) demonstrated that many temporal gaps exist in the depositional sequences of many of these caves, which makes correlations based on matching of apparent climatic cycles tenuous at best. Nevertheless, the deposition of limestone blocks and clasts is likely tied to colder conditions (Couchoud, 2003), although it is difficult to state more than that unequivocally.

10.4.2 Allochthonous Deposits

Geogenic materials are also transported into caves and rockshelters from outside. These items can include soils that emanate from cracks in the roof that are reworked by gravity and water (runoff) (Goldberg and Bar-Yosef, 1998; Macphail and Goldberg, 2000), fluvial deposits (Sherwood and Goldberg, 2001; Texier et al., 2006), and aeolian inputs (Karkanas et al., 2015; Karkanas and Goldberg, 2010), for example (Figure 10.3). The latter are particularly prominent in prehistoric sites along littoral parts of the Mediterranean climatic zone. Massive accumulations of sand – along with interspersed occupations – can be seen in Gorham's and Vanguard Caves, Gibraltar (Macphail and Goldberg, 2000) (see Chapter 9); Tabun Cave in Israel, and in many sites along the South African coast, such as Blombos and Die Kelders Caves (Henshilwood et al., 2001; Karkanas et al., 2015; Karkanas et al., 2020; Marean et al., 2000); in these sites, the influx of sand is related to glacio-eustatic sea level changes and as such are useful in the reconstruction and correlation of regional landscapes. Finer grained silts can also be blown into caves in the form of atmospheric dust produced by dust storms; silty deposits at the Middle Paleolithic/Upper Paleolithic sites of Hohle Fels and Geißenklösterle in Swabia, Germany, are of such a loessial origin (Goldberg et al., 2019; Miller, 2015).

A recent study used *magnetic anisotropy* to evaluate the depositional history (source, transport, sedimentation) of sediments in Gran Dolina and Galería from Atapuerca (Spain) (Parés et al., 2020), a site noted for its time depth and hominin content. They studied the orientation and packing of the sediments using anisotropy of magnetic susceptibility to sort out deposition by gravity (e.g. colluvium) vs. water transport as the two transport and depositional agents, which in turn have implications for the integrity of the archaeological finds. Grain orientations are different in both cases for both larger (e.g. gravel) clasts (see Chapter 2) and smaller (e.g. sand and silt).

> It follows that under quiet sedimentary conditions, if gravity is the only acting torque, the magnetic fabric is dominantly oblate (foliated). Particle long axes are horizontal but with no preferred orientation within the depositional plane. Nevertheless, under a hydrodynamic force, elongated grains will tend to align parallel to the flow direction, whereas the platy minerals will be slightly imbricated upstream. Elongated particles will tend to roll and orient their long axes perpendicular to the flow direction for higher flow velocities (Parés et al., 2020: 4).

Their results not only confirmed previous interpretations of the environments of deposition (gravelly vs. slackwater facies) but also demonstrated that there was no evidence of bioturbation. Moreover, they showed that depositional environments and the physical integrity of the deposits (e.g. modification by bioturbation) can be detected even if field observations are not clear (Parés et al., 2020: 10). The approach would seem to have significant potential for evaluating the degree of integrity of archaeological materials, and whether the original sediments might have been modified by natural bioturbation or human actions such as trampling, etc.

10.4.3 Geochemical Processes and Diagenesis

Geochemical processes are common in caves, but slightly less so in rockshelters. They are usually expressed as the accumulation and oxidation of organic matter, and the precipitation and dissolution of various minerals in the sediments, including those from archaeological materials such as bones and shells. The most common process is precipitation of calcium carbonate, which occurs in various forms that include stalactites, stalagmites and other ornamental forms (Gillieson, 2009; White et al., 2019). Massive, bedded accumulations of calcium carbonate are generally known as flowstone or travertine and can cover or intercalate with archaeological deposits (Figure 10.5). In suitable instances, where levels of contamination by detrital components (e.g. silt and clay) are low, these carbonate accumulations can be dated with uranium-series methods. Hoffmann et al. (2018) used U-Th to date carbonate crusts coating paintings from Spanish caves and showed that Iberian art pre-dates 64.8 ka, suggesting that the paintings were made by Neanderthals. On a larger scale, *speleothems* and flowstones have been intensively studied from several caves in the Mossel Bay area, along the South African coast, including investigation of stable oxygen and carbon isotopes of speleothems coupled with U-series dating (Bar-Matthews et al., 2010; Braun et al., 2019; Karkanas et al., 2020). These studies allowed for the integration of the nature and dating of paleoenvironments – including sea level change and dune incursion, which blocked cave entrances – and their relationship to human occupational history of the caves. These studies were enhanced by the use of GIS and 3D models of the caves to provide a solid contextual basis for the results.

Figure 10.5 Pinnacle Point Cave 13b (South Africa) showing calcite-indurated archaeological deposits in excavated profile capped with thin layer of flowstone (arrow), which accumulated between ~92 and 39 ka using U-Th (see Karkanas and Goldberg, 2010; Marean et al., 2010 for details).

Carbonate-charged water dripping on detrital sediment can lead to partial or total impregnation and cementation of the underlying deposits whatever their composition (Figure 10.5). This cementation results in hard, concreted masses, commonly called "breccia" or "cave breccia"; such deposits are widespread around the globe (Bruxelles et al., 2019; O'Connor et al., 2017; Rosas et al., 2006; Vandermeersch, 1981; Villagran et al., 2017a). Animal and human bones, seeds and other organic materials tend to be well preserved in these breccias, and thus they serve as valuable stores of environmental and cultural information.

Calcium carbonate can also be dissolved and dissolution can stem from a number of factors. Among the most prominent is the acidic conditions produced by guano, the decay of organic matter, and carbonic acid derived from carbon dioxide dissolved in water as it passes through the soil (see also "weathering" in Chapter 4). Associated with carbonate dissolution, particularly in caves of the Mediterranean region (e.g. Mallol et al., 2010) but also common elsewhere (Goldberg et al., 2015; Miller et al., 2016; Shunkov et al., 2018; van Vliet-Lanoe, 1986), is the transformation and formation of a number of phosphate minerals (Table 10.4) (Karkanas and Goldberg, 2018a; Karkanas et al., 1999; Weiner et al., 2002) (Figure 10.6). The geochemistry of these interactions is quite complex although recent research on modern deposits has begun to reveal some of the major processes and conditions operating within the caves (Berna et al., 2004). Shahack-Gross et al. (2004a), for example, showed the rapidity of formation of certain minerals – under acidic conditions – within a period of decades. As organic matter degrades, the amount and availability of Al, K, and Fe increase, whereas nitrogen and sulfur decrease. Minerals, such as brushite, taranakite, nitratite, and gypsum (Table 10.4) form first, although they are transient. With time, as

Table 10.4 Some of the more *common* authigenic minerals associated with cave sediment diagenesis (modified from Karkanas et al., 2000; Karkanas and Goldberg, 2018a; see also Hill and Forti, 1997; Karkanas et al., 2002; Miller et al., 2016; Shahack-Gross et al., 2004a).

Mineral	Formula
Brushite	$CaHPO_4 \cdot 2H_2O$
Calcite	$CaCO_3$
Hydroxyapatite	$Ca_{10}(PO_4)_6(OH)_2$
Fluor-carbonated apatite (Francolite)	$Ca_{10}(PO_4)_5(CO)_3F_3$
Dahllite (carbonated apatite)	$Ca_{10}(PO_4,CO_3)_6(OH)_2$
Ca,Al-phosphate amorphous	Ca, Al, P-phase, non-stoichiometric
Crandallite	$CaAl_3(OH)_6[PO_3(O_{1/2}(OH)_{1/2})]_2$
Montgomeryite	$Ca_4Mg(H_2O)12[Al_4(OH)_4(PO_4)_6]$
Taranakite	$H_6K_3Al_5(PO_4)_8 \cdot 18H_2O$ or $K_x[Al_{2-y}(H_3)_y](OH)_2[Al_xP_{4-x-z}(H_3)_zO_{10}]$
Leucophosphite	$K_2[Fe_4^{3+}(OH)_2(H_2O)_2(PO_4)_4]\,2H_2O$
Nitratite	$NaNO_3$
Millisite	$NaCaAl_6(PO_4)_4(OH)_9 \cdot 3H_2O$
Newberyite	$Mg(PO_3OH) \cdot 3H_2O$
Strengite	$Fe^{3+}(PO_4) \cdot 2H_2O$
Vivianite	$(Fe^{2+})3(PO_4)_2 \cdot 8H_2O$
Variscite	$Al(PO_4)\,2H_2O$

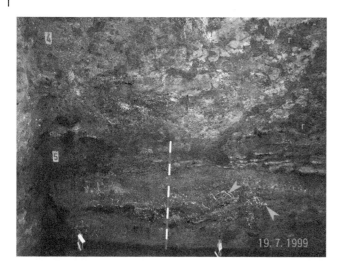

Figure 10.6 Middle Paleolithic deposits at Hayonim Cave (Israel). The geology here is quite complex (Weiner et al., 2002; Weiner et al., 1995b), but illustrated here are remnants of Neandertal fireplaces and significant post-depositional changes. The latter is shown by the precipitation of phosphate veins (green arrows) and a number of rodent burrows (r); these formed in antiquity whereby rodents burrowed into diagenetically altered clays and ashes (shown by the orange and whitish colors.

organic matter continues to decrease but still under acid conditions, clay minerals start to degrade, releasing Al and Fe, and possible formation of authigenic opal. Other phosphatic minerals (e.g. dahllite, crandallite, montgomeryite, taranakite, and leucophosphite; Table 10.4) also form under complex diagenetic pathways (Karkanas et al., 2000).

Implications for these diagenetic changes are many. The most apparent is the localized to total dissolution of bone in many cave deposits, particularly those in Mediterranean environments, such as Tabun, Hayonim, Kebara (Goldberg and Bar-Yosef, 1998; Weiner, 2010; Weiner et al., 2002). Thus, the occurrence (presence or absence) and distribution of bones, can be mostly a function of geochemical processes and not necessarily related solely to human activities. Berna et al. (2004) using a number of laboratory experiments showed that bone apatite (dahllite) is preserved in pH values above 8.1; in neutral and alkaline conditions bone will be preserved but the apatite will be recrystallized.

More subtle, however, are the diagenetic effects in which radioactive elements such as potassium can be concentrated. In turn, these localized concentrations can affect the dosimetry measurements needed to obtain chronometric dates using TL and ESR techniques (Weiner et al., 2002), thus yielding incorrect dates if they are not taken into account. Finally, results of mineralogical studies of diagenetically altered cave deposits suggest that the occurrence of the more insoluble minerals, such as leucophosphite, can be used to infer the presence of former fireplaces in cases where their actual presence is no longer visible (e.g. Schiegl et al., 1996).

In addition to chemical changes, physical diagenetic changes in caves are common as well. Primary among them is *bioturbation* at various scales. Most prominent is that done by rodents, which leave generally circular/elliptical shape features about 8–10 cm across (*krotovina*) (Figure 10.6), commonly filled with a contrasting sediment compared to that of the surrounding material (Goldberg and Bar-Yosef, 1998; Goldberg et al., 2007; Ponomarenko and Ponomarenko, 2019). One must be aware that in many cases, an individual visible burrow may only be the last of many burrowing events, in which large swaths of the deposits have been homogenized by these small mammals. Individual and repeatedly burrowed sediments can be inferred in the field, for example, by their color, generally loose consistency (although some burrows can be preferentially mineralized; Miller et al., 2016), or their artifact composition, which is anomalous and does not fit in with that as expected. In thin section, burrowed sediment is apparent by the granular nature of grains of mixed composition. Smaller burrows in caves (on the order of ~1 cm) are generally attributable to insects (Kooistra and Pulleman, 2018). These are more difficult to discern in the field but are visible in thin section (passage features – Stoops, 2020) as circular features or concentrations or fecal pellets.

The effects of bioturbation can be severe. It can totally obliterate original depositional signatures and at the same time mix archaeological materials (lithics, bones, charcoal) of different ages and cultural attribution; it can also lead to incorrect assessment of ages.

10.4.4 Biogenic Deposits

Biogenic inputs in caves are signified by bat and bird *guano*, as well as the remains of denning activities by carnivores and omnivores. Hyaenas, for example, can produce noteworthy accumulations of bones, coprolites (generally apatite), and organic matter (Goldberg, 2001b; Horwitz and Goldberg, 1989; Morley et al., 2019). Deposits linked to cave bear occupations have also been inferred from phosphate-rich sediments associated with both the bones of their prey and the bears themselves (Andrews et al., 1999; Miller, 2015; Schiegl et al., 2003). In Middle Eastern prehistoric caves, pigeons generate mm-size rounded stone gastroliths ("crop stones") that indirectly indicate a significant but now-invisible source of guano (Shahack-Gross et al., 2004a; Goldberg and Bar-Yosef 1998). Both anthropogenic and natural accumulations of plant remains (e.g. grass, wood, roots, seeds) can originate from vegetation growing at the cave entrance, which has been washed or blown into the cave; they can also be brought into the cave for use as bedding, fuel, or shelter, or even as nesting material (Goldberg et al., 2009b; Wadley et al., 2011; Wattez et al., 1990). Quantitative phytolith studies of cave sediments from Israel, France, and Spain have been instrumental in evaluating these biogenic/anthropogenic inputs (e.g. Albert et al., 1999; Albert and Weiner, 2001; Albert et al., 2000; Cabanes et al., 2010; Rodríguez-Cintas and Cabanes, 2017; Wroth et al., 2019)

10.4.5 Anthropogenic Deposits and Human Activities

Anthropogenic deposits are the result of human activities. Deposition can occur independently of – or pene-contemporaneously with – geogenic deposition, or it can be interspersed with geological depositional events over time. Furthermore, human activities such as trampling, sweeping, or digging can modify previous deposits creating new and complex sedimentological signatures. Meter-thick, often charred animal bedding and dung, recording centuries of pastoralism are a special case (Boschian and Montagnari-Kokelj, 2000; Macphail et al., 1997).

Human deposition can be purposeful or accidental. In the case of the latter, sedimentary grains can be brought into the cave attached to clothing, feet, carcasses, or plant remains. Although these processes surely happened in the past, it is difficult to identify them as say, "tracked in" sediment. On the other hand, other kinds of deposits and activities are more identifiable, either in the field or analytically.

More prominent features are fireplaces, *hearths*, and others associated with fire (Mallol and Goldberg, 2017; Mallol et al., 2017) (Figures 10.6 and 10.7). Although isolated hearths are not uncommon, striking combustion features are found in Paleolithic caves from the Middle East (e.g. Tabun, Kebara, Hayonim: Goldberg and Bar-Yosef, 1998; Meignen et al., 1989, 2007), Middle Stone Age

Figure 10.7 Anthropogenic deposits (a) Photo of anthropogenic remains in Middle Paleolithic Layer 8 at Pech de l'Azé IV (Dordogne, France), consisting of some remains of hearths, most of which have been redistributed by rake-out and some trampling; the dark layers are richer in fat-derived char, but not much charcoal (Dibble et al., 2018a) (photo courtesy of Old Stone Age and Pech de l'Azé Team). (b) Photo of interbedded sand and combustion features at Die Kelders Cave, South Africa. Both the sand and some of the burned materials have been partially modified by wind and humans. The former reworked the ashes and charcoal into bedded deposits, whereas the latter was responsible for displacing the combusted the materials as part of activities such as rake-out. The width of the profile is ca. 1.5 m.

sites in South Africa (Klasies River Mouth, Die Kelders, Blombos; Sibudu: Goldberg et al., 2009b; Mallol et al., 2017; Marean et al., 2000; Mentzer, 2016; Miller et al., 2013; Singer and Wymer, 1982). Those at Klasies River Mouth, Die Kelders, and Kebara (see Box 10.1) are particularly noteworthy for their degree of preservation and abundance. At Kebara, different types of combustion features could be recognized, including a massive accumulation of ashes that were dumped by the Neanderthal inhabitants (Meignen et al., 2007). These combustion features provide high resolution windows into human activities and technologies related to fire, including type of combustible, duration, and intensity of burning, cleaning, trampling and dumping, and their spatial distribution (Aldeias et al., 2012; Goldberg et al., 2012; Leierer et al., 2019; Mallol et al., 2013a, 2013b). Interestingly, two nearby sites in the Dordogne of south-western France indicate the presence of hearths in predominantly Last Interglacial deposits, with essentially little evidence of combustion in the overlying cooler deposits and a subject of considerable debate (Dibble et al., 2017, 2018b; Goldberg et al., 2018; Sandgathe, 2017; Sørensen, 2017).

On a similar note, caves figure into important issues such as the first uses of "controlled fire" (Sandgathe, 2017; Sandgathe et al., 2011). These include Swartkrans (South Africa) and Zhoukoudian, Locality 1 (China). Some geoarchaeological investigations have been directed to evaluating the nature of the latter, which roughly range in age from 250 to 600 ka. Micromorphological and FTIR analyses of the often-quoted "hearths" in Layer 10, are in fact poorly sorted interbedded accumulations of sand, silt and clay-sized mineral and organic matter (Figure 10.8) (Goldberg et al., 2001). Such fine-grained, laminated deposits more reasonably accumulated in standing water depressions; these were likely associated with the presence of the Zhoukou River, which at that time occasionally flowed into the cavern. Not all researchers, however, accept this interpretation, and remains of

(a)

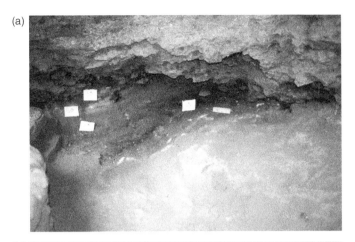

(b)

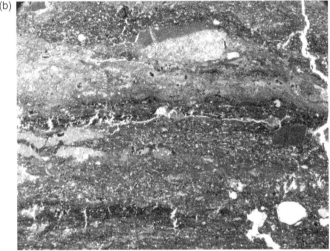

Figure 10.8   Zhoukoudian, Locality 1. (a) Putative hearth from Layer 10. The structured appearance of these finely laminated reddish and orange deposits was long thought to be a result of *in situ* burning activity associated with the controlled use of fire. Scale at right is 10 cm. (b) Photomicrograph of thin section from middle of the deposit showing finely inter-laminated mineral and organic material, only some of which is charred. These laminated sediments accumulated in standing water depressions within the cave early in its history (see Goldberg et al., 2001 for details). Plane-polarized light; width of field is ca. 6.4 mm.

hearths in the much later Layer 4 have been recently advanced (Gao et al., 2017). In any case, one has to wonder how many fires were needed to produce the accumulation of ~6–7 m of supposed ashes in Layer 4 (de Chardin and Young, 1929; Goldberg et al., 2001).

Among the first clear signs of repeated, purposeful combustion events comes from bedded hearth deposits from Qesem Cave, Israel, which date to between 300 and 400 kya (Karkanas et al., 2007; Shahack-Gross et al., 2014). Prior to that, evidence of ashes and heated rocks suggestive of burning at Wonderwerk Cave (South Africa) have been put forward by Berna et al. (2012), based on the occurrence of burned bones, pot-lidded stones, and possible ashes, although this is currently being re-investigated (Goldberg et al., 2015).

As discussed in further detail below (see also Chapter 17), new approaches are being developed to evaluate inorganic and organic evidence for the identification and use of fire

(Mentzer, 2014, 2016). One methodology that has been used for already a half decade in archaeology (Evershed, 2008) but now being applied to prehistoric settings is organic residue analysis (Brittingham et al., 2019; Leierer et al., 2019, 2020), which reveal plant and animal biomarkers. In turn, these reflect upon, for example, the length of occupation, fuel use and sources, and the controlled used of fire.

Constructions occur in caves and appear as early as the Upper Paleolithic. For example, Üçağızlı Cave (Turkey), a 2.5 m long alignment of rocks dating to ~ 35 kya appeared to create a repository, retaining dumped ashy sediments between the makeshift wall and bedrock wall; although unique for this time and place, its function remains unclear (Kühn et al., 2009). At Hayonim Cave (Israel), the Natufian Layer B (~12.3–13.4 kya cal) contained a number of constructed stone rooms and graves associated with calcareous ashy midden deposits (Stiner, 2005). Other examples include Iron Age and Medieval stone masonry walls enclosing cave entrances in the Dordogne (France) (Couchoud, 2003).

Smaller construction features consisting of prehistoric clay-lined pits and clay prepared surfaces have been investigated recently. In the oldest known example from Klisoura Cave (Greece), Karkanas et al. (2004) document the presence of intentionally clay-lined hearths dating to the Aurignacian (~34–23 kya). Using soil/sediment micromorphology, Fourier transform infrared spectrometry (FTIR; Chapter 17), and Differential Thermal Analysis (DTA), they were able to demonstrate an alluvial source of the clay that was brought to the site and used to line the structures, which were heated to between 400 and 600° C. In the New World, at Russell Cave (Griffin 1974) and at Dust Cave (Sherwood and Chapman, 2005), clay (Macphail, 2019b) surfaces were prepared for cooking activities starting in Late Paleoindian – Early Archaic times, at least 10 kya. In both sites, residual clay derived from deep within the cave system was used to create ~ 1 m 2–5 cm thick surfaces, that were then heated to temperatures between 400 and 900° C. Based on micromorphology and experimentation these surfaces are often stacked, suggesting repeated parching or cooking over thousands of years (Homsey and Sherwood, 2010; Sherwood and Chapman, 2005).

Similar intentionally-laid flat surfaces were prepared at the site of Yuchanyan, Hunan (China), one of the earliest sites with pottery and dating back to ~ 18,300 cal BP (Boaretto et al., 2009; Patania et al., 2020). Here, the clay was clearly brought in from outside to prepare the surfaces, which were used repeatedly as in the case of Dust Cave. Furthermore, results of FTIR analyses lead to the suggestion that the fires and pottery were used to boil bones and render grease from them. Possibly locally collected "wetland" clay was imported into Later Stone Age Taforalt (Morocco) for presumed artifact production.

Sediments can also accumulate through the action of waste *discard* and *dumping*. A good example of midden accumulation is the Late Stone Age (LSA) layers (late Holocene) at Die Kelders (South Africa). Here, sand is interbedded with midden material consisting of faunal remains (e.g. shell, fish, mammals, and birds), dumped ashy material, and organic matter. This dense midden accumulation abruptly overlies quartz sand aeolian deposits from the Middle Stone Age (MSA) and indicates that Holocene LSA midden formation took place when sea level was close to the cave (Marean et al., 2000).

Evidence of discard and dumping of ashes related to fireplace cleaning is documented in Kebara Cave (Israel) (see Box 10.1) where both bones and ashes are preserved in dense deposits from the Neanderthal occupation of the cave (~48–60 kya). Bones were repeatedly dumped and accumulated against the north wall of the cave (Speth and Tchernov, 2007; Speth et al., 2012), grading laterally into a ~80 cm accumulation of bedded calcitic ashes interspersed with sand-sized fragments of bone and terra rossa aggregates (Goldberg et al., 2007; Meignen et al., 2007). Such "house cleaning" activity is not normally associated with Neanderthals, and these deposits provide an insight into Neanderthal behavior.

| Box 10.1 | Kebara Cave, Israel |

The site of Kebara Cave, about 30 km south of Haifa, Israel, is developed within a Cretaceous limestone reef on Mt Carmel (Figure 10.9) and is one of the best studied prehistoric cave sites in the Old World. It represents an excellent example of a multidisciplinary excavation with a team from a variety of scientific backgrounds (e.g. prehistory, geoarchaeology, geophysics, geochemistry zooarchaeology, paleobotany), and countries (e.g. Israel, France, USA, Spain, UK). These investigators all focus their talents and expertise to elucidate a wide range of archaeological and environmental questions and issues (Bar-Yosef and Meignen, 2007; Bar-Yosef and Vandermeersch, 2007; Bar-Yosef et al., 1992; Meignen et al., 2017). Such investigations include dating the deposits and associated lithic, faunal, and human remains; evaluating past environments within and outside of the cave; reconstructing human behavior and evolution, and elucidating the use of space in the cave during the Middle Paleolithic and Upper Paleolithic (Figure 10.10a, b). From the geoarchaeological point of view, Kebara has a rich history and exhibits sediments of geological and human origin that have been modified by geogenic and anthropogenic agents. Only a few of these aspects are presented here (for details see Goldberg et al., 2007; Meignen et al., 2007).

One of the most striking aspects of the deposits at Kebara is the abundance of ashy burning features that provided new insights into Neandertal activities associated with fire (Meignen et al., 2007). Among others, these features include well-defined combustion structures, with shapes ranging from lenticular to tabular (Figure 10.11a, b, c), and a massive bedded ash dump (Figure 10.12a – e). Thus, specifically different types of activities associated with the Neanderthal occupants are evident by examining the sediments themselves, such as *in situ* burning events and the dumped ashes. The differentiation of these deposits and their interpretation constitute the fabric of the site's stratigraphy. These accumulations and associated activities reflect upon the length of occupation episodes and the localized use of space in the site.

Anthropogenic deposits constitute the majority of the Middle Paleolithic deposits at the site, but geogenic sediments also occur, and these are particularly marked in the Upper Paleolithic deposits (Goldberg et al., 2007) (Figure 10.13). Much of the sediment is ashy and

Figure 10.9 Kebara Cave, looking east toward the junction of Mt. Carmel the Coastal Plain. The cave is formed within a Cretaceous limestone reef during pre-Pleistocene times. At the mouth of the cave is a small rise that marks the accumulation of soil material that was washed down from the slopes above the cliff face.

(Continued)

Box 10.1 (Continued)

organic-rich, with some additions of aeolian quartz sand and silt; these deposits were emplaced by runoff, which emanated from the entrance of the cave beneath the drip line and flowed towards the rear of the cave. This deposition was promoted by the relief from the front to the back of the cave, which resulted from two factors. The first is that reworked soil materials derived from the slopes above the cave built up as a mound beneath the brow of the cave (cf. Figures 10.9 and 10.10b) and raised the height of the entrance section. In addition, at the end of the Middle Paleolithic and beginning of the Upper Paleolithic, the deposits in the direction of the rear of the cave subsided on the order of several meters (Laville and Goldberg, 1989). This collapse was a result of reorganization of sedimentary material that probably plugged cavities within the subsurface karstic cave network; similar types of subsidence can be seen today in many areas of active karst. In any case, with this newly created striking relief from front to back, deposits from the entrance were easily transported toward the back of the cave during heavy rainfall events, one of which we witnessed in October of 1984. Interestingly, when viewed in thin section, these well-bedded Upper Paleolithic deposits are composed of soil clasts of terra rossa reworked from soil above and outside the cave, and fragments of Middle Paleolithic combustion features, such as charcoal and ashes, that were being eroded from near the cave entrance and transported toward the rear. These thick, well-bedded Upper Paleolithic deposits (Figure 10.13a, b) indicate prolonged periods of deposition by runoff and furthermore suggest wetter climatic conditions during this time, roughly ca. 43 to 30 ka (Bar-Yosef et al., 1996). This last inference is supported by the presence of diatoms in some of the Upper Paleolithic layers whose analysis indicated the presence of a continually wet substrate (B. Winsborough, pers. Comm.).

(a)

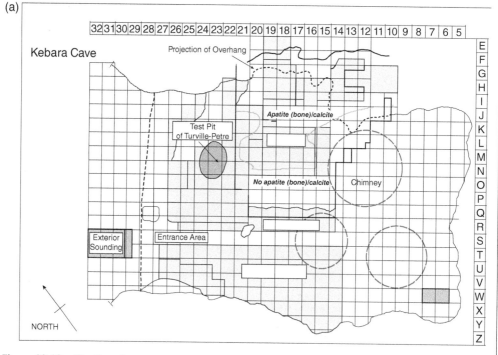

Figure 10.10 (Continued)

Box 10.1 (Continued)

(b)

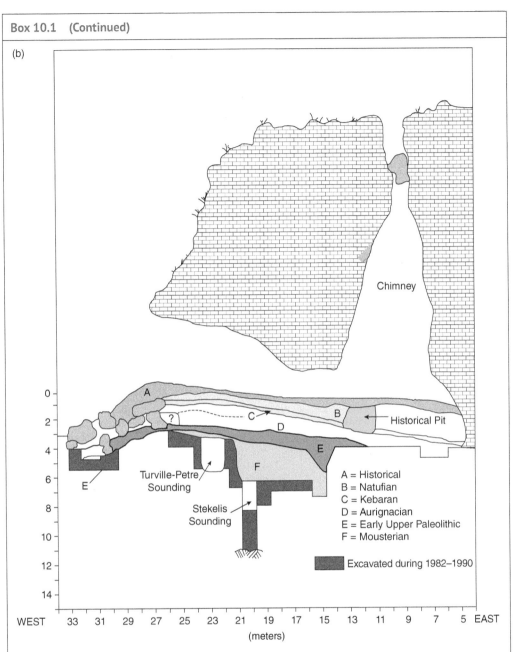

Figure 10.10 (a) Plan view of Kebara Cave showing excavations and location of profiles and areas shown in photos. The curved green line shows the boundary between layers that contain apatite (bone) and calcite (to north) from those in which bone has not been preserved but dissolved because of diagenesis. The recognition of the causes of such distribution patterns places constraints on how we interpret and date the archaeological record (Weiner, 2010; Weiner et al., 2007) (modified from Bar-Yosef et al., 1996). (b) Cross-section of Kebara Cave. The archaeological remains within the layers are as follows: A: Historical; B: Natufian; C: Kebaran; D: Aurignacian; E: Early Upper Paleolithic (Units I–III, IV); F: Mousterian. Note the build-up of roof fall and other sediment at the entrance, beneath the brow of the cave; this accumulation helps create a primary, depositional relief, with a natural slope toward the back of the cave (right). This slope was enhanced by a major subsidence episode that took place at the end of the accumulation of the Middle Paleolithic deposits, prior to Upper Paleolithic occupation (Laville and Goldberg, 1989) (modified from Bar-Yosef et al., 1996).

(Continued)

Box 10.1 (Continued)

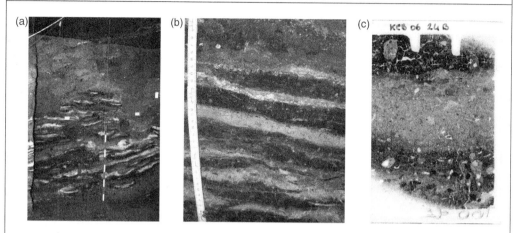

Figure 10.11 Features associated with *in situ* burning at Kebara. (a) The south profile of the Kebara excavations (Sq. O21) showing the abundance of combustion features marked by the black (charcoal-rich) and white (ashy) lenses. These features represent both burning events that took place within hollowed depressions (lenticular shapes), as well as fires made on flat surfaces (tabular/linear shapes). The patchiness or absence of distinct burned layers and homogenized grayish color in the upper part of the photograph is due to mixing by bioturbation of small and mid-size rodents. Scale is 1 m. (b) Detailed view of a number of tabular and elliptical hearths seen looking north in profile along M 19/20. These hearths are within the zone of extensive diagenesis (Figure 10.10a) so that the light ashy parts are no longer comprised of calcite but of various phosphatic minerals (see main text). (c) Scan of thin section from sample taken from this profile showing at least three different hearths comprised of charcoal, reddish terra rossa aggregates, and phosphatic ashes. Size of thin section is 50x75 mm.

Diagenetic alterations are widespread at Kebara and they have significant implications for interpreting the site's history and the use of the cave. The most striking feature is the lack of bones in more than half the excavated area of the cave (Figures 10.10a, 10.14). The immediate question that comes to mind, "is this lack of bones the result of primary factors, such as lack of discard or removal/cleaning by humans, or some other taphonomic factor?" The answer is revealed by detailed mineralogical analyses of the deposits (Weiner, 2010; Weiner et al., 1993, 1995, 2007), which indicate that the absence stems from the dissolution of bone by acids derived from organic matter (guano, human refuse). Such information clearly helps to demonstrate that bone distribution in many parts of the cave is due to post-depositional alteration and not to human activity. In turn, this knowledge influences the way we look at the zooarchaeological record and how it can be interpreted at the site. The second implication is that in many locations, not only have the bones disappeared, but also calcite (both as limestone clasts and as ashes from fires) has been dissolved or replaced by phosphate minerals that range from apatite to leucophosphite and taranakite (see text). These diagenetic alterations are responsible for concentrating radiogenic elements that in turn affect the dosimetry calculations, which constitute an essential part of thermoluminescence and electron spin resonance dating techniques. They are prominent techniques in dating deposits beyond the range of radiocarbon.

Box 10.1 (Continued)

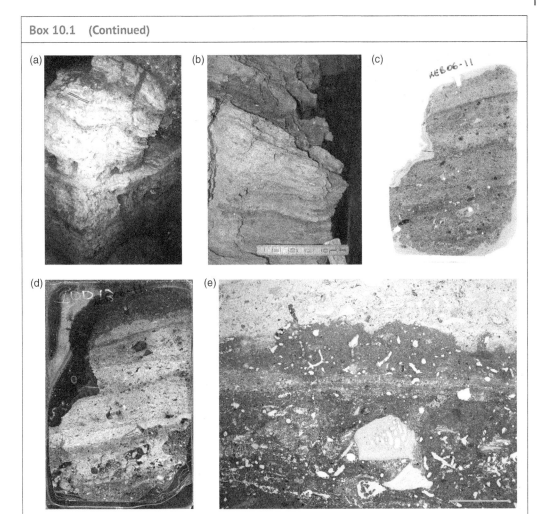

Figure 10.12 Dumped ash accumulations. (a) ~80 cm-thick accumulation of bedded ashy sediments against a bedrock wall in the northern part of the site (near Sq. I13). It provides insights into Neandertal behavior by signifying that this mass of deposit is a result of ash dumping. Scale is 20 cm long. (b) close-up of part of the ashes shown in (a) revealing fine bedding of the ashy sediment. Scale is 11 cm long. (c) thin section scan of sediment (b) above showing laminated nature of whitish ashy material alternating with reddish layers and bands rich in terra rossa clasts and bone. (d) Same thin section as in c but with the lid of the scanner left open (Goldberg and Aldeias, 2018), which serves to highlight the contrast between brighter ashy components and redder detrital terra rossa. (e) photomicrograph of upper part of thin section shown in (c) and (d). The yellow angular grains are bone fragments within porous ashes (yellow and gray); note the silty and clayey reddish lamination just above the bone, representing very low energy water flow. Scale bar is 5 mm.

(Continued)

Box 10.1   (Continued)

(a)

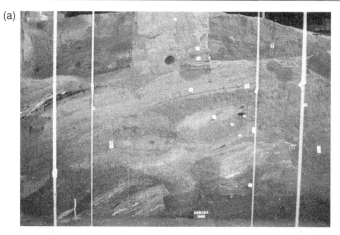

(b)

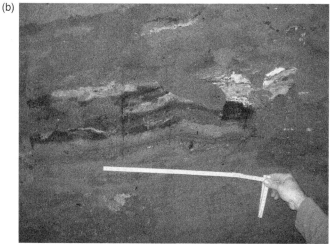

Figure 10.13   Laminated Middle and Upper Paleolithic deposits. (a) Shown here is the southern section of Kebara in 1986 (excavation squares Q15 through Q20), in which the lower half are tilted, collapsed, and burrowed hearths and other diagenetically altered sediments modified in part by the slumping of sediments into a karstic cavity below. Overlying them are well-bedded Upper Paleolithic sediments that also dip to the rear of the cave (left) deposited by sheetflow emanating from the mouth (right) toward the rear. Some burrowing is evident, as shown by the circular features in the center of the photograph. Upper Paleolithic hearths shown at the left-hand side continue on the adjacent perpendicular profile in (b). (b) Lateral continuation of burrowed Upper Paleolithic hearth complex in (a) along east face of square P15. The waterlain nature of the finely laminated underlying silts is evident below the ruler, which is 50 cm long. Diatoms were recovered from some of these laminated Upper Paleolithic deposits indicating damp conditions during this time.

Box 10.1   (Continued)

Figure 10.14   Post-depositional modifications. Remains of compressed hearths in sq. P20 are somewhat masked by the precipitation of whitish secondary phosphates, in this instance taranakite (see text). Additionally, some burrowing is evident just below the scale 'b'. Additionally, some burrowing is evident just below the scale; the large brown area behind the white string is a recent pit fill. Scale is 40 cm.

Some of the best preserved remains of Neanderthals (too numerous to name) have been found in caves. But more so, caves contribute strongly in the debate about whether Neanderthals intentionally *buried* their dead (Dibble et al., 2015; Rendu et al., 2014; Rendu et al., 2016). Interestingly, the discussion – in the past and up to the present – is generally centered on bones, objects, body position, etc.; ironically, little attention has been paid to helping resolve the issue from a geoarchaeological perspective (see Chapter 15). In the case of Roc de Marsal in France (Goldberg et al., 2017a), the geoarchaeological evidence points to *natural* formation processes involved with the initial deposition and later sediment covering of the Neanderthal infant, and not to anthropogenic ones.

Furthermore, caves more recently have served up specimens of Anatomically Modern Humans (AMH) in North Africa and the Near East (Hershkovitz et al., 2018; Richter et al., 2017). Both Final Paleolithic and Neolithic "*cemeteries*" have been found at Arene Candide, Liguria, Italy, with the latter including up to 40 burials, some in stone cists and some associated with the use of ochre (Maggi, 1997). Moreover, this cave is the archetypal example of Holocene anthropogenic deposits in the Mediterranean region (*fumiers*, in French), which display either a homogeneous or what appears to be an overall black and white layered appearance (Angelucci et al., 2009; Macphail et al., 1997). These layers are several meters thick and are best preserved from the Early Neolithic (ca. 5,500 cal BC) to the Late Neolithic/Chassey (ca. 4,000 cal BC). The homogeneous deposits represent a dominance of domestic occupation, which is reflected in the large amounts of associated eco- and artifacts. On the other hand, the layered sediments are often the *in situ* remains of burned *stabling* floors, reflecting periods when the cave was seasonally occupied by pastoralists who used tree leaf hay to fodder their animals (Courty et al., 1989, 1994); in many instances, they represent penning of sheep and goat. This model of sedimentation has been found to hold true in France, Switzerland, and across northern Italy into Spain (Binder et al., 1993; Boschian and Montagnari-Kokelji, 2000; Polo-Díaz et al., 2016), and even into Denisova in Siberia where the deposits are associated with diagenesis similar to that described above (Shunkov et al., 2018). In the rock shelters from Ramon Crater (Negev, Israel), occupations do not appear to have been seasonal but only just a few days in order to graze local pastures after spring rain (Macphail and Goldberg, 2018a: 358–362).

10.5 Environmental Reconstruction

Climate and environmental changes can be recognized in cave sediments, although it requires careful investigation to extract them from the sediments. As discussed above, the preserved coarse fraction of éboulis initially played a major role in climatic interpretation (Laville, 1964). More recently, however, the finer fraction of the sediments, as revealed by soil micromorphology in particular, can disclose some interesting paleoclimatic information. Micromorphological analysis of sediments at Hohle Fels Cave, south-western Germany, for example, showed a marked increase in *cool climate* cryoturbation during the Late Pleistocene in deposits that spanned the Gravettian through Magdalenian periods (Macphail et al., 1994a; Miller, 2015; Schiegl et al., 2003); the diagenetically altered Mousterian hearths at Grotte XVI, France have been strikingly deformed by cryoturbation (Karkanas et al., 2002), and the same can be said of hearths at Azilian Rocher de l'Impératrice, Brittany (France) (Naudinot et al., 2017). In the western Mediterranean area Courty and Vallverdú (2001) analyzed three caves and rockshelters, and asserted that warm conditions are associated with dripping water and biogenic precipitation of carbonates. Cold conditions can be associated with rapid deposition of exogenous sediments or the accumulation of debris associated with freezing activity. They noted, however, that the role of these detailed climatic inferences in human behaviors or activities remains to be demonstrated. At Roc de Marsal (SW France), large blocks appear to have been detached from the ceiling during a marked cold phase; prior to the collapse of these blocks the silty sand sediments were subjected to freeze-thaw that resulted in the formation banded fabrics (Aldeias et al., 2012; Couchoud, 2003). Concomitantly, the sediments show a distinct increase is mica-rich silt associated with inputs of aeolian dust.

In many instances, the geoarchaeological recording of climatic change in cave sediments can be eclectic. At Pech de l'Azé II, not far from Roc de Marsal, some layers visibly show the same banded fabrics as at the latter (Goldberg, 1979a). Yet, at the virtually adjacent site of Pech de l'Azé IV, deposits of comparable age lack these cold-induced features (Goldberg et al., 2018). Similarly, at Denisova cave in Siberia, microstratigraphic phenomena relating to cold in the deposits (Van Vliet-Lanoë and Fox, 2018) are muted or absent, although inputs of aeolian dust are possible though not micromorphologically distinctive (Derevianko et al., 2003; Morley et al., 2019). In many instances, more detailed and complete paleoenvironmental information is derived from other, non-geological data, such as micro- and macrofauna, pollen, and phytoliths, for example (Derevyanko, 2015; Madella et al., 2002; Rhodes et al., 2018; Weissbrod and Weinstein-Evron, 2020; Wroth et al., 2019).

10.6 Methods of Study of Cave Sediments

Laboratory techniques used in geoarchaeology are treated in some detail in Chapter 17. Here, however, we highlight some of those, which have played a role in the study of cave sediments, including those that have been applied more recently.

As in all scientific endeavors, as new instrumentation is developed and old standards are refined, our ability to evaluate the presence and type of anthropogenic sediments in caves has markedly improved recently. In the early years of geoarchaeological study, analysts relied on bulk samples (bags of dirt) collected from profiles representing lithostratigraphic units that could be identified in the field.

These bags were returned to the lab and underwent standardized sedimentological analyses such as particle size, pH (acidity), calcium carbonate and organic matter content, and possibly bulk chemistry. Thus, particle size analysis, for example, was and continues to be a basic technique used in cave sediment study (e.g. Bordes, 1972; Huckleberry and Fadem, 2007; Laville et al., 1980). But unlike sedimentology and soil science, where "natural" materials tend to follow processes and systems that have been studied and well known for decades, archaeological sediments are different. What are we actually analyzing in a grain size analysis of say, a trampled hearth, or a heterogeneous dumped deposit? Thus, although the use of bulk samples had some success in elucidating depositional and post-depositional processes that generally have been linked to paleoclimate and indirectly, chronology (Bordes, 1972), we are more aware now that archaeological sediments in general are complex and are usually comprised of a mixture of geogenic and anthropogenic components (Karkanas and Goldberg, 2018b). In this light, bulk samples commonly group multiple microstratigraphic units that can represent separate activities or sources and thus provide a "bulk result". Thus, they are hobbled to reveal many of the complex and subtle processes that occur in caves, with its typically heterogeneous deposits, often leading to bland or incorrect interpretations at best (Goldberg and Sherwood, 2006).

More recent research into cave and rockshelter deposits incorporates a microcontextual approach, which focusses on observing and noting mm- to cm-sized details of a deposit (Goldberg and Berna, 2010). It begins with careful field observations, followed by analyses that take into account such details. A much used technique nowadays that is enfolded in this approach is soil/sediment *micromorphology*, the study of *undisturbed* soils and sediments using primarily petrographic thin sections (see Chapter 17), which document composition (both mineral and organic), texture (size, sorting), and most importantly fabric (the geometric relationships among the constituents) (Courty et al., 1989; Macphail and Goldberg, 2018a; Nicosia and Stoops, 2017; Stoops, 2020; Stoops et al., 2018a). Within a single thin section it is possible to observe temporally ordered successions of microscopic features that reflect changes in depositional and post-depositional processes through time, whether produced by biological or geological agents. Thus, micromorphology can be viewed as a "methodological filter" and serves as a starting point from which further questions can be asked and analyses undertaken (Goldberg et al., 2017b).

As a consequence, micromorphology is now typically combined with *in situ* analysis of grains, features, cements, etc., on the thin sections themselves or from the cut slices from which they are made (Mentzer, 2014; Mentzer and Quade, 2013). Thus, the analytical results are contextualized with the slide and within the sample-space. Fourier Transform Infrared Spectrometry (FTIR) and micro-FTIR (Berna et al., 2012; Goldberg and Berna, 2010) can be carried out both in the laboratory (Weiner, 2010) and in the field (in real-time) (Weiner and Goldberg, 1990). Micro X-ray Florescence (XRF) and X-ray Diffraction (XRD) can also be used on embedded blocks from which the thin sections are made (see also Chapter 17 for other methods applied directly to the thin section; SEM/ EDS – Energy-Dispersive X-ray Spectrometry). The ability to map the micro-distribution of minerals in caves, revealing significant information about diagenesis, bone preservation, and evolution of sediments has significantly advanced the way in which we interpret the archaeological context.

At Elands Bay Cave (South Africa), for example, Miller et al. (2016) undertook a geoarchaeological study of the Middle Stone Age and Later Stone Age deposits using micromorphology, FTIR, x-ray diffraction (XRD), and organic petrology to evaluate the site formation processes at the site, including extensive diagenesis. Among other things, the site is particularly interesting for its wealth of anthropogenic deposits, and their preservation. They were able to document specific diagenetic minerals (mostly phosphates and silicates) that appear to map on to local zones of sediment moisture and to relate them to the absence of carbonates and bone, a situation similar to that of Kebara cited above. They also showed with micromorphology that ". . .anthropogenic depositional events were often separated by periods of more natural accumulation, possibly implying a more sporadic use of the site" (Miller et al., 2016: 124).

A technique that follows along the lines of the microcontextual approach is that of organic petrology, which uses reflected light microscopy to study coal and organic-rich components in rocks and deposits, and now "char" (Ligouis, 2017). The technique has its origins in the study of coal and other organic-rich rocks (Taylor et al., 1998), but its application to archaeological sites – both caves and open-air sites – has occurred only within the past 20 years or so (Clark and Ligouis, 2010; Goldberg et al., 2009b; Ligouis, 2017; Miller et al., 2016; Stahlschmidt et al., 2015b). Essentially, the technique examines finely polished surfaces of sliced rocks, thin sections, or blocks of impregnated sediments using reflectance and incident fluorescence microscopy. For prehistoric sites, the technique has been applied to evaluate the presence and degree of burning of both bones and organic matter. For example, Sibudu is a site noted for its abundant hearths, its microstratigraphic record and bedding that are about 60 ka old. Clark and Ligouis (2010) evaluated the effects of burning of bone by combining heating experiments with archaeological samples from Sibudu. The results showed that

> the reflectance of the bone matrix increases with increasing temperature, and that the fluorescence intensity decreases with increasing reflectance as a result of mineralogical transformations and the loss of organic matter. . .Bones which were produced at temperatures around 300°C were characterized in reflected light by heterogeneous colors with reddish internal reflections, and showed fissures, cracks, and numerous vesicles. White (calcined) bones formed at 800°C showed few fissures and cracks and were non-fluorescing (Clark and Ligouis, 2010: 2653).

These results reinforced the use of bone color as an indicator of burning, which in turn allowed for confident evaluation of bone burning (e.g. use as fuel) at the site where it appears to be related to site-maintenance activities.

More commonly, however, organic petrology, has been focused on organic matter to evaluate burning and the presence of hearths, and in the case of the open-air site of Schöningen, to clarify the environment of deposition of organic-rich deposits (Ligouis, 2017; Stahlschmidt et al., 2015a, 2015b). More specifically, it can aid not only in the identification of combustibles, but also in the evaluation of the conditions of combustion, including temperature, oxygen supply, and dryness and freshness of fuel; the last can be shown by evidence of fungal attack, for example (Ligouis, 2017).

At Sibudu, where extensive micromorphological analyses were carried out, organic constituents were studied by organic petrology, revealing the nature of fine microscopic particles not identifiable in thin section (Goldberg et al., 2009b). Organic petrography confirmed woody and herbaceous charcoal as a major organic element with identifiable tissues from leaves, stems, and roots. Remains of unburned and possibly burned plant resin were also identified.

Klasies River Mouth (South Africa) is a coastal cave noted for its thick sequence of deposits including many hearths, human remains, and use of ochre (Larbey et al., 2019). Organic petrology results showed poor plant tissue preservation due to permineralization, humification, as well as fungal attack that occurred prior to burning. In Cave 1A, various plant remains could be identified as stems, leaves, and fibrous and woody tissues. In cave 1, reflectivity measurement indicated temperatures of burning between 150–200° C, indicative of fuel that burned to completion. Fat-derived char was also observed in some cases. The identified plant remains, coupled with other analyses (micromorphology, scanning electron microscopy – SEM, and micro-FTIR) suggest the cooking of a variety of foods.

Finally, we must continually remind ourselves that the results of any of these analytical techniques are still only as reliable as the fieldwork that places the samples into the correct stratigraphic and archaeological context. The integrity of the fieldwork and precision of the field observations in the context of the depositional environment and the archaeological stratigraphy provide the framework in which to interpret the geoarchaeological laboratory data, and other artifacts.

10.7 Conclusions

Caves and rockshelters are special geoarchaeological situations, as they have the ability to accurately record "documents" of geological and human histories and activities. This statement is particularly true for prehistoric periods, such as the Lower and Middle Paleolithic, where anthropogenic deposits – and their inferred human activities and behaviors – are overall much better preserved in caves than in their open-air sites. Records of pastoralism are particularly well preserved, possibly because of their young, Holocene age. The detailed study of cave deposits by micromorphology and other, recently developed micro-techniques has revealed many new insights into interpreting the geological and archaeological records for both prehistoric and later periods. Issues of bone preservation and associated diagenesis, for example, not only have ramifications about interpreting the archaeological record and what people "appeared" to be doing at the site. They also provide necessary assistance, for example, in evaluating dates that are derived from objects and the deposits that enclose them, especially their stratigraphic integrity.

11

Human Impact: Changes to the Landscape

11.1   Introduction

The recognition of human impact on the environment is important to both investigations of modern and past landscapes. Although there are caveats, the past can be the key to understanding the present and possible future changes to the environment. This was a main motive for the meetings and volumes on "Past and Present Soil Erosion" and "Traditional Arid lands Agriculture. Understanding the Past for the Future", for example (Bell and Boardman, 1992; Ingram and Hunt, 2015). In the case of soil erosion and the formation of colluvium, few effects were considered to be of early prehistoric date by some authors (e.g. Fechner et al., 2014; Van Vliet-Lanoë et al., 1992). However, the question could be asked – what was the effect on the vegetated landscape of (a) Neolithic peoples in the Near East who manufactured burned lime to produce plaster floors (Garfinkel, 1987; Rollefson, 1990) or (b) early wet rice cultivation in China, when previously millet had been the chief crop (Gyoung-Ah Lee, 2007; Rosen et al., 2017)? Equally, Zuni arid-land agriculture employing runoff fields utilized colluvial soils (Sandor and Homburg, 2015). Mesolithic sites also sometimes show clear evidence of, albeit localized, erosion and colluviation (cf. Star Carr, North Yorkshire; Haakonshellaveien and Sotrasambandet, Bergen; Macphail, 2019c; Milner et al., 2012). We also note that not all clearances were for cultivation (see section 11.4)

For a long time, ancient edaphic – soil and vegetation – regimes were investigated only through palynology or land snails, for example (Allen, 2000, 2017; Dimbleby, 1962, 1976; Evans, 1972, 1975, 1999). These approaches attempted to distinguish between human impact and natural variations in climate, sea level, and tectonics, for instance. As soil science, especially soil micromorphology, was mainly employed in the study of paleosols and only a few archaeological sites had come under scrutiny, human impact was little considered. The first major suggestions concerning soil evidence for clearance by fire and the formation of early agricultural soils came from Romans and Robertson (Romans and Robertson, 1975a, 1983a, b, c). Such studies were followed up by experiments in *use of fire* and *ancient cultivation*, which were then discussed in the light of a number archaeological type-sites, such as camp site deposits and impact on soils by different cultivation implements and regimes (Courty, 1984; Courty and Fedoroff, 1982; Gebhardt, 1992; Macphail et al., 1990; Macphail and Goldberg, 1990). At the same time, the formation of occupation floors and possible use of space were being presented in the literature (Cammas et al., 1996b; Gé et al., 1993). These findings were not universally accepted (Canti et al., 2006; Carter and Davidson, 1998; Macphail, 1998; Macphail et al., 2006). Since then numerous other workers have extended these studies, validating these earlier suggestions (Deák et al., 2017; Mallol et al., 2017; Rentzel et al., 2017; Shahack-Gross, 2017). In this

Practical and Theoretical Geoarchaeology, Second Edition. Paul Goldberg and Richard I. Macphail.
© 2022 John Wiley & Sons Ltd. Published 2022 by John Wiley & Sons Ltd.

section, we explore some of the ways that past human activities can be detected, using soils and geomorphological records.

Past impacts can be assessed in a number of ways. These can range from (a) the slightest indication of an early human presence (e.g. forest clearance and associated minor soil disruption) to (b) major effects, such as landscape and soil modifications. The latter could include the formation of over-thickened cultosols such as plaggen soils, infilling of dry valleys with colluvium (see Chapter 5), and the construction of terraced and/or paddy field landscapes. As noted in Chapter 4, pedogenic processes are governed by the five soil-forming factors, and the factor *organisms* can sometimes be overly influenced by humans. Humans can directly affect the plant cover, through forest clearance, maintenance of grazing by fire and animal husbandry, and by the growing of "crops", which can include woodland exploitation through to arable farming (Deák et al., 2017). This chapter describes human impact from the least obvious effects on the soil, starting with forest and scrub management and *clearance* activities and how this can effect soil fertility, both positively and negatively. Such geoarchaeological studies have often been complemented by a suite of supportive paleoenvironmental techniques – macrofossils (e.g. cereal grains and seeds of weeds indicating manuring and field management) and microfossils (e.g. cereal pollen, and rice and cereal phytoliths) (see below for details).

11.2 Forest and Woodland Clearance

The terms "forest" and "woodland" are used differently around the globe. For example, forest is a legal term in England for land that was set aside for hunting that is not necessarily totally wooded, such as the Medieval "New Forest" (Hampshire/Dorset). The terms "virgin forest", "old growth woodland", and "primary woodland" are discussed in Peterken (1996: 16 *et seq.*). For our purposes, human populations first cleared "virgin forests" and "primary woodland", but there has also been clearance of secondary woodland and scrub. In Europe especially, woodlands have been managed in various ways (coppicing for charcoal, tree-shredding for leaf hay fodder), all of which may have impacted on soils.

The major implied effect of humans on the edaphic environment in British prehistory was recognized by G. Dimbleby, a pioneer in *environmental archaeology* working from the 1950s (Dimbleby, 1962). Through pollen analysis of barrow-buried soils, he showed that woodland had been replaced by heath plants, which have shallow rooting systems compared to broadleaved trees. Thus, heath plants do not aid in soil nutrient recycling; moreover, they produce chemicals (e.g. tannins) that slow down organic matter breakdown and mobilize sesquioxides (see Chapter 4, Figure 4.9). A French contemporary of Dimbleby, P. Duchaufour, categorized human effects in both soil degradation and amelioration (Duchaufour, 1982: 137–141). Using present-day edaphic studies, buried soils, pollen, and ^{14}C dating, these workers were able to recognize one of the first links between woodland clearance activity, the maintenance of open areas by fire and grazing and the development of leached soils (e.g. albic horizons, Spodosols) formed on poor geological substrates. Duchaufour also identified the connection between montane terracing, and the expansion of alpine grass (hay) meadows at the expense of conifer (e.g. larch) forest, and the replacement of *podzols* by brown soils with a mull horizon, a form of soil amelioration (Duchaufour, 1982: 140; Courty et al., 1989: 315). Their pioneering insights are an encouragement to present-day workers who are discovering anomalous soil characteristics that are difficult to explain simply through natural pedogenic processes and which may well be human induced. It can be added that management of vegetation by fire was not only a Mesolithic practice (e.g. Lewis et al., 1992; Lewis and

Rackham, 2011), but also one employed for millennia to the present-day in Australia in order to maximize potential animal and plant food resources (Cool burns: Key to Aboriginal fire management – Creative Spirits).

In the case of buried soils, the most important first step is to be able to differentiate ancient features from those caused by the process of burial (as studied by experiments – Chapter 14) and by other post-depositional effects (e.g. bioturbation; Johnson, 2002). One way to achieve this is through soil micromorphology, by observing the "hierarchy of the features", i.e., identifying the temporal ordering of soil formation and post-depositional pedofeatures according to their geometrical arrangement in the thin section (Courty et al., 1989: chapter 8) (see Chapter 17). For example, sesquioxidic coatings in the lower and least leached part of the spodic (Bh/Bs) horizon can be found superimposed on clay coatings. This may provide a proxy history of increasing soil acidity, with primary clay translocation (under probable broadleaved woodland) being followed by podzolization that perhaps developed under an ensuing vegetation dominated by ericaceous species. At some sites, podzolization followed clearance, and was maintained by grazing and burning (Dimbleby, 1962, 1985). This occurred on poor sandy substrates across Europe and dates mainly to the Bronze Age in the UK and Denmark, for example (Mikkelsen et al., 2006). However, lowland and upland podzolization can be traced to both Mesolithic (e.g. Star Carr – see below) and Neolithic sites, and through to Iron Age and medieval periods; in the last case it is usually due to large-scale monastic sheep grazing (Romans and Robertson, 1975b; Simmons, 1975) (see Box 4.2). The onset of podzolization was therefore asynchronous across Europe because of these human impacts, and thus the presence of podzols should never be employed as a single "marker horizon" or be used to provide some kind of proxy dating – as was attempted in the past (Macphail, 1986; Macphail et al., 1987). It is also proven that European Holocene podzols can equally form *naturally* under broadleaved woodland by late prehistoric times (e.g. Bronze Age/Iron Age) (Dimbleby and Gill, 1955; Mackney, 1961; Macphail, 1992b).

Throughout the present-day boreal region of North America, northern Asia, and Scandinavia, podzols occur naturally under conifers. In north Sweden these podzols can be anomalously shallow, being only 30 cm thick at some Late Iron Age (here dating into the first millennium AD) and Viking sites. These "post-glacial" soils have not experienced any kind of erosion, however, but have undergone only 2,000 years of development (instead of 10,000) because they were only recently exposed by *isostatic uplift* (Linderholm, 2003). Here then, is an example of an unexpected environmental situation specific to a particular regional and cultural setting that would need to be understood by everybody if they were to interpret correctly human activities for a given locality.

At Star Carr (North Yorkshire, UK), leached sands interdigitate with lake edge sediments (Milner et al., 2012). However, at lesser well-known Mesolithic sites on upland Exmoor (Farley Water), and in lowland Leicestershire (Asfordby) and Kent (Dartford River Valley), stone tools occur within colluvia likely dating to occupations that disturbed the local woodland cover and Holocene soils (Carey et al., 2020; Macphail and Goldberg, 2018a; Simmonds et al., 2011). Woodland clearance has also been inferred from wood charcoal in buried topsoil horizons, and possible "slash and burn" agriculture has been suggested in Scotland from such findings (Romans and Robertson, 1975a). This has seemingly been replicated by Swedish experimental "slash and burn" cultivated podzols where again coarse wood charcoal was concentrated in the uppermost few cm of the soil (Macphail, 1998). Similarly, the Forchtenberg-Experiment, which included clearances and ancient tillage regimes, was carried out on luvisols on the Hohenlohe Plain, southwest Germany (Schulz et al., 2011; see also reviews in Deák et al., 2017).

Where *treethrow* subsoil holes contain charcoal of mainly one tree species, it has been suggested that the fallen tree that formed this feature, was burned *in situ* – especially when burned soil is also present (Barclay et al., 2003; Campbell and Robinson, 2007; Macphail, 2011a; Macphail and Goldberg, 1990; Mark Robinson, pers. comm.). Such findings are not unequivocal evidence of purposeful clearance, however, because natural blowdowns occur during storms, and such opening up of the woodland landscape may have been taken advantage of by human groups. The presence of artifacts in treethrow holes is also not conclusive evidence of large-scale forest clearance, because again the use of such hollows could be opportunistic, and artifacts in the soil could also have fallen into the hollow as it became infilled by extant soils (Crombé, 1993; Crombé et al., 2015; Newell, 1980); they could have also provided shelter.

Treethrow is a common natural occurrence (Stephens, 1956), and treethrow formations can give clues to the nature of past woodland (Langohr, 1993) (see Figures 11.1–11.7). It is commonly thought that one method of forest clearance took the form of simply burning down trees. Deciduous trees – unlike conifers – do not readily burn, however, and other methods of tree removal must be envisaged. Recent studies have suggested ring-barking as a technique (where the bark is removed in a ring from around the base of the tree, for example with a stone axe; Figure 11.7); once dead, the tree is pulled down (Barclay et al., 2003; Romans, pers. comm.). In addition, clear soil micromorphological and bulk data show that trees, once pulled down (or blown over), were burned *in situ*; Stewart (1995: 37–38) has illustrated methods of using controlled burning for cutting down, and cutting through, cedar in NW America. The reader is referred to the many classic and modern studies on the character of ancient forests and woodland (Peterken, 1996; Tansley, 1939) and the effects of human impact upon them (e.g. Europe; Berglund, 1986; Huntley, 1990) and the Near East (Zeist and Bottema, 1982). Soil evidence of forest clearance and woodland management is presented below. More recently, both human and climatic impacts on woodland have been assessed through tree ring studies (Büntgen et al., 2011), and it is worth noting that increasingly nuanced studies of land snails (mollusks) have shown that some chalkland areas were never fully wooded and did not need to be cleared for pastoralism, for example (Allen, 2017).

Clearance effects have recently been reviewed and include both experimental and archaeological examples, including tumulus-buried soil evidence of Neolithic clearance by fire at Er Grah (Deák et al., 2017). This excellent example of early clearance and woodland management was identified for Megalithic Brittany, France (Gebhardt, 1993; Gebhardt and Marguerie, 2006). The authors were able to integrate soil and micropedological findings with pollen and charcoal data to identify clearance, management by fire, and leaching of soils at classic Neolithic buried soil sites such as megalithic Er Grah. Even on the acid granitic substrates of Brittany, there was an instance of calcareous ashes being preserved by rapid burial through monument construction.

Similarly, the first effects of podzolization could tentatively be dated to Neolithic occupation of the same granite in Cornwall (south-west England) at the outer rampart of the Carn Brea "fort" (Macphail, 1990a; Mercer, 1981). Here, a highly leached topsoil (Ah and A2 horizon) had apparently developed during only 1–2 centuries of local land management (see Figure 4.6a) that probably resembled that recorded by Gebhardt (1993). Only a weakly formed spodic horizon had begun to form at Carn Brea, however, and was superimposed on the acidic cambic Bw horizon that exhibited earthworm-worked soil. In order to construct the rampart, this soil was probably cleared of shrubs, and burning had led to the present bAh/A2 (Albic) horizon becoming rich in fine charcoal and phytoliths. Trampling, probably by people as opposed to animals, had compacted the surface soil and formed dusty clay coatings in closed vughs, seemingly immediately ahead of rampart construction (Rentzel et al., 2017; see also Figures 11.4a and 11.4b; Macphail, 1990b). On the same Cornish granite, however, at the Bronze Age kerb cairn-buried site at Chysauster, the contemporary Cambisol had formed under open oak/hazel woodland with local (possibly on-site) cereal cultivation. Whereas the Carn Brea soil had been completely sealed, at Chysauster the "open" ca. 0.50 m thick kerb cairn had allowed post-Bronze Age podzolization to alter much of the buried soil even under a large kerb stone (Smith et al., 1996). Similar studies have been undertaken in Scotland (Mercer and Midgley, 1997); at Hørdalen (Norway), clearance of conifer vegetated podzols led to a concentration of burned stones and charred root remains at the base of a cultural soil, which formed subsequently; fire effects were also consistent with an enhanced magnetic susceptibility (Viklund et al., 2013) (see below). The danger of mistaking post-burial podzolization for ancient podzolization has been recognized for a long time, and the continuation of the chemical effects of podzolization was demonstrated at the 33-year-old Wareham Experimental Earthwork (see Chapter 14) (Macphail et al., 2003b; see also Runia, 1988).

Clearance and its effect on other soil types and environments have been less well studied, but the likely clues to clearance activity are the same (see Arroyo-Kalin, 2012 and below). In areas of tropical Oxisols, the rainforest produces a highly nutrient-rich (N-P-bases) humus, which can be utilized for a few years for shifting agriculture by slash and burn, a practice which can eventually lead to secondary savannah and the loss of this fertile humus layer (Duchaufour, 1982: 412–413). Pre-Columbian occupations enriched and transformed soils producing anomalous dark "garden soils" once termed *terra preta*, but now more commonly called *Amazonian dark earths* (ADEs) (Arroyo-Kalin, 2017; Arroyo-Kalin et al., 2008; Glaser and Woods, 2004).

11.2.1 Forest and Woodland Clearance Features

Soil indicators of clearance of trees and scrub woodland and growth of shallower rooting shrubs, have been found at a number of archaeological sites. This evidence included the identification of disrupted soils with possible inclusion of charcoal, phytoliths, and burned soil (Figures 11.1–11.7). These findings have been employed to create some general models of vegetation clearance that can be utilized during soil investigations (Courty et al., 1994: table 4; Deák et al., 2017; Macphail, 1992a: table 18.2; Macphail and Goldberg, 2018a: table 9.1; Macphail et al., 1990: table 1).

Figure 11.1 Tree-throw: modern example from the "hurricane" of 1987 that produced "blowdowns" of mature beech trees that were shallow rooted in thin decalcified drift (Clay-with-Flints) over chalk (Chiltern Hills, Hertfordshire, UK).

Figure 11.2   Tree-throw subsoil hollow (section and part plan) formed by a "blowdown" during the Atlantic Period, ca. 5,000 years ago as indicated by mollusks in the subsoil fill; Balksbury Camp, Hampshire, UK (Allen, 1995).

Figure 11.3a   A section through Hazleton long cairn and its buried soil, showing the Neolithic old ground surface (A), the remains of an earlier ("Atlantic Period" – mollusks) tree-throw subsoil hollow (B), up-throw of marl (C) and deeper rock (D) from the Jurassic Oolitic limestone geology (Saville, 1990).

These models were also influenced by classic studies of American woodland soils and geology (Denny and Goodlett, 1956; Lutz and Griswold, 1939). In brief, shallower (0–20 cm) soil disturbance is caused if small, shallow-rooting plants are pulled or grubbed up, in contrast to deeper rooting trees and scrub (20/40–100 cm); some modern pit-and-mound (or hummock and hollow; Figure 11.1) topography reaches up to 4 m high (Peterken, 1996). In the case of shallower rooting shrubs, only A and A2 horizon material is likely to become mixed, whereas if deeper rooting trees are cleared or pulled over, subsoil horizon fragments (e.g. Bt, Bw, C) become mixed with topsoil material (see Box 4.1). Often this disruption is accompanied by loose soil being washed into fissures within the disturbed soil hollow, and orientation of the resulting textural pedofeatures is "right way

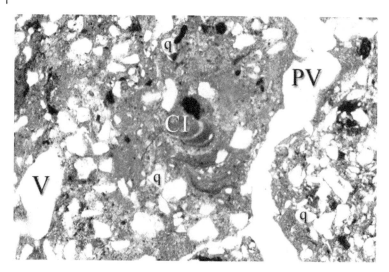

Figure 11.3b Photomicrograph of the "Atlantic Period" subsoil at Hazleton (Figure 11.3a Area B), showing: quartz sand (q) in the clay loam soil; an open vugh (V); crescent infills (CI) of dusty clay; and dusty clay coated planar void (PV). These infills and coatings resulted from tree-throw soil disruption. Plane polarized light (PPL), frame is ~1.35 mm. (Note: this is technically a Bt horizon formed under woodland, but here most of the textural pedofeatures – clay coatings – were formed by tree-throw not *lessivage*.)

Figure 11.4a Field photo of charcoal-rich Neolithic old land surface (OLS) termed the "midden area" at Hazleton (Figure 11.3a). The Neolithic occupation soil contained flint artifacts, pottery, domestic animal bones, and charred wheat and hazel nuts.

up" (Figures 11.3a and 11.3b). In contrast, fragments of relict mixed-in Bt horizon material may show intra-ped textural features that are not "right way up" (see Deák et al., 2017: Figure 28.19; see Figures 4.3c and 4.3d). It can be suggested that secondary clearances will produce mainly only small charcoal, such as when brash is burned. Along the Neolithic Stanwell Cursus (near Terminal 5, Heathrow Airport, UK) the side ditches were characterized by fine charcoal and burned sands embedded in clayey infills that also showed rubification (cf. Figure 11.6b), the disturbed clearance

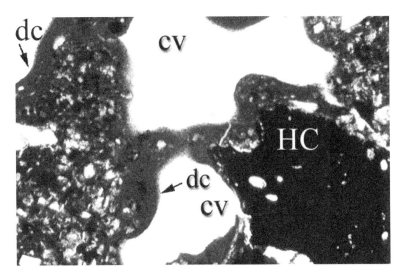

Figure 11.4b   Photomicrograph of Neolithic old land surface (Figure 9.4 OLS) at the "midden area" at Hazleton, showing: very dusty soil containing microscopic and coarse (HC – hazelnut shell) charcoal, the occupation (cultivation, trampled) disturbed soil characterized by closed vughs (CV) formed by very dusty clay (DC) void coatings (Macphail, 1990b). PPL, frame is 1.35 mm.

Figure 11.5   Prehistoric (alluvium-buried) paleo-landscape at Raunds, Nene valley, Northamptonshire, UK; a series of semi-inter-cutting tree-throw holes are outlined by curved "humus" infilled boundaries (see Figure 11.7). Most tree holes date from the Late Mesolithic/Early Neolithic to the Beaker/Early Bronze Age Period (Harding and Healy, 2011; Healy and Harding, 2007).

soil being preferentially slaked because of the release of K (potassium) from ashes (Macphail and Crowther, 2010; see also Courty and Fedoroff, 1982; Mallol et al., 2017); it can also be noted that pollen data from the site indicated only secondary woodland was present (Lewis et al., 2010). Lastly, the long cairn at Hazleton buried a whole series of different soil surfaces, including trampled turf soils, another with possible evidence of shrub clearance and also a charcoal-rich midden area where burned inclusions had no clearance origins but recorded a probable occupation surface (Figures 11.4a and 11.4b).

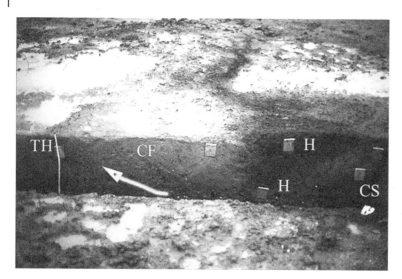

Figure 11.6a   Raunds: section through 2.5–3.0 m wide tree-throw hole (see Figure 9.7) showing the wide humus (H) infill on the right, and thin humus (TH) on the left. The humus is now almost totally replaced by iron and manganese staining. The central fill (CF) is less humic and is characterized by dusty clay void infills. The control soil (CS) profile (i.e., the local natural *argillic* Bt subsoil soil) outside the tree-hole is both poorly humic and contains few textural pedofeatures compared to the tree-hole (Macphail, 2011a).

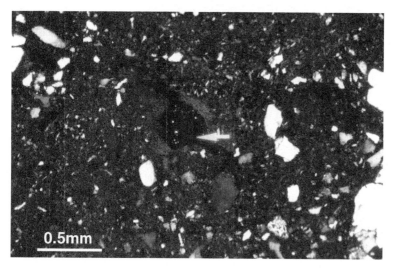

Figure 11.6b   Photomicrograph of burned soil in tree-throw pit, where the soil has a very high magnetic susceptibility due to this burning (e.g. $\chi = 894 \times 10^{-8}$ m^3 kg^{-1}), as identified in post clearance soil in Figure 11.7 (below). Soil, which was clearance disturbed, produced dusty void coatings, before being baked/rubified; note thin iron-;manganese void hypocoating to this void, which is a record of post-depositional waterlogging associated with rising water tables and alluviation. Crossed polarized light (XPL), scale bar = 0.5mm (photo by Rob Kemp at Royal Holloway, University of London).

The study of clearance features has also benefitted background investigations of recently (World War II) cleared soil in the Italian Apennines, and numerous ancient examples of treethrow holes (Courty et al., 1989: 286–290). Furthermore, natural treethrow holes dating to the Atlantic period in England (Balksbury Camp and Hazleton long barrow; Figures 11.2 and 11.3) have been differentiated

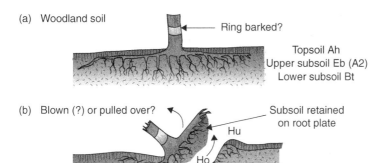

(a) Woodland soil

— Ring barked?

Topsoil Ah
Upper subsoil Eb (A2)
Lower subsoil Bt

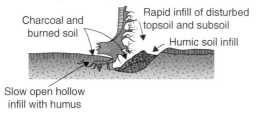

(b) Blown (?) or pulled over?

Subsoil retained
on root plate

Hu

Ho
Broken roots

Hummock (Hu) and Hollow (Ho) formation

(c) Tree trunk cut by fire or simply burned?

Charcoal and
burned soil

Rapid infill of disturbed
topsoil and subsoil

Humic soil infill

Slow open hollow
infill with humus

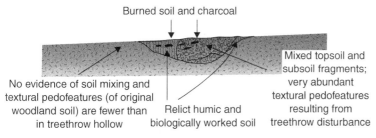

(d) Buried old landsurface (here truncated)

Burned soil and charcoal

Mixed topsoil and
subsoil fragments;
very abundant
textural pedofeatures
resulting from
treethrow disturbance

No evidence of soil mixing and
textural pedofeatures (of original
woodland soil) are fewer than
in treethrow hollow

Relict humic and
biologically worked soil

Figure 11.7 Model of tree-throw field soil features; (a) broadleaved woodland soil where clay is translocated into the lower subsoil Bt horizon by *lessivage* (see Figure 4.2a, 4.2c, 4.3a) – trees can be killed by ring-barking and are thus more easily pulled down and burned; (b) clearance may have taken place after trees were blown down or pulled down after ring-barking, and produced hummock (Hu) and hollow (Ho) landscapes; (c) *in situ* burning of the tree produced charcoal and burned soil (with an enhanced magnetic susceptibility), and while mainly subsoil fell off the root plate, humic topsoil infilled beneath the tree trunk and around the edges of the treethrow pit; (d) treethrow disturbance produced a mixed soil infill and very abundant textural pedofeatures compared to undisturbed soils between treethrow features (after Barclay et al., 2003; Courty et al., 1989: 127; Langohr, 1993; Macphail, 2011a; Macphail and Goldberg, 1990).

from those that are associated with human activity (e.g. Rounds; Figures 11.5 and 11.6). Pedological studies as well as personal observations (RIM) of pit-and-mound ground formed by past treethrow – including that caused by the hurricane that hit southern England in 1987 – show that treethrow holes receive (and sometimes hold) more water and humus than the surrounding soil (Peterken, 1996; Veneman et al., 1984) (Figures 11.1, 11.5–11.6a). The remains of an Atlantic woodland molluskan fauna were recovered from within the subsoil hollow at Hazelton (Figure 11.3), despite decalcification of the (later) Neolithic ("top") soil, and at Balksbury Camp (Figure 11.2) carnivorous mollusks were included in the species list (Allen, 1995; Saville, 1990; see also Allen, 2000, 2017).

That subsoil hollows formed by trees become infilled with organic-rich soil has long been recognized in the archaeological record (Limbrey, 1975). In plan (after the modern topsoil and over-burden has been removed), these features (3–5 m across) have a banana-shaped dark infill on one side, which is due to the concentration of relict organic matter here. In some cases, the latter is strongly replaced by iron and manganese mottling where water tables have risen (as in the flood-plains of the Upper Thames and Nene, UK, for example). However, measurements of organic car-bon and LOI can still be twice as much as those found in the rest of the soil fill or in the soil profile outside the hollow (see Table 16.4). At one site at Raunds, Northamptonshire, (UK), where a fallen tree was burned *in situ*, *magnetic susceptibility enhancement* was very high: with values (χ 10^{-8} m^3 kg^{-1}) of 894 units (concentrated rubified soil – burned "clay"; Figure 11.6b) and 96 units (fragmented rubified soil) (Campbell and Robinson, 2007; Macphail, 2011a; Macphail and Goldberg, 1990). These high values of χ can be compared to other areas of the fill (8–10 units), including the dark iron and manganese mottled soil fill (22–44 units) and the overlying alluvium (11–26 units).

As discussed in Chapter 4, activities of vegetation clearance on slopes can lead to erosion and colluviation. This was documented at Brean Down, (Somerset, UK; Bell, 1990) (see Box 4.1) and in the Italian Apennines, where it is possible that mineralogenic infilling of low ground contributed to peat initiation (Cruise, 1990). A major impact on woodland and the landscape in general is contemporary with colluviation that is dated to the Chalcolithic in Italy and the Beaker period in southern England (Macphail, 1992a) (see below).

11.3 Agricultural Practices

11.3.1 Cultivation

Sites of ancient cultivation have been identified worldwide, from the Americas (Jacob, 1995; Minnis and Whalen, 2015; Sandor, 1992; Sandor and Homburg, 2015; Sandor et al., 2021) and Europe (Deák et al., 2017; Gebhardt, 1990, 1992, 1995; Lewis, 2012) to Asia (Lee et al., 2014; Matsui et al., 1996; Rosen et al., 2017; Sunaga et al., 2003; Zhuang et al., 2013). The most obvious signs of cultivation are field features, such as *ard marks*, plough marks, and spade marks (Figures 11.10a and 11.10b; Box 11.1). These, however, must all be treated with caution, and not confused with drag lines, bases of wheel ruts, linear periglacial features, mole drains and other modern features (Gebhardt and Langohr, 2015). Other larger field features (e.g. lynchets; cultivation ridges and terraces) can also often be a clear means to identify *in situ* areas of cultivation (Barker, 1985; Evans, 1999; Fowler and Evans, 1967; Minnis and Whalen, 2015; Sandor, 1992; Sandor and Homburg, 2015). The presence of cereal pollen, phytoliths, charred cereal grains, and soil micromorphological features, which may be indicative of cultivation, also need to be interpreted carefully. For example, cereal pollen is often strongly associated with areas of crop processing as well as in cultivated soils (Bakels, 1988; Behre, 1981; Macphail and Linderholm, 2017; Segerström, 1991) and cereal phytoliths may occur through manuring (Devos et al., 2009, 2020a; Vrydaghs and Devos, 2018); soils can also include seeds of nitrogenous weeds that have been encouraged by *manuring* (Viklund et al., 2013). Equally, soil micromorphology data must be treated with caution, as a number of processes can produce features suggestive of cultivation, following the law of equifinality (see Chapter 16.6). It is therefore crucial that all human and environmental factors be taken into consideration when employing soil evidence to locate cultivation (Adderley et al., 2006; Arroyo-Kalin, 2017; Carter and Davidson, 1998; Courty et al., 1989; Deák et al., 2017; Macphail, 1998; Viklund et al., 2013). Some generalizations can be listed as follows:

Table 11.1   Some generalizations on global cultivation arranged by period and location.

Period	Location	Activities	Data support	References
Medieval, post-medieval	Northern Europe	Mixed farming, manured cultivation, plaggens	Nitrogenous weeds, exotic inclusions, anomalous LOI, P, PQ, and MS. Cereal pollen. phytoliths Tables 11.2–11.3; see Figure 4.8	Adderley et al., 2018; Blume and Leinweber, 2004; Hicks and Houlistan, 2018; Viklund et al., 2013
		Use of heavy wheeled plough		Henning, 2009)
Roman	Europe	Urban horticulture Low level arable manuring, iron ploughs	As above; moldboard ploughmarks (Box 11.1)	Deák et al., 2017; Gebhardt and Langohr, 2015
Prehistory	Europe	Ard ploughing	Ard marks	Fowler and Evans, 1967; Gebhardt, 1992; Lewis, 2012
Later prehistory to recent	Polynesia, Australia etc.	Use of digging stick		
Later prehistory to recent	Asia	Dry millet cultivation Wet rice cultivation	Millet seeds Charcoal Bunds and other landscaping; rice phytoliths and seeds; Figure 11.11a–11.11b	Lee et al., 2002–2004, 2014; Rosen et al., 2017; Zhuang, 2018
Prehistory	Near East/ Mesopotamia	Manuring around settlements		Wilkinson, 1990, 1997
e.g. Mayan	Mesoamerica	Maize		Minnis and Whalen, 2015
Paleo-native american	South-west USA	Irrigated fields; dry mulches, runoff fields	Figure 11.12a–11.12b	Sandor and Homburg, 2015

The only ways to tackle such complicated issues is to utilize well-studied cultivation sites as archaeological analogues of various ages and character, and to carry out and learn from experiments (Deák et al., 2017) (see Chapter 14). Data from slash-and-burn, non-manured tilled soils, and soils which have been strongly amended have been investigated for their field, soil micromorphological, and geochemical characteristics.

11.3.2   Manuring

There have been various suggestions for early manuring, including isotope studies on cereals that indicate early manuring, as well as other plant remains and pollen; e.g. nitrogenous seeds (Bakels, 1997; Bogaard et al., 2013). The geoarchaeological case for what happens to cultivated soils, in terms of soil stability, without "evident" manuring has also been discussed (Macphail and Goldberg, 2018a: 303–316). Examples of archaeological analogue studies are given in Box 11.1 for the site of Oakley (Suffolk, UK) (Macphail et al., 2014) and Medieval Malham, Yorkshire

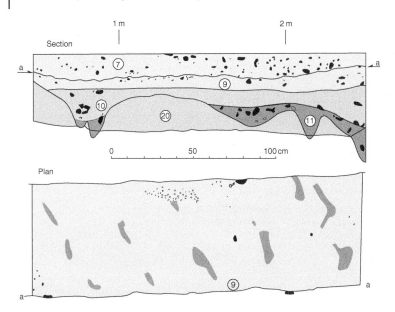

Figure 11.8a Beaker cultivation at Ashcombe Bottom: plan and section of possible ard marks at the Ashcombe Bottom site, Lewes, East Sussex, redrawn from Allen, 1994: 80; a machine trench cut across the dry valley (Sussex Chalk Downs); layer numbers are as follows: 20 – valley bottom periglacial solifluction deposits; 11 – truncated early Holocene soil (argillic brown earth; Allen, 2005; Macphail, 1992a); 10 – Late Neolithic/pre-Beaker colluvium; 9 – putative prehistoric old land surface formed of colluvium containing abraded Beaker pottery (consistent with ploughing) and featuring at its surface some 14 parallel grooves – possible ard marks; 7 – later prehistoric colluvium (see Chapter 4).

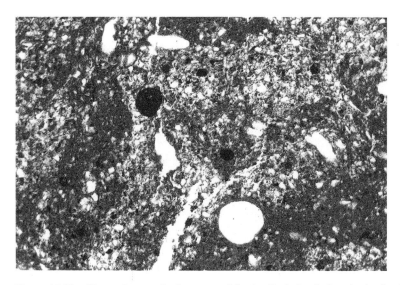

Figure 11.8b Photomicrograph of suggested Beaker Period ard-ploughed colluvial soils at Ashcombe Bottom (Layer 9, Figure 11.10a); within silt loam colluvium are gravel-size soil clasts, which are fragments of tillage crusts, with slaking separating loessic silts and clayey fine soil; note closed vughs and vesicles. PPL, ~3.3mm.

(Figures 11.9a–11.9e). Here, Late Roman/Saxon cultivation – first indicated by field features (moldboard ploughmarks) at field edge and within field locations – was examined using chemistry and micromorphology. An important issue here are the effects of cultivation and manuring on Spodosols; this practice in The Netherlands eventually led to the formation of a *plaggen* (a very thick cultosol; Table 4.1; Figure 4.8) during medieval times and earlier times (pp 316–322 in Blume and Leinweber, 2004; Macphail and Goldberg, 2018a; van de Westeringh, 1988). Notable at Oakley is the development of a fine tilth composed of phytoliths, "clay", and fine organic matter that forms a loose fill between sand grains, and grain and voids coatings (Figure 11.9a–11.9e). Current amounts of phosphate in the Ap horizons are not high, but around three times that in the bA2 horizon. Moreover, the presence of earthworm burrows (and a biogenic earthworm granule formed out of calcite) is totally anomalous in Spodosols, as is the presence of "exotic" inclusions (Adderley et al., 2018) including amorphous nodules of presumed "nightsoil" (cf. Karkanas and Goldberg, 2018; Nicosia et al., 2017). Phosphatic "nightsoil" nodules have a characteristic appearance, and sometimes exhibit vivianite (crystalline iron-phosphate) within an amorphous (Fe-Ca-P-Mn) matrix (Figures 11.9b–11.9e). In addition, arable fields ("infields") on the outskirts of Roman Canterbury (Kent, UK) (now sealed by the late 3rd century AD earth-based town ramparts) also received small amounts of town waste that included "nightsoil", bone, and other coarse anthropogenic ("exotic"/allocthonous) inclusions. The inclusions are comprised of gravel-size flint, burned flint, wood charcoal, an example of articulated phytoliths (cereal waste?), and fragments of amorphous organic matter. Here, in contrast to the naturally more fertile Luvisols, ploughsoils were found to have 1.4 to 1.9 times as much P as the underlying natural subsoil (Bt horizon formed in fine loam), and also show rather higher values for magnetic susceptibility and heavy metal concentrations (Cu, Pb, Zn) (Macphail and Crowther, 2002a) (Table 11.2; CW12). Dark Age to medieval cultivated and manured colluvial loessic soils at the Büraberg (North Hessen, Germany), were found to contain 2.5 times as much phosphate and twice the magnetic susceptibility signal (χ_{max}; Chapter 16.4) as the substrate (Henning and Macphail, 2004). Similarly, Sunaga et al. (2003) found at two sites in central Japan that volcanic mudflow-buried Edo Period cultivated soils had a higher available phosphate content and bulk density compared to buried uncultivated soils; the latter in contrast were characterized by more total carbon, total nitrogen and "easily decomposable organic matter". Additionaly in the Medieval Friary horticultural soils at Whitefriars Canterbury, manuring levels are many times richer in organic matter, phosphate, and general settlement waste inputs, as recorded by magnetic susceptibility and heavy metals Pb, Zn, and Cu (Table 11.2; CW64). Settlement waste is also conveniently geochemically documented for the Late Saxon town in road deposits from two streets (Table 11.2; CW21).

There are rare examples of plaggen-like soil profiles where the basal layer records clearance that involved burning (charred woody roots, burned rock and enhanced magnetic susceptibility) and overlying "plaggen" soil accumulation through manuring with both dung (raised LOI, P, and organic P – PQuota) and settlement waste (burned clay, sands, and enhanced magnetic susceptibility). Figures 11.10a–11.10e illustrate this from the Iron Age to Migration Period settlement at Hørdalsåsen, Sandefjord Kommune, Vestfold, Norway (Viklund et al., 2013).

11.3.3 Paddy Fields

Deliberate modification of wild habitats to grow rice began very early in the Far East (China, Japan and Korea) (see Figure 7.12). One of the earliest examples of rice-growing sites is found at the Neolithic site of Huizui, Henan Province (Rosen et al., 2017) (see also https://www.globaltimes.cn/page/202012/1210011.shtml). The early modified rice fields were of diverse morphologies

Table 11.2 Whitefriars, Canterbury, Kent, UK; chemical (excluding phosphate fractionation) and magnetic susceptibility data and comparing low level manuring of Roman town infields (CW12), Late Saxon town road fills and Medieval heavily manured horticultural soils at the Whitefriars, Monastery (data supplied by John Crowther; Hicks, 2015; Hicks and Houlistan, 2018; Macphail and Crowther, 2007a).

Bulk sample	Context[a]	Description[a]	LOI[b] (%)	CO_3^{2-}[c] (%)	pH (1:2.5 water)	Phosphate-P[d] (mg/g)	χ (10^{-8} m³)	χ_{max} (10^{-8} m³)	χ_{conv}[e] (10^{-8} m³)	Pb[f] (μg/g)	Zn[f] (μg/g)	Cu[f] (μg/g)
CW12		**Roman manured soils**										
BS1	?	Roman rampart	1.34	0.224	7.9	1.02	26.1	1200	2.18	6.9	10.4	5.2
BS2	1212	Buried topsoil	1.68	0.629	7.8	1.96	38.2	795	4.81	11.5	18.2	10.5
BS3	1212	Buried topsoil	1.67	0.776	7.8	1.59	52.1	729	7.15*	9.5	17.7	9.2
BS4	1212	Buried topsoil	1.71	0.738	7.8	1.38	34.3	722	4.75	9.4	17.2	8.6
BS5	?	Natural subsoil	1.27	0.738	7.9	0.962	20.0	795	2.52	7.5	9.5	4.5
BS6	1212	Buried topsoil	1.76	0.751	8.0	1.27	47.0	801	5.87*	9.8	16.1	8.4
CW21		**Saxon road fills**										
BS7	4207	Saxon road silts	4.21*	4.22	8.7	8.64**	228	1110	20.5***	99.6*	71.2*	28.2
BS8	4207	Saxon road silts	4.93*	4.61	8.5	8.43***	258	838	30.8****	84.8*	92.2*	30.5
BS9	5135	Saxon dark earth	2.95*	0.723	8.3	3.85*	86.3	1030	8.38*	36.0	56.9*	26.1
BS10	5235	Upper brickearth	2.15	1.01	8.1	2.63*	49.2	1000	4.92	20.4	34.7	17.2
BS11	6666	Brickearth	1.64	0.843	7.8	1.82	31.1	955	3.26	10.4	16.4	9.6
BS12	2113	Dk soil over Roman b'earth	3.24*	1.12	8.1	5.07**	124	1220	10.2**	62.2*	66.8*	37.3
BS13	1986	Saxon road silts	5.49**	8.85*	8.1	9.74***	183	864	21.2***	83.4*	86.6*	33.9
BS14	1986	Saxon road silts	5.83**	8.60*	8.1	12.7***	1050	2110	49.8***	140**	96.6*	50.7*
CW64		**Whitefriars horticulture**										
BS39	457	Medieval soil	4.11*	10**	7.6	9.39**	130	1490	8.72*	121**	89.0*	42.0
BS40	457	Medieval soil	3.61*	5*	7.6	10.4***	115	1520	7.57*	144**	75.8*	38.0

BS41	437/47	Medieval-Post Med soil	3.89*	5*	7.4	8.97**	180	1490	12.1**	144**	85.9*	41.3
BS42	2581	Med Friary soil	4.13*	5*	7.5	9.99**	237	1400	16.9**	271**	107**	47.5
BS43	2581/73	Med Friary soil	4.06*	5*	7.6	10.2***	218	1390	15.7**	309**	102**	49.6
BS44	2573	Med Friary soil	4.03*	2	7.6	9.69**	228	1320	17.3**	317**	98.8*	47.2
BS45	2577	Medieval soil	3.85*	2	7.6	9.44**	179	1340	13.4**	184**	101**	40.3

[a] Context/Description: Samples highlighted show signs of enrichment/enhancement in two or more of the key anthropogenic indicators: phosphate-P, magnetic susceptibility (as reflected in χ_{conv}), Pb, Zn, and Cu.

[b] LOI: Samples highlighted in bold have higher organic matter concentrations: * = 2.50–4.99%, ** = 5.00–6.39%

[c] CO_3^{2-}: Values presented to more that 1 d.p. are based on calcimeter measurements, whereas others are estimates, with 10% representing ≥10%. Samples highlighted in bold have higher carbonate concentrations: * = 5.00–9.99%, ** ≥10%.

[d] Phosphate-P: Figures highlighted in bold show likely phosphate enrichment: * = enriched (2.50–4.99 mg/g), ** = strongly enriched (5.00–9.99 mg/g), *** = very strongly enriched (> 10.0 mg/g)

[e] χ_{conv}: Figures highlighted in bold show signs of magnetic susceptibility enhancement: * = enhanced (χ_{conv} = 5.00–9.99%), ** = strongly enhanced (χ_{conv} = 10.0–19.9%), *** = very strongly enhanced (χ_{conv} ≥ 20.0%)

[f] Pb, Zn, and Cu: Figures highlighted in bold show signs of enrichment: * = enriched (50.0–99.9 µg/g), ** = strongly enriched (> 100 µg/g)

Box 11.1   Cultivation and manuring at Late Roman/Saxon Oakley, Suffolk, UK (and other European sites)

At the rural Roman site of Oakley (Suffolk), just outside and across the River Waveney from the small Roman town of Scole, Norfolk, a number of archaeological contexts were investigated (Ashwin and Tester, 2014). This included a late Roman/Saxon cross-ploughed field, where bulk and six thin section samples were taken to study the interior and field edge soils. Associated pollen samples were taken from two of the Kubiëna boxes, before crystic resin impregnation was carried out.

Modern and ancient soil cover

It can be argued that the dominant Roman and Saxon natural soils were Aquods (typical gley podzols and humo-ferric gley podzols; e.g. pH 4–5), as found today (Hodge et al., 1983); full details in Macphail et al. (2014). The buried Ap horizon (thin sections 1 and 2; Figure 11.9a), in which the pollen of cereals, grasses and herbs, including weeds of fallow ground were found, has more organic matter (1.37–1.53% LOI) and phosphate (190–230 ppm P) compared to the field's buried A2g horizon (0.49% LOI and 50 ppm P), and local Roman buried A2 horizons at Scole (0.44–0.59% LOI, 40–50 ppm P). The buried illuvial Spodic B horizons can also be compared chemically (bBsg – 1.43% LOI, 210 ppm P; bBh 0.89% LOI, 120 ppm P). Two areas of the ploughsoil were examined in thin section. Their subsoils showed two types of amorphous pedofeatures, (1) a spodic monomorphic sesquioxidic type of grain coating and (2) ferruginous nodular formations of hydromorphic (gley) origin (thin section 3), consistent with their past and present Aquod character. Ploughmarked soils (thin sections 1 and 2) contain frequent anthropogenic inclusions and features that are anomalous in a natural podzol, including burned flint, coarse wood charcoal, brickearth (constructional material?), clay (raw material?), a single biogenic calcite (earthworm) granule and relict earthworm burrows. An anomalous nodule contains rare calcite ash embedded with charcoal; it is probably an amorphous phosphatic iron nodule, as indicated by *in situ* vivianite (crystalline iron phosphate – e.g. $Fe_3(PO_4)_2$ $8H_2O$) (Karkanas and Goldberg, 2018) (see Figures 11.9a–11.9e). These additions are clearly evidence of manuring and at the early medieval Büraberg colluvial soil site similar nodules were encountered where phosphate levels had been enhanced (see Table 11.3) (Henning and Macphail, 2004). When embedding charcoal and ash these can be reasonably interpreted as evidence of nightsoiling, where the contents of the toilet bucket and/or cess pit were first "sweetened" by the addition of ashes. SEM/EDS analysis of this kind of nodular material in manured medieval soils used for ground raising at Malham Church, North Yorkshire (M. Roberts, pers. comm.) found a typical make-up of 32.1% Fe, 15.3% P, and 5.70% Ca., compared to the background soil (e.g. 4.33% Fe, 0.48% P, and 2.45% Ca) (Figures 11.9b–11.9e). The medieval cultivated soils at Whitefriars, Canterbury, UK are similarly phosphate enriched (Table 11.2). Experimental pig management also produced soils with seemingly *in situ*-formed phosphate nodules (Macphail and Crowther, 2011a)

Box 11.1 (Continued)

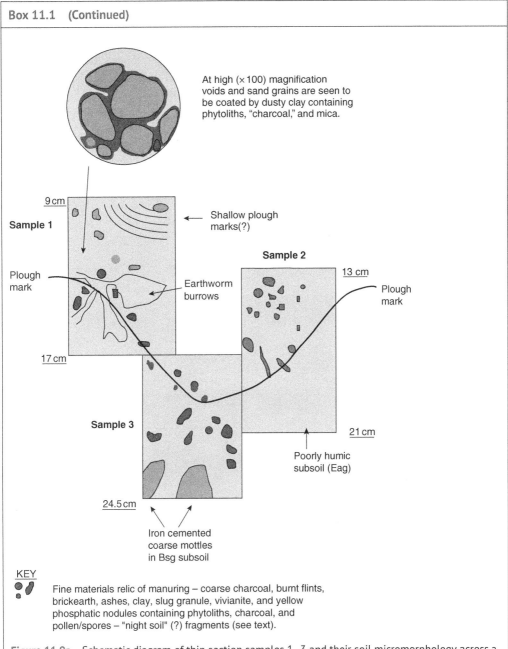

At high (×100) magnification voids and sand grains are seen to be coated by dusty clay containing phytoliths, "charcoal," and mica.

Sample 1

9 cm

Shallow plough marks(?)

Sample 2

13 cm

Plough mark

Plough mark

Earthworm burrows

17 cm

Sample 3

21 cm

Poorly humic subsoil (Eag)

24.5 cm

Iron cemented coarse mottles in Bsg subsoil

KEY

Fine materials relic of manuring – coarse charcoal, burnt flints, brickearth, ashes, clay, slug granule, vivianite, and yellow phosphatic nodules containing phytoliths, charcoal, and pollen/spores – "night soil" (?) fragments (see text).

Figure 11.9a Schematic diagram of thin section samples 1–3 and their soil micromorphology across a late Roman/Saxon moldboard plough mark at Oakley, Suffolk, UK (modified from Macphail et al., 2014).

(Continued)

Box 11.1 (Continued)

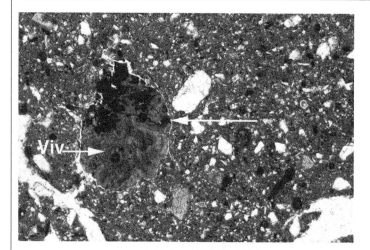

Figure 11.9b Photo micrograph of vivianite (Viv) phosphate nodule in manured medieval soil at Malham, North Yorkshire, UK. PPL, frame width is ~2.38mm.

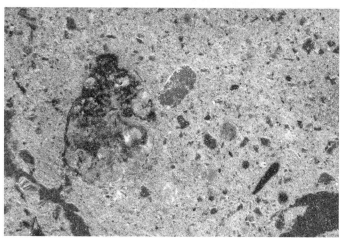

Figure 11.9c As Figure 9b, under OIL.

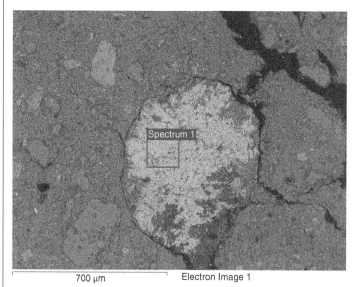

Figure 11.9d SEM/EDS X-ray backscatter image of phosphate nodule in Figure 9b.

700 µm   Electron Image 1

Box 11.1   (Continued)

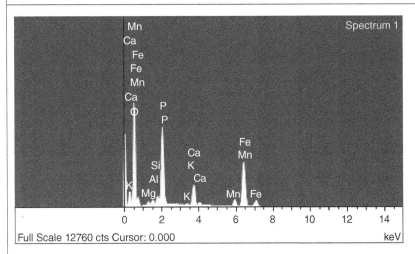

Figure 11.9e   X-ray spectrum of phosphate nodule in Figure 9b; typically nodule is an iron-phosphate-calcium compound (e.g. 32.1% Fe, 15.3% P, with 5.70% Ca and 3.82% Mn).

(Zhuang, 2018). For instance, the rice fields at the Tianluoshan site (7000–6500 BP) were of relatively simple form with notable absence of bunds and many irrigation features in the field (Zheng et al., 2009). It was during the late Majiabang culture (ca. 6300–6000 BP) that typical paddy fields with field bunds started to be built, being equipped with a wide range of irrigation facilities in, and surrounding fields (Zhuang, 2018). Such a form of paddy fields continued to develop and expand in the prehistoric Lower Yangtze River region, culminating in the enormous scale of fields at the Maoshan site (see section 6.6.4), and its spread to other regions of prehistoric and historic East Asia (Mizoguchi, 2013).

In paddy soils, the use of bunds to hold water seasonally can be suggested by generic microfeatures indicative of gleying (hydromorphism), such as iron and manganese nodules with soil impregnation especially concentrated along root channels and residues in very pale gleyed soils (Bouma et al., 1990; Lee et al., 2014; Matsui et al., 1996; Vepraskas et al., 2018). Some microfabrics show extreme depletion features due to repeated long-term water saturation, and associated slaking and mixing that result from ploughing, and trampling by humans and water buffalo at times (Liu et al., 2004; Zhuang et al., 2014).

Apart from extreme depletion and likely clay breakdown, microprobe and SEM/EDS studies have also shown that iron and phosphate (from manuring additions such as nightsoil) can be mobilized and redeposited within the paddysols (Figures 11.11a and 11.11b). Rice phytoliths are rarely visible in thin section, compared to bulk sample investigations. More commonly, the presence of fine charcoal is consistent with the burning of crop waste, and fine fragmentation caused by cultivation and trampling. At the Huizui site (China), dumped rice straw waste in ash deposits also included rice phytoliths. At Nara (Japan), a large thin section from the horizontal plane in an ancient paddy soil exhibited 2 mm wide hollow circles of iron and manganese impregnation grouped into five to seven of these circles. These have been interpreted as being relict of rice planting, when small bunches of rice were planted at a time

Table 11.3 Chemical and magnetic susceptibility data from the site of the Dark Age Carolingian (8th century) fortress of Büraburg, Nordhessen, Germany (Crowther in (Henning and Macphail, 2004).

Context/sample no.	LOI (%)	Carbonate estimate* (%)	Phosphate-P_o (mg g^{-1})	Phosphate-P_i (mg g^{-1})	Phosphate-P (mg g^{-1})	Phosphate-P_o:P (%)	Phosphate-P_i:P (%)	χ (10^{-8} m^3 kg^{-1})	χ_{max} (10^{-8} m^3 kg^{-1})	χ_{conv} (%)
4/4	n.d.§	1	0.392	2.12	2.51	15.5	84.5	41.3	393	10.5
3/3	n.d.	2	0.358	2.25	2.61	13.8	86.2	41.2	417	9.88
3/3a	1.69	2	0.407	2.34	2.75	14.9	85.1	41.2	483	8.53
2/2	n.d.	0.1	0.295	1.20	1.50	20.0	80.0	44.9	346	13.0
1/1	n.d.	0.5	0.312	0.696	1.01	31.1	68.9	39.9	774	5.16

Figure 11.10a Field photo of example of (plaggen-like) cultural soil profile Hørdalsåsen, Sandefjord Kommune, Vestfold, Norway; sampling tins in place, with lower tin recording clearance and the upper tin within the manured cultivated soil. The uppermost layer is the modern Ah horizon.

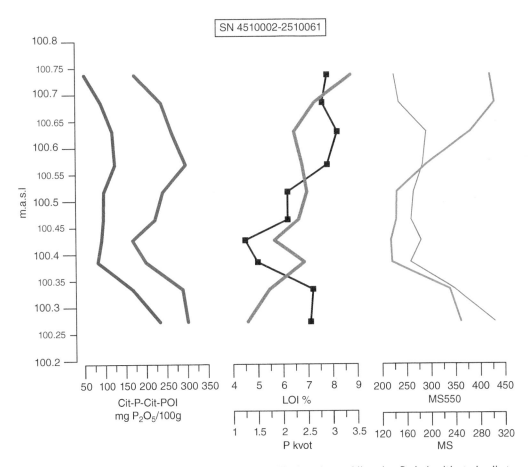

Figure 11.10b Geochemical profile through plaggen-like Iron Age to Migration Period cultivated soil at Hørdalsåsen, Sandefjord Kommune, Vestfold, Norway (Location: SN 4510002-2510061); enhanced MS at the base (100.25 m OD) records clearance by fire (cal 390 BC–135 AD; see Figures 11.10c and 11.10d), with upwards increased LOI, P and magnetic susceptibility values, around 100.6 m OD; are the result of manuring with dung (cal 220–420 AD; see Figure 11.11e) and burned settlement waste (Mjærum, 2012, 2020; Viklund et al., 2013) (diagram by kind permission of Johan Linderholm, University of Umeå, Sweden).

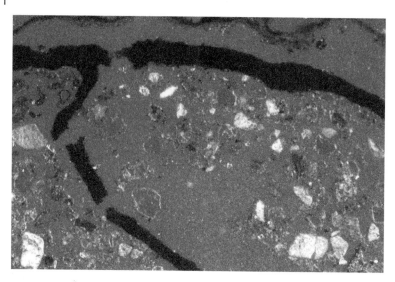

Figure 11.10c As Figure 11.10a, photomicrograph of clearance soil, characterized by charred roots, burned fine soil and sand, and natural acidic very thin organo-mineral excrements. OIL, frame width is ~4.62mm.

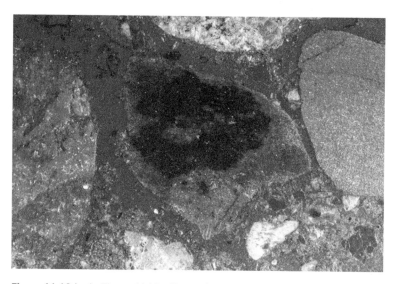

Figure 11.10d As Figure 11.10c, illustrating concentration of burned sands and gravels at the base of the soil profile, consistent with the enhanced magnetic susceptibility here (Figure 11.10b). OIL, frame width is ~4.62mm.

(Matsui et al., 1996). The adding of manure/night soil and the trampling of water-saturated soil by large animals (e.g. traction buffalo) and humans has produced distinctive dark, mottled, and massive structured soils. Fish can also be harvested from waterlogged paddy fields, as noted near Tonle Sap Lake (Cambodia).

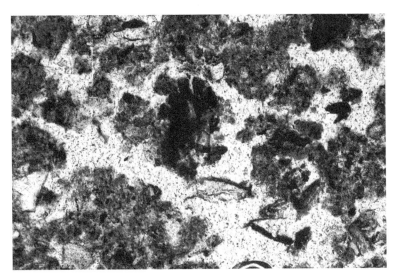

Figure 11.10e   As Figure 11.10a, photomicrograph of manured soil at around 100.2 m OD, where peaks of LOI, P and PQuota record high amounts of organic phosphate (see Figure 11.10b). Thin organo-mineral excrements indicate an increase in soil fertility (see Chapter 14) and are probably linked to the presence of humified organic materials (center) indicative of manuring with raw dung (Viklund et al., 2013).

11.4   Other Managed Changes to the Landscape

11.4.1   Development of Pastures

A major component of how settlements function is through *animal management* (see Chapters 12 and 13), with byres and stables within the settlement, and enclosures nearby, and with pastures in the settlement's wider landscape (Macphail et al., 2017a). Often the purposeful development of *grazing* lands is given less interest compared to cultivation, but it is just as an important land management practice. Although there are regional examples of the formation of pastures without clearance because a fully wooded landscape had not developed by the mid-Holocene (Allen, 2017), expansion of grazing land in other areas was linked to clearances, for example in the Ligurian Apennines, Italy, especially during the Chalcolithic (Cruise, 1990; Cruise et al., 2009). In addition, low alpine terracing and clearance of the *Larix* forest was linked to the development of pastures, and podzols were converted to brown soils (Courty et al., 1989: 309–318). The types of soils employed for grazing often have a Mull topsoil horizon (Table 4.1), and include Mollisols (Table 4.2) that once characterized the prairies and pampas. Modern and Neolithic lowland pasture soils are illustrated in Figures 14.2–14.6. In upland areas pastures were also kept open by grazing intensity, in some cases following Bronze Age clearances and sometimes exacerbated by medieval monastic flock management (see Box 4.2).

11.4.2   Mining and Quarrying

Woodland clearances may have little to do with agriculture and may be the result of disturbances linked to mining and industrial processes where wood is required, e.g. for fuel, before coppicing can be organized for a sustainable broad leaved wood crop (Cowgill, 2003). Conifers cannot be

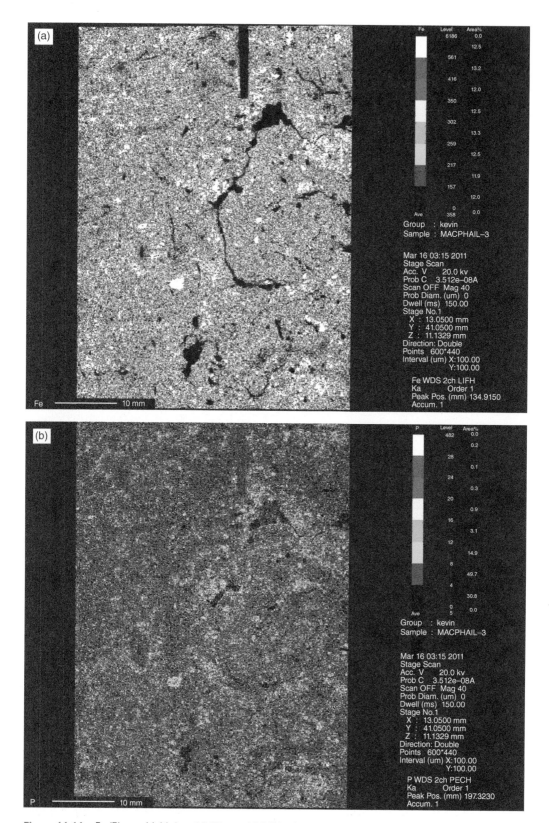

Figure 11.11 Fe (Figure 11.11a) and P (Figure 11.11b) microprobe element maps from Bronze Age (10th–4th Centuries BC) paddysol layer, at Gulhwa, located in the very south South Korean peninsula; waterlogged iron depleted matrix soil occurs alongside fine iron nodules ("white") and iron impregnations and channel hypocoatings (mottling; "pink"). Note coincidence of Fe and P–P is the result of manuring, and has become stabilized through iron mottling (Lee et al. 2014/with permission of John Wiley & Sons).

coppiced in the same way. Massive medieval (cal AD 1200–1400) coarse colluvium with included soil clasts, buried the edges of Bargone Lake at 831 m asl in the Italian Apennines; a markedly enhanced MS is consistent with the presence of burned inclusions and a sharp decrease in tree cover are consistent with mining disturbance (Cruise et al., 2009). Further effects of mining activity also include local heavy metal (Cu, Pb, Zn) concentrations of waterlaid tailings containing fine charcoal in "settling" ponds, and where processing produced markedly enhanced magnetic susceptibility values (see Chapter 12.4).

In some cases, mining impact has been slight while in others it has altered the whole landscape. At the 1st century AD site of Las Médulas (León, Spain) the Romans used hydraulic mining to extract gold from alluvial deposits, regolith, fluvial terraces, moraines, and fluvio-glacial fans; they removed about 300 million m^3 of material (Pérez-García et al., 2000) (see Figure 11.14). Similarly at the Talayotic (1st and 2nd millennia BC) site of Torre d'en Galmés (Menorca, Balearic Islands, Spain) – as well as others on the Island – reef limestone was quarried for the construction of houses, Taula (uprights with T-shaped horizontal rock slabs), and Talayots (massive rock structures) (Pérez-Juez, 2011; Pérez-Juez, 2013; Pérez-Juez and Goldberg, 2018). The effect of such extensive and widespread quarrying on agricultural soils and colluvia, which had to be removed in order to extract the blocks (some measuring meters wide and high) has yet to be fully evaluated. As is well-published now, the famous bluestones of Stonehenge (UK), were quarried in Wales, put into a circle, then removed and transported to England, and reused (Parker Pearson et al., 2015; Parker Pearson et al., 2021).

11.4.3 Irrigation, Water Management, and Canals

In arid areas, soils used for agriculture need to be irrigated, and fields have associated channel systems (Sandor et al., 2021; Sulas, 2018). When dry soils are wetted through irrigation they may become water-saturated. In thin section, these soils may exhibit a collapsed structure, which is manifested as a void type known as *vesicles*, where air bubbles were once trapped (Courty et al., 1989; Lee et al., 2014) (see Chapter 7.6). Two types of pedofeatures together may infer irrigated cultivation in thin section: biological features (e.g. root channels and excrements of mesofauna) that indicate anomalous high levels of biological activity in "arid" soils, and relict vesicles.

At the Early Agricultural Period (800–1200 BC) Las Capas site (Tucson Basin, Arizona, USA; Nials, 2011; Vint and Nials, 2015), canal sediments are clayey, and irrigated soils are affected by clayey inwash, compared to the sandy sediments of active natural streams (Figure 11.12a). Here also, planting holes were employed, and concentrated inwash microfeatures and traces of now-oxidized organic matter indicate that watering-in plants and use of amended soils was restricted to these planting holes (Figure 11.12b) rather than attempting to irrigate and manures whole fields; hence, this demonstrates the clear management of a restricted water resource. Runoff-fed open reservoirs and rock-cut cisterns were also recorded in the Negev Highlands (Israel) (Junge et al., 2018).

In other parts of the world, such as Protohistoric Lattes (southern France) and Sassanid Merv (Turkmenistan) water was managed using canals, with sluggish flow being recorded by clayey sediments, and occasional desiccation features of iron mottling and minor biological working. At early Islamic Merv, the deep brick-lined Madjān canal was constructed for urban water supply; its fill was analyzed employing 16 thin section and sub-sampled bulk samples (Macphail and Crowther, 2016; Williams, 2007, 2018). When freely flowing the canal was characterized by massive and very finely bedded clean silts and fine sands, while periodic clean standing water episodes produced upward-fining laminae (Figures 11.13a and 11.13b). Drying out and re-wetting could produce sediments with locally eroded clay clasts and collapsed channels, now forming polyconcave vughs. Free-flowing and standing water sediment fills are mirrored by inputs of few

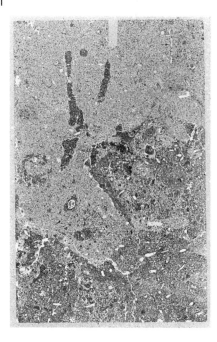

Figure 11.12a Las Capas, Arizona, USA. Scan of LC-M2, showing erosive and now-burrowed junction between fine sands (Canada del Oro drainage) and earlier-formed biologically worked cultivation soil, with clayey fine soil elements due to irrigation with low-energy canal water. Frame width is ~50 mm.

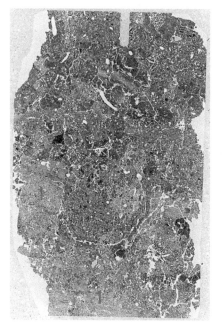

Figure 11.12b As Figure 11.12a. Scan of LC-M9 (planting hole) and showing three layers; (1) alluvial sediments, (2) cultivated mixed soil ("Ap") and (3) "canal" clay sealing the cultivated soil; this clay has also washed into voids down-profile (3a). Frame width is ~50 mm.

contaminants (phosphate-P in Figure 11.13c), although there is background "soot" and fine burned mineral material of urban occupation origin that are also recorded by LOI and % χ_{conv} (see Chapter 17). Muddy, stagnant water conditions led to iron staining microfeatures. The top three samples in the sequence, which seem to date to the Mongol invasion, show mainly stagnant muddy water sedimentary conditions and much waste disposal, as also reflected by the P, LOI and magnetic susceptibility (%$\chi_{conv.}$) (Figure 11.13c).

Figure 11.13a Madjān canal, Merv, Turkmenistan. Scan of 028–56; fine sediments include calcareous clay clasts (CC), evidence of muddy conditions and collapse, including a polyconcave vugh (pv); upwards standing water is recorded by microlaminated sands and silts (ML). Frame width is ~50 mm (2010 sampling by Rosemary Hoshino).

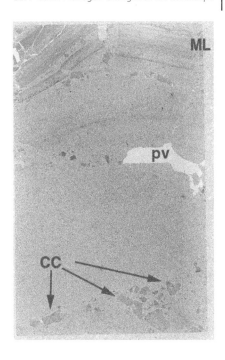

Figure 11.13b As Fig 11.13a. Photomicrograph of 028–56; upward-fining laminae indicate a period of standing water. PPL, frame height is ~4.56mm.

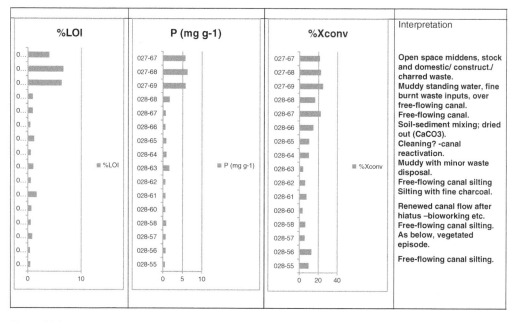

Figure 11.13c Datalog (% LOI, phosphate-P, and magnetic susceptibility % χ_{conv}) through ~4m of the Madjān canal, Merv, Turkmenistan (Williams 2007, 2018), with associated sediment micromorphology interpretations. Clean freely flowing and standing water conditions with episodes of sluggish muddy flow characterize the early Islamic fill, before the Mongol invasion when the canal was much more affected by waste disposal and poor flow.

Figure 11.14 Eroded landscape at Las Médulas León, Spain brought about by Roman hydraulic mining of gold from alluvial deposits (Wikimedia Commons).

11.5 Conclusions

The role of the geoarchaeologist is to attempt to differentiate human actions from natural phenomena, such as climatic change, and relationships between societies and climatic change. Workers in other environmental disciplines have the same task, and can provide vital complementary information (see Bargone Lake example, above). Changes to vegetation cover may be due to climatic deterioration and/or land use, as previously suggested in Boxes 4.1 and 4.2, but geoarchaeology can often clearly identify clearance impact, and a change to a cultivated land-scape, which can include paddyfields. This can be done through fieldwork allied to laboratory analyzes (eg. soil micromorphology and geochemistry, magnetic susceptibility, fractionated P). Although geoarchaeology can be applied to both site and landscape scales, palynology, and tree ring data can provide wider overviews of human impacts and climatic change. As shown in the Dilling (Norway) example, cooler and wetter conditions after AD 250 produced increased soil wetness in low ground areas that led to the spread of the settlement upslope (Box 18.4). We also note that clearances and other disturbances that trigger erosion and colluviation may stem from mining activities (Figure 11.14). In arid marginal environments manipulation of the landscape to concentrate runoff to irrigate fields or create canals for both agriculture and populations, are further examples of human impact.

12

Human Use of Materials

12.1 Introduction

The archaeological record traditionally consists of artifacts, features, and ecofacts; as we discussed in Chapters 2, 3, and 13, we also need to include archaeological sediments on equal footing with these typical archaeological phenomena. In antiquity, people used a variety of materials either in their raw form (e.g. lithics made from chert) or as processed materials such as mudbricks, mortars, ceramics to name a few. Study of these materials, provides a variety of insights into sources of these materials, how they were processed, transformed, and used, and what they can tell us about specific activities and behaviors in the past.

Among artifacts for prehistoric periods, we typically think of lithics (e.g. basalt and other volcanic rocks, limestone, and especially flint and chert), and processed/collected materials such as ochre. In later prehistory and historical periods these rock types are supplanted by materials generally made with more sophisticated technologies, which include metals (copper, bronze, iron), ceramics, and construction materials such as plasters and earth-based adobe and daub. The analytical study of many of these materials is usually subsumed under "archaeometry". However, since this book is not an archaeometry text, we have decided to limit our discussion to materials that humans actually make, particularly those that are employed in construction materials. The recycling of these materials in archaeological sites as fills and dumps is considered in Chapter 13. A further aspect of human materials and occupation deposits is what happens to them when a site is abandoned – how do they weather and become transformed by soil processes. Examples of this from Europe (dark earth) and South America (*terra preta*) are given in Box 13.1.

12.2 Lithics and Ochre

12.2.1 Lithics

It is important to be able to recognize the presence and orientation of small lithics, debitage, and microartefacts in archaeological deposits, because they provide information on taphonomy and site formation processes (Sherwood, 2001). Sometimes through excavation errors or loss of an archive, lithic debitage fragments (Angelucci, 2017) found in thin section may be the only evidence of occupation/tool making activity. Such findings in Paleolithic (Figure 12.1) and Mesolithic sites (Figures 12.2–12.4) can also confirm the presence of chipping floors, or colluvium derived from such occupational soils now lost to erosion (Harding et al., 2014). Pit houses and hut

Practical and Theoretical Geoarchaeology, Second Edition. Paul Goldberg and Richard I. Macphail.
© 2022 John Wiley & Sons Ltd. Published 2022 by John Wiley & Sons Ltd.

Figure 12.1 Thin section scan of hearth complex from the Middle Paleolithic site of Roc de Marsal, Dordogne, France, showing angular flakes of chert (arrows) within a lower combustion feature. The fragment at the right shows the effect of spalling from having been heated in the fire. Width of thin section is 75 mm.

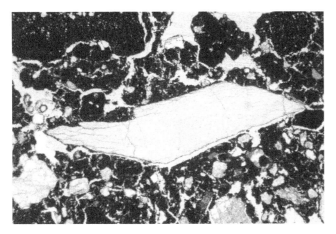

Figure 12.2–4: Mesolithic Haakonshellaveien, Bergen, Norway; photomicrographs of microdebitage within hut occupation floor deposits (Context 1026); all images with frame width of ~4.62 mm. **12.2:** fine charcoal-rich floor deposits with quartz schist microdebitage, PPL.

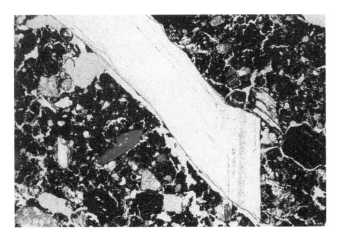

Figure 12.3 Example of colorless feldspathic microdebitage, within floor deposits, and featuring brownish fine calcined bone fragment, which also characterize this context.

Figure 12.4   As Figure 12.3, under OIL; note white calcined bone fragment.

locations of hunters and gatherers are sometimes difficult to identify unequivocally (Crombé, 1993; Crombé et al., 2015; Newell, 1980), but concentrations of diffusely stratified microartifacts such as micro-debitage, can provide convincing evidence (Cooper et al., 2017; Darmark, in prep.). In addition, at Boxgrove, a suggested animal/hominid trail soil-sediment included an "out of context" horizontally oriented flint flake, sparking some controversy (Macphail and Goldberg, 2018b; Roberts and Parfitt, 1999), while a basalt tool and bones were found embedded in tufa at Olduvai Gorge (Tanzania) (de la Torre et al., 2018). In some cases, especially in pedologically and sedimentologically active environments, the only stone-size material may be artifacts; at Chongokni (South Korea), large Acheulian-like Upper Paleolithic hand axes provided small patches of protected soil-sediments that allowed the identification of possible dry season occupation of highly active alluvial plains (Macphail and Goldberg, 2018a: 157–165; Yi, 2015).

12.2.2   Ochre

This red ferruginous and hematite-rich material is best known in rock art, burials and as painted floors (Hilton, 2002). At Arene Candide (Liguria, Italy) ochre was used in both Upper Paleolithic and Neolithic burials, and was composed of natural geological materials especially associated with limestone weathering and Quaternary terra rossa soils (Chromic Luvisols or Rhodustalfs (FAO, 2015; USDA, 2014) (Ferraris, 1997). In Scandinavia, some ochre has been recovered and at the once-coastal site of Holtebråtan-Tusse ochre has a chemistry of up to 98.3% FeO as measured by EDS. At early Mesolithic Lillsjön, Pengsjö i Ångermanland, Sweden, hunters and gatherers presumably following reindeer and elk herds that were colonizing the newly vegetated landscape exposed by post-glacial uplift, conceivably "manufactured" ochre from other iron minerals such as goethite possibly in hearths (Johan Linderholm, University of Umeå, pers. comm.). Here, relatively large fragments of ochre were analyzed (Figures 12.5–12.7; Fe = 62.0%; – 82.3% FeO); suspected ochre fragments (max 7 mm) with a 75% Fe content were also found at Mesolithic Sotrasambandet, west of Bergen (Norway) (Tøssebro et al., In prep). At the La Prele Mammoth site ochre nodules and stains could be traced to a quarry 100 km distant, demonstrating not only that Paleoindians transported such materials (as well as chert) over long distances, but also that it was important in Paleoindian networks (Zarzycka et al., 2019). Engraved pieces of ochre dating to ~100 kya has been recovered from Blombos Cave (South Africa) (Haaland et al., 2020; Henshilwood et al., 2009).

Figures 12.5–12.7: Early Mesolithic Lillsjön, Pengsjö i Ångermanland, Sweden; scan of thin section LL-R556-1; sand-dominated Mesolithic pit fill, with coarse areas of very dark red ochre. Frame width is ~50 mm.

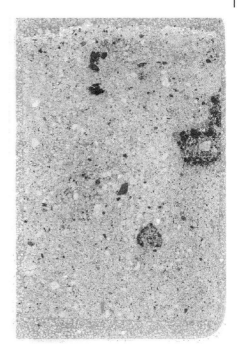

Figure 12.6 Photomicrograph of sands and opaque ochre. PPL, frame width is ~4.62 mm.

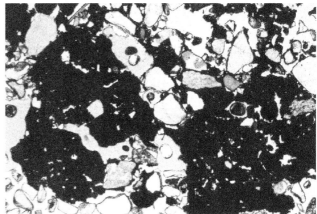

Figure 12.7 As Figure 12.6, under OIL, illustrating typical orange-red ochre color, and showing compound nature of ochre ingredients.

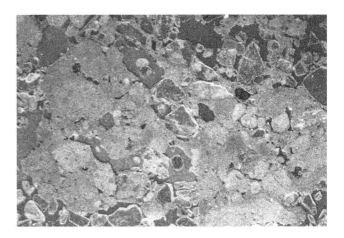

12.3 Constructional Materials

In this section we discuss a variety of building materials that include intact soil (turf, sod blocks), soil and sediment (regolith) ground-raising layers, disaggregated and manufactured soil materials (adobe, daub, mudbrick), and manufactured lime- (and gypsum-) based plaster/mortar. Plasters, mortars, and adobe/daub commonly have both a *matrix* (fine binder) and a *temper* (coarse component). The latter can include plant (straw, reeds) or geological/anthropogenic (rock, ceramics, and/or brick fragments) material. The terms "matrix" and "temper" are used henceforward.

12.3.1 Intact Earth-based Constructional Materials

The amount of earth-based structures in archaeology is often grossly underestimated apart from the most obvious upstanding ramparts, earth mounds (Old World barrows and tumuli, New World mounds; Sherwood and Kidder, 2011; Van Nest et al., 2001) and tells, for example (see Chapter 13; Table 13.1). Adobe, mud brick, rammed earth, daub, clay, turf, and local variants have been used for construction across the globe. Even famous stone-built linear features such as Hadrian's Wall (England) and the Great Wall of China, have lengths once composed of earth-based ramparts. Furthermore, although the remains of Roman masonry structures are admired today, Roman Period populations across Europe employed local, imported, and processed soil and geological materials for constructional purposes (Blake, 1947; Macphail, 1994a, 2003b); lengths of Hadrian's Wall were constructed from local turf and/or rock quarried adjacent to the wall's route (see Wang et al., 2020 for construction at prehistoric Liangzhu City, China). Of particular interest are soil micromorphological investigations since 2008 of four turf rampart locations at Oxford (UK), where various turf components record soils evidencing animal trampling and pastures. The interpretations are consistent with two sets of thin section-correlated chemical data, including LOI and P. Turf in ramparts can be dated by context and artifacts, but this is not always the case. At Oxford, Roman, Saxon, and 17th century English Civil War rampart constructions occur. At one location, however, pasture turves were dated by four OSL assays to a likely Saxon defense structure (the Oxford burgh?) (Graham Spur and Michael Smith (Museum of London Archaeology – MOLA) and J.-L. Schwenninger (Luminescence Dating Laboratory unpublished reports).

It is therefore important to understand both the geology and soil types of an area for sourcing both quarriable rock and earth-based building materials that have been used, and are currently in use in the world. Stoops (2017a) also notes that laterite when quarried quickly hardens on exposure producing a brick-like material for construction used both in southern India and for example at the sites of Angkor Wat and Angor Thom in Cambodia (Figures 12.8–12.10; see Chapter 4.4.5). Chemically, these porous and pisolitic laterites are composed of 20–50% SiO_2, 12–22% Al_2O_3, and 23–50% Fe_2O_3, and because goethite and hematite iron minerals dominate these laterites they have a very low magnetic susceptibility (Uchida et al., 1999). In addition, the Maya in Mesoamerica quarried and processed a soft (weathered?) limestone called "sascab" (Littmann, 1958). In modern experiments at Xunantunich, Belize, sascab was mixed with water and a small amount of lime to produce useful blocks on drying out (Jaime Awe, Northern Arizona University, pers. comm.). Slabs of sascab could also be used for floors, which were then plaster covered (E. Graham, UCL, pers. comm.).

Figure 12.8a Field photo of Angkor Thom city wall, Cambodia, constructed of laterite blocks. The late 12th century Khmer Empire capital city is moated and enclosed by 3 km-long walls that are 8 m high.

(a)

Figure 12.8b Close-up of Figure 12.8a, showing laterite blocks with presumed "lifting holes".

(b)

Figure 12.9 Field photo of Angkor Thom Temple. There is a core of laterite blocks faced by limestone.

Figure 12.10 Traditional use of mud brick in rural areas of abandoned house Covarrubias Castilla, Spain. Note the additions of angular rock fragments (photo: Amalia Pérez-Juez).

12.3.1.1 Turf and Sod Blocks

"Turf" and "sod" blocks are broad terms for soil used for construction (Figures 12.11–12.13). They can simply be composed of unconsolidated sediments and soil *sensu lato* (Gokstad Ship Mound, Norway; Shiloh Mound, USA), well-drained mineralogenic topsoils (Easton Down long barrow and Experimental Earthworks, UK), naturally carbonate-rich turves (Khokhlova et al., In review), and wetland pasture and peaty soils (Cannell et al., 2020; Huisman and Milek, 2017; Milek, 2006; Nicolaysen, 1882; Sherwood and Kidder, 2011; Whittle et al., 1993). In general, topsoil turf involves living grass-covered and grass-rooted mull humus ("prairie") Ah soil horizons (see Table 4.1 and Box 12.1). A variety of topsoils occur and include, for example, "laminated mull" horizons where soils are poorly drained, such as Viking Age wetland pastures at the Gokstad Mound and early parts of Hadrian's Wall, Cumbria, UK (Macphail et al., 2013a). Calcium carbonate-rich humic topsoils of chernozems were employed in a 4th millennium BC Kurgan (complex mound) at Essentuki (Stavropol Krai, Ciscaucasia, Russia); similarly constructed kurgans have also been investigated (Khokhlova et al., In review). Peat (Histosols), and acid moder and mor humus horizons can be used, for example, in ramparts and walls, but their high organic content can be very susceptible to oxidation, leading to shrinkage. They are often thought to be totally unsuitable for turf roofs where they are exposed to oxidation (see below). It has been shown, however, that wetland turf can become naturally transformed by the dying off of the wetland flora and its replacement by "dryland species" (Sigurðardòttir, 2008: 7). The Experimental Earthwork Project investigated the post-depositional transformations affecting turf mound constructions on both base-rich (rendols) and acid (Spodosols) soils; findings are described in Chapter 14.

Box 12.1   Turf and sod blocks

Turf was widely employed as a building material, and until the 20th century was used regularly in Scotland and Ireland (sod-house), and by early settlers of the prairie regions of northwest America –"soddies" (Evans, 1957; Fenton, 1968; Huisman and Milek, 2017). It is still used in Scandinavia and Iceland, and its use studied experimentally (Figure 12.12). It is worth noting that a wide variety of turf types can be chosen and cut-to-shape in order to fulfil different constructional functions, including layers between stone courses; the construction of arches and turf "mattresses" in houses have also been recorded (Figure 12.11) (Milek, 2006; Sigurðardòttir, 2008; see below).

Some archaeological examples A large number of turf barrows and associated environmental studies have been reviewed for England (Macphail, 1987). A more recent study is the Romano-British kingly burial at Folly Lane, St. Albans, where the burial chamber is sealed by a putative collapsed turf tumulus (Niblett, 1999). Turf used for construction was recognized on the basis of its soil micromorphology and microchemical (microprobe) characteristics. These data suggested the use of topsoils (Ah horizon) from natural "woodland soils" (acidic argillic brown earths – alfisols) and dung-enriched topsoils that had been likely associated with animal management (Macphail et al., 1998). This finding was consistent with pollen analysis of the same samples and macrofossil remains at the site (Murphy and Fryer, 1999; Wiltshire, 1999).

Experimental turf roof Pollen analysis and soil micromorphology were also combined in order to characterize the experimental turf roof used on a wooden structure built by Roger Engelmark on the experimental "ancient farm" at Umeå (Bagböle) in northern Sweden (University of Umeå; Chapter 14). Analyses of two roof turves showed that the characteristics of the turf used were still recognizable, although somewhat transformed by its use. The 140 mm-thick turf roof comprised two turves over a birch bark roof liner, with the bottom turf being face down and top turf facing upwards and displaying a living grass sward (Figures 12.12 and 12.13) (Cruise and Macphail, 2000: figures 22.1 and 22.2). Furthermore, the successful turf roof at Umeå has only a very gentle pitch; the turf roof constructed on a steeply pitched roundhouse roof at Butser failed (Reynolds, 1979). If turf is cut to the right shape and laid as overlying strips (and or diamond-shaped blocks), relatively steep roofs can be successfully turfed (Sigurðardòttir, 2008: 7, 14–15).

Figure 12.11   Field photo at Skálholt Ecclesiastical Center, Iceland; reconstructed subterranean way between church and medieval settlement, with herringbone turf layers below wooden church floor.

(Continued)

Box 12.1 (Continued)

Figure 12.12 Umeå University ancient farm (Bagböle), north Sweden, 1994; turf roofed wooden house constructed by Roger Engelmark and Karin Viklund in birch and pine woodland. Note grass-covered "living" turf (mull humus) roof. Turves were cut from a local grassland that had been ameliorated during the period 1850–1950; boreal iron Spodosols are the natural soils (Cruise and Macphail, 2000; Engelmark, pers. comm.).

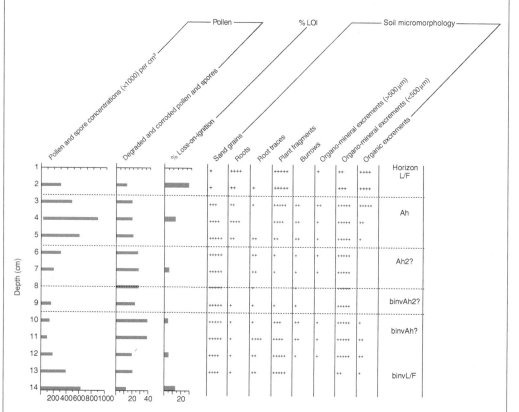

Figure 12.13 Diagrammatic section through the Umeå experimental turf roof (courtesy of Roger Engelmark and Johan Linderholm), showing an inverted (face down) turf at the base (binvAh – buried inverted Ah horizon) and "right way up" upper turf (e.g. Ah). Note pollen concentrations are highest in the binvL/F (Litter and Fermentation layer) and the Ah horizons, with LOI being highest in the "modern" L/F. Although organic excrements are concentrated in the "modern" L/F, much of the micromorphology of the upper turf is mirrored in the buried inverted turf (from Cruise and Macphail 2000). The way-up of turves in earth mounds is also often deduced from the pollen (Dimbleby, 1962; Wiltshire, 1999), although these data can be combined with measurements of organic matter and micromorphology (Macphail et al., 2013a).

Box 12.1 (Continued)

Ancient turf This has been identified through pollen and macrofossil analysis. However, even when strongly transformed by the effects of burial, the biological traits of the employed turf should still be recognizable through soil micromorphology (e.g. turf mounds, turf roofs, turf ramparts; Babel, 1975; Kooistra and Pulleman, 2018). As already noted, such turf may yield land use data contemporary with the construction. Lastly, mineralogenic peat and the humic turf of Spodosols (podzols) and gley soil have been - and still are - used for fuel. In the archaeological record, "peat ash" accumulations have been found at settlements, and as residues of manuring in soils (Adderley et al., 2006, 2018; Carter, 1998). In such leached soils, the rubification of burned mineral material may be sparse because the soils may well be strongly iron-depleted (Vepraskas et al., 2018).

12.3.1.2 Building Clay

The use of intact "building clay" can also be mentioned here. Coherent subsoil and regolith "clay" could be used in an unaltered way as slabs for ground-raising and floor preparation layers. As a result of its wide geographical cover, till was broadly used across northern Europe, and employed for floor construction in East Anglia (UK) and southern Sweden (e.g. ~5th century AD longhouse floor at Uppåkra, near Lund, Scania, Sweden; Macphail et al., 2017a). A common choice for floor material in western Europe was Pleistocene silt loams and fine sandy silt loams, and some Roman and medieval cities are characterized by deep quarry pits ("silt pits") that were later used as loci for dumping (e.g. the Roman and Medieval Louvre site, Paris; Ciezar et al., 1994). Sometimes ground-raising deposits can be differentiated from *in situ* subsoils because argillic microfabrics may not be oriented to "way-up" (Kühn et al., 2018; Macphail and Goldberg, 2018b). A building material termed "brickearth" (derived from loess deposits) was commonly utilized in Roman London and Canterbury, for example, and wooden framed buildings with brickearth "clay" walls from the 1st–2nd centuries AD commonly had painted lime plastered walls (Figures 12.14 and 12.15) (Macphail, 2003b). Well-sorted river sands were employed as temper in lime plaster, and in Roman London employed to create the arena floors (see Figure 20.21). Intact intertidal clay was utilized in the making of presumed bellows plates employed in salt working at Stanford Wharf (Biddulph et al., 2012) (see Box 12.2).

Figure 12.14 Field photo; Courage Brewery site, Southwark, London; upstanding remains of painted plaster-coated brickearth "clay" wall, with sampling tin in place.

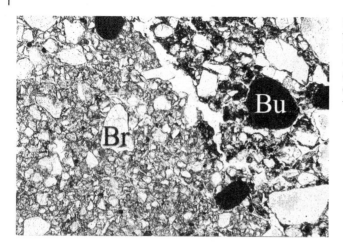

Figure 12.15 Photomicrograph of typical compact fine sandy silt loam brickearth (Br) wall and humic "dark earth" burrow (Bu), illustrating how earth-based building materials break down into soil. PPL, frame width is ~5.5 mm.

12.3.2 Adobe, Daub, and Mud Brick (Worked and Transformed Earth-based Building Materials)

Adobe *sensu lato*, and including daub and mud brick are soil-sediment based materials, which are commonly mixed with organic matter, (e.g. straw and reeds) or other organic binding agents, such as dung as a form of a temper (Cammas, 2018; Friesem et al., 2017; Karkanas and Goldberg, 2018b; Macphail and Goldberg, 2018b: 132–135); pottery and rock clasts are also common components in the Near East (Figure 12.10). The soil/sediment and organic matter are mixed with water, then puddled (e.g. by trampling in a puddling pit, or in a mold; see below) (Levine et al., 2004). In this process, the mixture loses all structure and most of the air is removed, so that on drying the mixture is massive (i.e., without structure), with no gas voids, and internal slaking is visible in thin section (Karkanas and Goldberg, 2018b). Non-calcareous mud plasters used for sealing floors and walls have also been encountered in Neolithic to medieval and modern contexts (Boivin, 1999; Macphail and Goldberg, 2018b).

Experimental and archaeological examples of earth-based building materials have been systematically characterized employing soil micromorphology in order to achieve more highly refined identifications from Roman Morocco and Iron Age France, including the well-known site of *Lattara* (Cammas, 2018). Mud daub structures slake under rainsplash impact if not roof-protected, and daub quickly returns to a soil state in abandoned buildings; daub is mainly preserved by having been burned, which essentially "fossilizes" it (Figures 12.16a and 12.16b) (Kruger, 2015). Mediterranean Bronze Age sunken house fills are commonly colluvial in character and formed by rainwash from weathering clay walls (e.g. Bronze Age Crete, Cyprus, and Minorca; Carpentier, 2015).

Earth-based structures, such as hearths, can also be strongly preserved because of heating (Mallol et al., 2017) or protected by cementation of calcareous ashes; hearths associated with long periods of salt-making are of particular interest (Sordoillet et al., 2018). In the case of coastal salt making, the use of intertidal sediments as a naturally concentrated salt source, has led to burned sediments characterizing both Mayan and Romano-British sites, the latter being well known for their "redhills" (Biddulph et al., 2012; Graham et al., 2017; Macphail et al., 2012a) (see Chapters 13 and 14).

Mud bricks and adobe may be shaped in a wooden "former", and when the former is removed, they are left to dry and harden in the sun. Daub is more commonly applied to, and built up on, wattle walls and trellis-like frameworks (see Figure 12.16b), which themselves are supported by more substantial wooden structures (Friesem et al., 2017; Wauchope, 1938). Different forms of rammed earth (e.g. pisé) are built up within wooden frames, sometimes with wooden

Figure 12.16a An example of plant-tempered burned daub from Ecsegfalva, Early Neolithic Hungary (Körös culture); daub contains alluvial clay clasts (C) and void pseudomorphs of plant (P) temper; note rubified edge (RE). Length of thin section scan is ~55 mm.

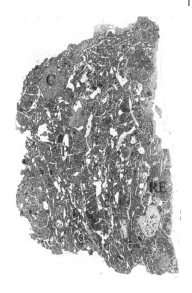

Figure 12.16b Burned adobe from an Iron Age house from el Cerro de la Gavia, Madrid, with wattle impressions (photo with permission from Amalia Pérez-Juez).

"reinforcing" supports and sometimes with an oblique horizontal boundary to the next horizontal layer, in order to impart structural strength (Pierre Poupet, CNRS, pers. comm.). One of the early geoarchaeological investigations into the study of tells and mounds, including pisé structures is that of Bullard (1970) at Tell Gezer, Israel.

Other variants of daub constructions include local English names such as cob, where the daub contains a high amount of chalky material, such as from East Anglian Chalky Boulder Clay; it is seen at the reconstructed Anglo-Saxon Village at West Stow, Suffolk, England (West, 1985). In comparison, clunch is a hand-puddled chalky soil that is employed to build up a thick (20–30 cm) freestanding wall (e.g. Butser Ancient Farm, England). In the latter case, the idea is again to remove all the air. This process may be surprising to those who have examined ancient adobe, daub, and mud brick, because these materials from archaeological sites appear to contain many voids. These voids, however, are in fact the spaces left by the oxidized plant temper; long articulated phytoliths and possible humic staining are sometimes the only remains visible (Courty et al., 1989) (Figure 12.16a). Plant material can also sometimes be identified from molds of these voids, or plant impressions (see Friesem et al., 2017 for wattle and daub) (Figure 12.16b). It can be noted that when small amounts of pounded chalk or other limestone, are added, when the mixture dries a small amount of secondary micrite forms and this helps the "cementation" process.

Adobe is the common term applied to earth-based building materials in Africa, the Far East, and the New World. In the USA it is defined as an unburned brick dried in the sun, the term being adopted during expeditions to ("Spanish") Mexico during the early 19th century. The term can also be employed to describe a "clay" used as a "cement". Both adobe and rammed earth (e.g. as used in China; Jing et al., 1995) preserve well in dry climates, but may collapse during a particularly wet season in some countries; plant-tempered adobe layers were constructed as a leveling preparation layers for tufa floors at Middle Neolithic (Yangshao) Huizui, Henan Province, China (Figure 12.17) (Macphail and Crowther, 2007a). Adobe/rammed earth walls therefore may have a thatch or tile top for protection against rain, and/or, like many turf walls, are built upon a masonry foundation (as protection from ground water) to offset slumping (Pierre Poupet, CNRS, pers. comm.) (Sigurðardòttir, 2008: 16).

Morris (1944) conducted one of the earliest studies of adobe buildings in pre-Conquest North America, and compared mud brick building techniques of the New World with those of 3500 BC Sumeria. He also made one of the first identifications of an adobe mud brick kiva wall at a Pueblo site near Aztec, New Mexico. Here the site, dated to AD 1090–1110, featured hand-formed bricks around 5–12 inches long, 3–5 inches wide, and 2–3 inches thick, but without temper. It was estimated that the construction of Huaca del Sol (Peru) (ca. 28 m high by 342 × 159 m) required more than 143 million adobes (mud bricks), with the Luna Platforms requiring more than 50 million (Hastings and Moseley, 1975). The adobe bricks, were manufactured variously from local water lain silt, more coarse desert soils, and more organic "sumps" within the valley. They were formed

Figure 12.17 Scan of 15 cm-long impregnated block (M18) at Huizui, Henan Province, China, that sampled Layer 5 – a Middle Neolithic (Yangshao) plant-tempered adobe preparation surface (APS), and Layer 6 – a series of fossiliferous tufa floor layers (TFL) composed of quarried slabs of tufa; the basal slab and overlying thicker slabs showing natural horizontal splitting. Tufa is a type of limestone formed in calcareous springs. Before soil micromorphology was carried out at the site, the calcareous floors (now tufa slabs) were assumed to be lime plaster floors. Above (BDD) is a layer of burned daub debris produced after the adobe-walled structure was razed by fire.

in molds and varied in size according to stage of construction, changing from 8–11 cm thick to 12–18 cm; a study of the size range, mold markings, and maker's marks also revealed chronological implications. In the Near East, geoarchaeological studies of adobe constructions from tells revealed information on the source of mud bricks, as well as decay processes acting at the sites, such as Tell Lachish (Goldberg, 1979a; Rosen, 1986) and Çatalhöyük (Turkey) (Love, 2012) (see Table 13.1).

12.3.3 Lime- and Gypsum- Based Building Materials

For convenience and as a very broad generalization for European studies, coarse-tempered materials are termed "mortar", such as that used to cement building stones (Rentzel, 1998) or as Roman arriccio (a coarse layer applied directly to the face of a wall), and generally finer-tempered coatings are called "plaster" (e.g. Roman intonaco: fine surface layer applied on top of arriccio; Mora et al., 1984: 10; Sue Wright and MOLA staff, pers. comm.). A painted, fine "plaster" over a coarse-tempered ("mortar") wall covering is well illustrated in Pye (2000/2001: fig. 1) (Figure 12.18). Plaster essentially differs from mortar by containing a temper dominated by sand rather than gravel-size material (Courty et al., 1989: 121).

There are numerous types of lime- (gypsum-) based constructional materials, some of which have been in use for around 9,000 years (Rollefson, 1990; Stoops et al., 2017a; Stoops et al., 2017b). The first use of lime plaster dates to the end of the Epi-Paleolithic and Neolithic (Friesem et al., 2019; Kingery et al., 1988); gypsum plaster was utilized within the Neolithic as early as 5000 years ago by the Egyptians (Stoops et al., 2017b). The Siloam Tunnel, Jerusalem for instance was lined with plaster, which has now been dated to around 700 BC (Amos et al., 2003). Lime plasters have also been recorded from ancient Roman and Greek sites, along with examples from Mesoamerica and China (135–137 (Karkanas, 2007; Karkanas and Goldberg, 2018b; Stoops et al., 2017a; Villaseñor and Graham, 2010).

In order to produce a lime-based mortar or plaster, limestone (including chalk) would be burned so that $CaCO_3$ was converted into quick lime (CaO); slaked lime is formed from this through the addition of water ($Ca(OH)_2$). The hydrated lime then reacts with carbon dioxide (CO_2) in the atmosphere to produce $CaCO_3 + H_2O$; lime was created from burned shells, including conch shells in Maya Belize (Blake, 1947; Graham et al., 2017; Stoops, 1984; Stoops et al., 2017a). Essentially, therefore, the composition of a mortar or plaster is the same as the original material. It occurs either as fine limey material, but more commonly with added coarse material, which serves as an aggregate (temper).

Gypsum is a ubiquitous mineral found in both marine and lake deposits. It also forms major geological strata, as for example the gypsum mines in Germany and around Paris where "Plaster of Paris" is manufactured by heating gypsum. As a cautionary tale, the presence of gypsum crystals in some early Medieval London occupation deposits confounded workers in conservation, but was easily explained as the result of the importation to the site of estuarine Thames plants and associated sediment that naturally contain gypsum (Kooistra, 1978). At the Roman city of Segobriga in Spain (Gómez and di Monti, 2009), sheets of selenite (gypsum) (*lapis specularis*) were mined and used extensively in windows because of their lucid properties (Figure 12.19).

12.3.3.1 Mortar
The term "mortar" refers to a mixture of quicklime with sand and water, used as a bedding and adhesive between adjacent pieces of stone, brick, or other material in masonry construction.

(a)
(b)
(c)
(d)
(e)
(f)

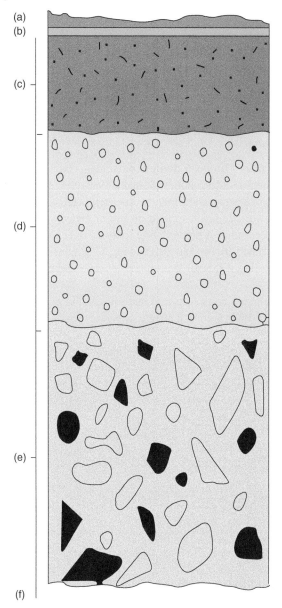

Figure 12.18 Schematic section through wall mortar, plaster and painted surface of the Roman Empire (from Pye, 2000/2001); diagram to show layered structure of painted wall plaster; (a) the painted surface; (b) layer of pure lime usually 1–2 mm thick, but which is not always present; (c) fine plaster, often less than 10 mm thick and sometimes mixed with ground marble or other white filler (temper); (d) coarse plaster, often more than 10 mm thick; (e) preparation layer of plaster of varying thickness, mixed with coarse filler (temper) and applied direct to wall; (f) wall below the plaster layers.

It commonly consists of one volume of well-slaked lime to three or four volumes of sand, thoroughly mixed with sufficient water to make a uniform paste easily handled on a trowel (www.infoplease.com/encyclopedia/science/tech/terms/mortar). There are variations, however, in this composition, which are delineated in thin sections of the hardened material according to their coarse/fine ratio (sand temper:fine matrix) (Macphail, 2003a; Stoops, 1984). As noted above, lime mortar hardens by absorption of carbon dioxide from the air (Blake, 1947). Various forms of mortar have been recorded from as early as the Pre-Pottery Neolithic in the Near East (Garfinkel, 1987; Goren and Goldberg, 1991; Rollefson, 1990).

The term "mortar" is used here in a generic sense, for a coarse-tempered "cement" that can be applied to walls, and make solid floors. At Roman sites the latter is exemplified by *opus signinum*

Figure 12.19 Photo of translucent Roman windows composed of selenite (gypsum) from the Roman forum at Cartagena (Spain). The frame of each pane is ~20 cm across.

Figure 12.20 Field photo; Colchester House (PEP89), City of London. In this profile, soil micromorphology samples were taken through the natural buried soil, the overlying "early dark earth" and into the massive concrete and tiled post AD 350 floor of a substantial aisled building (possible basilica; Sankey, 1998). The sealed "early dark earth" records soil formation in weathered earth-based structures (Macphail and Linderholm, 2004b), with artifacts and coins associated with busy 1st–2nd C London acting as an entreport (Sankey, pers. comm.). The dark earth above the floor formed between the 4th C and medieval times.

that often would have had a fine plastered surface, or be tessellated (with "coarse" ca. 2 cm size ceramic tessera bricks), or covered with a mosaic (using "fine" ca. 1 cm size stones etc). At Roman Colchester House, London, an *opus signinum* floor was tiled (Figure 12.20).

At the Courage Brewery site, Southwark, London (Cowan, 2003), a reference fragment of 1st century AD *opus signinum* was found to be composed of pure microcrystalline (micritic) calcite cement with dominant gravel-size (>2 mm) clasts (temper) that include flint, burned brickearth, limestone, chalk, and pottery (Courty et al., 1989). It also has a fine and medium well-rounded sand-size temper likely derived from Thames alluvium ("river sand"; Blake, 1947). At the London Arena site a thick mortar floor marks a late phase of use, below which are the arena sands

(Figures 12.21 and 12.22). Chalk fragments and other small limestone pieces may be relict of incompletely burned lime, as found in building waste for the reconstruction of a Romano-British "Sparsholt" villa, at *Butser Ancient Farm* (Figures 12.23 and 12.24) (Butser Ancient Farm, 2009; Karkanas, 2007). Mortar from the façade of the Alamo, San Antonio, Texas, USA, was also found to be characterized by moderately sorted micritic grains and pisolites, many showing a burned (lime-making?) history; these coarse materials were likely sourced from the local San Antonio River (Figures 12.25–12.26) (Goldberg, 1996).

On the other hand, coarse limestone and pounded pottery were typical tempers added for strength, and in the case of the latter produced a strong water-resistant "hydraulic" mortar

Figure 12.21   Field photo; Roman London Arena, London Guildhall; dark earth over mortar floor. A whole series of laid arena sands underlies the mortar floor (Figure 12.21).

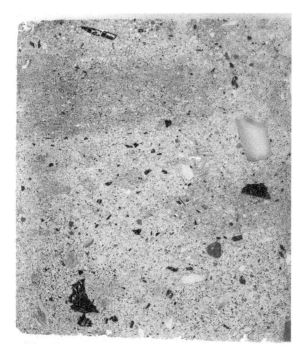

Figure 12.22   Scan of Roman London Arena sand layers (Figure 12.20), London Guildhall; sands are diffusely layered with examples of flint and charcoal, with areas and clasts of poorly cemented mortar, perhaps recording earlier "floors" and/or repairs (cf. Rentzel, 2009). Washed and raked sands are artificially layered. Frame width is ~65 mm.

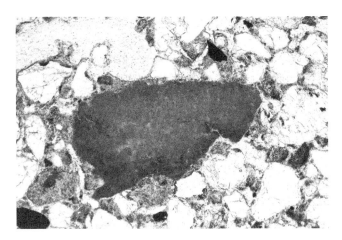

Figure 12.23 Photomicrograph of mortar building waste from the reconstruction of the Romano-British Sparsholt Villa at Butser Ancient Farm, Hampshire, UK. Incompletely burned chalk is a relict of lime-making, within the weathering sand-tempered mortar – hence decalcified voids. Chalk is weakly rubified and very finely fissured. PPL, frame width is ~2.38 mm.

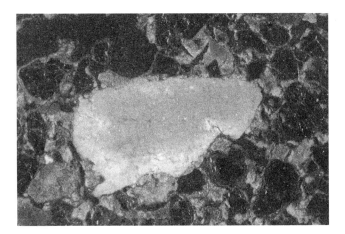

Figure 12.24 As Figure 12.23, under OIL; rubified sands, grayish yellowish lime binder remains and whitened chalk are in evidence.

(Blake 1947: 322–323). Here SiO_2 has been released and strengthens the cementation process. Hydraulic mortar is especially important for water basins and aqueducts (Malinowski, 1979; Rentzel, 1998). The manufacturing process for the famous pozzolanic cements used volcanic rocks as temper, because again the silicate minerals reacted with the lime to produce a hard material (Ca-Al-silicates) that even hardens under water (Brown, 1990; Lechtman and Hobbs, 1983; Malinowski, 1979); the Maya use of volcanic materials to produce pozzolanic plaster has already been noted (Villaseñor and Graham, 2010). Although low amounts of organic matter in the cement may originate from accidental contamination of material from the mixing trough, a small addition (10%) of impurities was said to add to the mortar's hardness (Blake 1947). A typical ratio of *coarse*

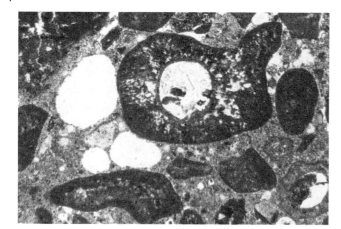

Figure 12.25 Photomicrograph of mortar from the façade of the Alamo, San Antonio, Texas, USA; moderately sorted coarse sand and gravel-size micritic pisolites (temper) with fine micritic cement. PPL, frame width is ~4.1 mm.

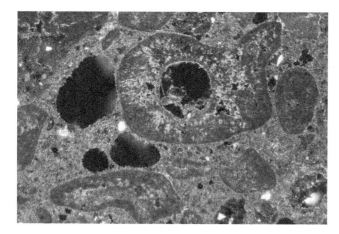

Figure 12.26 As Figure 12.25, under XPL; note rare quartz silt component and probable voids recording decalcification.

(gravel and sand) to *fine* (cement) material at 60:40 found in Roman London sites, is comparable to the soil micromorphological findings of Stoops (1984) from 15 examples of mortar from Roman Pessinus, Turkey.

Weathering can break down the lime-based cements (Figures 12.23 and 12.24), hence the need for "waterproof" hydraulic mortar in some constructions. Cements that were used for building foundations standing below water level had to be particularly resistant to water. One way to offset this problem was achieved during the mid-13th century building of the (short-lived) barbican of Henry III at the Tower of London (Keevill, 2004). Here, the stones were "cemented" with molten lead, hence the anomalously high levels of Pb found during heavy metal analysis of the juxtaposed and newly formed moat sediments (Macphail and Crowther, 2004).

12.3.3.2 Mortar Floors

Mortar floors can have a variable lime plaster component and may be surfaced by a fine plaster (Karkanas and Efstratiou, 2009; Karkanas and Van de Moortel, 2014) (see Figures 12.20–12.21). A well-made mortar floor with a flint gravel temper at the high status Dragons Hall, Norwich, UK was capped by a fine plaster screed (Macphail, 2003a; Shelley, 2005). Microprobe mapping picked out the siliceous (Si) nature of the flint gravel and quartz sands in the mortar and plaster screed surface, as well as the dominance of calcium (Ca) in the fine matrix of micritic calcium carbonate. Higher amounts of siliceous sands are present in the occupation floor and humic post-abandonment (post-razing) deposits, which include charred thatched roof remains and likely associated examples of guano. At the Ottonian (10th century) Church site, Magdeburg (Germany) a number of "mortar floors" were separated by beaten floor deposits. However, the mortar appeared to be building work residues, rather than floors, *sensu stricto* (Macphail et al., 2007a). The term "floor" also has to be nuanced in some contexts. For example, the lime plaster floors of Maya Marco Gonzalez (Belize), are in fact constructed surfaces for salt production (see below). Sometimes, mortar/plaster floors in North-West Europe are difficult to recognize, because of cool moist weathering conditions encouraging the dissolution of the micritic matrix (see Chapter 4) (cf. 135–137, Karkanas and Goldberg, 2018a; Stoops et al., 2017a). Very thin ghosts of the micritic cement – now without any high interference colors of calcite – may occur and these can still be white under oblique incident light (OIL). Equally, a lime plaster floor may only occur as an anomalous floor layer of well-sorted sand or have the appearance of a sandy clay loam, and it can be difficult to find any traces of the original calcitic cement. Medieval examples of this were found at Odense, Funen, Denmark (Figures 12.27–12.29).

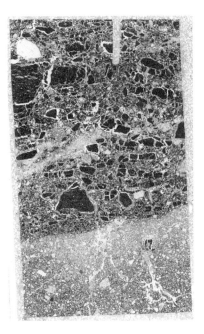

Figure 12.27 Scan of medieval floor layers at Odense, Funen, Denmark, comprising charcoal-rich fuel ash waste spreads including vesicular and isotropic silica slags (LOI = 23.7%), over pale yellow-brown remains of a suggested lime plaster floor (Figures 12.28 and 12.29).

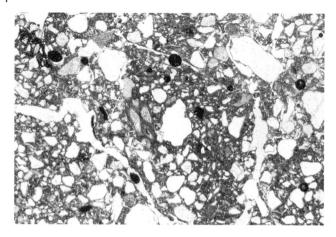

Figure 12.28 Photomicrograph of suggested weathered lime plaster floor at Odense, Funen, Denmark (Figure 12.27); now a sandy clay loam with darker areas of residual weakly calcitic character (enhanced birefringence consistent with the presence of calcite under XPL – not illustrated). PPL, frame width is ~4.62 mm.

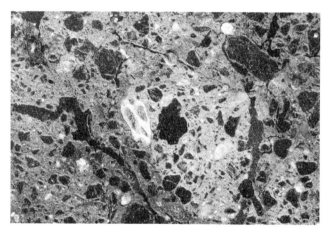

Figure 12.29 As Figure 12.28, under OIL; note yellowish gray colors and small area of grayish white suggested lime plaster residues.

12.4 Metal Working

Increasingly, ferrous and non-ferrous metal working indications have been found in geoarchaeological deposits. In the first instance, *iron-working* debris can be found in loose bulk samples by the simple use of a hand magnet (J. Cowgill, pers. comm.). Such industrial debris and installations can have a very strongly enhanced magnetic susceptibility (e.g. 8001 χlf 10^{-8} m^3 kg^{-1}; metalworking smithy oven with hammerscale present at Åker gård, Hamark, Hedmark, Norway; see Chapter 13).

Iron working and slags In thin section, hammerscale, the material that is produced when the hot iron is hit with a hammer, is opaque, but under oblique incident light can show multiple fine layering, picked out by metallic and reddish brown colors according to preservation. EDS analysis of some iron fragments at the Åker gård smithy records 100% FeO (see Figures 13.4a–13.4d).

Iron slag is often globular with a vesicular porosity, and is a mixture of both newly formed silicate minerals (olivine group iron silicate – fayalite) and iron that shows a dendritic pattern, with a grayish metallic luster under OIL (Angelini et al., 2017; Macphail, 2003a). Often such materials are associated with strongly heated rocks and other minerogenic materials, as well as fuel ash waste (Berna et al., 2007; Röpke and Dietl, 2017). In addition to the use of instrumentation (SEM/EDS, microprobe) further work may be carried out by specialists using a metallurgical incident microscope. Occupation floor deposits may record anomalously high magnetic susceptibility simply because *iron slag* has been trampled in from local artisan working areas (Late Saxon domestic floors with a wide variation in magnetic susceptibility; range 153–3278 χlf 10^{-8} m^3 kg^{-1}; Macphail et al., 2004). If iron slag undergoes hydromorphic weathering ("rusting") at wet sites, the iron loses much of its magnetic susceptibility (cf. Crowther, 2003).

Non-ferrous metal working Clues to non-ferrous metal ore processing may simply involve standard petrology, for example in the identification of artificially compounded aggregates of tin ore – *cassiterite* – characterized by high relief and marked cleavage. Such aggregates from the processing of local placer deposits (river sands) were found in a Bronze Age pit, near Bodmin Moor, Cornwall, UK, where bulk chemistry measured very strong enrichment of tin, which is normally in trace element amounts (7070 µg g^{-1}; ca. 0.0071% Sn and microprobe established the high purity of the aggregates (94–98% Sn by microprobe count; T. Rehren, UCL, pers. comm.) (Macphail and Crowther, 2008). Use of furnaces, even cremations, can produce "furnace prills" – spherical to subspherical aerosols of coarse silt to sand size. These vary from weakly ferruginous aluminum-silicate compounds to ferruginous examples, likely originating from iron working. In rare cases, such prills are typical of non-ferrous metal ore processing (J. Merkel, UCL, pers. comm.). Examples of ore residues and prills from the medieval district of Torvet, Trondheim are given in Figures 12.30–12.33. Although copper dominates, significant amounts of tin and lead also occur; such ores were also a source of arsenic and antimony. Vesicular, isotropic siliceous glassy slags (18.8% Si) also occur, that are pale green to pink under OIL, and are tin- and lead-rich (12.0% Sn; 8.12% Pb); other non-ferrous metals are recorded – copper and zinc (3.13 Cu; 1.55% Zn). At 11–15 Borough High Street, London Borough of Southwark, Roman artisan/industrial non-ferrous metal working waste, including clearly greenish (PPL and OIL) copper and tin corrosion products (79–93% CuO). These included possible copper sulfate and carbonate associated with background ashy deposits; high tin copper alloy also occurs as demonstrated by SEM/EDS analyses (71–72% SnO_2) (Macphail and Crowther, 2015b). It can also be noted that 12th century bell production at Magdeburg Cathedral,

Figure 12.30 Photomicrograph of medieval urban deposits in the Torvet district of Trondheim, Norway (M14133; C14240); compact upper Layer 5354 contains non-ferrous metal spherule ("droplet" or "prill") within with a metal salt "cemented nodule". PPL, frame width is ~4.62 mm.

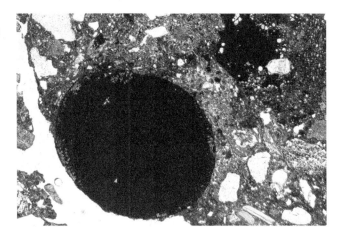

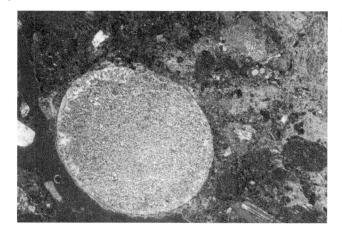

Figure 12.31 As Figure 12.30, under OIL.

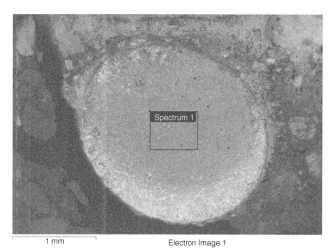

1 mm   Electron Image 1

Figure 12.32 As Figure 12.31; SEM/ EDS X-ray backscatter image. Spheroidal "prill" – copper-dominated furnace slag/aerosol. Scale = 1 mm.

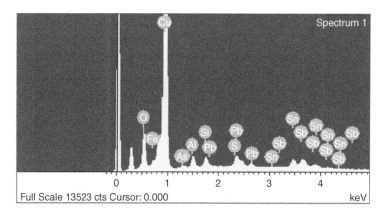

Figure 12.33 As Figure 12.32, X-ray spectrum of spheroidal prill in 13.32, recording presence of sulfur (0.96% S), copper (51.7% Cu), arsenic (1.43% As), tin (6.04% Sn), antimony (4.44% Sb) and lead (5.95% Pb).

Germany produced contaminated deposits (bulk analysis: 156 µg g^{-1} (0.000156%) Pb, 40.8 µg g^{-1} (0.0000408%) Zn and 441 µg g^{-1} (0.000441%) Cu), with partially corroded opaque to greenish bronze casting droplets also being present (microprobe: 29.3% CuO, 17.8% SnO$_2$, 0.97% PbO; $n = 4$) (Macphail et al., 2007a). Further examples can be found in Angelini et al. (2017).

Non-ferrous metal contamination is often not visually obvious but can be found employing heavy metal analysis (Cu, Pb, and Zn). Medieval and post-medieval Tower of London moat deposits were characterized by small amounts of heavy metals probably due to the Royal Ordnance and Mint being located in the Tower of London (Macphail and Crowther, 2004). At the Late Roman site of Stanford Wharf, anomalous staining of muddy clay floors in a structure suspected to be linked to salt working were analyzed employing SEM/EDS, and concentrations of lead were found (max 54.1% Pb); floors were thus more securely interpreted as the location of lead vessels employed in salt working, which had already been suspected (Biddulph et al., 2012; Macphail et al., 2012a). At the medieval Swedish site of Ingarvet waterlaid mining tailings are characterized by burned mineral inclusions (MS lf = 172.4–217.8 χlf 10^{-8} m^3 kg^{-1}), and heavy metal accumulations (1.12% Cu, 0.155% Pb and 0.303% Zn; XRF data), correlating with a peak in iron (2.53% Fe), observed as iron staining.

12.5 Fuels and Fuel Waste

Many sites are characterized by charcoal, with concentrated coarse round wood material being interpreted as fuel ash waste, even whilst no calcitic ash may be found, due to leaching (see Figure 12.23), winnowing, and/or water sorting (Goldberg and Macphail, 2012) (see Chapter 13 for other fuel materials; Mallol et al., 2017). When deposits include large amounts of charred grain, for example, this more likely records waste from burned grain stores as for example, at the Viking settlement of Heimdalsjordet (Norway) (Macphail et al., 2017). If *in situ* fire installations are investigated the make-up of the fuel components may be recognizable. Some Norwegian Iron Age fire pits have fuel waste in them with twig wood charcoal below more coarse round wood remains, indicating the fire pit was lined with twig wood; probable one-year-old willow (*Salix* sp.) twig wood was found lining two pits at Unnerstvedt and Ragnhidrød, Vestfold, Norway, for example (Viklund et al., 2013). Anomalously high amounts of chlorine (Cl) in wood charcoal in hearths at a number of Norwegian coastal sites infers the use of driftwood (e.g. Linderholm et al., 2019). The ashy remains of low temperature fires employing twig wood of local mangrove origin occur as stratigraphic components over heated lime plaster "floors" in Maya salt working levels at Marco Gonzalez (Belize) (Graham et al., 2017). The use of turf and/or peat as a fuel has already been noted (12.3.1.1). Coal and clinker remains of burned coal (Canti, 2017) are more likely to be medieval and post-medieval in age, with such waste being found in Elizabethan (16th century) manured ridge-and-furrow soils near Wolverhampton (UK) (Hewitson et al., 2010).

In contrast, high-temperature minerogenic fuel ash slags, which are colorless to gray, isotropic, and vesicular, have been noted in Weiner (2010), and microprobe and EDS studies have found these to be Si dominated. If linked to earth-based furnace make ups, quartz and feldspars can be melted and fused into these plant residues (cf. glassy slags; Figures 12.34 and 12.35; Berna et al., 2007; Röpke and Dietl, 2017). Such vesicular slags should not be confused with ashed monocotyledonous plant remains (cf. Canti and Brochier, 2017), where the vessel system is only slightly altered and fused stems are preserved (Figure 12.36). An abundant, probable example of this was found in redhills linked to salt working on the Essex coast in the UK (Stanford Wharf), where wetland plants, possibly including sea rush leaves (*Juncus maritimus*), had been utilized as a low temperature fuel (Biddulph et al., 2012; Hunter, 2012; Macphail et al., 2012a). Bellows plates from this site are an unusual form of *briquetage* because they appear to be ceramic like, although in fact they are composed of little altered intertidal clay, and unusually have a "green glaze" (Box 12.2, Figures 12.36–12.40).

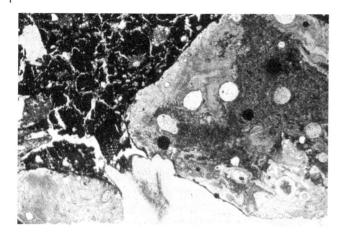

Figure 12.34 Photomicrograph of medieval urban deposits in the Torvet district of Trondheim, Norway (M8592; Smithy Profile West, Layer 28083 lower); charcoal-rich fuel ash waste and included melted vesicular glassy slag. PPL, frame width is ~4.62 mm.

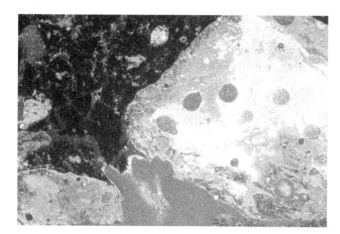

Figure 12.35 As Figure 12.34, under OIL, showing red and greenish colors. Glassy ash has an X-ray spectrum recording a tin- and lead-rich (12.0% Sn; 8.12% Pb) siliceous glassy slag (18.8% Si), with other non-ferrous metals recorded – copper and zinc (3.13 Cu; 1.55% Zn).

Box 12.2   Enigmatic green glaze 'ceramic' briquetage linked to salt working

At the Stanford Wharf coastal salt working site on the tidal River Thames, Essex, UK, an enigmatic late Roman ceramic-like material with a green glaze was investigated, because it was completely different from the site's pottery, normal briquetage, and burnt clay deposits (Biddulph et al., 2012; Macphail et al., 2012). Soil micromorphology, FTIR (F. Berna; Simon Fraser University, Vancouver, Canada), SEM/EDS and microprobe methods were applied. Salt working waste deposits forming 'redhills' of burnt estuarine sediments used in the 'sleaching' method of salt working, also included white nodules that are siliceous burnt plant stem debris (Figure 12.36). The briquetage is "manufactured" from estuarine sediments with a mica-smectite clay component, as established by FTIR. Sections through the green glaze briquetage record the decreasing effects of heat downward, forming (1) a surface green glaze composed of strongly heated silicate glass (minimum 700– 800°C to around 1000°C), (2) below an upper blackened part being more strongly heated (>700°C) than the underlying rubefied layer (400°C– 700°C). Microprobe elemental mapping and Line analyses were also carried out

Box 12.2 (Continued)

(Figures 12.37–12.40). The dominantly siliceous glaze contains statistically significant (Mann-Whitney U test; $p < 0.05$) greater quantities of Na, P, and Fe compared to the briquetage. Fe is possibly responsible for its greenish colour, and P is probably derived from burned fuel. It is possible that Zn may also have a similar origin (marine plants and sediment may have concentrated Zn from sea water). Amounts of Na are anomalously high but diminish away from the glaze surface; this trend can be viewed as supporting a salt (NaCl) making origin for this green-glazed briquetage.

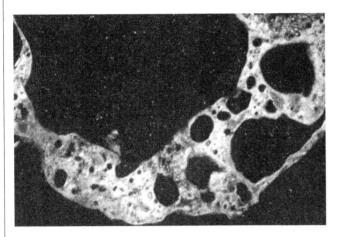

Figure 12.36 Stanford Wharf, Thames Estuary, Essex, UK. Photomicrograph of 'white nodule' (probable ashed plant stem), showing relict vascular structure which has a pseudo-vesicular appearance. SEM/EDS analyses: one area revealed that Si dominates, with values of Si=28.2% (60.2% SiO2); also present are: Na=2.47%, Al=7.38%, P=1.89%, K=4.08%, Ca=3.62% and Fe=4.18%. Oblique incident light (OIL), frame width is ~4.62 mm.

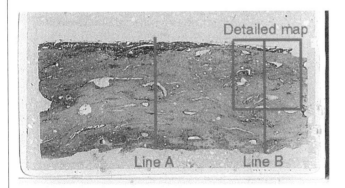

Figure 12.37 Scan of M4240A1 (large 'Green Glaze' fragment) composed of plant tempered *briquetage* coated with vesicular glass. Sample underwent both SEM/EDS and microprobe analyses. Whole thin section and smaller (detailed map in blue) were mapped for elements. Two quantitative line analyses (×100 100×100 μm areas), Lines A and B were carried out. FTIR was employed on parallel thin section M4240A2. Frame width ~48 mm.

(Continued)

Box 12.2 (Continued)

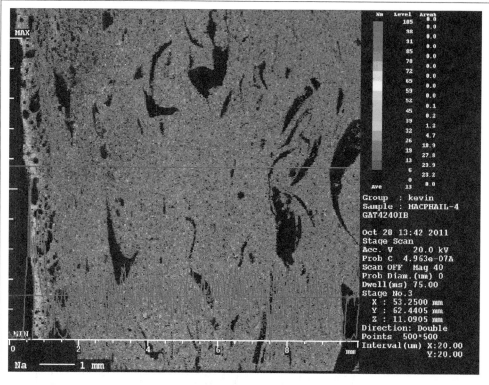

Figure 12.38 Detailed microprobe map area (see Fig 12.37); Na (sodium) is specifically concentrated in the green glaze layer (left edge; Na=12% n=4). Scale=1 mm.

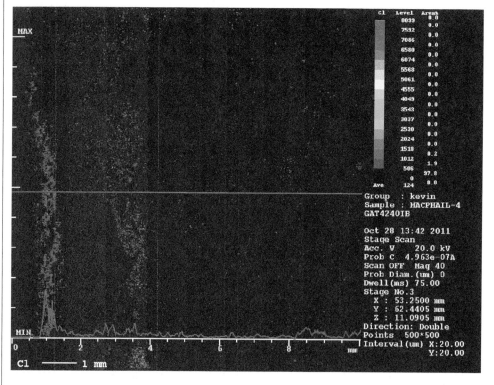

Figure 12.39 As Fig 12.37; Cl – chlorine also appears to be focused on the surface, but as Cl is a major component of the resin impregnating the thin section, Cl in the form of salt (NaCl) sensu stricto cannot be proven to be focused in the green glaze, but given the surface concentration, it seems plausible.

Box 12.2 (Continued)

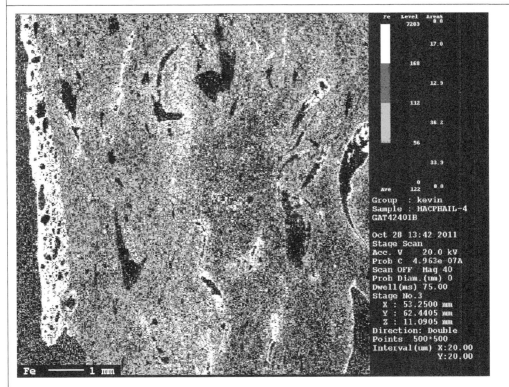

Figure 12.40 As Fig 12.37; Fe – iron is most strongly concentrated along the green glaze surface (left edge), but also highlights the laminated structure and relict plant temper material (void pseudomorphs). Scale=1 mm.

12.6 Concluding Remarks

When working in geoarchaeology it is important to realize how much of the excavated context is composed of anthropogenic deposits; this is discussed in Chapter 13. In this chapter we have looked at some of the individual components that make up a site, in terms of the human use of materials. First, the identification of natural materials manipulated into stone tools and ochre are briefly noted for early prehistoric sites. Secondly we show that the utilization of earth- and geology-based materials varies greatly, and includes the simple use of turf and soil for constructions, the manipulation of soils and regolith into forms of adobe, and finally full processing and transformation of natural materials into lime/gypsum plasters and mortars. We also focus on some other industrial activities, and examine the differences between iron production and working, and non-ferrous metal pre-processing and processing, as well as fuel wastes and how these products can be recognized and best analyzed.

13

Anthropogenic Deposits

13.1 Introduction

In other chapters, we described and discussed natural sediments and soils (Chapters 2 and 4), human effects on soils (Chapter 11), and human materials that can be found on sites (Chapter 12). This chapter examines how the wide variety of anthropogenic deposits, which may include natural soil and sediment components as well as the human use of materials, are understood in terms of site formation processes and the interpretation of sites with noticeable occupation, such as found in tells (Table 13.1). Moreover, the investigation of anthropogenic deposits gives us insights into constructions (and destructions), how sites functioned, and fundamentals such as use of space within buildings, and in settlements and the landscape as a whole. Earth and turf may have been deliberately used to build a mound, but soils and sediments may be tracked in and accumulate on floors and in hollow ways. Equally, discarded human use of materials may characterize an activity area, contribute to *middens* or be deliberately employed for *manuring* or even used for ground-raising constructions. In this light, anthropogenic deposits must be considered as integral parts of the archaeological record on the same footing as material remains such as lithic remains, ceramics, architecture, etc. (Karkanas and Goldberg, 2018b; Miller, 2011).

This chapter will introduce some typical anthropogenic deposits and associated site formation processes, components within the settled landscape such as monumental constructions, and make up of small (farmsteads) to large settlements (towns) (see Table 13.2). It will also deal with variations in intensity and longevity of occupation. For example, both *tells* and urban sites reveal major changes in anthropogenic deposition over centuries (or millennia) representing intensive occupation: Borduşani-Popină Tell (Romania), 1st century AD Roman to Medieval London and Canterbury, UK, and 3rd millenium BC to Medieval Byblos, Lebanon (Figures 13.1a–13.1c and 13.2a–13.2g). Clear differences are noted between the natural sediment of the Borcea River alluvium and anthropogenic deposits making up the tell. Important finer scale studies that differentiate use of space, and which are all broadly termed "*occupation surfaces*" but which may be of domestic, stabling or pathway origin, are also presented (see Tables 13.1, 13.2). Also important is the recognition and evaluation of destruction levels in urban areas, when reconstructing use of space and settlement morphology through time because destruction deposits may also contribute ground-raising constructions (Forget and Shahack-Gross, 2016; Kruger, 2015); burning also changes the morphology, magnetic susceptibility, and chemistry of byre deposits, for example (Macphail et al., 2004) (see Figures 13.2e and 13.6).

Practical and Theoretical Geoarchaeology, Second Edition. Paul Goldberg and Richard I. Macphail.
© 2022 John Wiley & Sons Ltd. Published 2022 by John Wiley & Sons Ltd.

Table 13.1 Micromorphological and macromorphological attributes from Near and Middle Eastern (Çatalhöyük, Turkey; Tel Brak, Abu Salabik, Iraq; Saar, Bharain) and Romanian tell sites (Borduşani-Popină and examples from Hârşova) (modified from Matthews et al., 1997 – see also Cammas et al., 1996; Courty et al., 1994; Haită, 2003, 2012; Shahack-Gross, 2011; Courty, pers. comm.).

Context	Prepared or unprepared surfaces	Accumulated deposits	Post-depositional modifications
Roofed structures			
Food preparation	Loam plastered surface Silt loam sediment slabs; phytolith-rich mat remains	Discrete strong parallel oriented lenses of organo-mineral deposits with grindstone fragments; vegetal pseudomorphs and siliceous graminae plant fragments; fish residues	Organic staining, bioturbation, salt formation immediately below surface
Food cooking adjacent to ovens and hearths	Plastered and compacted surfaces	Multiple layers of moderate horizontally oriented loam with organo-mineral material, burnt fuel, oven fragments, bone, flint/obsidian flakes, charred grain, tubers, date palm phytoliths	Subhorizontal cracking in plaster floors; salts; bioturbation; organic staining
Storage in small rooms or bins	Clayey plaster with included organic matter; gypsum plaster	Charred cereal grains, either pure or mixed within building debris;	Bioturbation
Reception/"clean activities"	Well prepared plasters often with finishing coats and impressions of mats/rugs	Thin lenses of charred and siliceous plant remains; sterile silty clay with strong parallel orientation	Organic staining and horizontal cracks
Ritual (associated with altars, sculptures and wall paintings	Multiple plaster layers, often with fine coats; occasionally painted; impressions of mats/rugs	Burnt remains; waterlain crusts; red ochre remains	Organic staining, salts and bioturbation
Probably roofed			
Stables	Very few prepared surfaces; overall undulating; silt loam sediment slabs	Interbedded lenses of fragmented dung pellets; oxidized and ferruginized weakly humic dung spherulite concentrations	Organic staining, salts, and bioturbation; phosphate-stained
Razed buildings			
Roof remains and ground-raising	Blackened and weakly rubified surfaces	Ash, coarse wood charcoal (timbers) and semi-melted phytoliths (thatch), with rubified mud brick (walls)	Minor bioturbation when sealed with ground raising layers
Unroofed			
Domestic courtyards and streets	Few plasters; aggregate hard-core surfaces; unprepared surfaces; bitumen pathways	Layers or unoriented deposits with cultural references; reworked wind- and water-lain sediment clasts; uncompacted refuse deposits; undisturbed wind and waterlain deposits; dung; dung spherulites, bone and phosphate-rich laminae, fish residues	Salts; bioturbation; wind and water reworking; phosphate staining

(Continued)

Table 13.1 (Continued)

Context	Prepared or unprepared surfaces	Accumulated deposits	Post-depositional modifications
Civic, administrative, and ritual courtyards	Mudbrick foundations with baked brick, lime plaster and plastered surfaces; few unprepared surfaces	Mineral-rich remains with some burnt and cultural refuse; thin layer of ash	Salts; bioturbation; wind and water reworking
Middens	Few prepared surfaces; mainly unprepared surfaces with different depositional episodes and in situ burning	Unoriented massive deposits; some wind and water-laid deposits and in situ burning; ashed-dung; bone and phosphate-rich laminae, fish residues	Bioturbation; settling and compaction; organic staining; phosphate staining

Table 13.2 A suggested guide to settlement composition including soil micromorphological examples.

The settlement

Within the settlement	Peripheral to the settlement	The settlement's wider landscape
Constructions		
Remains and residues of occupation surfaces and use of manufactured lime floors, mortar, daub, adobe/mudbrick etc.; storage pits, wells (also below), 'pit house' (e.g. Grubenhäuser), and post hole, and ditch fills that demarcate long houses (roof ditches) and settlement plots. Also present are hearth deposits, urban gardens; roof remains (roof collapse and razed buildings).	Ramparts and walls (earth and turf constructions; and buried soils); moats, ditches, millponds, paddy fields (potential fish source), grain dryers, baking ovens and cooking pits; landscaped gardens; fairs and markets.	Monuments – tumuli, grave mounds and other features – cursuses (see below); road systems, animal enclosures (see below); arable fields, and associated constructions- ridge and furrow; soft and hard rock quarries, field and forest boundaries.
Trackways, roads and paths		
'Rural' signatures – dung traces, minor P concentrations. Tracks linking byres and field. 'Urban' signatures – food, food preparation and domestic waste, with hearth and industrial residues, and concentrated fecal materials.	Chiefly 'rural' in character, in and out of settlement, linking local waterholes, 'infields', stockyards and pastures ('cattle paths' of Norway); some spillage of organic and settlement waste manures; stock and vehicle movements. Slipways, waterfront and harbors - shallow water sediments sometimes rich in refuse.	Chiefly 'rural' in character, linking major settlements (Bronze and Iron Age precursors of British Roman road system in places); accessing wider landscape – woodland resources and arable fields, with animal passage along droveways (e.g. transhumance) and pastures.

Table 13.2 (Continued)

The settlement

Within the settlement	Peripheral to the settlement	The settlement's wider landscape
Other transport (harbors and waterfronts)		
Animal management		
Stabling activity and byres, including tri-partite longhouses; specialized features from pig and bird husbandry (e.g. dovecotes).	Enclosures and corrals, and associated shallow waterholes.	Pasture soils and effects of stock concentrations (grazing, browsing and woodland pig husbandry); rock shelter stabling; accessing ponds and wetland.
Water management		
Sediments of well use and disuse; other features – moats and ditches; lead pipes (Pb traces).	Aqueducts, canals, ponds/fish ponds, reservoirs and millponds; irrigated fields and drainage ditches.	(For examples of hunting and gathering around natural water sources – see Borduşani-Popină tell)
Waste disposal 1; middening		
Street-side and street dumping; middening, pit and other feature fills and 'farm mound' creation; expedient refuse disposal within structures.	Major manuring with dung and settlement waste of 'infield' horticulture, with spillage along trackways.	
Waste disposal 2: latrines and cess pits		
Occurrence of concentrated human waste in cess pits and outflows (~sewers).	Human waste as manure.	
Specialist domestic and industrial activity		
High temperature burned sediments and fuel waste, ferrous and non-ferrous metal working, with leather and wood crafts, and use of lead (e.g. alloys and construction); storage and pitfills, food processing (cereal, fish and meat 'butchery', smoking, blubber boiling) and salt working.	Furnaces, lime kilns, grain dryers, and salt working; industrial waste in minerogenic manures and as recycled for constructions (temper in lime plasters and mortars, and daub – 'clay walls').	Fuel and raw material gathering and management of resources (woodland coppicing; iron, limestone, flint, clay for ceramics and floor constructions; intertidal plants and sediments for low temperature fuels/ salt-processing)
Mortuary practices		
Burials not normally found *within* Roman settlements, but Late Roman graves do occur; graveyards in ecclesiastical space; excarnation features in Iron Age settlements.Tomb re-use	Graves, cremations and excarnation features; grave mounds juxtaposed to settlements, including boat graves and ship burials.	Tumuli, grave mounds, mortuary houses, ship burials and excarnations.

(a)

(b)

(c)

Figure 13.1 Byblos, Lebanon. (a) An example of 3rd millennium BC(?) tell deposits (sample M5A) preserved in front of and below the 12th century AD crusader castle of Gibelet; (b) sampling monolith M5A; (c) sample M5A – undisturbed stratified occupation sediments (courtesy of M. Beydoun, American University of Beirut).

Figure 13.2a Field photo; Chalcolithic (4700–4000 Cal BC) tell site of Borduşani-Popină, Borcea River, Ialomiţei County, Romania, looking down from top of the approximately 9 m-high tell across recently reclaimed wetland formed on Borcea River alluvium (Figure 13.12b); the river and wetland provided food resources – Mallard duck and fish (Figure 13.2f), such as carp.

Figure 13.2b   Field photo; Borcea River terrace alluvium forming island upon which the Borduşani-Popină tell is located (Haită, 2003, 2012). These are natural sediments. Alluvium was sampled employing 80 to 200 mm size monoliths. Monolith 2 is at the top of the alluvial sequence.

Figure 13.2c   As Figure 13.2b, scan of M2 illustrating massive bedded alluvium, affected only by minor rooting and burrowing, and impregnative iron staining due to ground water (water table) fluctuations. Frame width is ~50 mm.

Figure 13.2d   As Figure 13.2c, photomicrograph of M2; burrowed alluvial coarse silts and very fine sands of loessic origin. Slabs of alluvium were employed for ground raising, making floors, and in mud brick manufacture. PPL, frame width is ~4.62 mm.

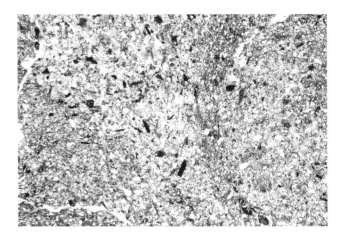

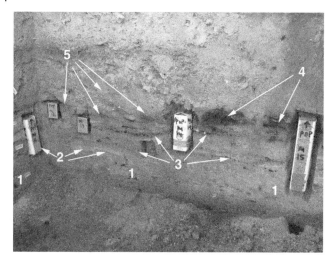

Figure 13.2e Field photo of anthropogenic deposits at Borduşani-Popină, Area 4, North side monoliths 14-15-16 and South side monoliths 11-12-13. 1: ground-raising mixed loess building material and remains of relict floors; 2: intact floors; 3: open area trampled debris spreads – middening – including latrine and fish processing waste (thin section sample M15C; Figure 13.2f); 4: blackened burned remains of byre floor deposits, mudbrick, and thatch (thin section sample 15A; Figure 13.2g); 5: rubified burned mudbrick, which has also been used for ground-raising.

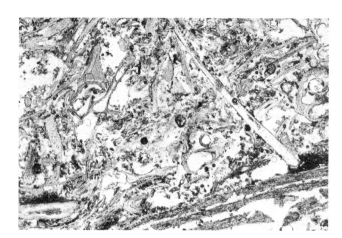

Figure 13.2f As Figure 13.2e (Deposits Number 3), photomicrograph of M15C (see Figure 13.2e). Here there is a marked concentration of fish bone and unknown yellow material, both of which are highly autofluorescent under blue light (BL). This is a presumed fish processing deposit (fish bones from freshwater pond fish such as carp are common at the site; Popovici et al., 2003; Radu 2003). There is also much reddish amorphous material and staining of bone; other butchery deposits can show reddish stained bone remains. PPL, frame width is ~4.62 mm.

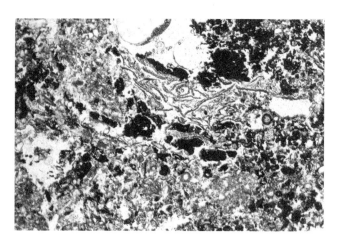

Figure 13.2g As Figure 13.2e (Deposit Number 4); photomicrograph of burned-down (razed) structure (M15A, 30022); the floor deposits include charred and ashed dung below coarse wood and twigwood charcoal, ash, and the vesicular siliceous remains of melted phytoliths (thatch). PPL, frame width is ~4.62 mm.

13.2 Concepts and Aspects of Anthropogenic Deposits

13.2.1 Modeling Anthropogenic Deposits

As noted above, one approach to the interpretation of anthropogenic deposits is to attempt to distinguish primary *waste*, primary transposed waste, and secondary-use waste (Renfrew and Bahn, 2001; Schiffer, 1987). Such distinctions are critical to interpreting correctly past human actions and behaviors.

Primary waste and activity areas In prehistoric contexts, this may simply be composed of a single material such as flint flakes on a Lower Paleolithic chipping floor where accumulation represents *in situ* reduction of a flint nodule to make a hand axe. Major examples occur at Boxgrove, West Sussex, UK, where hand axes were made on both a temperate marine mudflat (Unit 4b) and on an unstable cool climate interstadial sloping soil (Pope et al., 2020; Roberts and Parfitt, 1999) (Unit 8; see Figures 4.11a–4.11d; 9.1d–9.1f). Moreover, within Unit 4b, another activity area (GPT17) records both bone and flint artifacts at a horse butchery site.

Cooking and food preparation (burned bone, fish residues, oyster shell, and eggshells; Figure 13.2f; see Figure 13.13b), storage (cleaned grain), and stabling (*dung* accumulations; – see Figures 13.3a–13.3f and 13.4a–13.4b) all produce primary waste. Pottery production areas are characterized by kilns, the remains of kilns and pottery waste, while use of grain dryers may be indicated by deposits of burned daub, charred grain, and fuel ash waste of rake-out origin. Metal working produces both fuel ash and furnace fragments that record high temperatures compared to domestic hearths and drying ovens, for example (see Chapter 12) (Angelini et al., 2017; Berna et al., 2007; Dammers and Joergensen, 1996; Röpke and Dietl, 2017). Iron slag can be highly residual and does not necessarily indicate *contemporary* iron working. For example, it is a ubiquitous component in 4th century dark earth at Worcester, UK although iron working took place most intensively in the mid-3rd century (Dalwood and Edwards, 2004). The identification of co-occurring remains of high-temperature furnaces, iron slag,

Figures 13.3a–13.3e Figs 13.3a–3e: Ashed and oxidised stabling remains from rock shelters. Figure 13.3a: Neolithic pastoralism at Arene Candide cave, Liguria, Italy; an example of both "primary use related" (in situ stabling floor deposits), and "secondary transposed" refuse deposition (locally dumped stabling waste), best differentiated using soil micromorphology. (See Experimental data in Chapter 14). (a) Primary residues - rapidly accumulated Middle to Late Neolithic (Chassey) stabling layers (5,000 to 5,700 BP) composed of alternating brown ("stabling crust") and gray (ashed leaf hay fodder, dung, and wood) layers. The thickest brown layers however, are dumped ashed stabling waste.' Layer Cake' ashed and oxidized stabling remains from caves and rock shelters, as detailed in Figures 13.3b and 13.3c.

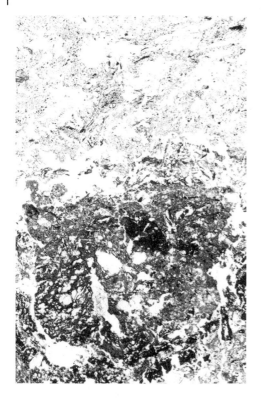

Figures 13.3b–13.3c Photomicrographs of junction between stable floor "crust" layer where bedding and fodder composed of leaf hay (e.g. *Quercus ilex* leaves and twigs), which has been compacted by animal trampling, and stained by phosphate-enriched dung and liquid waste. This material was burned after use, probably in springtime after over-wintering and the birth of lambs; the "wet" crust resisted total combustion compared to the gray ashed fodder and dung remains above (Maggi, 1997). (b and c, PPL and OIL, respectively; frame height is ~5.5 mm.)

Figures 13.3d–13.3e Oxidized Atzmaut rock shelter stabling deposits originating from "over-nighting", rather than foddered over-wintering stabling episodes, with probable goat grazing of areas of herbage linked to spring rains; it is assumed that herds were moved from rock shelter to rock shelter according to availability of grazing (Rosen, 1988; Rosen et al., 2005).

Figure 13.3d Atzmaut rock shelter: Early Bronze Age levels are composed of unburned but humified organic and horizontally layered "compressed dung" (herding floor crust); individual pellets cannot be discerned and birefringent plant fragments (cellulose) are rare, indicating humification. Deposits are less organic (13.7–14.0% LOI) but more phosphate rich (4.32–7.17 mg g^{-1} phosphate-P) compared to recent dung deposits. PPL, frame width is ~5.5 mm. (Macphail and Crowther, 2008).

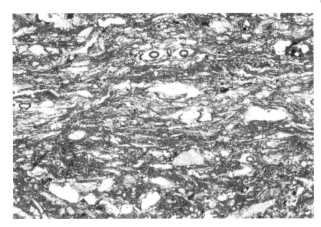

Figure 13.3e Detail of Figure 13.3d, under XPL: note abundant calcite "dung" spherulites within amorphous organic matter. When totally oxidized, this dung layer will become mainly characterized by spherulite concentrations. Frame width is ~0.47 mm.

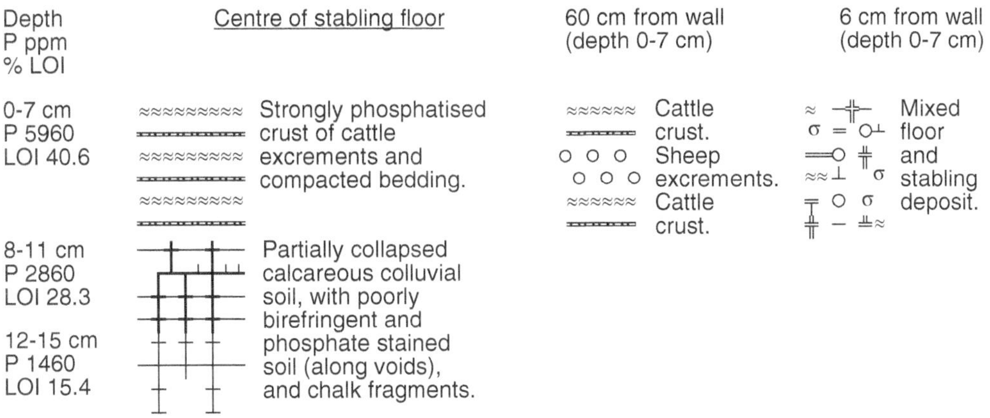

Depth P ppm % LOI	Centre of stabling floor	60 cm from wall (depth 0-7 cm)	6 cm from wall (depth 0-7 cm)
0-7 cm P 5960 LOI 40.6	≈≈≈≈≈≈≈≈ Strongly phosphatised ▬▬▬▬▬▬ crust of cattle ≈≈≈≈≈≈≈≈ excrements and ▬▬▬▬▬▬ compacted bedding. ≈≈≈≈≈≈≈≈ ▬▬▬▬▬▬	≈≈≈≈≈ Cattle ▬▬▬▬▬ crust. ○ ○ ○ Sheep ○ ○ ○ excrements. ≈≈≈≈≈ Cattle ▬▬▬▬▬ crust.	≈ ─┼─ Mixed σ = ○⊥ floor ═○ ╫ and ≈≈⊥ σ stabling ┬ ○ σ deposit. ╫ ─ ⊥≈
8-11 cm P 2860 LOI 28.3 12-15 cm P 1460 LOI 15.4	┼┼┼ ┼ ┴ ┼ ┼ ┼ ┼ ┼ ┼ ┼	Partially collapsed calcareous colluvial soil, with poorly birefringent and phosphate stained soil (along voids), and chalk fragments.	

Figure 13.4a Diagrammatic representation of site formation process at the Moel-y-Gar byre/stabling roundhouse, Butser Ancient Farm, showing different microfacies across the floor that "record" stabling between 1977 and 1990; phosphate (P), loss-on-ignition (LOI), and soil micromorphology (Macphail and Goldberg, 1995; Macphail et al., 2004).

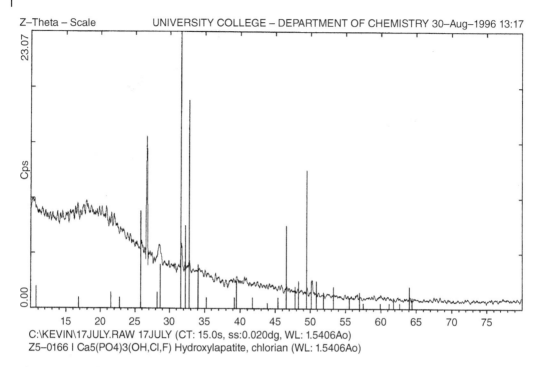

Z–Theta – Scale UNIVERSITY COLLEGE – DEPARTMENT OF CHEMISTRY 30–Aug–1996 13:17

C:\KEVIN\17JULY.RAW 17JULY (CT: 15.0s, ss:0.020dg, WL: 1.5406Ao)
Z5–0166 I Ca5(PO4)3(OH,Cl,F) Hydroxylapatite, chlorian (WL: 1.5406Ao)

Figure 13.4b XRD (X-ray diffraction) analysis of the phosphate-cemented organic (40.9% LOI) Moel-y-gar byre floor crust (Butser Ancient Farm, Hampshire), supporting the view that the high amounts of Ca (mean 2.5% Ca) and P (mean 0.5% P) as measured by microprobe, were at least in part in the form of hydroxylapatite (which was typically autofluorescent under blue light); strangely this seemed controversial to some other workers (Canti et al., 2006; Macphail et al., 2004, 2006).

fuel ash, and ore in one location is more indicative of a production site. Hammerscale is also strongly residual, but when found embedded in a hearth, *in situ* iron working/smithying seems highly likely, as at Norwegian Late Iron Age (Migration Period) Åker gård, Hamar, Norway (Macphail et al., 2017d) (Figures 13.5a–13.5d).

Salt production (see Chapter 9.4) has long been associated with "redhills" in England, which are accumulations of fragmented briquetage and fuel waste, especially on the coast where brine has been boiled; at Stanford Wharf, Essex on the Thames Estuary Romano-British use of briquetage was replaced by the use of lead vessels in the Late Roman Period (Biddulph et al., 2012). In the latter case, bulk chemical and optical soil micromorphology supported by SEM/EDS found high lead concentrations (54% Pb) (Macphail et al., 2012a). At Marco Gonzalez, Belize Late Classic (ca. AD 600–750) Maya salt working produced low temperature heated (rubified) plaster floor hearths, and alternating waste layers, which include clasts of burned marine sediment used as a source of brine, fuel waste composed of ash and charcoal from local mangrove woodland, for example (Graham et al., 2017; Macphail et al., 2017b). Weathering and soil formation associated with the development of anomalously rich broadleaved woodland has led to the development of an atypical *terra preta*, termed Maya dark earth (Graham, 2006; Graham et al., 2016) (Box 13.1; Figures 13.6a and 13.6b).

Figure 13.5a   Åker gård, Hamar, Norway; Norwegian Iron Age (6th-8th C AD "Migration Period") multi-phase long house, with a smithy hearth (Context A40347); here magnetic susceptibility is the highest at the site ($8001\ \chi lf\ 10^{-8}\ m^3\ kg^{-1}$) (Macphail et al., 2017d) (Lars Erik Gerpe and Jessica Leigh McGraw, Cultural History Museum, University of Oslo, pers. comm). a) Photomicrograph of M41703, showing thin iron fragments – hammerscale (2) and example of siliceous spherule (1: "furnace prill"). PPL, frame width is ~2.38 mm.

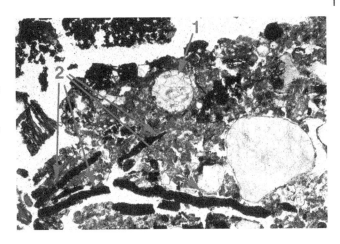

Figure 13.5b   As 13.5a, under OIL; note siliceous (colorless) spherule (1) and metallic iron flakes (2).

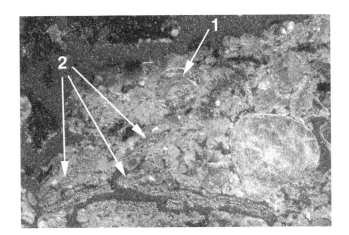

Figure 13.5c   SEM X-ray backscatter image of iron flakes – hammerscale. Scale bar = 1 mm.

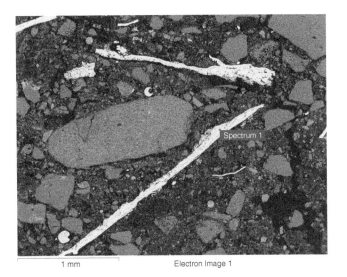

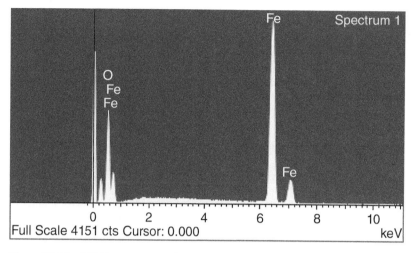

Figure 13.5d   EDS X-ray spectrum 1 shown in 13.5c; 100% FeO.

Box 13.1   *Terra preta* (ADEs) and European dark earth

Amazonian dark earths These Amazonian hortic Anthrosols, which have gained the acronym of "ADEs", were originally termed *terra preta* (dark earths). They are associated with pre-contact (pre-Columbian) settlement and occupation deposits and have been intriguing archaeologists and soil scientists for nearly 60 years after being first recorded by explorers in the 19th century (Arroyo-Kalin et al., 2008; Graham, 1998; Holliday, 2004: 320–323; Smith, 1980; Sombroek, 1966) (see reviews in Arroyo-Kalin, 2017; Lehmann et al., 2004; Sombroek et al., 2002). "Indian black earth", or *terra preta do Índio*, vary in color from a dark earth that is rich in cultural artifacts (mainly ceramics with some lithic material) to a brown soil, the latter termed *terra mulata* (Sombroek, 1966); "occurrences" varying in area from 0.5 ha to >120 ha, and up to 2 m in thickness (McCann et al., 2001; Woods and McCann, 1999). The present-day importance of *terra preta* is that they are much more fertile than the surrounding soils (e.g. Ferralsols and Acrisols), and are much more capable of sustainable agriculture, because they may contain higher amounts of organic matter, Ca, Mg, Zn, and Mn; they are even "mined" by local people to improve land for horticulture (Graham, 2006; Sombroek et al., 2002) (Bill Woods, pers. comm.). The two main types of ADEs differ both in color and character. *Terra preta* contains much more Ca and P compared to *terra mulata* (McCann et al., 2001; Woods and McCann, 1999), and although containing similar amounts of organic matter, they record a "near absence of artifacts" (Sombroek et al., 2002). This important research also showed that sites are not simply the result of universal kitchen middening and/or relict habitation but also developed from attempts at soil improvement. Rather, and as a generalization, the smaller areas of *terra preta* reflect Amerindian "habitation", whereas the more widespread *terra mulata* have resulted from "agriculture", with soil amelioration having been induced by slash and burn, mulching, and composting (Arroyo-Kalin, 2017; Lehmann et al., 2004; Woods and McCann, 1999, Figure 3).

   More recently a specific form of "habitation" type of ADE has been identified at Marco Gonzalez, Ambergris Caye, Belize – a small ~300m × 150m island, which is now surrounded by mangrove swamp that has developed since Maya times (i.e., probably after the 1500s AD) (Figures 13.6a-13.6b). Importantly, the "habitation" history, which at least dates from 300 BC, is

Box 13.1 (Continued)

extremely well documented through excavation including deep sondage (to −0.050m asl) studies since the 1980s (Graham et al., 2017; Pendergast and Graham, 1987). Unlike typical ADEs, this site is carbonate rich, with a substrate formed from coral and Pleistocene limestone of the Belize Barrier Reef (Gischler and Hudson, 2004). Moreover, occupation and especially salt processing carried out using low temperature fires on lime plaster hearth floors during the Late Classic (AD 600−750/800), produced much charcoal and calcareous ash (see Chapter 12) (Figures 13.6c-13.6g). Dark earth developed in the weathered, rooted, and burrowed remains of originally stratified deposits, which were also disturbed by caches, burials, and the more recent activities of land crab burrowing and soil removal for gardening on the caye (Figures 13.7a−13.7c). Characteristically, pH values are alkaline (between pH 8.5 and 9.1), and Specific Conductance (~salinity) is as high as 5070 (µS; see Chapter 17) due to the additional presence of salt working residues (Graham et al., 2017). Despite being blackish in color and containing high amounts of charcoal, these are atypical ADEs because they record the weathering and biological working of carbonate-rich occupation deposits, with relict lime floors occurring as pale "ghost" features in these "Maya dark earths" (Macphail et al., 2017; Arroyo-Kalin, pers. comm.). The island has currently an unusually species-rich "broadleaved caye woodland", which is apparently associated with modern surface soils that are the most organic (highest LOI at 26.9−28.0% LOI) and mercury (Hg)-rich (max 467 ng g^{-1} Hg) they are the least alkaline and most leached layer type found at Marco Gonzalez (pH = 7.9; carbonate = 35.9%; specific conductance = 455−477 µS) (Macphail et al., 2017). Thin section observations show a Mull humus type with a leaf litter surface layer (Figure 13.7d), aggregated amorphous organic matter here forms broad excrements (granules/crumbs) which embed fine mineral grains and remains of lime floors, shell, and reefstone fossils. The amorphous organic matter component probably includes the remains of termite nests. Significantly, and unusually for an ADE, these surface soils and litter layers include very little fine or very fine charcoal, and essentially result from the development of the broadleaved woodland that has rooted into the underlying Maya dark earth from which it has gained its nutrients.

European dark earth

As urbanism declined in what had been the Roman or Classical world, a number of processes and events were recorded in deposits called dark earth (Figures 13.9a-13.9b). The interpretation of these dark earth deposits has not been straightforward, however (Cammas, 2004; Galinié, 2000; Nicosia et al., 2012) (see reviews in Macphail, 1981; Macphail and Linderholm, 2004b; Nicosia et al., 2017). It has also increasingly been argued that dark earth did not develop because of the total abandonment of towns and cities, but rather it resulted from a change in the use of urban space, and the once-stratified Roman archaeology became obscured or totally lost (Galinié, 2004; Macphail et al., 2003a; Yule, 1990). As oxidizing conditions normally allow few organic remains to persist, investigations have tended to focus on geoarchaeological methods. Occasionally, however, other independent techniques have also been employed to improve interpretations of use of space and include palynology, phytoliths, and small artifact recovery (Borderie et al., 2014a; Devos et al., 2009, 2013; Laurent and Fondrillon, 2010; Sidell, 2000). The chief thing to remember is that the term "dark earth" is simply a convenient field name, and is not an interpretation or identification (Galinié, 2000, 2004).

(Continued)

Box 13.1   (Continued)

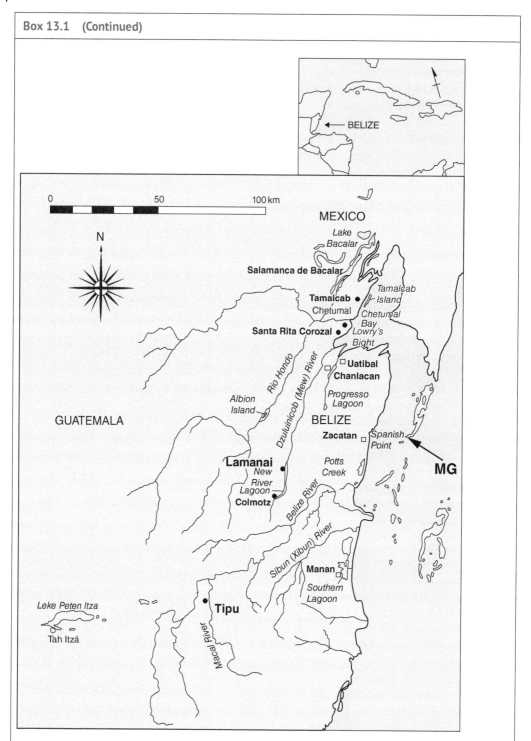

Figure 13.6a   Regional map showing location of Belize and Spanish Period sites; the island of Marco Gonzalez (MG) is at the southern end of Ambergris Caye (Graham et al., 2016, 2017; Graham and Pendergast, 1989; Macphail et al., 2017b/with permission of Elsevier).

Box 13.1 (Continued)

Figure 13.6b Aerial photograph of Marco Gonzalez, looking south-west, showing broadleaved caye wooded island area within mangrove swamps. The "small pool" which was cored is arrowed.

Figure 13.6c Field photo of Op 13-1 (Str 14), east face, showing monolith samples 1–4 and MG Contexts: MG 359 ("Maya dark earth"); MG 364, 367, 369 and 371 (remains of weathering lime floors and salt processing residues and burrow-mixed "Maya dark earth" (Box 13.1); MG 374 and 377 (mainly intact lime floors and ash layers of salt processing origin); MG 382 (lime floors and trample, sealing cached Early Classic basal-flange bowls – see fragments on the board by the ladder and Figure 13.6d); MG 383 (ash and fine bone-rich waterlaid deposit; Figures 13.6e and 13.6f). On the west face, monoliths 5 and 7 sampled a lateral example of salt processing layers/ floors (MG 377); monolith 6 sampled MG 383 to a depth of −0.230 m asl.

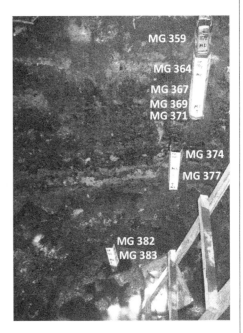

Figure 13.6d Reconstructed cached Early Classic basal-flange bowl (390/1) found below lime plaster floors (see Figure 13.6a).

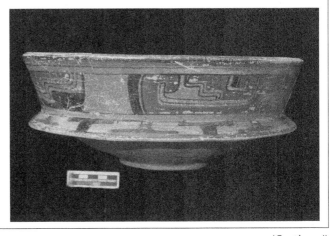

Box 13.1 (Continued)

Figure 13.6e Scan (M4D); Early Classic levels (MG 383) are composed of a weakly humic (3.66% LOI), fine sediment (93.6% silt and clay, with ~6% sand), which is strongly calcareous (70.2% carbonate), alkaline (pH 8.9) and strongly saline (3540 specific conductance) with a very strongly enriched phosphate content (23.2 mg g^{-1} phosphate-P). It is essentially an anthropogenic deposit, however, that was waterlaid originally, although now affected by some burrowing by small mesofauna. Anthropogenic inputs are chiefly calcitic ash (including much wood ash?), charcoal, human fecal material/dietary waste (including fish bone), with background shell, pot and few sand (including geological bioclasts). This deposit can be interpreted as a footslope colluvium/ash and kitchen waste dumped into standing water that surrounded the island/occupation mound (E. Graham, UCL, pers. comm.). The deposits record both hearth ash deposition and populations producing both kitchen waste and fecal material.

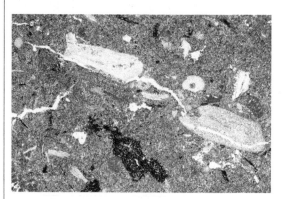

Figure 13.6f Photomicrograph of M4D (Figure 13.6e), illustrating waterlaid ash, containing charcoal, and semi-leached and rubified fine and very fine bone. This kitchen hearth discard deposit probably includes human fecal material. It is also iron-depleted and has possible iron hydroxide staining of iron pyrite, suggesting anaerobic conditions at times. PPL, frame width is ~4.62 mm.

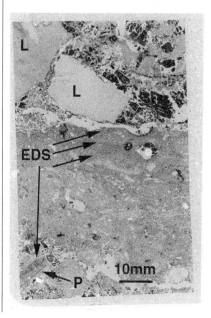

Figure 13.6g Scan of thin section M3B (Str 14, Op 13-1; west face; Figure 13.6c); here a Late Classic heated lime plaster floor is overlain by chaotic dumps of coarse limestone clasts (L), charcoal (fuel residues) and pink colored plaster and sediment fragments, including isotropic sediment containing sponge spicules. The floor also seals a loose deposit that includes a sediment-coated Coconut Walk ware potsherd (P). Energy-Dispersive X-Ray spectrometry (EDS) studies were carried out on pot examples and the plaster floor (EDS). Sponge spicule-rich sediment coating to pot sherds indicate that saline mudflat sediment-enriched brine was heated and not simply sea water. Weathered and bioworked lime plaster floors merge upwards into dark earth at this site (see Figure 13.6c and Box 13.1)
Scale bar = 10 mm.

Box 13.1   (Continued)

Figure 13.7a   Field photo of labeled Op 13-3 (Str 8) showing monoliths and Contexts. Monolith 8 sampled lime floor remains within 'Maya dark earth' (MG 366, 276 and 379) and weathered lime floor remains (MG 380) of the town. Monolith 9 also covered weathered lime floor remains (MG 384). A large lump of sediment from MG 385 (Ref 2) included an intact lime floor over 'sands'. Monolith 10 sampled burrowed disturbed Late Classic deposits, which also included a thick intact ash layer (MG 387) (Graham et al., 2017; Macphail et al., 2017).

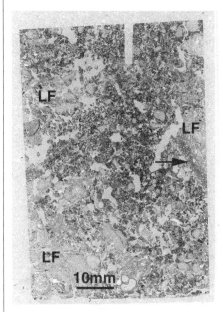

Figure 13.7b   Scan of M2A (Str 14, Op 13-1; west face; uppermost MG 364); weathering lime plaster floor fragments (LF) in 'Maya dark earth' – here shown as a very broad burrow fill of dark soil composed of thin and broad organo-mineral excrements from invertebrate soil mesofauna activity. Note use of potsherd as temper in floor plaster (arrow). Scale bar = 10 mm.

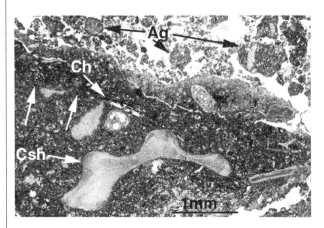

Figure 13.7c   Photomicrograph of M1A (Str 14, Op 13-1; west face; MG 359); lime plaster floor fragment within 'Maya dark earth'. Fragment retains example of plastered surface (arrows) and inclusion of charcoal (Ch). Tempering material include small gastropods, fossils and large example of probable conch shell (Csh). Soil is composed of fine pellets (very thin excrements) and thin to broad aggregated granular organo-mineral excrements (Ag). PPL. Scale bar = 1 mm.

(Continued)

Box 13.1   (Continued)

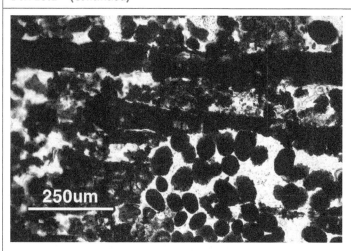

Figure 13.7d   Photomicrograph of MG14-M2 (surface soil with Litter layer at Str 14, Op 13-1; 28.1% LOI); mainly burrowed hollow leaves partially infilled with very thin organic excrements typical of litter layers. Some residual cellular leaf material remains. Plant material has undergone both 'browning' and 'blackening' decomposition (Babel in Bullock et al., 1985). Note very thin oval-shape organic excrements of Litter-dwelling small invertebrate mesofauna. Reddening due to iron staining (chemical measurements of surface soils elsewhere found 0.85–0.92% Fe, where mercury was concentrated – 241–467 ng g^{-1} Hg). It is presumed that rooting by current broad leaved woodland is bringing various nutrients, such as phosphate (11.4 mg g^{-1} P), to the surface, making such dark earth soils a source for creating gardens (Graham, 2006; Graham et al., 2016). Such ferruginization also testifies to localized and episodic hydromorphism (waterlogging). PPL. Scale bar = 250 μm.

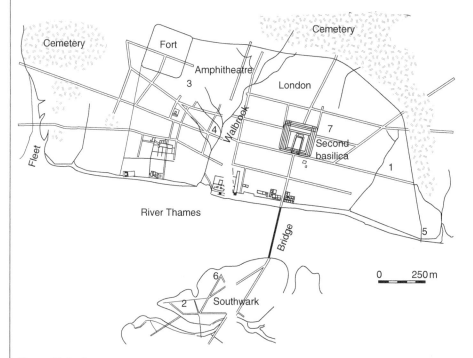

Figure 13.8   Roman (ca AD 50–180) London and Southwark showing locations of the extra mural cemeteries, the fort, amphitheater, second basilica, and sites mentioned in the book: 1. Colchester House; 2. Courages Brewery; 3. London Guildhall (Amphitheater); 4. No 1 Poultry; 5. Tower of London; 6. Winchester Palace; 7. Whittington Ave. (modified from Watson, 1998).

Box 13.1   (Continued)

Figure 13.9a   Field photo; the London Guildhall site, showing the junction of the arena surface of the amphitheater abandonment ca. AD 250 or 364, and the overlying dark earth formed between the 3rd–4th century and 11th century farmstead (Bateman, 2000; Bateman et al., 2008; Bowsher et al., 2007).

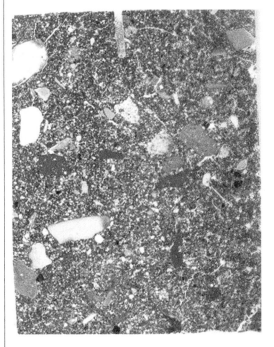

Figure 13.9b   As Figure 13.9a, scan of typically homogeneous decarbonated 'secondary dark earth' or 'mature dark earth' formed over centuries; soil has a total excremental microfabric due mainly to earthworm working although smaller organo-mineral excrements are present from smaller invertebrate mesofauna, such as Enchytraeids. Note large amounts of gravel size residual inclusions, brick, flints, and mortar. Frame width is ~65 mm.

(Continued)

Box 13.1   (Continued)

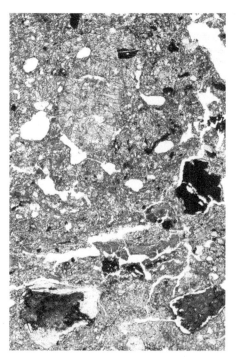

Figure 13.9c   Photomicrograph of dark earth at Tarquimpol (Decempagi), La Moselle, France. Diffusely layered post hole fill of structure contemporary with short-lived dark earth formed in the weathered remains of the Late Antique ~350–450 AD rampart-surrounded settlement that replaced the Roman town. Fill includes inputs of humic soil, fine soil inwash (coated vughs), and yellow orange phosphate deposition (3.25% LOI, 3.99 mg g^{-1} P), recording occupation. Note weakly dark gray earthworm granule testifying to both bioworking and weathering of dark earth. Tarquimpol was a trading and transport hub (Henning et al., 2012). PPL, frame height is ~4.62 mm.

Figure 13.9d   As Figure 13.9c, under XPL, showing non-calcitic fine fabric and weathering biogenic calcite of earthworm granule.

Box 13.1 (Continued)

Identification of its components and site formation features are then investigated in order to record of use of urban space through time.

Through the 1980s and 1990s investigations focused upon the role of abandoned and weathered Roman structures (cf. Friesem et al., 2011, 2014a, 2014b), and their residues contributed to the formation of dark earth (Yule, 1990). For example, abandoned buildings constructed of brickearth floors and plaster-coated brickearth walls (see Chapter 12.1.1), began to collapse as biological agencies destroyed supporting wood beams, or structures were robbed. In fact, a wide variety of mechanisms have been identified that could lead to dark earth forming out of constructional materials and occupation deposits (Tables 13.3 and 13.4). Some of these processes have been modeled from Second World War blitzed areas of London and Berlin (Macphail, 1994a; Sukopp et al., 1979; Wrighton, 1951).

Once-stratified building debris and occupation deposits developed as a homogenized dark earth soil, which can reflect the original constructional materials of an abandoned building, forming in the first decades a very thin humic calcareous soil or pararendzina (Entisol; see Chapter 4). The breakdown of earth-based materials such as daub, brickearth, and loamy cob and rammed earth yielded mainly clays, silts and sands (Cammas, 2018; Macphail, 1994a). The weathering (decarbonation) of lime-based mortar and plasters is exactly the same as soil formation on limestone, where pararendzinas form. This breakdown occurs through attack by humic acid from plants, and under moist conditions when calcium bicarbonate ($Ca(HCO_3)_2$) is removed in solution by rain water containing more or less dissolved CO_2 – carbonic acid (HCO_3) (Duchaufour, 1982: 74). Weathering in the form of decarbonation occurs and dark earth is thus generally strongly decarbonated (see above, for Maya dark earth). As evidence of this, dark earth that has taken 400–600 years to form often contains biogenic calcite (e.g. earthworm granules) from biological activity, but many of these granules show incipient decalcification. It can be termed "secondary dark earth". A current example of this has been found 1.5 m below the medieval cloisters of Exeter Cathedral, here reddish brown "primary dark earth" records demolition of Roman earth-based buildings, above which mainly decarbonated "secondary dark earth" is present which features decalcifying earthworm granules. Where there has been a change in use of space and once-disused ground is reoccupied for middening purposes for instance during the Late Roman (4th–5th centuries AD) this weathering of the calcareous anthrosol and associated soil biogenic calcite can be shown to have occurred over a shorter period – perhaps 100–200 years (e.g. Canterbury and Winchester). In the same way, new middening can include higher amounts of organic matter. The sparse pollen that was indicative of waste ground and characterizes primary dark earth formation was replaced by Late Roman deposits with a pollen spectra consistent with dumping of byre waste for example (Macphail et al., 2007). Dark earth soil development (when not truncated) can be traced upwards into Late Saxon and early medieval times (~9th–11th centuries AD), when "greenfield" sites (grazed pastures) were reoccupied as Saxon burghs (fortified towns) and newly established medieval settlements (Macphail, 2014) (for a medieval cultivation land use in

(Continued)

Box 13.1   (Continued)

Table 13.3   Components of urban stratigraphy and their potential for weathering (updated from Macphail, 1994a).

Urban stratigraphy (and land-use)	Materials and features	Components
(a) Clay and timber buildings and structures	Brickearth, manufactured daub, cob, and rammed earth	Clay and silt, with sands
	Lime-based mortar	Calcite cement and gravel
	Lime-based "plaster" (also including Maya lime floor hearths)	Calcite cement, silt, and sand
	Timbers	Wood (charcoal if burned)
	Roofs	Tile, thatch or turf (the last containing organic matter and phytoliths, and a focus of biological activity)
(b) Backyard middens and manure heaps	Middens	Human and dog coprolites, bone, ash (calcium carbonate), organic matter, phytoliths, oyster shell, wood and Gramineae charcoal (focus of biological activity)
	Manure	Herbivore/omnivore (e.g. pig) coprolites, organic matter, phytoliths, diatoms, Gramineae charcoal (focus of biological activity)
(c) Cultivated plots	Cultivated soil	Finely fragmented "urban stratigraphy" of building debris, midden and manure material, plant organic matter, some times mixed with natural soils (focus of biological activity)
	Gardens	As above, and including ground-raising activity and soil importation.
(d) Pit areas and dumps	Pit fills, dumps, and cess pits	Mixed natural soils and underlying geology with all kinds of urban materials of mineral and organic origin, including industrial waste, cess and nightsoil (sometimes the focus of biological activity)
(e) Destruction leveling and fires	Dumps, collapses, and razing	Mixed, sometimes burned) building materials, often with charcoal and other urban materials (see above)
Ditto	Robber trenches	Infills over wall foundations, etc.
(f) Abandonment and fires	Unoccupied buildings and building shells	Accumulated collapsed and weathered building debris; calcitic ash dumping and organic waste from visiting humans, animals and nesting birds (focus of biological activity)
(g) Rural	Grassland, shrubs, and woodland	Originally heterogeneous components (as above) becoming increasingly homogenized into dark earth

Box 13.1   (Continued)

Table 13.4   Phases of dark earth formation elucidated from an analogue (Post-War Berlin) and three selected Roman/early medieval sites (Macphail et al., 2003a/with permission of Cambridge University Press; see also Macphail, 2014).

Site	Dark earth phase	Interpretation
Berlin, Germany Monitored pedogenesis in post-war (1945–1978) Berlin's wastelands	Pararendzina formed in poorly weathered concrete and brick-rich building debris.	*Total site abandonment* Very shallow (100 mm) base-rich soil formation after 20–30 years (closed woodland canopy formed after 28 years)
Courage Brewery, Southwark, London, UK 1st/2nd century – Late 5th century Roman London; dark earth below post-medieval dumps	Dark earth Phase 3: calcareous brown earth development.	*3 – Essentially abandoned waste ground ca. AD400–1050* (Mature 300+ mm thick decarbonated but still base-rich soil formation)
	Dark earth Phase 2: calcareous brown earth developing in accreting ash-rich midden deposits.	*2 – low intensity urban activity – middening/shrub and ruderal vegetation in abandoned house plots during 200–300 years, developing into wasteland (late Roman cemetery)*
	Dark earth Phase 1: pararendzina formed in brickearth and lime-based mortar/plaster debris over (sometimes partially robbed out) brickearth floors	*1 – Extremely short-lived (1–5 years?) total site abandonment* (Extremely thin (10–50 mm) base-rich soil formation with likely moss/lichens/fungi and ruderal vegetation cover.
London Guildhall, UK – Roman Arena Mid-3rd century – Late 5th Roman/mid-11th century Saxon and early medieval; dark earth sealed by early medieval occupation deposits and floors	Dark Earth Phase 4: immature brown calcareous earth formation	*4 – High intensity middening* (Increasing Late Saxon urban use of adjacent land) (150 mm poorly weathered midden dumps)
	Dark earth Phase 3: Mature pararendzina development.	*3 – Essentially abandoned waste ground ca. AD400–1050* (40 mm mainly decarbonated base-rich soil)
	Dark earth Phase 2: Pararendzina forming in ash-rich midden/butchery/nightsoil dumps.	*2 – low intensity urban activity (middening) (50–100 years?); ca. AD364–400* (50 mm A/C soil profile)
	Dark earth Phase 1: pararendzina formed in and over weathering Arena mortar floor.	*1 – Short-lived (10–100 years?) total site abandonment ca. AD 250 or 364*
Deansway, Worcester, UK Late Roman to Saxon (4th–9th century); dark earth sealed by Saxon deposits and rampart of burh (ca. AD 890)	Dark earth phase 2: mature brown earth development.	*2 – Very low intensity occupation (grazing?) during ca. 300–500 years (8th century Saxon hearth)*
	Dark earth phase 1: brown earth forming in accreting dung-rich midden deposits (stock area?)	*1 – Short-lived (100 years?) non-typical urban land-use, developing into wasteland (late Roman cemetery)* (250–300 mm Ah horizon)

(Continued)

Box 13.1 (Continued)

Brussels, Belgium, see Devos et al., 2009, 2013). In short, a number of British Roman cities record: the weathering and reworking of 1st–2nd century AD urban structures and occupation deposits (primary dark earth); Late Roman urban settlement which is rural in character (secondary dark earth); Saxon and medieval occupation, which often led to phosphate contamination of the earlier-formed dark soil (tertiary dark earth) (see Borderie et al., 2014b).

Two sites of note have been studied in France – at Tours (Centre) and Tarquimpol (La Moselle). The remains of earth-based structures and associated human activities were of primary concern. At Prosper-Mérimée Square, Tours, the density of small artifacts through dark earth dating to the 7th–8th centuries indicated a high population density consistent with heavily compacted surfaces and concentrated latrine waste disposal found within the dark earth layers (Fondrillon, 2007; Galinié et al., 2007). From the 8th century onwards urban land use changed, and the dark earth accumulated as a heavily manured horticultural soil, finally being employed for viticulture by the 12th century according to historical records. Tarquimpol is a rare example of short-lived dark earth formation (cf. Colchester House, London; Macphail and Linderholm, 2004b) (Figures 13.9b-13.9d). Here, the original Late Antique city was destroyed – possibly by the Roman authorities themselves rather than by the Huns (Henning et al., 2012; Henning, pers. comm.), and a circular earth rampart-protected settlement replaced it for approximately 100 years (AD 350–450), during which time it flourished as a road system hub associated with transport, artisan activity, and probably the salt trade. Fragments of preserved charred turf and burned daub testify to the presence of earth-based structures, and enhanced magnetic susceptibility and phosphate concentrations, along with mineralized human waste also record typical occupation features. One post hole fill, with chemical and soil micromorphological characteristics of dung inputs may testify to the presence of animal management as would be expected for a trading and transport hub (Henning et al., 2012; Macphail, 2014).

Further examples of perhaps hidden evidence of higher than expected populations comes in the form of dark earth having a history linked to soil development in urban middening deposits. These dark earth soils with high very fine charcoal contents and sometimes relict calcitic ash, together with small concentrations of bone, coprolitic bone, burned bone, and both probable human and dog coprolites, seem to record hearth ash dumping along with food and latrine waste discard. Such waste ground ash disposal in Roman Southwark mirrors the practice of within house ash dumping in AD 79 Pompeii, for example (Fulford and Wallace-Hadrill, 1995–1996, 1998). The Roman dark earth at Exeter is an accreted accumulation with traces of ash residues and a marked concentration of bone waste, including much fish bone and at least one example of fish bone embedded in a probable dog coprolite (Figures 13.10a-13.10c); clearly Exeter had not been abandoned and the population was exploiting the resources of the River Exe and nearby sea.

See Table 4.1 for soil types (Bowsher et al., 2007; Cowan, 2003; Dalwood and Edwards, 2004; Sukopp et al., 1979).

Box 13.1   (Continued)

Figure 13.10a   Field photo of Exeter Cathedral Cloisters excavation, with Roman demolition (2033, 2029, 2034) including both earth- and lime-based constructional debris, dark earth (2008) and medieval (2007) levels. (Unearthed: Updates from the Cloisters Archaeological Investigation – Exeter Cathedral (exeter-cathedral.org.uk).)

Figure 13.10b   As Figure 13.10a, scan of M3A, bioworked dark earth with residual gravel size sandstone (SSt), basalt (Bas; some used in puzzolanic mortars), fish bone (FB) and mineralized coprolites (Cop). Frame width is ~50 mm.

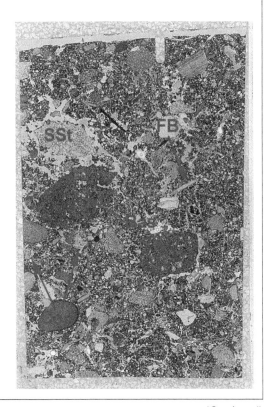

(Continued)

Box 13.1 (Continued)

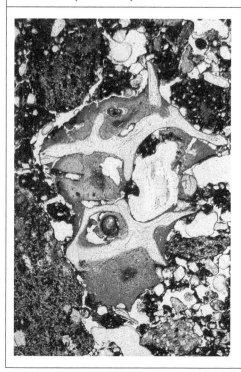

Figure 13.10c   As Figure 13.10a, photomicrograph of M2A, basalt-rich dark earth with mineralized (dog?) coprolite embedding probable fish bone, evidencing middening by a local population utilizing local resources – fish from the River Exe and sea. PPL, frame height is ~4.62 mm.

13.2.1.1 Primary Transposed Waste

The most obvious primary transposed waste deposits are middens, especially "kitchen" and shell middens where waste is juxtaposed to an activity area where food was processed (Corrêa et al., 2013; Goldberg, 2000b; Stein, 1992; Villagran et al., 2009). These can obviously be shell rich, providing information on range of exploitation and the contemporary marine environment; equally, if fish bones are present these may be from shallow and deep-sea waters. Shells have been a regular source of evidence for identifying seasonality, as in Mesolithic cave deposits at Uzzo Cave, Sicily (Mannino and Thomas, 2004–2006; Mannino et al., 2007). At the Uzzo Cave, there an important junction between hunter and gather occupation, where anthropogenic remains are diluted by natural geological and animal (e.g. guano) inputs, and overlying Early Neolithic layers of wholly anthropogenic character associated with use by pastoralists (charcoal, ash, charred dung). In such cases, the deposits are both the context and important element of the material culture (Karkanas and Goldberg, 2018b).

Stein's (1992) seminal work on shell mounds cited above, helped alert the geoarchaeological community to their value as records of the geological and archaeological past. Although there are earlier instances of detailed paleoenvironmental midden studies (Balaam et al., 1987) it was Stein's (1992) lead in the USA that sparked increased research into unraveling their complex stratigraphy and site formation including experimental research (Aldeias et al., 2016; Balbo et al., 2011; Gutiérrez-Zugasti et al., 2011; Villagran et al., 2011a, 2011b) (see Chapter 9.5 and Figure 9.9). Aldeias and Bicho (2016) studied the Mesolithic mound of Cabeço da Amoreira along the Tagus River in Portugal. Using micromorphology and microfacies analysis, they demonstrated the complexity of the Mesolithic occupations, which involved activities such as dumping tossing, and the spreading out of material below occupied surfaces. Moreover, they demonstrate hiatuses in sedimentation resulting from total abandonment of the mound.

Other primary discard deposits can include ash from domestic hearths, and as hearth cleaning was regularly practiced in Roman houses, ash deposits can be found in disused rooms (e.g. in the House of *Amarantus*, Pompeii; Fulford and Wallace-Hadrill, 1995–1996, 1998; www.bsr.ac.uk/research/archaeology/completed-projects/pompeii-project/pompeii-excavations). Weathering ash deposits dumped into house shells and other disused properties is a common background material for late Roman dark earth formation, e.g. in Southwark, London (Cowan, 2003) (Figure 13.5; Box 13.1). Several studies of Mousterian (Middle Paleolithic) cave sites have suggested that accumulations of fuel ash waste, that have included char, are also the result of purposeful hearth cleaning and ash dumping (Goldberg et al., 2009b; Meignen et al., 2007); char is commonly a residue from cooking meat for example (Goldberg et al., 2012; Ligouis, 2017; Mallol et al., 2017; see Table 13.2). Significant studies have also been undertaken to evaluate the use of bone as fuel (e.g. Costamagno et al., 2005; Théry-Parisot et al., 2005).

13.2.1.2 Secondary-use Waste

A chief example of the secondary use of waste, is manuring of cultivated fields. This has already been discussed in Chapter 11.3.2. Organic manures collected from the byre (see below 13.5.2) raise levels of organic matter and phosphate measured in these fields, and also may produce an anomalously high magnetic susceptibility because manure *sensu lato* has included "settlement waste" such as hearth debris (see Adderley et al., 2018). Burned daub, burned flint, and pot sherds can occur in these contexts, and the spread of pot sherds outside urban limits in the Middle East testified to this kind of manuring (Wilkinson, 1990, 1997). Nodules of nightsoil with a Fe-Ca-P chemistry may also be included, and night soil manuring was regularly practiced in Europe up to the 20th century (see Figures 11.9a-11.9e). For example, in Norway when the railway connected Dilling and Oslo in 1879, trains brought *pudrette* (*latrine* waste mixed with peat) to Dilling railway station from where it was spread on the fields (Berggren, 1990); amorphous iron-cemented sands, can contain 0.13% P, 0.15% S, 0.13–0.15% Mn, and 15.5–16.3% Fe (EDS data; Macphail et al., forthcoming). Usually an "*infield*" cultivated plot would contain more residual manuring materials compared to an "*outfield*". Another example of reuse is when debris is employed in ground raising during construction, and this includes examples from Roman London (Figures 13.8, 13.9a, 13.9b, 13.10a, 11.9a) and the Chalcolithic tell site of Borduşani-Popină, Borcea River, Romania, where debris of burned mud brick and loess slabs from earlier-razed buildings was regularly employed for ground raising (Haită, 2003, 2012; Macphail et al., 2017a; Popovici et al., 2003; Radu, 2003) (asma/Online_Video_11.1.md at master · geoarchaeology/asma · GitHub) (see Figure 13.2e).

13.2.1.3 Caveats to This Model

Experience has shown us that when we have a holistic approach to archaeological context and material culture studies, the above model – like most models – signifies a preliminary and limited approach; for example fitting in cremation, excarnation, and inhumation *funerary* deposits is probably not appropriate for such models. This is because geoarchaeology has added so much more information to the cultural archaeology database, and this has to be integrated fully if a site is to be properly understood; archaeological sediments are not simply a neutral background containing artifacts that alone are studied (Goldberg and Berna, 2010; Karkanas and Goldberg, 2018b; Miller, 2011; Stein, 1987). The deposit matrix is sometimes more informative than these artifacts for site reconstruction, especially when studied using soil micromorphology, and chemical (organic matter, phosphate, heavy metals), physical (particle size, magnetic susceptibility) and microartefact characteristics (Chapter 17). This holistic approach to understanding a deposit is generally known as the study of site formation processes.

13.2.2 Site Formation Processes

The limitations of the discard models can be illustrated by how regularly occurring archaeological features lose much of their interpretational worth if only fitted into such schemes. For example, *in situ* floor coverings in domestic space and bedding layers in sleeping areas (Goldberg et al., 2009b; Macphail et al., 1997) *could* be construed as an activity deposit. In the same way, the thick accumulation of plant remains in waterlogged medieval houses of Trondheim and Oslo, Norway (Macphail and Goldberg, 2018a: 229), which is interpreted as accumulated *floor coverings*, is a primary waste deposit. In these flooring examples, it may be better to employ aspects of "occupation surface" models, which includes a wide variety of scenarios associated with construction and use (Cammas et al., 1996b; Courty et al., 1994; Gé et al., 1993). In the specific case of the urban medieval Norwegian floor deposits, these often include sub-horizontally oriented fish bones, which in high concentrations likely point to fish processing, and thus represent both a "primary" and "secondary" use deposit (see Korsmyra, Norway; Beijersbergen et al., 2018; Bryn and Sauvage, 2018).

Another important instance is construction deposits (see 13.3). What were once natural soils and sediments could be used for ground raising, making "clay floors", and building small mounds (grave mounds, *tumuli*) or enormous monumental structures; some of the best studied examples are in the USA (e.g. Hopeton, Moundville and Shiloh; Gage, 2000; Mandel et al., 2003; Sherwood and Kidder, 2011) (see section 13.3). The Gokstad Burial Mound, in Vestfold, Norway, although enclosing a whole Viking Longship is small in comparison (Cannell et al., 2020; Macphail et al., 2013a; Nicolaysen, 1882). Tells themselves are settlement-accumulations composed of a variety of occupation deposits, mudbrick, and associated earth-based building materials (see section 13.4.2; Table 13.1).

There are also whole sites that are termed middens (see above and Chapter 9.5): Potterne and Chisenbury, Wiltshire, UK (Late Bronze Age/Early Iron Age) are midden sites 2–4 ha in size and 1–2 m thick (Lawson, 2000; McOmish et al., 2010). The multidisciplinary study of these sites, based upon a few excavated areas and sondages, found that in fact that these were settlements and not just large middens formed by rubbish disposal, although the exact nature of these sites is still under debate. Soil micromorphology and associated microchemical studies (e.g. microprobe) found that animal management was probably responsible for deposit accumulation including probable inputs from dogs, pigs and cattle for example (Macphail, 2000). At Chisenbury, more alkaline chalky soils allowed calcitic dung spherulite concentrations – of probable ovicaprid origin for the most part – to be preserved (see Canti, 1999; Shahack-Gross, 2011; Macphail unpublished report) (see also Shahack-Gross, 2017). Deposits at both of these sites also included mineralized *cess* (human waste) with a typical hydroxylapatite chemistry and ingested cereal phytolith remains (Macphail, 2016b). Although cess pits and latrine outflows display concentrated cess, much mineralized human waste is in the form of phosphatic nodules (Brönnimann et al., 2017b; Karkanas and Goldberg, 2018a). These also are accidentally spread across occupation floors, along passage ways and are often present in trampled soils between houses in settlements. Moreover, road and *trackway* deposits within and between settlements are also likely to contain fecal waste. These originate, for example, from the passage of stock and draft animals, and in some urban areas where the toilet bucket has been emptied into the street, quite high concentrations of human waste can occur (Macphail and Goldberg, 2018a: figure 11.14; Macphail et al., 2017a) (see section 13.4). At the Borduşani-Popină tell (Figure 13.2e), mineralized cess was analyzed by EDS: 17.4–17.8% P, 0.0–0.36% S, 38.3–39.2% Ca, 1.14–1.49% Fe.

13.3 New World Mounds and Monumental Earthen Architecture

As stated by Sarah Sherwood (The University of the South, Sewanee, TN, USA, pers. comm.), mounded, massive anthropogenic accumulations are not merely confined to the Old World but are also widespread in North and Central America. Although shell middens in mounds are common, massive earthen constructions are less widespread and tend to date to the last two millennia or so. One of the earliest excavations in America was conducted by Thomas Jefferson on a small (ca. 3.5 m high) mound in Virginia (Sharer and Ashmore, 2003). More typical and striking examples of earthen mounds, however, can be found in the Mississippi Valley (e.g. Cahokia; Poverty Point; Watson Brake), Ohio Valley (Hopeton), South-East USA (Etowah, Moundville), and throughout Central and South America. The Mississippian Period (ca 900–1600 AD) consisted of farmers (primarily maize producers) organized in chiefdoms that concentrated throughout the Mississippi River drainage in eastern North America and traded and shared exotic goods and iconography (citations). Their villages were typically organized around a central plaza with earthen mounds.

As in the Old World, these constructions are complex stratigraphically and socially, and overall, the same general principles of site formation occur. Again, the geoarchaeological task is to use our knowledge of the deposits and depositional/post-depositional processes to gain access to the *detailed* activities of the people who lived there and constructed them. In the New World, geoarchaeological work on mounds – with the exception of a few shell mounds and middens (Stein, 1992: xix), and ceremonial mounds (Van Nest et al., 2001) – is immature and tends to concentrate on landscape scales, such as land use, soil fertility, etc.; smaller-scale examination of individual deposits, as emphasized below, is not common. When one considers the size of these mounds and their abundance it is mystifying that so little effort has gone into the detailed study of their stratigraphic make-up.

Different geoarchaeological techniques have been applied to the study of mounds, ranging from larger-scale aspects and site survey, down to smaller scale studies of mound formation. The former types of investigations have employed geophysical methods (Herz and Garrison, 1998), whilst in the latter case, physical and chemical characterization (grain size, soil chemistry, and micromorphology) have been used (e.g. Cremeens et al., 1997).

Geophysical techniques have been applied to several areas of the New World. In eastern North America, for example, Dalan and Bevan (2002) raised the issue of how one distinguishes natural from human-made accumulations without having to probe the sediments directly by coring, for example. They were interested in the "lumpiness" of the geophysical data resulting from anthropogenic mixing of materials; it differs from most geophysical investigations, which concentrate on-site survey and the isolation of culturally-induced anomalies. They studied directionality data using seismic and magnetic susceptibility techniques at Cahokia Mounds State Historic Site, Illinois; at the Hopeton Earthwork in southern Ohio, magnetic susceptibility and seismic studies were also carried out.

While results at Cahokia were not diagnostic, those at the Hopeton Earthwork were more useful. They concluded that different earthworks have variable geophysical signatures, and that different methods and techniques are successful at different sites. Valuable recognition of anthropogenic signatures and the appropriate method to recognize them are dependent on the type of earthwork, the types of construction materials and processes employed, and on post-depositional processes. In fact, a combination of different techniques is needed to obtain significant results. It is also quite clear that the results of detailed studies of earth mounds and experimental versions in Europe, have still much to offer to New World investigations (Bell et al., 1996; Gebhardt and Langohr, 1999; Langohr, 1991; Macphail et al., 1987).

Moundville Some initial and very promising work was recently conducted at Moundville, a late Mississippian site in central Alabama. The site rests upon an alluvial terrace composed of sands, silts, clays, and gravels, situated ca. 45 m above the Black Warrior River. The geoarchaeological investigations aimed to explore aspects of the site structure, construction, and site formation employing relatively low impact and invasive techniques (Gage, 1999; Gage and Jones, 1999; Gage, pers. comm.). As ground-penetrating radar (GPR) can penetrate to several meters below the surface, it was preferred to shallower techniques such as resistivity, gradiometry, and magnetometry (see Chapter 15). Furthermore, a program of coring was undertaken in order to corroborate the geophysical results. The 4 inch diameter cores were split and logged.

One of several mounds was chosen, Mound R (AD 1350 and 1450), which is the third largest structure at the Moundville complex. It sits about 6 m above the surrounding flat area, and covers an area of ca. 6375 m². "Mound R is not considered to be a burial mound, but rather a domiciliary platform which speculatively would have supported various elite residences, of *wattle* and daub construction" (Gage, 1999: 307). The geogenic sediments constituting the matrix of the mound are derived from both the alluvium as well as red, gray, and purple clays of the locally outcropping Coker and Gordo Formations. These materials are thought to have been derived from local quarries termed "borrow areas", transported with baskets, and dumped on the mound.

Both the GPR survey and coring revealed a sequence of stratified deposits that point to discrete phases of construction. Calibration of the results was achieved by comparing surface deposits with previously excavated areas. It was concluded that burned organic horizons formed marker horizons that represent the end of an occupation phase. Where no organic material was present, it appeared that the mound summit at the time was leveled in order to produce a stable, flat surface. This leveling was accomplished by dumping and spreading basket-loads followed by compaction by the site's builders and occupiers.

Interestingly, Gage related these types of data to dynamics of the population at the site (Gage, submitted):

> Mound R's patterns of occupation mirror the changes in social organization proposed for the site. With the changing population, successive episodes of decreased volumes of construction become apparent until probably the end of Moundville III or early Moundville IV when Mound R is abandoned.
>
> For whatever reason. . .successive episodes of construction over years, decades, or centuries was undertaken by groups of laborers in an effort to modify the existing communal landscape. The commonality of building techniques and methods leads to a similarity, notable in almost all such earthen structures. Basketload upon basketload was excavated from secondary locations, transported to a common locale, and deposited to form mounded earthen structures. As a result, profiles of these structures reveal individual piles of sediment as well as summit surfaces buried by subsequent additions of mound fill.

Other North American mounds include Cotiga Mound (Cremeens, 1995; Cremeens et al., 1997) and Hopeton (Mandel et al., 2003; Price and Feinman, 2005; Van Nest et al., 2001), and as noted above, more recent investigations have taken place at Shiloh Mound, Cahokia, and Poverty Point (Sherwood and Kidder, 2011).

Meso- and South American mounds In Central and South America, geoarchaeological investigations associated with mounds and occupational deposits have tended to focus on the regional scale. In Belize for example, much of the work has focused on issues of soil, landscape development and settlement change (e.g. Dunning et al., 1998; Pope and Dahlin, 1989), although many studies link soil and sediment analysis to architecture and site function (Smyth et al., 1995). Remote sensing, satellite imagery, and geophysics are the common tools.

In Brazil, Bevan and Roosevelt (2003) used geophysical techniques and excavation to elucidate patterning of sediments and features, and to understand how they reflect on overall site history of the organization of its occupants. They employed several types of geophysical techniques, but the magnetic data were instrumental in revealing a number of cooking hearths and burial urns, and enabling the estimation of "the number of cooking hearths within the mound without being able to isolate or separately detect any one of them" (Bevan and Roosevelt, 2003: 328).

Furthermore, the geophysical analysis was accompanied by excavations, which demonstrated that baked clay hearths and their associated burned floors are the cause of the magnetic and conductivity anomalies. Before these excavations were carried out, the geophysical survey suggested that there could be as many as 8000 hearths in the mound, if these hearths were the only magnetic objects in the mound. A re-evaluation of the magnetic survey, based on the findings of the excavations, now suggests that there may be about 2200 hearths within the mound.

13.3.1 Maya Gonzalez, Belize

At the Maya site of Marco Gonzalez, Ambergris Caye, Belize (Figure 13.6a), which dates back to at least 300 BC, continued occupation led to the growth of a >3.5 m-thick mound site that forms a small island within modern-day mangrove swamp (Figure 13.6b). Here also there are numerous hearths, but these are associated with salt working and use of Coconut Walk ware; some 2 m of lime plaster hearth floors were recorded and these are increasingly weathered upwards (Graham et al., 2017; Graham and Pendergast, 1989; Macphail et al., 2017b) (see Box 13.1). Soil formation models linked to Maya settlements have been under scrutiny for many years, including the development of fertile *terra preta* (Maya dark earth) (Evans et al., 2021; Graham, 2006; Isendahl, 2012.)

Even after many years, sondage excavations have exposed very little of the full extent of the site, with investigations being also greatly slowed by the presence of numerous inserted inhumations and cached objects, which have to be recovered for museum studies (Figures 13.6c and 13.6d). In addition, rises in sea level and associated on-site water tables (Dunn and Mazzullo, 1993) have not only led to the island becoming surrounded by mangrove, but have also made the excavation and intact sampling of the lowest deposits very challenging. The lowest intact layer (MG 383) sampled for soil micromorphology dates to the Early Classic (ca. AD 250 to 550/600), and are typically pale gray in color. The earliest deposits are Protoclassic-Terminal Preclassic (ca. AD 100 to 250?) (Graham and Pendergast, 1989; Pendergast and Graham, 1987), but were inaccessible at this sondage. The Early Classic levels are composed of a weakly humic (3.66% LOI), fine sediment (93.6% silt and clay, with ~6% sand), which is strongly calcareous (70.2% carbonate), alkaline (pH 8.9), and strongly saline (3540 specific conductance) with a very strongly enriched phosphate content (23.2 mg g^{-1} phosphate-P). It is an essentially anthropogenic deposit that was waterlaid originally, although now affected by some burrowing by small mesofauna. Anthropogenic inputs are chiefly calcitic ash (mostly wood ash?), charcoal, human fecal material/dietary waste (including fish bone), with background shell, pot and few sand (including geological bioclasts) (Figures 13.6e and 13.6f). This can be interpreted as a footslope colluvium/ash and kitchen waste dumped into standing water that surrounded the island/occupation mound. The deposits record both hearth ash deposition and populations producing both kitchen waste and fecal material. Overlying Late Classic (AD 600–750/800) deposits are characterized by lime floor hearths (Figure 13.6g; Contexts MG 364–377) and post-depositional Maya dark earth (Context MG 359; see Box 13.1).

13.4   Settlement Archaeology

Few occupation deposits were studied using geoarchaeological techniques 40 years ago. In Western Europe most deposits forming urban archaeology – including *dark earth* (see below and Box 13.1) – were simply machined away although their importance had been noted as early as the beginning of the twentieth century (Biddle et al., 1973; Carver, 1987; Grimes, 1968; Norman and Reader, 1912). Sporadic studies across Belgium, France, Italy and, the UK in the 1980s and 1990s (Brogioli et al., 1988; Ervynck et al., 1999; Gebhardt, 1997; Macphail, 1981, 1983; Macphail and Courty, 1985; Yule, 1990), produced models for interpreting urban deposits and the reconstruction of a settlement's use of space (Macphail, 1994a). Urban studies across Europe fully took off in the 2000s, especially in Brussels (Belgium), Paris and Tours (France), and in cities in northern Italy and Scandinavia (Cammas, 2004; Cammas et al., 1996a; Cremaschi and Nicosia, 2010; Devos et al., 2009, 2020a; Fondrillon et al., 2009; Galinié, 2004; Heimdahl, 2005; Macphail et al., 2003a; Macphail and Linderholm, 2004b; Nicosia et al., 2017; Sidell, 2000; Wiedner et al., 2015); studies in Greece have mushroomed within the past decade (Karkanas, 2007; Karkanas and Efstratiou, 2009; Karkanas and Van de Moortel, 2014; Karkanas et al., 2018, 2019). In Eastern Europe (e.g. Hungary and Romania), the Near and Middle East, and China, urban settlements are often in the form of tells (Matthews, 1996, 2005; Matthews et al., 1996, 1997, 2020; Rosen, 1986; Shahack-Gross, 2011; Shahack-Gross et al., 2005) (see Table 13.1). In the Far East and the New World structural remains in past cities have been more commonly investigated in terms of preservation of monuments and the effects of cyanobacteria for example at Maya cities across the Yucatán Peninsula in central America and at World Heritage Sites in Cambodia (Gaylarde et al., 2012; Straulino et al., 2013) (see Chapter 12). Water management as studied in Africa, for example (Sulas, 2018), is also well recorded at a number of World Heritage Sites. Few investigations of anthropogenic deposits to study past settlement activities have taken place, however, except for satellite and ground survey to investigate water management (e.g. Evans et al., 2007). At Angkor Wat and other Cambodian Khmer royal sites reservoirs and canals are extremely important features; at Merv, Turkmenistan a possible Sassanid canal seems to have been in use until the destruction of the city by the Mongols (i.e. ~3rd to 13th centuries AD), and shows both periods of running and stagnant water (Williams, 2007, 2018) (see Chapter 11.4.3). Water management is also linked to some of the earliest evidence of rice cultivation (Zhuang et al., 2014).

This section deals mainly with stratigraphy and deposits of complex societies; only occasional reference is made to upstanding structures, because they are often rare apart from stone-founded churches, temples, and palaces. Standing structures are in the area of specialists in building analysis where constructional material (e.g. stone) conservation and standing structure survey are key components; and an essential background to any geoarchaeological investigations of the same site (Clark, 2000; Keevill, 1995). At the Tower of London (13th–18th century moat fill) and the London Guildhall (e.g. Roman Amphitheater – early medieval accumulations), analyses of deposits included the collapsed barbican built by Henry III and the earth-based (brickearth) amphitheater and its sand and mortared floors, respectively (Figure 13.5; see Figure 13.9a) (Bateman et al., 2008; Keevill, 2004). The latter fell out of use in the 4th century AD towards the end of the Roman occupation (*sensu stricto*), and became infilled and the leveled location of a small farmstead by AD 1050. It can be compared to parallel studies examining constructional waste and building methods at Roman Basel, Switzerland (*Augusta Raurica*) (Rentzel, 1997, 2009).

In western Europe, the Classical World (in this chapter we focus on the Roman Empire,~50 BC–~400 AD), produced a number of materials that were specific to this civilization (see Chapter 12 for discussion of New World lime plasters). In addition to the stone-founded buildings of both Roman urban areas and villas, and their use of various types of lime-based mortar and plaster (*sensu lato*), occupation could produce significant amounts of roadside deposits, midden accumulations and debris from domestic, industrial, and stock management practices (see below for road and trackway deposits). Numerous Roman structures, as well those from the early Medieval period, were also constructed of timber, and earth-based materials, such as clay, daub, rammed earth, and cob (or clunch) (Cammas, 2018), which produced easily weathered deposits. As such, they became a major constituent of deposits termed "dark earth" (Box 13.1;Table 13.3). During the Roman period there was systematic use of different constructional materials, such as lime-based mortar and the use of waterproof hydraulic mortar (puzzolanic cements mixed with volcanic inclusions) (see also Villaseñor and Graham, 2010). At Roman Exeter, the medieval castle is built on a basalt plug, a handy source for the Romans here (see Figure 13.9).

Some Roman activities were carried out for the first time on an industrial scale (e.g. iron working at Worcester, UK, producing a legacy of highly residual iron slag; Dalwood and Edwards, 2004) (see Chapter 12). Roman use of space could also be highly structured (e.g. *insulae*, space organized within the street plan that are both visible in *Pompeii* and Roman Vienne or London). A settlement could comprise roads/tracks, structures/houses, and land used for dumping/cemeteries, the last usually on the contemporary town edge (see Table 13.2, and below). It should also be noted here that Roman towns in England were not defined from the outset by walls. For example, towns like London and Canterbury, were established in the first century AD, but were not enclosed within walls until the third century AD (Salway, 1993) (Figure 13.8). Moreover, land use during the history of a settlement could also radically change during the lifetime of an occupation. In Southwark, London and Deansway, Worcester, for example, late Roman (fourth century AD) cemeteries were found in waste areas that were once a residential part of the 1st–2nd century town (Cowan, 2003; Dalwood and Edwards, 2004). At Canterbury the 3rd century AD walls actually buried infield cultivated soils, which although manured were much less potentially fertile compared to medieval soils associated with the Augustinian Friary founded in 1325 (Whitefriars) (Hicks, 2015; Hicks and Houlistan, 2018; Macphail and Goldberg, 2018a: 325–326) (see Table 11.2).

In the urban city of Munigua (Sevilla, Spain) Gutiérrez-Rodríguez et al. (2019) used soil micromorphology and PXRF to document the stratigraphic evolution and site formation from pre-Roman (4th century BC) through abandonment in the 8th to 9th century AD. They documented local metal production (lead production to iron smelting) in furnaces followed by thermae ("baked clay") construction during a phase of urban planning, and ultimately by abandonment, recycling of metals, and the construction of pavements during Late Antiquity. Another Spanish example of protohistoric earth-based urban constructions was investigated at Alcanar, Tarragona (Mateu et al., 2013)

Both homogeneous deposits, dark earth *sensu lato* and stratified urban deposits that are difficult to interpret have been investigated to provide insights into little-understood periods of urban history (Box 3.1). Some medieval (10th–13th century) rampart-buried dark earth in Brussels has been shown through combined soil micromorphology and phytolith studies to be dung-manured culti-vated soils. At the Court of Hoogstraeten, on the other hand, these methods tentatively revealed pre-10th–12th century pastures before manured arable soils were developed during the 10th–12th century, and a 16th century dark earth formed from dumped stabling waste (see Figures 13.11a, 13.11b, 13.12c), and matted floors (Devos et al., 2009, 2013, 2020). Further integrated investigations by Devos et al. (2016) also utilized pollen and macrofossil data in poorly drained soils and sedi-ments in the Senne Valley, and were able to identify different areas with histories of early medieval (10th–11th century) wet grassland, later medieval cultivation and late medieval and post-medieval (15th century) waste disposal that reflected the increased population in the city. Such studies of open space accumulations are less often studied (e.g. 13th–16th century Magdeburg Cathedral precincts, Germany), but includes the Lier market place study in Belgium, where three series of dark earth layers recorded early medieval (12th century) agriculture, and increasingly intense dep-osition of occupation waste through the 14th century, when traffic and craft activities contributed to the accumulated deposits (Wouters et al., 2016).

As at the Senne Valley sites, Belgium, urban studies at the Uffizi, Florence, Italy found that a major component of the middening deposits was human waste (10th–11th century); another, but natural contributor, was clay inwash of alluvial origin produced by flooding – the flood of November 4th 1177 (Nicosia et al., 2012). A major small artifact study, with a soil micromorphology and chemistry component at the Prosper-Mérimée Square, Tours, France, was carried out and dark earth with features dating to the Late Roman (*Antique* Period; 4th century AD) to the 11th century was investigated (Fondrillon, 2007; Fondrillon et al., 2009; Galinié et al., 2007). A major discovery was that vestigial earth-based building remains were associated with latrine waste disposal and high concentrations of small finds indicated much higher populations in the 7th–8th centuries (Merovingian into Carolingian Dynasties), for example, than had been previously thought. By the 11th century, historical documents show that the dark earth soils supported vineyards. Urban land use had thus changed through time and space. At the Roman city of Tarquimpol, La Moselle, France, and after its destruction in the 4th century, the open settlement (with typical stone-founded theatre and street plan) was replaced by a small earth rampart-protected settlement composed of daub- and turf-built structures associated with Tarquimpol's brief 5th century role as a road hub (Henning et al., 2012; Macphail, 2014).

During the 9th–11th centuries raiders, traders, and settlers from Denmark, Norway and Sweden (sometimes grouped as the "Norse" or the "Vikings"), occupied and/or controlled towns and cities across northern Europe all the way to modern Ukraine and Russia. They even estab-lished new dynasties (e.g. the "Russe" or "Rus'" along the Dneiper River (Kyiv)) and the whole of northern and eastern England was "Danish" for a while – Danelaw. One famous King of Viking York (Jorvik) was Eric Bloodaxe. As organic occupation deposits accumulated in these coastal urban locations (e.g. Riga, Latvia; Oslo and Trondheim, Norway) or towns easily acces-sible from the sea via major rivers (York), UK; here, the local water table rose as ground levels rose, further aiding the preservation of these usually easily-oxidized materials (Addyman et al., 1976). At Coppergate (now the Jorvik Centre), for example, some of the first intensive studies of well-preserved organic remains were carried out. This study allowed detailed assess-ments of living conditions in general and the analysis of within-settlement stabling practices, in particular (Kenward and Hall, 1997; Kenward and Hall, 1995). Human waste (especially its para-site egg content) came under scrutiny; thousands of whip worm and round worm eggs were counted for example (Jones, 1985) (see also Pümpin et al., 2017); a wide-ranging diet was also recorded in thin section investigations (see Macphail and Goldberg, 2018a: table 7.5). Similarly, wooden buildings and organic deposits at Riga, that are of a specific period dating to the pre-Hansa town occupied by the indigenous "Liv" (12th–13th centuries AD) underwent environmental studies; these included geochemistry, which revealed various activities such as metal working, fish processing, refuse disposal and use of plants (Banerjea et al., 2016). Comparable early medieval Slavic towns, with a Viking (Norse) component (e.g. Brünkendorf on the North German Plain) associated with anomalous dark soils, have also been investigated and compared with *terra preta* (Wiedner et al., 2015) (Box 13.1; Figures 13.9 and 13.10). The soil chemical analysis of black carbon and organic chemistry (e.g. biomarkers) revealed manure inputs from omnivores and herbivores as well as human excrement.

Combined soil micromorphology and microchemical studies of waterlogged and semi-waterlogged medieval (11th–14th century) deposits within the Norwegian settlements of Oslo, Tønsberg, and Trondheim, were used to examine wooden house floors, wooden pavement, and exterior contexts. Macrofossil remains were sometimes difficult to identify definitively because partial humification had occurred; soil micromorphology of undisturbed deposits showed that humified monocotyledonous plant remains were relict plant-based floor coverings (Sofi Östman, University of Umeå, pers. comm.). These floor coverings were not often removed but just renewed, becoming thicker through the life of a house (Macphail, 2016a; see also Macphail, 2005). The ubiquity of sub-horizontally oriented fish bone fragments scattered in these deposits and the occurrence of fish bone concentrations in middening deposits indicated that fish processing was common (cf. Simpson et al., 2000). Macrofossil and thin section studies also confirmed the presence of latrines where sphagnum moss had been utilized, probably as "toilet paper"; despite ground water movements and leaching secondary iron-calcium-phosphate cementation often occurred in such contexts. Craft working debris layers included traces of silver (Ag) associated with other non-ferrous metals, could also be observed when thin sections also underwent SEM/EDS studies. In the same contexts furnace prills (a spherical aerosol) could occur; at one location at Torvet, Trondheim prills contain antimony, arsenic, copper, lead, and tin (see Human materials – Figures 12.30–12.32).

13.4.1 Some Issues in Settlement and Urban Archaeology

Many of the research objectives set out for the study of historic settlements in Europe can be applied worldwide. It has also been argued that the term "settlement" should not always be restricted to complex societies *sensu stricto*. Well organized hunters and gatherers with dwellings, will have areas for different processing and gathering activities according to the locality of natural resources, as suggested by the study of Mesolithic sites and modern-day Evenki in Siberia (Grøn and Kuznetsov, 2004). Distinct activity areas and features involve, for example:

- Outdoor food preparation
- Skin processing
- Waste disposal
- Toilet activities
- Storage platforms and larger more peripheral areas for collecting bark and gathering firewood

In long-lived occupations from complex society sites, there is always a need to ascertain the layout of a settlement in order to understand how the occupation functioned and to identify and understand how this may have changed and developed through time. Some areas may have always been open ground where stock were kept, middens accumulated, or where markets were held, whereas other parts may have had religious, commercial (warehouses, shops, and food preparation), industrial, and (high and low status) housing land use. Space could also be divided up within structures themselves (Tables 13.1 and 13.2) (Cammas et al., 1996b; Courty et al., 1994; Cruise and Macphail, 2000; Macphail et al., 2004, 2017a; Macphail and Cruise, 2001; Matthews, 1995, 2010; Matthews et al., 1996).

Another important aspect of settlement archaeology is continuity (Galinié, 2000; Leone, 1998); in fact at all occupation sites this is an important component of any study. There are important examples of brief changes to "urban" life at both Roman London and Colchester, Britannia's capital at the time, after these with Verulam (St Albans) were razed to the ground during the Boudiccan revolt (AD 59/60). The native tribes in the east of the country (the Iceni, Catuvellauni) had managed to defeat various Roman forces and destroy these major towns, leaving behind the typically fire-scorched remains of brickearth "clay" walled buildings (see 12.3.2). It was at this time (~AD 60–70) that a short-lived period of horticultural land use can be demonstrated at both Colchester and London, and it took some time before fully functioning Roman urban construction/life recommenced. At Whittington Ave, London horticultural soils and middening deposits were sealed below ground-raising brickearth levels forming the foundations for the (AD 70, AD 90–120) Roman basilica (Macphail, 1994a; Watson, 1998). One of the chief ways of investigating continuity of urban occupation has been through the study of road and trackway alignments, and their relationship to a settlement's contemporary land use. At Elms Farm, Essex, there appears to have been continuity between the town's Late Iron Age origins and the Roman settlement (Atkinson and Preston, 1998, 2015). In the same way, Roman roads have been found to postdate Iron Age roads, for example, at Sharpstone Hill, Shropshire, UK (Roman Margary Route 64) (Malim and Hayes, 2011). At Canterbury, in contrast, metaled lanes of the seventh century Saxon town show no aligned relationship to Roman roads, even though only 200–300 years separates them (e.g. at Whitefriars; Alison Hicks, and Mark Houlistan, personal communication; Hicks, 2015; Hicks and Houlistan, 2018). The Saxon roads themselves occur over post-Roman dark earth, and Early-Middle Saxon *grubenhäuser* or sunken feature buildings of broadly Germanic origin (see Box 13.2).

13.4.1.1 Deposit Formation

One of the characteristic of road and trackway deposits is their apparent homogenization caused by muddy trampling and churning by the passage of people, animals, and vehicles, although fragments of dung and broad layering may sometimes be preserved according to use (Gebhardt and Langohr, 2015; Macphail et al., 2017a; Rentzel et al., 2017) (asma/Online_Video_11.2.md at master · geoarchaeology/asma · GitHub). Norwegian Iron Age rural (Bamble, Vestfold) and Romano-British oppidum (Winchester, UK) trackways are composed of 0.40–0.80 m of semi-homogenized sediments, with dung or fine organic fragments of likely dung origin, and a closed vughy porosity and matrix intercalations and void infills of structural collapse and slaking origin characterizing them (Ford and Teague, 2011; Rødsrud, 2020). Other examples from English Iron Age settlements and pre-Roman road systems, and a Swedish Iron Age rutted roadway between farmsteads (Malmö) also show bulk and microchemical evidence of phosphate deposition, presumably linked to traffic by stock (Engelmark and Linderholm, 2008: 80–81; Macphail, 2011c).

Especially associated with Roman towns are the fills of roadside gullies and dumps made on open ground next to roads. Roadside gully fills are characteristically sand-dominated sediments, washed from the road surface; however, there is a ubiquitous minor content of small bone fragments and human coprolitic material, as well as secondary phosphate deposition. The last occurs as crystalline *vivianite* (e.g. $Fe_3(PO_4)_2$ $8H_2O$) and/or as amorphous iron-calcium-phosphate (Karkanas and Goldberg, 2018a; Macphail and Goldberg, 2018a: table 2.3), as for example in

Very rarely, dung spherulites (Shahack-Gross, 2011) are preserved in road deposits when organic dung remains have been lost to oxidation. One enigmatic type of roadside dumped sediment that accumulated in waste ground in Roman towns, is typically organic. Where well preserved (e.g. as a result of a high water table at No 1 Poultry), such deposits are characterized by high LOI (25.9% loss-on-ignition; Figures 13.11a and 13.11b) (Burch et al., 2011; Macphail, 2011b; Macphail et al., 2004). They are also rich in phosphate (e.g. ~4080 ppm P_2O_5). The layered nature of the organic remains and the highly humified character of some of them imply that these deposits are formed from dumped dung-rich stabling waste (Macphail et al., 2007a) (see below for stable floors). This geoarchaeological interpretation is consistent with the results of the macrobotanical study (Burch et al., 2011). These deposits seem to owe their origin to the presence of many horses (as evidenced by faunal remains studies) and other stock in early first century London (e.g. AD 40s–50s). At that time, the developing urban area of London acted as a frontier town on the edge of an expanding Roman Empire. Rowsome (2000) alludes to the American "western" towns of Wichita and Dodge City of the 1870s as suitable analogues. Equally, at 7–11, Bishopsgate, London, a room with a domestic brickearth floor was reused as a stable, the charred remains of which seem to date to the late Hadrianic fires (Macphail and Cruise, 1997; Macphail et al., 2004; Sankey and McKenzie, 1998).

Late Roman (4th century AD) deposits that were once dung-rich, and which represent stock concentrations were also found in waste ground areas of Deansway, a suburb of Roman Worcester (Dalwood and Edwards, 2004; Macphail, 1994a). Similar open-air deposits were found at the small Late Iron Age Roman town located near the Essex coast at Elms Farm, Heybridge. Here they are associated, with animal management relating to the role of this settlement as a market center for the surrounding area (Atkinson and Preston, 2015). The use of space and changing use of space in Roman structures has been under-studied, because much more emphasis has been given to architectural remains. Increasingly, however, a less simplistic approach has been undertaken to scrutinize a variety of lifestyles. At the House of Amarantus, Insula 9, Pompeii, not all rooms had properly constructed floors. Room 3 in House 12 had a beaten floor, while room 4 was used to stable a mule that succumbed to the effects of the eruption of Vesuvius in AD 79. Its skeleton rested upon compacted stabling floor deposits in which P. Wiltshire found pollen of cereals and grass of assumed fodder, hay, bedding, and dung origin (Fulford and Wallace-Hadrill, 1995–1996; Fulford and Wallace-Hadrill, 1998). Similarly, at Phoenician Tel Dor, dung residues were found within a monumental building (Shahack-Gross et al., 2005; see also Shahack-Gross, 2011 for other Near Eastern examples, and insights into destruction levels).

Figures 13.11a and 13.11b Dumped stabling waste at Roman No 1, Poultry, London, United Kingdom (Macphail, 2011b; Macphail et al., 2004), in which plant macrofossils point to open area dumping of waste from post-Boudiccan animal husbandry (Burch et al., 2011; Rowsome, 2000). Scan of M891, illustrating laminated organic tissue fragments, amorphous organic matter, and intercalated silts (neutral pH 6.6; high organic matter content (20% LOI), with high amounts of phosphate (3950 ppm P_2O_5), and very low magnetic susceptibility ($\chi = 20 \times 10^{-8}$ m^3 kg^{-1}). Scanned image of 13 cm-long thin section.

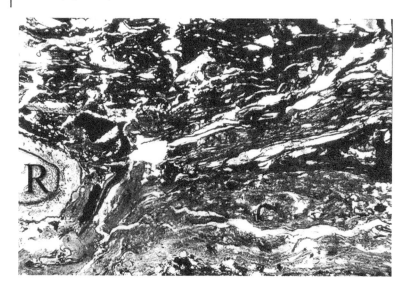

Figure 13.11b As Figure 13.11a. Photomicrograph of thin section M422 showing detail of well-preserved stabling waste deposits; a finely bedded organic deposit of plant tissues, amorphous organic matter, commonly intercalated with silt, which is extremely humic (34.6% LOI), neutral (pH 6.6), with very high amounts of phosphate (4170 ppm P_2O_5), and very low magnetic susceptibility ($\chi = 14 \times 10^{-8}$ m^3 kg^{-1}). Note fleshy rooting (R) of this wet deposit. PPL, frame width is ~5.5 mm.

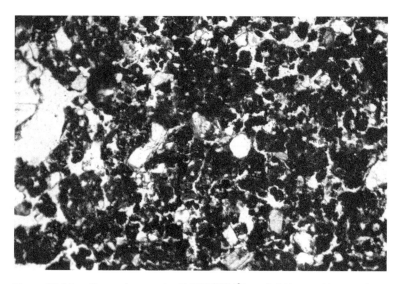

Figure 13.11c Photomicrograph of M41292B (Åker gård, Harmar, Norway; lower byre floor – slumped into top of pit A42080); pellety amorphous organic matter – bioworking of dung in "composting" byre floor. Note orange iron and phosphate staining from use of overlying byre floor layers as deposits accumulated. SEM/EDS: 0.49% S, 5.15% Ca; 2.10% P, 5.40% Fe (iron staining) (Table 13.5). PPL, frame width is ~2.38 mm.

13.5 Occupation Surface Deposits

Occupation surface deposits, which have come under increasing study over the last 30 years can include floor accumulations, both associated with use of space within structures – domestic, artisan, and animal management activities – as well as along passage ways within and between houses, for example (Cammas, 1994; Cammas et al., 1996b; Courty et al., 1994; Gé et al., 1993) (see Tables 13.1 and 13.2). Trackway and road deposits within and between settlements have already been discussed. Animal pounds where dung accumulates are also examples of occupation deposits (Shahack-Gross, 2011, 2017; Shahack-Gross et al., 2003, 2008), but being open areas they are often less well preserved apart from phosphate concentrations and minerogenic residues (*phytoliths*, *dung spherulites*) (Portillo et al., 2020). When enclosures are demarcated by a ring ditch, dung residues may become dumped and preserved in these, as proxy use of space indicators. Roadside dumps have already been noted.

When attempting to reconstruct the morphology of a settlement the investigation of occupation surface deposits is a key method for identifying use of space (Tables 13.1 and 13.2). Floors and surfaces, such as a "hard standing" pebbled surface, can be prepared by importing stones and/or clay ("clay floor") and also be constructed using mortar or plaster, as already described in Chapter 12.

13.5.1 Floor Deposits

Presented below, are some investigations of use of space in structures, which can include:

- rooms within stone-founded buildings and mud brick constructional remains in tells (see above), as well as,
- the main functional areas of tri-partite longhouses (Scandinavia) and Mediterranean rock shelters, and,
- variously constructed wooden buildings (e.g. plank-built stave building and wattle-walled byre at the London Guildhall), and the sunken byres (postals; Belgium), and individually built small (byres) and large (domestic living space) roundhouses (later prehistoric Europe, and Butser Ancient Farm).

For simplicity, the main characteristics of byre and stable layers are given separately from those of domestic floor deposits, although as can be imagined that unlike today, these spaces do not record exclusively stock or human activity traces.

13.5.2 Animal Stabling and Byre Deposits

Areas of animal management have been investigated from Kenya, North Africa, Eastern Europe, and the Near East, for example, and preservation of dung remains is essentially governed by rates of oxidation. Commonly, only minerogenic remains – phytoliths, dung spherulites, and calcium oxalates from partially digested plant remains – occur as "signals", but these have to be in high concentrations to be significant, because dung spherulites can be ubiquitous as a blown dust component. When buried, however, such as in tells and rock shelters, dung spherulites can be found embedded in partially oxidized compacted dung pellet deposits (e.g. Chalcolithic tell at Borduşani-Popină, Borcea River, Romania and Early Bronze Age rock shelters in the Ramon Crater, Mizpe Ramon, Negev Desert, Israel; see Figures 13.3e–13.3d) (Gur-Arieh et al., 2014; Haiță, 2003; Macphail et al., 2017a; Macphail and Crowther, 2008; Rosen, 1988; Rosen et al., 2005). Seemingly, the earliest and mainly oxidized stabling and byre deposits have been reported from Neolithic Mediterranean caves (Angelucci et al., 2009; Boschian and Montagnari-Kokelji, 2000; Brochier, 1983; Cremaschi et al., 1996; Wattez et al., 1990) (see Figures 13.2a–13.2c). Similar oxidized remains occur in dry southern European and African locations (Cammas, 1994), and arid Middle Eastern (Matthews, 1995; Matthews et al., 1997; Garcia-Suarez et al., 2020) open-air sites (Portillo et al., 2020; Shahack-Gross, 2011, 2017; Karkanas and Goldberg, 2018b: 117–120). These

Table 13.5 Åker gard, Hamark, Norway; stabling area of Iron Age long house, which had burnt down; bulk estimate of organic matter (loss-on-ignition), phosphate (P_2O_5), ratio between organic and inorganic phosphate (PQuota), magnetic susceptibility (MS – χ 10^{-8} m^3 kg^{-1}), and selected energy-dispersive X-ray spectrometry data (EDS).

Deposit	%LOI	P_2O_5 (ppm)	PQuota	MS	P (%EDS)	S (%EDS)
Burned layer	1.3	1350	1.0	845	2.2–5.3	0.0–1.3
Charred stable floor (dung)	4.7	2820	1.0	175	3.0–5.7	1.50–2.70

ashed and oxidized remains reflect grazing and foddering regimes, and dung management practices. For instance, ashed dung may also occur as fuel ash waste.

More organic byre and stabling remains are especially typical of northern and western Europe. To improve our understanding there have been experiments at Butser Ancient Farm and archaeological and ethnographic studies (Ismail-Meyer et al., 2013; Macphail et al., 2004; Reynolds, 1979; Reynolds and Shaw, 2000) (see Chapter 14). Of particular note at the Moel-y-gar round house stable was the lateral variation over just a short distance from the round house center to the wattle and daub wall (see Figure 14.27). The center recorded the best preserved laminated bedding and fodder refuse layer, in part preserved by being cemented with hydroxylapatite (see Figures 13.4a–13.4b). In Roman and medieval London, floor deposits and surface accumulations within and without post and wattle buildings were found to be highly organic and phosphate-rich, with low magnetic susceptibility values; here preservation is also due to the deposits being partially waterlogged (Figures 13.11 and 13.12). Open area byre waste dumps, which are assumed to derive from stable cleaning, have the same character of layered phosphate-embedded and/or stained monocotyledonous (grass and cereal stems) plant remains. Another example within the precincts of Winchester Cathedral was dominated by grass and herb pollen, and was characterized by a high LOI and phosphate (Goldberg and Macphail, 2016). *In situ* floors are rarer. One, almost unique example comes from a razed Iron Age long house at Åker gård, Hamark, where thick ash remains preserved the byre floor Table 13.5). The latter was made up of pebbled surface (stones embedded in an earth floor), above which lateral samples recorded 20–40mm-thick organic dung residues showed partial working by acidophile invertebrate mesofauna (Figure 13.11c). Some dung residues had become charred and organic phosphate had been partially mineralized to produce lower PQuotas (~1.0) compared to unburned dung layers (PQuota = ~2–3).

Typically, the low amounts of mineral material found in byre deposits are often in the form of silt, which probably reflects ingestion of silt during drinking and eating by stock (although animals can trample in soil clasts). Moreover, fragments of dung, insect-worked dung, and fragments of material have been identified. Thus, both the chemistry and micromorphology are consistent with the character of experimental stabling deposits found at Butser and the detailed investigations of early medieval London Guildhall, and elsewhere (Brönnimann et al., 2017a; Macphail et al., 2004; Macphail et al., 2007a; Shahack-Gross, 2017; Viklund et al., 1998). At the London Guildhall, a pilot study recorded the soil micromorphological, chemical and pollen characteristics of an early medieval animal byre floor deposit (455), the dung employed to wind-proof its wattle wall (456), and similarly organic deposits outside the pen (457), and compared them with ditch-dumped presumed animal bedding (251), and the contemporary and close by essentially domestic stave house floors (428, 429, and 430) (Figures 13.12a–13.12d). As found in the Butser Farm Experiment (Chapter 14), compared to organic and homogeneneous byre floor material, domestic use of space produces less organic and more diverse deposits.

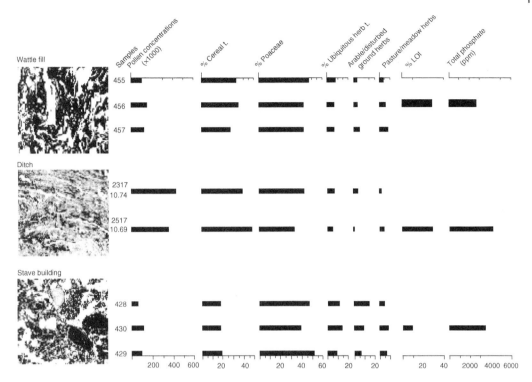

Figures 13.12a–13.12d Schematic comparison of early medieval settlement deposits associated with a small farm at the London Guildhall, employing seven thin sections, eight pollen, and three chemical analyses. **Figure 13.12a** Schematic microstratigraphical signatures at Early Medieval London Guildhall (GYE), United Kingdom (AD 1060–1140), soil micromorphology samples (Figures 13.12b–13.12d), selected chemistry and pollen data (t = type, for example, cereal type pollen) (Cruise and Macphail, 2000; Macphail et al., 2004, 2007a, 2007b).

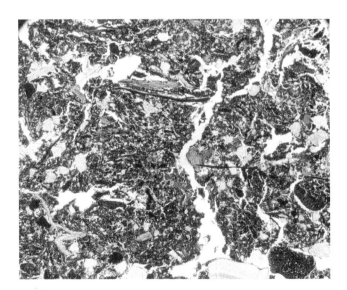

Figure 13.12b Photomicrograph of animal stable (GYE 456; top layer in Figure 13.12a) – wattle wall fill of small byre composed of highly humified probable cattle dung (28.9% LOI, 6800 ppm P). PPL, frame width is ~6.4 mm.

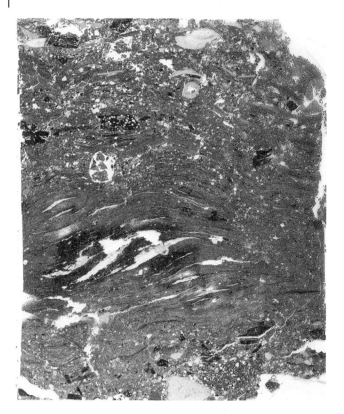

Figure 13.12c   Scan of ditch fill (GYE 251; middle layer in Figure 13.12a), with finely laminated organic remains – cereal-rich stabling refuse (28.6% LOI, 4250 ppm P_2O_5). Frame width is ~65 mm.

Figure 13.12d   Photomicrograph of domestic stave building (GYE 429; bottom layer in Figure 13.12a); heterogeneous minerogenic beaten floor deposits; with charcoal, burned bone, sands, anthropogenic soil clasts). PPL, frame width is ~6.4 mm.

Animal byres have also been investigated in the context of manuring generally, for example, and models concerning how dung (and other domestic refuse) reaches the fields (Bakels, 1988, 1997; Carter and Davidson, 1998; Macphail, 1998; Mücher et al., 1990), and how phosphate is distributed across field systems (Viklund et al., 2013). In Scandinavia, cool climates have led to better preservation of organic matter and organic phosphate compared to Europe and the Middle East, for example (see below), and humified dung residues can still be recognized within manured soils. In buildings, patterns of phosphate distribution have also been utilized to infer the location of animal stalls and byres in long houses for all periods (Conway, 1983; Crowther, 1996; Linderholm et al., 2019; Viklund et al., 1998). Roman period European examples of sunken byres (postals) in Belgium and possibly southern Norway (Mikkelsen et al., 2019), and use of dung heaps to age and compost dung (cf. Mücher et al., 1990); see Figure 13.11c) before adding it onto the fields, for example at Dilling, Østfold, seems to record a major step forward in the development of mixed farming (Mjærum, 2020). This seems to have allowed easier absorption into cultivated soils because it was already partially broken down (humified) compared to simply dumping "raw" byre waste onto the fields, and may also have concentrated such nutrients as phosphate (Macphail et al., in review; Macphail et al., forthcoming).

13.5.3 Beaten (Trampled) Floor Deposits

An occupation surface may simply form through trampling of the *in situ* topsoil or other substrate, and a beaten earth floor is developed (Karkanas and Goldberg, 2018b: 146–148; Macphail et al., 2004; Rentzel et al., 2017). It also should be noted that trampling by stock also affects byre floors, perhaps only compacting bedding layers (floor crust), but also coarsely fragmenting some areas as well, as noted when sheep and goats were housed at the Moel-y-gar roundhouse, Butser (see Figures 13.4 and 14.29–14.31). Several geochemical approaches to the study of floors have been carried out, and include bulk chemistry (e.g. LOI, P), magnetic susceptibility, and such instrumental methods as XRF, ICP and FTIR on bulk samples or *in situ* layers (hand held, pXRF) and micro-XRF, micro-FTIR, microprobe, and SEM/EDS on thin sections (Mentzer, 2017b; Mentzer and Quade, 2013; Wilson, 2017) providing multi-element data (Entwhistle et al., 1998; Middleton and Douglas-Price, 1996; Banerjea et al., 2015a) (see Chapter 17). In the view of the authors, however, the best and most easily interpretable results come first from soil micromorphology, especially when all optical methods are combined with instrumental analyses directly on the thin section. This is because floor layers can vary dramatically over the millimetric scale. In fact, beaten (trampled) floors are characterized by trampled-in layers that vary in thickness from 200 mm to 1 mm, with each layer potentially representing trampling-in from a different environment. They may also be separated by thin re-plastering layers.

This is what gives beaten floor deposits their essential heterogeneity, and often minerogenic character compared to stabling deposits (e.g. Figures 13.12a and 13.12d; see Table 12.4). For example, burned soil, ashes, and charred food waste can be derived from a local hearth and kitchen areas, while soil, dung, human coprolites, and earthworm granules can be tracked in from outside the structure. Equally, iron slag, leather, and bark can be trampled in from the neighborhoods' industrial and craft activities. These examples, taken from early Medieval London and Winchester, for example show that simple bulk analyses from a "floor" may not only average many microlayers of different use but also reflect trampling from different areas within a structure and outside that structure (Macphail, 2011b; Macphail et al., 2007a, 2011). This point has already been made by previous authors (Cammas et al., 1996b; Courty et al., 1989;

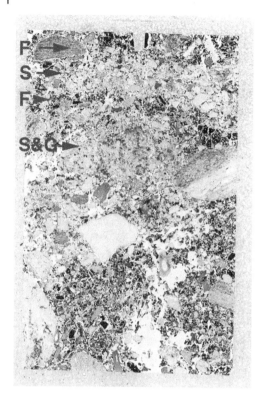

Figures 13.13a–13.13d Occupation floor deposits. **Figure 13.13a** Scan of M2267 (Lower Cultural layer, Profile 2275; early to late Neolithic dwelling site at Korsmyra in Bud, Møre og Romsdal, Norway; Bryn and Sauvage, 2018); occupation floor deposits, including possible diffusely layered charcoal dominated floors (F) between sands (S) and sands and gravel (S&G) layers, indicative of intermittent use. Chemical measurements on floors produced an occupation signal (16.4% LOI, 4870 ppm P); iron-phosphate stained charcoal (e.g. ~14–16% Fe, with ~6–7% Ca, ~6–7% P and ~1–5% Cl), which can be chlorine enriched indicating possible use of drift wood as noted at other coastal Norwegian sites. Frame width is ~50 mm.

Gé et al., 1993; Matthews et al., 1996). Less well weatherproofed areas such as entrance ways will exhibit muddier trampling features (Rentzel et al., 2017) compared to dry areas around hearths. This situation can become more complicated in long houses, which have byres, because entrances are also used by stock.

Some occupation floor deposits are illustrated. These include Neolithic Korsmyra on the west coast of Norway, which records a long period of intermittent occupation (3700–2300 BC; Bryn and Sauvage, 2018), where trampled humic and very fine charcoal-rich floors are interstratified with probable beach sands (Figure 13.13a). Large numbers of fish bones were found at the site (Beijersbergen et al., 2018) (Figure 13.13b). In strong contrast, clay and timber Roman buildings in Canterbury (Whitefriars), UK, were characterized by a series of constructed clay- and beaten-floor layers (Figure 13.13c). Lastly, anomalous microlaminated domestic floor deposits have been encountered, which are far too finely laminated to be trample spreads *sensu stricto*. These can occur below wooden floor remains in medieval Trondheim and Tønsberg, Norway, and possibly similar deposits were encountered in the air space ("pit") below a reconstructed wooden floored grubenhaus at West Stow, Suffolk (French and Milek, 2012). Such deposits are best expressed at the medieval hospital of St Mary Spital (Spitalfields, London; Harward et al., 2019). Here a whole series of floor deposits recorded cyclical (diurnal?) trampling across putative overlying plank floors; tracking-in from outside brought in fine soil, biogenic calcite, while in-house activity spread kitchen hearth residues consisting of fine burned bone, eggshell, ashes and fine burned mineral material (Figure 13.13d) (Macphail and Crowther, 2006b).

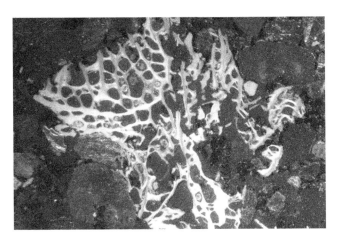

Figure 13.13b As Figure 13.13a; photomicrograph of M2267 highlighting probable calcined (burned) fishbone; large numbers of burned fish bone were found at the site (Beijersbergen et al. 2018/with permission of University Museum of Bergen). OIL, frame width is ~4.62 mm.

Figure 13.13c Scan of CW25 (CW46 Roman floor sequence, Whitefriars, Canterbury, UK); 2750 is a heated, mud-plastered (high status?) brickearth clay floor; 2734 is a beaten floor composed of ash-rich burned domestic soil debris; 2748 records a supposed lower status brickearth "clay" floor composed of Eb&Bt soil horizon material; 2747 is another domestic beaten floor deposit with hearth and cereal processing residues; B = burrow with in-mixing of Context 2746. Scan width is ~50 mm.

It should also be noted that simply because traces of dung or dung spherulites are encountered this does not "identify" a stable floor – such materials can easily be trampled in. Equally, the presence of examples of slag or heavy metal chemical traces does not necessarily mean industrial activity. Floor deposits merely need to be characterized, and then ascribed a functional use or uses.

Figure 13.13d Scan; Medieval Spitalfields, London, United Kingdom; a series of compact mineral-rich plank-sifted (domestic) floor deposits composed of brickearth flooring material, with coarse and fine black charcoal, fine bone, shell, ash, burned eggshell and earthworm granules (Harward et al., 2019). Note post-depositional burrow. Findings are consistent with experimental Pimperne house data from Butser Ancient Farm and London Guildhall floors (Macphail et al., 2004, 2007a). Frame width is ~50 mm.

13.5.4   Pit House Studies

The term "pit house" is broadly used to describe any form of sunken occupation structure in the Far East (Korea and Japan), USA, Canada, and Scandinavia, while the generic term SFB (sunken feature building) is employed in the UK; in the US southwest, they also include kivas (Van Keuren and Roos, 2013). In mainland Europe often the German term "*grubenhaus*" (or grub hut) is employed for a "sunken feature building" – SFB (see Box 13.2). In studies of First Nation sites in British Columbia, Canada (starting around 4800 BP at Keatley Creek for example), it was difficult to isolate unequivocally pit house floors because of very marked biological working of occupation deposits and collapsed constructional materials such as the roof, especially in the first instance by scavenging animals (Goldberg, 2000a; Hayden, 2004). In cooler climate sites along the northwest coast of Norway, Mesolithic to Neolithic pit house fills, wooden constructional materials can be preserved – sometimes as possible birch bark floors. In other cases, wood has become humified but because of post-depositional ferruginization it remains identifiable as wood. Compact occupation floors – sometimes of a rather muddy nature – may also be found, with concentrations of charcoal and sometimes with rich fish bone deposits (see Figures 13.13a and 13.13b). Lastly, some of these coastal pit house sites were associated with Mesolithic hunters and gatherers following game herds exploiting newly vegetated coastal landscapes after ice retreat and land exposure after uplift. Seasonal activity is recorded at one site by fine charcoal layers within accumulated Laminated Mull humus horizons, on newly wooded substrates recently exposed by post-glacial uplift. For further examples of geoarchaeological data from trackway, post hole and pit house studies see asma/Online_Appendix_11.md at master · geoarchaeology/asma · GitHub.

| Box 13.2 | Grubenhäuser |

Grubenhäuser are sunken-featured buildings of mainly Germanic origin and record settlement during 5th–8th centuries AD in northwest Europe – a timespan sometimes called the "Migration Period". Their exact use is still under debate but they are important settlement features in landscapes where remains of domestic rectangular houses have often been totally lost. Most studies have considered them to have once had a suspended plank floor (Gustavs, 1998) and that fills are simply "tertiary" in nature (Figures 13.14a and 13.14b) (Maslin, 2015; Thomas, 2010; Tipper, 2004). Too often the fills of grubenhäuser have been ignored in geoarchaeological studies but detailed analyses from mainland Europe, England and Scandinavia (Norway and Sweden), for example, have shown what unique information can be extracted to help identify habitation and settlement patterns during the so-called Dark Ages. The primary findings from this approach are that grubenhäuser fills can be grouped into two types (Macphail, 2016; Macphail et al., 2006):

1) *moderately homogeneous fills* that could relate more dominantly to *in situ* formation, construction/use and disuse history; and
2) *heterogeneous fills* that probably relate to secondary use and dumping, and which provide evidence of site activities.

Homogeneous fills

In England, at West Heslerton, Yorkshire, Stratton, Bedfordshire, and Svågertop, Malmö (Sweden), homogeneous fills gave evidence of possible infilling by mainly mineralogenic construction

Figure 13.14a Field photo, Lyminge, Thanet, Kent, UK. Field photo of typical Saxon grubenhaus (~560–660 AD; SFB 1), showing a pair of primary posts located on a longitudinal axis of the pit, both of which had been replaced during the lifetime of the structure, together with an additional six posts located at the corners and the mid-point of each of the longitudinal sides of the pit (photo courtesy of G. Thomas, University of Reading, UK; Thomas, 2010). This has a mainly tertiary fill (Maslin, 2015), at the base of which an iron plough coulter was found in suggested contact with the remains of a wooden suspended floor (see Figure 13.14b), some 30 mm above the chalk substrate exposed in the photo.

(Continued)

Box 13.2 (Continued)

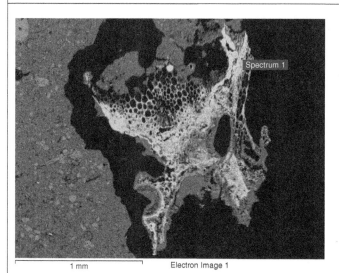

1 mm   Electron Image 1

Figure 13.14b As Figure 13.14a, SEM X-ray backscatter image of supposed remains of suspended wooden floor in grubenhaus – ferruginized wood. Note presence of probable iron- impregnated knot wood fibers; mean 2.68% Si, 2.40% P, 2.84% Ca, 65.6% Fe. Scale bar = 1 mm.

material used in the construction of the grubenhaus. This was either composed of (a) local turf (for low turf walls and roofing?) from natural soils that were little affected by settlement occupation, or (b) from turf (and daub) formed during the lifetime of an occupation. In the first case (a), this gave information on the natural environment, and when grubenhäuser fills are of this kind, it could imply that either these structures were constructed early in the life of the occupation or perhaps were constructed on the expanding margin of a settlement. In the second case (b), "turf" contains large amounts of small anthropogenic inclusions, bone, charcoal, human and animal coprolites, burned material, and secondary phosphate features. All of these components indicate that "soil" has formed in an occupation-"contaminated" substrate. This finding permits the inference that this structure is unlikely to have been constructed and used early in the history of the occupation. Where dating material is so sparse, as in this specific early medieval period, such interpretations based upon fills, could clearly aid phasing.

A 3D plotting of finds of this last type (b) at West Heslerton of an example, produced the archaeological interpretation of a grubenhaus with a tertiary fill (Tipper, 2001, 2004) (Figures 13.15a and 13.15b). It essentially amounts to a different way of explaining the same phenomenon. This anthropogenic material-rich grubenhaus fill also records values of organic matter (6.2–7.0% LOI; $n = 6$), P (3160–5040 ppm P) and magnetic susceptibility (χ 619–895 × 10^{-8} m^3 kg^{-1}) that are all high; in the case of P, it is 3–5 times higher than pre-Saxon levels. Similarly, χ is 6 times higher. The bulk analyses totally support the interpretation of the fill using soil micromorphology.

At Svågertop, a natural soil-infilled grubenhaus also contained little evidence of human activity (Grubenhaus 433: χ mean 38.6 × 10^{-8} m^3 kg^{-1}; P$_2$O$_5$ mean 170 ppm; $n = 9$), compared to three others (e.g. Grubenhaus 1954: χ 81.8 × 10^{-8} m^3 kg^{-1}; P$_2$O$_5$ mean 2200 ppm; $n = 20$). The presence of turf in Grubenhaus 433, was recognized on the basis of humic soil being present that is characterized by organic-rich excrements of soil mesofauna in a biological microfabric (Kooistra and Pulleman, 2018), as found in experimental turf roof samples in Sweden, for example (Cruise and Macphail, 2000) (Box 12.1)

Box 13.2 (Continued)

Figure 13.15a Field photo of Middle Saxon, grubenhaus; West Heslerton grubenhaus 12AC09507, North Yorkshire, United Kingdom. A typical tertiary fill is present (Tipper 2004 / with permission of Landscape Research Centre).

Figure 13.15b Field photo, as Figure 13.15a, West Heslerton grubenhaus 12AC09507; bulk data: lateral control sample WH124: 6.8% LOI; $\chi = 858 \times 108$ m^3 kg^{-1}; 4410 ppm P.
Bulk data of vertical sequence:
WH125: 6.2% LOI; $\chi = 895 \times 10^{-8}$ m^3 kg^{-1}; 3230 ppm P
WH123: 6.4% LOI; $\chi = 816 \times 10^{-8}$ m^3 kg^{-1}; 3160 ppm P
WH122: 6.7% LOI; $\chi = 863 \times 10^{-8}$ m^3 kg^{-1}; 3870 ppm P
WH121: 7.0% LOI; $\chi = 619 \times 10^{-8}$ m^3 kg^{-1}; 5040 ppm P
(LOI = loss on ignition; χ = magnetic susceptibility; P= phosphorus).

(*Continued*)

Box 13.2 (Continued)

Heterogeneous fills

These can provide specific information on events, activities, settlement morphology, discard practices, lifestyle, cultural, and domestic economy. In addition, it can clarify the origins and character of fragile objects such as unfired loom weights formed out of "soil". For example, at Svågertop the burned dung microfacies type is composed of a phytolith-rich matrix that includes articulated phytoliths of monocotyledonous plants, layered microfabrics (pseudomorphs of cattle dung?), charred amorphous (humified) organic matter (dung?), silt, diatoms, burned soil, and rooted and iron-depleted sediment. The effects of burning on magnetic susceptibility (χ) and phosphate concentrations induced by the burning of dung have been noted above (Macphail and Goldberg, 2018a: table 2.3). The presence of silt is consistent with the ingestion of mineral matter from grazing and drinking, diatoms, and gleyed sediment fragments, being indicative of fodder/grazing originating from wetland areas. These data contribute to the analysis of the integrated economy, which included the necessity of over-wintering cattle and the use of their manure for barley production (Engelmark, 1992; Engelmark and Linderholm, 1996; Viklund, 1998).

Various tertiary fills have been recorded across Europe (Guélat and Federici-Schenardi, 1999; Maslin, 2015), along with the remains of wooden remains of suspended floors (French and Milek, 2012; Gustavs, 1998; Macphail and Goldberg, 2018a: figure 11.7; Plates 12a–b), and examples of pit houses with corner-located hearths – but which are normally of an early medieval date (Wegener, 2009). Another variety is termed a podstal and the "cellared" area is believed to have been used as a byre (Mikkelsen et al., 2019).

It is often the case, that if grubenhäusser (or Migration Period pit houses) are investigated at a site, only one or two are studied. Exceptions were at the sites of Svågertop (Figures 13.16a and 13.16b) and West Heslerton, noted above, and when several are studied within Saxon

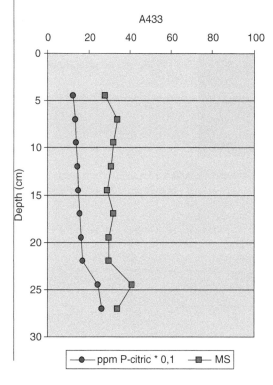

Figure 13.16a Phosphate and magnetic susceptibility diagram; Grubenhaus with "homogeneous" fill – of local soil/turf: Svågertop, Grophus 433: vertical sampling column through the grubenhaus fill and underlying soil (below 23 cm), showing measurements of χ (MS × 10^{-8} m^3 kg^{-1}) and P$_2$O$_5$ (ppm P citric, × 0.1). Note how little difference there is between the soil and the fill, supporting the soil micromorphological evidence of a natural soil-infilled grubenhaus. Such a finding implies the likelihood of little anthropogenic impact on this part of the site – possibly indicating that it is an area peripheral to a settlement and/or implying that it could be an *early* feature. It also may suggest that turf was a major component in its construction and that this has fallen in (figure by Johan Linderholm, Umeå University).

Box 13.2   (Continued)

Figure 13.16b Grubenhaus "heterogeneous" fill of dominantly burned (cattle?) dung: Svågertop, Grophus 1954: vertical sampling column through the grubenhaus fill and underlying soil (below 39 cm), showing measurements of χ (MS × 10^{-8} m^3 kg^{-1}) and P_2O_5 (ppm P citric, × 0.1). Here there is a clear contrast between the high values recorded in the anthropogenic fill, which results from dumping of occupation waste, and the natural low levels in the soil (figure by Johan Linderholm, Umeå University).

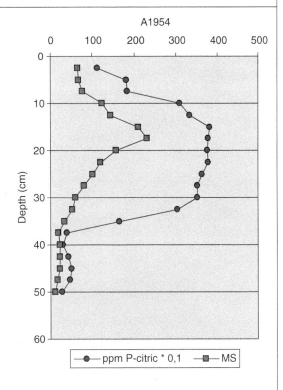

A1954

villages for example, they have provided clear indications of mixed farming (Macphail and Goldberg, 2018a: appendix 11.2, 8–11; https://geoarchaeology.github.io/asma/appendices/Online_Appendix_11/). A number have been investigated from East Anglia, a focus of migration broadly during AD 500–800, such as West Stow and Eye, Suffolk (West, 1985), the most famous coastal site being Sutton Hoo (Carver, 1998). A recent example is Kentford, Suffolk where a contribution to reconstructing the Saxon settlement included geoarchaeological study of six grubenhäusser, two buildings and two pits of note (Macphail, 2020). The chief finding here, was the clear proxy evidence of sheep/goat husbandry, in the form of concentrations of dung spherulites and phytoliths, the former preserved because of the buffering effect of calcitic ash deposits, including wood ash, where the regolith is chalky (Figures 13.17a and 13.17b). Other inputs involved earth-based constructional materials and artifacts (clayey and chalky daub and suggested loom weights), burned mineral material including concentrations of fire-cracked flints and charcoal, and phosphatic latrine waste, the last being another common component of such tertiary fills (Figures 13.17c). Overall, fill contexts are either moderately to moderately strongly enriched in phosphate (range 870–3080 ppm P), while magnetic susceptibility is also weakly to moderately enhanced throughout (MSlf = 25–52 χlf 10^{-8} m^3 kg^{-1}), likely reflecting the presence of burned weakly ferruginous soils. Microchemistry directly on thin sections (SEM/EDS; Chapter 17) helped identify possible furnace aerosols (prills; 12.0–18.7% Fe), mineralized human waste (e.g. 12.6% P, 28.0% Ca) and charred dung pellets (e.g. 3.89% P, 10.6% Ca). Common to many such settlement sites, including Norway, is the importation of wetland plant material (Figure 13.17d) into the site for animal bedding, thatching etc., with the Kentford village probably exploiting the nearby River Kennet valley; phytolith analysis at Eye produced similar conjectures (Macphail and Crowther, 2015b).

(Continued)

Box 13.2   (Continued)

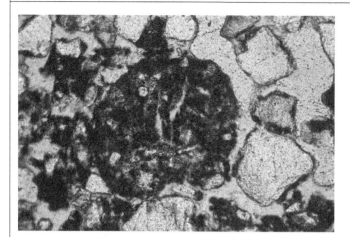

Figure 13.17a   Photomicrograph of M15 (SFB 0145, 0230–0242; Early Saxon SFB; grubenhaus, at Kentford, Suffolk, UK); possible ashed dung remains – possible from a sheep/goat pellet. PPL, frame width is ~0.90 mm.

Figure 13.17b   As Figure 13.17a, SEM X-ray backscatter image M215B (Kentford Pit 0358; 0362); ashed dung with high concentrations of dung spherulites. Various spectra of: 8.61–9.45% Si, 3.82–3.89% P, 0.10–0.14% S, 0.18–0.21% Ca, 0.81–0.82% Fe; 10.7–12.3% Si, 0.84–1.09% P, 0.0–0.02%S, 0.28–0.42% K, 4.21–4.22% Ca, 0.08–0.23% Mn, 1.24–1.1.30% Fe. Frame width is ~2.4 mm.

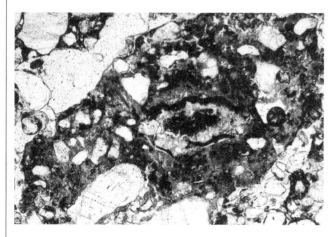

Figure 13.17c   Photomicrograph of M15 (Kentford SFB 0145, 0230–0242); nodule of calcium phosphate cemented sands and more iron phosphate filled center. (SEM/EDS: center with 8.17% Fe, 7.52% P and 1.4% Ca; sands cemented by 25.5% Ca, 11.9% P and 1.02% Fe). PPL, frame width is ~2.38 mm.

Box 13.2   (Continued)

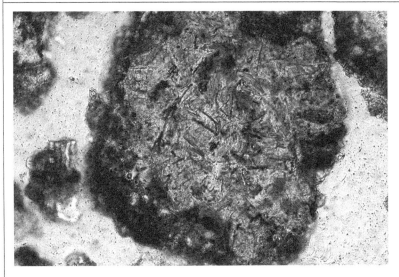

Figure 13.17d   Photomicrograph of M163 (Post-in trench Building 28, Kentford – ditch fill 0837); phytolith-rich sediment clast – imported wetland material(?). PPL, frame width is ~0.47 mm.

13.6   Final Remarks

Elsewhere in the book we have dealt with a whole variety of materials used and produced by humans (Chapter 12) as well as how humans actually impacted on the landscape (Chapter 11); the latter for example, is exemplified by manured cultivation, which has produced soils enriched with materials of human origin. However, in those chapters, anthropogenic deposits *per se* were not characterized systematically. This is done in this chapter, where we propose various models concerning anthropogenic material disposal practices and deposit characterization according to settlement frameworks. We review the model of describing object disposal as: "primary", "primary transposed" and "secondary use", with examples and see how such models like these fit into such phenomena as European dark earth and *terra preta* of Meso- and South America. We also note that funerary features discussed in Chapter 15 are difficult to fit into such models. Furthermore, we discuss the make-up of European and Near Eastern tells, and North American and Mesoamerican mounds, as special accumulations of anthropogenic deposits. Settlement morphology and urban archaeology studies are another way of modeling patterns of archaeological material dispersal across a site, and as a way of comprehending how a settlement functioned. With this in mind, we detail our knowledge of trackways, occupation floors, byres, etc., and also make a special note of how pit house fills can be variously interpreted. One of the most important aspects of anthropogenic deposits is that they are the result of a complex of human activities and behaviors, just as are the manufacture of stone tools, or the exploitation of plants and animals. They should be considered and studied as equals to these latter contributors (artifacts, bone and plant remains) to the Archaeological Record.

14

Experimental and Ethno-Geoarchaeology

14.1 Introduction

A major issue in geoarchaeology is how well is the past being interpreted? It is often asked, how can morphological and micromorphological features and numerical data be understood accurately? In the sciences, empirical data from field and laboratory observations and experiments are commonly employed to help make interpretations, often based on comparisons with present-day environments. For example, numerous modern sedimentary environments have been studied, and these can be utilized to elucidate depositional environments of fossil deposits (Reineck and Singh, 1986) (see Chapter 2). Similarly, the very many soil types that are present on the earth have been well documented (Chapter 4), and they can be used to infer past conditions of soil formation.

In archaeology, ethnoarchaeology tends to focus on the formation of deposits and the history of objects such as bones and lithics found in them (Brochier et al., 1992; Gifford, 1978; Grøn and Kuznetsov, 2004; Shahack-Gross et al., 2003). On the other hand, experimental archaeology was also employed to "calibrate" archaeological findings, often linked to artifact taphonomy and preservation (Armour-Chelu and Andrews, 1994; Bell, 2009; Bell et al., 1996; Breuning-Madsen et al., 2001; Breuning-Madsen et al., 2003; Macphail et al., 2010; Mallol et al., 2013; Newcomer and Sieveking, 1980; Nielsen, 1991; Sherwood, 2001). The scope of experiments in archaeology and ethnoarchaeological experiments has recently widened to encompass a variety of topics. These include animal management, different types of floors and occupation surfaces (domestic occupation and byre floors, threshing floors; Banerjea et al., 2015a; Gebhardt, 1995; Macphail et al., 2004; Milek, 2012; Shahack-Gross et al., 2009); sweeping, trampling, and trackways have also come under scrutiny (Miller et al., 2010; Rentzel and Narten, 2000; Rentzel et al., 2017). Experiments in cultivation have received more attention (Deák et al., 2017; Gebhardt, 1992; Gebhardt and Langohr, 2015; Lewis, 2012; Schulz et al., 2011). There has been a focus of attention on the deterioration and collapse of earth-based buildings (Friesem et al., 2011, 2014b; Milek, 2006) and Paleolithic pyrotechnology. The latter includes how hearths and other combustion features form and degrade, and conditions of combustion (Aldeias et al., 2016; Mallol and Henry, 2017; Mallol et al., 2007, 2017). Lastly, many workers have found the opportunity to create reference materials within their own projects, and this can be as relatively straight forward as adding to an archaeological database by burning rocks at controlled temperatures for example (Berna et al., 2007; Canti, 2017; Karkanas, 2007; Röpke and Dietl, 2017), or creatively producing lime mortar (Karkanas and Goldberg, 2018b: 115–116, 125–127) (see Chapter 12).

The problem most frequently encountered in geoarchaeology, which like geology and archaeology is an historical science, is that there are no modern analogues for many past situations. For example,

Practical and Theoretical Geoarchaeology, Second Edition. Paul Goldberg and Richard I. Macphail.
© 2022 John Wiley & Sons Ltd. Published 2022 by John Wiley & Sons Ltd.

where does one find "virgin" soils on which to carry out paleoagricultural experiments (Macphail, 1990d; Rösch et al., 2004; Schulz et al., 2011). Many situations have not yet been fully studied or investigated at all. In addition, there is the effect of burial, and all kinds of "aging" (i.e., organic matter oxidation, natural compaction, phosphate-loss), and other post-depositional processes that alter or delete the original components, thus frustrating attempts to accurately understand the past. There are some researchers, of course, who would give up at this stage. We know, however, that geologists have successfully understood past environments where soils and loose sediments have been completely transformed into rock (Wright, 1986, 1992). There is no reason, therefore, why geoarchaeologists should not attempt to elucidate ancient landscapes and human activity from archaeological soils and deposits. On the other hand, the challenge for geoarchaeologists could be seen as being more complex than that for geologists who deal solely with natural processes. The human activities that the geoarchaeologist attempts to recognize can be extremely diverse and are sometimes unknown to the modern academic mind or experience (e.g. Neanderthals were not modern humans and how many geoarchaeologists have had farming experience, for example?). Previous practices may simply be too counter-intuitive to us to appreciate, nowadays sheep farmers keep ewes because today's money is in lambs, while in the past, wethers (castrated rams), were husbanded because they produced the finest wool. Human actions also commonly interact with geological/pedological processes, making the situation even more complex (Adderley et al., 2018) (N. Fedoroff, pers. comm., 1990).

Experimental archaeology is a thriving field, and numerous experiments having been devised over the years to understand a variety of problems crucial to archaeology. Outside the realm of lithic technology where replication experiments abound (e.g. Amick et al., 1989; Dibble and Rezek, 2009; McPherron et al., 2014; Stafford, 1977), issues mostly center around the techniques and production of stone tools, and the integrity of archaeological assemblages and taphonomy. They include trampling studies as a means to interpret artifact distributions (Gifford-Gonzalez et al., 1985; Miller et al., 2010; Nielsen, 1991; Villa, 1982; Villa and Courtin, 1983), or evaluations of artifact movements in natural settings (Bertran et al., 2019; Rick, 1976; Shackley, 1978). Most recently, researchers have increased their efforts to understand Paleolithic pyrotechnology turning to experiments in order to discern processes associated with burning of wood and bone (Costamagno et al., 2005; Stiner et al., 1995; Théry-Parisot, 2002) and to understand the processes of combustion and its effect on substrates (Aldeias et al., 2016). Great progress has been made recently in terms of site formation processes and their effect on microstratigraphy (Mallol et al., 2013, 2017; Mentzer, 2014).

In the context of soils *sensu stricto*, although there is a strong experimental and empirical database for the extant soil cover of the earth (Brady and Weil, 2008; Duchaufour, 1982; FAO, 2015; USDA, 2014), we have had to carry out experiments in order to understand better the archaeological soil archive. The first section in this chapter therefore deals with the effects of burial and aging of soils. Secondly, we give examples of the processes involved in anthropogenic deposit and feature formation, and how these effects can be identified in the archaeological record. The premises underpinning these experiments are firmly based upon archaeological analogues and/or reference analogue or ethnoarchaeological observations, which are our other chief sources of information for developing and checking interpretations, especially if they involve as many independent disciplines as possible. Such a holistic approach has been deemed controversial because they are complicated and necessarily produce results and nuanced interpretations, even though every single technique is individually scientifically sound and tested (Canti et al., 2006; Macphail and Cruise, 2001; Macphail et al., 2004, 2006). When it comes to elucidating the complexities of sites, this latter approach has proven itself time and again (Banerjea et al., 2015a; Deák et al., 2017; Macphail et al., 2007a; Mallol et al., 2017; Rentzel et al., 2017).

14.1.1 Some Experimental Opportunities

Investigations into the effects of burial on archaeological materials and old land surfaces were initiated during the early 1960s with a number of experimental earthworks (see below 14.2). Equally innovative, and in many ways more challenging as it requires a full time commitment, was the construction of some "Ancient Farms" and settlements (see section 14.3). Furthermore, sometimes, unique experimental opportunities need to be taken up, for example the multidisciplinary characterization of a newly marine-flooded coastal landscape at Wallasea Island (Essex, UK) (see below). Although we note the results from experimental industrial activities, such as charcoal making (Gebhardt, 2007), construction material manufacture that produces anthropogenic sediments (Karkanas and Goldberg, 2018b: 99–148), and use of furnaces (Berna et al., 2007; Röpke and Dietl, 2017), we focus here on reconstructions of occupations and settlement structures (e.g. hearths, occupation floors and surfaces), and paleoagriculture (cultivation and stock management). Some results from the Experimental Earthworks Project and related research, and flooding of coastal soils at Wallasea Island, are dealt with first.

14.2 Effects of Burial and Aging

It is absolutely crucial that any interpretation of buried soils and structures is based upon exact knowledge of preservation conditions once they are buried or abandoned. Major transformations affecting ancient soils and sediments are chemical and biochemical ones, such as oxidation of the organic content, and physical changes, as for example, compaction and bioturbation. In order to replicate these processes, geoarchaeological experimentation was deemed essential. Thus, the Experimental Earthworks Project was set up after the Charles Darwin centenary meeting of the British Association for the Advancement of Science in 1958 (Bell et al., 1996). Two British experimental earthworks were constructed in 1960 (Overton Down, Wiltshire) and in 1963 (Wareham, Dorset), and these have been monitored every 1, 2, 4, 8, 16 and 32/33 years ever since, with projected final excavations in 2024/26 (Bell et al., 1996) (Figure 14.1). Both experiments were located

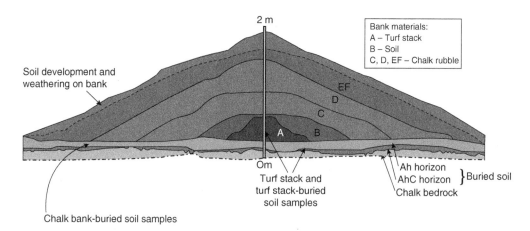

Figure 14.1 The Experimental Earthwork at Overton Down, Wiltshire, UK, 1992 (Bell et al., 1996); section drawing and soil sampling points: the chalk bank-buried soil (2 thin sections and ~10 bulk samples from two profiles) and the turf stack-buried soil (9 thin sections and ~18 bulk samples from two profiles) (modified from Crowther et al., 1996).

in areas of ancient barrows (earth mounds) that involved base-rich Typic Rendolls (rendzina) at Overton Down (Chalk) with typical A/C horizon profiles, and strongly contrasting acid Orthic Spodosols (humo-ferric and ferric podzols) at Wareham (Tertiary sand) with fully developed Mor-Moder Humus, Ah, A2, Bh, and Bs horizons (Tables 4.1 and 4.2; Figures 4.2b and 4.9); comparative control samples from modern soils were used at both locations. Earth mound (*sensu lato*) studies are not confined to England of course, and different constructional and environmental situations have been encountered in Belgium (Gebhardt and Langohr, 1999; Langohr, 1991), Denmark (Breuning-Madsen et al., 2012; Breuning-Madsen et al., 2001), at Shiloh, USA (Sherwood and Kidder, 2011), and in eastern European and Near Eastern tells, although these are characterized by much drier conditions (Haită, 2003, 2012; Rosen, 1986; Shahack-Gross et al., 2005) (see Chapter 13).

14.2.1 Overton Down

Background control soils The control Rendzina (Rendol) soil comprises a 1.5–2.0 cm-thick root mat, an 18 cm-thick Ah horizon (range 15–22 cm), and a very stony AhC horizon (Crowther et al., 1996). The Ah horizon is rich in organic C and N (~18.9% organic matter). It has a low C/N range of 8.07–10.7 indicative of a well-humified organic fraction (see Chapter 4). In thin section (Sample 15: 10–80 mm) the soil displays a total excremental fabric, with roots concentrated in the top 2 cm. Here the soil has a very open compound packing porosity around crumb structures (porous or "spongy" soil aggregates produced by earthworms) and other finer organomineral excrements, and an average void space of 55%. Below the rooted layer, between 20 and 80 mm, coarser crumbs and finer subangular blocky structures become more common as a result of coalescence through wetting and drying, and the mean void space is reduced to 32%. As is typical of mull horizons (Babel, 1975) and of other grazed rendzinas and high base status brown earths (e.g. Martin Down, Hampshire (Carter, 1987; Carter, 1990); Maiden Castle, Dorset (Macphail, 1991)), the soil is dominantly earthworm-worked. The micro- and meso-size organo-mineral excrements noted above are present throughout the thin section studied, and many of these are probably derived from enchytraeid activity, in particular through the reworking of earthworm excremental fabric. While this is a typical feature of mull humus rendzinas (Babel, 1975; Bal, 1982), the actual biomass of enchytraeids is likely to be very small compared with that of earthworms (Wallwork, 1983).

Buried soils At Overton Down (excavated 1992) the thickness of the Ah horizon had decreased from ca. 180 mm to ca. 90–100 mm; this took place in part through organic matter loss (ca. 11.0% to ca. 7.59–7.88% organic C) (Crowther et al., 1996). An open microfabric characterized by 55% void space (Figures 14.2 and 14.3), and showing signs of earthworm-working and grass rooting had become compacted (minimum 14% void space at 1 cm depth). Moderately broad (>500 μm) mamil-lated earthworm excrements were replaced by aggregated excrements and a spongy microfabric composed of thin (50–500 μm) cylindrical organo-mineral excrements (Figures 14.4 and 14.5). In addition, soil acidity increased under the turf stack (from pH 6.9 to 5.6). Where the soil was buried by chalk rubble, the soil became more alkaline (pH 7.9) because here, earthworms mixed the buried soil with the overlying chalk. *Lycopodium* spores (which had been laid on the old ground surface at the time of monument construction) were mixed 90 mm upwards into the overlying chalk bank (Crabtree, 1996) (the stratigraphy of land snail populations can also be affected by burial; Carter, 1990). Micro- to meso-sized nodules of Fe and Mn had also formed after burial at Overton Down. The formation of iron pans and pseudomorphic iron-replacement of plant mate-rial are both typical manifestations of buried soils that result from localized gleying (Crowther et al., 1996; Limbrey, 1975). Interestingly, the buried soil at the nearby Neolithic (e.g. ca. 3400–3600 cal BC) long barrow at Easton Down (8.5 km distant) exhibited very similar soil microfabric types

Figure 14.2 As Figure 14.1 – the humic rendzina control profile (Soil Pit 1) within the enclosure protecting the Overton Earthwork from grazing animals. This protection has resulted in the formation of a 2 cm-thick grass mat and incipient surface gleying; terrestrial diatoms are also present in this moist surface grass mat; the control profiles have an Ah horizon that varies in thickness from 15–22 cm. Trowel for scale.

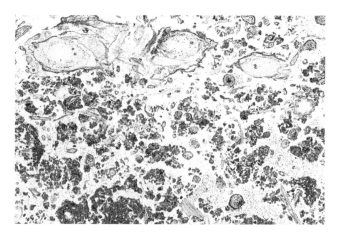

Figure 14.3 Photomicrograph of the Ah horizon of control profile (Soil Pit 2) outside the Earthwork enclosure at 1.5 cm depth, showing a heavily rooted topsoil (note cross-sections through living grass roots) and a very open compound packing porosity (mean 55% voids), and very abundant humic and organo-mineral excrements. Plane polarized light (PPL), frame width is ~9 mm.

to those at Overton Down (Whittle et al., 1993) (Figures 14.6 and 14.7). This evidence showed how quickly long-lasting changes occurred in Rendolls once they were buried, i.e., changes that began over the first 32 years after burial are clearly recognized in soils some 5,000 years old.

The ditch sediments have also been studied through both granulometry and chemistry (Bell et al., 1996: 90–95). Thin sections revealed that chalk and soil banding currently visible in the ditch fills is probably an ephemeral seasonally-formed feature, where chalk gravels are deposited over winter; in contrast, more humic soils accumulate during the summer. Such layering has often disappeared from UK archaeological ditch fills on similar soils where bioworking predominates (Macphail and Cruise, 1996: 106). By comparison, alternating stony and organic ditch fills at Early Bronze Age, Rorkoll nordre (Vestfold, Norway), could be recording coarse minerogenic layers from winter talus deposition and spring time byre cleaning episodes (Viklund et al., 2013).

Figure 14.4 As Figure 14.1 – chalk bank-buried soil and Kubiëna box samples; the buried soil had become compressed (~11 cm thick), although overall, the chalk bank-buried soil ranges in thickness from 5 to 18 cm.

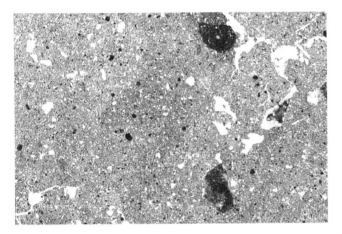

Figure 14.5 As Figure 14.1, photomicrograph of the chalk bank-buried rendzina showing changes to soil structure. Its structure had become more massive through the aggregation of earthworm excrements, with concomitant loss of porosity (22.7% to 48.4% less void space compared to the control profile). Organic carbon has been reduced from 11.0% to 7.9% (mean values), and earthworms have mixed chalky soil from the bank with the original decalcified soil, raising the pH (from pH 6.9 to pH 7.9). Plane polarized light (PPL), frame width is ~9 mm.

14.2.2 Wareham

14.2.2.1 Background Control Soils

The soils at Wareham, which are generally well drained, are either humo-ferric podzols of the Shirrell Heath series or humus podzols of the Leziate series, the latter reflecting the very low amounts of sesquioxides present (Avery, 1990; Findlay et al., 1984; Macphail et al., 2003b). They are typical of podzols found on lowland heaths and show the high degree of natural soil variation present on the Eocene sands of southern England. Four control profiles were studied (Figure 14.8). The upper horizons of the control soils typically comprise a 40–70 mm thick litter (L) and fermentation (F) layer, sometimes with a discontinuous humus (H) horizon. These and all topsoils included in this study were classified according to humus form after Babel (1975: 446).

Figure 14.6 Neolithic Easton Down Long Barrow, Bishops Cannings, Wiltshire, UK (Whittle et al., 1993). Field photo of humic soil and coarse chalk-constructed long mound over thin (max 80 mm) buried rendzina Mull A1h horizon (thin section sample D). Soils are broadly calcareous down-profile (pH = 8.1; carbonate = 2–5% of fine earth – <2 mm) (Crowther, 1996). Only few small patches of the aggregated soil are composed of decalcified soil. Soil chemistry is similar to that at Overton Down: organic C = 4.75–5.31%, %N = 0.440–0.480; C:N = 10.3–11.1), but with higher phosphate concentrations (P = 4.25–4.55 mg g^{-1} (0.425–0.455 %P) – the latter is consistent with the presence of fine burned bone occupation relicts.

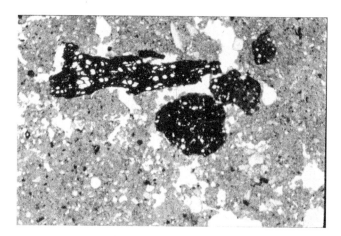

Figure 14.7 As Figure 14.6, photomicrograph of decalcified buried Mull A1h horizon, with micro-inclusions of charcoal and fine rubified – burned – soil clasts consistent with measured magnetic susceptibility enhancement (χ = 2.11–2.24 μm^3 kg^{-1}; 13.1–13.3% χ-conversion). Plane polarized light (PPL), frame width is ~5.36 mm.

At Wareham, the black humose Ah horizon is very variable in thickness (range 40–170 mm). In thin section, the LFH can be seen to have an overall blocky structure that is strongly characterized by broadly laminated coarse plant residues (max. 20%) including *in situ* ericaceous roots, and increasing proportions of dark reddish brown organic, excrement-dominated fine organic matter down-profile (optical counting data). This reflects the activity of arthropods and Enchytraeids in such soils (cf. Van der Drift, 1964; Zachariae, 1964). In contrast, the medium sand-size quartz-rich Ah horizon contains much fewer plant fragments and is massive with a microaggregate structure and a brownish black microfabric attributable to Enchytraeids reworking arthropod feces and microbiological breakdown of humus.

Figure 14.8   Wareham Experimental Earthwork, Dorset, UK (1963 construction of experimental bank and ditch). It is an example of one of three sampled control profiles – a podzol formed on Tertiary sands. Typical of the other control soils there is:

- a ~40 mm thick Ao/LF(H) superficial organic layer (LF) – heather (*Calluna*) roots are visible from the cleared vegetation;
- a humus-enriched minerogenic topsoil (Ah horizon);
- a strongly leached and iron-depleted upper subsoil Ea(A2) horizon, showing thin humic downwash laminae;
- an illuvial Bh(s) subsoil horizon (high amounts of humus but smaller concentrations of sesquioxides – Fe_2O_3 and Al_2O_3), and C horizon, Tertiary sands and thin pipe clay layers (Hillson, 1996; Macphail et al., 2003b).

Averaged over the four profiles, the L layer has a higher organic C concentration and a higher C/N ratio than the F layer. It is also less acidic, with a mean pH of 4·0. The Ah horizon has a mean organic C concentration of 6–8%. The Ah horizon is also very acidic and has a low nutrient status, with mean concentrations of available phosphate-P and K of 18·1 and 43·3 mg $^{-1}$g, respectively. Despite having iron-deficient topsoils, magnetic susceptibility values are relatively high especially in the top 20 mm of the Ah horizon, and this presumably reflects a combination of natural fermentation processes and burning (Gimingham, 1975). The lack of a well-developed H horizon below the L and F layers results from humus immaturity and is also a likely result of burning and recent disturbance of these layers.

14.2.2.2   Buried Soils

The buried Spodosol at Wareham, changed dramatically from 1962 to 1980, with a reduction in thickness of the LF(H) horizon from ca. 70 mm to 1.6–3.6 mm (Figures 14.9–14.14). However, by 1996 in the LFH superficial organic matter horizon an (amorphous: H_r) humus form was still developing from an (identifiable excrement and plant fragment-rich: F_m) humus form (after Babel, 1975). This development took place despite continued "ferruginization" resulting from both gleying (induced by the poorly draining nature of basal turf stack), and continued *podzolization* (Table 14.1) (Babel, 1975; Macphail et al., 2003b). For example, there were increases in the pH of the buried sand, in alkali-soluble humus, and in the pyrophosphate-extractable C, but marked decrease in available K. Image analysis and soil micromorphology showed that while

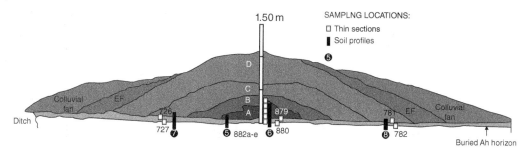

Figure 14.9 Wareham Experimental Earthwork. Simplified cross-section through the bank and ditch showing the 1996 excavated section, from where turf (Layer A) and subsoil sands (Layers B, C, D, E and F) were removed to construct the bank. Bank constructed of different soil materials: Layer A = turfstack; B = Ah and Ea horizons; C = B, Bh and Bf horizons; D = BCu and other deeper subsoil horizons; EF = material derived from process of adjusting ditch sides to correct angle (after Evans & Limbrey, 1974). Bank erosion has formed a colluvial fan. Turf bank and turf-buried soils (879, 880 and 882a-e) and sand-buried (726, 727; 781–782) soils were sampled. (Drawing supplied by John Crowther; Macphail et al., 2003b).

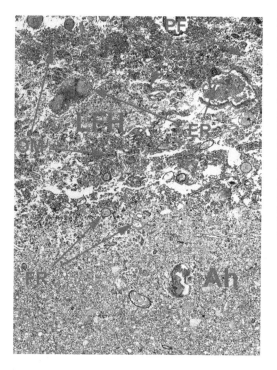

Figure 14.10 Wareham Experimental Earthwork. Digital flatbed scan of thin section 681a (LFH/Ah of control soil profile 4; Figure 14.8;) showing some 40 mm of the LFH horizon, composed of 20.6% total plant residues, including 13.4% organ residues (mainly woody ericaceous roots; ER), with some 26.0% total organic excrements (in passage features; PF), and with a dark reddish brown organic microfabric (OM) (Macphail et al., 2003b). The underlying Ah horizon also contains 6.8% organ residues (mainly roots; Fine Roots – FR), 27.0% passage features (commonly excrement-filled burrows), but only 4.4% organic excrements compared to 18.0% organo-mineral excrements, and with a dominant intergrain (enaulic) microaggregate related distribution/microstructure. Frame width is ~65 mm.

there was an increase in the proportion of organo-mineral excrements, plant residues, and organic excrements decreased. Microprobe also showed that the buried soil had been enriched in iron in some places, confirming the theory of continued podzolization as indicated by the bulk organic chemistry.

Such monitoring of buried soils has been crucial to the understanding of processes not only affecting monument-buried soils, but also those which result for instance, from the weathering of

Figure 14.11 As Figure 14.9; thin section 781 (sand-buried soil – sand Layer D/bLFH/bAh) illustrating minerogenic nature of sand Layer D, the very thin (2–3 mm thick) nature of the relict bLFH horizon (MF3, Fm/Hr), the strongly burrowed (30.1% passage features PF) nature of the bAh horizon, and paucity of plant remains (5.5% total plant residues) in a sand-dominated (53.4%) massive structured, intergrain (enaulic) and coated (chitonic) microaggregated/related distribution and microstructure (MF7, Ah). Sesquioxides appear to be weakly concentrated in the relic bLFH (0.28% Fe, 0.66% Al). Frame width is ~65 mm.

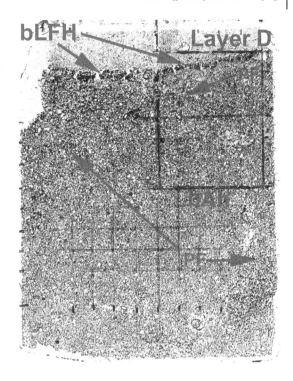

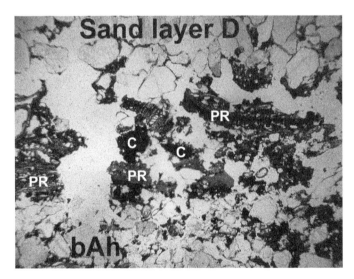

Figure 14.12 As Figure, 14.11; photomicrograph of thin section 781 (sand bank-buried soil) showing junction of buried LFH and sands of bank, with plant remains (PR) and an artificial concentration of charcoal (C) that has been resistant to oxidation; organic matter from the bLFH has been worked by mesofauna into the overlying sand layer D. PPL, frame width is ~2.6 mm.

fills in pits, ditches, tree-throw holes and pit houses (Goldberg, 2000a; Macphail, 2016c; Macphail and Goldberg, 1990; Maslin, 2015; Veneman et al., 1984). Equally, in alluvial soil sequences (cumulic soils), the diffusion of ephemeral land surface topsoils into soils forming in the overlying alluvium, after alluviation, follows the same model of organic matter aging and bio-mixing

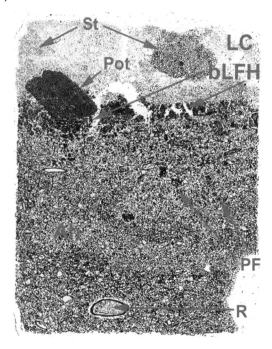

Figure 14.13 scan of thin section 726 (sand-buried soil – sand Layer C/bLFH/bAh; Figure 14.9) showing sand Layer C containing clean sand, subsoil sandstone (St) and at the junction of sand layer C/bLFH, a piece of pottery (Pot). The bLFH horizon (MF3, Fm/Hr) is only 3–4 mm thick and is strongly Fe stained (3.49% Fe) below the piece of pottery. The bAh horizon here displays more organic fragments (2.1%) and dark organic matter (17.9%) compared to 781 (0.9% and 15.4%, respectively), and can be seen to feature more coarse (>500 μm) passage features (PF – burrows) (cf. 14.4% with 4.6%); woody Ericaceous roots (R) also occur. Frame width is ~65mm.

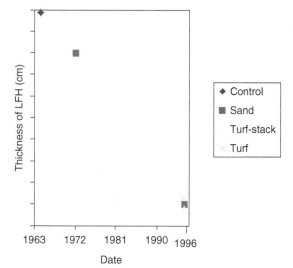

Figure 14.14 Wareham Experimental Earthwork 1963–1996: LFH thickness in 1963 (assumed thickness from 1996 control profile), 1972 (after 9 years burial), 1980 (17 years burial), and 1996 (33 years burial) (Bell et al., 1996; Evans and Limbrey, 1974; Macphail et al., 2003b).

(Catt, 1986, 1990) as that identified at Overton Down. One major finding at Wareham was the concentration of charcoal along the "old land surface" produced by periodic heath fires and now found within the 70 mm thick LFH (Figures 14.11 and 14.12). Such a concentration results from the fact that charcoal is more stable compared to the other forms of organic matter present. A concentration of charcoal along an old ground surface, when found under an archaeological monument, could easily be interpreted as indicating pre-construction clearance, however. The results from the experiment at Wareham now point to the possibility that a "taphonomic" concentration of charcoal has to be envisioned at some sites. (Sedimentary concentrations of charcoal also have to be considered, for example as linked to slow marine inundation or freshwater flooding of

Table 14.1 Wareham Experimental Earthwork; changes to the buried soil (Macphail et al., 2003b/with permission of Elsevier).

Measurement	Turf-buried	Sand-buried
Field		
LFH (thickness)	− − −	− − −
Ah (thickness)	0	0
Chemistry		
Organic C and N	− ?	− ?
C/N	− −	− −
pH	0	+ +
Alkali soluble humus	+ +	+ +
Pyrophosphate ext. C	+ +	+ +
Pyrophosphate ext. Fe and Al	0	0
χ (magnetic susceptibility)	−	−
Phosphate-P and available P	0	0
Available K	− − −	− − −
Soil micromorphology image analysis		
Void space	−	− − −
Organic fragments	− − −	− −
Dark organic matter	− −	− − −
Mineral	+	+ + +
Large ~2mm size voids	−	− − −
Medium ~1–2 mm size voids	+ +	+ + (Layer C – 726)
		− − (Layer D – 781)
Small ~<1mm size voids	+ +	+ + +
Void shape factor (irregular/elongate)	−	− − −
Soil micromorphology optical counts		
Plant residues and passage features	− −	− − −
Organic excrements (total and variety)	− −	− −
Organo-mineral excrements (total)	+ +	+ +
Organo-mineral excrements (aggregated)	+ ?	+ ?
Soil micromorphology description		
LFH (laminae thickness)	− − −	− − −
LFH (Iron concentration) (Microprobe)	+ +	+ + +
Ah (microstructure and related distribution)	Small change	Change
Ah (amorphous organic matter)	0	0

Key: + + + marked increase, + + increase, + slight increase, 0 unchanged;
− slight decrease, − − decrease, − − − marked decrease.

sites where soil charcoal is liberated (e.g. Mesolithic coastal Goldcliff, Gwent, Wales, see Chapter 9; Alluvial Iron Age Gudbrandsdalen, Uppland, Norway, see Chapter 6 and Figure 17c).)

Magnetic susceptibility (see 17.2.6) is low at Overton Down ($\chi = 0.357$–$455\ \mu m^3\ kg^{-1}$; 8.18–9.42% χ-conversion), but given the likely background levels in chalk soils (0.357–$0.455\ \mu m^3\ kg^{-1}$), where $>0.42\ \mu m^3\ kg^{-1}$ is deemed indicative of burning in these weakly ferruginous and thus poorly responsive soils, the Overton Down soils likely record some burning (Allen, 1988; Crowther et al., 1996). In this case, magnetic susceptibility at the Neolithic soils at Easton Down clearly testify to inputs of burned soil, as found in thin section (Figure 14.7; $\chi = 2.11$–$2.24\ \mu m^3\ kg^{-1}$; 13.1–13.3% χ-conversion). At Wareham, magnetic susceptibility is naturally low in Ah horizons of podzols ($\chi = 0.5$–1.4; 5.7–8.8% χ-conversion), with turf-burial leading to a weaker signal ($\chi = 0.2$; 2.4–2.7% χ-conversion), probably due to continued leaching after burial (Macphail et al., 2003b). The sand-buried soil, however, recorded a slightly higher magnetic susceptibility ($\chi = 1.5$–1.8; 9.0–12.2% χ-conversion), probably due to the influence of the overlying sand layers that contain non-leached material, and possibly because of greater biological activity here in the more porous sandy deposits.

The preservation and movement of buried artifacts, including organic remains, were monitored in the Experimental Earthworks Project. Bone, for instance, was preserved at base-rich Overton Down, whereas most had been lost at acidic Wareham (Bell et al., 1996). At the Historical-Archaeological Experimental Centre in Leyre, Denmark, the sealing of buried pigs in wooden coffins, either through the use of grass turves (sods) or by compacting wet soil, led to the rapid establishment of anaerobic conditions and the cessation of meat decay (Breuning-Madsen et al., 2001a, 2003). Although these experiments were of only short term (ca. 3 years), the results demonstrate that the authors were able to replicate the likely conditions of burial of some well-preserved Bronze Age burials.

Moreover, such experiments contribute to our understanding of unique wooden Viking ship burials at Gokstad and Oseburg (Vestfold, Norway) (Bonde and Christensen, 1993). At Gokstad, anaerobic burial conditions were induced by sealing much of the ship with marine clay (intertidal mudflat sediments), below a thick cover of turf (Cannell et al., 2020; Macphail et al., 2013b; Nicolaysen, 1882). The currently ~5m-high mound was first investigated using a motorized screw auger, followed up by 100 mm cylinder coring; the latter provided continuous cores for sub-sampling (Figures 14.15 and 14.16). The turf came from poorly drained pasture soils with a Laminated Mull (Barrat, 1964) topsoil produced by a sedge grassland vegetation as suggested by macro- and microfossil studies exactly correlated with geochemical and soil micromorphological studies (Figures 14.17 and 14.18). In addition, concentrations of iron phosphate (vivianite) along plant litter layers produced by phosphate-rich ground water migration within the waterlogged mound, wood splinters associated with wood-working, and the likelihood of stored turf becoming vegetated (rooting evidence unrelated to original soil formation) were found. Whilst mound studies are often useful for investigating the mound-buried soils, at Gokstad the ground had been dug out to allow the ship, which had been probably dragged up a shallow coastal watercourse, to be more easily buried within and by sediments from the marine regolith – the land had only emerged during the Iron Age. In fact, only one small intact buried soil area was found. It had a muddy surface and conjecturally was ground left in order to access the ship (by a gangplank?) so that the chieftain's body and his grave goods could be deposited (Rentzel and Narten, 2000; Rentzel et al., 2017). Data on ditch silting from both the Overton Down and Wareham experiments allowed the suggestion that the Viking Age robber trench that accessed the ship burial was simply left open, while recent excavation trenches were back-filled.

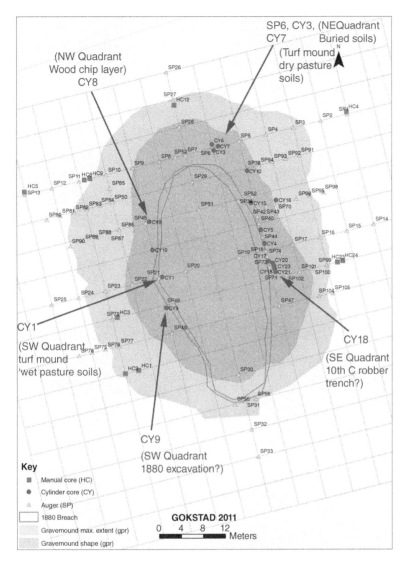

Figure 14.15 Gokstad Mound: Location of spiral coring (SP) and extracted cylinder cores (CY), showing central area of ship burial and 1880 excavation (now back-filled). Examples of spiral coring and cores mentioned in the text: clockwise – North-East Quadrant CY3: location of turf mound and buried soil/ intertidal sediment/till sequence, and hypothetical access point to ship burial during its AD 900 construction (Figures 14.19); SP 16–18 (Figure 14.16); CY21: sample sequence through 10th century robber trench (Figure 14.20); South-West Quadrant CY9: sampled location of 1880 trench; CY1: turf mound; examples of sedge grassland pollen in poorly drained "laminated mull" turves (Figures 14.17 and 14.18) (after Cannell, 2012; Cannell et al., 2020/Cambridge University Press / CC BY 4.0).

14.2.3 Wallasea Island, Essex, UK

In order to create more coastal wetland for migrating birds, an experiment was carried out by the Environment Agency, which involved the breaching of sea walls in the estuary of the River Crouch (Figures 14.21 and 14.22). Both grassland and arable land were flooded in 2006, and in 2008 soil pits

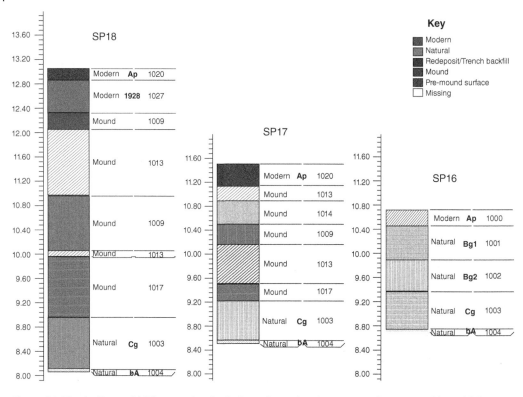

Key
- Modern
- Natural
- Redeposit/Trench backfill
- Mound
- Pre-mound surface
- Missing

Figure 14.16   As Figure 14.15; example of spiral core logs, showing truncated pre-mound intertidal sediments, which were analyzed in core samples along with mound materials (Macphail, 2012; Macphail et al., 2013a) (figure by Cannell, 2012).

Figure 14.17   As Figure 14.15; thin section scan of MB2 (CY1, Context 1024 turf mound); compact "turf 2" over sedge litter surface layer of "turf 1" (arrow) (laminated mull horizon; Figure 14.18). Frame width is ~50 mm.

Figure 14.18   As Figure 14.17; photomicrograph of laminated mull horizon characterized by layered plant litter and amorphous (humified) organic matter, with associated vivianite (e.g. $Fe_3(PO_4)_2H_2O$) formation (V) where phosphate in this mound has been mobilized and redeposited within (OM); EDS data for vivianite: Fe = 45.6% (58.7% FeO) and P = 16.7% (38.2% P_2O_5). "Blue" vivianite formed under anaerobic conditions, which seemingly permitted preservation of the Gokstad Ship and grave goods. PPL, frame height is ~4.62 mm.

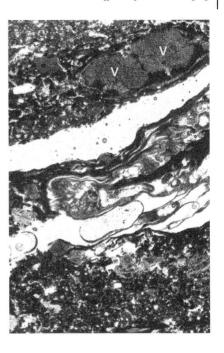

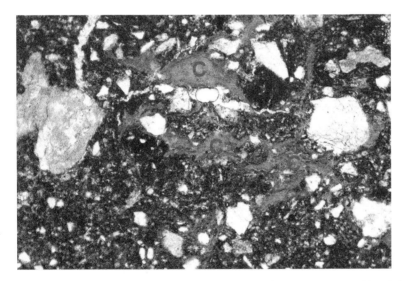

Figure 14.19   As Figure 14.15; photomicrograph of MD2 (CY3; Context 1015 bAg); humic soil with anomalous yellowish brown clay void infills (C) as possible evidence of human trampling on exposed surface or other disturbance associated with accessing the ship burial from the north-east quadrant during grave mound construction. PPL, frame width is ~2.38 mm.

were dug in both unflooded (control profiles) and flooded areas, the last at very low tide, to see what changes had occurred over this short time span (Figures 14.23–14.26). This was an important investigation given the large amounts of archaeology found in the intertidal zone in the UK and elsewhere (Belgium, Israel), and the major and increasing risk of coastal flooding on the dryland

Figure 14.20 As Figure 14.15; flatbed scan of thin section MG2 (CY18, Context 1050, Robber Trench fill); strongly heterogeneous trench deposit with reddish brown areas of microlaminated silt and clay void infills testifying to muddy soil disturbance during the tenth century break-in into the ship's burial chamber (between AD 953 and AD 975; Bill and Daly, 2012). Frame width is ~50 mm.

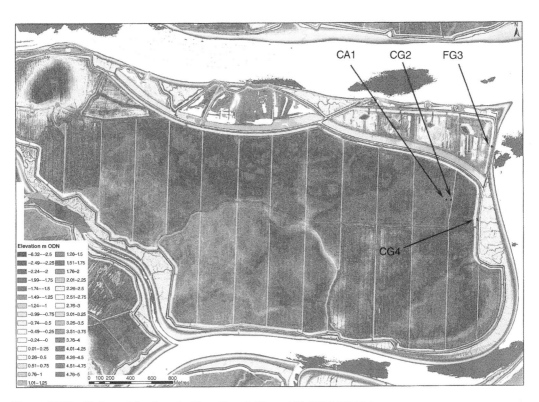

Figure 14.21 Wallasea Island on the River Crouch (Essex, UK). RSPB LIDAR base elevations and locations of soil sampling sites: CA1: Profile 1 – Control Arable Pit 1 CG2: Profile 2 – Control Grassland Pit 2; FG3: Profile 3 – Flooded Grassland Pit 3; and CG4: Profile 4 – Control Grassland Pit 4.

(a)

FG3

15. 7. 2008

(b)   Breach in old sea wall

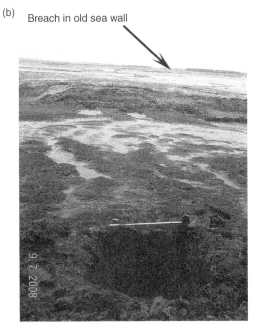

9. 7. 2008

Figure 14.22   As Figure 14.21. (a) Newly flooded (since 2006) areas of arable and grassland within breached sea walls. FG3 (Pit 3) is located on grassland adjacent to the "borrow" ditch, River Crouch in background. (b) Seaweed-covered estuarine muds over grassland soil profile (Pit 3). Borrow ditch, flooded arable fields and the sea wall in background (purposely breached by the Environment Agency).

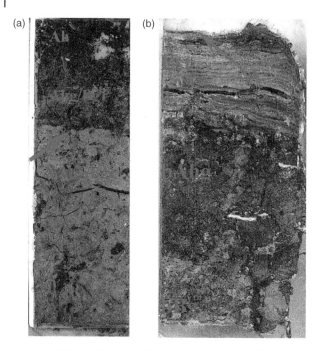

Figure 14.23 As Figure 14.21. (a) Control Grassland Pit 4 (CG4): ~16 cm-long resin-impregnated block. Humic Ah horizon, a mixed/dumped Ahg/Bg1, a buried "surface" (arrow), and buried bAhg/Bg1 horizon. Mottling and blackened organic matter reflect the low elevation/wet environment. (b) Flooded Grassland Pit 3 (FG3): ~11 cm-long resin-impregnated block, showing algae (A) coated estuarine clay laminae (Est) over buried Ahg and Bg horizons.

archaeological sites. Some Paleolithic sites have experienced very marked environmental and landscape change, the Acheulian coastal site of Boxgrove is now 22 km from the current sea shore.

The main impacts can be summarized as (Table 14.2):

- burial of terrestrial soils by typical laminated intertidal mudflat sediments (marine alluvium; Wallasea soil series) with silty clay infilling some voids in the buried soil (Figures 14.22 and 14.23) (Avery, 1990; Reineck and Singh, 1986);
- dramatic increase in salinity (from 83–115 to max 8590 µmho) and associated concentrations of Na and Cl in the laminated sediments, and increased amounts of Ca, compared to the buried soil (Figure 14.24), with concomitant and
- new foraminifera fauna of salt marsh and tidal flat character (Figures 14.25 and 14.26).

These findings have been particularly relevant to investigations of salt working coastal sites from:

1) Romano-British to Late Roman Stanford Wharf (Essex, UK), where inundated Neolithic soils (e.g. 129–169 µmho) were affected by Na^{++} induced soil dispersion and structural collapse before being sealed by marine sediments. These continued to accumulate through late prehistory, also burying "red hills" of salt working site residue origin (max. 2080 µmho) (Biddulph et al., 2012; Macphail and Crowther, 2012); and

2) Maya (Late Classic – AD 600–750/800) Marco Gonzalez, Ambergris Caye (Belize), a site recording numerous low temperature lime plaster hearths, fuel ash waste, and rubified marine sediments (a chief source of enriched brine), which retain their saline character (e.g. 2500–5700 µmho). This is in contrast to the overlying weathered "Maya dark earth" soils that have a much reduced salt content (min. 184 µmho) (Graham et al., 2017; Macphail et al., 2017b).

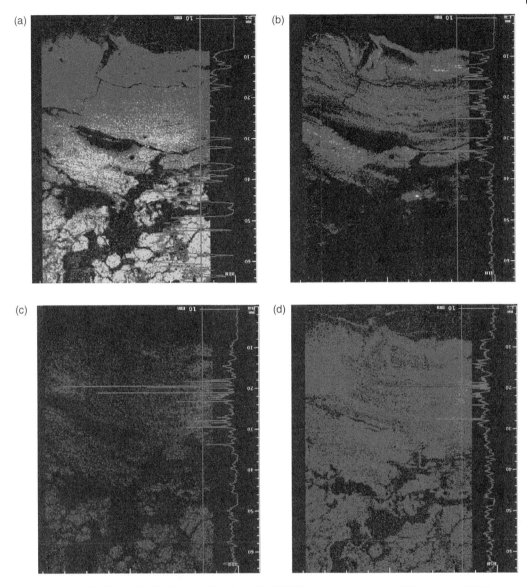

Figure 14.24　As Figure 14.21. Flooded Grassland Pit 3 (FG3): microprobe maps of thin section M5A1 (junction of buried Ahg horizon and overlying estuarine mudflat laminae) showing elemental distributions (a) Si; (b) Ca; (c) Na, which occurs both as saline (NaCl) salts and as sodium carbonate; and (d) Cl. Width of thin section is ~50 mm.

14.3　Experimental "Ancient Farms" and Settlements

During the 1970s–1990s, "ancient farms" such as Bagböle, Umeå (northern Sweden) (Scandinavian Iron Age; ca. BC 500–1000 AD) and Butser (Hampshire, UK) (Iron Age and Romano-British; ca. BC 1000–400 AD), were established to mainly test ancient farming methods (Reynolds, 1979; Viklund et al., 1998), such as:

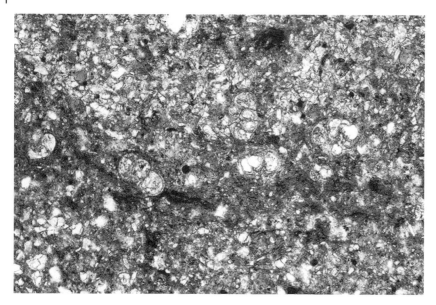

Figure 14.25   As Figure 14.23b. Photomicrograph of laminated silty clay marine alluvium, with iron staining probably linked to original detrital organic matter such as seaweed. Concentrations of "circular" bioclasts (foraminifera) are of tidal mudflat and high saltmarsh origin (Table 14.2). PPL, frame width is ~2.38 mm.

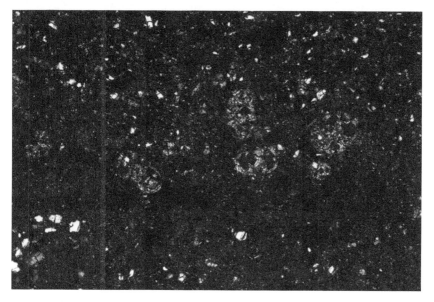

Figure 14.26   As Figure 14.25, under XPL illustrating calcitic sediment and forams (pH = 8.2; specific conductance = 8950 μmho; Table 14.2; see also Box 13.1).

- slash-and-burn agriculture and the growing of barley and rye at Bagböle; and
- mixed farming of wheat and other crops, with animal management of goats, sheep, and cattle, supplying dung for manuring, and with associated reconstructed round houses in the first instance; a Roman villa was more recently built to the pattern of the excavated Sparsholt villa, UK (see Figures 12.23 and 12.24).

Table 14.2 Wallasea Island (River Crouch), Essex, UK; basic soil characterization and magnetic susceptibility, foraminifera, and pollen data (Macphail et al., 2010/with permission of Elsevier) (see Figure 14.21 for locations).

Horizon	Depth (cm)	LOI (%)	pH (water)	Spec conductivity (μmho)	$(10^{-8}\ m^3\ kg^{-1})$	χ_{max} $(10^{-8}\ m^3\ kg^{-1})$	χ_{conv} (%)	Foraminifera summary	Pollen summary
Control arable pit 1 (CA1)									
Apg	0–4	4.52	7.6	115	16.7	1190	1.40	Very poorly preserved relict saltmarsh and tidal flat types	Cereal, "rape", very few tress
Apg	4–20	4.32	8.0	83.0	18.6	1240	1.50	Ditto	
Apg	20–40	4.03	8.1	98.0	16.9	1210	1.40	Relict low-mid and tidal flat, and mid-high saltmarsh types	
Flooded grassland pit 3 (FG3); 2006–2008									
Estuarine clay	0–5	6.95	8.2	8950	18.4	321	5.73	Super-abundant low-mid and tidal flat, with mid-high saltmarsh types	"Saltmarsh" with 19–44% trees
bAh	5–10	14.2	7.0	4750	18.6	185	10.1	As above including inwashed fossils and relict wet grassland types	Grass with cereal
bBg	20–40	2.91	8.3	1740	13.2	783	1.69	Relict low-mid and tidal flat types	and "saltmarsh"
Control grassland pit 4 (CG4)									
Ah	0–5	26.7	7.4	5580	15.0	725	2.07	Common low-mid and tidal flat, and mid-high saltmarsh types	Grass and sedge,
Bg?	20–37	2.21	8.5	1610	13.6	776	1.75	Relict mixed saltmarsh and tidal flat spectra	With trace of trees and "saltmarsh"

Other "ancient settlements" of note are (Viking) Leyre, (Denmark), (medieval) Melrand (Morbihan, France), and West Stow (Suffolk, UK) (Anglo-Saxon/Migration Period). "Folk Villages" and open-air museums have also been utilized, including St Fagans Open Air Museum (Cardiff, Wales, UK) and data recovered from excavating Roman Silchester, UK (Banerjea et al., 2015a, 2015b; Bell, 2009).

14.3.1 Butser Ancient Farm

Founded in 1975, Butser Ancient Farm, near Petersfield, Hampshire, UK was set up by Peter Reynolds to investigate some archaeological questions on farming and rural settlement during the Iron Age and Romano-British Periods (Reynolds, 1979, 1981, 1987). For example, Iron Age life-styles often have had to be reconstructed on the basis of truncated/eroded, dry land sites, where postholes and pits are the only field features. Most importantly for geoarchaeologists, the project tackled theories concerning roundhouse construction/use and *paleoagriculture*. It is beyond the scope of this book to discuss the yields from ancient crop production, or how grain was successfully stored in pits. Yet, other feature-fills, such as postholes (Engelmark, 1985; Reynolds, 1995), have supplied additional data used for settlement reconstruction that are crucial to geoarchaeological studies. Two important experiments concerned the study of different floor deposits: (1) the stabling of animals in a roundhouse, which was subsequently burned down as part of the experiment when the site had to be relocated, and (2) the development of "beaten" floors in domestic roundhouses (Figures 14.27 and 14.28).

The now abandoned (1990) Old Demonstration Area at Butser Ancient Farm was located on a lower downland (short calcareous grassland) slope and dry valley site on Chalk, some 5 km south-west of Petersfield, Hampshire, UK. The site had been under grassland for some 100 years prior to the Demonstration Area being established. The location is broadly mapped by the Soil Survey of England and Wales as having a cover of brown rendzinas – Rendolls (Andover 2 soil association; Jarvis

Figure 14.27 Butser Ancient Farm, Hampshire, UK, 1990: the 1977–1990 razed Moel-y-gar stabling roundhouse, just before sampling of the stable floor and the burned (rubified) and scorched daub wall material. The daub wall was pushed over and the site leveled (a frequent occurrence in Roman London, e.g. after the AD 59–60 Boudiccan revolt razed the town (Rowsome, 2000); the buried stable floor was re-sampled in 1995.

Figure 14.28 Butser Ancient Farm, 1994: the New Demonstration area, with the newly reconstructed Moel-y-gar roundhouse (foreground) and the new large domestic Longbridge Deveril Cowdown round house (background) from which newly forming beaten floor deposits were sampled. Note small door to Moel-y-gar roundhouse – suitable for the short stature of Dexter cattle used for ploughing at Butser.

et al., 1983, 1984), although the valley bottom, where the houses and arable fields were situated, has a typical (colluvial) brown calcareous earth soil cover (Coombe 1 soil association/Millington soil series; Gebhardt, 1990; Jarvis et al., 1983). The large domestic (Longbridge Deveril) roundhouse was reconstructed in the New Demonstration Area, 3 km to the south on non-calcareous and siltier (loessic) drift soils (Figure 14.28) (paleoargillic brown earths, Carstens soil association; Jarvis et al., 1983, 1984, 116).

14.3.1.1 Experimental Floors at Butser (and Trampling Experiments)

Whereas for the Paleolithic context, the notion of "floors" and "occupation surfaces" is a difficult matter to recognize and interpret, it is more clear-cut when dealing with architecture and complex societies of later time periods. Experiments have been useful in interpreting occupation surface (e.g. floor) deposits (see below), which were originally based upon case studies and modeling (Cammas et al., 1996b; Gé et al., 1993; Matthews, 2005; Matthews et al., 1997). At Butser, this was carried out in order to differentiate domestic "beaten" floors from animal stables (Heathcote, 2002; Macphail and Cruise, 2001; Macphail et al., 2004). This approach was, however, deemed "unscientific" (Canti et al., 2006; Macphail et al., 2006) (see section 14.4 for ethnoarchaeological research; Boivin, 1999; Goldberg and Whitbread, 1993; Matthews et al., 2000).

The chief findings from the Butser 1990, 1994, and 1995 studies, which provide much of the basis for modeling occupation surface (floor and floor deposit) formation can be summarized as follows:

Stable/Byre Floor (Figures 14.29–14.31, Table 14.3)

Although the most detailed results derive from the Moel-y-gar stable at Butser, where three areas were studied across the floor, other stables holding cattle, sheep and goats, and horses have also been studied (Heathcote, 2002; Macphail et al., 2004; see also Brönnimann et al., 2017a; Shahack-Gross, 2017). The chief characteristics found at the Moel-y-gar have been found clearly replicated in a series of Roman and early Medieval sites in London (see the London Guildhall below; Macphail et al., 2007a), but site conditions (desiccating, waterlogged, charred/ashed, mineralizing) and cultural practices have also to be considered when studying archaeological sites. For example, only very rare phytoliths, but many woody remains would be present if leaf hay was employed as fodder (Akeret and Rentzel, 2001; Ismail-Meyer and Rentzel, 2004; Ismail-Meyer et al., 2013; Rasmussen, 1993; Robinson and Rasmussen, 1989); different fodder and grazing plants, combined with variations in environmental conditions, will result in a number of dung residue types (Brönnimann et al., 2017a; Macphail et al., 2007a; Shahack-Gross, 2011; Viklund et al., 2013).

It was noted above that phosphate derived from wastes reacted with the underlying buried chalky rendol (rendzina) (Figure 14.29) substrate. To examine this more closely the stable floor was sampled from three locations, and three different types of strata were found – (a) a stabling (cattle?) crust, (b) layered sheep/goat dung, and (c) a highly turbated and mixed layer of fragmented dung, crust, and daub, essentially from the same "period" and "context" of the experiment (Figures 14.29–14.31).

Numerous Neolithic to medieval stabling and byre deposits have been investigated alongside reference dung and ethno-geoarchaeological investigations (see 14.3). Case studies indicate that when uncemented and humified, dung layers may become pelletized (cf. Moder humus) and may have a typically high Inorganic P:Organic P ratio (mineralized phosphate has a low Inorganic P:Organic P ratio).

Domestic "Beaten" Floor

Beaten floors were studied from both the Pimperne round house (dismantled in 1990) (Figure 14.32) and the newly built Longbridge Deveril Cowdown roundhouse of 1994, revealing how rapidly such a beaten floor formed. In fact, its massive surface structure started to develop almost as soon as the roof was finished, the grass cover died, and trampling began to affect the exposed surface

Figure 14.29 Butser Ancient Farm, Hampshire, UK, 1990; 13 cm-long thin section scan of the stabling floor and buried soil, from center of Moel-y-gar stabling roundhouse. Frame length is ~130 mm.

Figure 14.30 Moel-y-Gar 0.6 m from wall. Layers of compact crust and sheep/goat excrements. 65 mm wide thin section scan.

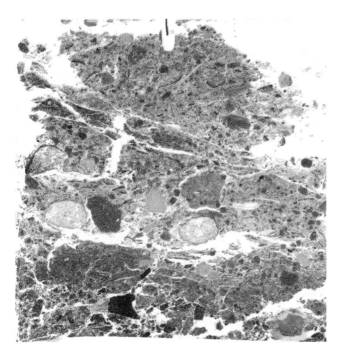

Figure 14.31 Moel-y-Gar next to wall; fragmented and mixed dung, bedding/fodder, crust, daub (from wall). 65 mm wide thin section scan.

Table 14.3 Bulk analysis of organic matter (LOI), P (2N nitric acid), phosphate (P_2O_5; 2% citric acid) and magnetic susceptibility (χ) on Butser experimental and London archaeological stabling and domestic beaten floors (Table 14.4). Included are control sample data from grassland outside the round houses, animal pasture, manured soils from Butser, and non-manured arable fields (Table 14.5). (Measured on <500 μm soil fraction; * measured on <2 mm soil fraction) (Macphail et al., 2004/with permission of Elsevier).

Site and context	Depth mm or Sample No.	%LOI	P (ppm)	P_2O_5 (ppm)	χ ($\times$ 10^{-8} m^3 kg^{-1})
BUTSER 1990					
Demonstration area grassland	0–80	16.6	1770		
Animal pasture	0–80	32.6	1600		16*
Arable field (manured)	0–80	18.1	2820		17*
Arable field (manured)	90–170	17.5	2220		26*
Arable field (non-manured)	0–80	19.0	2010		23*
Arable field (non-manured)	0–80	15.8	1990		
STABLING					
Moel-y-gar 1990					
Stable crust	0–60	40.9	5960	4730	27
Stable floor	60–100	32.3	2840		18
Buried soil	100–150	22.8	1460		16
7–11, Bishopsgate					
(charred Roman stable)	Sample 3	9.7	9350	4080	128
	Sample 5	4.3	9220	4160	79
No 1, Poultry					
(Roman stabling)	432	28.3		4300	14
Stabling	422	34.6		4170	14
Stabling	437	36.5		4170	16
Stabling	468	16.2		3920	27
Stabling	891b	20.9		3960	25
Stabling	891a	18.9		3950	16
Stabling	Mean *n* = 6	*25.9*		*4080*	*19*
Stabling	Std. Dev.	*8.51*		*155*	*5.78*
Below stabling	968-1a	4.8		2450	86
Saxon stabling pit fills					
	385-8	27.4		4330	75
	385-9	13.1		4140	241
BEATEN FLOORS					
Pimperne House 1990					
Compact surface	0–30	20.2	2430		47
Buried soil	30–80	19.9	2310		28

Table 14.3 (Continued)

Site and context	Depth mm or Sample No.	%LOI	P (ppm)	P$_2$O$_5$ (ppm)	χ (× 10^{-8} m^3 kg^{-1})
Longbridge Deverel Cowdown 1994					
Main area	0–0.5	12.0	1360		
	0.5–4	8.5	1220		
	4–8	8.1	1240		
Rear area	0–0.5	12.0	1880		
	0.5–4	8.7	1270		
	4–8	8.1	1290		
No 1, Poultry (Roman beaten surfaces)					
	887a	4.3		1750	91
Beaten surfaces	968-3	3.6		1300	59
Beaten surfaces	968-4a	7.0		660	114
Beaten surfaces	968-4b	7.4		1050	173
Beaten surfaces	968-7a	3.9		1320	156
Beaten surfaces	973-2	13.9		3270	227
Beaten surfaces	Mean *n* = 6	*6.7*		*1560*	*137*
Beaten surfaces	Std. Dev.	*3.89*		*911*	*60.8*
No 1, Poultry (Saxon beaten surfaces)					
	234-4a	8.1		3340	785
Beaten surfaces	234-5b	9.6		3860	779
Beaten surfaces	234-6a	17.8		3020	231
Beaten surfaces	234-6b	15.6		3110	153
Beaten surfaces	234-7a	11.5		3070	3278
Beaten surfaces	234-7b	20.7		2470	2894
Beaten surfaces	385a	5.9		2680	174
Beaten surfaces	Mean *n* = 7	*12.7*		*3080*	*1185*
Beaten surfaces	Std. Dev.	*5.43*		*449*	*1330*
CASE STUDY					
London Guildhall (Early medieval)					
Wattle fill	456	28.9	6800	2740	54
Drain fill	251	28.6		4250	61
Stave building beaten surface	429	9.0		2190	84

(Reynolds, pers. comm. 1994). The crumb structures were not being renewed by biological activity, nor were the blocky structures being reformed by wetting and drying (more recent investigations have been carried out; Bell, 2009). It is possible that this immature beaten floor was recording mainly natural surface soil amounts of organic matter and phosphate in 1994; minor phosphate

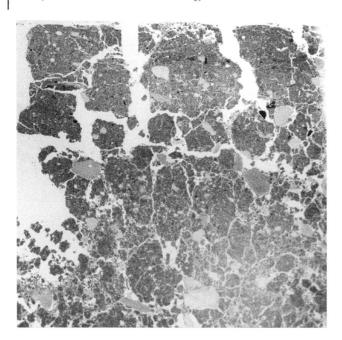

Figure 14.32 Butser Ancient Farm, Hampshire, UK, 1990. Domestic Pimperne house floor near hearth 1975–1990 (Frame width is 6.5 cm).

0–30 mm: beaten floor; 2430 ppm P, 20.2% LOI, χ = 47 × 10^{-8} m^3 kg^{-1} (note massive structure and typical vertical fissures of dried-out sample)
30–80 mm: Buried soil; 2310 ppm P, 19.9% LOI, χ = 28 × 10^{-8} m^3 kg^{-1} (note remains of blocky structure of original Mollisols Ah horizon).

accumulation in the back area was perhaps being best preserved from visitor erosion (0.5 cm depth: LOI = 12%; 1360–1880 ppm P; 0.5–8 cm depth: LOI = 8.1–8.7%; 1220–1290 ppm P; Table 14.3) (Macphail et al., 2006).

Sampling of the 1990 Pimperne house was also used to examine a straw matted area (Figure 14.33). There are many examples of archaeological floors being covered by some kind of plant "matting" which subsequently have given rise to compact floor deposits that are occasionally associated with planar voids relict of monocotyledonous plant "mats", long articulated phytoliths, and iron staining (Cammas, 1994; Macphail et al., 1997; Matthews, 2010; Matthews et al., 1997).

In archaeological sites beaten floors can be even more heterogeneous and contain far more bone and coprolitic waste than is allowed at sites such as Butser! Hence archaeological beaten floors may contain higher amounts of phosphate than their experimental counter parts (Table 12.3). Also, magnetic susceptibilities can be very high if buildings have been subsequently burned (see below and Figure 14.34) or when industrial activities such as metalworking are recorded in the trampled deposits (e.g. χ = 2894 and 3278 × 10^{-8} m^3 kg^{-1}).

The taphonomic effects of trampling on artifact distribution has long been of concern in archaeology (Barton, 1992; Gifford, 1985; Nielsen, 1991). Deleterious "poaching" effects (breakdown of soil structure due to animal traffic) and crust formation in soils were also recognized around water holes, for example (Schofield and Hall, 1985; Valentin, 1983), and investigators attempted to replicate them experimentally (Beckman and Smith, 1974). As a consequence, micromorphology and chemistry were used to identify Neolithic and Bronze Age cattle pounds at Raunds (Northamptonshire, UK) (Harding and Healy, 2011; Healy and Harding, 2007; Macphail, 2011a) in ways similar to the characterization of

Maasai animal enclosures (Shahack-Gross, 2017; Shahack-Gross et al., 2003, 2004b, 2008b) (see secction 14.3). Muddy trackways and road deposits also have some similar features, as shown from case studies and experiments in human trampling (Macphail et al., 2017a; Rentzel and Narten, 2000; Rentzel et al., 2017).

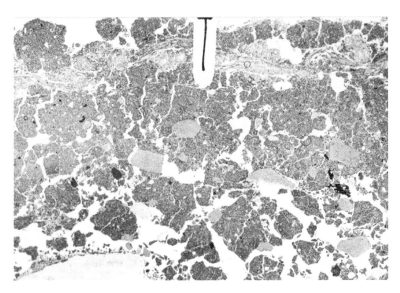

Figure 14.33 As Figure 14.32, Pimperne house floor; matted area 1975–90 (Frame width is 6.5 cm).

Detail of mat (arrow) and massive structured beaten floor.
Fragmented remains of pre-matted soil floor.

Table 14.4 Stables (byres) and domestic space: a summary of characteristics based upon experiments and case studies (see Figures 13.4a and 13.4b).

Floor type	Characteristics
Stable floors	Typically, homogeneous where high concentrations of organic matter, phosphate, and pollen grains may be preserved. Organic matter may occur in the form of layered plant fragments which, depending on pH, are either cemented or stained by phosphate (e.g. hydroxyapatite). If preservation conditions permit the survival of pollen grains, they are likely to be abundant and possibly highly anomalous with respect to the surrounding area
Domestic floors	Typically heterogeneous; floor deposits are comparatively minerogenic, massive structured, and contain abundant anthropogenic and allocthonous inclusions such as burned soil, charcoal, and ash. Plant fragments are less common and may occur in single layers of organic mat remains (or as phytolith-rich residues). The palynology of domestic floors is an under-investigated subject, but available data suggest the likelihood of far more diverse weed pollen assemblages (reflective of settlement flora), lower concentrations and poorer preservation than in stabling deposits. Magnetic susceptibility is likely to be enhanced and P and LOI will reflect *in situ* activities and trampled-in materials

14.3.1.2 Post Holes

Posthole Fills

Although feature-fills are known to include charred grain (Engelmark, 1985) and animal bones as well as artifacts, they also contain recoverable geoarchaeological information. When the Pimperne roundhouse, a large domestic structure with a diameter of 12.8 m, was dismantled in 1990 after a 15 year life, Reynolds (1995) found that most of the posts had rotted in the ground, sometimes leaving a post-pipe lined with bark. The postholes had also begun to silt up and already contained artifacts from the use of the round house by visitors. Such a finding is crucial to archaeologists and environmentalists wishing to determine use of space from patterns of posthole fills. In fact, the tripartite division of Scandinavian Iron Age longhouses is based upon the recovery of macrobotanical fossils from postholes. In general, the houses are thought to have been divided up into areas for: (1) fodder storage and stabling, (2) cooking, and (3) grain storage (Engelmark and Viklund, 1986; Myhre, 2004; Viklund et al., 2013). It is also considered more useful for understanding the chronology of a settlement by dating charcoal from post holes and post-prints, than by including all dates from a site, such as from hearths, pits, especially if driftwood was utilized as a fuel (J. Linderholm, pers. comm.; Macphail et al., forthcoming).

Reynolds (1995) demonstrated that posthole fills actually recorded the use of the site during its lifetime (or first 10–15 years), and thus any sediment within these postholes recorded the structure's use, and not post-occupation fill. On this basis, 21 posthole fills and "wall gully fills" from a rare example of a British Early Neolithic longhouse at White Horse Stone, Kent, UK were analyzed for loss on ignition (LOI – organic matter), phosphorous (P), and magnetic susceptibility (χ), a bulk study complimented by three thin sections (Macphail and Crowther, 2006a). This investigation was carried out to reveal distribution patterns of these parameters as post Neolithic erosion had removed all "floor" soils and other surface occupation deposits.

The results suggested that both enhanced levels of magnetic susceptibility and phosphate enrichment related to a predominant domestic use of the long house throughout its inhabitation. This was indicated by (a) generally enhanced magnetic susceptibility levels and (b) only moderate phosphate enrichment, which could all simply be explained by the presence of fine burned soil (and fine charcoal) and bone; these results are consistent with the inclusion of trace amounts of pottery and flint flakes found in thin sections. This domestic use model was consistent with the recovered finds, pottery, and flints. It can also be noted that the bulk and soil micromorphological character of the fills was a close match to experimental domestic floors formed at Butser. These were soil dominated (i.e., minerogenic) by inclusions of fine charcoal and fine burned soil; they had no resemblance to stabling deposits (see below). Further examples of post hole fill types are available (asma/Online_Appendix_11.md at master · geoarchaeology/ asma · GitHub)

Reynolds (1995) also found that the ground surface between posts of the wattle wall of the Pimperne House was burrowed by rodents (rats, mice, and voles), forming a gully. This feature could easily be misidentified as a drip gully or wall constructional feature in the archaeological record. We can note, therefore, that at White Horse Stone (see above) the longhouse "drip gully" had soil micromorphological characteristics consistent with both burrowing (biologically worked natural soil) and infilling with soil from the beaten earth floor (fine charcoal-rich soil). Recent analyses of some Norwegian long houses at Dilling, Østfold, reveal clay-based structures and flues, which are interpreted as house heating installations. The presence of abundant burned small mammal bones in them apparently testifies to house vermin living in the flues during the summer when houses were not heated (Macphail et al., forthcoming).

14.3.1.3 Experimental Wattle and Daub Buildings and Their Destruction by Fire

Experimental results from occupation surfaces at Butser came from the large domestic Pimperne House (1975) and the smaller (7.6 m diameter) Moel-y-Gar roundhouse (1979) when the former was dismantled and the latter burned down in 1990 (Figure 14.27). The fire that burned down the thatched roofed Moel-y-Gar house lasted only around 20 minutes. Major scorching and rubification of the wattle and daub walls (and magnetic susceptibility enhancement) took place above the wooden door lintel (Reynolds, pers. comm., 1990). This produced an order of magnitude higher χ value ($\chi = 190 \times 10^{-8}\,\mathrm{m}^3\,\mathrm{kg}^{-1}$) compared to floor deposits and soils outside ($\chi = 12$–$17 \times 10^{-8}\,\mathrm{m}^3\,\mathrm{kg}^{-1}$). The daub walls were then knocked over and the debris used to bury the stable floor, which was excavated five years later in 1995 (Heathcote, 2002).

Similar observations have been carried out on Mesoamerican sites and ethnoarchaeological examples investigating the breakdown of unburned daub and spread of sintered daub (Kruger, 2015). Experimental plant-tempered mudbricks were burned in a furnace to see how long it took them to burn in order to estimate that during the destruction of Megiddo (Iron Age 1, Israel), fires would only have lasted 2–3 hours (Forget and Shahack-Gross, 2016). It should be noted that it is not at all unusual to find a stratigraphic sequence composed of a scorched surface, overlain by a jumble of burned and rubified daub wall material. This stratigraphy occurs after a timber and daub wall construction has been burned and the site either collapsed or was leveled on purpose (Karkanas and Goldberg, 2018b). Furthermore, this has been recognized at three sites in Roman London, including structures seemingly razed during the Boudiccan revolt of AD 59–60 at No 1 Poultry and at Whittington Ave, London (Brown, 1988; Burch et al., 2011; Macphail, 1994a, 2011b; Rowsome, 2000). The characterization of different burned materials employing controlled heating experiments has also been carried out (Berna et al., 2007; Canti, 2017; Röpke and Dietl, 2017).

The distribution pattern of magnetic susceptibility was also mapped across the Pimperne house floor by Peter Reynolds and Mike Allen (unpublished; Figure 14.34). It is important to note that much information has come from *failures* in experimental reconstruction, so that the number

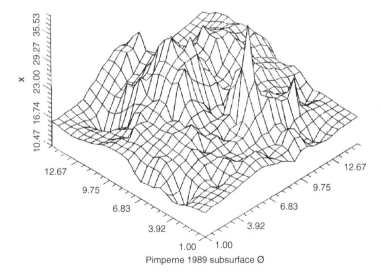

Figure 14.34 Butser Ancient Farm, Hampshire, UK, 1990: magnetic susceptibility distribution map of the domestic Pimperne roundhouse floor, showing the "high point" of the hearth area. Map scale is in meters; vertical scale magnetic susceptibility is ~10-36 × 10^{-8} m^3 kg^{-1}. (M. Allen and P. Reynolds, unpublished data).

of realistic models concerning the fabrication of Iron Age structures can be reduced, a great achievement (Reynolds, 1979). One example of particular interest to geoarchaeologists, is how *not* to construct a turf roofed house (see Chapter 12).

14.3.1.4 Reference Materials

Research at Butser Ancient Farm also provided a wealth of reference material, including dung from Dexter cattle (traditional draft animals used on the farm for ploughing), goats, and ancient mouflon sheep; pasture soils and plough soils with varying manuring regimes provided key background material for thin section and bulk analysis studies (Gebhardt, 1990; Heathcote, 2002; Macphail et al., 2004; Macphail and Goldberg, 2018b). Lastly, the site now contains a reconstructed small Roman villa with hypocaust, based upon the Sparsholt villa (Winchester, Hampshire) (Reynolds and Shaw, 2000), which provided reference building materials such as burned daub from walls and an oven (see 14.2.5.3). They also supplied examples of different "Roman" building materials, such as lime plaster, mortar, and tesserae for thin section study in relationship to Roman dark earth formation (see Box 13.1; Figures 12.23 and 12.24).

14.3.1.5 Ploughing and Manuring

The fields and pastures at Butser have been available for many studies of plough soils and tillage method experiments (Gebhardt, 1990, 1992; Reynolds, 1979, 1981, 1987); Lewis (2012) also made parallel studies of "Viking" soils at Lejre (Zealand, Denmark) and experimental soils at Rothamsted (Hertfordshire, UK) (see also reviews in Deák et al., 2017). Butser soils were mainly analyzed for their LOI (organic matter) and phosphate content, as ways to elucidate their fertility and the effects of different manuring regimes employing dung from the Moel-y-gar stable for example (Table 14.3). Chemical traces or "biomarkers" (5ß-stigmastanol and related 5ß-stanols) of inputs from animal dung in the manured fields were also investigated (Evershed et al., 1997).

Parallel *paleoagricultural* studies were carried out at Bagböle (Umeå, north Sweden), where, like Butser, the aim was to interpret more accurately weed and crop remains at Iron Age sites. In contrast to the brown calcareous (chalky) colluvial soils of Butser, agriculture on acid Spodosols (iron podzols) formed on outwash sands was carried out (Viklund et al., 1998). The soil micromorphological results demonstrated tillage and biological homogenization of manured soils with sometimes clear evidence of relict stable manure being present, although it is often best preserved in the soils of north Sweden where oxidation is less pronounced compared to southern England (Figures 14.35 and 14.36) (Gebhardt, 1990, 1992; Macphail and Goldberg, 2018a: 308–312). Results from slash-and-burn cultivation, which affects only the topmost few cm compared to ploughing (Macphail, 1998), were compared with fields undergoing various manuring regimes, as at Butser (see also Rösch et al., 2004). A pasture at Butser was not unexpectedly characterized by the highest LOI (36.2%) but by a relatively low phosphate concentration (1600 ppm P), compared to one manured field (2820 ppm P; 18.1% LOI), where this practice had seemingly led to this high phosphate concentration (Table 14.3).

At Bagböle, permanently manured soils accumulated the most organic matter (LOI = 10.5%) and highest concentrations of *organic P* with a $P_{total}/P_{inorganic}$ ratio of 4.9. These findings permitted the modeling of the circulation of phosphate and organic matter from the byre to the fields (Figure 14.37) (Engelmark and Linderholm, 1996). Iron Age fields, which were also identified from the cereal pollen in soils (Segerström, 1991), had high amounts of organic P preserved when compared to long house occupation soils, a finding common across Scandinavia (Viklund et al., 2013). This model has proven useful when large excavations of Iron Age (~500 BC – AD 1000) villages have taken place and settlement morphology reconstruction is attempted; the interdisciplinary study of Ørlandet (near Trondheim, Norway), is a case in point (Linderholm et al., 2019) (see Chapter 13).

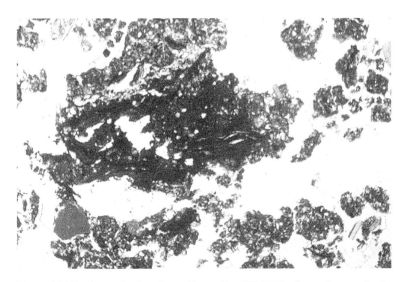

Figure 14.35 Butser Ancient Farm, Hampshire, UK, 1990: photomicrograph of experimental manured Ap horizon (formed on a colluvial Rendol1/rendzina) showing inclusion of relict dung fragments in chalky soil. PPL, frame width is ~3.35 mm.

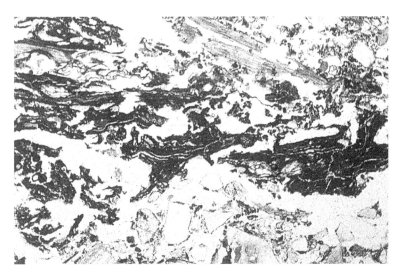

Figure 14.36 Umeå Ancient Farm, N. Sweden: photomicrograph of experimental manured Ap horizon (formed in a shallow iron Spodosol/podzol) showing inclusion of relict dung fragments in acidic sands. PPL, frame width is ~3.35 mm.

Such findings are consistent with enhanced phosphate levels characterizing manured archaeological soils that can also contain settlement waste (burned flint and daub, shell, grindstone, and charred coarse plant fragments and fuel ash residues) (Adderley et al., 2006; Carter and Davidson, 1998; Lewis, 2012; Macphail et al., 2000; Wilkinson, 1997). Other "farming" experiments included extemporized studies of medieval and Saxon pig husbandry, which showed surface soil disturbance in pig pastures and phosphate and fecal residue concentrations along pig pathways in an enclosure (Gebhardt, 1995; Macphail and Crowther, 2011).

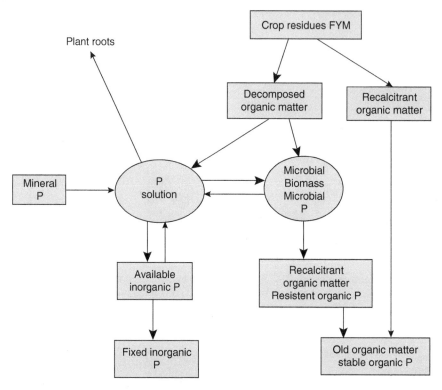

Figure 14.37 A model of the turnover of soil organic matter constituents and the mineralization-immobilization of phosphate in Swedish cultivated soils (FYM=farmyard manure). The main flow is indicated by dense arrows (redrawn from Engelmark and Linderholm, 1996).

Table 14.5 Swedish experimental soil data (Macphail et al., 2000); mean values of % LOI (550°C); 2% citric acid soluble P (P_{inorg}); 2% citric acid soluble P after ignition at (550°C) that converts organic P to inorganic P (unpublished data provided by Roger Engelmark and Johan Linderholm, Environmental Archaeology Centre, Department of Archaeology and Sami Studies, Umeå University, Sweden).

Context	%LOI	P_{inorg} ppm	P_{tot} ppm	P ratio (P_{tot}/P_{inorg})
Slash and burn	5.3	94	423	4.4
Modern field	6.4	105	436	4.2
Permanent field (manured)	10.5	139	684	4.9
Two-field system	5.3	161	493	3.1
Three-field system	6.4	153	541	3.5

14.4   Ethno-geoarchaeology

Increasingly, a large number of ethno-geoarchaeology experiments and studies have been carried out recently. The study of current burned-down earth- and wood-based structures from Mesoamerica has already been noted (Kruger, 2015). The decay and collapse of mud brick and wooden rooved buildings from the eastern Mediterranean have also been undertaken to aid the understanding of such sites as mud brick tells (Friesem et al., 2011, 2014a, 2014b). Turf-walled and turf-rooved houses, which are mainly associated with the Nordic world, but which were also in use in any location where turf, in all its forms, was available (e.g. soddies of Ireland and prairies of the USA), have also been reconstructed (Huisman and Milek, 2017; Milek, 2006) (see 12.2.1; Figures 12.11 and 12.13).

There have also been many studies of recent and historic structures, such as byres where stabling deposits have accumulated (Brönnimann et al., 2017a; Karkanas and Goldberg, 2018b; Lisá et al., 2020: 117–124; see Figures 14.29–14.31). Recent stabling deposits in Mediterranean and Near East have also come under scrutiny. These once-organic deposits can be preserved as minerogenic residues such as phytoliths (opal silica), calcium oxalates (e.g. $CaC_2O_4 \cdot H_2O$), and dung spherulites ($CaCO_3$), because of aging (oxidation) and ashing (Brochier, 1996; Brochier et al., 1992; Canti and Brochier, 2017; Macphail and Crowther, 2008). As noted in Chapter 13, microstratigraphic analysis of structured deposits is necessary if we are to elucidate the use of space and cyclical (seasonal) activities (Angelucci et al., 2009; Boschian and Montagnari-Kokelji, 2000; Macphail et al., 1997). Open-air animal enclosures studied from Kenya were found to include soils retaining both organic and poorly organic remains from animal stocking, with fossil material being linked to diets of mainly either monocotyledonous and/or dicotyledonous plants, with preservation being affected by factors such as waterlogging and ashing, for example (Shahack-Gross, 2017). Both ethno-geoarchaeological and geoarchaeological data were consistent in this review, and they parallel the findings from Arene Candide, Butser, and multiple prehistoric stock areas at Raunds (UK) (Macphail and Goldberg, 2018a: 436–455).

The taphonomic effects of trampling on artifact distribution and occupation deposit formation has long been of concern in archaeology (Barton, 1992; Gifford, 1985; Nielsen, 1991; Rentzel et al., 2017). Surface soil structure breakdown ("poaching") has been linked to livestock concentrations for example (Schofield and Hall, 1985; Valentin, 1983, 1991), and such soil changes and deposit formation associated with the pounding of livestock has benefitted from both pedological and ethnoarchaeological investigations (Macphail, 2011a; Shahack-Gross, 2011, 2017; Shahack-Gross et al., 2003, 2004b). Muddy trackways and road deposits also exhibit some related features, as shown from case studies and experiments in human trampling including laboratory studies on dry and moist loamy sands (Macphail et al., 2017a; Rentzel and Narten, 2000; Rentzel et al., 2017). In the realm of interpreting ancient hearths and combustion zone formation (Mallol et al., 2017; Mentzer, 2014), both experimental and ethnoarchaeological approaches have been utilized, for example, to detail the cleaning of hearths (Mallol et al., 2007, 2013). Interestingly, although threshing floors have been examined, apparently no evidence for this activity was recorded in thin section (Gebhardt, 1995; Shahack-Gross et al., 2009). Woodland management – coppicing – for fuel production in the form of charcoal is well known and produces soils with embedded coarse charcoal (e.g. at Butser); past charcoal working locations were also investigated from the Paimpont forest (Ille-et-Vilaine, France), the Zonien forest (Belgium) and at the Écomusée de Haute-Alsace (Ungersheim, France) (Gebhardt, 2007). The chief findings relate to operations related to the local deforestation for wood exploitation, the preparation and management of the plots before, during, and after the production phase and the domestic activities of the charcoal burners, along with the identification of turves used in the charcoal clamp.

14.5 Conclusions

The above description and discussion of geoarchaeological experiments and ethno-geoarchaeological investigations, and most importantly their application to real sites, has shown that there are several innovative ways to understand sites more fully and correctly. Firstly, there are many specific well controlled, albeit restricted in scope, laboratory-based experiments examining the effect of heating on geological, man-made, and organic materials. Secondly, there has been the experimental manufacture of building materials, especially lime mortar. Thirdly, we owe a great debt to those archaeologists who organized and carried out long-term experiments, involving earthworks and "ancient farms", where geoarchaeology was only one part of multidisciplinary investigations. These holistic efforts, although producing complicated results, allowed the development of consensus interpretations, and ways with which not only layers and features could be identified, but also allowing insights into how entire settlements functioned – a real aim in archaeology. Ethno-geoarchaeological studies, such as those studying animal management in Africa, and historic-recent stables in Europe, also have contributed to the accuracy of identifying use of space. Lastly, a few extemporized "experiments" – e.g. a trampled muddy path formed during an excavation and an Environmental Agency flooding of coastal lowlands – were utilized to produce important data on the effects of flooding and inundation on archaeological soils and artifact distribution. Although there is always a need to make specific experiments in the laboratory, and ethno-geoarchaeological opportunities may arise, research from long-term experimental/ancient settlement is still urgently needed.

15

Geoarchaeology in Forensic Science and Mortuary Archaeology

15.1 Introduction

At one time forensic science teams commonly consisted of a pathologist, a toxicologist, and a fingerprint expert. However, as crime scenes – such as clandestine graves and weapons caches – have been investigated more thoroughly, the techniques of archaeological excavation have been applied more frequently; a study of insect remains (e.g. fly pupae) is often a first step in suggesting timelines in a murder (Hunter et al., 1997). More recently, the pioneering use of palynology at crime sites by Patricia Wiltshire has led her to develop a completely new applied discipline, that of environmental profiling (Wiltshire, 2019; Wiltshire, 2002). This approach entails the application of plant and animal ecology, geology, and pedology to produce circumstantial evidence to help support the case, either for the prosecution or defense.

There is a very real and important role for the forensic geoarchaeologist to play in modern forensic science (Pye and Croft, 2004). This participation may involve simply suggesting to search teams where a clandestine grave may *not* be located, as for example where the soil is too shallow (e.g. rendzinas/Rendols on chalk). Or, it might help link a suspect to a crime site, when the mud on the suspect's shoe matches that found at the grave site (Pirrie et al., 2004; Pirrie et al., 2013). Environmental profiling is, however, a holistic approach, where both seeds (and other macro plant remains) and palynomorphs (pollen, plant spores, fungal material, and other microscopic entities) from specific ecologies could only come from a specific environment with particular soil types and geology (Horrocks et al., 1999; Wiltshire, 2019). Equally, if certain plant remains have been found on a suspect's clothing and/or vehicle, they could link a suspect to a specific environment, which could then be searched for a clandestine grave (Pringle et al., 2012).

Allied to these innovations is the increasing interest in the geoarchaeological study of graveyards. This can also include finding ancient unmarked *cemeteries* (Cannell et al., 2018; Faul and Smith, 1980) – as well as investigating various mortuary practices, such as inhumations, excarnations ("sky burials") and cremations. The identification of these practices may provide important cultural information and can be considered just as important as the investigation of other cultural features – houses, pits, waste deposits, cultivated fields, for instance – that make up complex society settlements (Macphail et al., 2017a). Recently, geoarchaeology viewpoints have contributed to the debate about whether Neanderthals intentionally buried their dead (Dibble et al., 2015; Goldberg et al., 2017; Rendu et al., 2014; Rendu et al., 2016); this and the views of Upper Paleolithic people on cave paintings have been discussed previously (Mel Brooks, *History of the World, Part I*).

Practical and Theoretical Geoarchaeology, Second Edition. Paul Goldberg and Richard I. Macphail.
© 2022 John Wiley & Sons Ltd. Published 2022 by John Wiley & Sons Ltd.

It is well established that skeletal and clothing remains, and grave goods such as deposits in pots and weapon remains, can provide unique data on populations and their lifestyles; one example is the presence of sickles in cremation graves at Dilling (Vestfold, Norway), suggesting new developments in farming (L-E Gerpe, Cultural History Museum, Oslo, pers. comm.) (Box 18.4). Such studies may also tell us something about clandestine graves. More widely, analyses of excavated settlements and their associated cemeteries have been successfully undertaken, for example at "rural" Middle Saxon (Anglian – 4th–7th century AD) West Heslerton (North Yorkshire, UK) (Haughton and Powlesland, 1999; www.intarch.ac.uk/journal/issue5/westhes_toc.html) and "urban" medieval Spitalfields, London (Connell et al., 2012; Harward et al., 2019); as found in this volume, both settlement sites were rigorously investigated employing geoarchaeological techniques within a multiproxy framework (see Section 13.5.3).

15.2 Mortuary Practices

When Professor Jan Bill, Director of the Viking Museum, Oslo, was asked "What was the Viking mortuary practice – inhumations, excarnations, or cremations?", the answer was "All three". In the UK, some Early Iron Age settlements (ca BC 800–400) were well known for their excarnation platforms (e.g. Oxley Park; Brown et al., 2009), while a chief Saxon (AD 400–800) milestone is a change from "pagan" cremations (contemporary with famous boat graves at Sutton Hoo) to "Christian" burials at this Suffolk, site in the UK (Carver, 1998; Fern, 2015).

15.2.1 Excarnations, Ship, and Boat Burials

Geoarchaeological identification of an excarnation is probably the most problematic – and reflects the same difficulties encountered when searching for more recent clandestine graves (see below). This is because the body is completely exposed to weathering, insect infestation, and the ravages of raptors and other scavenging animals (this is the equivalent of Tibetan sky burials and Native American platform burials). At these sites, bones and other body parts become scattered, as found in archaeological experiments in taphonomy (Armour-Chelu and Andrews, 1994). In this way, the practice resembles what happens when murder victims are simply disposed of in the undergrowth; and it may be weeks/months later before a dog walker finds a skull or other bones (Wiltshire, 2019). More unusually, "burials" in peats (Histosols) led to loss of bone but preservation of soft tissues, best known as "bog bodies" including Grauballe Man in Nebelgård Mose (Jutland, Denmark) in 1952 and Lindow Man in Lindow Moss (Cheshire, UK) in the 1980s (Brothwell, 1986, 1995); one of the present authors spent much time sharing a bench with the remains of Lindow Man's last meals (Holden, 1995).

Archaeologists may also find anomalous skulls in association with small post hole groupings, to suggest an excarnation platform was present. At the Iron Age site of Oxley Park (Milton Keynes, UK), the selection of a sub-sample from a post hole fill located an unexpectedly high phosphate concentration (10.8 mg g^{-1} P (1.08% P) in bulk sample), associated with a fragmented bone concentration (Brown et al., 2009; Macphail and Crowther, 2008). While the thin section study tentatively identified a neo-natal tooth, bone alteration was consistent with that found in raptor guano layers present in caves (Figure 17.8a and 17.8b; see Chapter 10) (Macphail and Goldberg, 1999). Archaeologists had also found other human remains, and so an interpretation of a child's sky burial and associated bird scavenging was acceptable.

Another form of excarnation was practiced during the Viking Period: exposure of the body in a wooden chamber (Box 15.1A). There are detailed examples of grave chambers at Hesby (Vestfold, Norway) (Gollwitzer, 2012). Of particular interest are the mapped concentrations of phosphate and body stains (see Box 15.1) that record the location of the bodies, while the remains of one wooden chamber also indicate exposure of the chamber (Viklund et al., 2013). Geochemical characterization of body stains through trace element studies has a long history (Keeley et al., 1977).

Box 15.1   Norwegian graves, excarnations, boat burials, and cremations

Box 15.1A Viking graves and chamber burials (excarnations); Hesby, Vestfold, Norway

Several grave/grave chambers were excavated at Hesby, Vestfold, Norway (Gollwitzer, 2012). Phosphate mapping of the area in and around excavated robbed Grave 1130946 (Figure 15.1) showed a complicated pattern of phosphate concentration although the highest amounts centered around the main grave (Viklund et al., 2013). Here, there was a marked contrast

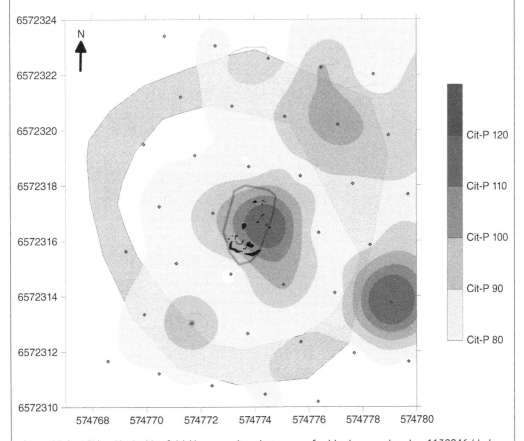

Figure 15.1   Viking Hesby, Vestfold, Norway; phosphate map of robbed grave chamber 1130946 (dark blue equals highest phosphate concentrations), which included a soapstone vessel; presence of charred plant materials may indicate presence of plant/food offerings, with associated soil also showing a slightly enhanced magnetic susceptibility (Viklund et al., 2013) Figure by kind permission of Johan Linderholm, University of Umeå.

(Continued)

Box 15.1 (Continued)

between the grave fill and the surrounding backfill, and this junction was also investigated employing soil micromorphology and SEM/EDS on the thin section (Figure 15.2). The presumed body stain can be clearly differentiated from the backfill, the body stain being characterized by neo-formed iron microfeatures (amorphous impregnative staining and void hypocoatings) possibly also recording high levels of mesofauna activity, and now showing concentrated amounts of iron and phosphate especially (SEM/EDS data: 44.9% Fe, 14.4% P; Figures 15.3–15.8). At the base of another grave (wooden grave chamber S1130513) located on the gentle slope downhill of the presumed settlement's location, the voids between moderately well-preserved fissured wooden fibers were infilled with fine colluvium originating from ploughing of manured soils upslope (Figure 15.9). The bodies were thus exposed in chambers whilst normal life carried on. In this area, this was agricultural activity, which included ploughing with concomitant soil erosion and colluviation, eventually sealing this form of excarnation.

Boat and ship graves are a special case, with both plank and nails preserving ship outlines at 7th century AD Sutton Hoo, on the Deben River (Suffolk, UK); the site is also well known for its body stains and its cultural links with Vendel southern Sweden (Carver, 1998). Although there are exceptional examples of the Gokstad and Oseberg Viking Ship burials where sealing of the grave mound with estuarine clay has both preserved the intact wooden ships and skeletal remains (www.khm.uio.no/english/visit-us/viking-ship-museum/) (Bill and Daly, 2012; Cannell et al., 2020; Macphail et al., 2013a; Nicolaysen, 1882). The much more recent (2018) discovery of the Gjellestad Viking ship burial by georadar in Østfold, Norway, associated with other burial mounds and long houses (www.niku.no/en/2019/08/finner-de-rester-av-gjellestadskipet/),

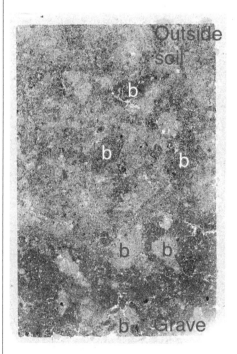

Figure 15.2 As Figure 15.1, Scan of M2131140 (Grave S1130946); note diffuse and burrowed (b) boundary between the dark fine silty clay grave fill and the outside sandy soil. Frame width is ~50 mm.

Box 15.1   (Continued)

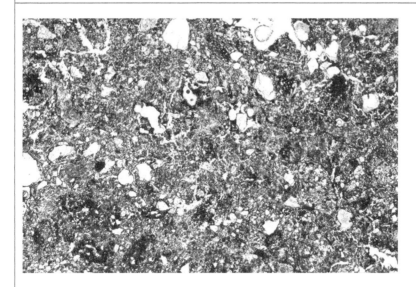

Figure 15.3   As Figure 15.1, photomicrograph of M2131140 (Grave S1130946); fine silty clay fill inside grave, with very abundant iron staining and void hypocoating. PPL, frame width is ~4.62 mm.

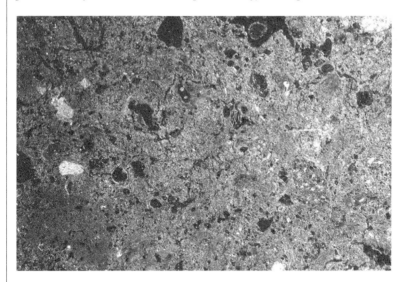

Figure 15.4   As Figure 15.3, under OIL.

found that this long ship was not well sealed in the mound. It had undergone major fungal attack; the iron-poor surrounding soils had also not permitted iron staining of the wood (Macphail et al., 2020a). It is also worth noting here that experiments in Denmark where an anaerobic burial environment was created by strongly compacting the turf mound, seems to have replicated well-preserved body remains in log coffins recorded in tumuli (Breuning-Madsen et al., 2001; Breuning-Madsen et al., 2003).

(Continued)

Box 15.1 (Continued)

Figure 15.5 Detail of Figure 15.3; note fibrous secondary iron in voids. Iron also impregnates a relict organic content. PPL, frame width is ~0.90 mm.

Figure 15.6 As Figure 15.5, under OIL. Note also red burned mineral grain which shows the anthropogenic nature of the fine soil fill, consistent with enhanced magnetic susceptibility.

Boat burials may be a special case, and these are typical of Scandinavian settlement sites and include Viking age examples (Box 15.1B). Often these may be seen in the field as narrow feature-fills, with only iron staining picking out wood remains – especially for example where nails have oxidized and ferruginized wood fibers (Figures 15.12 – 15.15) (Macphail et al., 2017a). At this Heimdalsjordet site the pelvic region where a sword had been laid was particularly phosphate rich. Both calcium phosphate (possibly hydroxylapatite) and iron-calcium-phosphate remains occur. As at Gjellestad (above), oxidized woody fibers may be the only traces of the

Box 15.1 (Continued)

Figure 15.7 As Figure 15.1; SEM, X-ray backscatter image of iron void hypocoatings and impregnations (as in Figures 15.3–15.6). Scale bar = 300 μm.

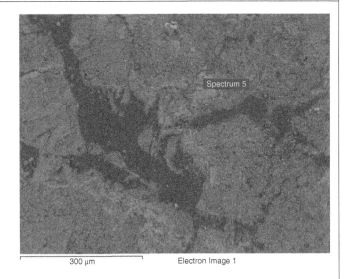

Figure 15.8 X-ray spectrum from Figure 15.7; mean 33.8% Fe, max 48.1% Fe; mean 9.34% P, max 14.4% P.

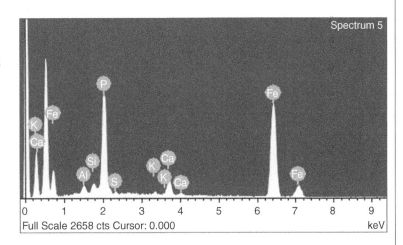

boat's wooden construction, and these may not necessarily be iron stained, but may be phosphate enriched (e.g. 2.84% P) as at another boat burial at Røddekrysset (Trøndelag, Norway). There is also evidence of double boat burials, where the same location was used ~100 years later to insert another boat burial (www.lifeinnorway.net/a-viking-grave-mystery-in-norway/)

Box 15.1B Viking boat burial, Heimdalsjordet, near Sandefjord, Vestfold, Norway

The 9th–10th century coastal settlement site of Heimdalsjordet in Vestfold, Norway is only 500 m away from the Gokstad Ship Mound, and is famous for rivalling nearby Kaupang as a Viking trading center, for example in the number of Middle Eastern dirhams found there

(Continued)

Box 15.1   (Continued)

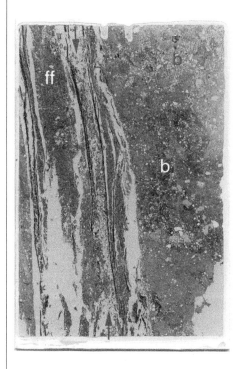

Figure 15.9   Scan of M213124A; base of Chamber Grave S1130513, showing wooden fiber remains of coffin; some fissures between fibers have a fine soil fill (ff), others show organic excrements of mesofauna comminuting the wood (arrows). Coarse sandy soil has become mixed-in due to burrowing (b). Frame width is ~50 mm.

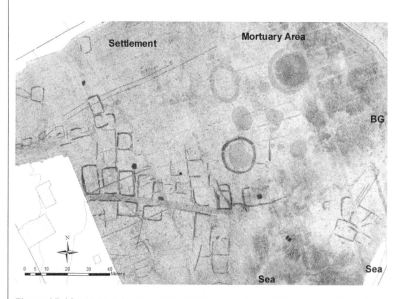

Figure 15.10   Heimdalsjordet, Vestfold, Norway; 9th–10th century Viking Age settlement (ditch-separated property parcels) grave mounds and boat grave (BG – approximate position) in mortuary area, with the seashore to the south (Sea). Graphic representation of GPR (Ground-Penetrating Radar Data) (Bill and Rødsrud, 2017: fig. 11.3).

Box 15.1 (Continued)

Figure 15.11a Heimdalsjordet,
boat grave fill 5529; excavated
surface.

Figure 15.11b As Figure 15.10, excavated boat grave fill, with sword, and micromorphological samples
of boat fill (M4) and body stain in the suspected pelvic region by the sword (M3).

(Bill and Rødsrud, 2017). Located on the Viking Age seashore, it is believed to have been
seasonally occupied (Macphail et al., 2013a, 2017a). There are ditched property parcels,
and a mortuary area, which includes a boat grave (Figure 15.9). The long narrow boat grave
emplaced in beach sands included a moderately well preserved sword and associated body
stain of a presumed Viking warrior; and traces of the boat's wooden keel were sampled
(Figures 15.10–15.12 (C. L. Rødsrud, Cultural History Museum, University of Oslo, pers. comm.).
As found in many such boat graves, the wood has become strongly pelletized by biological
activity, with sometimes iron contamination associated with boat nails preserving some of
these organic microfabrics; locations of "nails" in the field may in fact be iron mineralized

(Continued)

Box 15.1 (Continued)

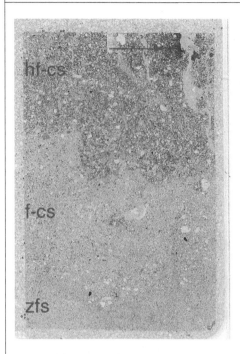

Figure 15.12 Scan of M9574 (vertical sample through Heimdalsjordet boat grave 5529); possible boat keel. The boat grave was inserted into beach silty fine sands (zfs), which are overlain by poorly sorted fine, medium, coarse and very coarse sands (f-cs), and few fine gravel. Grave soil is composed of humic fine to very coarse sands (hf-cs), with a pellety humus formed of wood residues worked by an acidophile invertebrate mesofauna. Iron staining and very poor pseudomorphic wood preservation may mark the location of an iron nail (arrow). Frame width is ~50 mm.

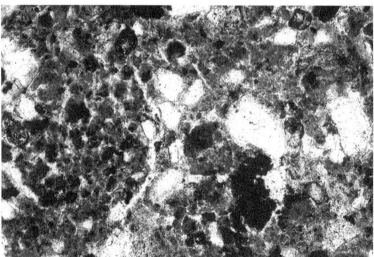

Figure 15.13 Photomicrograph of M9574 (Heimdalsjordet boat grave 5529); possible boat keel; pellety humus – wood residues – and secondary iron staining from weathering boat nails. PPL, frame width is ~0.90 mm.

wood fibers, where a nail had been located (Figures 15.12–15.15). The Sutton Hoo ships are famous for their preserved ship nail outlines (Carver, 1998). At the Heimdalsjordet boat grave, sampling in the pelvic region by the sword (Figure 15.11) revealed concentrations anomalous reddish and yellowish amorphous material, the latter having the isotropic appearance of mineralized cess containing microfossils such as phytoliths (see Chapter 11). SEM/EDS

Box 15.1 (Continued)

Figure 15.14 Photomicrograph of M9574 (Heimdalsjordet boat grave 5529); possible boat keel; iron staining of fibrous wood from boat nail location (see Figure 15.12). PPL, frame width is ~2.38 mm.

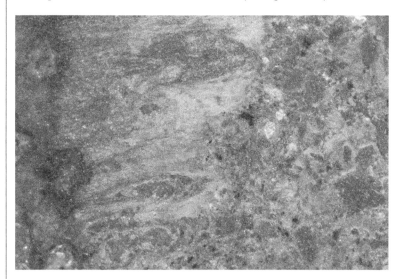

Figure 15.15 As Figure 15.14, under OIL.

analyses identified the reddish material as a Fe-P-Ca complex, while the yellowish material is probably a form of hydroxylapatite (CaP) (Figures 15.15–15.18). Such probable mineralized gut remains can be compared to that found at Voldskogen, Rygge, Norway (Iron Age, inhumation Grave A132; Figure 15.19) (K. Orvik and S. Solheim (Cultural History Museum, University of Oslo), pers. comm.).

Box 15.1C Inhumations and cremations

Bodies in coffins may leave very little evidence apart from sometimes a scatter of human bones if conditions are not acid (see Figure 15.28). When soils are poorly drained coffin wood

Box 15.1 (Continued)

Figure 15.16 Photomicrograph of M7652 (horizontal sample through Heimdalsjordet boat grave and pelvic region of body stain; Figure 15.11); pale yellow amorphous calcium-phosphate (weathered carbonate hydroxyapatite) relict of fecal material within gut (cf. "cess"). PPL, frame width is ~2.38 mm.

Figure 15.17 SEM/EDS X-ray backscatter image of M7652; there are areas of (1) amorphous ~hydroxyapatite (15.5% P, 35.5% P_2O_5; 20.3% Ca, 28.4% CaO) of probable mineralized gut fecal remains origin; some secondary iron staining has also occurred in places (8.43% Fe, 10.8% FeO). Area 2 is bioworked mineralized fecal remains with a similar CaP chemistry. Area 3 is formed of secondary iron phosphate. Scale = 300 μm.

Box 15.1 (Continued)

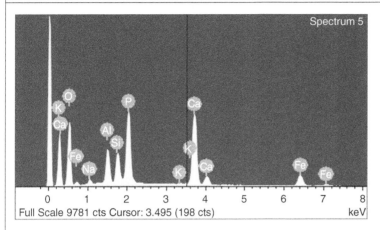

Figure 15.18 As Figure 15.17, X-Ray Spectrum of ~hydroxyapatite – note peaks of P and Ca.

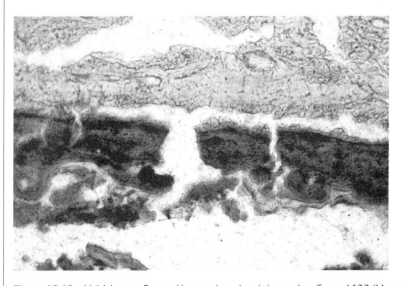

Figure 15.19 Voldskogen, Rygge, Norway, Iron Age inhumation Grave A132 (Mound A100); photomicrograph of M5044; mineralized (phosphatized) gut/coprolitic material. PPL, frame width is ~0.90 mm. EDS: mean 25.8% Al, 17.6% P, 0.40% S, 0.47% Ca, 2.38% Fe, 0.47% Ba, n = 7; max 19.6% P, 0.55% S, 0.64% Ba).

may preserve. Several coffin remains were found at Iron Age Sarpsborg, Østfold, Eastern Norway (Rødsrud, 2007). At inhumation grave F551, a trace of the coffin noted in the field was sampled. In fact, several very thin layers were found: a trace of thin probable amorphous phosphate ("FeP"), was mirrored by a layer of wood fibers, below which phosphate (vivianite) was identified – again associated with wood traces (Figures 15.20–15.24). Cremation burials are more likely to be associated with burned mineral and organic remains. At the central Grave A106, in

Box 15.1   (Continued)

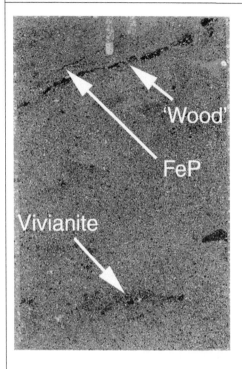

Figure 15.20   Flatbed scan of inhumation Grave F551 (Bjørnstad), illustrating "wood" layer (from the coffin) and iron-phosphate layer (from body?), and an underlying thin layer where vivianite (crystalline Fe-P) is present. Frame width is ~50 mm.

Figure 15.21   Photomicrograph of F551 (Figure 15.20), showing two thin layers in the grave; a top layer of amorphous iron-phosphate ("FeP"; from the inhumation?) and a lower layer of wood remains (from the coffin). PPL, frame height is ~4.6 mm.

Box 15.1 (Continued)

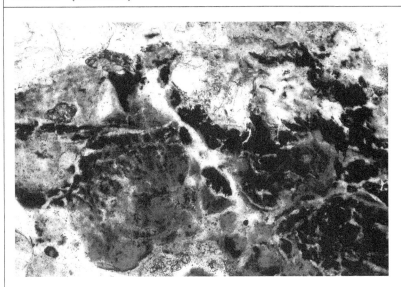

Figure 15.22 Detail of Figure 15.21, showing wood fibers, part cemented by amorphous iron phosphate – also likely preserved by acid and sometimes anaerobic conditions. PPL, frame width is ~0.90 mm.

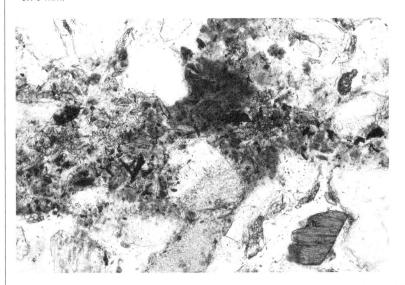

Figure 15.23 As Figure 15.20, photomicrograph of vivianite layer; a detail of a thin layer composed of very fine wood fibers and crystalline iron phosphate (vivianite). PPL, frame width is 0.90 mm.

mound 1, at Botshaug, Akerhus, Norway (K. E. Sæther, K. Orvik and F. Iversen, Cultural History Museum, University of Oslo, pers. comm.), calcined bone, with a typical bone chemistry (Figures 15.25–15.27) was found alongside vesicular char. The latter cemented with anomalously high amounts of presumed barium sulfate (EDS data: e.g. 37.9% Ba; 17.0% S).

(Continued)

Box 15.1 (Continued)

Figure 15.24 As Figure 15.23, but under XPL showing crystals of vivianite which are indicative of anaerobic conditions.

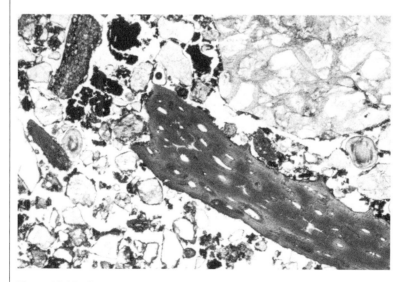

Figure 15.25 Photomicrograph of M805 (Botshaug, central grave A106, Mound 1); cremation residues composed of charcoal and calcined bone, in weakly humic and podzolic sands. PPL, frame width is ~4.62 mm.

Box 15.1   (Continued)

Figure 15.26   As Figure 15.25, under OIL.

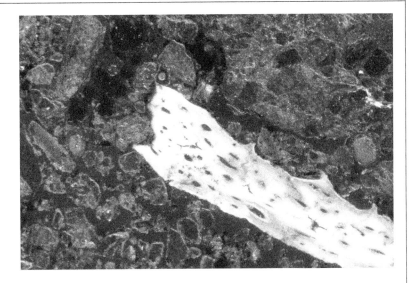

Figure 15.27   As Figure 15.25, BSE X-Ray backscatter image of bone and charcoal. Scale = 1 mm. EDS Spectrum data: 19.5% P and 39.6% Ca.

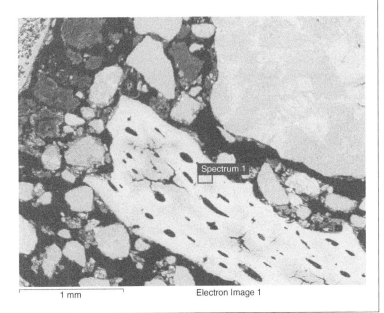

1 mm   Electron Image 1

15.2.2   Inhumations

A combination of soil micromorphology, associated instrumental analyses (microprobe, EDS), and chemical and organic matter fractionation (gas chromatography, liquid chromatography and mass spectrometry) have been employed to investigate and understand better inhumation remains (Burns et al., 2017; Pickering et al., 2018; Usai et al., 2014); a program initiated by Don Brothwell and Raimonda Usai at York University. By using this sophisticated combination of methods, the

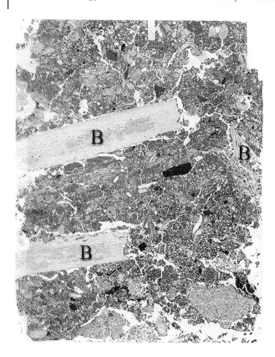

Figure 15.28 Buried bones (B) – Dark Age nunnery soil, Bath, UK. Dark earth soil formed in post-Roman building debris (Jurassic Oolitic Limestone building stone fragments, mortar tempered with limestone clasts and charcoal). "Ancient" bones in this cemetery are set in mature earthworm-worked and homogenized soil. Thin section scan; height is ~80 mm.

authors were able to investigate the chemical relationship between human remains and grave fills, and spatial variations associated with different parts of the body, human tissues, gut contents, and organic chemistry, along with chemical signatures indicative of a wooden coffin, and in one instance suggesting the use of non-native wood. They suggest that this could have been an import or use of driftwood; wood charcoal found in hearths of probable driftwood origin has also been found in coastal sites in Norway, as suggested by their anomalously high chlorine content.

As in boat burials, woody residues may be preserved but in graves are most often construed as the remains of wooden coffins. At the Iron Age site of Bjørnstad (Sarpsborg, Østfold, Eastern Norway), two grave fills were examined (Rødsrud, 2007). In one grave (F551) in a gley soil, a trace of the coffin seen in the field was found to be composed of a very thin line of wood residues associated with vivianite, with 2–3mm above, a thin layer of amorphous phosphate (CaP?) is present (Box 15.1C). Presumably, these are the remains of both the coffin and phosphate derived from the corpse. EDS chemistry measured concentrations of P, S, Ca, and Fe for example. At grave F13, coffin wood and possible body tissues were identified – the latter being autofluorescent under BL; again these elements seems to have been concentrated. At Voldskogen (Rygge, Norway; Kile-Vesik and Orvik, 2020), Grave A132 a likely mineralized human tissue fragment was identified and currently has an anomalous Al-P chemistry, presumably due to leaching of the original calcium content (mean 25.8% Al, 17.6% P, 0.40% S, 0.47% Ca, 2.38% Fe, 0.47% Ba, $n = 7$; max 19.6% P, 0.55% S, 0.64% Ba) (see Box 15.1A, Figure 15.9).

Grave fills can also provide much background on soil environments contemporary with the inhumation. At Bath, UK, the Dark Age burials had diffused into the dark earth (Figure 15.29); the same phenomenon had occurred on the peripheries of Late Roman Southwark (London) and Worcester, and the upper grave cut had been lost through natural soil working (Macphail and Goldberg, 1995). It also appears that the AD 1050–1150 cemetery at Copenhagen's City Hall

Square (Denmark), did not occupy an area of pre-urban arable land. Instead, graves had been dug into waste ground used for intensive urban middening (Macphail et al., 2020b; Stafseth, 2021).

15.2.3 Cremations

Although the presence of calcined bone remains are the most obvious indicators of cremations, such *burned bone* can occur in kitchen hearths, rake-out and industrial wastes without any relationship to mortuary practices (Mallol et al., 2017; Villagran et al., 2017b). Context is important (Angelucci, 2008). Other indicators may include fine size (~0.4 mm) spherules, as at Late Roman Winchester, in deposits with enhanced magnetic susceptibility (Booth et al., 2010). Some of the most obvious cremation contexts occur as secondary "burials" in grave mounds and are calcined bone-rich (Macphail and Goldberg, 2018a: section 11.9.5). Other and more recent investigations of Norwegian Iron Age grave mounds and their surroundings have been carried out.

Sometimes, the thin section may have missed the bone in a suspected cremation deposit, or it has been lost to leaching; however, burned organic and mineral debris were found suggesting cremations at Sebbetåa, Vest-Agder and Voldskogen (Rygge, Norway), while colluvium derived from a grave mound at Blaker kirkegard, Sørum, Akerhus, included two calcined bone fragments. A well-preserved cremation at the grave mound of Botshaug, Eidsvoll, Akerhus, however, shows both calcined bone and vesicular char occur (Box 15.1C). The bone has a typical CaP composition (~35% Ca and 19% P), while the vesicular char/charcoal slag has an unexpected chemistry (5.92–28.6% Al, 0–5.92% Si, 13.9–17.0% S, 1.15–7.52% Ca, 0–2.53% Fe and 33.4–44.4% Ba) – recording an anomalous dominance of relict barium sulfate/barite(?).

15.3 Soils and Clandestine Graves

In any given landscape, vegetation and soil patterns reflect geology and geomorphology (e.g. the catena; Figure 4.5), which, when understood, can provide useful guides to where a clandestine grave may be located. An essential tool is the soil map (see Chapter 15). The legend that goes with the map will also yield valuable information. For instance, some soil types that are, or have been, affected by freezing and thawing (e.g. through periglacial activity) are described as having "patterned ground". The presence of stone stripes, soil stripes, ice wedges etc., needs to be known before ground is probed for graves, because on some shallow soils, these features provide deep soils that could be mistaken for clandestine graves. Ancient tree-throw subsoil features and solution hollows in limestone are other natural features that need to be differentiated from *bona fide* archaeological features or clandestine graves. In addition to soil maps a good idea about landscape morphology can be gained by studying stereo pairs of air photographs, which provides a 3–D view (see Chapter 16). Although these also give information on the vegetation cover to be expected (Wiltshire, 2019), unfortunately most aerial photographs are "out of date" in terms of the vegetation, which could have changed dramatically since the photos were taken, but again this may be important if "cold" cases are reviewed.

Soil maps and guides provide information on soil types, soil depths, and soil horizons that can be expected in any one location or region. Areas with shallow soils can therefore be eliminated from any searches. On the other hand, areas covered by colluvium can be targeted because of greater soil depth. In addition, disturbed ground can be identified by the presence of anomalous subsurface, subsoil material, on the surface. For example, in areas of Spodosols (podzols) where there are very

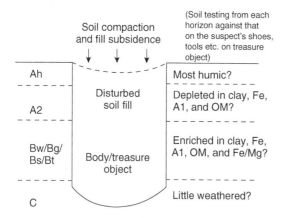

Figure 15.29 Variables affecting testing of clandestine burial/excavation sites. Different soil horizons have different characteristics such as chemistry, clay mineralogy and microfossil remains (e.g. pollen) – a simple "soil sample" may not be specific enough to link soil on the suspect's shoes/tools or soil on a treasure object, to the clandestine grave of treasure site.

strong color contrasts between dark humic topsoils (Ah horizons), gray leached A2 horizons, and ochreous illuvial Bs horizons, the presence of bright reddish and yellow Bs material on the surface is anomalous (see Chapter 4). Equally, the identification of mottled reddish brown and gray gleyed subsoils (Bg horizons) and stony and rocky parent material should be an obvious clue to soil disturbance if these are found on the surface.

A correct understanding of pedogenesis and soil horizons at a crime scene is therefore extremely useful, basic information for the forensic scientist and their "scene of crime officer" (UK – SOCO; USA – CSI) colleagues. If a spade belonging to the suspect exhibits traces of soil, and a link to the crime scene grave is sought, it is critical that more than one soil sample is collected. Someone untrained in soil science may not realize that soils are commonly made up of different soil horizons, and that sometimes these have strongly contrasting chemical and grain size characteristics. A suite of soil analyses can be expensive, but if only a topsoil sample is tested against soil on the spade that originates from digging into subsoil, it is money wasted. Again, the local soil guide will provide information on the types of soil horizon to be found, and the basic chemistry and grain size of these. For example, the Chiltern Hills (Hertfordshire, UK) are formed on Cretaceous Chalk. Their soil cover has been mapped as paleoargillic brown earths and these are bi-sequal in character, i.e. they have formed out of two parent materials – silty loess over clayey drift (Avery, 1964; Jarvis et al., 1983, 1984). The deep subsoil (Ctg horizon) parent material that rests on the chalk, sometimes in solution features, is called Clay-with-Flints, a stony clay partly derived from long-weathered Tertiary Reading Beds. Topsoils and the upper subsoils, however, are formed in a fine loamy loess (aeolian coarse silt and very fine sand – ca. 40–100 μm in size). Thus, the Ah, A2 (Eb) and Bt horizons have strongly contrasting grain size and chemistry compared to that of the Ctg lower subsoil horizon. It can be seen then that sampling of all these horizons is needed (both individual samples and a bulk sample of all horizons) if the soil on the spade is to be properly compared to that of the crime site (Figure 15.30). The same principle applies to studies of valuable artifacts (treasure), which are from unknown provenances. Any soil attached to them has to be compared to possible clandestine excavations just as thoughtfully.

Soil type and its relationship to drainage and groundwater levels, for example as modeled in a catena, can also give some insights into likely body preservation (Burns et al., 2017; Pickering et al., 2018; Usai et al., 2014). Base-rich, well aerated soils generally encourage rapid decomposition of a corpse through the activity of earthworms, snails, slugs and other invertebrates such as beetles and flies (e.g. coffin flies) (Hopkins et al., 2000; Turner, 1987). Poorly drained, cool, clay-rich soils, in contrast, may permit human remains to undergo less rapid decomposition, especially

Figure 15.30 Soils, burials, and canine sensing. Natural waterlogging and gleying can produce foul smells (sulfides and alcohols) that may indicate to a dog that a rotting corpse is present. On the other hand, a real clandestine burial may yield no identifiable smell either to dogs or foxes if a sealing soil layer is present – buried pig experiments; Wiltshire, UK forensic palynologist, pers. comm.; Modified from Wiltshire and Turner, 1999).

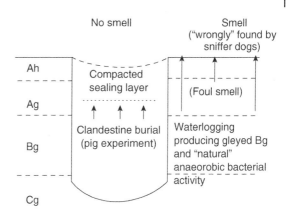

Figure 15.31 How to sample a clandestine grave for geoarchaeological evidence. Five potential sampling locations identified; 1 – control profile; 2–4 – soil micromorphology sample locations to help reconstruct "time" and "climatic" conditions of burial, evidence of trampling, post-burial weathering, and to differentiate "exotic" materials brought in by "suspect"; 5 – surface soil materials picked up on footwear or on trousers (kneeled-upon surface). All samples to be tested (palynology, macrofossils, mineralogy and chemistry) against tools and clothing etc., in possession of suspect (Wilson, 2004; Wiltshire, 2019; Wiltshire, pers. comm.).

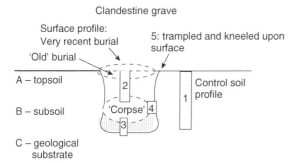

if well sealed. As such they are less liable to be dug up by scavenging animals such as foxes (Turner and Wiltshire, 1999). Dogs are commonly employed by search teams seeking clandestine graves. Likely locations are probed, for example using a narrow-gauge auger, to produce a small opening in the soil. These are left open so that a dog can locate a grave through the smell of decomposition that is generated by a corpse (Figure 15.30).

The presence of poorly drained soils can have a detrimental influence on the effectiveness of dogs searching for clandestine graves (Wiltshire, UK forensic palynologist, pers. comm. 2002; see also Wiltshire, 2019). Heavy textured (fine silty and clayey) soils with poorly draining characteristics can be seasonally waterlogged. Under such conditions, animals and plants can die and then may well undergo anoxic decomposition by obligate anaerobic bacteria, and these give rise to a suite of chemicals (e.g. sulfides and alcohols) that are both foul-smelling and indicative to canines that rotting corpses are present (Turner and Wiltshire, 1999). The study of a soil map will readily help identify locations where seasonal waterlogging may occur, as for example, with the presence of stagnogleys (UK) or pseudogley/gleysols (FAO/Europe) (see Chapter 4). In the American soil classification system, waterlogging is suggested by the formative element "Aqu". Aquepts, for example, can include periodically wet brown soils (Soil Survey Staff, 1999). It is worth remembering that these soils have horizons with the subscript "g" (Ag, Bg, Cg), which indicates that soils are mottled due to the effect of "hydromorphisim" (waterlogging), and ochreous, gray to blue-green colors may be encountered. The strange chemical and biochemical effects of acidic peat water on bog bodies has already been noted. Ground-penetrating radar (GPR; see Chapter 16), as used to carry out pre-excavation mapping, has also been employed to examine supposed cemetery areas, and to locate simulated burials (Cannell et al., 2018; Pringle et al., 2021).

It also must be remembered that the excavation of a clandestine grave has to be carried out meticulously, as in archaeology. Both samples and finds should be even more carefully recorded, as they become potential "exhibits" in any ensuing trial. The proper documentation of bulk and monolith samples is therefore absolutely essential and should follow the protocols of the police force involved.

15.4 Provenancing and Obtaining Geoarchaeological Information from Crime Scenes: Methods and Approaches

The analysis of soil materials has been used as circumstantial evidence in the investigation of crimes. It is difficult to link a suspect absolutely to a crime scene through soil analysis, and more commonly, soil data is employed to demonstrate that a suspect is lying. For example, the suspect may claim that the mud on a shoe or in the wheel arch of the getaway car came from a certain location, but soil analysis may show that this is unlikely to be true. Numerous geoarchaeological methods are utilized in this kind of work. One common major constraint is the small quantity of sample that may be available to work on. In the case of palynology, this is not a problem because palynomorphs may be present as thousands of grains per gram of soil. On the other hand, a muddy stain on the trouser knee of the suspect does not yield 40 gm of soil for standard grain size analysis or 10 gm for bulk chemistry (see Chapter 17). Techniques that require very small amounts of sample (e.g. 0.5 gm or less) that permit essential replicate analyses to be carried out are therefore much more commonly employed (Wilson, 2004). Replicates are required to offset both natural heterogeneity and allow statistical testing, as well as ultimately providing persuasive and defendable evidence in the witness box (Jarvis et al., 2004).

Methods include many that examine non-labile/stabile characteristics in soils. For example, calcium carbonate content ($CaCO_3$) is liable to change because of weathering, but a total chemical analysis of elements (including calcium) may well yield reproducible results. Jarvis et al. (Jarvis et al., 2004) recommend the combined use of two different types of ICP analysis (see Chapter 17), Inductively Coupled Plasma-Atomic Emission Spectrometry (ICP-AES), and Inductively Coupled Plasma-Mass emission Spectrometry (ICP-MS). ICP-AES is efficient in measuring oxides (SiO_2, Al_2O_3, MgO, CaO, etc) whilst ICP-MS is able to measure trace elements at extremely low concentrations, i.e., in parts per million (ppm) rather than as percent. Individual mineral grains can be identified according to their mineralogy, as done traditionally using the petrological microscope, or just as often, now employing X-ray techniques (EDAX, microprobe, *QemSCAN*; Pirrie et al., 2004, 2013) (Chapter 17). Grains can thus be provenanced in the same way as pollen suites.

Grain size analysis on very small samples is carried out using Coulter counting or laser diffraction methods that now can provide amounts of clay- ($<2\ \mu m$) to sand- ($<2000\ \mu m$) size material. Equally, the study of surface texture of quartz grains under the SEM has been successful, because both very small quantities of sample are needed, and because assemblages of surface texture types can be unique to different geological and soil environments (Bull and Goldberg, 1985) (Chapter 2). *QemSCAN* has the advantage of both identifying the mineralogical composition of grains (and soil particles) and recording their shape and size (Pirrie et al., 2004; Pirrie et al., 2013)

15.5 Additional Potential Methods

As yet, soil micromorphology, the study of thin sections of undisturbed soil (see Chapter 16), has been underused in forensic geoarchaeology. This probably stems both from the conservative nature of police and their forensic advisers (who have to justify funding of such work) and the common paucity of sufficient amounts of soil material to be analyzed, especially undisturbed samples. Some interpretations of thin sections can also be equivocal and perhaps not stand up in a court of law, but the clear identification of micro-clues and/or sequencing of grave fills would probably be sustainable as circumstantial evidence (see below). A gravesite can be investigated in the same way as an archaeological feature-fill is studied. Anomalous soil heterogeneity and poorly sorted textural pedofeatures (e.g. silt and dusty clay void infills) are indicative of disturbed ground, as investigated from tree-throw holes (Macphail, 2018: 295–297; Macphail and Goldberg, 1990). Equally, the effects of burial on soils have been investigated experimentally on both base-rich and acid soil types (rendzinas and podzols) from southern England and it is now known how long it takes for both meso- and micro- soil fauna to begin to transform their buried environment at these specific locations and soil types

Lastly, soil micromorphology in archaeology has been successful in identifying trampling of microscopic materials ("micro-clues") into "beaten floors", from a wide variety of anthropogenic and natural sources, and this could be used as a form of provenancing. Examples from Butser Ancient Farm (Hampshire, UK) and late Saxon and Norman London sites, show how fragments of industrial (e.g. iron slag, burned soil), domestic (e.g. burned bone, cereal processing and toilet waste), and animal management activities (e.g. dung and stabling crust) can all be recorded. Many crimes occur in environments strongly influenced by human activity, and just as in archaeological soil micromorphological investigations, there is the potential for these environments to be reconstructed. This is especially the case when carried out in parallel with other techniques employed in environmental profiling such as palynology.

15.6 Practical Approaches to Forensic Soil Sampling and Potential for Soil Micromorphology

If, for example, a grave site was being investigated, there are a number of locations and features that could be usefully investigated through soil micromorphology and other techniques (bulk chemistry, mineralogy, grain size and palynology) (Jarvis et al., 2004; Wiltshire, 2002). Potential pathways to the crime scene site could be sampled. Here, only the surface 1 mm or so could be of interest, if the ground was hard. A sample of loose soil crumbs would be collected from each sampling location, to complement soil for pollen, chemistry, and macrofossils. This loose soil would then be dried and impregnated and then made into a thin section, so that intact soil could be examined at the micro scale and used to try and match intact soil material gathered from the suspects clothing, shoes, car, and tools. If the crime scene is muddy, however, it may be worthwhile taking an undisturbed sample through the uppermost, homogenized trampled layer, including shoe print impressions. This would be the same as taking a Kubiëna box through a tire track where material from the uppermost 10–30 mm track deposit could conceivably have been collected in the suspect's car – on the tires, in the wheel arches, and/or even trampled by the suspect into the car's foot wells (Figure 15.31). Obviously, these samples would also need to be compared to samples from the suspect's normal home surroundings and areas of movement.

At the gravesite itself, loose soil samples should be taken from both the grave as well as the surrounding margins of the grave cut, the probably most trampled part of the crime scene (see Bodziac, 2000). During the excavation of the grave, the most important locations to sample, are (Figure 15.30):

1) the complete soil profile including the topsoil (A horizon), subsoil (B horizon) and parent material (C horizon). This can be accomplished by taking an undisturbed soil monolith, to provide a control profile of the undisturbed soil at the site (Goldberg and Macphail, 2003);
2) the grave fill, from one or more locations;
3) at the junction between the fill and the undisturbed subsoil/parent material. This could permit tool-marked soil to be compared to material on tools used to excavate the grave. It would also allow estimations of weather conditions at the time of grave cutting (e.g. soil wash would be greater if dug when raining). Finally such sampling would help establish the condition of the corpse when buried and possible age of the burial, because body fluids would have likely produced soil staining (e.g. phosphate, iron and manganese features) (Breuning-Madsen et al., 2003; Burns et al., 2017; Pickering et al., 2018); and
4) at the top of the grave fill/new soil surface, where trampling is likely to have occurred.

15.7 Conclusions

Although there has always been an interest in searching out unmapped cemeteries through phosphate mapping and the chemistry of body stains and information that ship burials may contain, there has been very little systematic study of these features until quite recently. This chapter examines the kinds of geoarchaeological signatures that help differentiate excarnations from inhumations and cremations. Of special interest are the associated remains of coffins and other wooden remains such as in exposed grave chambers and boat burials. Identifiable residues associated with the pelvic region of bodies have been of special interest. Such remains, and an understanding of how exposed bodies in excarnations may be scattered and reworked by fauna can be applied to forensic investigations of unburied bodies. and in fact, geosciences are increasingly being employed in forensic investigations of crime sites. If bulk and soil micromorphological analyses are to be useful in this context, however, there are two essential caveats to be remembered. Firstly, it is important to have a grounding in the types of natural geological, pedological and anthropogenic materials of the crime site and its environs, and secondly, because findings have to stand up in court – hence the term "forensic" – sufficient replicates and comparisons need to be made. The new integrated science of environmental profiling is also introduced.

Acknowledgments

The authors thank PEJ Wiltshire, E Wilson, and Hertfordshire Constabulary for their comments on an earlier version of this Chapter, along with discussions with the late Don Brothwell and late Raimonda Usai.

16

Geoarchaeological Field Methods

16.1 Introduction

Both archaeologists and geoarchaeologists face the same fundamental question whenever they investigate any archaeological site or landscape: "where could possible archaeological deposits have formed, what kind of deposits are there, how did they get there, and how can they be found?" Whilst this is a simple question to ask, it is a pivotal question for any geoarchaeological and archaeological investigation, and the computer and field methods used to address this question will ultimately dictate the outcomes from any research or commercial project. As the previous chapters have shown, the distinction between a "natural sediment" and an "archaeological sediment" is not a helpful concept (see Chapters 2, 3, and 13). For example, alluvium can form through deforestation and encase an archaeological site; both the alluvium and the archaeological site contain sediments that have formed as a consequence of human activity. However, the two different types of sediment have developed along different scales, both spatially and temporally, although they intersect at the point of the archaeological site. Therefore, how do geoarchaeologists and archaeologists determine where sediments occur within a landscape or archaeological site? And how do geoarchaeologists and archaeologists reconcile the investigation of sediment units across sites and landscapes that form and exist over different scales?

Ultimately, the deposition of sediments and formation of soils, provide the foundation of all geoarchaeological data. Therefore, in order to investigate any aspect of past human activity, the first step is to identify areas of landscapes and archaeological sites that are likely to contain sediments and features of interest for further study. Today, this first step of investigation is undertaken through desktop computer research, using techniques such as high-resolution topographic data, multispectral data, and aerial photographs. Following this, there is likely to be initial field investigations, such as using cores (boreholes) and/or gouge cores and/or geophysical survey, to define the spatial extent and stratigraphy of major sediment units or archaeological features across a site or a landscape. Some sites then undergo a trial trench study, typically between 1–5% of the total area, for ground-truthing, prior to a more detailed excavation phase. Finally, with identified areas of high potential, the process moves into excavation, whereby the recording of features and the recovery of artifacts is intimately related to the removal of sediments and deposit sequences for laboratory analysis. In many ways, the scales of investigation become more targeted and smaller in extent, as we move from larger datasets generated on a computer, such as topographic maps, through to smaller areas of geophysical survey/limited ground intrusion, e.g. cores, through to targeted excavations.

Practical and Theoretical Geoarchaeology, Second Edition. Paul Goldberg and Richard I. Macphail.
© 2022 John Wiley & Sons Ltd. Published 2022 by John Wiley & Sons Ltd.

It is worth noting that the "modern-day" geoarchaeologist has to master, or at least be conversant with, many sub-disciplines. Understanding how to use GIS systems to map spatial data, selecting the right type of geophysical survey to model sediments, and being able to record and sample sediments in the field are all key skills we need to understand. It is important that the geoarchaeologist heads out into the "field" with appropriate strategies for the project they are undertaking. Such strategies may entail regional mapping of sites or features at a macro scale, modeling alluvial deposits on a cross-section over a valley at a meso scale or they may focus on documenting and sampling the stratigraphy of a site or feature on the meso to micro scale (see Chapters 5–7). These different scales of operations will of course require different field protocols, with different recording and sampling methods employed. Below we consider some of the most common methods used to carry out geoarchaeological fieldwork. This chapter is not designed as a recipe for all field investigations, but rather it provides a flavor of the suite of desktop and field techniques that the modern geoarchaeologist can employ. As always, every landscape, every site, and every feature presents a unique geoarchaeological question and will need a considered and tailored approach to realize its full significance.

16.2 Sediment Stratigraphy Across Time and Space (Deposit Models)

The preceding chapters have demonstrated the huge range of variability that the geoarchaeologist can encounter in terms of *sediment sequences*. However, the formation of sediment sequences, those that vertically accrete will not occur across all parts of a landscape as studied at the macro-landscape scale. Indeed, within any landscape, certain areas will become "sediment receptors", areas where sediments build up over time, even if this process is discontinuous, e.g. a river valley, base of a slope, an archaeological ditch. This can be contrasted to areas of the landscape where deposits are not forming; in these areas, relatively static topographic surfaces are provided by bedrock (e.g. a limestone). This static topographic surface might have archaeological features cut into it or placed on top of it, and surface scatters could become incorporated during soil profile development (Courty et al., 1989). The distinction, then, between sediment receptors and lateral accretion of sediments versus static topographic surfaces is a useful one to start with to categorize a landscape or wider area, as these two different zones provide different geoarchaeological potentials and are suited to investigation using different techniques.

At this point, it is necessary to tie together some concepts of sediments and stratigraphy that we covered earlier (Chapters 2 and 3) and integrate these with the real-world complexity of geoarchaeological sediments. Different scales of geoarchaeological investigation is a key theme and this will provide a central component to the discussion of field methods below (cf. Courty et al., 1989). However, it is also important to consider the size of the sediment receptor and how this relates to the scale of the geoarchaeological investigation, including paleoenvironmental and climatic proxies.

Larger sediment receptors, such as river valleys and associated alluvial deposits, can contain sediments that are linked to human activities in a *catchment* or reach scale (a reach is a section of river with similar hydrologic conditions). In such an example, landscape changes, e.g. human-induced deforestation, can be the trigger for the onset of alluviation, providing a wider perspective of human-environment interaction (Chapters 5–7). Samples taken from such a deposit sequence can link together the analysis of sediments formed through soil erosion caused by tree cover removal, with paleoenvironmental data, such as pollen or mollusk analysis, identifying changes in the vegetation structure (e.g. Fyfe et al., 2003). This can be contrasted with much smaller sediment receptors, such as a posthole from a house on an archaeological site (Reynolds, 1995). In this example, the posthole provides a highly localized deposit sequence, related to intra-site human activities in the roundhouse and its immediate environs. Analysis of this deposit sequence can identify activities undertaken within the house, such as the use of fire, cooking, metalworking, livestock byres, etc. Ideally, geoarchaeologists will analyze a range of sediment receptors across a landscape providing a sliding scale of spatial analyses, linking together regional and site-specific narratives (Macphail et al., 2017a).

Secondly, because geoarchaeologists now employ a range of techniques to investigate wider aspects of sediment deposition across landscapes and sites, there is a requirement to analyze and interrogate the sediment sequences. This facilitates a visualization and assessment of different areas of a landscape or site for its geoarchaeological and archaeological potential. The correlation and stratigraphic interpretation of these sediment units comes under the umbrella term of *deposit modeling* (Carey et al., 2019). Deposit models provide a visual representation of the vertical and lateral distribution of sediment units beneath the modern ground surface whereby, major lithostratigraphic units are correlated in three dimensions to provide a model of the deposit

(a)

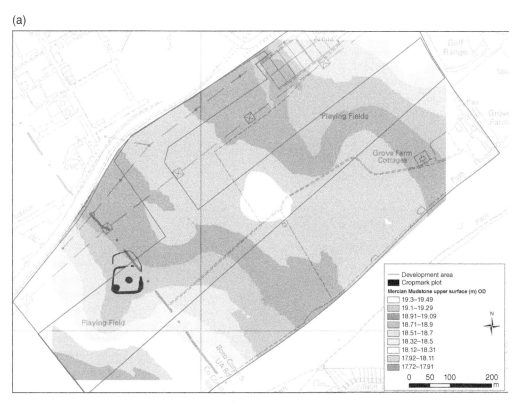

Figure 16.1   (Continued)

(b)

(c)

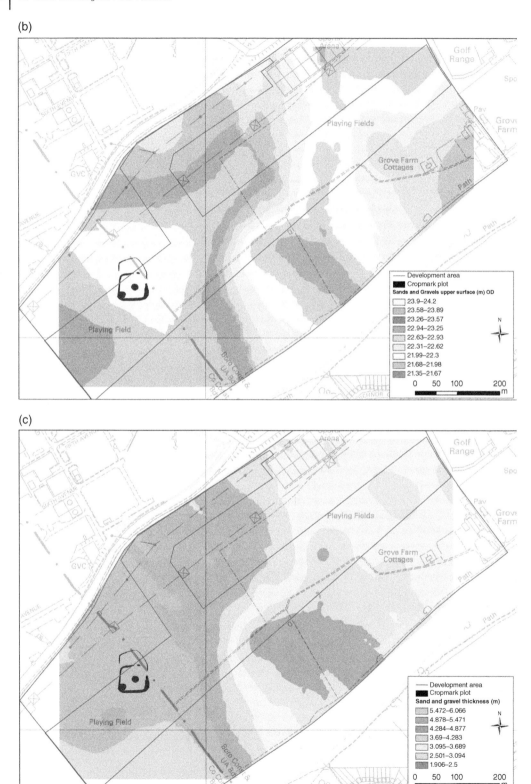

Figure 16.1 (Continued)

(d)

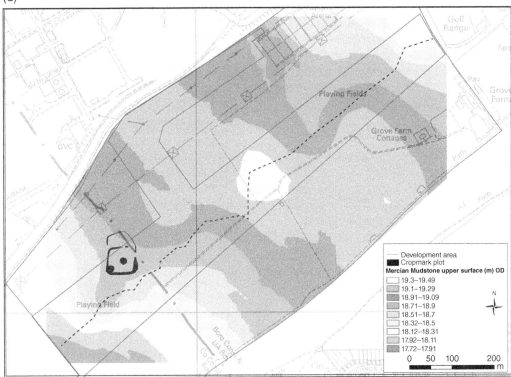

Borehole Cross Section (transect marked as dashed line above)

Southwest   Northeast

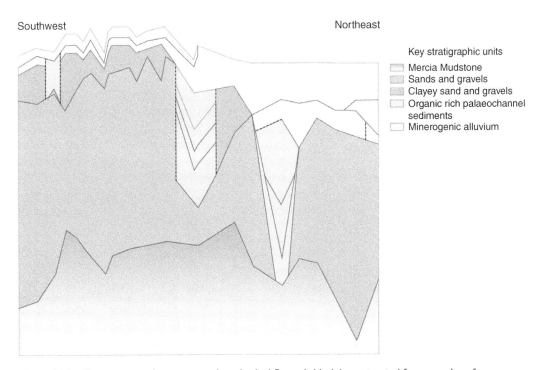

Key stratigraphic units

Mercia Mudstone
Sands and gravels
Clayey sand and gravels
Organic rich palaeochannel
sediments
Minerogenic alluvium

Figure 16.1 Some outputs from a geoarchaeological Deposit Model constructed from a series of purposive boreholes on the River Trent, UK: (a) The modeled surface of the bedrock, beneath the alluvial deposit sequence; (b) the modeled surface of the sand and gravel alluvial deposits; (c) the thickness of the sand and gravel alluvial deposits; and (d) a simplified cross-section through the deposit sequence, defining the major lithostratigraphic units. Note how the enclosure defined from the cropmarks is located on a topographically higher area of sand and gravel close to a paleochannel.

sequences. Such deposit models can be visualized through 2D views of the data, such as plan view interpolated images that map the surface topography or thickness of a key deposit, e.g. a buried land surface on a river terrace or a paleochannel, etc., or by using cross-sections that provide a temporal stratigraphic interpretation of the sediment sequence (Fechner et al., 2014: fig. 12) (Figure 16.1). In addition to landscape sediment modeling, smaller scale examples include tracing various occupation surfaces across a tell or characterizing posthole fills in a long house (see Figures 13.13a–13.13e and 18.19–18.21).

Often a deposit model is constructed prior to the excavation phase of a project, being used to target excavation in areas of high potential (see Carey et al., 2018 for examples). However, the data derived from the excavation can be incorporated into the deposit model, allowing its refinement and increased accuracy. The data used to construct a deposit model can come from older excavations, boreholes/coring, remote sensing, geophysical survey, and more recent project excavations. Consequently, a Geographical Information System (GIS; see section 16.2.1) is often used to house this multidimensional/multitechnique data and to produce a deposit model as a set of surfaces, although alternatives software exists for sediment modeling derived from the geological sciences, such as Rockworks™ (https://historicengland.org.uk/images-books/publications/deposit-modelling-and-archaeology/).

It is now time to turn our attention to the methods geoarchaeologists can employ to investigate landscapes and areas, to maximize the retrieval of geoarchaeological data. This discussion will be framed around the scales these techniques are commonly employed at, working from the macro through to the meso scale, with the micro scale covered in the next chapter (Chapter 17).

16.3 Macro-scale Methods

Multiple methods and data sources can be used to investigate the geoarchaeological record on a landscape scale. Such methods tend to work on a spatial dimension (a XY set of coordinates), although one or more Z dimensions of data are collected (a Z dimension is a measurement of a physical phenomena, e.g. elevation, spectral reflectance, electrical capacity of the ground, etc). Some data sets have multiple Z dimensions of data, such as the collection of *multispectral* data with multiple bands. This "macro-scale" geoarchaeological research has significantly developed during the last 20 years, aided by the adoption of GIS within geoarchaeological research programs. These spatial database management systems, coupled with higher resolution topographic, multi- and hyperspectral data sets, digital regional geological and soil maps, and access to digitally recorded archaeological sites and find spots, means the geoarchaeologist now has a powerful set of data with which to start their investigation.

16.3.1 Geographical Information Systems

Geographical Information Systems (GIS) are used extensively for geospatial analysis in heritage management (Conolly and Lake, 2006) and are powerful software packages that enable the creation of a geodatabase to combine information from vector and raster datasets. They are used to collate, manage, and analyze information of an archaeological site or landscape by combining multiple sources of spatially distributed information with modern and historic mapping. As computing power has increased, it has become possible to incorporate large volumes of data, with a variety of software tools that are now easily accessible. GIS also provides a way to undertake image processing, statistical operations, and subsequently a variety of spatial analyses and theoretical modeling.

Many early applications of GIS in archaeology were concerned with developing predictive models for the distribution of archaeological sites (e.g. Allen et al., 1990). These spatial analyses can provide various insights into the behavior, social organization, and cognitive structures of past human societies. Such techniques have been used to investigate routes of travel, cost functions, site catchment, and resource distributions, as well as the inter-visibility of archaeological sites within a landscape (see Gillings et al., 2020 for a thorough overview). More significantly, for geoarchaeological investigations, these digital and quantitative methods can also provide geomorphological classification of landscapes, facilitating investigation of past human-environmental interactions and the reconstruction of paleolandscapes (Siart et al., 2018). The ability of GIS to integrate archaeological spatial datasets with topographic, remotely sensed, and geophysical data, means these systems have become an essential component of archaeological and geoarchaeological practice (see Box 18.4 for use of *Shape Files* supplied by Cultural History Museum, University of Oslo).

16.3.2 Topographic Maps

Topographic maps are routinely used in geoarchaeological and archaeology projects. They can provide a context for the understanding of site distributions, proximity to certain landscape features, such as plateaux, sheltered areas (e.g. canyons), water sources, types of vegetation, and minerals resources (information available from US Geological Survey (USGS – http://erg.usgs.gov/isb/pubs/booklets/topo/topo.html; http://erg.usgs.gov/isb/pubs/booklets/usgsmaps/usgsmaps.html#Topographic%20Maps).

Generally, the most useful aspect of topographic maps is the elevational data. This is expressed as individual spot elevations (benchmarks), but usually with contour lines of equal elevation. Their spacing reflects the type of terrain, whether steep, with closely spaced lines, or more gently sloping, with large spaces between contours. Latitude and longitude, as well as UTM (Universal Transverse Mercator) coordinates can be read off the map directly or interpolated for other areas. Depending on scale of the map (see below) it can also portray a number of useful features that could be of geoarchaeological interest. These for example include, location of mines, buildings, and roads, as well as important surface features (e.g. levees, dunes, moraines, channels, beaches, reefs, and shipwrecks). Swamps and types of rivers are also indicated, whether they are perennial or intermittent. The location of springs in arid environments can be particularly valuable for locating potential sites.

Maps are available at a variety of scales. In the US, they can be found at scales of 1:24,000 (7.5-minute maps; 1 inch = 2,000 feet; occasionally at 1:25,000 in metric units), the 15-minute maps (1 inch = 63,360; 1 inch = 1 mile), 1:100,000, and 1:250,000 (www.usgs.gov/products/maps/topo-maps). The larger the scale, the less detailed information is portrayed, but the trade off is in viewing a larger area at one time. Also available in the US are other types of maps, such as Orthophotomaps, which are multicolored, distortion-free, photographic image maps. They are produced in standard 7.5-minute quadrangle format from aerial photographs, and they show a limited number of the names, symbols, and patterns found on 7.5-minute topographic quadrangle maps. They are particularly useful in showing areas of low relief.

16.3.3 Digital Elevation Models

Digital Elevation Models (DEM) are visualizations of the land surface derived from a distribution of regularly spaced or gridded elevation values. They typically relate to points recorded in a three-dimensional (XYZ) coordinate system. They are produced using GIS software and are extremely valuable in providing a visualization of topography, most commonly through an interpolated surface displayed with a color ramp. They can be derived from elevation data captured using a total station and GPS or by digitizing topographic maps (Hageman and Bennett, 1999, 2003). However, they are now most commonly derived from high-resolution remote sensed data.

There are two sources of satellite derived elevation data available, which provide near-global coverage and can be freely downloaded electronically from the U.S. Geological Survey Earth Explorer (https://earthexplorer.usgs.gov/). The Shuttle Radar Topography Mission (SRTM) utilizes interferometric radar, which compares two radar images or signals taken at slightly different angles, allowing for the calculation of surface elevation (www.usgs.gov/centers/eros/science/usgs-eros-archive-digital-elevation-shuttle-radar-topography-mission-srtm-1-arc?qt-science_center_objects=0#qt-science_center_objects). This provides data at a spatial resolution of ~30 m for most areas. The Advanced Spaceborne Thermal Emission and Reflection Radiometer (ASTER) also provides a DEM at a spatial resolution of ~30 m, though this is derived from stereo-pair images and is not considered to be as accurate as SRTM data. Nonetheless, both datasets are likely to be too coarse to be of use for the identification of specific archaeological features. However, they can be useful for larger-scale morphological information (e.g. slope, slope azimuth, and curvature), which is fundamental for identifying and mapping larger-scale geomorphological landforms that may be of geoarchaeological importance (Napieralski et al., 2013).

In addition to satellite remotely sensed data, Light Detection and Ranging (*LiDAR*, also referred to as Airborne Laser Scanning (ALS)) is another dataset that is often available for the creation of DEMs (see Figure 14.21). It is an active remote sensing technique, which involves transmitting pulses of infrared or near-infrared light at the ground from an aircraft (Hannon, 2018). The precise location of the sensor is known from an onboard Global Navigation Satellite System (GNSS) and by calculating the length of time the pulse takes to reach an object and be reflected back to the sensor, it is possible to record the location of points on the ground with a high degree of accuracy (White, 2013). These data can then be used to create three-dimensional elevation models, and using the first and last returns, it is possible to create a Digital Surface Model (DSM) and a Digital Terrain Model (DTM) (Davis, 2012).

A DSM can be used when the elevation of buildings or vegetation may be of interest, but for most geoarchaeological applications, a DTM is more appropriate as it provides a "bare-earth" model of the land surface. As such, most archaeological applications of LiDAR are focused on the generation and visualization of DTMs as elevation rasters (Fernandez-Diaz et al., 2014). Such DTMs have been widely used to identify archaeological features from a diverse range of environments and are a baseline data source for many macro-scale geoarchaeological investigations. This is in part due to the increasing amount of publicly available LiDAR resources across Europe and many other regions. This increasing availability has also led to a renewed interest in the study of the topographic context of archaeological sites and landscapes, as well as an integration of geomorphological studies in archaeological projects (Opitz and Herrmann, 2018).

LiDAR is highly effective at mapping geomorphological features through extant topographic variations. In most geoarchaeological applications, a simple diverging color scale differentiates between values at the higher and lower end of the recorded elevation. This enables the LiDAR data to be displayed in such a way that subtle, low-relief variation is more meaningfully represented, allowing the delineation of different archaeologically relevant landforms (Challis et al., 2011). For example, river terraces, paleochannels, and other alluvial landforms are often clearly identifiable within a constrained DEM, even if they are too subtle to be identified by the naked eye. Despite this, more deeply buried landforms, or those that do not display any significant surface expression, will not be identifiable in LiDAR data or may be very subtle.

16.3.4 Remote Sensing

The term "remote sensing" refers to a group of techniques that can be used to observe and provide information regarding the earth's surface. Within archaeology, it is largely concerned with the identification of surface variations that can be related to the presence of subsurface remains. Remote sensing has previously been focused on aerial photographs, but a wide range of other techniques are now becoming more accessible. This includes optical methods such as satellite and airborne multi/hyperspectral imaging and "active" techniques such as LiDAR, mentioned above, and Synthetic Aperture Radar (SAR).

16.3.5 Aerial Photographs

In contrast to satellite images, aerial photographs are taken at lower altitudes, either from airplanes, but also formerly from small balloons or from photo towers constructed at the site (Whittlesey et al., 1977) and more recently through the use of UAVs (unmanned aerial vehicles or *drones*) (Campana, 2017) (Figures 16.2a, 16.2b). Aerial photographs have been used for over a century to depict landscapes, and these early photographs have been valuable in documenting recent landscape modifications and their rates of change, even if they are over relatively short time spans.

Aerial photographs can be classed as either oblique or vertical (Figure 16.2, 16.3 and 16.4; see also Figure 13.6b, Marco Gonzalez, Belize). In *oblique* photographs the axis of the camera is tilted from the vertical, generally 20° or more (Lattman and Ray, 1965). This provides a somewhat artistic feeling for the landscape and relief (the difference between high and low points on a map) of

Figure 16.2a Drone photo of southern part of the Talayotic site (Iron Age) of Torre d'en Galmés, Menorca, Spain (buildings date to 5th century BC–2nd century BC). Visible are different circular domestic structures (forming groups of houses), as well as cisterns and storage pits. The aerial view allows the identification of built areas and open spaces that might have served for circulation, animals. North is to the right (photo courtesy of A. Pérez-Juez).

Figure 16.2b Drone photo showing the juxtaposition of structures from different time periods: the circular building date to the Talayotic period (5th–2nd centuries BC); the rectangular structures were built during the Muslim control of the island (al-Andalus), between the 12th and 13th centuries AD using stones from the Iron Age houses. North is at the top (photo courtesy of A. Pérez-Juez).

Figure 16.3 An oblique aerial photograph of Gibraltar showing a marine Pleistocene abrasion platform (A) and Gorham's and Vanguard Caves (B) along the eastern side of the Peninsula (see Chapter 9).

landscape features, both cultural and natural, are enhanced; the downside is that features in the foreground appear larger than those in the background, with concomitant differences in scale (Figure 16.3).

Vertical aerial or georectified photographs are generally more useful in archaeology and particularly in disciplines, such as geology (Hamblin, 1996; Lattman and Ray, 1965; Marcolongo and Mantovani, 1997; Mekel, 1978; Ray, 1960) and botany, where field mapping is done. This is due to the stereographic properties of overlapping vertical photographs, which can be used to view the ground surface in three (XYZ) dimensions. By using multiple overlapping two-dimensional images, very high-resolution three-dimensional topographic maps of sites and landscapes can be produced (Campana, 2017). An airplane flying at a specific altitude will take a series of photographs so that there is about 60% overlap from one photograph to the next, and about 25% sidelap of photographs from successive adjacent flight paths (Figure 16.4). When two of the overlapping pairs are viewed with a stereoscope, a 3D image of the landscape can be seen.

These stereo images provide a much clearer view of the landscape, including the vertical and spatial distribution of rocks and strata, vegetation, springs, and other important features. It should be remembered that images in simple stereoscopic viewing are vertically exaggerated by an amount that is a function of the degree of overlapping of the photo pairs. The overall scale of the photo depends on the height of the aircraft above the ground, as well as the focal length of the camera. In addition, it should be remembered that within each photo the scale can be somewhat variable since objects at higher elevations will occur at a different scale than those at lower elevations.

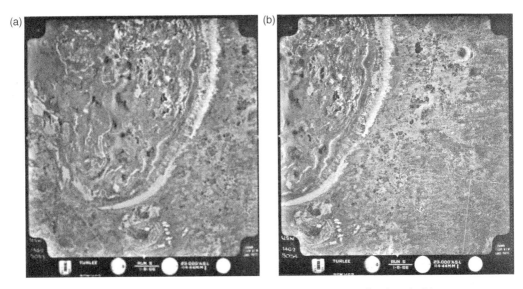

Figure 16.4 Stereo-pair of aerial photographs (a) and (b). These are normally viewed with a stereoscope but it is possible to train oneself to view the photographs stereoscopically without the device. Shown here is a portion of the eastern side of Lake Mungo in SE Australia. The banana-shaped ridge on the right is called a lunette and is composed of a sequence of aeolian sands (both quartz grains and sand-sized pellets of clay blown up from the lake bottom) and paleosols that formed during the late Pleistocene. In the right-hand side of photo (b) weakly developed, vegetated linear dunes can be seed downwind of the lunettes; left of the lunette, the floor present-day lake surface is hummocky. Within the lunette deposits are prehistoric sites, including the earliest use of ritual ochre in a burial (Mungo III), and a cremation (Mungo I); dating is controversial, but recent estimates place both burials at about 40,000 years ago (Bowler et al., 2003) (aerial photograph used with permission © Department of Lands, Panorama Avenue, Bathurst 2795, NSW, Australia, www.Lands.nsw.gov.au).

This variability is of only slight consequence in areas of low relief but can be significant in highly mountainous terrains with great relief. Thus, because of the lateral displacement of objects in an aerial photo, depictions of objects differ from those of maps, which are true-to-scale representations of distances. Consequently, linear distances on aerial photos are not equal as would be the case for maps (Lillesand and Kiefer, 1994).

Interpretation of photographs translates the visual patterns on the photograph into information that becomes useable to the observer. The typical characteristics that are observed include the following (Lillesand and Kiefer, 1994):

- Shape – morphology of objects, some are characteristic, some not.
- Size – both absolute and relative, particularly with regard to the scale of the photograph.
- Pattern – the spatial arrangement of individual entities, including repetition, layout, and ordering.
- Tone and hue – relative brightness or color of objects or materials; different types of soils will have different tones.
- Texture – frequency of change of tones and their arrangement on the photograph, typically by objects too small to be seen individually, as for example a smooth lawn versus a coarse forested canopy.
- Shadows – provide outline and relief, particularly if the photograph is taken at a low sun angle.
- Site – topographic or geographic location, reflecting for example, peaty soils in lowland locations, vs. coniferous vegetation on well-drained upland acidic sands.

Awareness, knowledge, and experience with the above can be applied to geoarchaeological situations, with particular emphasis on geology and soils. In geological and geomorphological mapping, for example, landforms (e.g. alluvial fans), lithology (e.g. basalt vs. rhyolite), and structure (folds and faults) can be identified, as well as mineral resources (Rowan and Mars, 2003). Similarly, drainage networks (both on the surface and partly buried as relics) and topography can be clearly seen, the latter often a reflection of structure and lithology. As shown in Table 16.1, a horizontally bedded limestone may have the following characteristics in an aerial photograph (Lillesand and Kiefer, 1994):

Similarly, deposits associated with glacial, aeolian, lacustrine, etc. landforms can be identified by associations of characteristics. Those for loess deposits, for example are shown in Table 16.2.

Soil mapping (see 16.3.12 below) can be made more efficient and cover larger areas with aerial photos. Most soil survey maps in the US are presented on an aerial photo base. In addition, the use of photos provides insights into moisture characteristics of the soil. With the use of stereo pairs (see above) topographic boundaries such as breaks in slope can then be drawn directly onto the

Table 16.1 Characteristics of horizontally bedded limestone (modified from (Lillesand and Kiefer, 1994: 251).

Topography	gently rolling with sinkholes varying form 3 to 15 m deep by 5 to 50 m wide
Drainage	centripetal drainage into sinkholes; few surface streams; sinking streams
Erosion	gullies with gently rounded cross-sections are developed into clayey residual soils
Photo tone	mottled due to numerous sinkholes
Vegetation and land use	typically farmed except for sinkholes which are usually inundated with standing water

Table 16.2 Characteristics of loess (modified from (Lillesand and Kiefer, 1994), p. 261.

Topography	Undulating mantle over bedrock or unconsolidated materials; tends to fill in pre-existing relief
Drainage and Erosion	Gullies with dendritic pattern and broad, flat bottoms, and steep sides with piping
Photo tone	Light, resulting from good internal drainage; tonal contrast demarcated by gullies and associated vegetation.
Vegetation and land use	Farming

photograph, and plateaux, slopes, and valleys demarcated. The vegetation cover also seems to "stand up", and be identifiable. Such a desktop reconnaissance can be a vital step before working "on the ground", whether it be for an archaeological survey or for a search for a clandestine burial (see forensic chapter).

16.3.5.1 Air Photos in Archaeology

Aerial photos are often useful for seeing things that are simply invisible on the ground. Aerial photography in archaeology has a long history, going back to photographs in England of the well-known site of Stonehenge which was photographed from a military balloon already in 1906 (Bradford and González, 1960). A classical example of early work in archaeology is that of J. Bradford, who mapped over 2000 Etruscan graves at the site of Tarquinia in Central Italy; most of these tombs were not visible at the surface, although some had been previously discovered centuries before (Bradford and González, 1960).

Aerial photographs can be used to discover archaeological sites and to map, measure, and interpret these remains, whilst placing these within their wider context. For sites with architectural features, a number of features show up on photos that aid in their recognition. *Shadow marks*, for example, are produced by differences in surface relief that are augmented by oblique illumination of the sun's rays and are particularly enhanced with low sun angle; they result in shadows being cast on the surface. Such marks can be produced by ditches or embankments, or the construction of monuments, such as the Poverty Point site in Louisiana (Kidder, 2002) or other mound sites in the central and SE USA.

Crop marks and *soil marks* (Evans and Jones, 1977; Jones and Evans, 1975; Riley, 1987) pick out crop responses to soil variations relating to archaeological features. They are best developed:

- in arable areas where there can be a marked summer water deficit (e.g. in the UK, eastern and southern England and eastern Scotland);
- where bare soil marks relate to the occurrence of light colored subsoils (e.g. on chalk);
- where shallow loamy soils have rooting depths between 30–60 cm and crop marks are caused by soil moisture deficit acting with nutrient supply and soil depth.

Generally, *crop marks* are better recorded in cereals than grasses, appearing first in shallow soils when the potential moisture deficit, which occurs when water transpired by a crop exceeds precipitation, is greater than the amount of water in the soil available to the plant. More faint crop marks also occur in wet years, probably because of waterlogging, which may lead to nitrogen shortage, or archaeological features may cut through drainage impedance levels, also producing a crop mark (Bellhouse, 1982). Soil conditions, crop marks, and potential soil moisture deficit have been linked to better plant growth over a ditch, because of deeper soil allowed plants to reach greater available water (Evans and Jones, 1977; Jones and Evans, 1975).

Crop marks are also produced by differences in vegetation related to soil moisture conditions and controlled by subsurface features. Ditches filled with organic-rich material, for example, will tend to favor vegetation growth, resulting in slightly more rigorous vegetation growth than outside the ditch area, which may be rocky and with less available moisture. Similarly, above a buried wall, soil cover may be thinner and consequently, vegetation will be slightly stunted in comparison to adjoining areas where soil cover is thicker. Vegetation may be grass, crops, weeds, or woods.

Soil marks show up on photographs as differences in tone, texture and color and moisture of the soils resulting from human features such as ditches, or heaping of soil. A ditch, for example, may promote the accumulation of darker humus within the depression, serving as a contrast for the surrounding, lighter, less humus-rich soils (Bradford and González, 1960). This combination of crop and soil marks has been applied extensively within the UK to map archaeological remains and features, such as the National Mapping Programme (NMP) (https://historicengland.org.uk/research/methods/airborne-remote-sensing/aerial-investigation/). Finally, other types of constructional features resulting in changes of the landscape or surface coatings on rocks can be visible. Among the most visible and striking are the famous Nazca lines in Peru.

16.3.6 Satellite Imagery

Satellite images are acquired from spacecraft at elevations greater than 60,000 feet (USGS: http://erg.usgs.gov/isb/pubs/booklets/aerial/aerial.html), and as such, they generally provide less detail than that acquired from an aerial photograph taken at a lower altitude. They employ diverse types of sensors that are responsive to different parts of the electromagnetic (EM) spectrum. The successful application of these instruments is dependent on several factors and their application can be hampered by issues surrounding the resolving power of the remote sensing instruments. For example, the scale and spatial resolution of the data, the specific portions and amount of the EM spectrum covered, and the time of year when the image is collected, has a significant impact on the utility of the data and consideration of this is therefore paramount.

Data in digital images are represented as small equi-sized areas (picture elements or *pixels*) in which the relative brightness values of each pixel are indicated by a number (Canada, 2004). Thus, the smaller the pixel size, the greater is the spatial resolution and our ability to distinguish individual objects on the ground surface. It is essentially the size of the smallest object detectable by the sensor. The electromagnetic spectrum embodies types of energy that are characterized by differences in wavelength measured on distance in meters or nanometers (nm) and frequency (waves per second or GHz) (Figure 16.5a). It spans the longer wavelengths of radio waves (e.g. 10^3 m),

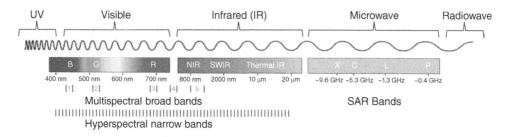

Figure 16.5a Regions of the EM spectrum covered by common remote sensing techniques (not to scale) (drawn by Nicholas Crabb).

down to very short wavelengths of "hard x-rays" (10^{-10} m) used in x-ray diffraction techniques, for example; visible light, infrared (IR) and ultraviolet light (UV) are somewhat in the middle of the spectrum with wavelengths centering between 10^{-4} to 10^{-7} meters.

Many satellite-borne remote sensing instruments use energy derived from the sun, which travels through the atmosphere and strikes the earth's surface or target (Figure 16.5b) (Canada, 2004). Much of this energy is re-emitted to the atmosphere in different wavelengths (and frequencies) where it is collected by different sensors within a satellite (or by different satellites) that detect the different types of electromagnetic radiation. Energy collected by the sensor is transmitted electronically to a

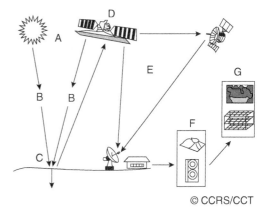

© CCRS/CCT

Figure 16.5b Generalized scheme of remote sensing detection using a satellite (from 2004 National Resources Canada: *Fundamentals of Remote Sensing*).

facility (E) that processes the data into an image that can be obtained as digital data or a hardcopy printout (F). These untreated images can be used or enhanced in order to make observations and interpretations about the objects being recorded (G). It should be kept in mind that certain factors influence this path of energy transmittal and detection. These aspects include, for example, the energy source, scattering and absorption in the atmosphere, energy/matter interactions at the earth's surface (e.g. scattering and reflection), and the type of sensor (e.g. whether passive or active) (Canada, 2004; Lillesand and Kiefer, 1994).

16.3.7 Satellite Multi and Hyperspectral Data

In contrast to aerial photography or panchromatic imagery, which only detects the visual component of the electromagnetic spectrum, multi/hyperspectral sensors are sensitive to emitted or reflected radiation over different wavelengths (Beck, 2011). This can be advantageous as vegetation health/stress variations that may relate to cropmarks and indicate the presence of subsurface archaeological features and geomorphological landforms, are sometimes better expressed in non-visible NIR (near-infrared) parts of the EM spectrum.

Multispectral sensors are "passive" instruments that separate electromagnetic radiation into a small number of broad regions of the spectrum (bands), normally covering the visible and NIR wavelengths, although there are sensors that include the thermal and Short-wave Infrared (SWIR) range. Hyperspectral sensors, on the other hand, co-collect data in many (100+) narrow bands, which can help to improve classification and contrast of features (Figure 16.5a). The theory is that if more of the spectrum is sampled, the detection of features is not limited to those which express contrasts in a smaller group of wavelengths (Beck, 2011).

Despite the potential, there has been relatively limited application of satellite multi and hyperspectral imagery for archaeological and geoarchaeological research. This is largely due to the cost of deploying systems that can provide suitable spatial resolution for the definition of individual features. However, there are now a wide range of satellite multispectral datasets available (see Table 16.3). Some of these are low spatial resolution satellite platforms, such as LANDSAT, ASTER, Sentinel-2, SPOT, and IKONOS, which provide 4–30 m imagery. Many of these systems also provide a higher resolution panchromatic band that enables pan-sharpening of the multispectral

Table 16.3 Specifications of various multispectral and hyperspectral sensors that have been deployed for archaeological research (various sources).

Sensor	Type	Platform	Spatial Resolution	No. of Spectral bands
LANDSAT	Multispectral	Satellite	*LANDSAT 4,5* – 30 m	7
			LANDSAT 7 – 30–60 m (Panchromatic band – 15 m)	8
			LANDSAT 8, 9 – 30 m (Panchromatic band – 15 m)	9
ASTER		Low spatial resolution	15–90 m	14
Sentinel-2			10–60 m	13
Hyperion	Hyperspectral		30 m	220
SPOT	Multispectral		5–20 m (Panchromatic band – 0.8 m)	4
IKONOS			4 m (Panchromatic band – 1 m)	4
QuickBird	Multispectral	Satellite High spatial resolution	2.4 m (Panchromatic band– 0.6 m	4
GeoEye			1.65 m (Panchromatic band 0.41 m)	4
Pléiades			2 m (Panchromatic band 0.5 m)	5
WorldView			WorldView 2 – 2 m (Panchromatic band 0.5 m)	8
			WorldView 3 – 1.24 m (Panchromatic band 0.31 m)	8
Daedalus ATM		Airborne	1.5 m	11
MIVIS	Hyperspectral		3 m	102
AVIRIS			1–4 m	≤224
AISA Eagle			1–3 m	≤244
AISA Hawk			2–6 m	≤254
CASI			0.5–10 m	≤288

bands, though this imagery is still often too coarse to be of use for the detection of individual archaeological features. However, if remains contrast sharply with their surroundings such as for relatively large, linear features (e.g. roads, earthworks and a variety of settlement and structural remains) they are often identifiable. Moreover, broad geomorphological units and variations in geology can also be apparent.

Low spatial resolution satellite multispectral imagery has been used for geoarchaeological data collection, with their low cost, worldwide coverage, and historic value (availability of images dating back to the 1970s) being attractive aspects (Parcak, 2009). Applications include archaeological site prospection in England in which satellite data provided complementary information to that found in aerial photos (Fowler, 2002), the detection and surveying of former patterns of land-use in Spain (Montufo, 1997), and the investigation of geomorphological, paleoecological, and archaeological evidence of mid-Holocene environments and land-use in Yemen (Harrower et al., 2002).

One of the main advantages of low spatial resolution multispectral sensors is that they regularly cover the same area at different times of the year. This is significant as the visibility of archaeological crop or soil marks vary over time. In most temperate regions, cropmarks will be visible only when plants become stressed during dry periods, so the specific time and date that data are collected is of importance. More recently, the Sentinel-2 constellation, which comprises two identical satellites, is designed to give high revisit times of approximately 5 days. Though this is often constrained by cloud coverage, it has been recently used for a range of archaeological applications focused on identifying optimum periods for the detection of cropmarks (Agapiou et al., 2014). It also allows for multitemporal and quantitative analyses of different image enhancement techniques (e.g. Abate et al., 2020). Thus, despite the low spatial resolution, this technology can still be valuable for geoarchaeological research.

As mentioned earlier the costs of higher spatial resolution (0.5–2.5 m) multispectral sensors such as the QuickBird, GeoEye, Pléiades, and WorldView constellations have been a significant barrier to their wider application. However, their costs have been reduced in recent years and are now more affordable, particularly if archival datasets are available. They generally provide a small number of spectral bands (4–8), with at least one covering the NIR. Most studies have, therefore, focused on the utility of the NIR band(s) to emphasize moisture and vegetation changes linked to the presence of small-scale archaeological features (Lasaponara and Masini, 2011). This has shown to be advantageous, even where aerial imagery has been extensively deployed previously (Keay et al., 2014). As such, these systems have significant potential in regions where access is difficult and where aerial imagery is limited or unavailable. Additionally, they can be used to study geomorphological landform assemblages to provide insights into landscape evolution, which can subsequently be related to archaeological and geoarchaeological potential (see Box 16.1).

The only satellite-borne hyperspectral system is Hyperion, which was decommissioned in 2017. It provides 220 spectral bands, each at 30 m resolution. Although the spatial coverage is patchy, it can be used to examine more surficial spectral characteristics. For example, it has been used to search for occurrences of ancient raw materials in Oman (Sivitskis et al., 2018) and a copper mining district in Jordan (Savage et al., 2012). It has also demonstrated some potential to detect sites of Neolithic settlements in Greece (Alexakis et al., 2009, 2011), but its archaeological capability is significantly constrained by the low spatial resolution of the imagery.

16.3.8 Airborne Multi and Hyperspectral Data

The geoarchaeological application of airborne multi and hyperspectral sensors is still relatively uncommon, but as is the case with aerial photography, the lower altitude enables high-resolution imagery. However, these datasets are generally very costly, and their availability can also be quite restricted. In the UK, the Natural Environment Research Council (NERC) Airborne Research & Survey Facility (ARSF, formerly Airborne Remote Sensing Facility) provided the means to obtain data from a variety of sensors, many of which have been used for archaeological research. For example, the Daedalus Airborne Thematic Mapper (ATM) provided 11 bands that cover the visible, NIR, SWIR, and thermal regions of the EM spectrum. The high resolution, and the presence of a thermal band, have been particularly advantageous for investigation in areas where traditional aerial photographs have not always been successful. For example, it has helped better delimit areas of archaeological activity and aeolian sands (see Chapter 6) in the Vale of Pickering, Yorkshire, UK and agricultural landscapes in Norway (Powlesland et al., 1997; Grøn et al., 2005). In addition, there are now a range of other hyperspectral instruments that can be programmed to provide data in the order of 100s of bands at a spatial resolution of ~1 m. This includes the MIVIS (Multispectral Infrared Visible Imaging Spectrometer), AVIRIS (Airborne Visible/Infrared Imaging Spectrometer), CASI (Compact Airborne Spectrographic Imager), and the AISA Eagle and Hawk systems.

The majority of archaeological applications of airborne hyperspectral sensors have focused on the detection of cropmarks through a detailed analysis of vegetation or by trying to identify specific signatures of archaeological features (Luo et al., 2019). However, the amount of technical expertise required to collect and process this data is significant, mainly due to the quantity of data (Doneus et al., 2014). Identifying the most effective band combinations or spectral indices is particularly challenging and numerous computational, or statistical techniques are often required to reduce this (Lasaponara and Masini 2012). Despite this, Challis et al. (2009) have explored the use of CASI data for the geoarchaeological investigation of river valley floors to identify a range of alluvial landforms.

16.3.9 Unmanned Aerial Systems

Some of the shortcomings of satellite and airborne multi and hyperspectral imagery can be overcome with lightweight multispectral sensors that can be mounted on small Unmanned Aerial Systems (UAS) or drones. There are now numerous multispectral sensors available for use on UAS that typically provide data at a spatial resolution in the order of 0.1 m, meaning that imagery can be provided at very high spatial resolution. This heightened sophistication has largely been driven by developments in precision agriculture and ecology, but their low cost and increased flexibility in choosing the timing of data collection is a significant advantage over many satellite and airborne systems (Cowley et al., 2017). However, despite the low cost and increased mobility of these sensors, they generally provide a more limited overall coverage of the EM spectrum as the focus is on a small number of targeted narrow bands. This lower level analysis is less problematic with the development of UAS mounted hyperspectral systems, but this approach has yet to be fully explored in either a geoarchaeological or archaeological context, despite a wide range of compatible sensors now available. As the cost of these systems comes down, there will likely be an increase in the number of investigations that deploy this technology.

16.3.10 Active Remote Sensing Techniques

There are a small number of other remote sensing datasets that can be advantageous to geoarchaeological investigations. In contrast to multi and hyperspectral sensors, these are "active" techniques, which transmit a pulse of energy towards the ground surface and record the amount of energy reflected or returned to a receiver. This comprises LiDAR and SAR, and their capability to produce highly detailed topographic elevations models has already been highlighted (section 16.3.3). However, a further advantage of these sensors is that they also provide a measure of the return amplitude (intensity) of each received (backscattered) echo, which can be related to the reflective characteristics of the ground surface (Höfle and Pfeifer, 2007).

Research undertaken by Challis et al. (2006) explored the capability of LiDAR intensity data to investigate whether the relationship between increased soil moisture, which is known to reduce reflectivity and can be associated with paleochannels and organic-rich deposits, would result in a reduced reflection of the laser pulse. Although it was concluded that the use of this data to predict preservation potential was problematic, there was a correlation between lidar intensity and sediment character. It was also suggested that vegetation cover was the principal cause of the weak response, so it may have significant potential in other non-vegetated areas. Although other applications of this data are rare, further exploration is warranted, as the examination of intensity data will likely assist in the identification of key sedimentary units.

In a similar manner to LiDAR intensity data, SAR also records the amplitude of the reflected (radar) signal. This "backscatter" data can be used to define surface conditions such as texture, soil moisture content, and compactness. Investigations incorporating SAR systems were initially focused on detecting shallow subsurface features, predominantly in desert regions, exploiting the capability to penetrate dry sand and detect relatively large, buried features (Wiseman and El-Baz, 2007). However, despite this potential, the approach is not as commonly used as other remotely sensed datasets, largely because of perceived difficulties in accessing and processing this data, a lack of expertise amongst archaeologists, and because the spatial resolution is generally considered too coarse (Tapete and Cigna, 2017).

In recent years, higher resolution (X-band) systems such as TerraSAR-X and COSMO-SkyMed have become available and are better suited for the study of archaeological sites, despite their limited surface penetration. They are particularly useful for the study of geomorphological landforms and have been used to identify paleochannels, water management systems and other landforms in coastal/intertidal areas (e.g. Bachofer et al., 2014; Gade et al., 2017). Although the advantage of subsurface imaging is significantly reduced where dense vegetation cover is present, the short revisiting times can enable the detection of changes in certain features and be extremely useful for monitoring the condition, or destruction of existing cultural heritage sites through interferometry (Chen et al., 2017). This time series is also of use in a geoarchaeological context as consistently "wetter" or "drier" areas recurring in the same location over a long period can be revealed, which would not be identifiable from a single image (see Box 16.1).

16.3.11 Geological Maps

The use of geological maps is an essential part of geoarchaeological investigations, and as such, should be consulted before, during, and after fieldwork. A remarkable amount of information is available on geological maps that is useful to the geoarchaeologist and many are now available as digital files that can be loaded straight into a GIS, to be used alongside other data types. A geological map displays the types of rocks and sometimes sediments exposed at the surface (although it is noteworthy that some geological maps do not include superficial deposits; these sediments might be recorded in *soil maps*; see 16.3.12 and Chapter 4). However, many geology maps do include both lithified (consolidated) materials, such as bedrock, and looser, normally younger, sediments such as alluvium, colluvium, or aeolian deposits. Information about existing rock and sediment types can be useful in locating possible sources of raw materials for lithic (e.g. chert, volcanic rocks) or ceramic production (e.g. clays for paste and coarser materials for temper). In turn, knowledge of their distribution can be important in determining distances and energies of expense of procuring such resources, and be instrumental in guiding the collection of samples of materials for instrumental analysis (e.g. X-ray fluorescence (XRF); see Chapter 17), as used in resource fingerprinting for example. In the case of the very large collection of obsidian (volcanic glass, commonly black) tools at Arene Candide, Liguria, Italy, neutron activation analysis showed that during the important Mediterranean "colonization" period of the Early Neolithic people relied on obsidian from Sardinia and Palmarola, whereas by the Late Neolithic there was shift to Lipari (Ammerman and Polgase, 1997). Likewise, pXRF of the large sarsen stones at Stonehenge fingerprinted the source of the sarsen stones to West Woods, 25 km away from Stonehenge, allowing the postulation of transportation routes (Nash et al., 2020).

Outcrop patterns of different bedrocks and sediments also provide information about geological structures, such as folds and faults, reflecting the tectonic history of an area. The presence of faults or lithological discontinuities (breaks), coupled with the lithological distributions, can indicate the location of water sources (springs) or zones of mineralization, which could be used, for example, to determine the provenance of ores used in smelting. Furthermore, the orientation of strata and structures, indicated with strike and dip symbols, can be used to project the occurrence of strata with depth. Similarly, patterns of more recent Quaternary sediments, when used in conjunction with soil maps (see section 16.3.12) can reflect relative ages of the deposits and the landscapes associated with them. Most geological maps provide structural cross-sections that portray the earth's fabric beneath the surface.

Included in many maps is auxiliary information such as a booklet that provides a variety of geological details. These items include a geological overview of the area, complete descriptions of the rocks and deposits, earthquake history, tectonic structure and history, regional raw materials, hydrology, and soils. Geological maps are obtainable from national and local geological surveys at a variety of scales that typically range from 1:25,000 to 1:1,000,000. Some are available on the web, including Spain (Instituto Geológico y Minero de España: www.igme.es/), France (Bureau de Recherches Géologiques et Minières: www.brgm.fr/en/website/infoterre), and the UK (the British Geological Society: https://mapapps.bgs.ac.uk/geologyofbritain/home.html). Another example is the Geological Survey of Norway (NGU; www.ngu.no/), providing both geological and sea level information for the Dilling project (see Box 18.4). However, in some countries, geological maps have to be purchased.

It is worth recording here that, although geological maps are an extremely valuable source of information for the geoarchaeologist, they are constructed and utilized at a scale that might not contain the high-resolution data that geoarchaeologists require. For example, Quaternary alluvium might be recorded and mapped as a blanket (undifferentiated) deposit (e.g. "Qal") by a geological survey team, whereas for the geoarchaeologist that alluvial sediment body might contain peat dominated sediments, fine-grained sediments, and paleosols. The details of these different sediment units within the alluvium is essential for the geoarchaeologist in understanding the likely distribution of archaeological sites and features, alongside recreating the landscape history for that area.

16.3.12   Soil Survey Maps and Soil Mapping

Like geological maps, soil survey maps are also an important resource for the geoarchaeologist and should be consulted before, during and after fieldwork, and at the reporting stage. As with geological maps, many soil maps are available as GIS shapefiles and can be readily added to a project database for integration with other data. As already described different types of soils develop in response to the five soil-forming factors (see Chapter 4). Although soil maps can be drawn from remotely sensed data and previously studied areas, they need to be checked and refined in the field. These can be national, regional or agricultural agencies (e.g. British Forestry Commission, US Department of Agriculture) (Countries with their own soil classifications have been listed (Avery, 1990; Bridges, 1990). Other countries have used guidelines from the FAO (Food and Agriculture Organization) or *Soil Taxonomy* (USA; http://soils.usda.gov/technical/manual/; http://soils.usda.gov/technical/fieldbook/) (FAO, 2015; USDA, 2014), although some countries have had soil maps drawn up by external commercial companies (King et al., 1992).

Soil maps are produced using the techniques of soil profile ("pedon") description, sampling and laboratory analysis, in order to produce a soil classification. Soils are then recorded through an auger survey (see section 16.5.1 for augering). A pedon is the smallest identifiable three-dimensional soil unit and a polypedon is composed of one or more contiguous pedons, i.e., a three-dimensional soil body that makes up a specific "soil series" (Buol et al., 1973). Once a soil series has been identified and classified it is then mapped by surveying with auger or trenching. It can be noted here that because of financial and time constraints, a soil type may have only been mapped by two auger borings per square km. As a result, many soil maps use a large amount of extrapolation. For example, the 1983 (and still current) 1:250,000 soil map of England and Wales (e.g. Jarvis et al., 1983) was compiled as a combination of many earlier detailed maps (1:63,360 –"1 inch map") which gave a 20% coverage of these two regions, and new information and testing based upon small pit and auger borings at an average frequency of 250 per 100 km^2 (Avery, 1990: 45).

In both the UK and the USA, soil series are named after the locality where it was first studied or where it is most extensively present (Avery, 1990), and are defined according to "intrinsic properties of practical significance" (Jarvis et al., 1984: 54). Apart from being defined according to soil type (Table 4.2), they can also be defined in relation to substrate type (geology, including organic soils), texture (clay, silt, or sand, or sand over silt, etc.), and mineralogy, e.g. "ferruginous" or "saline", containing weakly or strongly swelling clay, and if stony, composition of the stones, as used for the present soil map(s) of England and Wales (Clayden and Hollis, 1984).

According to the detail required (e.g. local vs. regional study), different scale soil maps can be used. In the first instance, the soil surveyor would probably use a 1:10,000 base map. The detail from this would be transcribed (and simplified) onto a lower scale map, for example 1:50,000 for a "county" soil map. In North America the USDA has an online Web Soil Survey that can be queried to produce a soil map of a particular area (https://www.nrcs.usda.gov/wps/portal/nrcs/main/soils/survey/). Information from individual soil survey reports can be used to create global soils maps at a scale of 1:5,000,000, and only some 18 sheets are needed to cover most of the world (FAO-Unesco, 1988; FAO, 2003, 2015).

The amount of detail and accuracy required to produce a soil map for a geoarchaeological survey is, of course, much greater, and is more likely to be 10 to 20 borings/km^2. Two types of survey can be carried out, namely *free survey* and *grid survey* (Avery, 1990: 37). The first is employed by experienced soil scientists who use variations in topography and vegetation from aerial photography (see section 16.3.5) to infer the location of different soil types that can then be later checked by augering. Soil map boundaries can be sketched in along a break of slope, for example, and then inspected by augering. Within this kind of survey, it may be important to know the variation in depth of the soil profile from plateau top to valley bottom, i.e., a catena profile (see Chapter 4), and an auger transect can then be carried out. The second type of survey, the grid survey, is useful when recording a single parameter, such as soil depth. This can be as detailed as required, from 1 auger per 10 m, 50 m, or 100 m, etc.

As a note of caution, modern soils, their depth, chemistry, and physical properties, may have very little in common with the ancient soil cover, since many of the soil-forming factors may have been different in the past for an area. For example, on Dartmoor, UK, the present-day podzol cover only dates to the post-Bronze Age or Iron Age, before which the soils had the attributes of bioactive brown soils (see Chapter 4). Furthermore, if the modern topsoil appears to be highly fertile and produces good crops of strawberries, such information is only extrapolated to past landscapes and cultures at great peril to the worker's reputation!

One of the major uses of soil surveys is its predictive value in evaluating site distributions and site survey. As an example, we used county soil maps to help guide the survey of a small valley/river system in northern Alabama, USA. The occurrence of Inceptisols (soils with scant profile differentiation) on the flood plain of tributaries to the Tennessee River indicated that older sites (e.g. Archaic or Paleoindian, i.e., > 2 ka) would not be present on the soil surface, but could possibly be found at depth, buried by the young alluvium on which this soil developed. Similarly, remnants of more developed fluvial soils in the valley (mapped as Ultisols and alfisols) indicate older soils and underlying deposits, suggesting that older sites might be found within these fluvial deposits; relatively young sites could naturally occur on the surfaces of these older soils).

In addition, state and county soil surveys, in cooperation with US Natural Resources Conservation Service (NRCS, as part of the US Dept of Agriculture) provide a wealth of potentially useful information including: temperature and precipitation data; lengths of the growing season; wildlife habitats; and chemical, physical, and engineering properties of the soils. Thus, the occurrence of different types of soils can point to certain subsurface conditions, which in the past might have influenced where former settlements might have occurred (Fechner et al., 2004). For example, the presence of aquic soils indicates waterlogging for at least part of the year; therefore, such soils are an unlikely locus for past occupation. Conversely, soil units with good drainage would be much more favorable locales for both past and present settlement. A relationship between sands and loamy sandy soil survey map units and Neolithic settlements has been noted in Belgium (Ampe and Langohr, 1996). Nevertheless, one must be aware that present-day drainage characteristics might not necessarily represent those in the past: modifications can come about by past climate changes, or tree clearance, which has the effect of reducing evapotranspiration, and more recent drainage, can drastically influence groundwater levels (see Box 18.4).

16.4   Meso-scale Methods

This section describes a number of methods for geoarchaeological data collection at the meso-scale. These techniques require the deployment of field teams to collect data, whether through non-intrusive methods, such as geophysical survey or limited intrusive methods, such as gouge coring, and boreholes. The deployment of field teams to collect such data require high-resolution survey systems, typically differential or corrected GPS providing sub-meter accuracy for the recording of fieldwork. These techniques are routinely employed at a scale between the wider (macro) landscape and local (meso-micro) archaeological site.

Meso-scale techniques of investigation include field data collected through boreholes/cores, such as transects across a floodplain providing reach wide sedimentation models and/or geophysical survey across parts of a landscape to identify archaeological sites or areas of local sediment deposition. The deployment of field techniques at this meso-scale is usually guided by the results of the macro-scale methods, identifying areas of interest within a wider landscape that can be targeted for higher resolution data collection.

Often, from a geoarchaeological perspective, techniques applied at this meso-scale aim to provide a picture of sediment deposit sequences (deposit models) at a variety of locales, and linking these together into a more unified picture. Such an example would be to record multiple transects down stream through a river system, documenting the deposition of alluvium across time and space (see Figure 6.23; Gladfelter, 2001). Ultimately, this meso-scale of investigation allows resources to be most effectively deployed at the meso to micro scale, whereby sites and features can be excavated, sampled, and analyzed in the laboratory through techniques such as thin section, chemical, and particle size analyses (see Chapter 17).

16.4.1 Geophysical Survey Methods

Geophysical survey is a non-destructive method used to examine and reveal variations in subsurface sediment architectures, through a variety of measurements of the physical properties of the ground. Within a geoarchaeological context, such data are used to map different sediment units and relate these to geomorphological features, e.g. the extent of buried Quaternary deposits that can contain Pleistocene hominin evidence (Bates et al., 2007). From a more archaeological context, surveys can be undertaken to identify anomalies that have been caused through anthropogenic activities, e.g. pits, ditches, use of fire, etc, allowing the identification and mapping of archaeological sites.

Geophysical techniques can be subdivided in a number of ways, with a common division between those techniques that are active or passive (Table 16.4). However, from a geoarchaeological viewpoint, it is more useful to consider the application of techniques from a perspective of those that have the capacity to model subsurface variations in deeper deposits (>1 m) and those that are routinely used to look at subsurface variation in shallow deposits (<1 m). This division has a direct influence on the reasons for undertaking a survey and the interpretations of the data produced. As a rule of thumb, shallower geophysical techniques are used to map archaeological sites and features, whereas deeper geophysical methods are used to model sediment deposits.

Geophysical techniques that are routinely used to analyze deeper deposit sequences include ground-penetrating radar (lower antenna frequencies) (GPR) and electrical resistivity imaging (ERI); these techniques are used to model larger-scale variations in subsurface sediment architectures and relate these to larger-scale lithostratigraphic sediment units. They can be a useful way to

Table 16.4 Geophysical techniques commonly used in both geoarchaeological and archaeological projects.

Deeper geophysical methods	Depth range	Application	Active or passive
Electrical Resistivity survey (ERI)	Up to ca. 25 m on an 4 m electrode spacing	Create sections through the deposit sequence to model sediments. Using multiple sections to create a 3D solid cube.	Active
Ground-Penetrating Radar (GPR – lower frequency)	Up to ca. 4–10 m with a 50–200 MHz antennas	Create sections through the deposit sequence to model sediments. Using multiple sections to create a 3D solid cube.	Active
Electromagnetic conductivity survey	Up to ca. 3.5 m with a coil spacing of 5 m	Create a map of electrical conductivity contrast across an area, to identify large changes in sediment composition.	Active
Shallower geophysical methods			
Magnetometry	Up to ca. 1.0 m BGL with a sensor separation of 1 m.	Produce a map of magnetic variation across an area to identify archaeological features	Passive
Earth resistance survey	Up to 0.67 m with a 0.5 m electrode separation	Produce a map of relative resistance contrast across an area to identify archaeological features	Active
Ground-penetrating radar (GPR – higher frequency)	Up to ca. 1–2 m with a 400–1000 Mhz antennas	Create a 3D solid cube model to identify and analyze archaeological sites and features	Active

measure depth to bedrock and the morphology of the bedrock surface. Such techniques are frequently undertaken to produce sections through a deposit sequence using transects, although multiple transects can be welded together to provide 3D volumes of data that can be sliced into multiple depths or sections. In addition, electromagnetic survey can be used to provide deeper bulk measurements of conductivity to model sediment compositions. In all cases, it is suggested that the use of boreholes or gouge core data is collected to ground-truth the models constructed.

In comparison, shallower techniques such as GPR (higher frequency antennas) earth resistance survey and magnetometer/gradiometer survey are used to map the distribution of anomalies across a surface and identify discrete archaeological features such as pits, ditches, banks, etc. Such techniques are ideally suited to areas of static topographic surfaces buried by a relatively shallow soil profile and are ineffective for mapping features across deeper sediment deposits. These techniques are undertaken in area (plan) survey, and in the case of earth resistance and gradiometer, no depth dimension is provided, just a bulk measure of the phenomena for the top <1 m of sediment or soil.

It should be noted that these deeper and shallower techniques are not mutually exclusive. ERI has been utilized to produce 3D models of large archaeological deposit sequences such as tells (e.g. Berge and Drahor, 2011), whilst gradiometer survey has been successfully used to map the interface between terraces and paleochannels (Carey and Knight, 2018). However, geoarchaeologists should always consider the reason for undertaking the survey and the sorts of subsurface conditions they expect to encounter. If mapping archaeological features is the goal of the survey, using electrical resistivity with a wider electrode spacing is unlikely to yield results. Conversely, using earth resistance survey to identify features buried under 10 m of peat deposits will also be a waste of time and effort. As discussed by Kvamme (2001), not every technique is appropriate for all situations, and commonly more than one is used in a project or survey, such as the Stonehenge landscapes project (Gaffney et al., 2018). Details on the these techniques can be found in several books and web sites, including Clark (2000), Conyers and Goodman (1997), Gaffney and Gater (2003), Garrison (2003), Herz and Garrison (1998), and Kvamme (2001); the journal *Archaeological Prospection* contains numerous examples of applications of these techniques within archaeological projects (https://onlinelibrary.wiley.com/journal/10990763). In addition, Schmidt et al. (2015) provide an extremely useful resource derived from Historic England guidelines, that includes a detailed overview of the issues to be considered when undertaking or commissioning a geophysical survey.

16.4.2   Deeper Methods Electrical Resistivity Imaging

Electrical Resistivity Imaging (ERI) or Electrical Resistance Tomography (ERT) injects an electrical current into the ground between combinations of four electrodes. Electrodes are placed in transects at set intervals, routinely spaced 1 m, 2 m, and up to 5 m apart, with multiple electrodes set out in any one transect. For example, in the case of the Iris Syscal system, a line of 72 electrodes is used. Along this transect, different combinations of four electrodes are selected by a switching unit or field computer; by selecting electrodes further apart, a great degree of depth penetration can be achieved compared to selecting combinations of electrodes closer together. By ascribing each apparent resistivity measurement to a location beneath the center position of the four electrodes and a depth proportional to their horizontal separation a "pseudosection" can be produced (Schmidt, 2013). These vertical sections of measurement points can then be constructed down through the subsurface, thus illustrating a sediment profile (Figure 16.6a). As the separation between all electrodes is placed at a set distance, e.g. 1 m, the measurement is apparent resistivity,

(a)

(b)
Electrical resistivity section of an alluvial sequence on the River Trent, UK.

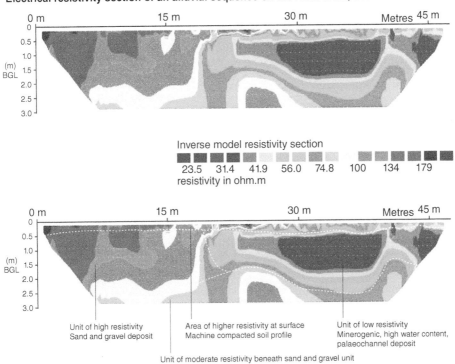

Figure 16.6   (a) A resistivity transect being collected in the field, showing a line of electrodes, spaced at a regular interval, with the switching unit visible, and (b) a resistivity section through an alluvial sediment profile on the River Trent, UK. The difference between the high resistivity sand and gravel deposits and the minerogenic, waterlogged fill of the lower resistivity paleochannel are clearly visible. In addition, the relative depth of the complete gravel sequence is recorded here, demonstrating a relatively thin layer of sand and gravel overlying the Mudstone bedrock.

which is resistance over a set distance, Ohms.m. Like all geophysical data, electrical resistivity data requires processing, specifically inversion, requiring some skill and experience to be able to correctly interpret the data. The parameters of the particular inversion can be altered to suit the data and array being processed and can also incorporate topographic data.

The type of array, the number of electrodes used, and the separation between them dictate the maximum depth of investigation of the survey. The profile allows areas of higher and lower resistivity to be shown in section. Areas of higher resistivity can be related to sediment architectures that have a lower capacity to contain water and conduct electricity, such as sands and gravels. In contrast, sediment architectures that have a higher capacity to retain water and conduct an electric current, e.g. clays and fine silts, are visible as units of lower resistivity. Through producing a section, different sediment architectures and sometimes geomorphological features, e.g. paleochannels, can be interpreted. Often, for geoarchaeological research, it is sufficient to undertake key transects to define variation in sediment deposits across an area (Figure 16.6b).

The acquisition of electrical resistivity data is a time-consuming process and often single transects provide enough information to model large sediment contrasts, especially when the method is complemented with other data, e.g. lidar topographic maps. However, if transects are collected on a regular grid the individual 2D transects can also be combined and processed to give a 3D output. Electrical resistivity has been more commonly employed in geoarchaeological projects in recent years, including the 3D modeling of the Quaternary deposits in southern England (Bates et al., 2007), and modeling alluvial deposit sequences on the River Lugg (Carey et al., 2017).

16.4.3 Deeper Methods Ground-Penetrating Radar Survey (Lower Frequency Antennas)

Ground-Penetrating Radar (GPR) surveys using lower frequency antennas (particularly 50–200 Mhz) have been increasingly used in geoarchaeological research over the last 20 years. Antennas in this frequency range emit a continual pulse of electromagnetic energy (radiowaves) which are transmitted down into the ground profile. The speed of the radar pulse is affected by the types of material it travels through, with different sediments having different RDPs (relative dielectric permittivity). As a consequence, when a boundary is encountered between two sediments with contrasting RDPs, a bounce effect occurs, reflecting some of the radar pulse energy to the receiver. Likewise, when sediment units are encountered with relatively little variation in their RDP, these sediment units are visible in section. The lower the frequency of the antennas the greater the depth of penetration of the radar pulse into the ground, although the actual depth of penetration is governed by the subsurface sediments encountered. Sediments with a high RDP, such as clay, rapidly attenuate the radar pulse, meaning depth penetration and data resolution can be limited. Lower frequency antennas also work at a lower resolution, minimizing their ability to identify archaeological features such as pits and ditches.

GPR surveys also use the collection of transects of data, although frequently multiple transects of data are combined to produce a 3D cube of the survey area. The data can be viewed as either a series of sections, or converted into horizontal depth slices through a velocity correction (see Section 16.4.7). Although radar antennas can be either dragged or towed, survey speeds can be relatively slow. For many geoarchaeological purposes, targeted transects can be used to identify larger-stratigraphic changes beneath the surface and identify subsurface surface geomorphological features, such as land surfaces or paleochannels (Figure 16.7a).

GPR has been very effectively used to reveal the extent of sites buried by masses of volcanic ash, such as at Ceren, in El Salvador (Conyers, 1995; Conyers and Goodman, 1997). The application of

(a)
Ground penetrating radar transect (200 MHz), River Trent, UK.

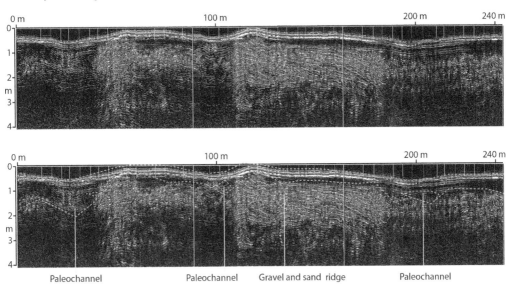

Paleochannel   Paleochannel   Gravel and sand ridge   Paleochannel

Gouge core transect sediment boundary   ------

(b)

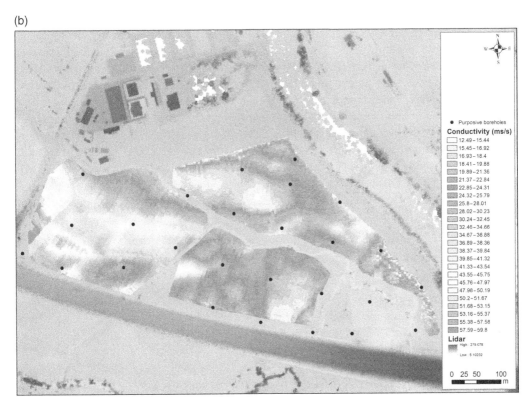

Figure 16.7 (a) A GPR transect and (b) an electrical conductivity image of an alluvial corridor.

GPR for geoarchaeological research is extensively covered by Conyers (2016), showing the application of this technology within multiple types of depositional environment.

16.4.4 Deeper Methods Electromagnetic Survey

Whereas both GPR and ERI require direct contact or insertion into the ground surface, Electromagnetic Induction (EMI) surveys collect indirect readings and are significantly faster to collect. The instruments comprise a pair (or pairs) of coils including a transmitter and receiver coil at a fixed distance from one another. The transmitter coil generates a low-frequency electromagnetic field, which causes eddy currents and in turn induces a secondary magnetic field, which is then detected by the receiver coil (Tabbagh, 1986). This enables the measurement of an in-phase component (or magnetic susceptibility), which represents the ratio between the primary and secondary magnetic fields, and a quadrature component (or apparent conductivity), which represents an average of the conductivity of all components of the subsoil in the measured volume. The depth of investigation is dependent on the orientation and distance between the pair(s) of coils. The further the distance between the transmitter and receiver coils, the deeper the penetration, although this also results in a larger volume being measured, thus reducing sensitivity to smaller objects and thinner sediment deposits.

EMI surveys have been used in geoarchaeological investigations to map aspects of geomorphological variation in relation to paleolandscapes (Figure 16.7b). Relatively wide transects (2–5 m) are often sufficient to reveal differences in soil magnetic susceptibility and ground conductivity relating to changes in sediment composition. In such instances, the use of instruments with a relatively wide coil separation (2–5 m) is more appropriate, allowing for the detection of features and landforms that may be deeply buried. For example, such systems have been used to map a moated site alongside a large marine tidal channel (Simpson et al., 2008), as well areas of deeply stratified sediment sequences containing potential archaeological sites (Bates and Bates, 2000). However, there is an increasingly wide range of instruments that comprise several coil separations that allow data to be recorded from multiple nominal depths. These systems can enable the reconstruction of the morphology of specific landforms such as paleochannels and provide lateral connections between other intrusive borehole investigations (De Smedt et al., 2011). Additionally, data from closer coil separations (and transect intervals) can be used to identify specific archaeological features (e.g. pit/ditches). This versatility of multi-receiver EMI systems allows for the most significant layers from the dataset to be selected, which can be highly advantageous for geoarchaeological research (De Smedt et al., 2014).

16.4.5 Shallower Methods Magnetometry

Magnetometry techniques are commonly applied in archaeology and measure subtle variations in the Earth's magnetic field across the ground surface, with part of this magnetic variation being caused by archaeological remains, such as kilns, pits, ditches, etc. By traversing a surface with a magnetometer, irregularities in the earth's magnetic field (anomalies) produced by these features can be detected (Gaffney and Gaiter 2003).

Magnetometers respond to a wide variety of archaeological features and offer quicker ground coverage, compared to the other shallower techniques (Schmidt et al., 2015). They detect archaeological features through both magnetic susceptibility contrast and through the identification of thermoremanantly magnetized features (evidence of heating events) (Clark, 2003). Magnetometer survey is routinely used to identify cut features such as ditches and pits that have infills (and hence have an increased magnetic susceptibility), alongside inwashed anthropogenic materials such as ashes, wastes from domestic and industrial use. All of these sediment types change the magnetic contrast compared to the surrounding material into which they are cut. Correspondingly, upstanding features, such as mounds and banks, are often identifiable due to the heaping up of sediment or soil, which again provides a contrast in magnetic susceptibility, compared to the surrounding soil. However, structural remains such as walls are normally poorly defined by this technique (mudbrick and decayed mudbrick are the same material), unless they are constructed using material that is magnetically different from the surrounding soil, such as ceramic building materials. Sediments and/or features that have been exposed to heating provide strong magnetic signals, due to the magnetization of Fe within the fired sediments. Features that remain *in situ*, such as a kiln or a furnace, provide an alignment to magnetic north at the point of firing, whilst material that has been removed from its location of firing, e.g. a brick, tile, or rock, will still provide a strong magnetic signal, although the orientation toward magnetic north at the point of firing will have been lost (Gaffney and Gater, 2003; Gose, 2000).

A range of magnetometers is available for archaeological prospection, but the most commonly utilized is the fluxgate gradiometer. With this instrument, two magnetometer sensors are mounted vertically above one another at a set distance (between 0.5–1.0 m apart) (Gaffney and Gater, 2003). Both the upper and lower sensors measure the Earth's magnetic field but will provide different readings relating to the vertical component of this. As the lower sensor is closer to the ground surface, the magnetic signal will be influenced by buried features and objects to a greater extent (Kvamme, 2001), so the difference or "gradient" between the two sensors can be used to detect variation in the subsurface, which may be caused by the presence of archaeological features.

There are several lightweight handheld instruments available that typically comprise two gradiometers separated by a distance of 1 m, thus allowing relatively large area coverage in a short period (3–4 ha per day). Data using these instruments are normally collected as a series of traverses in regular (20–30 m) grids. However, more recently even larger areas can be surveyed in single sessions through using cart-based systems with real-time GPS corrections. These can also be towed by an all-terrain vehicle, allowing for multiple sensor arrays, which can comprise 16 sensors, separated by a distance of 0.25 m. As a result, it is now possible to collect multiple hectares of data per day at a very high spatial resolution. This capability, combined with their ability to detect a wide range of archaeological features, means that is common for commercial investigations to map anomalies related to archaeological features and thereby provide maps of sites and wider areas. There are numerous surveys reported in journals such as *Archaeological Prospection*; examples include mapping buried tombs in Egypt (Odah et al., 2005) to Mesopotamian cities in Turkey (Creekmore 2010), and pretty much everything else in between (Figure 16.8a)!

16.4.6 Earth Resistance Survey

Earth resistance survey injects an electrical current into the ground and the resistance is measured in Ohms. Commonly, four electrodes are used (two voltage and two current) and as the separation between electrodes is variable, only a relative resistance is measured (as opposed to resistivity above). Resistance is mapped across an area, with contrasts in resistance produced by features that provide a greater or lesser resistance to flow of the current (Gaffney and Gaiter 2003; Schimdt 2013). Cut features such as ditches and pits that have infilled with sediments, provide a higher moisture content in comparison to surrounding bedrocks, whereas walls, cists, and features constructed of stone will have a much lower capacity to conduct an electrical current, providing high resistance features.

(a)

(b)

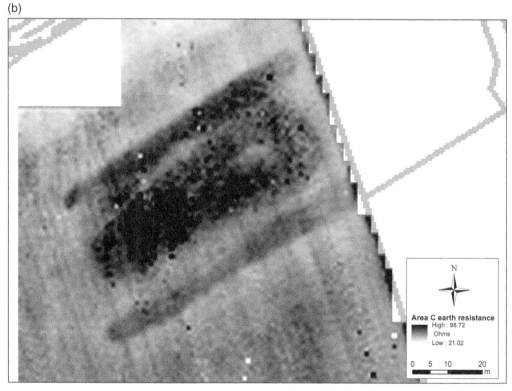

Figure 16.8 (a) A processed gradiometer plot of the Hill of Ward, Ireland, showing an extensive prehistoric and historic archaeological landscape, with a myriad of features, overlain on a lidar topographic hillshade (Image used with permission, Steve Davis, University College Dublin.); and (b) an earth resistance survey of a Neolithic long barrow, at Damerham, Hampshire, UK, clearly showing the mound and ditches (chalk filled, free draining features) as higher resistance features compared to the surrounding chalk bedrock covered by a thin Rendzina soil profile. The identification of ditches as higher resistance features is the opposite of what might normally be expected; however, the survey was conducted during a very wet August and consequently the soils surrounding the ditch were waterlogged, compared to the more freely draining barrow mound and ditches, containing chalk rubble (Data used courtesy of Steve Davis, UCD).

Resistance surveys are best conducted in wetter seasons to enhance contrast among objects and features. A common separation of the mobile electrodes in an earth resistance survey is 0.5 m, which allows an approximate depth penetration of 0.7 m beneath the ground surface. The rate of survey for earth resistance survey is still relatively slow compared to magnetometer survey, so often earth resistance surveys are targeted over smaller areas to define specific features of interest. Numerous earth resistance surveys have been applied to map archaeological features across the world, although they are often undertaken as part of wider campaigns using a suite of geophysical survey methods, due to their slower rate of coverage. Examples range from defining Neolithic causeway enclosures (Schofield et al., 2020) to mapping Victorian Park features (Parkyn 2010) (Figure 16.8b).

16.4.7   Ground-Penetrating Radar (Higher Frequency Antennas)

GPR surveys using higher frequency antennas (ca. 400–900 Mhz) are now widely used to provide a detailed characterization of subsurface archaeological remains. The principles of survey are much the same as for lower frequency antennas (see section 16.4.3), but the higher frequency, or shorter wavelength, provides increased vertical resolution of data at the expense of depth penetration. This provides more accurate detection and definition of archaeological features within the near-surface (0–1 m BGL). However, for higher frequency GPR surveys, often the most valuable information is not contained within the vertical profile of the individual transects, but from the generation of images that connect anomalies from closely spaced transects to produce multiple horizontal "time slices" through the subsurface (Goodman and Piru, 2013). These images not only provide a plan view of an archaeological site at various depths but also enable the visualization of the data as a three-dimensional volume, which can facilitate detailed characterization of specific features (Leckebusch, 2003).

The effective definition of small-scale archaeological features is not only governed by the frequency of the antenna, but also by the horizontal resolution, which is a factor of the density of data points taken along, and distance between, each transect. A higher number of readings distributed along the profile (e.g. 0.05 m) with a close interval spacing of < 0.5 m will result in a high spatial resolution, which allows for the more accurate interpretation of complex archaeological remains such as settlement and structural features (Neubauer et al., 2002). As such, most GPR surveys using higher frequency antennas are undertaken as area surveys over a grid comprising a series of regularly spaced transects.

GPR survey is generally more time-consuming than the application of magnetometry or earth resistance and is often used to target areas of interest at an archaeological site where there may be known archaeological remains. A good example of the advantage of such an approach has been shown by geophysical surveys in the Stonehenge World Heritage Site, where a hitherto discovered long barrow was revealed through a magnetometer survey (Roberts et al., 2018). The addition of a targeted GPR survey over this monument, however, identified clear internal features comprising an oval arrangement of small-scale post holes and pits.

A more recent development in GPR survey has been the application of multichannel radar systems, which enable the simultaneous collection of multiple transects of GPR data. Owing to their larger size, these are normally towed by an all-terrain vehicle, thus vastly increasing the amount of data that can be collected in a single survey session. The system also normally incorporates a Global Navigation Satellite System (GNSS) or a self-tracking total station that provides real-time positioning and large-scale coverage without the need to set up individual grids or survey areas (Trinks et al., 2018). These instruments contain pairs of very closely spaced transmitter and receiver antennas with the same central frequency, typically providing a horizontal profile spacing of <0.1 m. Such rapid data collection has allowed entire archaeological sites and landscapes to be mapped at an unprecedented level of detail. For example, it has enabled the interpretation of entire Roman Cities, such as at Falerii Novi in Italy (Verdonck et al., 2020) and Carnuntum in Austria (Neubauer et al., 2013), and detailed the extensive funerary landscapes at Gjellestad in Norway (Gustavsen et al., 2020) and parts of the Stonehenge World Heritage site (Gaffney et al., 2018).

16.4.8 Interpreting and Using Geophysical Data Sets

It is worth briefly considering how geophysical survey data needs to be processed and how the data from geophysical prospection surveys can be integrated with other geoarchaeological data. Processing of any geophysical survey data is a highly skilled process, with each individual survey technique requiring a set of processing steps, often in bespoke software. Whilst there is no intention here of describing the processing steps for each type of survey instrument and survey type, it is important that the processing of the data provides the best identification of the features and units being investigated. However, after processing, the data still requires interpretation. The old adage, that "the data does not speak for itself" has never been truer than when describing geophysical data sets. Archaeologists need to know how those data sets can contribute to their project, and how the findings can enhance the understanding of a site.

The interpretation of geophysical survey data can be neatly divided between those that require an interpretation as a section and those that require interpretation as a surface or plan view. The interpretation of sections is primarily for ERI or GPR survey and from a geoarchaeological perspective; they are used to investigate sediment sequences, key interfaces between different sediment units and geomorphological features. Through integration with accurate topographic maps, the depths of these interfaces and sediment units can be calculated as below ground level, and point values can be added into deposit models to depict these surfaces, a process referred to as surface picking.

The interpretation of features in plan view is more easily achieved through importing the processed geophysical surface/s into a GIS. These can be depth slices of ERI or GPR data, as well as the surfaces produced from electromagnetic conductivity, earth resistance, and magnetometer survey. For plan view interpretations, a polygon shapefile can be constructed that records the attribute information of the interpreted features, allowing color coded symbology to display the results. These anomalies can record zonation of geomorphological units using EM data or individual archaeological features from earth resistance survey, gradiometer survey, etc.

For all interpretation schemes, there are basic procedures that should be adhered too. Each individual anomaly, whether defining a sediment unit, archaeological feature or object, requires a unique identifier. Often anomalies can be grouped together to define a larger feature, i.e. several sediments filling and defining a paleochannel or multiple anomalies associated with a furnace, etc. Ultimately, the aim of any interpretation is to convey the geoarchaeological and archaeological information from the data, so that a non-specialist can relate a set of interpretations back to a set of anomalies on the processed data plots. Like all aspects of geoarchaeological research, this requires some skill and a lot of practice!

16.5   Coring and Trenching Techniques

A further suite of techniques for the meso-scale analysis of landscapes and archaeological sites is the application of limited intrusion methods into the ground surface. These techniques can be summarized as auguring and coring (including CPT), and trenching techniques. These methods provide a window into the sediment deposit sequences and potentially the types of archaeological remains that reside within a given survey area. These techniques, particularly coring, are often undertaken alongside the deployment of deeper geophysical survey methods. Such techniques provide an invaluable ground control mechanism for the interpretation of these geophysical data sets, and occasionally to ground-truth remote sensing multispectral data sets and identification of topographic features (see Box 16.1). Likewise, coring/boreholes are often undertaken as a survey method in its own right, either to construct a deposit model or to provide cross-sections to allow targeted sampling for paleoenvironment and geoarchaeological purposes.

By undertaking limited ground intrusion fieldwork such as coring and trenching, the vertical and lateral extents, and thicknesses of sediments and archaeological features can be directly observed. Furthermore, information can be retrieved on the texture, composition, and fabric of the sediment units, detail that cannot be achieved by just using the macro-scale techniques. It should be noted that understanding the data gathered from coring is facilitated by having a notion of the types of soils/sediments/features to be expected. This information can be already collected from roadcuts, river/gully exposures, or archaeological excavations.

16.5.1 Augering Techniques

Augering techniques involve inserting a screw-like or gouge-like auger into the ground and bringing material to the surface (Rowell, 1994). The material is generally unconsolidated and accumulates in the auger head; in the case of a screw auger, it adheres to the blades of the auger. Parameters such as color, composition, and texture can be recorded in stratigraphic order (see section 16.6) by removing the sediment retained on the auger blade (Figure 16.9a). Each sediment sample can either be laid out, in the order it was removed from the ground, or put into a labeled sample bag. Moreover, since it is possible to measure the depth that the auger has reached, either by inserting a ruler to the bottom of the auger hole or by marking the depth reached on the shaft of the auger, it is possible to make accurate recordings of the depth beneath the ground surface of a sediment and its thickness. If a sample is required for thin sectioning techniques, ideally it is better to remove these sediments as intact blocks, from a trench section or thicker core (borehole) (see Box 18.1 and below).

Several types of augers are commonly used. The most common, and cheapest, is a posthole digger or hand auger, usually a bucket auger (or open "Jarrett"/posthole auger) (Figures 16.9b and 16.9c), which can be purchased as part of a soil sampling kit. In this case, the auger head is attached to a vertical pipe-like stem with a horizontal handle. The vertical part is commonly composed of threaded steel pipes that extend the augering to depths of several meters. Hand augers are usually 40–100 mm (1.5 to 4 inches) in diameter. Screw-type augers are handy, but can be difficult to remove from the ground (Stein, 1986). In heavy clay soils and sediments it is preferable to try to take a smaller, yet retrievable, sample, compared to more loose sandy soils and sediments, where larger samples are more easily taken (Appendix 1.15.1).

Gouge-type augers can be as narrow as 10 mm or 60 mm for deeper coring. The gouge auger is inserted into the ground and after being rotated, it will remove an intact profile, as long as the deposits are not too loose, wet, or stony. This profile can then be recorded (photographed, drawn), and the material can be used to sub-sample for bulk chemical analysis. This is a common method used when carrying out a phosphate survey, for example, especially when the subsoil horizon is the focus of archaeological soil phosphate (Linderholm, 2003). Conversely, because there is a risk of contamination, gouge augering is seldom used in palynological studies.

A more robust type of augering involves a larger mechanically driven corkscrew-type blade (up to 900 mm/36 inches) mounted on the back of a truck (Figure 16.9e). With this technique much larger volumes of material can be brought up to the surface, and greater depths can be achieved.

(a)

(b)

(c)

(d)

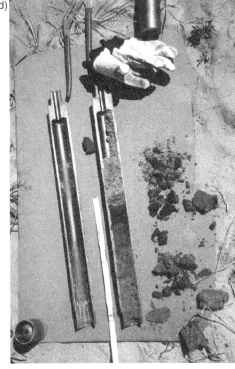

(e)

Figure 16.9 Different types of augering and coring equipment. (a) A large, truck-mounted auger during field work at Zilker Park, Austin Texas; the auger has a diameter of about 1 m. Material is scraped off the blade, described, sampled, and processed by sieving for artifacts or ecofacts. (b) A post hole digger is illustrated in the foreground. This is useful in collecting samples relatively near the surface (~ upper 25–30 cm. (c) A hand bucket soil auger used for examining surface material or for starting deeper boreholes. (d) A split-spoon core is laid open showing sandy material at the top overlying a darker, organic-rich buried horizon associated with the archaeological material dating to the late prehistoric period. (e) A mechanical corer (GeoProbe®) taking a sample next to the Tennessee River in Knoxville (photo courtesy of S.C. Sherwood).

Furthermore, material can be sieved for artifacts and ecofacts, permitting the documentation of potential sites at given depths. This larger auger is more costly to operate, and is usually rented by the day. Again, its use is a balance between budget, as well as surface area and depth to be evaluated. At the Gokstad Viking Ship Mound screw reconnaissance augering was followed up with cylinder coring, the latter providing intact cores for laboratory analyses (Figures 14.15 and 14.16) (Cannell et al., 2020; Macphail et al., 2013a).

16.5.2 Coring (Boreholes)

Coring is a common technique for obtaining intact subsurface information, and like augering, hand and machine coring devices are common. Either technique involves pushing a hollow pipe beneath the surface to a measured depth, and then removing the corer and extracting the material from the pipe. To achieve greater depths, a weighted slide hammer is used to pound the core into the ground. Usually the pipe is lined with a plastic tube, enabling it to be capped at both ends and stored for later analysis (e.g. granulometry, mineralogy, pollen, diatoms, soil micromorphology). Depending on the type of device, core lengths can be up to 1 m; the longer the core, however, the more difficult it is to bring it up from the subsurface.

When sampling peat bogs piston corers or a "Russian" corer can be used. The latter has a half circle section and a thin blade to hold it in position in a soft sediment such as peat. The meter or half-meter-long closed chamber is inserted, then the shaft rotated 180 degrees. This ingenious tool then opens and cuts a half-section out of the soft sediment at the same time as the cutting edge re-closes the chamber. Hence when the corer is withdrawn there is minimal contamination of the sample from nearer surface deposits. This method is very helpful when striving to avoid contaminating pieces of peat that may contain thousands to millions of pollen grains. The half-section of retrieved soft sediment core is then slid into a labeled length of semi-circular drainpipe. The soft sediments of lakes are sometimes very sloppy and difficult to core. In the boreal region, these are best cored when frozen in winter.

Several types of corers are available, depending on type of sediment/soil, intended depth of sampling, and budget. A hand corer is usually 25–50 mm (1 to 2 inches) in diameter, and the coring head can fit on to the same pipe stems used in hand augering. The coring head can be of several types depending on type of sediment. A common type is the split-core sampler, which in which the pipe of the coring head is split in half and held together by a ring at the tip of the sampler. After removal from the ground, the ring is unscrewed from the core barrel and the plastic liner can be lifted out (Figure 16.9d).

More ambitious, and expensive, coring strategies involve mechanical coring rather than manual coring. The smallest is probably a one/two person held machine driven by a two-stroke motor that "vibrates/rams" the corer into the ground. As with augering, the coring device can also be part of a self propelled vehicle, such as those manufactured by GeoProbe® (https://geoprobe.com/), or mounted on the back of a small truck (Giddings corer) or larger drilling rig (Figure 16.9e). In these cases, hydraulic pressure is used to either push or drill the core into the ground, reaching depths of tens of meters, and thus alleviating the difficulty of removing manual cores from several meters' depth. Mechanical coring enables a large area or transect to be sampled during the course of a day. In addition, it may justify the expense of renting a corer.

16.5.3 Cone Penetration Testing (CPT)

A further method that requires consideration for identification of subsurface sediments is CPT. Unlike the gouge and auger methods described above, CPT does not primarily sample the deposit sequence, but instead inserts a conus into the ground using hydraulic pressure from a mounted vehicle. A constant speed is maintained, normally around 2 cm/s, and pressure gauges measure the mechanical tip resistance and sleeve friction. Different sediment architectures provide different resistances to the insertion of the conus; for example an organic peat deposit will have a very low resistance. As data recording is continuous, a very good estimation of sediment properties through the profile can be constructed. CPT is often done by commercial geotechnical companies not for archaeological purposes, but this information can be subsequently used for geoarchaeological research. Recently, Verhegge et al. (2021) mapped the Late Glacial coversand landscape paleotopography from CPT data in Northwest Belgium. They found three organic-rich, stabilization surfaces within the coversands, which were dated to the Late Glacial period, identifying a high potential for the recovery of Late Upper Paleolithic archaeological remains.

16.5.4   After Coring

In the case of using augers and corers, the first step after extracting the sediments is to "log" the cores, i.e., identify the different sediment units, measure their thickness/asl limits and describe them (see section 16.6.5). In addition, the depth of any artifacts recovered should be recorded and located within a specific sediment unit. The cores are often logged in the field, but they can also be removed to the laboratory for logging. Cores that are between 50 and 100 mm wide are also suitable for selective or continuous sampling for thin sections and soil/sediment micromorphology.

After recording the characteristics and dimensions of the sediments recovered, data can be exported into a deposit model to provide a plan view of the depth of a specific sediment unit that is recognized in multiple cores, e.g. a paleosol, or a transect cross-section drawn of the deposit sequence (see Box 16.1). Often, the majority of auger and core samples taken and recorded will be not be used for further analysis. However, typically, cores that have importance, i.e. they are an excellent representation of the deposit sequence encountered, or, they contain a significant deposit sequence, such as paleosol, land surface, or discrete anthropogenic deposit, can be taken to the Geoarchaeological Laboratory for further analysis. When samples are taken back to the laboratory, the cores require cleaning, a photographic record using a digital camera and a detailed log made of the core corresponding to the field recording.

After cleaning the core, but prior to sub-sampling, further analyses of the core can be undertaken prior, such as:

- Running the core through a magnetic susceptibility meter coil, recording χ (Chapter 17) at regular intervals or at the location of visible stratigraphic units.
- Using X-radiography to record details of the fine-scale stratigraphy resulting from deposits rich in "opaques", such as alluvium (Barham, 1995). Measuring of χ and making X-radiographs can only be carried out on cores within plastic containers. Sometimes metal monoliths are used to collect samples, but these can have plastic inserts that are then employed for these techniques (see section 16.7).

After cleaning the core can be sub-sampled at regular intervals (routinely 1 cm), and each individual sub-sample can be analyzed (Chapter 17) for:

- Particle size analysis using a laser particle sizer to provide a quantification of the sediment texture.
- Magnetic susceptibility using a lidded pots with a magnetic susceptibility meter.
- Geochemical contamination and composition of the sediment using techniques such as traditional wet chemistry, ICP-MS, or ICP-AES.
- Calculation of organic and carbonate contents through the use of a muffle furnace.

16.5.5 Trenching

Trenching follows along the lines of augering and coring in revealing the detailed aspects of sub-surface soils and sediments, whilst allowing archaeological features and artifacts to be sampled and recorded. Trenching can be part of test excavations before full excavations are started, or part of an exploratory program to evaluate site limits, or to record the extent of certain types of deposits, or to provide sections for sediment sampling for geoarchaeological or paleoenvironmental analysis.

The simplest trenching involves hand digging test pits that vary in size; a simple test pit is commonly 0.50 × 0.50 × 0.50 m. The material that is removed can be systematically sieved for recovered artifacts, and the exposed profiles can be studied, although their small size might limit the degree of detailed observations. Larger trenches of different sizes can also be undertaken, such as the 1 × 1 m "telephone booth" or 2.0 × 0.50 m "slit trench". For safety concerns, these types of trenches can only reach depths of about 1.2 m without needing shoring up or through stepping the trench sides.

In mitigation or commercial archaeology as part of planning requirements (CRM in the US) where time is critical, mechanical trenching is the method of choice, and many trenches can be dug in a day or two, thus justifying the costs in renting the heavy machinery. A backhoe or 360° excavator are most commonly employed for trenching (Figure 16.10). In the USA, often the machine can be borrowed from the agency that is contracting the work, such as a State Department of Transportation, which has access to this equipment, or hired in with a skilled driver. In all cases, the machine excavation of a trench needs to be done under the supervision of a trained archaeologist. Mechanical trenches can range in length from a few to tens of meters in length, and can reach depths of up to 3 to 4 m. However, with greater depths, the trench must either be stepped for safety reasons to avoid collapse or shuttered, thus adding expense and time to the trenching effort. In restricted areas, such as on urban sites, or in sensitive areas where wide excavations are not permitted, shuttering is required to minimize the limit of the excavation; at Maiden Castle, Dorset, UK, for example, deep trenches were stabilized by steel shuttering, with the 7 m deep ditch sequence accessed through gaps between the shuttering (Sharples, 1991).

Figure 16.10 A backhoe digging a trench at the western end of the mound of Titris Höyük, Turkey. Excavation director Dr. Guillermo Algaze on the left oversees the operation that attempted to establish the stratigraphic relationship of the anthropogenic mound fill on the left with the flat alluvial floodplain sediments at the right and background. Part of the mound is overlapped by the alluvium.

During commercial programs, archaeology can be evaluated through evaluation (trial) trenching, with the protocol typically being 2–10% of the area trenched, using trenches 30 m long by 2 m wide, which are then hand cleaned. However, simple area calculations for evaluation trenching can be misleading, where the trenches are placed over deeper deposit sequences, multiple steps are required to stop the sections collapsing, in order to access the deeper parts of the sediment profile. Such stepping increases the area investigated and the impact on the site, whilst also being more time-consuming and hence expensive to excavate. In comparison, trenches excavated over static topographic surfaces can be quickly and efficiently stripped.

Although trenching is a more intrusive technique, and one which leads to a greater degree of disturbance to the landscape or site being investigated, trenching provides clear and relatively large windows into the subsurface environments. Whilst coring, which is spatially punctuated, can provide lithostratigraphic correlations over large areas, trenching provides a much higher resolution of continuous stratigraphy and greater detail into the nature and variation of the sediments and possible soils. Ideally, the data from trenching is integrated with earlier data sets, such as cores/boreholes and geophysical data within a deposit model, to provide a more holistic interpretation of site and landscape history.

By using trenching it is possible to directly observe lateral and vertical changes of deposits, soils, and archaeological features and place them in their proper stratigraphic context, as is done in site excavation. It is perhaps the best way to obtain large scale, yet detailed stratigraphic information from several places over large areas. The intact section edges provides access to undisturbed sediment sequences, allowing detailed column samples using tins or drainpipes to be collected for laboratory analyses. In some cases, trenching might be the only excavation undertaken on a site or area and consequently it will provide the only window of opportunity to obtain samples for further analysis.

16.6   Describing Soils and Sediments

Although the recording of sediments and soils has been frequently commented on during this chapter, it is worth now considering how such information is documented for cores and sediment sections exposed through trenching and excavation. A large degree of variability exists in the way that stratigraphic information is recorded (Courty et al., 1989: fig. 3.3; Fechner et al., 2004, 2014), and this arises from fundamental differences in ways that profiles are approached (see Figure 3.1).

Geoarchaeologists (and geologists) are trained to observe lithological differences, recording lithostratigraphic units on the basis of color, texture, composition, and sedimentary structures, for example (see Chapters 2 and 3, Stratigraphy and Sediments). Emphasis is placed on depositional criteria, although post-depositional effects (e.g. mottling, rooting, diagenesis, pedogenesis) are included. They tend to be less concerned with soil genesis, although many of the same descriptive criteria are employed. Geoarchaeologists may also tend to look at lateral facies or microfacies, which are critical in understanding site histories (Courty, 2001; Karkanas and Goldberg, 2018b). When dealing with non-open-air sites (e.g. caves, buildings, features) a pedo-stratigraphic approach may be less practical and a simple lithostratigraphic description may be more appropriate and pragmatic.

Soil scientists, on the other hand, tend to view profiles in soil stratigraphic units that center around soil horizons, whether they are actually forming or are relict features that have been buried but are still preserved. They see the section in terms of smaller scale pedological processes, which

Ap

Ep

2Ab

2Bw1b

2Bw2b

3Btb

Figure 16.11 Field photo of profile 2, Hopeton Earthworks site, Ohio (USA), by Rolfe Mandel (see Table 16.6).

for example, strongly affect "surfaces"/topsoils by subaerial weathering or subsoils by down-profile translocation of particles and solutes, or by groundwater movement (see Chapter 4). It is worth noting that the description of sediments or lithostratigraphic units and soils are not mutually exclusive. Paleosols are frequently encountered during geoarchaeological research, constituting a pedogenic profile now preserved as a series of sediments.

The most important point in the documenting of the profile is the stratigraphic coherence of the recording, i.e. although multiple soil horizons might exist within the paleosol they can be covered as one lithological unit, as they represent an extant soil prior to burial, although the individual soil horizons can be recorded within the wider sedimentary unit. The important point is to record such complexities in site formation during the documentation process, so that the descriptions allow a holistic interpretation of the deposit sequence to be made and avoid artificial or nonsensical stratigraphic divisions.

The Hopeton, Earthworks site, Ohio (USA) is a Mississippian period site (ca. 200 AD) in which soil material was mounded above the floodplain. Soil stratigraphic descriptions were done by Rolfe Mandel, Kansas Geological Survey (Mandel et al., 2003). In this recording (Figure 16.11), which is representative of soil stratigraphers, the stratigraphic breakdown (shown in the photo and accompanying table) tends to focus on soil horizons. With this type of descriptive approach, there is a distinct genetic flavor, such the use of "3Btb" that indicates that this indeed is a "buried illuvial horizon."

Note that in Table 16.5, the following criteria are systematically entered (see also Fechner et al., 2004; Hodgson, 1997):

- color, with Munsell designation (the standard is to record the moist color, but dry colors may also be recorded if useful contrasts are noted);
- texture, using soil nomenclatural terms (a geologist might use "clayey silt" for "silt loam" or similar term where appropriate);
- moisture content (e.g. moist);
- structure (e.g. platy; subangular blocky);
- cohesion, compactness, firmness (e.g. soft, friable);
- composition with sizes and abundance (e.g. "common siliceous pebbles");
- roots;
- pores and porosity;
- lower boundary, with degree of transition and morphology (e.g. "abrupt wavy").

At the same time that the deposits are described, key profiles are photographed at various scale (wide shots through to close-ups) and drawn (Hester et al., 1997; Hodgson, 1997). Sediment color and a broad outline of the fine fabric component and its relative abundance should be included, such as silty clay or sandy silt, etc. Inclusions should be included such as clasts, chalk inclusions,

Table 16.5 Profile description (below) of Southern end of wall Profile 2 Hopeton Earthworks Site, Ohio (USA). Note the soil nomenclature, type of information included in the description, and the inclusion of archaeological "stratigraphic units" (see Figure 16.11; photo and description furnished by Rolfe Mandel; see Mandel et al., 2003 for details).

Depth (cm)	Stratigraphic unit	Soil horizon	Description
0–10	Wall Fill B2	Ap	Light yellowish brown (10YR 6/4) silt loam, yellowish brown to dark yellowish brown (10YR 5/4 to 4/4) moist; weak medium and fine platy structure parting to moderate fine and medium granular; soft, very friable; many fine and very fine roots; abrupt wavy boundary.
10–30	Wall Fill B1	Ep	Very pale brown (10YR 7/4) to light yellowish brown (10YR 6/4) silt loam, yellowish brown (10YR 5/4) moist; weak medium platy structure parting to weak fine and very fine subangular blocky; soft, very friable; common siliceous granules; common siliceous pebbles 30–40 mm in diameter; common fine and very fine roots; common fine, medium, and coarse pores; abrupt wavy boundary.
30–55	Wall Fill A2	2Ab	Brown to dark yellowish brown (10YR 4/3 to 4/4) silt loam, dark brown to dark yellowish brown (10YR 3/3 to 3/4) moist; weak fine subangular blocky parting to weak fine and very fine granular; slightly hard, friable; many fine, medium, and coarse pores; common worm casts and open worm burrows; common siliceous granules; common siliceous pebbles 30–40 mm in diameter; common fine and very fine and few medium roots; common fine, medium, and coarse pores; gradual smooth boundary.
55–83	Wall Fill A2	2Bw1b	Brown (7.5YR 5/3) loam, brown (7.5YR 4/3) moist; weak fine subangular blocky structure; slightly hard, friable; common pale brown to very pale brown (10YR 6/3 to 7/3) silt coats on ped faces and in pores; common worm casts and open worm burrows; common siliceous granules; common siliceous pebbles 30–40 mm in diameter; few fine and very fine roots; common fine, medium, and coarse pores; clear smooth boundary.
83–103	Wall Fill A1	2Bw2b	Brown to dark yellowish brown (10YR 4/3 to 4/4) loam, dark yellowish brown (10YR 3/3 to 3/4) moist; weak fine subangular blocky structure; slightly hard, friable; common pale brown to very pale brown (10YR 6/3 to 7/3) silt coats on ped faces and in pores; common worm casts and open worm burrows; many siliceous granules; many siliceous pebbles 2–70 mm in diameter; few fine and very fine roots; common fine, medium, and coarse pores; abrupt wavy boundary.
103–120	Sub-Wall Soil	3Btb	Yellowish brown (10YR 5/4) silty clay, dark yellowish brown (10YR 4/4) moist; few fine prominent red (2.5YR 4/6) mottles; weak medium and fine prismatic structure parting to moderate fine and very fine subangular blocky; very hard, extremely firm; common distinct brown (10YR 5/3) clay films on ped faces; few round pebbles 30–50 mm in diameter in lower 5 cm of horizon.

and artifact positions, etc. Here key stratigraphic boundaries, unit descriptions and the position of sub-samples can be included (Figure 16.12). For exposed sections, which are more common in CRM/survey work, sections are hand drawn again including information on key units, although in some cases data recording sheets, which permit rapid sketches of key strata and features might be used. All units should be assigned individual numbers for inclusion within stratigraphic matrices

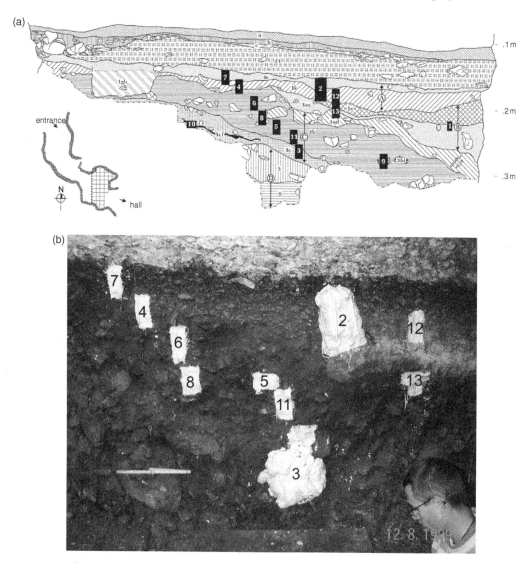

Figure 16.12 Profile drawing (a) and photograph (b) of Profile 2 from Hohle Fels Cave, Germany showing deposits and locations of micromorphological samples. Due to the stony and crumbly nature of the deposits, it was necessary to jacket them in plaster in order to remove intact samples (see Table 16.6 for systematic descriptions of deposit). Shown are Gravettian and Magdalenian silts and clays rich in éboulis (the lighter layer at the top dates from the 1940s). The white squares and blocks are pieces of gauze soaked in Plaster of Paris. In samples 4, 6, and 8 for example, a swath of gauze is placed on the section to stabilize it. After it dries, the sediment is partly excavated around it, leaving a boss of sediment which is then covered with more gauze as shown in samples 2 and 3. This results in an area of sediment that is stabilized on the front and sides. It is then removed from the back and the exposed part is sealed with a final layer of plaster/gauze; for other sampling methods see Goldberg and Macphail (2003).

Table 16.6 Sediment and profile description of Profile 2 from Hohle Fels Cave (Schelklingen, Southern Germany) (cf. Figure 16.12) (modified from Goldberg et aL, 2003).

Geol. units	Munsell color	Field observations	Arch. layers	Archaeological features and comments	Uncalibrated ^{14}C age (AMS in italics)
1. Modern stratigraphic complex					
a	2.5Y 7/4	Cave pavement (Höhlenfestschichten) consisting of alternating banded layers up to 2 cm in diameter; coarse yellow, rectangular gravel, and blackish fine sediment-bearing pebbles.	0		*Post-1958*
b	5 Y 2.5/1	Slag-bearing layer containing pebble chips and few 3–4 cm size pebbles mixed with some finer clayey sediment	0		*1944*
C	7.5YR 3/4	Thin gravel "leveling layer" with rounded pebbles up to 2cm across.	0		*1944*
D1	2.5Y 5/6	Sandy deposit containing limestone gravel with boulders up to 45 cm in diameter. Gravel displays preferred sub-horizontal orientation. Yellow-ochre.			
2. Stratigraphic A – complex (Holocene; below Unit 1 k, pure and typical Magdalenian)					
0c	5Y 2.5/1	Slightly rounded to well-rounded 2–5 cm size limestone fragments lacking preferred orientation and occurring within a dark gray/black argillaceous silt with fine- grained calcareous sand.	0/1	Mixed horizon: containing metal age and Neolithic ceramics and Magdalenian artifacts; disturbed Holocene surface caused by clearing in 1944	
Magdalenian					
1k	2.5Y 2/0	Clayey silt with 1–2 mm size irregularly to slightly rounded limestone gravel, slightly elongated and without recognizable orientation.	I	Magdalenian lithic and organic artifacts	*13,240 ± 110* *13,085 ± 95* *12,770 ± 220*
Pit filling of Qu 76/77/86/87					
1gb	10YR 5/2	Medium gray, very dense clay, with homogeneous distribution of rather big-charcoal fragments. Inclusions of medium sized, angular to slightly rounded limestone.	0	Contains black fired ceramics	

(Continued)

Table 16.6 (Continued)

Geol. units	Munsell color	Field observations	Arch. layers	Archaeological features and comments	Uncalibrated ^{14}C age (AMS in *italics*)
Stratigraphic B-Complex (moonmilk-sediments and limestone gravel)					
1s	10YR 8/2	White, fine calcitic sand with slight amount of interstitial clayey silt; subangular limestone inclusions, 2–5 cm, mostly without orientation.		essentially sterile	
3as	2.5Y 5/6	Very heterogeneous layer. In the northern part, a marked unconformity separates a matrix-rich area from a matrix-poor one.			
		Matrix-rich area: Matrix is pale yellow/brown, pale gray/brown, partly darker brown silt with little clay and fine sand. High abundance of 1–8 cm, angular limestone gravel inclusions displaying irregular orientation, occurring along with very well-rounded, 1cm diameter clasts.			
		Matrix-poor area: Matrix is virtually lacking, and clasts consist mostly of 2–8 cm, and some 10–13 cm of limestone fragments/gravel; these are angular to rounded, and irregular to slightly elongated shape, without orientation.			
Stratigraphic C-Complex (mainly Gravettian)					
3ad	10YR3/2-4/4	Heterogeneous layer with pockets of variable brightness (generated by washing out of finer components) and containing different types of sediments; the darker sediment possibly exhibit an anthropogenic component. The lighter matrix consists of dark gray-brown silt with small amounts of clay and calcareous sand. The dark matrix is blackish silt with small amounts of calcareous sand and a somewhat higher clay component compared to areas of the bright matrix. Inclusions (2–7 cm) are relatively abundant, mostly angular, but some partly rounded; they dip northwards.	I/IIb	Mixed Magdalenian/Gravettian-horizon (reworked and displaced Gravettian)	
3b-complex	7.5YR4/6	Reddish silt with little clay and fine calcareous sand. Inclusions of angular to slightly rounded limestone gravel occur in variable amounts without orientation.	II b	Gravettian lithic and organic artifacts	*27,150 ± 600*

Unit	Munsell color	Description	Layer	Archaeology	¹⁴C ages
3bt	2.5Y2/3	Dark gray, partly black silt, with an abundance of sand-size burned bone (*Knochenkohle*) and a small amount of clay. Inclusions consist of some stones, bones and bone ash (>1cm).	II b	Gravettian lithic and organic artifacts	
3c	10YR5/6	Very moist clayey silt with many limestone clasts of variable size. They range from angular to rounded, and dip towards the NE.	II c	Gravettian lithic and organic artifacts	29,550 ± 650
3cf	5Y 2.5/1	Black clayey silt with a high proportion of bone ash that underlies a limestone-rich gravel layer.	II c	Gravettian lithic and organic artifacts	28,920 ± 400

Stratigraphic D-Complex (mainly Aurignacian)

Unit	Munsell color	Description	Layer	Archaeology	¹⁴C ages
3d	10YR 4/4	Red-brown argillaceous silt with ca. 60% of 4–8cm, slightly rounded limestone gravel inclusions. Matrix only slightly compacted, contains abundant small pores and cavities.	II d	Aurignacian?	29,560 + 240/-230 30,010 ± 220
5	7.5YR 4/4	Generally loose mixture of angular limestone gravel and moist clayey silt with a minor component of coarse calcareous sand.	II e	Aurignacian lithic and organic artifacts	
6	7.5YR 5/6	Rounded limestone gravel with little interstitial matrix. Mostly coarse-grained calcareous sand, downward-facing edges or fracture planes of limestone fragments frequently bear gray clay crusts.	III	Aurignacian lithic and organic artifacts	30,550 ± 550 31,100 ± 600

Notes on descriptions:

1) These are provisional descriptions based on unpublished work by J. Hahn (12.09.1994); Waiblinger (1997); N. J. Conard and T. Prindiville (02.09.1997); and P. Russell and N. J. Conard (09.98). The geological units (GH in original publication) and archaeological layers (AH in original publication) serve as preliminary field descriptions and are subject to modification during future excavations.

2) Colors were determined using the Munsell Soil Color Chart on moist samples, both in the cave under artificial light conditions and outside the cave in natural light.

3) Differentiation of individual sediment units on the basis of their calcium carbonate contents was not possible due to the large number of limestone clasts.

4) ¹⁴C ages are uncalibrated and compiled from Conard and Bolus (2003), Conard and Floss (2000), Hofreiter et al. (2002), and Housley et al. (1997).

and when the recording is being done in conjunction with archaeological excavation, the context numbers assigned by the archaeological excavators need to be applied to the sections.

In addition, the positions of sampling locations for paleoenvironmental proxies, sediments, and dating need to be clearly marked on the section. Often a project will have a set protocol for field recording, although in other circumstances it is down to the individual geoarchaeologist to decide on the recording process. In all cases, the more information that is recorded at the time, the better. Of course, like always, there is no substitute for experience; however, basics concepts of sediment description, pedogenic processes and stratigraphic coherence must be adhered to.

16.7 Collecting Samples

After the sections have been described, it is possible to proceed with sampling. There are many different strategies for the collection of samples; the method employed will depend on the overall goals of the project, the specific questions being asked, what types of analyses are envisioned (grain size or DNA) and on the budget(s). However, two types of strategies are commonly used: (i) the collection of bulk samples from sediment units, and (ii) the collection of intact sediment deposits using plaster of Paris bandages or tins.

The collection of bulk samples allows sediments to be analyzed for a variety of physical, chemical, and biological parameters. Many of these samples are for classical pedo-geological analyses, including granulometry (grain size analysis), pH, carbonate content, heavy minerals, organic content, magnetic susceptibility and geochemical analysis (see Chapter 17; Courty et al., 1989). To collect these bulk samples, both random and systematic sampling techniques can be employed.

For example, it is common to systematically sample an exposed section from bottom to top, collecting representative samples from the principal stratigraphic units. Samples are taken from the bottom upwards, which avoids cross contamination of samples, as any disturbed material will fall downwards onto the previously sampled section, not onto the section above that is yet to be sampled.

Bulk sampling can also be undertaken when deposits or features are scattered throughout a site or area, for instance taking spot samples from the major units. When taking bulk samples, contacts between stratigraphic units are to be avoided to preclude mixing of deposits of different lithologies and depositional events. This is particularly important when collecting samples of microfossils, heavy minerals or for chemical elemental analysis, where just a trace of "soil contamination" may produce misleading results. In addition, the sampling trowel or other tool must be carefully cleaned for each sample. In some cases, it is recommended that plastic trowel be used for sampling for magnetic properties.

The other sampling strategy involves collecting intact, undisturbed samples for continuous subsampling and micromorphological or microprobe analysis (for details see Courty et al., 1989; Goldberg and Macphail, 2003). Such samples can be problematic to collect, as commonly archaeological sediments are loose, poorly sorted mixtures of coarse and fine material; imagine trying to sample a shell midden in one piece!

Fortunately, years of experience and the benefit of getting older(!) have led to several successful approaches in the removal of undisturbed samples. Strategies of sampling vary according to type of sediment, but considerations of leaving unsightly and unstable large holes in witness sections should be considered. The easiest sampling involves compact sediments or soils, such as loess. In this case, blocks of sediment can be cut out of the profile using a knife and trowel, and then wrapped with toilet paper/paper toweling and tape, labeling the sample number and way-up with an indelible marker (Figure 16.12b and Figure 16.13a).

(a)

(b)

Figure 16.13 Strategies for collecting micromorphological samples. (a) Consolidated or firm samples can be simply cut out of an exposed profile and wrapped with paper and packaging tape. A column of stratified burned material (bones, charcoal, ashes) is being wrapped up by P Goldberg with paper towel and tape; cave of Pech de l'Azé IV, (Dordogne, SW France)(Goldberg et al., 2018/ with permission of HM Government of Gibraltar). (b) Sampling of colluvial sediments and paleosol on Dartmoor, UK, using a drainpipe. The sample is marked on the section and material on the side of it is removed, before placing the drainpipe over the sample and removing the sediment sample.

Similarly, metal or plastic boxes (so-called Kubiëna tins) and plastic downpipe (sometimes with its back removed) can be inserted into the profile, removed and covered to ensure that they remain intact (Figure 16.13b). It may look neat to have a carefully lidded box, but it may better just to have the lid on the outside face (where it can photographed *in situ*), and plenty of soil/sediment on the inside face after digging out, as this can then be sub-sampled later for bulk samples. The Kubiëna-box sample, with one lid on, and protruding soil on the "back" can easily be wrapped with plastic film. Sometimes, an unexpected large stone or artifact may also protrude out of the "back" of the sample, and any attempt to remove this will destroy the sample. Again such an unwieldy sample can be wrapped, and then supported by additional wrapping in a couple of plastic bags. Obviously, such samples have to be transported and stored carefully.

If fine-grained cohesive sediments are being sampled, such as alluvium, overlapping metal monoliths can be hammered into the section. If these samples are then going to be used for X-radiography, however, they will require plastic inserts to collect the sample. A low-cost alternative to metal monoliths for the sampling of sediments is to use square plastic drain pipe, with one side sawn off. A column of sediment is first exposed in the section and the drainpipe is eased onto the sediment (Figure 16.13b; asma/Online_Video_1.1.md at master · geoarchaeology/asma · GitHub). This low-cost technique provides high quality samples and provides a cheaper alternative to the metal tins described above, being successfully used on sandy deposits in caves in Gibraltar and South African (Goldberg, 2000b; Goldberg and Macphail, 2003; Macphail and Goldberg, 2000), to colluvial and paleosol deposits on Exmoor (Carey et al., 2021).

However, more demanding sediments to sample are coarse deposits. Here, the best strategy is to encase the sample in a plaster jacket (Goldberg and Macphail, 2003) (Figure 16.12b). Strips of gauze infused with plaster of Paris (commonly available in pharmacies or online) are dipped in water and applied to the exposed surface to stabilize it. Once hardened deposits can excavated from around the sides, leaving a raised bulge above the profile surface, which can also be covered with plaster bandage. The sample, now surrounded on three sides, is detached from the profile, and the remaining surface is covered, and when dry, can be transported.

For all methods of sampling, the sediment unit boundaries, the sample number, site name and context numbers can be written on the sample container, before extraction and wrapping in clingfilm and black plastic (bin bags). Samples should be photographed *in-situ* before extraction to provide the context, and a record of the sampling location and process; detailed notes should be made at the same time on the sample location, the sediments sampled and the reasons for taking the sample.

16.8   Integrating Samples and Proxies

It is crucial to correlate all geoarchaeological sampling with other paleoenvironmental sampling, the recovery of artifacts, and the relationship between the geoarchaeological samples to the archaeological contexts (Macphail and Cruise, 2001). It is useful to repeat the fact here, that if samples are too far apart they can tell different stories even though they come from the *same context*. Although this might be appreciated better on occupation sites, because of the heterogeneity and numerous lateral changes in microfacies (Courty, 2001), it can be true of many sediments and soils, which will all demonstrate some degree of lateral variability. For example, pedogenesis is influenced by small-scale differences in drainage or vegetation cover, and by subtle, localized anthropogenic impacts, and by variations in rates and types of sedimentation.

When considering the retrieval of geoarchaeological samples, it is often the case that a single sample will be sub-sampled and split between multiple researchers from different sub-disciplines. Thus, a core or monolith sample can provide samples for microfossil investigation (pollen, diatoms, ostracods, mollusks), bulk analyses (physical and chemical studies), with enough remaining undisturbed material for micromorphology thin section analysis. When separate samples are taken for different proxies and analyses, it is imperative that the samples are accurately recorded, with appropriate contextual information. If such an approach is carried out, information from all the sub-disciplines can be viewed together and enhance a *real* interdisciplinary investigation (see Box 18.1). Some current research projects are using micro-contextualized samples from impregnated thin section blocks to study organic residues, for example, (Jambrina-Enríquez et al., 2019; Lambrecht et al., 2021; Leierer et al., 2020).

When setting up an interdisciplinary program, it may be worthwhile to carry out further analytical methods in addition to those done routinely. The palynologist, for example, may wish to know the ratios between organic and mineral deposition in a peat deposit, and so high-resolution organic matter analysis could be carried out to match the pollen sampling intervals. In turn, a palynologist may be willing to look at minerogenic layers that are of more interest to the geoarchaeologist, rather than simply concentrating on the most organic and probably most polleniferous horizons (Champness et al., 2012; Macphail et al., 2007a). In some cases, the potential of samples for an interdisciplinary study is not realized until a thin section is observed and fossils are found, overcoming to some extent the absence of a bulk sample for macro- or microfossil analysis, which was not collected through some oversight.

In fact, charcoal, diatoms, nematode egg, and palynological data have been readily revealed from the thin sections themselves (Goldberg et al., 1994; Macphail et al., 1998; see also various chapters in Nicosia and Stoops (2017)). Although such observations of fossil remains commonly have no statistical value *sensu stricto*, they nevertheless can provide valuable data on the sedimentary environment represented by the specific fossil materials. In traditional sedimentology *all* fossil assemblages would be analyzed in terms of their depositional environment (Reineck and Singh, 1986). This happens less often in archaeology, where fossils are often recovered (along with artifacts) and analyzed without close regard to the geoarchaeological origin of the sediments in which they were deposited. Thin section analysis early on in a project can therefore often guide ensuing environmental studies.

Figure 16.14 provides an example of recording at North Farm, Avebury, UK. Here the digging of test pits along the corridor of the River Kennet was being used to sample sediments and date the sequence of alluviation in the valley and colluviation at the floodplain edge. In this sequence, key

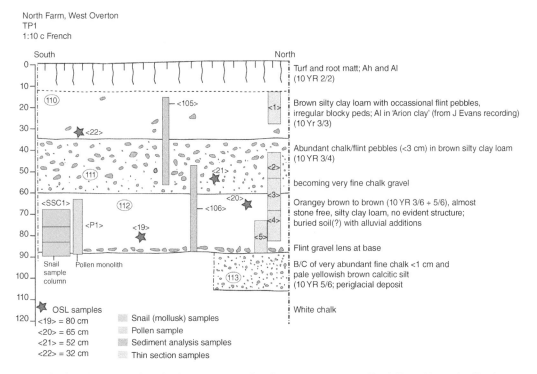

Figure 16.14 The geoarchaeological recording of sediment sequences at North Farm (drawn by Charly French). Here a range of samples and proxies are being taken from the same section. The detailed stratigraphic recording, combined with the precise positioning of the samples for different proxies, has provided a coherent analysis and interpretation of this sequence.

sediment units are drawn and described, alongside the positioning of sampling tins for further sediment analyses, paleomolluskan samples for environmental reconstruction, small tins for thin section analyses and optically stimulated luminescence samples, to date the sequence of sedimentation. It is also worth highlighting that layer (112) is a paleosol recorded at the base of the sequence as a single unit, although the removal of soil micromorphology sections has allowed confirmation of the soil structure and type and definition of the individual soil horizons. Above this, a predominantly colluvial unit (111) and alluvial unit (110) have been deposited. This example shows the coherent recording of multiple sampling types for geoarchaeological and paleoenvironmental analyses coupled with the pedogenic and sediment features of the section.

Finally, during our geoarchaeological experience over the past decades, various issues and problems have been met with; some general points from these experiences is worth emphasizing (see also Goldberg and Aldeias, 2018):

1) Apart from of what technique is actually used, the *exact* location of the sample must be noted: its position in the field should be indicated on a photograph, as well on a section drawing. If total station/GPS recording of artifacts and features is being done at the site, the sample should be shot in, and the ID no. of the sample (or equivalent unique identifier) should be noted by the sampler *and* the archaeologist.

2) Micromorphological samples are always given a site code. However, when a specialist has a cooler filled with samples from many sites over several years awaiting the go-ahead for analysis, such site codes can be difficult to sort through. It is therefore useful to at least append a recognizable site name to samples and maintain a digital database of site codes and names, sample numbers, and *the reasons for taking each sample.*

3) During sampling, good communication is essential with the director of the site and area supervisors in order to ensure that samples will address archaeological issues. Such communication also ensures that sampling for radiocarbon dating, snails, pollen, seeds, etc, will be integrated with the geoarchaeological micromorphological or bulk samples.

4) When visiting a site, the geoarchaeologist should err on the side of caution, even if it is the somewhat pessimistic assumption that this will be his/her last visit to the site. Thus, every effort should be made to collect samples from the major stratigraphic units, features, off-site soils, or any other specific deposits, with the notion that with these samples, a basic understanding of site formation history is attainable. There may not be a tomorrow or next season, to collect further samples. While working in Gibraltar, we had planned to collect samples from a critical profile that spanned Middle Paleolithic and Upper Paleolithic deposits, but decided to put it off until following year. That winter recorded the highest rains in a century: the profile had slumped during these heavy rains, and it took several field seasons to get back to the original profile.

5) It should be kept in mind that samples represent an invaluable resource and archive. Depending on the project and expenses of getting to the field area, they can have in effect high price tags attached to them. A field project in South Africa, for example, involves a costly plane ticket, as well as field subsistence costs. In addition, samples are often collected from exposures that will be excavated, and thus represent an archive of deposits that are no longer visible or available for study or observation. Thus, on the advice of a close colleague (the late O. Bar-Yosef), where and when possible, it is best to collect as many samples as possible, even if there is no current budget to analyze these materials. They can always sit on a shelf until the money or scientific questions arise, but at least they are there and will remain an archive of the archaeological record for the time being.

6) Impregnated samples have a virtually indefinite shelf-life, and in fact can help preserve DNA in samples (Slon et al., 2017; Massilani, D., Morley, M.W., Mentzer, S.M., Aldeias, V., Vernot, B., Miller, C., Stahlschmidt, M., Kozlikin, M.B., Shunkov, M.V., Derevianko, A.P., 2022. Microstratigraphic preservation of ancient faunal and hominin DNA in Pleistocene cave sediments, Proceedings of the National Academy of Sciences 119, no1). Air-dried bulk samples also possess good longevity. Undisturbed samples, however, if stored for a long time can suffer deterioration. Metal tins may rust, calcium carbonate or other salts in the sample can migrate towards the outer part of the sample; the sample may also be affected by biological activity. The best storage method is in a cooler, at about 2–3° C. This is particularly true for organic deposits. Freezing imparts artifacts on soils and sediments, and is normally only used for ice cores, or for lake sediments taken during winter in boreal areas.

Box 16.1   Remote sensing in alluvial environments: the Lower Lugg Valley, Herefordshire, UK

Alluvial floodplains are challenging environments for the detection of archaeological remains through remote sensing techniques. Often, alluvial corridors contain a thick blanket of fine-grained alluvial sediments that cover the valley floor. This alluvium can preserve significant archaeological remains, including organic materials, but this preserving alluvium also renders most prospection techniques, such as shallow geophysical survey and aerial photography relatively ineffective. However, the analyses of landform assemblages (see Chapter 5) from satellite and airborne remotely sensed data can provide insights into landscape evolution, which can subsequently be related to archaeological and geoarchaeological potential. For example, at the Lower Lugg Valley in the UK, Airborne LiDAR, Satellite multispectral and SAR data has been used to analyze the surface of the river valley floor and relate to this to the subsurface environment.

The Lower Lugg Valley is located close to the Welsh Border in Herefordshire, UK. It is dominated by a broad (ca. 1.5 km wide), level floodplain and the present course of the river meanders through the valley on a north-south orientation and is highly sinuous. The alluvium is characterized by 2–3 m of fine-grained silty clay alluvial deposits that cover the width of the valley floor. Previous archaeological excavations in the wider area have shown that the alluvial sediment sequence is interspersed with significant and complex archaeological deposits and features, and also sediments deposited within paleochannels.

Although displaying localized variability, the main alluvial sequence can be subdivided into three distinct units, which vary according to the underlying subsurface topography (summarized in Table 16.7). This alluvial sequence is most extensive where the underlying gravel surface is lowest, with shallower accumulations occupying higher ground. Significantly the higher, drier areas have been shown to provide the foci for archaeological activity, particularly during prehistory, whereas the lowland flood-prone areas were likely unattractive locations for archaeological activity (Jackson and Miller, 2011). As such, it is possible to model the distribution of archaeological remains in relation to the subsurface topography. The starting point for this model is the application of multiple remote sensing methods, which are used to subdivide the valley floor into different landform units, such as paleochannels and areas of higher, drier floodplain. Such interpretations can then be subsequently ground-truthed through archaeological cores/boreholes and excavation.

(Continued)

Box 16.1 (Continued)

Table 16.7 Summary of major alluvial units, geomorphological features and associated archaeological remains at Wellington Quarry (summarized from Dinn and Roseff, Jackson et al., 2010 and Carey et al., 2017).

	Description	Extent and relationship with associated archaeological material
	Topsoil	
UNIT 1	Red-brown silty clay. Typically, 0.5–0.75 m thick, but varies significantly.	Fills former channels and deep cut features. Seals all Romano-British and earlier deposits.
UNIT 2	Yellow-brown clay silt. Often exceeds 1 m where underlying gravel is lowest. Shallow towards the floodplain edges.	Long-lived deposition commencing during the early Holocene – Iron Age. The presence of thin bands of organic/humic material also indicate periods of stasis. Romano-British and Iron Age features cut into the surface. Some prehistoric features are sealed by it and some cut it.
UNIT 3	Red-brown silty clay, with subrounded gravel clasts. Rarely exceeds 0.3 m in thickness.	Widely recorded across the floodplain at the interface with the late Devensian sand and gravel.
	Devensian (Pleistocene) sand and gravel deposits	

In the UK, and many European countries, Open Government License LiDAR data can be freely downloaded. By 2021, the Environment Agency National LIDAR Programme aims to provide 1 m spatial resolution elevation data for all of England. Use of this technology is, therefore, extremely cost-effective and it provides an excellent starting point for any investigation concerned with visualizing topographic features. For this study, a 1 m DTM composite was downloaded as a series of ASCII files and subsequently converted to a raster format (Davis, 2012). Owing to the relatively flat nature of the floodplain, the data are displayed as a tightly constrained color scale to enhance any subtle, low-relief variations (Figure 16.15a). This clearly highlights the central, vertically accreting portion, of the alluvial corridor and highlights the generally flat nature of the floodplain, as well as a series of more discrete alluvial landforms. The clearest of these is the highly sinuous meandering former river channel, located directly east of the present river course (L1), though it is probable that this is comprised of multiple channels and deposits. To the west of the river, there is also a notable high point, which may relate to a gravel ridge or island (L2). Despite this, potentially deeply buried landforms, or those that do not display any significant surface expression, will not be identifiable or may be very difficult to visualize and it is advantageous to compare this against other remotely sensed datasets.

Box 16.1 (Continued)

Geomorphological landforms and the nature and depth of alluvial deposition can affect vegetation health. As such, the observation of vegetation health using multispectral sensors can be beneficial, relating surface conditions to subsurface sediments. Theoretically, specific landforms such as paleochannels or areas of organic-rich deposits, increase soil moisture, allowing for increased vegetation health during dry periods. Conversely, the presence of river terraces and gravels islands that are more freely draining, may appear as areas of less vigorous vegetation. However, plant health is not a direct indicator of soil/sediment depth, and numerous factors can affect the specific nature of the response. For example, if deeper sediments are in low-lying areas and are exposed to prolonged flooding, vegetation health will be reduced. Moreover, different land use and agricultural regimes will also affect the detectability of features and it will not always be possible to identify expressions of subsurface remains.

As has been highlighted in Section 16.3.7., there are numerous satellite multispectral sensors that can be used for geoarchaeological investigation. These each have different advantages and limitations, but to allow for the characterization of specific alluvial landforms a relatively high spatial resolution is required. As such, multispectral data from the WorldView-2 constellation was obtained from a notably dry period during May 2020. This data includes eight spectral bands at 2 m resolution and a single panchromatic band at 0.5 m resolution. The higher resolution panchromatic image was used to pan-sharpen the coarser bands, producing an eight-band image of the area at 0.5 m resolution. A single False Color Composite multispectral image, combining NIR, red and green bands is reproduced here, as these are known to highlight crop and soil marks. However, it is possible to perform various image enhancements and calculate spectral ratios, indices, and other statistical methods to increase the contrast of these features.

Within the image presented in Figure 16.15b it is possible to identify the course of a former channel to the east of the River Lugg (M1), which is most clearly visible in the southern portion of the area. The response is more complex than the LiDAR data, owing to variation in the composition of the channel deposits, with more coarse material probably being located at the channel edges. Paleochannel M1 is less clear to the north of the area, as the fields were recently ploughed at the time the image was collected reducing vegetation cover, although it is still visible as more poorly defined soil marks. To the west of the River Lugg, a gravel ridge or island has been interpreted. This was identified as a high point in the LiDAR data, but is characterized by an area of healthier vegetation surrounded by more sparse vegetation/bare soil (M2). Whilst this is the inverse of the anticipated response, the lower-lying areas surrounding were subject to extensive flooding in the preceding winter months, which has stunted the growth of the vegetation in this area.

Furthermore, an extensive time series of radar images were acquired via the COSMO-SkyMed (CSK) mission, which is the largest constellation operating in the X-band, comprising four spacecraft, allowing short revisit times. The data comprised 74 images distributed between 2014 and 2020 that were collected in "StripMap" mode, providing a spatial resolution of approximately 3 m. Within floodplain environments, which are typically covered with open pasture or crop, the subsurface imaging capability of these sensors is negligible. However, a time series can be used to study consistently wetter/drier areas, particularly in relation to

(Continued)

Box 16.1 (Continued)

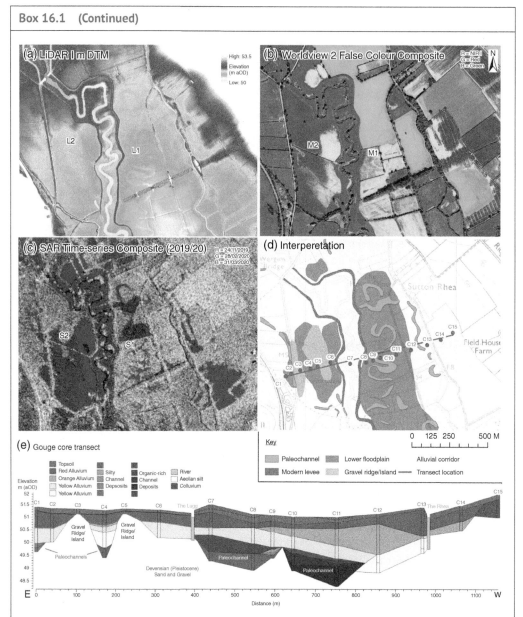

Figures 16.15a–e Comparison of remote sensing techniques deployed at the Lower Lugg Valley, Herefordshire, UK. Within each image it is possible to identify a series of paleochannels and other landforms such as a gravel island, which have subsequently been confirmed by a gouge core transect located across the floodplain. There is a clear difference visible in the alluvial deposit sequence between the west and east side of the current river course, allowing a model of the subsurface environment to be constructed from the surface (image by kind permission of Nicholas Crabb; Crabb).

modern flooding episodes. For example, in Figure 16.15c three SAR images from the winter of 2019/2020 are presented in a false color composite. The flooding is characterized by a specular response from a smooth water surface, resulting in no signal being scattered back to the sensor, hence when the valley is in flood larger areas appear as dark areas in the radar images.

Box 16.1 (Continued)

The extent of areas that consistently flood, were used to identify areas of slightly lower surface topography. This is significant as lower-lying, frequently flooded areas are likely to have been unfavorable for past human activity, but their consistently wet nature likely contributes to the preservation of important paleoenvironmental remains. Moreover, several morphological aspects of the floodplain, including the former channels (S1) and gravel islands (S2) are also very well defined.

The analysis of each of these remote sensing datasets has identified landforms that can be interpreted in terms of their archaeological and paleoenvironmental potential (Figure 16.15d, e). However, a more meaningful interpretation has been achieved through the combination multiple data sources, each measuring different ground surface properties that can be used as a proxy indicator of subsurface sediment architectures (Figure 16.15e). Despite this, it is essential to undertake at least some "ground-truthing" through borehole sampling/sediment logging to confirm these interpretations. A single gouge-core transect has been undertaken for this area, and much of the complexity identified corresponds with the recorded subsurface deposition. Moreover, it is noteworthy that although many of the landforms are deeply buried (>2 m), they are still identifiable as surficial responses to some degree. (Nicholas Crabb, University of Brighton, is kindly acknowledged for contributing to this section).

16.9 Conclusions

In this chapter we have attempted to acquaint the reader with some of the basic methods that are needed to carry out geoarchaeological fieldwork. Not all the techniques will be applicable or suitable in all instances, and in many cases decisions will need to be made about which technique is to be employed over any other. In any situation, the conditions within a project area, coupled with the experience of the site team, will to a large account dictate the strategy that is employed. Unexpected results and new situations will always arise, and not every eventuality can be planned for in advance. The key issue is to select the right suite of techniques to be able to give the team the best chance of acquiring the highest quality data, no matter what scale of project is undertaken.

The reader should be able to follow the general strategy of how to tackle a project, regardless of the specific strategies and tactics employed. This pathway starts with gathering data at the macro-scale, with aerial photos, geological maps, multispectral, and topographic data, down to the meso-scale of boreholes, coring, augering, and geophysical surveys, through to the meso-micro scale interface of recording sediment exposures and sections and the taking of bulk and continuous intact samples.

We cannot emphasize strongly enough how sampling for a project has to be integrated with all of the project team members and types of data acquisition. Without accurate field observations and proper contextualization of samples, profiles, and landscapes, any analytical work carried out subsequently in the laboratory may be of limited use, be totally uninterpretable, or even provide inaccurate results, regardless of the degree of laboratory precision.

17

Laboratory Techniques

17.1   Introduction

Geoarchaeological research is comprised of three essential parts, and each is necessary in order to achieve the best results. So while early planning in the office and detailed fieldwork can often represent a major part of a geoarchaeological study, work in the laboratory is essential in order to clarify issues and test hypotheses that have arisen. It is worth remembering here that any model or updated hypothesis has to be built on data (see Chapter 1, Introduction). Below we provide the basics for some of the most important and common laboratory techniques that are used in geoarchaeology, which have evolved greatly since our 1st edition of 2006. We concentrate on why a technique should be employed, rather than the details of the laboratory procedures themselves, due to obvious limitations of space. The latter can be found in general texts, for example (Artioli, 2010; Carter, 1993; Gale and Hoare, 2011; Garrison, 2010; Weiner, 2010), as well as books and articles on soil science and sedimentary petrography that are cited throughout the text; geological surveys and soil organizations publish some methods on their websites.

17.1.1   General Overview and Considerations

The decision concerning which laboratory studies to undertake is based on the questions one wishes to answer. These can be general ones or site-specific ones, such as what is the origin and significance of a particular layer; is this a refuse spread, a trackway, remains of a floor or manured agricultural field? Such questions can be posed prior to excavation or during fieldwork, and they are particularly needed in the case of grant proposals and bids for funding and contracts when methods need to be outlined at the planning stage (see Chapter 16). This is an important point, since having these questions in mind and the possible methods to solve them, obviates the possibility that the wrong type or quantity of sample will be taken during the excavation. Also, as more data become available, for example from thin sections or other analyses, additional procedures may be deemed necessary; this is especially true of multi-season studies, but less likely for mitigation archaeology. Most important, however, is to remember that quality is better than quantity. As noted in Chapter 16, small, well-focused samples (e.g. for organic matter, phosphate and magnetic studies, ca. 10–20 gm) are better than large "mixed" samples that include diverse components that lack microcontextual associations (Courty et al., 1989a; Goldberg and Berna, 2010; Macphail and Goldberg, 2018a). Any data from these bulk samples will be far more difficult to interpret, or to

Practical and Theoretical Geoarchaeology, Second Edition. Paul Goldberg and Richard I. Macphail.
© 2022 John Wiley & Sons Ltd. Published 2022 by John Wiley & Sons Ltd.

relate to microstratigraphic details in the thin sections. It is crucial when deciding which methods to employ, that the following are clearly thought through:

a) All techniques are time consuming, and many are expensive. It is therefore crucial to ask: is this analysis really necessary, what questions are we trying to answer, and what will the results add to the site's interpretive dataset? Simply, are the costs of time, labor and cash worth it?

b) In cases when a laboratory always carries out a suite of analyses, are they all appropriate to the study, and will all the data be useful? On the other hand, if laboratory protocols are well established, numerous samples can be analyzed rapidly if automatized machines are employed.

c) When dealing with ancient soils that are either on the surface or buried, is the property that is being measured now applicable to the understanding of past conditions? This is especially important when comparing buried soils to so-called "control" profiles, particularly if the latter are present-day soils. Similarly, does the measurement of soil acidity (pH) for a paleosol really represent the conditions at the time of its formation and burial? (See Chapter 14).

The amount of required "accuracy" and "precision" must also be thought through. Accuracy, the closeness of the measurement to true value, is constrained by the equipment being employed. For example, if a balance only measures to three decimal places, then it may be better to report only two decimal places, by rounding up. If a mean value is calculated, the answer may be given by the calculator, to six decimal places, which of course is nonsense, and "inaccurate", when the balance only weighs only up to three decimal places. Again, rounding up to two decimal places is necessary. More importantly, "precision" refers to how reproducible are the results. If several measurements of pH are taken, the time the sample is in water can affect pH, and the pH reading can theoretically shift (although not demonstrated in practice; Gale and Hoare, 2011). Pragmatic decisions need to be taken concerning how representative the sample is, how many replicate analyses/measurements are required to ensure that the results are robust. Statistical testing can be unreliable when carried out on small sample populations. In this context, rapid screening methods, such as using a magnetic susceptibility probe (see below) and/or NIR (Near-Infrared Spectroscopy) imaging on cores or even on archaeological sections, can be very efficient (Linderholm and Geladi, 2015).

When carrying out laboratory analyses, one should strive to produce data for each sample that is effective in answering a common problem. Thus, if we are trying to establish the presence of a buried argillic paleosol, organic matter estimates, phosphate, and clay content might be suitable sets of analyses to choose. In any case, all material should be analyzed from a split of the homogenized sample (Appendix A1.17.1). Every report should cite which methods were used, naming an authority and detailing any variations that were carried out. Different bulk analytical methods, for any given technique, will give different results (Holliday, 2004: appendix 3). If studies are aimed at making comparisons with work by others, or replicating investigations, exactly the same methods need to be undertaken. With regard to these points, when European dark earth (Chapter 13) was first investigated in detail, a multiple analytical approach was used (11 methods, Macphail, 1981). Some methods proved more useful than others in answering the specific questions on the dark earth, and these methods, with a further six techniques were applied on subsequent sites (Macphail, 1994; Macphail and Courty, 1985). Techniques to study dark earth are selected on a case-by-case basis, and several more analytical methods have recently been used. The characterization of the phosphate-cemented stabling crust at the Moel-y-gar stable (Butser Ancient Farm, Chapter 14) also underwent a mixed geoarchaeological approach, which involved field identification, pollen and macrobotanical studies, bulk chemistry (organic matter, magnetic susceptibility, and various phosphate analyses), soil micromorphology, microprobe, and XRD (Macphail et al., 2004). Consistent with microprobe measurements of Ca and P, XRD confirmed the identification of the hydroxyapatite cemented stabling floor crust (Macphail et al., 2006) (see Figure 13.4b). We advocate a flexible approach that provides both continuity with earlier studies, and that also takes advantage of useful new techniques as they emerge. XRD (X-ray Diffraction) is a traditional method for identifying minerals, and has been supplemented with FTIR in a number of laboratories (see below).

17.2 Physical Techniques

17.2.1 Grain Size Analysis

Grain size or particle size analysis is a routine study in geoarchaeology and other earth sciences (e.g. sedimentology, pedology). When studying sediments it is carried out in order to determine, for instance, energy of transport/deposition and environment of deposition (see Chapter 2). The uniformity of grain size (sorting) (Figure 2.2) may provide information on the type of sediment source; is a deposit totally aeolian in character or are there admixtures of fluvial material, as when differentiating between a loess and a brickearth (see Chapter 8)? Alternatively, in soil science, there may be more than one parent material, producing a soil profile formed in two sequential sediments; similarly, the movement of clay within a profile (translocation) can be inferred from grain size data.

As discussed in Chapter 2, grain sizes form a size range continuum, from fine (clay, <2 or 4 µm), fine to coarse silts, fine to coarse sands, through granules (gravel and stones), coarse cobbles, and boulders (see Figure 2.2, Tables 2.2 and 2.4).

Any particles <2 mm in size are classed as the fine fraction or "fine earth", and are usually estimated as clay, and silt- and sand-size fractions. The word "estimation" is employed because it is difficult to measure exactly all the different size ranges. Material >2 mm is termed the coarse fraction or stones, and is often determined less frequently. If the deposit contains large amounts of material that is <63 µm in size (i.e., silt and clay) for example loams, and/or organic matter, the deposit/soil needs to be pre-treated (Avery and Bascomb, 1974; Carver, 1971; Soil Conservation Service, 1994). The first pre-treatment removes all non-mineralogenic material (organic matter), and separates the non-sandy elements, usually by "wet sieving" (see next). It may also be desirable to eliminate calcium carbonate, to produce carbonate-free grain size estimations (Catt, 1990).

The most basic method for estimating the sand and stone size ranges is through sieving. In pedology, each horizon is described in the field according to its "texture" (grain size) through "finger texturing", and its stone content or stoniness through visual estimation (see Chapter 16). The precise nature of this stoniness may be very important, and stones can be individually measured or sieved into 2–4 mm, 4–8 mm, 8–16 mm, 16–32 mm, etc., fractions (see Tables 2.2 and 2.4). If sediments are totally dominated by sand, the different size ranges are analyzed by dry sieving the <2 mm fraction, normally employing a minimum of 63 µm, 125 µm, 250 µm, 500 µm, and 1,000 µm sieves (Tables 17.1a–17.1b). In sedimentology, a settling tube is used in which sediment is introduced into a column of water and the rate of sedimentation is measured, with coarser grains settling first. This method more accurately depicts the sedimentary dynamics associated with transport and sedimentation. The equipment is not standardized, and relatively expensive, however, so sieving is a much more straightforward and cost-effective procedure, hence its popularity.

Since the finer silt and clay sizes are too small to measure directly by sieving, their size and abundance must be measured indirectly. There are many ways to estimate clay and silt fractions. The first includes wet sieving (to exclude the sand component), the use of a sedimentation column, and bulk instrumental methods of using a laser counter or SediGraph (Malvern Instruments, 2007) (Figures 17.1a and 17.1b). With mixtures of sand, silt, and clay (e.g. loams, Figure 2.2), silts and

Table 17.1a Example of traditional grain size analyses – from different sedimentary environments at Lower Palaeolithic Boxgrove, UK: % CaCO₃, clay, silt and sand; grain size on a decalcified basis (from (Catt, 1999)

Sample	CaCO$_3$	Clay < 2 μm	Silt 2-4 μm	Silt 4-8 μm	Silt 8-16 μm	Silt 16-31 μm	Silt 31-63 μm	Sand 63-125 μm	Sand 125-250 μm	Sand 250-500 μm	Sand 500-1000 μm	Sand 1000-2000 μm
Unit 6	51.9	44.4	4.0	5.8	7.2	17.3	13.0	3.0	0.7	0.1	0.1	0.1
Unit 4c	1.7	40.0	5.1	6.0	9.0	12.1	22.6	5.0	0.2	0.1	0.1	0.1
Unit 4b	25.7	39.2	4.4	4.3	7.0	10.2	25.6	6.8	0.2	0.1	0.1	0.1
Unit 3	13.0	3.4	2.0	0.4	0.3	1.2	22.7	50.6	14.2	1.1	0.1	0.1

GTP 13: Unit 3 = Beach sand; Unit 4b = Estuarine silt; Unit 4c = Old land surface (ripened 'soil' formed in partially decalcified uppermost Unit 4b).
GTP 10: Unit 6 = Colluvium (brickearth colluvium)

Table 17.1b Example of traditional grain size analyses – from different soil horizons at prehistoric Raunds, and including the overlying medieval alluvium; data from Turf Mound and Barrow 5 Raunds, UK; % clay, silt and sand (From Macphail, 2011)

Sample and soil horizon	Clay < 2 μm	Silt 2-16 μm	Silt 16-31 μm	Silt 31-63 μm	Sand 63-125 μm	Sand 125-250 μm	Sand 250-500 μm	Sand 500-1000 μm	Sand 1000-2000 μm	Texture
22 (Ap/alluvium)	36	12	10	11	4	10	14	2	1	Clay
24 (bA)	9	6	5	15	6	17	37	4	1	Sandy loam
26 (bB2)	12	3	2	13	9	25	32	3	1	Loamy sand
27 (bBt)	24	2	3	9	8	17	27	7	3	Silty clay loam
28 (bC)	3	5	2	10	7	20	33	10	10	Loamy sand

NB: bC = Nene river terrace sands and gravels; bBt = clay-enriched argillic horizon; bB2 = upper subsoil which also includes anomalous argillic features; bA = barrow buried topsoil which also includes anomalous argillic features; Ap = medieval ploughsoil formed in post-prehistoric fine alluvium which buries the barrow

clay materials are pre-treated to remove organic matter and to disaggregate the soil and disperse clay particles. (The importance of "clasts" of mixed grain sized material is noted below.)

The standard procedures outlined here broadly follow those of Avery and Bascomb (1974). Organic matter is removed from the sample using hydrogen peroxide (H_2O_2) mixed with de-ionized water. (This loss of material needs to be deducted from the original soil total weight, and it should be consistent with the measured organic matter.) In addition, carbonate can be removed using HCl and results presented on a carbonate-free basis. The procedure then continues with the now

Figure 17.1a Example of grain size analyses (phi scale) from different soils, sediments, building materials, and the dark earth at Courages Brewery, Southwark, London, UK. Dark earth includes high amounts of local sandy alluvium (used in all kinds of building activities, mortar manufacture, etc; see Chapters 12 and 13), with contributions from brickearth building "clay" and anthropogenic dumping. Such confirmed findings were possible only through parallel soil micromorphology analyses.

mineralogenic part of the sample that should be disaggregated using a disaggregating agent such as sodium hexametaphosphate (formerly the main ingredient in Calgon, which now uses sodium citrate as an active ingredient). The sample is then mechanically stirred or shaken, and when totally disaggregated the suspended sediment can be used in a sedigraph or laser counter to measure grain size, which nowadays includes all of the <2 mm fraction. Traditionally, however, the sample is washed through a 63 μm sieve into a 500 or 1000 ml cylinder, in order to separate silt and clay from the sand content (>63 μm). The sample then undergoes testing for silt and clay content, by either the pipette or the hydrometer methods. Essentially the "sediment column" is "paddled" until all the particles are evenly distributed, and the particles begin to settle according to Stoke's Law, at a given rate that is temperature-dependent. Coarse silt falls quickest, while clay stays in

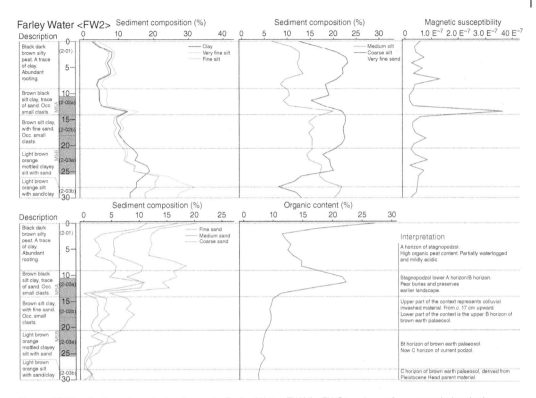

Figure 17.1b Sediment analysis of sample Farley Water FW16 <FW2>, using a 1 cm sample/analysis interval; data are complemented by monolith log, context depths, %LOI (estimated organic matter) and magnetic susceptibility (Carey et al., 2020/John Wiley & Sons / CC BY 4.0).

suspension for the longest time. At specific time intervals (e.g. 30 seconds to 8 hours), therefore, sediment is drawn off from a given depth, and the amount of sediment present is dried and weighed. This permits the calculation of the different percentages of different grain sizes. The hydrometer method allows the measurement of the density of sediment, again giving information on the nature of the sediment in suspension at different times. In both methods, how much fine (<1 μm) and coarse (<2 μm) clay, fine silt, medium silt and coarse silt is present can be calculated (Table 17.1; see Appendix A1.17.2).

The measured amounts of the different fractions are then plotted to produce a grain size curve. The different grain sizes have already been presented (Table 2.1). In sedimentology, grain sizes are normally expressed on the phi (ø) scale (e.g. 0.25 mm = 2.00 ø; 0.50 mm = 1.00 ø; 1.00 mm = 0.00 ø; 2.00 mm = −1.00 ø, etc.). A number of examples of grain size curves, and estimated amounts of the different size factions, are shown here (Figure 17.1a; see Chapter 2). As an example, it is useful to compare a loess with a brickearth: the latter is overall less well sorted because the well-sorted coarse silt and very fine sand-size aeolian fractions are mixed with medium size sand of fluvial origin. Both have been utilized for building clay and bricks (Chapters 12 and 13). In the Chiltern Hills, a range of chalk hills in southern England, clay-rich subsoils were covered by a more silty loessic drift (Avery, 1964; Avery, 1990). This was strongly eroded in prehistoric times after clearance, and now is mainly found in subsoil throw hollows and early Neolithic ditchfills (Chapter 11). Across many parts of the USA and Eurasia, Alfisols (Luvisols) formed under woodland are differentiated in part from Mollisols developed under

Table 17.1c   Example of grain size analyses – from different soils, sediments, building materials and the dark earth at Courages Brewery, Southwark, London, UK : % clay, silt and sand (from Macphail, 2003) (see Fig 17.2a)

Sample	Clay < 2 μm	Silt 2-16 μm	Silt 16-31 μm	Silt 31-63 μm	Sand 63-125 μm	Sand 125-250 μm	Sand 250-500 μm	Sand 500-1000 μm	Sand 1000-2000 μm	Texture
Southwark Roman										
1. Brickearth (clay wall)	14	3	11	27	17	14	12	3	<1	Fine sandy silt loam
2. Sandy Thames alluvium (Pre-Roman subsoil)	5	1	2	5	10	28	40	8	1	Loamy sand
3. Fine Thames alluvium (Roman)	24	11	31	9	6	5	7	1	1	Clay loam
4. Roman fine dumps ('sweepings')	29	5	13	11	5	8	13	5	2	Clay loam
5. Dark Earth (*n=4*)	17	4	9	10	12	14	23	8	3	Sandy loam

NB: 1 = Brickearth (a loess-like fine sandy alluvium that was quarried for building clay); 2 = Thames river sands (early Holocene alluvium); 3 = Roman fine alluvium; 4. Roman fine dumps (contains abundant fine anthropogenic inclusions); 5 = Dark earth (a mixture of brickearth (1) and sands (2) derived from constructional raw materials (e.g., sand-tempered mortar), as well as finely fragmented building debris (mortar and brick), and occupation dumps (4) that include ashes, phytoliths, fine bone, shell and coprolites etc.)

prairie vegetation, by measured increased amounts of clay in their subsoil (Bt) horizons (Table 17.1b) (see Courty et al., 1989b: fig. 8.6). In coastal areas, there are enormous differences between sandy beach sediments, and silt- and clay-rich estuarine deposits (Table 17.1a; see Chapter 9). Lastly, anthropogenic deposits such as dark earth are very poorly sorted because of dumping and the common presence of earlier buildings (Table 17.1c; Figure 17.2). Particle size analysis clearly shows that fine sandy silt loam brickearth (building "clay"), early Holocene river sand, and dumped material have contributed to the overall grain size of dark earth (Macphail, 2003). Additionally, the grain size of Roman alluvium is comparable to that of dumps of silty Roman street sweepings. Hence, when carrying out granulometry on anthropogenic deposits the resulting data may well be uninterpretable without complementary soil micromorphology. Grain size analysis was also carried out on the Late Bronze Age/Early Iron Age site of Potterne and produced essentially bimodal distribution curves. The sand-size material originated from the local Greensand soils, but the silt-size material included many phytoliths, which characterized this midden (Macphail, 2000). Hence, grain size data from occupation deposits need to be interpreted with a large measure of caution.

Like any technique, grain size has its limitations and pitfalls. For example, when a sediment contains aggregates composed of sand silt and clay, disaggregation of these clasts will not reflect the depositional energies or conditions of the *clast itself*, because the clast is the grain size that was transported (Table 17.1a – Boxgrove Unit 6). Also, there may be problems when deposits are rich in mica, because lath-shaped silt-size mica takes longer to settle than round silt particles. Deposits rich in diatoms or phytoliths also settle at anomalous rates. Finally, the use of

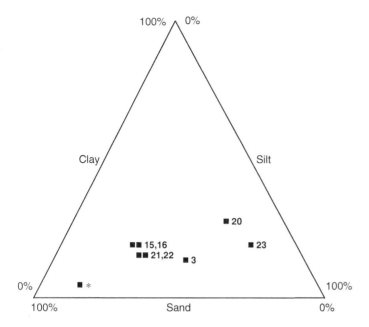

Figure 17.2a   Triangular Grain Size Diagram (see Figure 2.1; Table 17.1c); example of dark earth, Courages Brewery, Southwark, London, UK: particle size classes for local sand (*), alluvium (23), brickearth building clay (3), "silty" dumps (20), and dark earth (15, 16, 21 and 22) (data after Macphail, 2003). Dark earth includes high amounts of local sandy alluvium (used in all kinds of building activities, mortar manufacture, etc.; see Chapter 12), as well as contributions from brickearth building "clay" and anthropogenic dumping. Such confirmations were possible only through parallel soil micromorphology analyses.

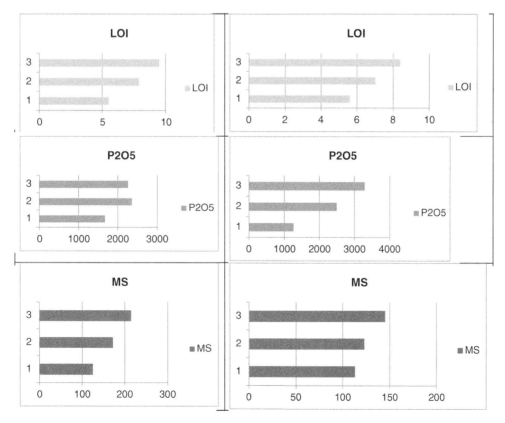

Figure 17.2b   As Figure 17.2b; chemical (LOI and Phosphate – P_2O_5) and magnetic susceptibility (MS) log diagrams through two dark earth profiles, showing generally increased signals upwards, which testify to increased middening inputs through time, and not stagnation/city abandonment (data after Macphail and Linderholm, 2004).

grain size analysis to characterize anthropogenic deposits should be strongly questioned for many of the reasons mentioned above: what is its value when studying a sediment composed of aggregates of clay, brick, ashes, bone fragments, and organic matter (see Table 17.1c; Figure 17.2a)? Certainly the sedimentary dynamics cannot be reliably determined as is in the case of geogenic sediments.

A relationship between grain size data and soil micromorphology needs to be emphasized (Appendix A1.17.3). In general, it is not a bad idea to conduct grain size analysis after thin sections have been examined or if it becomes clear that such results will be truly meaningful in terms of sedimentary process. In fact, micromorphology should be thought of as a methodological filter used to direct subsequent analyses (Goldberg et al., 2017). Mixed size populations and their geometries are generally evident under the microscope.

17.2.2 Soil Characterization

There is insufficient space here to itemize and explain all the many soil analyses that can be carried out, as well as describe all the specific different methods employed. We have therefore chosen to present a short guide to the most commonly used techniques. The chief methods that are available in most laboratories, especially those in soil or earth science departments, are:

- water content
- total organic matter (e.g. by loss on ignition, LOI)
- pH
- organic carbon
- nitrogen, carbonate
- iron (Fe)
- aluminum (Al).

Other elements, such as heavy metals (e.g. copper (Cu), lead (Pb), tin (Sn) and zinc (Zn)) may also be analyzed; however, now with the greater availability of X-ray fluorescence (XRF), and Inductively Coupled Plasma-Mass Spectrometry (ICP-MS), most elements can be measured easily and in bulk (see 17.6 below – and Figure 18.9; see Table 17.2 for example of analyses carried out at Marco Gonzalez – Belize). But again, as above, there are caveats to the indiscriminate use of these methods; for example, XRF can struggle to accurately measure lighter elements such as Na (J. Linderholm, University of Umeå, pers. comm.). A whole series of techniques for chemical analyses on archaeological materials are published in Nicosia and Stoops (2017).

The measurement of water content at 105° C is useful as samples are all analyzed on a water-free weight basis. All that is needed is a simple drying oven and a scale/balance to measure weights before and after heating.

Low temperature LOI (loss on ignition) is an important measure of the amount of organic matter in the soil or sediment (e.g. Ball, 1964). Its measurement can be an indicator of intensity of human activity or the degree of oxidation associated with soil formation (see Chapter 4 and Figure 17.2b). LOI is carried out in a furnace by heating an oven-dried sample in a crucible; a desiccator and desiccant are used to make sure that samples do not take up moisture between heating and weighing (for further details on LOI and associated soil techniques see Appendix A1.17.4).

Table 17.2 Example of multi-method bulk sample analyses (Maya salt working deposits and post-depositional dark earth formation at Marco Gonzalez, Ambergris Caye, Belize; Macphail et al., 2017) LOI, carbonate, pH, conductance, phosphate-P, magnetic susceptibility, selected XRF (Ni, Cu, Zn and Pb) and Hg data.

Bulk sample	LOI[a] (%)	Carbonate[b] (CaCO$_3$ equiv, %)	pH[c]	Specific conductance[d] (µS)	Phosphate-P[e] (mg g^{-1})	χ^f (10^{-8} m³ kg^{-1})	χ_{max}^f (10^{-8} m³ kg^{-1})	χ_{conv}^f (%)	Hg[g] (ng g^{-1})	Ni (µg g^{-1})	Cu (µg g^{-1})	Zn (µg g^{-1})	Pb (µg g^{-1})
x0-5cm Str8	28.1***	35.1	7.9	455	11.4**	70.3	188	37.4	467.5	19.8	30.1	48.2	7.7
x8a	12.0**	50.5*	8.6*	1900*	7.25*	87.3	167	52.3	204.6	27	19.1	39.8	7.7
x8b	8.67*	54.8*	8.7*	2220*	6.26*	88.1	157	56.1	189.6	27	23.8	34.9	8.4
x8c	6.24*	59.1*	8.8*	2240*	3.66	127*	145	87.6	82.2	24.5	16.1	21.6	7.6
xMRef3	4.41	58.8*	9.1*	1090*	2.32	212*	225	94.2	53.4	26.7	26	15.7	7
x9a	5.41*	57.9*	8.9*	2620**	4.50	139*	158	88.0	128.8	19.7	18.6	33.8	6.8
x9b	5.36*	59.3*	8.7*	2880**	2.42	209*	209	100	54.2	33.7	18.3	20.5	7.3
x9c	5.34*	60.9*	8.6*	4080**	1.88	206*	191	>100	27.3	33.5	19	20.8	8.6
xMRef2	2.02	75.0**	9.0*	2340*	1.15	59.4	57.3	>100	29.6	27.3	12.6	10.7	5.2
x10a	5.24*	57.6*	nd	nd	1.82	203*	197	>100	74.9	32.2	17.6	21.1	7.2
x10b	3.28	61.7*	nd	nd	1.92	164*	155	>100	36.2	36.3	17.3	15.3	6.1
x1a Str. 14	7.22*	52.1*	8.9*	697	5.42*	197*	214	92.1	213.5	21.7	19.3	35.6	6.9
x1b	6.51*	43.4	8.8*	1420*	2.00[†]	444*	440	>100	44.3	52.9	22.5	30.7	11.4
x2a	2.61	60.1*	nd	nd	3.12[†]	77.8	92.3	84.3	32.8	2.2	21.3	33.1	3.6
x2b	12.6**	47.7	8.7*	2500**	2.00[†]	144*	166	86.7	64.3	14.9	27.6	40.6	6.6
x2c	8.12*	44.0	8.7*	2930**	3.65[†]	362*	348	>100	38.9	50.3	23.9	32	11.1
x2d	18.4**	45.4	nd	nd	1.62	105*	144	72.9	34.5	10.4	22.2	33.6	6
x3a	19.9**	49.7	8.5*	5260***	1.09	163*	244	66.8	31.6	12.1	23.3	36.2	5.1
x3b	5.89*	33.5	8.7*	5220***	2.10[†]	641*	714	89.8	40.1	49.8	21.8	38	12.2
x3c	14.4**	56.0*	8.6*	5700***	1.19[†]	110*	198	55.6	26.1	13.1	18.5	44.8	8.9
x3d	7.28*	53.0*	8.8*	4900**	6.67*	257*	303	84.8	50.6	23.7	21.7	38.6	7.7
x3e	3.72	63.8*	8.8*	3340**	1.37	96.4	189	51.0	44.1	14.2	7.5	21.9	7

(Continued)

Table 17.2 (Continued)

Bulk sample	LOI[a] (%)	Carbonate[b] (CaCO₃ equiv, %)	pH[c]	Specific conductance[d] (μS)	Phosphate-P[e] (mg g⁻¹)	χ[f] (10⁻⁸ m³ kg⁻¹)	χmax[f] (10⁻⁸ m³ kg⁻¹)	χconv[f] (%)	Hg[g] (ng g⁻¹)	Ni (μg g⁻¹)	Cu (μg g⁻¹)	Zn (μg g⁻¹)	Pb (μg g⁻¹)
x3f	6.99*	48.6	8.8	5070***	5.66*	374*	402	93.0	42.9	35.8	21.8	33.5	7.5
x5a	5.39*	48.9	8.9	5010***	5.52*	286*	323	88.5	63.5	7.6	26.5	47.5	3.8
x5b	5.84*	54.9*	8.9	3980**	18.7*	85.4	159	53.7	48.8	<2.0	20.9	56.5	5
x5c	9.21*	55.1*	nd	nd	5.43*†	158*	205	77.1	37.6	21.7	19.3	35.6	6.9
x6a	4.65	65.8*	nd	nd	21.2***	4.8		>100	34.6	52.9	22.5	30.7	11.4
x4a	5.51*	60.5*	8.7	5580***	22.5***	121*	111	>100	56.4	2.2	21.3	33.1	3.6
x4b	3.66	70.2*	8.9	3540**	23.2***	8.6	14.8	58.1	40.0	14.9	27.6	40.6	6.6
x0-5cmStr19	26.9***	39.4	8.0	477	8.09*	120*	194	61.9	**241.4**	12.8	26.6	52.8	8.3
x12a	5.42*	58.3*	8.4	184	3.91	86.2	137	62.9	**104.1**	2.3	18	24.2	4.2
x12b	8.65*	50.0*	nd	nd	3.26†	132*	175	75.4	75.7	17.9	16	27.7	6.9
x13a	5.87*	48.0	8.5	5020***	7.03*	388*	451	86.0	28.6	29.6	19.6	31.9	8.4
x13b	2.70	59.4*	8.8	3620***	36.5***	17.8	16.3	>100	14.6	<2.1	23.9	74.3	4.8
x13c	4.02	58.8*	8.7	3560***	25.3***	76.3	99.4	76.8	45.7	<2.1	32.6	60.5	4.3
x14a	3.85	56.6*	nd	nd	26.9***	62.4	81.3	76.8	50.4	4.9	36.5	72.2	4.8
x14b	3.91	63.4*	8.8	3680**	22.4***	16.5	21.0	78.6	29.7	<2.0	29.9	68.1	6.1
x14c	3.30	67.1*	8.8	3040**	24.5***	16.1	26.8	60.1	43.5	<2.1	38.5	81.2	7.1
x14d	3.36	65.0*	8.8	3480**	28.1***	13.1	20.4	64.2	47.5	<2.2	37.4	80.6	4.3

[a] LOI: values in bold indicate notably higher LOI values, which reflects the amount of organic matter and/or charcoal present: * = 5.00–9.99%, ** = 10.0–19.9%, *** ≥ 20. 0%.

[b] Carbonate: values highlighted indicate higher carbonate concentrations: * = 50.0–74.9%, ** ≥ 75.0%.

[c] pH: values highlighted indicate pH ≥ 8.5; nd = not determined because of insufficient sample.

[d] Specific conductance: values highlighted indicate higher values: * = 1000–2440 μS, ** = 2500–4990 μS, *** ≥ 5000 μS; nd = not determined because of insufficient sample.

[e] Phosphate-P: † indicates that phosphate-P was determined on residual samples from the LOI analysis; values highlighted indicate likely phosphate-P enrichment:
* = 'enriched' (5.00–9.99 mg g⁻¹), ** = 'strongly enriched' (10.0–19.9 mg g⁻¹), *** = 'very strongly enriched' (20.0–39.9 mg g⁻¹).

[f] Magnetic susceptibility: data are difficult to interpret (Graham et al., 2017); χ values ≥ 100×10^{-8} m³ kg⁻¹ are highlighted.

[g] Mercury (Hg): CV-AFS data suggest particular enrichment of surface and near-surface dark earth layers.

Organic matter is also often determined by the analyses of C and N, and these are measured by acid digestion. This analysis is normally carried out only in properly equipped soil laboratories, but it provides very useful information on bioactivity and biodegradation, in the form of the C:N ratio. Similarly, this kind of "wet chemistry" is used to analyze different important pedological fractions of Fe and Al (Bascomb, 1968; Bruckert, 1982; Macphail et al., 2003b). For example, it may be important to know the proportions of "mobile" and "crystalline" forms of these components in a soil. Spodic/podzolic subsoils (Bh, Bs) have accumulated mobile Fe and Al (sesquioxides), whereas brown soils mainly feature crystalline forms of these compounds. CEC – cation exchange capacity – is another useful fundamental test, measuring the amounts of Ca^{++}, K^+, Mg^+ and Na^+, along with H^+. The amounts of these ions are related to pH and degree of leaching (see Figure 4.9). In the same way, C:N results from the degree of bioactivity and bio-decay in soils, with low C:N (~5–10) being found in base-rich bioactive soils (e.g. high earthworm or termite populations), and values of 10–20 in acid mor humus horizons of Spodosols (see Macphail and Courty (1985) for CEC and C:N ratio data in dark earth).

Other analyses that can be undertaken are measures of plant-available K and available P. These measurements are in reality mainly used to determine present-day soil fertility, and are not suitable for archaeological deposits. They have usefully been carried out, however, at Experimental Earthworks, for example (Crowther, 1996c; Macphail et al., 2003b) because these studies are engaged in monitoring modern or short-term soil changes after burial (Chapter 14). In addition, a common question when studying soils of historic gardens is, how fertile are the garden soils and do they require amendments to the soil when being reconstructed, as was undertaken at Hampton Court Privy Garden (London, UK) (Macphail et al., 1995; Vissac, 2002)?

When studying ancient soils, and under normal temperate conditions, a number of characteristics are labile or easily transformed. This was also demonstrated at the two Experimental Earthworks in the UK (see Chapter 14). Firstly, pH can be affected by cations leaching out of the overburden, or through down wash of unstable fine material, or by organic matter decay. Organic matter itself is highly affected by post-depositional oxidation, and frequently cannot be compared to amounts of organic matter in present-day "control" soils. C:N ratios may also change through time, and it is worth noting that archaeological deposits with high amounts of charred organic matter, have anomalously high C:N ratios, despite being once bioactive (Courty and Fedoroff, 1982).

17.2.3 Phosphate Analysis

This technique is one of the most commonly employed in geoarchaeology, having an archaeological pedigree going back at least to the 1930s (Arrhenius, 1934) but more common since the 1950s (Arrhenius, 1955; Proudfoot, 1976). Interestingly, Holliday (2004: Appendix 2) notes at least 50 extraction methods, some 30 of which have been used in archaeology. From the outset it is useful to differentiate between the various forms of phosphate such as organic, and inorganic iron and calcium compounds, which are reported as an oxide (P_2O_5), or elemental phosphorus (P) (Schlezinger and Howes, 2000). The mineral dahllite ($Ca_{10}(PO_4,CO_3)_6(OH)_2$) is a major component of bone and occurs as a diagenetic product in many prehistoric cave settings (Goldberg and Nathan, 1975; Karkanas et al., 2000; Karkanas and Goldberg, 2018a) in addition to several other phosphates (Weiner, 2010; Weiner et al., 2002). A well-known example of a phosphate mineral in later periods is vivianite (e.g. $Fe_3(PO_4)_2$ $8H_2O$), which commonly occurs in waterlogged environments associated with organic remains (Rothe et al., 2016). Both *phosphate* and *P* can be measured by wet chemistry, ICP-OES (Inductively Coupled Plasma-Atomic Emission Spectroscopy) and/or through microprobe, SEM/EDS, micro-XRF, and micro-FTIR (Fourier Transform Infrared

Spectrometry) analyses, but amounts of measured phosphate will always be *more* than reported P. The ratio between P_2O_5 and P is 2.2916, and this figure is used to calculate P from P_2O_5. Amounts of P can be reported as ppm (parts per million), mg g^{-1}, or as percent: 1,000 ppm P = 1.00 mg g^{-1} = 0.10%; 10,000 ppm P = 10.0 mg g^{-1} or 1.00% P.

Phosphate analysis is important because it can provide not only an indication of diagenesis (particularly bone preservation), but also an important measure (amongst several) of the intensity of human occupation. As noted above, there are many ways of extracting phosphate, and it is up to the researcher to choose the most appropriate method (Bethell and Máté, 1989). Rudimentary measures such as spot testing (Eidt, 1973) and plant-available phosphate (Brady and Weil, 2008; Cornwall, 1953; see also Eidt, 1977) are not useful, because the former is a non-quantitative test, and on most sites phosphate will be present, just as a contaminant; in the latter case, the extraction method is usually far too weak to provide a useful signal. Even in impoverished Spodosol topsoils Crowther found the presence of phosphate (Ah horizons: 0.827 mg g^{-1} P; 21.2 mg l^{-1} available P), but with much less available phosphate compared to the Privy Garden soil at Hampton Court (plate-bande fills: 0.942 mg g^{-1} P; 95.8 mg l^{-1} available P) (Macphail and Crowther, 2002b; Macphail et al., 2003b).

Two principal approaches for phosphate analyses are preferred. The first relates to available-phosphate testing as is commonly used in agronomy; a sample is subjected to a mild leaching in which the more easily solubilized phosphate is removed. The second employs measurement of *total phosphate* in which the entire sample is dissolved. In the latter case, it may be misleading to dissolve the whole sample in a very strong acid (e.g. hydrofluoric) before measuring phosphate, since geogenic P (e.g. apatite within rock fragments such as chalk) will be included, thus giving confusing, irrelevant, and probably regionally specific data. The study of the Hampton Court soils, however, is a special case where the measurement of available phosphate was important to the reconstruction and re-planting of the garden in order to record present-day soil fertility (Thurley, 1995).

Nevertheless, in geoarchaeology generally, a total phosphate method is required to extract phosphate from (a) organic matter, including dung and immobilized recalcitrant/aged organic material, (b) bone and coprolites (inorganic "apatite") and (c) inorganic secondary minerals, such as vivianite and other finely crystalline and amorphous Ca-Fe forms, as suggested by *SEM/EDS* and mineralogically characterized by *micro-FTIR*, for example (Karkanas and Goldberg, 2018a: table 1). It became obvious that it was useful to differentiate or fractionate organic and inorganic phosphate, which was Eidt's strategy (Eidt, 1984): he attempted to infer soil use by measuring the various types of *fractionated phosphate*. Determination of phosphate fractions is commonly carried out by testing a sample twice. The first, measures inorganic phosphate (*before* ignition/oxidation), whilst the second measures any residual inorganic phosphate (*after* ignition/oxidation) that had previously been in an organic form (yielding a measurement of Total Phosphate) (see Figure 17.3).

Two approaches can be identified. The first uses a relatively weak acid (2% citric acid) in order to be able to *gently* differentiate inorganic from organic phosphate. This method is used across Europe since the 1950s and has a long track record for finding archaeological sites in Scandinavia (Arrhenius, 1934, 1955), and identifying anthrosols/plaggen soils (Driessen et al., 2001; Pape, 1970). It is still used successfully for Nordic sites (see below). It was, however, designed for acid and leached soils of this boreal region, which are generally carbonate-free and rather naturally low in phosphate (Engelmark and Linderholm, 1996; Linderholm, 2007) (see Linderholm et al., 2019). It therefore has the disadvantage that its extraction capacity is weakened if large amounts of free calcium carbonate are present, as in calcareous soils. This effect, however, can be offset if a few drops of HCl is added during the procedure as this removes the buffering effect of the carbonate (Linderholm, University of Umeå, pers. comm.). Generally speaking, this method has produced good results for samples containing up to around 4500 ppm P_2O_5 (Viklund et al., 2013) (see Figures 18.14 and 18.20).

A second approach is to use stronger acid extraction (e.g. 2N nitric acid, again pre-treating the sample with HCl to offset any calcium carbonate), or oxidation methods (Bethell and Máté, 1989; Dick and Tabatabai, 1977), and as employed at Butser Ancient Farm by C. Bloomfield (ex-Soil Survey of England Wales; Macphail et al., 2004). It can be argued that these acids are less discriminating when separating organic and inorganic phosphate. Eidt (1984) differentiated several types of P in his work in South America. Kerr (Kerr, 1995; Schuldenrein, 1995), and Holliday (2004) reviewed the subject of phosphate analysis, with Kerr and Schuldenrein also carrying out their own fractionation methods on prehistoric North American sites. The presence of P at sites has also been linked to other elemental concentrations, for example in body stains (Keeley et al., 1977), floors (Banerjea et al., 2016; Entwhistle et al., 1998; Middleton and Douglas-Price, 1996) and campsites (Linderholm and Lundberg, 1994) (see Table 17.2).

Movement of P in archaeological sites is a key issue. Considerable amounts of P can be moved down-profile through pedogenic effects, perhaps to a depth of 1 m over 1,000 years (Baker, 1976); redeposition, for example by groundwater, also occurs (Thirly et al., 2006). Hence when studying the phosphate distribution in an area of boreal Spodosols in Sweden, the illuvial Bs horizon was sampled because surface phosphate dating to the lifetime of the site, has been translocated (illuviated) down-profile (Linderholm, 2007). On the other hand, Romans and Robertson working at Strathallan (Scotland), suggested that P could move significantly within archaeological time, i.e., 10 mm per 20–40 years (Barclay, 1983). At Raunds (Northamptonshire, UK), P that likely originated from prehistoric livestock concentrations was carried down-profile through clay translocation during the life of the sites (Macphail, 2011c). Movement of P and neoformation of various phosphate minerals (identified by x-ray diffraction and FTIR) in caves through time is well documented (Courty et al., 1989b; Jenkins, 1994; Karkanas and Goldberg, 2018a, 2018b; Karkanas et al., 1999, 2000; Macphail and Goldberg, 1999; Macphail et al., 2012b; Schiegl et al., 1996; Weiner et al., 2002). However, short-term movement has also been reported within an experimental stable (during 15 year use) (Heathcote, 2002; Macphail et al., 2004) and where bone was buried at the Experimental Earthwork at Overton Down (Crowther, 1996a, c). Certainly at Worcester (UK, Saxon), cess pits led to the phosphate contamination of the earlier-formed dark earth; at the London Guildhall and Pevensey Castle, marked phosphate staining of dark earth from medieval *in situ* stabling of animals and latrines occurred (Macphail, 2011d; Macphail et al., 2008; Macphail and Linderholm, 2004) (see Figure 17.2b). Much of the phosphate was clearly secondary, occurring as both amorphous and crystalline (vivianite) void infills; at the London Guildhall, P was *statistically linked* to concentrations of heavy metals by J. Crowther (Macphail et al., 2007a, b). There, both P and heavy metals were statistically associated with LOI (organic matter) and thus of organic origin (probably dung and human waste).

The analysis of total phosphate has been applied in many ways. It varies from simply measuring amounts of phosphate in an archaeological section or soil profile, to identifying different areas of phosphate distribution through transects, or mapping phosphate distribution using samples taken on a grid (Crowther, 1997). (Phosphate has also been commonly mapped alongside geophysical measurements, such as magnetic susceptibility (e.g. Maiden Castle (UK) and Ørlandet (Norway) – (Balaam et al., 1991; Linderholm et al., 2019); see Figures 18.17, 18.20, and 18.21.) Historically, the earliest chemical studies recorded high phosphate levels in association with anthropic agricultural soils (Barnes, 1990; Eidt, 1984; Liversage et al., 1987; Pape, 1970; Sandor, 1992), cemeteries (Faul and Smith, 1980), assumed stock areas and drove-ways, structures housing animals (Conway, 1983; Crowther, 1996b), and mounds (Kerr, 1995), for example.

Differential distributions of mainly organic phosphate and predominantly inorganic phosphate have been commonly found in central Sweden. The values coincide with the remains of long house structures (large amounts of dominantly inorganic P) as identified archaeologically, and arable fields (small concentrations of mainly organic P) as also indicated by the presence of cereal pollen (Engelmark, 1992; Engelmark and Linderholm, 1996; Segerström, 1991). Years of experimental agriculture at Umeå Ancient Farm (Chapter 14) established that manure (organic P) was used to promote the growth of barley in late prehistoric (AD 0–1000) Sweden, where impoverished boreal podzols are common; such fields develop an organic P signature. On the other hand, in settlement/long house areas, inputs of inorganic P from bone and human waste in the form of now-mineralized cess predominated. Overall and through time, amounts of P can also give proxy estimates of numbers of humans present for given periods under some circumstances (Linderholm, 2007). Other Iron Age to Viking Period fields and settlements from Denmark and Norway have been similarly investigated through LOI and P, alongside magnetic susceptibility and soil micromorphological investigations. *PQuota*, i.e. the ratio between organic and inorganic phosphate, can be plotted against LOI, in order to help identify potential manuring associated with stock management intensity in settlements, through time and space, as at Ørlandet (Norway) (Figure 17.3a; see also Figures 18.19–18.21).

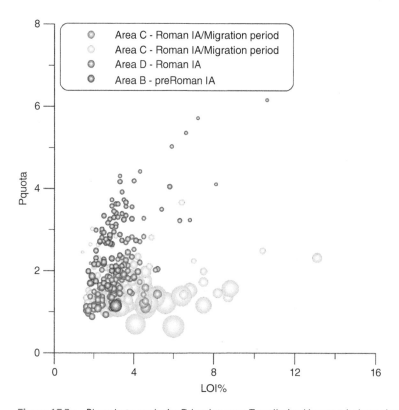

Figure 17.3a Phosphate analysis: Ørlandet, near Trondheim, Norway: A chronological overview on the Pquota and loss on ignition relationship based solely on posthole fills from house areas B, C, and D. Circle sizes are relative to the amount of CitP (citrate-extractable P) (Linderholm et al., 2019).

One specific study in Belgium found phosphate concentrations in Roman Period sunken byres, or postals (Mikkelsen et al., 2019).

17.2.4 Multi-element Geochemical Analysis

Phosphate analysis (section 17.2.3) has been used in multiple ways in geoarchaeological and geo-prospection research, but one of its earliest uses was to investigate areas of wider landscapes to identify archaeological sites (Arrhenius, 1955; Craddock et al., 1985). Whilst the results of interpreting such surveys were fraught with difficultly, the application of these techniques opened up the ideas of hidden landscapes and site histories (Heron 2001). The idea is that human activities in the past might not be identifiable at the excavation level, but they could be recognized through geochemical analysis. Since this early pioneering work, multi-elemental geochemical surveys have been to investigate past landscapes, sites and features, to elucidate past human activities.

Previously, multi-element data sets were constructed by undertaking multiple single element analyses, through using instrumentation such as Atomic Absorption Spectrophotometry (AAS). These analyses were time-consuming, requiring each element to be measured individually. More recently, the advent of Inductively Coupled Plasma-Mass Spectrometry (ICP-MS), Inductively Coupled Plasma-Atomic Emission Spectrometry (ICP-AES) and Inductively Coupled Plasma Optical Emission Spectrometry (ICP-OES), has allowed the acquisition of a wide range of metal concentrations per analyzed sample, greatly reducing laboratory time. These methods are commonly referred to as "wet chemistry" methods, requiring a destructive acid digestion of the sample, normally in a water bath or on a hotplate, before subsequent dilution. In comparison, other methods such as portable X-Ray Florescence (*pXRF*), whilst offering lower precision and detection levels to the "wet digestion" methods, offer the ability to produce multi-element data sets through a non-destructive analysis, which is important when analyzing more sensitive archaeological materials, such as artifacts. However, data sets produced by pXRF often have numerous "no detection" values, where an element concentration was below detection limits, making subsequent data analysis more complicated.

Multi-element analyses of archaeological sediments and samples have been applied on a variety of scales, ranging from the landscape scale (Bintliff *et. al.* 1992) to the landscape or wider site scale (Entwistle et al., 2000; James, 1999; Linderholm and Lundberg 1994). They have been used to identify activity areas within houses (Middleton and Price 1996), analyze intra-site organization of space (Misarti et al., 2011; Vittori Antisari et al., 2013), intra-site analysis of metalworking residues (Carey and Juleff, 2013; Cook et al., 2005), intra-feature analysis (Cook et al., 2010) and intra-feature analysis combined with soil micromorphology (Macphail and Crowther, 2008). For example, Milek and Roberts (2013) applied ICP-AES, coupled with other geoarchaeological methods such as micromorphology to identify different activity areas within a Viking house, Iceland, whilst Wilson et al. (2008) used ICP-AES to define different activity areas within recently abandoned historic farming sites in the UK. Furthermore, a reoccurring theme within archaeological and environmental geochemical work has been the identification of metalworking residues through geochemical survey, such as the identification prehistoric copper contamination (Jenkins et al., 2001), copper and lead mining (Mighall et al., 2009), tin extraction (Thorndycroft et al., 1999), and the mapping of metalworking residues at site-specific scales (e.g. Cook *et. al.* 2005; Carey and Juleff 2013). In addition, non-destructive analyses such as pXRF are commonly being applied to analyze artifacts for provenancing (see section 17.5); this method has recently been applied to geochemically fingerprint the megalithic sarsen stones at Stonehenge, enabling their original location to be identified (Nash et al., 2020).

The power of using multi-element analyses is that individually, single elements can produce conflicting and contrasting results that can make interpretations difficult. Multiple processes, both related to past human activities and also post-depositional effects, can cause variations in element concentrations. However, the acquisition of larger multi-element data sets removes this ambiguity; the entire data set can be modeled using multivariate techniques, moving the discussion away from single elements, into signatures related to past human activities. Often, multi-element geochemical data will contain several signatures, some of which can be attributable to processes in the past. This approach was used on the reconstructed roundhouse at Trewortha, where multivariate analysis of the geochemical dataset were spatially plotted to identify areas of recent "experiential Bronze Age" metalworking. In this case the plotting of factor scores modeled geochemical signatures, with the factor scores showing the relationship between an individual sample and the formation of the new principal component. At Trewortha, PCII effectively defined a metalworking signature with a high level of association to Sn, Cu, and Pb, with the plotting of the factor scores effective for the mapping of the recollected working areas (Figure 17.3b; see Carey et al., 2014). As a final point, the application of using multi-element

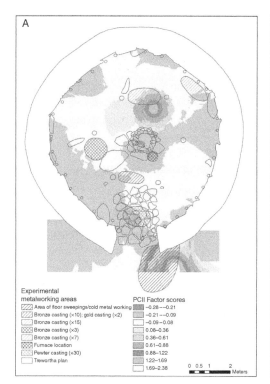
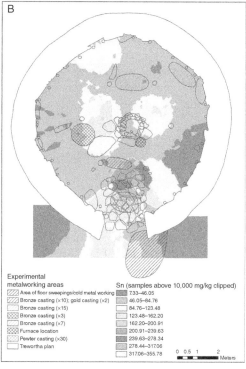

Figure 17.3b The results of Principal Components analysis of the multi-element data set at a reconstructed roundhouse, Trewortha, Bodmin Moor (Cornwall, UK), showing (a) the analysis of the geochemical data produced PCII that was strongly associated with Cu, Sn and Pb. This defined a metalworking signature, with the PCII factor scores for each sample plotted spatially and interpolated through krigging. This analysis was used to identify areas containing metalworking residues and these were related to the recollection of events within the roundhouse and (b) the plotting of an individual element Sn, a key metal for the alloy bronze. The spatial definition of this element bears a strong relationship to the PCII factor scores, strengthening the interpretation of metalworking residues (Cu also demonstrated a similar spatial relationship) (Carey et al., 2014/with permission of Elsevier).

geochemical analyses can only realize its full potential when the context of the sample is understood. This is of course easier, when the analysis is integrated with the excavation of a site, and the contexts and relationships of features, sediments and deposits is known. However, when used on wider scales, the appreciation of the context of the sample becomes more difficult to understand, producing additional variance into the data set, sometimes complicating the interpretation of the data. Like all proxies, context specificity and understanding is key to its successful deployment. Control thin sections are able to record the micro-contexts where multiple floor layers, or trampled deposits, including tracked-in materials, could complicate floor interpretations, and also allow the potential for SEM/EDS, micro-FTIR, and micro-XRF to exactly identify non-ferrous metal working materials (see Figures 17.6c–17.6f).

17.2.5 Magnetic Susceptibility

Together with analyses of organic matter and phosphate, low-frequency mass-specific magnetic susceptibility (expressed as χ_{LF}) is a third key laboratory technique commonly employed to provide bulk data at archaeological sites on soils and pyrotechnical activities (Clark, 2000; Crowther, 2003; Crowther and Barker, 1995) (Figure 17.2b; Table 17.2). The magnetic susceptibility of a soil or sediment is generated by the amount of magnetic minerals that are present (Gale and Hoare, 1991: 201 *et seq.*). The prime influence on magnetic susceptibility enhancement that is of interest to archaeologists is biological activity, which is responsible for the formation of the magnetic mineral maghemite; this occurs in topsoils through the "fermentation processes" associated with alternating oxidation/reduction conditions (Longworth et al., 1979; Tite and Mullins, 1971). The second is burning, which also mainly effects topsoils (Oldfield and Crowther, 2007), and this enhances χ because iron minerals become aligned (a phenomenon also used for paleomagnetic dating of hearths; Clark, 2000). It is therefore clear that if a soil/sediment is low in iron (Fe) it is likely that it will not develop a high χ value, and Fe content is dependent on geology and soil factors (Tite, 1972). Examples of deposits and soils that are low in Fe are (non-ferruginous) peats, or an albic horizon.

There are several different types of measurements and analytical treatments used in the study of magnetic susceptibility. These include raw measurements of samples, samples that have been treated/heated, and recalculated values (Clark, 2000). As noted above, field sections and cores can be rapidly assessed as a sample screening process, before fully quantitative standardized laboratory analysis of dry weighed samples where magnetic susceptibility measurements are made on a known mass of soil. Homogenized fine soil is normally weighed in a plastic pot, with often ~10 gm being used. The pot is then being placed in the magnetic susceptibility meter, and a series of readings are taken. Data are normally recorded as SI units ($\chi - \times 10^{-8}\,\mathrm{m^3\,kg^{-1}}$).

Measurement of χ_{LF} is the most common one carried out in archaeology. It is also valuable to know what the maximum potential value for χ is in a given sample (i.e., χ_{max}; Crowther and Barker, 1995; Crowther, 2003). A simple, qualitative measure can be gained by testing the sample a second time after it has been ignited to estimate LOI (Viklund et al., 2013). Such measurements also provide proxy information on amounts of iron present. Waterlogged sediments often have a very low χ, but when ignited at 550°C they can produce very high values (χ_{550}) because of the presence of bog iron (cf. Roman peat at Warren Villas, Bedfordshire, UK): $\chi = 5 \times 10^{-8}\,\mathrm{m^3\,kg^{-1}}$; χ_{550} = $6098 \times 10^{-8}\,\mathrm{m^3\,kg^{-1}}$). Equally, deposits that become waterlogged are affected by gleying (hydromorphic leaching) processes and the formation of mobile ferrous iron. This can lead to a decrease in the value of χ.

As levels of magnetic susceptibility (MS) often relate to geological background, ways of producing MS data that can be compared within and between sites have been suggested. This has been done by comparing MS and potential maximum MS of a sample; the latter is achieved by heating the sample. There are both quantitative and semi-quantitative methods. Quantitative potential susceptibility (χ_{max}) can be determined by first heating a sample under reducing conditions, followed by oxidizing ones. This is carried out by mixing samples with household flour and adding lids to create reducing conditions (Crowther, 2003; Crowther and Barker, 1995; Graham and Scollar, 1976). Crowther (2003) found that he could measure fractional conversion – $\chi_{conv}(\chi_{LF}/\chi_{max})$ this way and has created a database from 30 sites. Some of his major findings are that magnetic susceptibility results may be problematic where the χ_{max} is naturally low (e.g. in poorly ferruginous soils) and where the present χ_{max} is no longer representative of the original circumstances in which the soil/sediment formed (e.g. through gleying); leaching under conditions of podzolization has a similar effect. He also demonstrated that χ_{conv} does provide a measure of enhancement in samples that are known to have been affected by heating/burning, with values of $\geq 30\% \ \chi_{conv}$ being recorded in certain contexts (e.g. burned daub from Early Neolithic Ecsegfalva (Hungary): $\chi_{LF} = 717 \times 10^{-8} \ m^3 \ kg^{-1}$; $\chi_{conv} = 48.1\%$). Applications of magnetic susceptibility include profile logs (see Figure 17.2b), grid mapping across sites (see Figures 18.17, 18.20, and 18.21) and features in long houses, such as the location of hearths (Viklund et al., 2013) (see Tables 11.2 and 11.3 for presentation of χ, χ_{max}, and χ_{conv}). At Vanguard Cave, Gibraltar magnetic susceptibility was a component in grouping microfacies types, with combustion zones having the strongest enhancement ($\chi = 128$–$234 \times 10^{-8} \ m^3 \ kg^{-1}$), compared to silt beds ($\chi = 22$–$107 \times 10^{-8} \ m^3 \ kg^{-1}$) and blown sands having the lowest signal ($\chi = 14$–$90 \times 10^{-8} \ m^3 \ kg^{-1}$) (Macphail et al., 2012b).

Magnetic susceptibility has also been used in larger-scale endeavors, for example examining lake cores, alluvium, and other long sequences (Needham and Macklin, 1992: e.g. fig. 2.4; Oldfield et al., 1986). In China, for example, the technique has been employed extensively to recognize paleosols and associated paleoclimates (Tang et al., 2003). It appears that the amounts of magnetite and maghemite are enhanced during times of pedogenesis. Thus, in thick loess sections, fluctuations in magnetic susceptibility values are clearly higher in paleosols than in the intervening values for aeolian dust (loess). Although the exact mechanisms involved in the enhancement process are not exactly known, they do clearly point to environmental changes, which can be linked with oxygen isotope stages found in deep-sea sediments. Consequently, many loess sections can be correlated based on their stratigraphies and magnetic susceptibility changes.

Similar findings have been reported from Huizui (Henan Province, China), where the presence of eroded topsoils and rolled pottery sherds in alluvial sequences is linked to enhanced magnetic susceptibility levels (Liu et al., 2002–2004; Rosen, 2008; Rosen et al., 2017). It was clearly confirmed when loessic alluvium and natural subsoils (2.79–4.41 %$\chi_{conv.}$) provided control measurements for comparing Late Neolithic (Longshan) ash pit fills (9.16–23.4 %$\chi_{conv.}$) (Macphail and Crowther, 2007a). The same study found that supposed lime plaster floors recorded negligible magnetic susceptibility enhancement (1.80–2.45 %$\chi_{conv.}$), supporting the view that such "floors" were composed of tufa slabs as identified in thin sections of four tufa-constructed floors.

As noted above, techniques such as these, while important, should be supplemented with other analytical data and should not be taken at face value. Other measurements of χ, including χ_{550}, χ_{max}, and ratios between χ and χ_{550}, χ and χ_{max} (%$\chi_{conv.}$), as well as amounts of total iron present, are all useful additional data when employing magnetic susceptibility to reconstruct past environments (Allen and Macphail, 1987; Clark, 2000; Crowther, 2003; Crowther and Barker, 1995; Longworth et al., 1979).

17.2.6 Sourcing (Provenancing)

A common first step in site investigations involves sourcing raw materials of manufactured and constructional materials, establishing the provenance of sediments and archaeological deposits, and evaluating the origin of a soil's parent material. Methods to accomplish these goals are discussed below where we deal with Instrumentation Techniques (17.6), and the analysis of minerals (including "heavy minerals") through the optical microscope, X-ray methods (XRD), and chemical and microchemical studies (XRF, NAA, ICP-MS, microprobe, FTIR and micro-FTIR, SEM-EDS, "Quemscan") (e.g. Weiner, 2010). These methods also allow the investigation of sediment diagenesis, and type and maturity of soil formation.

We present an example where SEM-EDS on thin sections was combined with XRD (*X-ray Diffraction*) on correlated bulk samples from the Late Pleistocene, Upper Paleolithic site of Chongokni (Korea), where some 8 m of soil-sediments overlie a basalt flow; Acheulian-like hand axes and quartzite core artifacts are concentrated in the Red Clay (Unit II) (Yi, 1988, 2005; Yoo, 2007) (Figure 17.4). Soil micromorphology results suggested red soil microfabric development in brown soil-sediments, giving the unit an overall red color, but SEM/EDS analyses showed that difference was unrelated to total iron content. XRD was thus carried out on the sequence (Figures 17.4a–17.4c). Hematite (αFe_2O_3), which was absent from yellow alluvial Units II and IV and was identified in the soil-sediments of Unit II (Figure 17.4d), is formed through cycles of dehydration-hydration causing *natural* rubifaction. This is consistent with a highly seasonal climate (Courty et al., 1989b, 165– 168; Duchaufour, 1982: 179, 374–375; Brady and Weil, 2008: 327; see also Stoops and Marcelino, 2018: 695–698). The findings from Chongokni may suggest dry season occupation of the floodplain (Macphail, 2010; Yi, 2015).

Figure 17.4a Field photo of Late Pleistocene Upper Paleolithic site of Chongokni, Korea, on the Hantan River (a tributary the Imjin River). Yi (2005, 2015) excavated ~8 m-thick deposits overlying an uneven basalt bedrock (~51 m asl), which is a typical geological sequence for the area. This sequence (from the base upward) is composed of: Lacustrine Clay (LC: Unit VIII, ~0.5– 1.0 m), Clastic Sands (CS: Unit VI, >2 m-thick coarse cross- bedded clastic sands containing vesicular basalt), Yellow Sands (YS: Unit V, >1 m), Yellow Silt (YZ: Unit IV, ~1 m), Yellow Red Clay (YRC: Unit III, ~1 m), Red Clay (RC: Unit II, ~2 m of artifact-bearing soil sediments with "Acheulian- like" stone tools; see Figure 17.4b), and Red Brown Clay (RBC: Unit I, ~1 m- thick sediments) below the modern (truncated?) excavated soil surface at ~59– 61 m asl. Artifacts, such as quartzite hand axes and cores, including the famous Chongokni hand axe found in 1978, are typical occurrences in the Red Clay (Yi, 1988; Yoo, 2007).

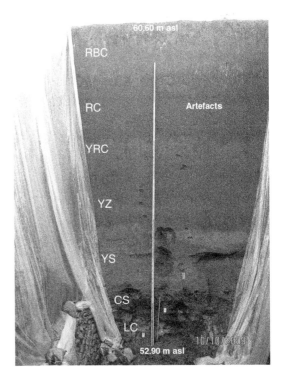

Figure 17.4b As Figure 17.4a; quartzite core sample; core-buried and protected soil sample M28 in Unit II, Red Clay, and unburied soil surrounding quartzite core sample M27 (NE 07A). Core is still in-place.

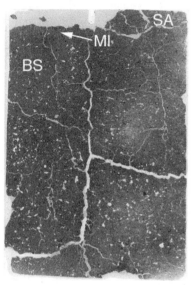

Figure 17.4c Scan of thin section M29 hand axe-buried soil (Unit 60.20 m asl); curved void space at top of thin section above 'MI' is due to removal of hand axe. It is assumed that the hand axe was 'dropped' onto the biologically homogenised soil-sediment (buried soil BS); burrowing led to a soil accumulation (SA), which started to bury the hand axe. Ensuing flood inundation produced muddy inwash (MI) into thin contemporary burrows below the hand axe. Soil micromorphology of buried soils in Unit II could differentiate a reddish soil microfabric layer over a brown microfabric, but SEM/EDS found no clear difference in iron content (e.g. red microfabric: 7.29–7.66% Fe; brown microfabric: 6.93–7.75%. XRD was therefore carried out on a number of bulk samples in order to identify broad mineralogical trends at the site (Fig 17.4d) (Macphail, 2010). Frame width is ~50 mm.

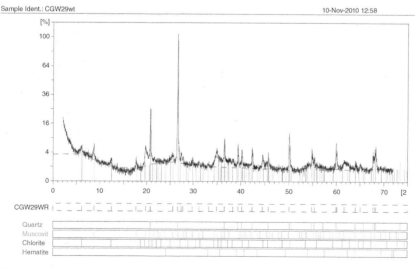

Figure 17.4d As Figure 17.4a and 4c, XRD mineralogical analysis on bulk Sample 29 (Unit II). In both Unit II samples (28, 29) red soil colours appear to be due to small amounts of hematite. Samples from the Yellow Sands (Unit V) and Yellow Silt (Unit IV) contained no hematite, although some was present in a sample from the Red Brown Clay (Unit I), which overlies Unit II and is probably partially derived from it (data from Steve Hirons, Geology, UCL).

17.3   Microscopic Methods and Mineralogy

17.3.1   Binocular and Other Macro-methods

Binocular examinations can be made on loose samples, thin sections, and polished blocks, with magnifications usually ranging from x4 to x20. These lower magnifications provide information about the larger-scale features of the sample that might not be evident with higher magnifications alone. In the case of thin sections, useful macroscopic investigations can be carried out using binocular microscopes with both plane- and crossed-polarized light, as in normal petrological microscopy (see below 17.3.2). Even if binocular microscopes only have an incident light source, polished blocks (i.e., impregnated blocks with a resin-coated or spray coated "cover slip") can be studied to view the fabric, including macrostructure, inclusions, coarse voids, and bedding. Anomalous and enigmatic organo-minerals, possibly from extraterrestrial impact origin, have been identified employing routine binocular observations (Courty, 2012; Courty et al., 2008). Such minerals were found concentrated in a charcoal-rich alluvial layer of wildfire origin that was widespread over kilometers along the Gudbrandsalen (Lågen River, Norway) (Macphail et al., 2016)

The coarse fraction (>2 mm) of loose samples can be analyzed for composition, texture, shape, and roundness. This technique is often used on coarse natural sediments, such as river gravels. It has also been applied to the study of microartifacts (e.g. lithics, charcoal) in anthropogenic deposits to examine spatial patterning of activities (Rosen, 1991; Sherwood et al., 1995) or to dark earth, in which stratigraphic units are difficult to determine, but where features and units may be indicated by concentrations of coarse inclusions. In the latter case, this approach has been carried out to establish amounts of Roman mortar, tile and brick, stones, bone, charcoal, and food waste in the form of mussel shells in deposits from sites in London and more recently in France (St Julien, Tours, France) (Fondrillon, 2007; Fondrillon et al., 2009; Macphail, 1981; Sidell, 2000). Middens have also been studied in this way to identify stones, botanical (seeds, wood and charcoal fragments) and faunal (shells, animal and fish bones) remains (Balaam et al., 1987; Stein, 1992; Villagran et al., 2009) (see Figure 7.5). Lastly, "invisible" stratigraphy has been elucidated through the use of X-radiography, where undisturbed cores are collected in 450 mm long 100 mm diameter plastic tubes, as standard U4/U100 samples (see Chapter 16) and then X-rayed (Barham, 1995). This technique has been commonly applied to marine sediment cores, along with other logging techniques already noted, such as magnetic susceptibility, and elemental studies (Rothwell and Rack, 2006) (see Box 18.1). Trace fossils left after bioworking were examined by X-radiography in order to help record changes from fluvial to estuarine and marine depositional conditions in the fill of the Red River paleochannel during the Pleistocene in the Gulf of Tonkin, South China Sea (Wetzel and Unverricht, 2020)

17.3.2   Petrography – Soil Micromorphology

Soil micromorphology is considered by some to be one of the "new techniques" in geoarchaeology. In fact, the employment of soil micromorphology in archaeology has a history almost as old as the subject itself: Kubiëna first developed the subject from geological petrography in the mid-20th century (1930s–40s) in Germany. It was almost immediately taken up by Cornwall at the Institute of Archaeology, University of London (Cornwall, 1953; Kubiëna, 1938, 1953), who produced reports up to the 1960s (in Evans, 1971, 1972; Macphail, 1987a). The Scottish soil scientists Roman and Robertson also applied this pedological technique to archaeological sites during the 1970s and early 1980s (Romans and Robertson, 1975, 1983a, 1983b). The subject also evolved internationally during this period (Courty and Fedoroff, 1982; Courty et al., 1989b; Courty and Nørnberg, 1985; Goldberg, 1979b, 1980, 1981).

Since these early efforts, archaeological soil micromorphology has taken off as a mainstay of geoarchaeological investigations, especially in Europe (International Soil Micromorphology Working Group, initiated in 1990 by Macphail), with numerous articles in conference proceedings, in several journals, such as *Geoarchaeology, Journal of Archaeological Science, Archaeological and Anthropological Science*, and *Quaternary International* (Macphail, 2013). There has always been a strong tradition in the USA of applying soil micromorphology in the realm of soils, especially for soil classification and even for the study of NASA's moon rock (Douglas, 1990; Soil Survey Staff, 2014). Despite the long-term major contribution made by Goldberg (Universities of Texas and Boston), however, this technique is still completely underused by North American geoarchaeologists (Collins et al., 1995; Holliday, 1992) compared to Europe. Here, the working group under the umbrella of International Union of Soil Sciences, Commission 1.1 Soil Morphology and Micromorphology, has been holding one to two international workshops annually (e.g. Universities of Barcelona and Lleida, Spain, 2012; Macphail, 2013, 2019d).

17.3.2.1 Protocols in Soil Micromorphology

A number of protocols have been formulated for the scientific description of thin sections that can be applied to archaeological soils. The earliest ones were introduced by Kubiëna and Brewer (Brewer, 1964; Kubiëna, 1938, 1953), and the most recent and most commonly used internationally, are by Bullock et al. with recent revisions by Stoops (Bullock et al., 1985; Stoops, 2003; Stoops, 2020). Much can also be learned from FitzPatrick (FitzPatrick, 1984, 1993), although much of his terminology is not in mainstream use, and now the most up-to-date textbook on the micromorphology of soils and regoliths is Stoops et al. (2018a). These descriptive systems, combined with those from sedimentary geology (e.g. Tucker, 2001) have been used to describe many kinds of archaeological materials and situations. For example, archaeological cave sediments, Pleistocene sediments, floor deposits, "manufactured" materials (e.g. daub and mortar), pit fills, and soils *sensu stricto* have been described (Courty et al., 1989b; Karkanas and Goldberg, 2018b; Macphail and Goldberg, 2018a; Nicosia and Stoops, 2017). These works have permitted the accurate characterization and interpretation of archaeological deposits. However, as more complex and completely novel materials are investigated, the use of the petrological microscope has had to be supplemented by a whole range of complementary techniques (e.g. fluorescence microscopy, cathodoluminescence, SEM-EDS, microprobe, micro-XRF, micro-FTIR – Fourier Transform Infrared Spectrometry) (e.g. Berna, 2017; Mentzer, 2017b; Stoops, 2003) (see below).

17.3.2.2 Sample Processing Methods

Essentially, micromorphology relies on the use of undisturbed samples in order to conserve their microcontextual information (Goldberg and Berna, 2010). Thus, the undisturbed sample that was retrieved in the field (Chapter 16) needs to remain undisturbed during processing. This entails the removal of any moisture before the sample is impregnated with a resin that will eventually produce a hard block after curing. Further details of sample processing methods for thin section manufacture, and some guidelines concerning the handling of thin sections are given in Appendix A1.17.5 and A1.17.6.

17.4 Thin Section Analysis

Essentially, petrological studies are carried out using plane polarized light (PPL), crossed polarized light (XPL), and oblique incident light (OIL). The use of the petrological microscope can be found in textbooks on petrology (e.g. Nesse, 2014), the internet (Nikon, 2021), and in soil micromorphology literature (Bullock et al., 1985; FitzPatrick, 1984; Stoops, 2020). Another commonly long employed technique is fluorescence microscopy (Stoops, 2017b; Van Vliet et al., 1983), which can be carried out using an attachment to the petrological microscope (see below).

The most important aspect of the procedure is to have a continuum of observation – starting with the field – and then from the lowest to the highest magnifications, i.e., from ×1 (hand held thin section), through ×25, ×40, ×100, and ×200/400. Thus, there will be no observational break between what has been seen in the field and what is observed under the microscope. Many suggestions are given in Courty et al. (1989b: 63–78) and Macphail and Goldberg (2018a: 66–96).

Macrostructure, field layers, and coarse inclusions cannot be recognized at the higher magnifications (>×10). Equally, the composition of the fine fabric cannot be fully elucidated at low magnifications (<×4). A study protocol should therefore include the following six steps:

1) *Preliminary examination* of the thin section by eye (×1), by using a hand lens, microfiche reader, or binocular microscope (×4–×20) – the last may also provide both PPL and XPL observation (see above). Photos or scans of the entire thin section using a flatbed scanner are important for seeing the overall fabric and organization of the components (Arpin et al., 2002; Carpentier and Vandermeulen, 2016; Gutiérrez-Rodríguez et al., 2018; Haaland et al., 2019); without this step, one risks not seeing the forest for the trees. The scanned image can be examined using software packages such as Adobe Photoshop® or Image J (https://en.wikipedia.org/wiki/ImageJ) whereby different levels of lighting and contrast, and color balance, can be utilized to see what may be present but not immediately obvious. To repeat: the chief role of this macro-study is to relate what was seen and recorded in the field (photos, images, section drawings, contexts sheets, notebook), to the laboratory study (whole contexts, layers within contexts, etc.) (Courty, 2001; Karkanas and Goldberg, 2018b). At this stage, macrostructure, internal organization of the components, and heterogeneity, for example, can be identified and examined more closely during stage 2.

2) *Perusal of the thin section* using the polarizing microscope can be carried out at ×25–×40 magnifications. Areas/features/inclusions of interest can be identified, and information on these can be related back to the field and low-power microscopic observations. In some cases, it may be desirable to summarily examine all the thin sections at this stage, to see if field and macromorphological identification of similar features is correct. It is important at this early stage to switch rapidly between PPL, XPL, and OIL and to link theses observations with those at the binocular scale. For example, variations in amounts of organic matter may show up well in PPL, whilst calcium carbonate-rich areas (e.g. patches of ash, secondary carbonate) become strongly evident under XPL because of their high interference colors. Equally, OIL will reveal iron and manganese stained features, while fragments of "black" charcoal and "red" burned (e.g. rubified) soil become obvious. This may also be a good time to employ blue light (BL; or UV filters), to reveal the presence of residual roots, bone and calcium phosphate-enriched materials (some coprolites; phosphatized soil/sediment), and other autofluorescent materials (see below (Stoops, 2017b)). Areas and features can be denoted on the slide by drawing on the cover slip with a marking pen, or on the back of the thin section, noting a two-dimensional coordinate; drawing on a printed version of a scanned image is also an effective way to keep track of observations.

3) *Detailed observation and description* of areas/features/inclusions should be carried out carefully *without* any interpretative weighting. If interpretation is introduced at this stage, it can undermine the systematic gathering of soil micromorphological data. It is recommended that any new worker or student *sensu lato* should employ Stoops (2020), but they should be also aware of slightly varying approaches in Bullock et al., (1985), while Courty et al. (1989) and

Macphail and Goldberg (2018) have a more thematic approach to geoarchaeological site studies. There are numerous specialized articles that can be sought, and a way into these is through selectively reading thematic chapters in Nicosia and Stoops (2017) and Stoops et al. (2018a). Although all new researchers are strongly urged to undergo proper training in soil micromorphology and to record all their information systematically, it is important that the results be presented in such a way as to be rapidly accessible to both specialist and lay colleagues alike. Examples of description are given by Bullock et al. (1985: 142–145) and Stoops (2003: 137) also includes a tabulated presentation. In addition, we present a suggested order of description and a number worked examples from a range of site types (Appendix 1, Tables A1.17.6.2–3). Ways to estimate approximate percentages of grains, voids, or other objects are repeated in these books, following FitzPatrick (1984). With experience and much time spent estimating percentages, great accuracy can be achieved with considerable less time than manual counting of such objects (e.g. Acott et al., 1997; Crowther et al., 1996). Individual characterization and identification of features and materials may also require other instrumental analyses (e.g. EDS, microprobe).

4) *Presentation of data* needs to be carried out in ways best suited to the findings, type of site and chosen audience (see Chapter 18). Although systematic presentation is useful, it must be remembered that each site and audience varies and the presentation of data should be flexible. Very few readers are going to wade through a mass of descriptive information for its own sake, and so presentation of material must be very carefully balanced to be both inclusive and relevant. In Appendix 1.17 we present a standard format (Table A1.17.6.1), and how it can be employed to describe a freshwater sediment (Table A1.17.6.2; see Figures 17.6e–17.6h), a barrow-buried soil (Table A1.17.6.3; see Figures. 17.6a–17.6d), and an anthropogenic deposit (Table A.1.17.6.4). An additional approach has been to present the semi-quantitative data in the form of count tables (Macphail and Cruise, 2001). These were taken direct from (Bullock et al., 1985), and have six "frequency" categories for common features and materials such as gravel, soil microfabric and ped types, and six "abundance" categories for pedofeatures – clay coating types, mesofauna excrements – and inclusions, for example: charcoal, dung spherulites, roots, ash, bone fragments. These tables provide a lot of detail in a log that can be correlated with macrofossil, chemical and pollen data for example. They also allow reliable checks and comparisons when numerous microstratigraphic units from large profiles and lateral samples. A more basic tabulation of only three categories for "present", "abundant" etc., is fine for quick assessments, as used by pollen and diatom scientists. We believe that a primary step in the interpretational process is the characterization and identification of soil/sediment microfabric types or soil microfacies types based upon description and semi-quantitative data (SMTs – Soil/Sediment Microfabric Types), as Table A1.17.6.1) (Courty, 2001; Macphail et al., 2007a; Macphail and Goldberg, 2018: table 3.2). An SMT is described in detail at all magnifications under PPL, XPL, OIL, and BL, and natural soil horizons and archaeological deposits could very have different characteristics. For example, a Bw horizon will not often have the very fine charred organic matter inclusions that can be found in a manured ploughsoil, for example. Also, a well-preserved hearth may well display a layer with a SMT composed solely of calcitic ash crystals.

5) *Interpretations* need to be carefully explained and justified. They should first follow "identifications" of SMTs, inclusions, grains/clasts, and (*pedo/soil*) features (i.e. to include all sedimentological, pedological, and anthropogenic processes). This strategy will allow for an initial, first level of interpretation (e.g. homogeneous material, mixed material, illuvial or gleyed character, high or low biological activity). At this time, *primary* materials such as sediments can be

differentiated from *secondary* materials such as soils. This stage is then followed by a second level of interpretation (e.g. identifying the presence of a specific types of deposit – e.g. colluvium, soil horizon, occupation layer, or cultivation soil; French, 2003: 47 *et seq.*). The third level of interpretation attempts to place the slide in a broader interpretive context (both at the regional and site level) such as overall soil type, mass-movement deposit, type of land use, stable, house floor, building collapse. It should be recognized, however, that it may not always be possible to reach the third level of interpretation on micromorphological data alone, and thus, one should avoid jumping to this level too soon during thin section analysis, as it can bias the observation process. Moreover, crucial identifications of anomalies may be erroneously disregarded if descriptions are carried out with an "agenda" that weighs descriptions in favor of a predetermined interpretation.

6) *Testing of interpretations* is done by reviewing the first five stages, and comparing findings with those from independent studies in other disciplines, such as bulk chemistry, magnetic studies, palynology, and archaeological data. Such combined datasets can also be presented together in tables, so that the skeptical reader may be persuaded by consistent findings from different methods. It should be emphasized, that as in all geoarchaeological studies, interpretations can also be reviewed against results from field excavation, as well as the interpretations presented from zooarchaeology, archaeobotany, artifact recovery and dating. In the examples presented in Appendix 1 (Tables A1.17.6.2–4), the interpretation of freshwater sediment sequence at Boxgrove was not only supported by thin section analysis of calcite root pseudomorphs, but by bulk counts of slug plates and *earthworm granules*, and the presence of relict saline foraminifera washed in from the eroded marine substrate (Pope et al., 2020; Roberts and Parfitt, 1999). At Raunds, the interpretation of soils becoming both phosphate-enriched and poached through stock concentrations was consistent with macrofossil (cattle bone, grass tuber, dung beetle) and archaeological findings (Harding and Healy, 2011; Healy and Harding, 2007; Macphail, 2011a). Cave sediment diagenesis was substantiated by detailed mineralogical analyses using FTIR (Karkanas et al., 2000; Karkanas et al., 1999; Weiner et al., 2002), now enhanced with much more focused micro-FTIR on individual objects (e.g. bone, soil aggregates) in thin sections (Berna, 2017; Goldberg et al., 2009) (see Chapter 10). At the London Guildhall, interpretations of soil micromorphology and complementary chemistry and pollen were based upon experiments (Macphail et al., 2004) as well as on well-preserved remains of timber buildings (Bowsher et al., 2007; Macphail et al., 2007a) (see Chapter 14). Many other examples are provided throughout this book.

Finally, the first to third level interpretations of thin sections can be based upon post-depositional features, some pedological and geogenic (see Tables A1.17.3 and A1.17.5). When these are observed and described in detail they often show a temporal sequence that parallels the "law of superimposition": original deposition→bioturbation→calcite cementation→phosphatization. As such, one has to evaluate which post-depositional features came first, second, third (etc.), and last.

For example, at the Fryasletta site (Gudbrandsalen, Lågen River, Norway) (see Box 17.1; Macphail et al., 2016), two examples of multi-method interpretative approaches are given:

1) a wild fire event:
   a) Chemistry, magnetic susceptibility, particle size, organic residue, pollen and SEM/EDS were combined to investigate a geographically widespread charcoal layer within an alluvial sequence (Figures 17.5a and 17.5b). The charcoal is unrelated to any *in situ*

burning, but results from local fluvial reworking and deposition of charcoal-rich wild-fire soils at sites where pioneer plants (e.g. *Epilobium angustifolium* t. – fireweed/rose-bay willowherb – and *Pteridium*) had invaded (Figures 17.5c and 17.5d; Tables 17.3a and 17.3b).

2) an alluvium-buried pasture soil:
    1) a humic, bioworked laminated Mull pasture soil (Layer 10) is
    2) sealed by fine (silty) flood alluvium (Layer 9) (Figures 17.5e and 17.5f),
    3) rooting and channel (void) formation occurred with
    4) the voids subsequently becoming affected by iron-phosphate charged groundwater (Thirly et al., 2006), possibly mobilized from the pasture soil; concomitantly, voids became infilled with secondary iron-phosphate (Landuydt, 1990) (Figures 17.5g and 17.5h);
    5) later a coarser alluvial episode occurred (Layer 8), which buried this 1–3 sequence.

In an example from the Quaternary site of Boxgrove (UK), meltwater flow and *sedimentation* produced polyconcave and closed vughy porosity and associated textural "pedofeatures" of inter-calations and void coatings, as final pedofeatures formation (see Figures 17.7a and 17.7b see Table A1.17.4).

Furthermore, it is crucial to continually bear in mind the principle of *equifinality* when inter-preting soil micromorphology. So many different natural and human-induced processes can lead to the similar microfabrics and pedofeature(s). Thus, caution needs to be exercised during descrip-tion, otherwise results may simply follow wishful thinking to attain a "finding".

Another example of equifinality is the presence of iron-phosphate nodules, which through EDS can often be shown to have Fe-P or Fe-Ca-P chemistry. Examples of these were found in soil tram-pled by pigs in an experimental pig enclosure (Macphail and Crowther, 2011), but when these nodules are found they cannot simply be employed to "identify" a pig sty for example. Such nod-ules occur in natural sediments through ground water movement (Karkanas and Goldberg, 2018a; Thirly et al., 2006), in trackway and stockyard deposits (probably derived from fecal waste-associated animal traffic) and in cultivated fields manured with "nightsoil" (e.g. mineralized human fecal waste) (Macphail, 2011c; Nicosia et al., 2017; Shahack-Gross, 2017).

We hope that neophytes do not find the above too daunting, but it is exactly what the authors have to do at each site, in order to make objective observations and the best possible interpretations according to site circumstances. In fact, if objective observations are made and integrated with other geoarchaeological data (bulk and other instrumental analyses), and the site's archaeological and geological context, reasonable interpretations should be forthcoming if the six-step protocol outlined above is adhered to.

17.4.1 Fluorescence Microscopy and Reflected Light

Standard petrography is usefully combined with the use of fluorescence microscopy by adding a fluorescence attachment to the petrological microscope. Observations can be made either using ultraviolet light (UV), blue violet light (BV), or blue light (BL). The attachment combines an excit-ing filter between the light source (sometimes as a form of incident light) and the thin section, and a suppression filter between the object being studied and the eyepiece (Stoops, 2017b). A number of identifications can be made using UV, BV, and BL, although in pedology it was also commonly

Box 17.1   Methodological approach to geoarchaeological investigations of alluvium along the Lågen River, Gudbrandsdalen, Oppland, Norway: (1) wildfire event (broadly dated to 3200–3600 BP) and (2) pasture soils (2800–2900 BP).

Figure 17.5a   Fryasletta: Profile 1a, Trench 1. Field sample P42 (thin sections M42A and M42B), across alluvial layers 1152, charcoal-rich 1150, and 1151 (see Figure 17.5b). Samples M9 and M10 were also collected across 1150 elsewhere in this trench.

Figure 17.5b   Fryasletta: Scan of thin section M42A, with sandy upper 1152a containing an example of twigwood charcoal (blue arrow) and showing relict silt loam layering. 1150 is composed of brown silt loam and a concentrated layer of wood charcoal and wood char. "1151b" is and upward extension of silt loam deposition associated with 1150 ponding (see Tables 17.4a, b for magnetic susceptibility, chemistry and pollen analysis). Overlying 1151a is erosive poorly sorted sand and gravel alluvium. Frame width is ~50 mm.

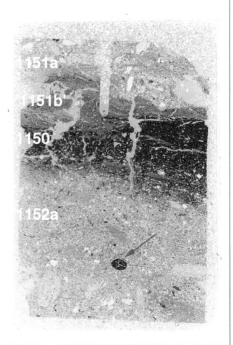

(Continued)

Box 17.1 (Continued)

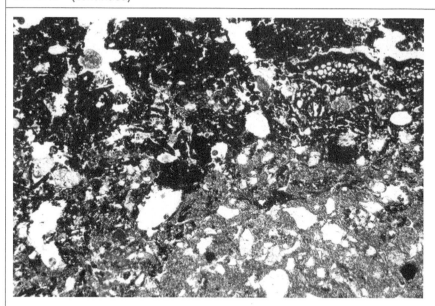

Figure 17.5c Fryasletta: Photomicrograph of M42A, contact between Context 1150 – lowermost laminated fine and coarse silts ("silt loam"; note weak iron staining – not rubification) and uppermost 1152a – biomixed humic sediment and fine charcoal in sandy alluvium. Iron stained silt loam with 5.92–7.06% Fe and 0.0–0.71% Mn (EDS – energy-dispersive X-ray spectrometry). PPL, frame width is ~4.62 mm.

Figure 17.5d Fryasletta: Photomicrograph of M9B; 1150 silt loam infill affecting sands and gravels of 1152. Orange colored (rubified) burned bone and fine charcoal (originally riverside small carnivore scat?). PPL, frame width is ~0.90 mm.

Box 17.1 (Continued)

Figure 17.5e Scan of M63A (Fryasletta, Gudbrandsdalen, Lågen River valley, Oppland, Norway); alluvial silt (L9) seals humic pasture soil, L10. L8 records later higher energy fluvial sands and gravel deposition; humic soil and charred dung (green arrow) have also been reworked. Red arrows (iron stained uppermost turf, Figures 17.5f and 17.5g) and green arrows mark areas of EDS analyses. Frame width is ~50 mm (Macphail et al., 2016).

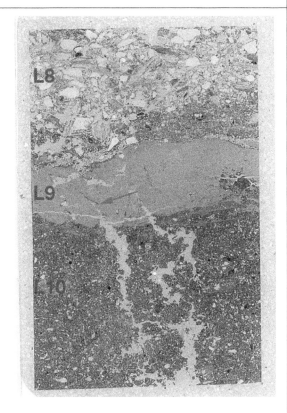

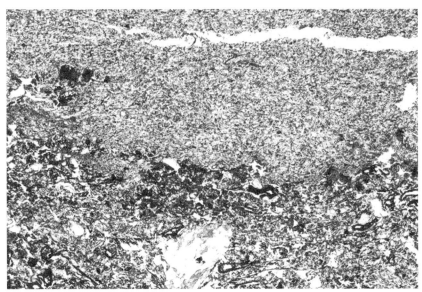

Figure 17.5f As Figure 17.5e; photomicrograph of M63A, uppermost Layer 10, a "laminated mull" humic pasture topsoil, with blackened/aged "grass" leaf litter, of poorly drained soil, sealed by alluvial flood silt. PPL, frame width is ~4.62 mm.

(Continued)

Box 17.1 (Continued)

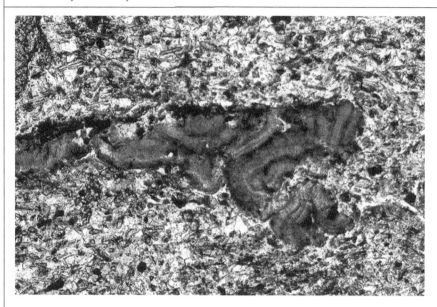

Figure 17.5g Detail of Figure 17.5b; Layer 9; minerogenic alluvial silts, with reddish brown secondary iron void infills ("ferrihydrite"?). PPL, frame width is 0.90 mm.

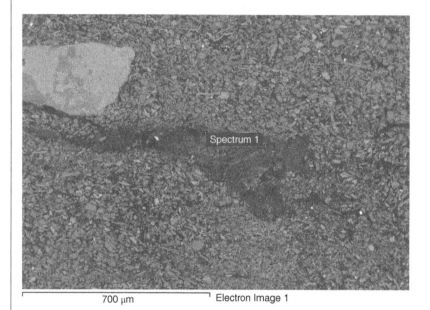

Figure 17.5h Backscatter image of alluvial angular silts and secondary iron void infill (Spectrum 1) – location of EDS analysis; iron is dominant (43.2% Fe), with 1.73% P, 6.33% Ca and 0.59% S (relict root material?) (local silts contain 5.54% Fe and 0.24% P).

Box 17.1   (Continued)

Table 17.3a   LOI (estimated organic matter), fractionated phosphate and magnetic susceptibility data from Fryasletta, Gudbrandsdalen, Norway (supplied by Johan Linderholm, MAL, University of Umeå)

Sample	MSlf	MS550lf	CitP	CitPOI	Pppm	PQuota	LOI
Fryasletta							
1151 (Dark Silt loam)	49	124	38.2	120.6	530	3.16	5.8
1150a (Charcoal)	71	156	26.2	103.6	450	3.95	7.1
1152a (Silt loam)	73	53	85.9	99.8	430	1.16	1.1
1152b (Sandy alluvium)	76	57	74	89.7	390	1.21	1.6

Table 17.3b   Pollen analysis of a sample (PMM42, 2.5cm, Layer 1150) from Fryasletta, Norway.

	No. counted	% count
Trees and tall shrubs		
Alnus	16	7
Betula	40	18
Corylus t.	5	2
Pinus	49	23
cf. Hedera	1	<1
Herbaceous taxa		
Cyperaceae	1	<1
Epilobium angustifolium t.	68	31
Fabaceae undiff.	2	1
cf. Geranium	2	1
Unknowns	2	1
Spores		
Polypodium vulgare	1	<1
Pteridium	27	12
Pteropsida (mon.) indet.	2	1
Pteropsida (trilete) indet.	1	<1
Total	217	100
Pollen preservation		
Normal	70	32
Crumpled	24	11
Corroded	2	1
Degraded	108	50
Split	13	6
Total	*217*	*100*

employed to study void space (through image analysis) made apparent by adding fluorescent dyes to the resin used for impregnation (Jongerius, 1983; Stoops, 2003: 24–27). "Free" aluminum can produce strong autofluorescence in some soils, such as spodic horizons (Van Vliet, 1980), and may cause some burned phytoliths to be autofluorescent because they also have become coated with "free" Al (Van Vliet-Lanoë, University of Lille, pers. comm.; see also Devos et al., 2020b). It is also likely that the chemical composition of amorphous organic matter in spodosols, andosols, and planosols (especially aluminum complexes) affects autofluorescence (Altemuller and Van Vliet-Lanoë, 1990; Van Vliet et al., 1983).

Organic material such as cellulose, roots, fungi, pollen, and spores often become more visible using fluorescence microscopy (Courty et al., 1989: 48–50, plate Id; Davidson et al., 1999). Lastly, the presence of calcium phosphate as various types of apatite can be inferred from their autofluorescence under blue light. Guano-induced phosphatized limestone (in caves), bone, and coprolites of birds, dogs, carnivores, and humans can be identified (Figures 17.5a and 17.5b); in the Middle Pleistocene levels at Westbury-sub-Mendip, Somerset, UK, phosphatic bird guano embedded with small mammal bones produced a "Yellow Breccia" (Andrews, 1990; Courty et al., 1994; Macphail and Goldberg, 1999; https://geoarchaeology.github.io/asma/appendices/plates/: plates IIe–g, IXg–h).

Although reflected light (or oblique incident light – OIL) should be employed as routinely as PPL and XPL in order to differentiate opaque materials under the petrological microscope, specific reflected light studies can also be carried out. Simple OIL observations employing a good flashlight or fiber optics are most commonly used to differentiate opaque minerals, grouped under "heavy minerals" such as "golden" pyrite and black metallic magnetite, for example; OIL is often employed for describing nodular impregnations of iron and iron-manganese, and possibly associated iron-depleted microfabrics. Occasionally, non-ferrous metal working can be identified, as for example, possible soldering with lead produced lead droplets embedded in ash found in Late Roman dark earth at Free School Lane (Leicester, UK) (Borderie et al., 2014a; Macphail and Crowther, 2007b). Special reflectance microscopes are used by archaeometallurgists, for example, to permit the identification of some oxides: at Free School Lane, Thilo Rehren (UCL, pers. comm.) suggested that lead was surrounded by red lead oxide and white lead carbonate (Figures 17.6c–17.6f; Table 17.4). In addition, reflected light methods which originated in coal and organic petrology have been applied to all kinds of organic remains, including various types of char – wood char from wild fires and hearths, and remains from burned bone and fat-derived char (Ligouis, 2017). The last has been found in both prehistoric caves and recent seal-hunting sites (Goldberg et al., 2009; Villagran et al., 2013).

17.4.2 Image Analysis of Impregnated Monoliths and Thin Sections

In soil science, for example, a fluorescent dye in the resin made voids "visible" and were studied by a system called "Quantimet" (Jongerius, 1983; Murphy et al., 1977). With the widespread use of PCs and quantification software, such measurements could be carried out quickly (McBratney et al., 1992; 2000). Image analysis techniques have evolved from the analysis of voids and the size, shape, orientation of constituents to three-dimensional studies of soil using X-ray computed tomography (Terrible et al., 1997; Tovey and Sokolov, 1997) as well as archaeological materials (Frahm, 2019).

Image analysis has been used in sedimentology and soil science for decades using digital cameras linked to computers, and often with the emphasis on analyzing of pore volume, shape, structure, and size and shape of grains for agricultural soil reasons, associated with drainage and crop

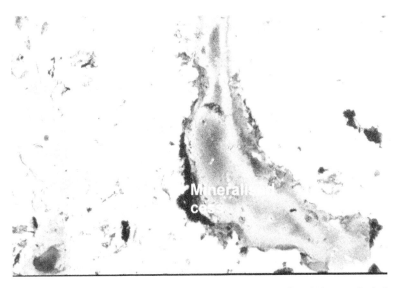

Figure 17.6a   Photomicrograph of little-weathered lower fill of Norman Period cesspit at Monkton, Kent; also present in this context are mineralized fly pupae and *Ascaris* parasite eggs. PPL, frame width is ~3.3 mm.

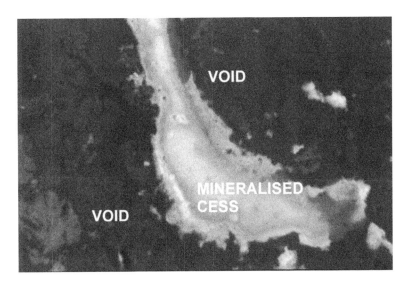

Figure 17.6b   As Figure 17.6a, under blue light (BL), showing strong autofluorescence consistent with an "apatite" mineralogy. Mineralization of organic remains in cess pits has commonly been encountered (Carruthers, 2000/with permission of Wessex Archaeology).

rooting, for example, and have also been employed to infer soil genesis and process, and sediment history relating to source, transport, deposition and diagenesis (Francus, 2004; Mertens and Elsen, 2006; Ringrose-Voase, 1990; Ringrose-Voase and Bullock, 1984; Shang et al., 2018). The method has also been applied to ethnoarchaeological and experimental archaeology (Acott et al., 1997; Goldberg and Whitbread, 1993; Macphail et al., 2003b).

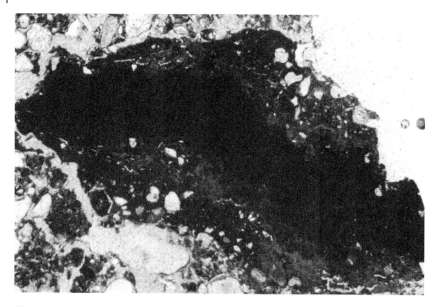

Figure 17.6c Photomicrograph of M11 (Late Roman dark earth, Free School Lane, Leicester); corroded lead droplet embedded in calcitic ashes, from artisan activity, such as lead pipe soldering. Bulk analyses measured 2560 µg g^{-1} Pb and an enhanced magnetic susceptibility (χ_{conv} = 7.41%) (Macphail and Crowther, 2007b). PPL, frame width is ~4.62 mm.

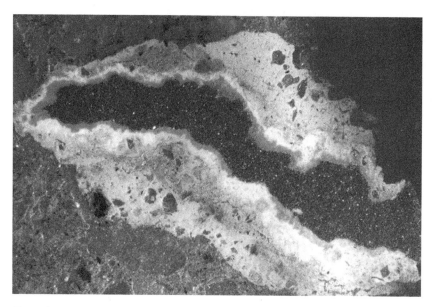

Figure 17.6d As Figure 17.6c, under OIL, showing zoning: (1) center of lead (max 77% PbO) containing small amounts of Al, Si, and Fe; (2) a border of pure red lead oxide (max 100% PbO; see Figures 17.6e and 17.6f); (3) an outer layer of white lead carbonate (max 78.2% PbO, and including Al, Si, P and Ca; linked to original ashes); and (4) outermost grayish brown contaminated ashes (46.3% PbO, with ~15% CaO (10.6% Ca) and ~18% P$_2$O$_5$ (7.8% P)) (original identifications by archaeometallurgist Thilo Rehren, University College London).

Figure 17.6e As Figure 17.6c, BSE X-ray backscatter image of lead droplet, with Spectrum 7 of red lead (17.6f). Scale bar = 1mm.

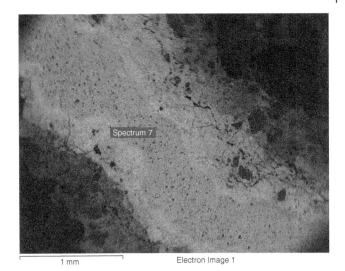

Figure 17.6f As Figure 17.6e, X-ray spectrum of red lead in 17.6e.

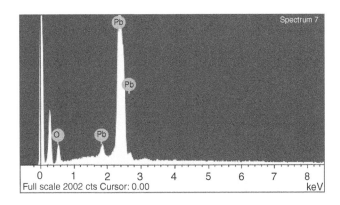

Some attempts have also been made in archaeology to quantify more accurately soil micromorphology components using image analysis (Bryant and Davidson, 1996; McBratney et al., 1992). Image analysis data cannot be interpreted alone, however, without standard soil micromorphological descriptions and accurate identifications. In the latter, manual semi-quantitative and quantitative estimates of features and inclusions that are difficult or near impossible to discriminate in practical terms by image analysis, can provide a data set on a par with those produced by image analysis. Some examples of combined studies are the investigation of a Bedouin floor and Experimental Earthworks Project (Acott et al., 1997; Goldberg and Whitbread, 1993; Macphail et al., 2003b). At the Experimental Earthwork at Wareham (Dorset, UK), a number of pilot studies were carried out, and eventually nine types of mesofauna excrements were manually counted; image analysis was employed mainly to accurately measure void space, organic fragments, dark (amorphous) organic matter and mineral grains of sampled areas at 1 mm intervals. The possible compression of the buried Ah horizon in comparison to unburied (control) soils was investigated by comparing void space. Image analysis data on amounts of "organic fragments" and "dark organic matter" was measured against bulk chemical estimations of organic matter (low temperature LOI and organic carbon). Results showed a consistent match between the two datasets.

Standard soil micromorphology was also combined successfully with image analysis of eight 130 mm-long thin sections through freshwater pond deposits at Boxgrove (Roberts et al., 1994). Here void space and void size were analyzed employing circular polarized light (Stoops, 2003: 21), in which voids are black compared to the predominantly anisotropic calcareous silt sediment. This accurate measurement was crucial in appreciating the amount of fine void space associated with packing voids (Unit 4*) compared to fewer but much more obvious and larger void spaces (closed vughs) in overlying Unit 4d1 (Figures 17.7a–17.7d; see Table A1.17.4) (Macphail et al., 2001).

Despite the above useful combined studies and the large number of cautionary tales and caveats regularly published by specialists in image analysis (Acott et al., 1997; Terrible et al., 1997), there has been a very worrying trend in research and grant applications, which play down the role of standard soil micromorphology in favor of techniques such as image analysis and X-ray tomography (Frahm, 2019). The thinking is that soil micromorphological identifications and interpretations are too difficult and *unreliable*, especially as they are "non-quantitative" (Canti, 1995). Features and materials in thin section have to be identified and clearly understood through standard soil micromorphology first, before image analysis can be used to quantify them. The suggestion that image analysis eliminates the bias of the observer and their interpretation may sound

Figures 17.7a–17.7d: Boxgrove waterhole thin sections 14c and 13c; whole 130 mm size thin section under circular polarised light (Figures 17a and 17c) and areas sampled for image analysis (binary images) of void space and void shape (Figures 17b and 17d, respectively) (Analyses by Dr Tim Acott, Greenwich University, UK).

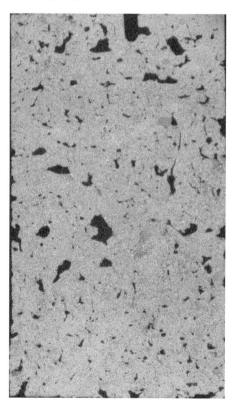 

Figures 17.7a and 17.7b: Soil Micromorphology: massive with dominant fine to medium, coarse meso to macro-channels, showing partial collapse, with polyconcave and closed vughs, and few complex packing voids in M13c. Frame height is ~130 mm.
Image Analysis: 8% voids; dominant fine and medium (61.9% – 101–200 μm^2), frequent coarse (24.1% – 201–500 μm^2) meso voids, few fine (6.5% – 501–1000μm^2) macro voids and very few medium and coarse macro and mega-voids (2.5% – 1000–2000 μm^2). View of center of Figure 17.7a; frame height is ~70 mm (Macphail and Cruise, 2001/with permission of Springer Nature).

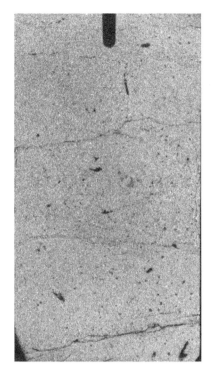
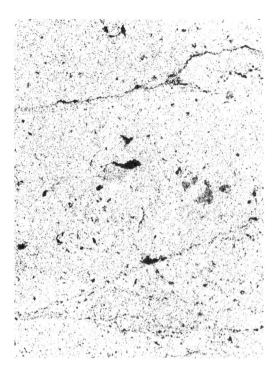

Figure 17.7c and d: Soil Micromorphology: *Structure and voids*: massive, burrow and channel microstructure; common very fine to fine (0.1–0.4 mm), frequent medium (0.8 mm) and coarse (1 mm), and few patches of very fine packing pores, in M14c. Frame height is ~130 mm. Image Analysis: 15% voids; very dominant fine and medium (78.5% – 101–200 μm^2), frequent coarse (20.3% – 201–500 μm^2) meso voids, very few fine (0.9% – 501–1000μm^2) macro voids and very few medium and coarse macro and mega-voids (0.3% – 1000–2000 μm^2). View of upper center of Figure 17.7c; frame height is ~70 mm (Macphail and Cruise, 2001/with permission of Springer Nature).

scientific but is in fact, nonsensical. Investigators need to *know* what they are "counting" (Crowther et al., 1996; Macphail et al., 2003b) (see Chapter 14). Blind use of instrumental methods is the very antithesis of holistic microfacies approaches that have contributed so much to our advances in geoarchaeology (Banerjea et al., 2016; Courty, 2001; Linderholm et al., 2019; Macphail et al., 2007a; Matthews et al., 1997).

17.5 Heavy Minerals and Mineral Provenancing

In sedimentary geology, the composition and texture of deposits can serve as an important indicator of source, transport, and depositional processes (see Chapter 2). For example, in large sedimentary basins that drain different types of bedrock, petrographic analysis of different grains can reflect the relative contributions and evolution of different source areas that furnish sediment into the basins (Pettijohn, 1975).

Similar strategies can be employed in geoarchaeology in different ways. Perhaps the foremost is the petrographic analysis of pottery in which mineralogy of pottery is examined. Information gleaned from the examination of mineralogy and texture, can be used to infer the source of ceramic temper (Quinn, 2013; Spataro, 2002); pottery use has also been investigated (Salquea et al., 2012).

In the New World (southwest USA), A.O. Shepard (Shepard, 1985) was a pioneer in ceramic petrography. Numerous other studies in North and Central America show the value of the technique (Iceland and Goldberg, 1999). Analyses of Roman ceramics (Peacock, 1982) were expanded by (Wieder and Adan-Bayewitz, 2002) who were able, by employing micromorphology of the local soils in Galilee (Israel), to identify specific raw material sources for Roman pottery (cf. Spatoro, 2002). Elsewhere, raw materials and trade routes have been located (Goldberg et al., 1986; Kapur et al., 1999).

In addition to thin sections, the mineralogy of an object or deposits, can be studied by analyzing individual grains under the microscope (Catt, 1986, 1999). Loose sand-size grains, for example, can be mounted on a glass slide and examined under the petrological microscope. Mineralogical composition can be determined optically using characteristics such as shape, cleavage, fracture, and birefringence. Specialists prefer to study grains that are not cemented in a mount, but which can be turned around to see their three-dimensional characteristics (J. Catt, ex-Rothamsted Experimental Station, pers. comm.).

Particularly useful in these types of single grain mounts are the so-called heavy minerals, which are those with specific gravities >2.2–2.9. These minerals are usually separated from the "light minerals" (e.g. quartz and feldspar) either with an electromagnetic separator or by using a heavy liquid. Concerning the latter the sediment is put into a test tube with the heavy liquid and the lighter grains float, while the heavy minerals sink to the bottom. Originally, bromoform or tetrabromethane were used as the liquid, but as these proved to be toxic, they were replaced by sodium polytungstate solution ($Na_6(H_2W_{12}O_{40}) \cdot H_2O$). Once the minerals are separated and mounted on the slide, they can be counted by scanning the slide systematically, using any of a number of point counting techniques. The percentages of the different minerals can be tallied and then interpreted (Catt, 1986; Catt, 1990).

Normally the suite of heavy minerals can point to a sediment source, i.e., its provenance, or help to identify possible locations of raw material used to manufacture ceramics. Alternatively, heavy minerals can be used as weathering indicators, because some minerals are more susceptible to dissolution than others, and this follows the opposite path to Bowen's reaction series (Pettijohn, 1975). Weathering indices have been used to estimate amounts of clay that can be formed in soils (Catt and Bronger, 1998). At Tabun Cave the lower two-thirds of the sedimentary column consisted almost entirely of zircon, tourmaline, and rutile, which are very resistant minerals (Goldberg, 1973; Goldberg and Nathan, 1975; Jelinek et al., 1973). The upper part (Layers B and C) contained a suite of minerals that matches those found in modern coastal sediments in Israel: hornblende, pyroxenes, staurolite, with relatively minor amounts of zircon, tourmaline, and rutile. The inference, corroborated by other analyses, is that these lower sediments had been subjected to intense diagenetic alteration, in which the less stable heavy minerals were removed, leaving a concentrate of the more stable ones. Traditional heavy mineral analysis through petrology is highly skilled. More commonly nowadays difficult grains will be placed in a SEM and so that its chemical composition can identified through EDS (X-Ray analysis), allowing quite rapid mineralogical identifications, as for example employed in forensic studies (Pirrie et al., 2004; Pirrie et al., 2013) (see Chapter 15).

17.6   Organic Residues

Lipid and fat analysis on ceramics have been employed to trace the possibility of stock husbandry (Bogaard et al., 2013; Salquea et al., 2012), while there is a long history of biochemical testing for fecal inputs, encompassing experimentally manured fields and Paleolithic cave deposits

(Evershed et al., 1997; Mallol et al., 2013b). For example, combustion features at Middle Paleolithic Lusakert Cave, Armenia, included hydrocarbons that could be differentiated from those produced by wild fires, implying the hominins were able to control their fires (Brittingham et al., 2019).

Analyzing lipid residues, coupled with experiments, C. Mallol and her research group (University of La Laguna, Canary Islands, Spain) have raised the bar over the last decade in the study of Middle Paleolithic site use and pyrotechnology from El Salt (Spain) (Jambrina-Enríquez et al., 2019; Mallol and Henry, 2017; Mallol et al., 2013a, 2013b, 2017). Some of their results show the construction of *in situ* hearths on exposed soil surfaces rich in herbivore excrements and the remains of angiosperms, suggesting only brief human occupations and possible seasonality (Leierer et al., 2019).

The interpretation of organic remains is not always clear, however, due to site penecontemporaneous activities. For instance, a biochemical pit fill study by Sven Isaksson (Archaeological Research Laboratory, Stockholm University) at the Iron Age "Sea-Kings Manor" (Avaldsnes, Karmøy, Rogaland, Norway), identified many components that yielded a mixed interpretive message. Non-heated fat residues were suggestive of meat processing, possibly by smoking, consistent with high concentrations of very fine charred organic matter ("soot") and phosphate (Macphail and Linderholm, 2017). In addition, given the nature of the site (essentially a farm), the presence of coprastanols is regarded as background contamination.

17.7   Instrumentation Methods: Scanning Electron Microscope (SEM), EDS, Microprobe, and Micro-FTIR

A number of additional instrumental methods are employed to analyse individual objects, as well as contextualized ones ensconced within thin sections (Courty et al., 1989: 50 *et seq.*). Of these, only Scanning Electron Methods (SEM), associated energy-dispersive X-ray spectrometry (EDS), similar electron microprobe and micro-FTIR (Fourier Transform Infrared Spectrometry), and micro-XRF are briefly explored here (Berna, 2017; Frahm, 2019; Mentzer, 2014; Wilson, 2017). These techniques are capable of examining materials that are too small (<2 μm) to be resolved by light microscopy, and also provide punctuated (spot) microchemical data on uniform grains, heterogeneous rock fragments, fine-grained sediments, soils, coatings and other post-depositional effects. It should not be forgotten that although the very highest magnifications can be used (e.g. up to ×60,000), it is useful to commence an SEM study at ×40 to provide a natural transition between observations with the light microscope and those with the SEM, especially when modern machines can take large thin sections. When two 75×50 mm-size thin sections can be loaded, numerous analyses can be carried out without time-wasting evacuations of the vacuum chamber between samples.

Different types of samples can be employed (see below). These can include an individual object, such as a piece of pottery, or several objects (e.g. mineral grains, plant or seed fragments). In soil work, it is common to examine loose aggregates of soil or sediment. In either case, the object(s) are mounted on a metal disc and coated with a conductive material such as gold or carbon. Alternatively, an uncovered thin section or impregnated block can be the object of study. These both must be polished, particularly if microchemical analyses are to be performed, either with SEM/EDS or the electron microprobe.

Scanning electron microscopy utilizes electrons to gather an image rather than light (Weiner, 2010). Essentially, an electron source is focused on and scans across the sample by using electromagnetic field lenses. When the electrons strike the object, both electrons and X-rays are given off, and each form of energy can be detected and each provides different sorts of information.

Secondary electrons are low-energy electrons produced close to the surface by inelastic collisions with the beam (Ponting, 2004). The images produced by secondary electrons are gray scale photographs with high definition and depth of field, and because the electron beam scans across the sample, it is possible to obtain three-dimensional view of the material. Visible in this mode, it is possible to observe dissolution or precipitation of crystals, as well as their relative formation history (e.g. gypsum crystals overlying calcite).

Backscattered electrons (BSE) are a result of elastic contact with the electron beam. With BSE, the energy emission is proportional to the atomic number of the element so that the higher the atomic number the brighter the image (see Figure 17.6e – e.g. Pb). The employment of BSE serves as a complement to secondary electron images for it is a relatively simple means to reveal variations of composition within a sample and thus disclose some of the very fine-scale fabric elements (Karkanas et al., 2000). It is typically utilized to examine pottery, metals, soils, and sediments. The exact elemental composition of materials cannot be judged from the BSE images alone, although shapes of components coupled with micromorphological data can often point to the correct mineralogical identification in the BSE image (see Figures 12.31, 12.32, 15.7, 15.8, and 15.28). To observe elemental abundance secondary electrons is employed. The BSE image is often used to identify areas for microprobe analysis (see below).

X-rays are produced in the process of secondary electron emission when it is replaced by an electron from a different part of the atom. What is important is that the X-rays that are emitted, have energies and wavelengths that are characteristic of each element and its energy shells (Ponting, 2004). Two techniques are used to measure the X-rays and thus infer elemental composition: energy-dispersive x-ray analysis (commonly written as EDS, EDX, EDAX, EDAXRA, EPMA) and wavelength dispersive X-ray analysis (WDX). EDS analyses can be performed on either loose grains or aggregates, or more efficiently in conjunction with micromorphological analysis, on a polished thin section (Weiner, 2010; Wilson, 2017). The latter approach allows the researcher to expand the continuum of observation from field through macro, micromorphological, and ultimately, elemental scale. An instrument that is similar to the SEM and is specifically oriented toward analysis of polished blocks or thin sections is the electron microprobe (see below).

With EDS it is possible to analyze the composition of specific grains, even those that are only a few micrometers in diameter because the electron beam can be focused to 1–2 μm across. This allows the determination of different elements, which in turn is an indicator of the mineral being observed. Such analyses are presented as spectra in which energies correspond to different elements. These spectra also indicate relative abundances of the different elements, which in turn can be utilized to infer the mineralogical composition of the material or sample and ultimately determine its history (Karkanas et al., 2000; Weiner et al., 2002).

SEM/Microprobe, which can have higher detection levels compared to SEM/EDS, and has been employed for

- elemental analysis;
- mapping elements (and their % distribution using image analysis); and
- mapping combinations of elements (and their % distribution using image analysis) (Wilson, 2017).

In this volume we give examples of microprobe analyses of guano in Mousterian Vanguard Cave (Gibraltar) (Figures 17.8a and 17.8b), a green glaze briquetage from Late Roman salt working at Stanford Wharf (Essex, UK) (Figures 12.37–12.39), and the effects of experimental marine inundation at Wallasea Island (Essex, UK) (Figure 14.24).

Other techniques have become more available in light of recent of technical advances. One that is being increasingly used in μ-XRF, which is similar to elemental mapping using SEM/microprobe

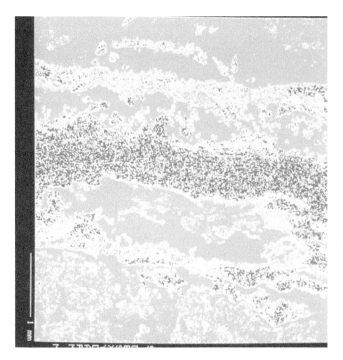

Figure 17.8a Microprobe elemental map (Van97-52, Layer 6a; Mousterian Vanguard Cave, Gibraltar); guano layer rich in P (pink) – Microfacies *Gii* (Macphail et al., 2012b). Frame width is ~4.5 mm.

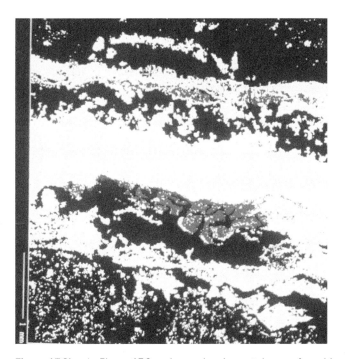

Figure 17.8b As Figure 17.8a, microprobe elemental map of combined Ca-P (yellow; 35.35% of the view by image analysis) in guano layer (max 36% Ca and 14% P). There are also concentrations of Fe-Mn (blue) and a scatter of mixed material at the base, composed of Al-Si-Fe-Mn. The occurrence of Fe-Mn is possibly linked to biologically worked organic residues.

but requires less preparation of samples, such as sample coating (see Mentzer and Quade, 2013 for details). They used the technique to investigate plaster floors from the Aceramic Neolithic site of Aşıklı Höyük (Turkey) showing different elemental compositions that could be eventually tied to different raw material sources.

FTIR studies on already described thin sections – micro-FTIR – have the advantage over bulk FTIR analyses of being able to produce mineralogical data for observed objects, grains, and features, just as in microprobe and SEM/EDS analyses. In general, hearth constituents and residues (ashes, transformed ashes, burned and unburned bones) and the temperature at which materials were burned, for example have come under special scrutiny (Berna, 2017; Berna et al., 2007; Goldberg et al., 2009; Mallol et al., 2017). Along with other evidence, FTIR analyses did not support the notion of early hearths at the Lower Paleolithic site of Schöningen (Germany) (Stahlschmidt et al., 2015).

A less typical example is Late Roman briquetage from coastal Essex, UK, where the possible bellows plate was made from local estuarine alluvium with a mica-smectite clay component (Biddulph et al., 2012; Macphail et al., 2012a). Sections through the green glaze briquetage record the increasing effects of heat upwards causing, 1) "rubification" (400°C – <700°C), with the 2) upper blackened part being more strongly heated (>700 °C). The green glaze itself is a strongly heated silicate glass (minimum 700–800°C to around/below 1000°C). Linked SEM/EDS and microprobe analyses indicated that the dominantly siliceous glaze contains statistically significant greater quantities of Na (sodium), P (phosphorus) and Fe (iron) compared to the briquetage (see Figures 12.36–12.39). It is commonly the case that micro-FTIR can be usefully combined with a series of associated analyses such as soil micromorphology, TL (thermoluminescence), and organic petrology (Mentzer, 2014; Mentzer et al., 2017).

17.8 Conclusions

This chapter presents some of the most commonly used laboratory techniques that are applied in geoarchaeology. Our chief concern here has not been to describe techniques *in detail* – as these can be researched from papers and textbooks – but to discuss *why* certain techniques are essential to address geoarchaeological questions. There are almost an infinite number of techniques that can be applied, and we have focused on the most important ones; additional ones appear in increasing frequency. We have outlined the methods for both bulk analyses and those employing instruments, and have provided some examples of results. We have also noted some important caveats so that techniques should not be used blindly. Thus, we have highlighted the fact that granulometry of anthropogenic deposits can produce potentially meaningless results leading to absurd interpretations, if other techniques such as soil micromorphology are not also carried out. Equally, soil micromorphological interpretations are undoubtedly enhanced if complemented by such approaches as SEM/EDS and microfossil analyses and bulk chemistry.

18

Reporting and Publishing

18.1 Introduction

Reporting the results from any geoarchaeological study is the single most critical step in any project. A student or professional researcher may be highly talented and have at their command a large database of geoarchaeological material, but if these data are not presented in a comprehensible way, with equally clear interpretations of the trends and the significance of the findings, then much of the work and effort is wasted and will ultimately have little impact. It is therefore important to make the report relevant to the project and the questions being asked. Similarly, it is not necessary to include all data that were collected, but only those data that are pertinent; additional tables, illustrations, raw data, and photo archive can be uploaded onto a designated website and/or physically stored on computer drives at the project headquarters.

Reports need to be concise, well illustrated and documented, with as little jargon as possible. The reader should be able to follow the logic of data presentation, interpretation and ultimately *its relevance to the archaeological goals of the project*. Therefore, good reporting skills need to be learned and practiced. Working to a timetable is also a necessary professional requirement, as is reliability. Writing a publication article for one's peers, will require further effort, partly because of the word limits on journal articles, coupled with the need to balance the requirements of an international readership and still cover a subject in a focused way. Few people can create more than a few such articles a year, despite academic management pressures to produce more, especially given that the investigation phase may have taken years to complete. The benefit of working in research teams is that different members can lead jointly authored papers through the lifetime of a project. It is also the experience of the authors that the sum is always greater than its component parts.

Outputs from geoarchaeological research can therefore take a variety of forms, depending on the scales of the investigation, the types of methods being applied and the proposed outcomes from the geoarchaeological investigation. The variation in these outputs can be summarized under three main themes:

1) geotechnical reports, often including geophysical survey data and GIS modeling, dealing with soft sediment geology that in many cases involves beds of alluvial sediments, colluvium, and buried soils, which are modeled from boreholes etc., and include possible archaeological and paleoenvironmental (e.g. peat) deposits,

2) archaeological reports and publications, where geoarchaeology is specifically focused on the site and its nearby surroundings; and

Practical and Theoretical Geoarchaeology, Second Edition. Paul Goldberg and Richard I. Macphail.
© 2022 John Wiley & Sons Ltd. Published 2022 by John Wiley & Sons Ltd.

3) most commonly, reports and publications that combine geotechnical overviews and specific focused investigations of sediments; these may include characterization by sediment micromorphology and laboratory analyses of samples.

In the following section, we make some suggestions about reporting and publishing based on our experience and that of our colleagues. We cannot provide a universal guide, however, as different organizations, countries, and journals have different requirements pertaining to national laws, local authority guidelines, and the specifications of a single contract.

18.2 Management, Reporting, and Publication of Archaeological Sites

After a few brief examples of site management types (Box 18.1), we describe some ways to construct geoarchaeological reports including some fieldwork report protocols and an example of fieldwork evaluations that led to sample assessment and their selection for analysis (Box 18.2). In terms of reporting and publication, we later suggest some of the ways data can be best presented,

---

Box 18.1 The Sørenga D1A borehole site, Søreng Inlet, Oslo: a combined geotechnical and laboratory investigation

In 2014 two borehole samples to ~17.00 m below sea level were collected in the challenging sediments found along the River Alna outlet into Oslo harbor at the head of Oslo Fjord, Norway (Bukkemoen, 2013; Linderholm et al., in preparation) (Figures 18.1–18.3). Only one core series was useful, and after careful evaluation even this one allowed only a small number of core sample studies (8 core subsamples from −4.8 m to −16.5 m asl). The chosen cores then underwent pre-sub-sampling visual assessment and detailed logging of sediment type employing a magnetic susceptibility probe and NIR scanner. These preliminary results permitted accurate sub-sampling and sharing of the chosen cores for undisturbed sediment micromorphology samples, and exactly correlated bulk analyses for particle size, magnetic susceptibility, chemistry (LOI, fractionated P and XRF), pollen, and macrofossil analyses. Contrary to expectation and sea level curve data (Sørensen et al., 2007), deep-water anaerobic sediments were found at the base of the core, most of the deposits had to be interpreted as shallow-water river delta sediments (Figures 18.4–18.8). The latter were increasingly rich in anthropogenic inclusions upwards, and were matched by phosphate and heavy metal enrichment (Figure 18.9). Moreover, these had C14 dates consistently grouped around the 1300s AD, although there are earlier medieval and later medieval outliers. It can be suggested that the best explanation for these anomalies of depth, sediment type, and dating involves slumping of the Alna River delta front, leading to these shallow-water sediments now anomalously occurring in the deep waters of the fjord. This caused some difficult discussions amongst the team after reporting the sediment micromorphology (Macphail et al., 2015), because the interpretation of shallow-water sediments did not fit the established sea level curve, but such slumping of fjord sediments is well documented (Holtedahl, 1975), and noted in Reineck and Singh (1986). These medieval sediments, in fact, recorded probable pro-delta, delta front, channel, levee, and marsh facies, which involved both shallow-water and subaerial environments. This characterization was clearly acceptable once the C14 dates became available to support the geochemical and sediment micromorphology findings, but of course it would have been easy at the reporting stage simply to "accept" a deep-water sediment interpretation based on the sea level curve rather than interpret the geoarchaeological evidence of shallow-water deposition.

Box 18.1 (Continued)

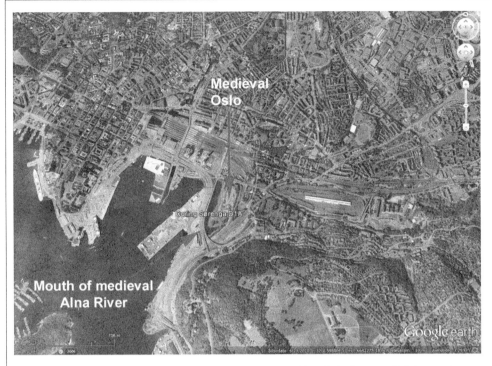

Figure 18.1 Aerial view of Oslo, Oslo harbor and location of borehole Sørenga D1A, just to the west of medieval Oslo (Google Earth image).

Figure 18.2 Field photo; borehole Sørenga D1A; drilling rig in action (Bukkemoen 2013/with permission of University of Oslo).

(Continued)

Box 18.1   (Continued)

Figure 18.3 Field photo; borehole Sørenga D1A; core extraction into plastic sleeve. Core sample was then taken to the laboratory at the Museum of Cultural History, Oslo, for visual, magnetic susceptibility and NIR (near infrared spectrometry) logging, ahead of sub-sampling for bulk and intact soil micromorphology analyses (Bukkemoen 2013/ with permission of University of Oslo).

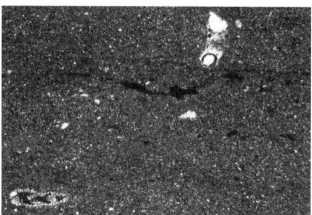

Figure 18.4 Photomicrograph of M23A (Core 2/23; −16.455 to −16.53 m); very compact diffusely laminated silty clay loam, with black pyrite void infills (Py) and channel infill that includes a coating of gypsum crystals (Gy). Plane polarized light (PPL), frame width is ~4.62 mm.

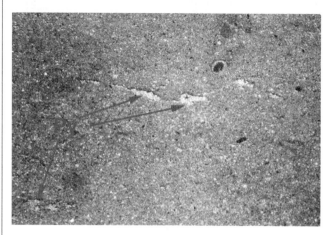

Figure 18.5 As Figure 18.4, under oblique incident light (OIL); pyrite framboids (Py) are brassy colored.

Box 18.1 (Continued)

Figure 18.6 Photomicrograph of M17A1 (Core 2/17, −13.355 to −13.43 m); this shows the junction between typical low-energy muddy and microlaminated silty clay loam sediments and overlying coarser, fine sandy loam deposits. PPL, frame height is ~4.62 mm.

Figure 18.7 Photomicrograph of M17A1 (depth as Figure 18.6); detail of junction between mudflat and more fine sandy sediments. PPL, frame width is ~2.38 mm.

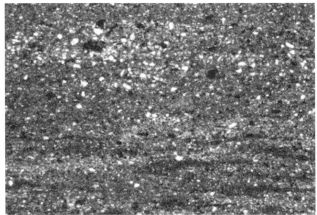

Figure 18.8 Photomicrograph of M17A1 (depth as Figure 18.6); the thin section records a broad burrow bringing down sands from beach/foreshore(?) sediments above; note presence of plant fragments. PPL, frame width is ~4.62 mm.

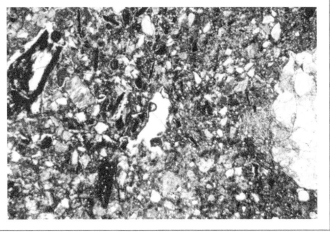

Box 18.1 (Continued)

XRF Sediment chemistry-Potentital contaminants

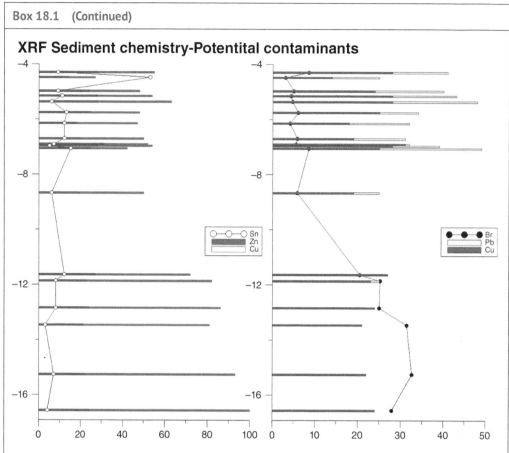

Figure 18.9 Example of geochemical log through Sørenga D1A core; XRF data and depth below sea level (~–4 to –16 m). Zn (zinc) and Br (bromine), which are linked to marine sedimentation, diminish upwards, while Cu (copper), Pb (lead) and Sn (tin) tend to peak upwards, probably reflecting medieval artisan activities and associated contamination (figure by Johan Linderholm, MAL, Umeå University; Linderholm et al., submitted) (with kind permission of Johan Linderholm).

including some illustrations using composite figures, statistical methods, and GIS (Box 18.3). Lastly, we provide some examples of how sites can be interpreted from the gathered data, to produce stand-alone and consensus interpretations, and the ways that these can be published.

18.2.1 Geotechnical Reports and Publications

In the UK, these works can follow various guidelines associated with modeling deposits that point to the identification of layers of archaeological interest, which in urban areas can involve meters of anthropogenic deposits (AGS, 2018; Carey et al., 2018; Historic England, 2020). For example, the new London Underground Railway ("Thameslink") led to combined geotechnical and archaeological publications (Ridgeway et al., 2019; Stafford and Teague, 2016), in which soil micromorphological reports (including Roman copper alloy working) was archived. High Speed 1 (Channel Tunnel RailLink) led to similar geotechnical publications, one with a sediment micromorphology component that examined the details of site formation processes during the Pleistocene (Bates and Stafford, 2013; Bates et al., 2014).

Box 18.2   How to write a report: a suggested fieldwork report protocol

Introduction

To contain: Name of site and exact geographical location; when visited; with whom was the site discussed, and who supplied information, including their affiliations (archaeological company, university); period(s) and type of site (urban, rural, tell); what type of exposure (open excavation, backhoe trench, sondage, building cellar); why the site was visited/aim of geoarchaeological study (geoarchaeological questions); what was done (field evaluation, section description, survey, sampling); associated work by other specialists and overall context of the work (ancient erosion, early tillage, urban morphology, human/animal occupation).

Methods

To include: List of techniques carried out and authorities employed (for field description, Munsell Color Chart), sampling, on-site tests (e.g. HCl for carbonate, finger texturing for grain size, coring, survey); any specific equipment employed (different augers, geophysical equipment, types and lengths of monoliths); follow-up laboratory techniques, e.g. pH (see Chapter 17).

Samples

This requires: A tabulated *list of samples* (type of sample: undisturbed monolith, with measurements of OD or relative depths); bulk samples, perhaps including size/weight; context number; key archaeological aspects; brief description (color, texture, compostion, etc.); GIS location, e.g. as *Shape* files). Further sampling requirements can be noted in the Discussion, below.

Results

Content: Present information on the site's location and known geology, soils and perhaps also modern land use (e.g. using Soil Survey of England and Wales for the UK, USDA guides for the USA); systematically present the field information, for example, logically describing the oldest rocks/parent material or earliest stratigraphy first; present each geological/archaeological section/soil profile in turn, referring to tabulated sample lists and descriptions, section drawings, field drawings, photographs/digital images and any other data (figures and tables) that have been generated (e.g. pH, HCl carbonate reaction; survey and coring data).

Discussion

Can be organized as follows: First set the scene by discussing the known information about the geoarchaeology of the site and what the fieldwork has revealed. Perhaps broadly discuss systematically the sequence in terms of oldest to youngest, or by area on the site. Secondly, discuss the geoarchaeological questions posed by the site, referring to why the site was visited/aim of geoarchaeological study and what are the broad aims and objectives of the wider investigation coupled to what information has been gained from the fieldwork. For example, this is the time to identify which samples have the potential (from a laboratory study) of answering the geoarchaeological questions (it may be necessary here to identify where further samples should be taken). Thirdly, suggest the various techniques that can be employed to answer these questions. Here, a proper literature-supported track record for each technique

(Continued)

Box 18.2   (Continued)

in providing useful data and interpretations has to be argued. For example, areas of supposed occupation could be isolated by measuring the LOI, P, and magnetic susceptibility of samples taken through a grid survey (see Chapter 16). Or, in order to seek out possible locations of Paleolithic hearths in caves, soil micromorphology should be combined with microprobe and FTIR analyses to identify ash transformations (see Chapter 10).

Future Work and Costs

Requirements: Here the number of samples and techniques to be applied are listed, based upon the preceding Discussion and the List of Samples. Estimated costs of analyses and reporting (see below) should be given; this should include the number of working days involved so that the work can be scheduled properly. It is also useful here to indicate how long procedures take: soil micromorphology, for example, requires lengthy processing times before thin sections can be made, usually several months (see Chapter 16). Also, a worker may have a backlog of work and may need to inform the "client" when the work can commence. Reports often have to fit a deadline, so that the archaeologist can write up his report with all contributions present. This is especially true of CRM/mitigation archaeology.

Bibliography

Present a reasonable list of cited publications; more are normally required for the post-excavation report/publication (see below).

Professional Responsibilities

Lastly, fieldwork may take a single day, and as such the "client" may expect a fieldwork report even the next day (by email) or very soon after, because the geoarchaeological information is required immediately to identify what are important contexts to sample further, or which areas of the site require further trenching or excavation. It may be that a number of questions were answered on site during the visit, but these answers need to be reinforced by a well thought out report. This is especially important when short, day to several days, field work is undertaken. During long periods of fieldwork, many "operational" questions can be resolved on site. If working internationally, flights and accommodation costs also need to be built into overall project costs.

In the United States, the Department of Defense document, *Final Governing Standards*, for example, provides guidelines for the management of both the environment and cultural heritage of sites. It is formulated individually for each site, but includes protective requirements of the USA and applicable international agreements (e.g. UNESCO conventions, International Charters and World Heritage Conventions; 1972 Paris Treaty on World Cultural and Natural Heritage; https://whc.unesco.org/en/conventiontext/).

In Spain, archaeological work falls under the competence/control of regional authorities (i.e., 17 different autonomous communities) that authorize, supervise, and assess final reports. Therefore, due to local legislation, site reports have to cover different aspects of archaeological work depending on where the site is located. There is a heavy emphasis, however, on the detailed reporting of material culture and findings. Although research teams conduct all kinds of scientific and laboratory analyses, reporting is not always required and is reserved for later publications (see Pérez-Juez, 2014).

INRAP (Institut National de Recherches Archéologiques Préventives/French National Institute for Preventive Archaeological Research; www.inrap.fr) is the organization in France that oversees salvage archaeology. Holocene paleoenvironment, regional soil and soil erosion studies and landscape reconstruction fall within this category. INRAP staff contributed data from a comprehensive investigation located in northern France, Belgium, and Luxembourg, comprising 74 colluvial and 26 alluvial settings (Rivers Scheldt, Meuse, Somme, and Rhine) (Fechner et al., 2014); the region of Lorraine was similarly investigated and reported on (Gebhardt et al., 2014). In Italy, the Soprintendenza, for example, combined with regional museum based monitoring and excavation can include paleoenvironmental studies (charcoal, phytoliths, bones, and sediment micromorphology) (Maggi, 1997).

Box 18.3   From field evaluation to sample assessment and selection for analysis at the A14 road scheme, UK (MOLA-Headland Infrastructure)

Here, archaeological monitoring was carried out by Cambridgeshire County Council archaeologists in the Historic Environment team. Works along a total of 34 km involved 40 separate excavations during 2017–2018. Some four site visits led to four site evaluation reports concerning mainly nine sites, which included a Bronze Age barrow, a Roman farmstead, Roman dark earth (Figure 18.10), Saxon villages, and background soils and alluvial soil-sediment sequences. A separate geotechnical study focused on the recent geology, and channels and paleochannels of the local rivers (e.g. River Great Ouse). The Soil Survey of England and Wales soil map was the primary reference for the soil study (Hodge et al., 1983). In 2018, 40 Kubiëna box samples were assessed visually for potential (Figures 18.11 and 18.12), and of these 33 were selected for their archaeological potential (based upon the field evaluations) and quality of sample (e.g. intactness). Archaeologists and paleoenvironmentalists discussed the sites and potential at a meeting in 2019, and sub-sampling of Kubiëna boxes for small bulk sample analyses was carried out in 2019, with processing of Kubiëna samples for thin section production and study commencing in 2020.

Figure 18.10   Field Photo – A14 road upgrade, Cambridgeshire, UK; dark earth area; soil profile an example of a series of gray, brown and blackish gray soil-sediments (alluvium) with extensive thin spreads of gravels. Earthworm-working, colluvial wash, high-energy fluvial activity and plough mixing could account for gravels within this sequence.

(Continued)

Box 18.3   (Continued)

Figure 18.11   Laboratory photo of opened Kubiëna Box sample A14 road upgrade, Cambridgeshire, UK; Example from Saxon village site with a pit fill containing fire installation waste. Here, there is just about enough sample to carefully extract a small bulk sample.

Figure 18.12   Laboratory photo of opened Kubiëna Box sample A14 road upgrade, Cambridgeshire, UK; Example of Roman dark earth, which has been poorly sampled, but which comes from a crucial context. A small bulk sample was collected from it, and a resin-impregnated thin section was successfully rescued from it.

In Norway, several organizations employ geotechnical methods, with an example of coring through the Gokstad Ship Burial Mound given below. Much of ancient Oslo harbor is now reclaimed land and knowledge of the morphology of the original fjord, associated river systems and wetland is an important aspect of various archaeological investigations, including those along the route of the new Metro "Follobanen". Here, for example, waterlogged organic remains associated with log-built structures were the subject of paleoenvironmental analyses (Buckland et al., 2017; Macphail, 2016). Medieval and post-medieval shipwrecks, harbor installations, and associated sediments were also studied using cofferdams. Particularly of interest is the borehole study of samples from the Sørenga D1A borehole, which involves the extraction of samples from 17 m below sea level at the deepest part (Bukkemoen, 2013), following the guidelines of Falck and Gundersen (2007) (Box 18.1; Figures 18.1–18.9). Some geophysical remote sensing techniques such as Georadar (ground-penetrating radar, GPR) can create clear models of sites ahead of trial trenching and excavation as at the Viking Age Heimdalsjordet settlement, Sandefjord, Vestfold, Norway (Bill and Rødsrud, 2017; Macphail et al., 2017a). This study method provided an image of the newly discovered Gjellestad Ship burial (Østfold) and was posted on the webpage before excavation began, although the latter has an ongoing newsfeed (www.niku.no/en/prosjekter/jellestad-skipet/; www.khm.uio.no/english/visit-us/viking-ship-museum/gjellestad-ship/index.html).

18.2.2 Archaeological Site Reports and Publications – Some Examples

As demonstrated throughout this book, archaeological sites come in all shapes and sizes, with varying levels of stratigraphic complexity and preservation. As the examples below show, these can vary significantly, from (1) a single site of open-air hunter and gatherer occupation through to (2) complex mounds, to (3) deeply stratified cave and urban sequences and multi-period occupations of a landscape, such as found along linear construction projects of major new routeways (4–5). Therefore, the reporting and publication of such projects can equally be as variable, targeted to different outputs, and use data from a variety of different methods, scales and complexities. Examples include:

1) An article on 3- to 4 m-thick cave sediments dating from the Middle and Late Pleistocene ("Denisovans") recording hominin presence along with cave dwelling carnivores and a chapter comparing different naturally formed and anthropogenic facies at Late Stone Age Taforalt (Cave of Pigeons), Morocco (Macphail, 2019b; Morley et al., 2019),
2) Papers on a Bronze Age complex kurgan mound (Ponto-Caspian Region, Russia) and a series of Native American Mounds in the Mississippi Basin that date as early as 3600–3000 BP; the latter may include the construction of artificial terraces and embankments even before the soil basket-loads and sod block mound was constructed (Khokhlova et al., 2021; Sherwood and Kidder, 2011).
3) The 12 m high Chalcolithic (~4500 cal BC) urban mud brick tell sequence of Popină Borduşani (Romania) and its wetland location between the Danube and Borcea Rivers, as well as the Romano-British, Roman dark earth and medieval stratigraphy at the London Guildhall, where the Roman arena, early medieval farmstead and medieval Guildhall were located (Bateman et al., 2008; Bowsher et al., 2007; Haită, 2012; Macphail et al., 2007a; Popovici et al., 2003).
4) A thematic publication focused on multi-site analysis of colluvia in Belgium, France, and Luxembourg along the high speed rail line, or the three year-long, multi-period (Bronze Age to Viking Period) site study along the E18 highway between Tønsberg and Sandefjord, Vestfold, Norway; in the latter case, seven major interdisciplinary paleoenvironmental investigations were focused along a 16 km stretch of the route (Fechner et al., 2014; Viklund et al., 2013).
5) The case where excavations have unearthed large areas of settlements that allowed for the development of models to show how sites functioned both internally and externally (hinterland, routeways, access to rivers and sea); some examples being coastal Heimdalsjordet, Ørlandet and Avaldsnes, Norway (Bill and Rødsrud, 2017; Linderholm et al., 2019; Macphail et al., 2017a; Skre, 2017).

Such settlement morphology studies have also commenced on Middle Saxon (Migration Period) villages across East Anglia, where Saxon peoples (e.g. 5th–6th century AD Sutton Hoo; Carver, 1998) established themselves during the so-called Dark Ages in the UK, with small halls and pit houses (sunken feature buildings – SFBs); the latter are archetypal negative features at the sites of Eye and Kentford, Suffolk, and with reconstructions at West Stow (West, 1985).

Other site-based thematic publications, for instance, have focused on the early use of fire at Qesem Cave, Israel, Middle Stone Age bedding deposits in Sibudu Cave, KwaZulu-Natal, South Africa, Viking grave fills in Iceland, occupation deposits from medieval Riga, Latvia, and urban dark earth across Brussels, which can be compared with atypical tropical dark earth at

Marco Gonzalez, Belize (Banerjea et al., 2016; Burns et al., 2017; Devos, 2018; Goldberg et al., 2009b; Karkanas et al., 2007; Macphail et al., 2017b; Shahack-Gross et al., 2014).

18.2.3 Combined Geotechnical and Archaeological Site Reports and Publications

In fact, many projects involve both geotechnical and archaeological site components. The High Speed 1 Rail Link through Royal Oak, London has already been noted (Bates et al., 2014). Numerous projects include both traditional excavation and geotechnical coring, which are not only employed for mapping and modeling deposits, but also for extracting samples from deeply buried stratigraphic sequences (see Gokstad Mound, below and Box 18.1). At Winchester, UK, borehole investigations through the River Itchen's alluvial soil-sediments involved microstratigraphic analyses of core samples (C14, pollen, soil micromorphology) in order to reconstruct the floodplain's paleoenvironment (Champness et al., 2012; Macphail et al., 2009). In addition to deep coring of alluvium recording middle Holocene tufa formation and hydoseral succession (woodland invasion and soil ripening of river peats), dumped medieval occupation deposits including lead-rich household refuse and phosphate-cemented stabling waste were within the then precincts of Winchester Cathedral.

As already noted (Box 18.1), background geology and sea level curves can play an important role in interpreting sites, especially where there has been post-glacial uplift, as for example in North America and Scandinavia (Kelley et al., 2010; Sørensen et al., 2007). Kaupang, near Larvik, Vestfold, Norway was an important Viking coastal settlement where knowledge of the sea levels of 1,000 years ago was crucial to the understanding of its settlement pattern (Milek and French, 2007; Skre, 2007). Similarly, the Sea King's Manor site of Avaldsnes, Karmøy, Rogaland includes turf-built boat houses, whose location and altitude are presumed to be governed by contemporary and local sea levels; one example had two phases of near-sea level construction dating to the Roman Iron Age and Migration Period, respectively (Bauer, 2017; Macphail and Linderholm, 2017b; Skre, 2017).

18.3 Management of Sites: Evaluation, Assessment and Reporting

Geoarchaeology is, however, normally only part of a holistic approach to archaeological investigation, which in addition to the study of features and artifacts will include a full arsenal of environmental techniques. For all projects, the selection of samples for laboratory analysis and reporting needs to be thoughtfully carried out to address the aims and scope of the original project design. Moreover, timetables and funding are finite, and the aim is to produce a report/publication to deadline and to budget. Also, during long projects, samples are continually being processed and studied so that later sampling becomes more focused. Some examples are provided below.

In China, the National Cultural Heritage Administration oversees all excavation and post-excavation works. Recently, they commissioned Peking University to write a protocol for excavation and report publications, which, in theory, is to be adopted by all field archaeologists in China. The protocol clearly states that publication reports should be objective, authentic, comprehensive, and systematic and that understanding natural environmental settings of a site should be an important part of the report along with other types of data (National Cultural Heritage Administration, 2009: 9). It also makes recommendations on how to survey and map the surrounding environment of the site, how to collect environmental and archaeological samples, and how to publish and archive results (National Cultural Heritage Administration, 2009).

In England, the ultimate authority in archaeology is Historic England, and city and county councils employ their standards and working practices, with similar processes followed in Wales

(CADW) and Scotland (Historic Scotland). Sampling strategies also come under the heading *Management of Archaeological Sites*, which has a moderately long history, commencing in urban areas and areas of industrial extraction (e.g. stone aggregate), with the formation of "Rescue". This organization (Rescue) attempted to stop developers simply machining "uninteresting" archaeological deposits away, such as dark earth (Barham and Macphail, 1995; Biddle et al., 1973). For example, when the Rose Theatre (Elizabethan and Shakespearean London) was discovered, a decision was made to preserve it, and it was sealed below a "terram", a durable plastic-based cover, and then monitored in order to try and keep the site waterlogged (Historic England, 2016b). A similar policy was enacted for Must Farm, Cambridgeshire, UK, an almost intact Bronze Age settlement, with waterlogging preserving a Bronze Age timber platform (Historic England, 2016a). General advice for managing projects with a geoarchaeological component is also given by Historic England (Historic England, 2015), although the authors would argue that it requires updating; the most recent cited reference to employing soil micromorphology is 2003. The more recent *Deposit Modelling* (Historic England, 2020) provides good advice on preparing a geoarchaeological report employing geotechnical data (see Chapter 16).

After excavation and sampling, many projects include screening of samples in order to prioritize and select the best samples based on context integrity, dating and the best means to address the specific archaeological questions of the site, including required justification of expenditures. This screening process is normally carried out as rapidly as possible, and never should become a time-wasting goal in itself. Screening and evaluation of samples includes findings from the fieldwork (see Box 18.1) and perhaps a preliminary laboratory study. In the UK, there may well be a separation of techniques designed to address broad landscape issues (geotechnical, geomorphological, and geological) from soils and microstratigraphic methods. In the context of the latter, issues at this stage can be: "Is this context phosphate rich?" or "Are there preserved microfeatures of occupation?". Landscape questions may include sea-level or base-level fluctuations and how they affect sites in a regional context. The geoarchaeologist may be asked to comment on what deposits, if any, could be suitable for macrofossil and microfossil recovery (see Table 4.1). Moreover, it is essential at this time that the research aims and objectives of the site be addressed, and that new information is fed back to augment and improve such research aims. A recent example of the above aims is illustrated by the A14 project road upgrade between Cambridge and Huntingdon, UK (see Box 18.3).

At the nationally iconic Norwegian site of the Gokstad Ship Burial Mound, Vestfold, Norway, a small trench excavation was carried out by the Museum of Cultural History ("Viking Ship Museum"), Oslo in order to reinstate the mound after recovery of the 1928 reburied remains of "Gokstad man" in 2007; this was over 100 years after the initial excavation and recovery of the Gokstad longship in 1880 (Nicolaysen, 1882). The Vikings who broke into the AD 895–903 burial chamber after AD 939, had no official permission of course (Bill and Daly, 2012). When a geoarchaeological coring exercise was organized to supply data for the Gokstad Revitalised Project (www.khm.uio.no/english/research/projects/gokstad/index.html), only a very small percentage of the site was allowed to be disturbed. Even when screw coring (for evaluation of mound and buried soil make-up) and cylinder coring (for removing samples) are added together, only 0.019% of the site was disturbed (Cannell, 2012). This coring and paleoenvironmental analyses of the cylinder cores produced extensive results on the make-up and late 9th century environment of the mound (Cannell et al., 2020; Macphail et al., 2013a).

At the Roman amphitheater and early medieval site of the London Guildhall, assessment included laboratory studies (bulk chemistry, soil micromorphology, and integrated pollen analysis), which were then discussed within the research context of parallel experimental investigations, thus helping to establishing the bona fides of the assessment (Macphail et al., 2004). Final results came out as selected findings within the archaeological volumes and as a thematic article on occupation deposit microfacies (Bateman et al., 2008; Bowsher et al., 2007; Macphail et al., 2007a).

18.4 Components of Geoarchaeological Reports and Publications

Geoarchaeological reports and publications will no doubt vary from investigation to investigation, depending on the specific goals and requirements of the project. Nevertheless, there are a certain number of elements that should be included irrespective of the exact nature of the project. A well-constructed text, with a clear and detailed summary, is essential for both reports and publications, because clients and readers need to be informed quickly and easily. The report itself can then be mined for details by the reader.

Good illustrations are crucial, with the following considerations:

- Include map location(s), of local and regional scale.
- Include field photos, annotated photos of profiles showing stratigraphy and sampling locations; profile drawings are useful if more detail of sampling is required.
- If soil or sediment micromorphology is involved, microphotographs should be large enough to be useful; a page of multiple postage stamp size photos is worse than useless.
- Supportive illustrations from use of instrumentation may also include spectrographs, and these have to be labeled.
- Tables of data can be useful, but if these can be replaced with a figure, this may be better for the reader; full tables can be archived or published as "supportive data".
- Some figures may be composites of photos, tables, instrumental data such as spectra.
- All figures require comprehensive captions explaining exactly what is being illustrated and why.
- Lastly, for easy access by a reader, major points of interest should be included in an extended summary, which also acts as a self-review of the data by the author, and how it has been interpreted.

Further suggestions on a reports content can be found in Historic England (2020).

18.4.1 Fieldwork and Assessment/Evaluation Reporting

An example of a field report protocol and suggested contents is given in Box 18.2. Most reports such as these end up as gray literature and may be used by a variety of persons other than the archaeologist involved (e.g. the "curator" for the town or district, or other planning officers), or be utilized in a review many years later. Organizations may have their own webpages where non-sensitive reports can be uploaded and accessed. Consequently, it is important to specify in the *Introduction* the details of the site visit: date and people involved; site details, period, specific finds, etc; reasons for visit/site questions being addressed, and tasks/methods carried out. The *Results* section should include a tabulated list of profiles/sections examined and samples taken, including type (e.g. undisturbed monolith, bulk sample, associated samples for pollen and macrofossil analysis), and their provenance. This table may also contain descriptive field information, and any laboratory data, such as pH. Aerial photos, field photos, section drawings, survey results etc. may also be included, along with references to descriptions and desktop studies (e.g. data from topographic, geological, and soil maps). The *Discussion* will vary according to the task, but it should specifically cover reasons for the site visit and the questions being addressed; it should logically employ the field data gathered and any supportive laboratory information, organized either by site chronology or by section/profile location. In support of these arguments, examples of analogue sites, similar situations, and experimental results should be cited.

It is here in the discussion that follow-up work suggestions can be made. These should be argued on the basis of the field findings, the quality of the samples taken, the importance of the site, and the questions being addressed. Here, appropriate laboratory techniques can be discussed along with their track record in achieving *answers* to questions specific to the site. This report section should also contain bibliographic support, for example, indicating why this suggested work would contribute to an established research agenda for a region or topic.

Follow-up work needs to be costed: cost per sample preparation, analysis, and per diem (day rate) of study and reporting. It is also important to give a timescale, for the length of time this work will take, and state when the work can begin. Do not forget that while a pollen sample can be processed in a day, soils studies, especially soil micromorphology, take a long time: sample processing takes weeks to months from monolith to thin section. Whereas a thin section may be described in a day, interpretation may not be instantaneous, and complementary micro-FTIR/EDS/microprobe studies have often to be planned well in advance because instrument time can be limited.

18.5 Post-excavation Reporting and Publication

The format of a full post-excavation report (and publication) follows the same plan as field reporting described above. In contrast, publication in a journal is likely to be much more thematically based, shorter and less detailed. On the other hand, it will be more focused, as for example, the integrated geoarchaeological and paleoenvironmental results from Butser Ancient Farm (Chapter 14) and the study of brown soils on Exmoor, Devon from archaeological sites (Carey et al., 2020; Macphail et al., 2004). Some journals allow supportive data to be presented on a linked webpage (Goldberg et al., 2009a; Gutiérrez-Rodríguez et al., 2018; Morley et al., 2019). Unlike many reports, which tend to be single-authored, a publication may be multi-authored, and contain inputs from a series of specialists. This type of publication will therefore need from each specialist a series of *Methods*, *Results*, and *Discussion* sections (and supporting sample tables and data presentations; see Appendix 1.17.2), before an overall *Discussion* and *Conclusion* can be written. Sometimes the geoarchaeology is presented in two parts, as (1) the geotechnical, field and associated phasing evidence and (2) findings from geochemistry and sediment micromorphology (Macphail et al., 2013b; Wenban-Smith et al., 2013).

It is often the case that the geoarchaeology (including possible integrated paleoenvironmental data) is subsumed within a volume or larger paper (e.g. London Guildhall), with possibly inclusion as an Appendix (Boxgrove) or as an online chapter (Stanford Wharf) (Biddulph et al., 2012; Bowsher et al., 2007; Macphail et al., 2012a; Pope et al., 2020). In the case of the Guildhall project, the most important findings were presented in a thematic paper (Macphail et al., 2007a). Similarly, a chapter in a project volume can be reworked into a thematic article, with project results also contributing to teaching guides; e.g. Swiss Neolithic Lake village site of Arbon/Bleiche 3 (Ismail-Meyer, 2017; Ismail-Meyer and Rentzel, 2004; Ismail-Meyer et al., 2013). Such large projects, with enormous datasets, are probably the most difficult to publish as a stand-alone article or volume chapter (but see, Ismail-Meyer et al., 2020).

In some of the earliest archaeological reports in China, environmental and geoarchaeological works already constituted an important component, partly because many early archaeologists were also trained geologists. For instance, as a young geologist, Yuan Fuli conducted a well-designed geological survey around the Yangshao site in Henan, the results of which were systematically published in the excavation report (Andersson, 1947). However, the publication of the Banpo excavation in 1950–1960s (Institute of Archaeology of Chinese Academy of Social Sciences and Banpo Museum of Xi'an, 1963) established a paradigm of publishing environmental studies as appendixes to archaeological reports, which was followed by numerous reports published after 1960s. The situation that environmental data is considered a non-essential part of an archaeological report has recently started to change in some major research projects. In the voluminous report of a 10-year and multidisciplinary survey of the prehistoric Eastern and Middle Luoyang Basin, for example, geoarchaeological results constituted a key element (e.g. chapters 5 and 9) for the discussion on the development of social complexity in the region (Institute of Archaeology of Chinese Academy of Social Sciences and Yi-Luo River Joint Archaeology Team, 2019).

Commonly, details from single sites are published, but few attempt to go further in their interpretation to see how a site or settlement functions. An exception for hunter and gatherer archaeology was produced by Grøn and Kuznetsov (2004). Modeling the make-up of complex society settlements and how they function in a landscape was suggested for sites in Romania (tell) and Norway, the latter including reconstruction of the coastline (Macphail et al., 2017a). Such an approach was tested for the Iron Age to Viking Period site of Ørlandet (west of Trondheim), where marine resources were identified along with sea level changes that allowed greater areas of grazing to be accessed through time (Linderholm et al., 2019) (see below). The large site of Dilling, Rygge, Østfold is also being tackled in the same way, with sea level, superficial geology, and climatic change (Büntgen et al., 2011) also requiring consideration (Box 18.4). Overall, phosphate levels are low to moderately low, which is typical of rural occupations (Linderholm, 2007; Macphail et al., 2000). The ubiquity of byre waste and use of organic manures is recorded by the predominance of organic phosphate (Engelmark and Linderholm, 1996) (see also Myhre, 2004).

Producing fully integrated interdisciplinary reports/papers can be more difficult. For example, the London Guildhall report recounts detailed findings from 50 thin sections (and selected

---

Box 18.4   Geoarchaeology and mixed farming settlement studies at Dilling, Rygge Municipality, Østfold, Norway (a Pre-Roman Iron Age/Roman Iron Age site ~200 BC-AD 400, with Migration Period (~AD 400–550) outlier)

The ca 5.5 hectare site of the Dobbeltspor Dilling Project, Østfold, Norway has produced a large database for improving the modeling of Iron Age settlement and farming in southern Norway. 137 houses/different house phases (e.g. three-aisled buildings), and other settlement features such as fields, trackways and pit houses were excavated with environmental archaeology samples undergoing geochemical, macrofossil, and soil micromorphology analysis. The geoarchaeology component involved 92 thin sections and some 600 bulk sample studies (LOI, MS, MS550, and P) (Macphail et al., 2021). Both survey and feature-specific analyses were undertaken. Phosphate fractionation was also carried out on a large proportion of these in order to produce PQuota figures (the ratio between organic and inorganic phosphate), and selected thin sections were additionally characterized by microchemistry using SEM/EDS on the uncovered thin sections.

The first task was to reconstruct sea levels and landscape for ca. AD 0, as many house phases spanned this period. The Geological Survey of Norway (NGU; www.ngu.no/) maps provide combined sea level and superficial geology information. The Iron Age settlement is located on the lowest ground covered by fine intertidal marine sediments (fine sands, loamy fine sands and fine sandy silt loams), whereas upslope, beach sands occur. The Migration Period houses spread upslope across the beach sands and onto higher ground occupied by an end-moraine

Box 18.4   (Continued)

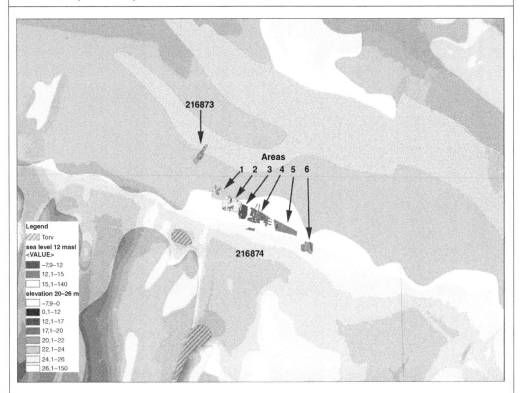

Figure 18.13 Geological map of Dilling settlements; 2 km wide, north oriented map of elevations and geology. Most Iron Age structures (Areas 1–6) occur on fine marine intertidal sediments (yellow), with beach sands upslope (green). The Migration Period site upslope occurs across the boundary between the beach sands and poorly sorted stony end-moraine (pink). Brown and green striped areas of peat ("torv") record the remnants of probably much larger areas of wetland (adapted by Johan Linderholm, MAL, University of Umeå, Sweden from The Geological Survey of Norway (NGU; www.ngu.no/) maps).

(Figure 18.13). Normally, due to post-glacial uplift, more recent settlements are found on lower ground, but it is the opposite here. Speculatively, this may be due to the low ground becoming wetter as the climate deteriorated during the Migration Period (Büntgen et al., 2011). Some peat area remnants ("torv") suggest that wetland was more extensive in the past, but this has diminished greatly due to recent drainage and agricultural practices.

Bulk analyses were employed to produce an overall geochemical characterization of the site. Phosphate concentrations are low to moderate, as is commonly the case for Scandinavian Iron Age rural settlement sites (Linderholm, 2007; Macphail et al., 2000). Such mixed farming communities, where animal management and manuring of fields is typical, also show a predominance of organic phosphate (PQuota >1.0; Figures 18.14a–18.14d) (Engelmark and Linderholm, 1996; Myhre, 2004). A naturally low iron content in the marine sediments is consistent with low magnetic susceptibility enhancement (often MS = 10–30 χlf 10^{-8} m^3 kg^{-1}), apart from localized areas of ovens and furnaces (e.g. >80 χlf 10^{-8} m^3 kg^{-1}). Large geochemical and geophysical databases allow statistical observations, such as through PCA (principal components analysis; Figure 18.15). It is useful to examine one area, Area 6, in terms of the

(Continued)

Box 18.4   (Continued)

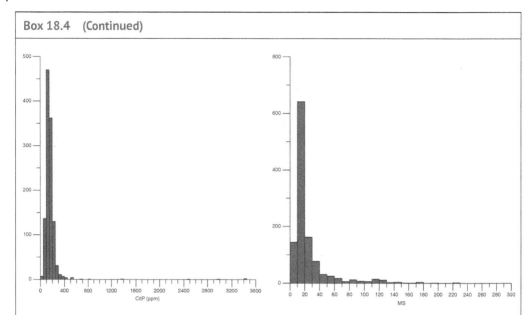

Figure 18.14a   Phosphate chemistry for nearly 500 samples, showing that most record ~50–300 ppm CitP (citric acid soluble P).

Figure 18.14b   Magnetic susceptibility for nearly >600 samples, showing that most record low to small amounts of enhancement (e.g. ~10–40 χlf 10^{-8} m^3 kg^{-1}), with few outliers, probably due the overall low iron content of the marine sediments.

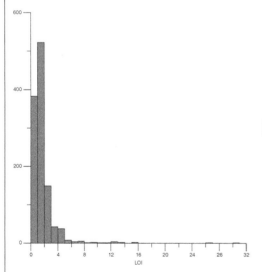

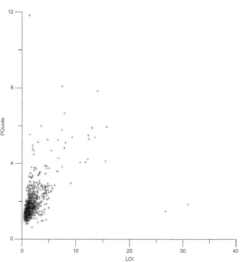

Figure 18.14c   LOI (% loss-on-ignition – proxy measurement of organic matter, including charcoal) for nearly 500 samples, with most showing a low to moderately low LOI, and indicating generally oxidized nature of the soils and deposits.

Figure 18.14d   PQuota (OrgP/InorgP) and %LOI scattergram clearly indicating that although soils have been oxidized, much organic phosphate is present (e.g. PQuota is >1.0); in turn, this high PQuota points to the likely presence of organic manure in the fields and byre waste in the houses and other features, including trackways.

Box 18.4 (Continued)

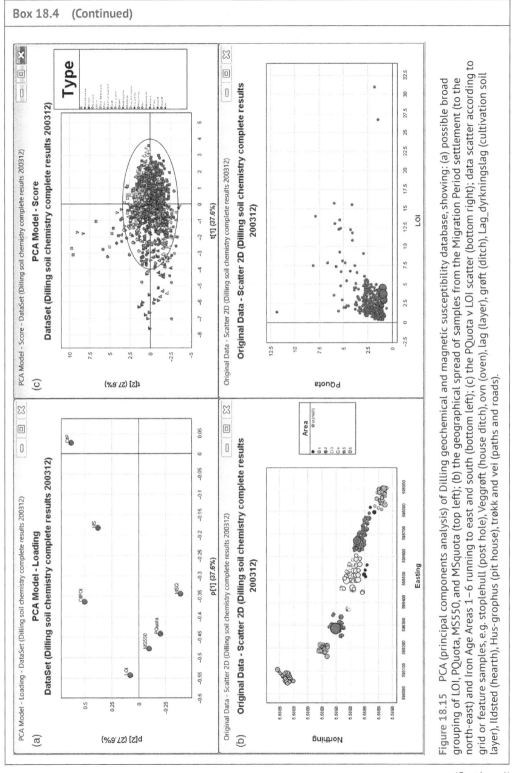

Figure 18.15 PCA (principal components analysis) of Dilling geochemical and magnetic susceptibility database, showing: (a) possible broad grouping of LOI, PQuota, MS550, and MSquota (top left); (b) the geographical spread of samples from the Migration Period settlement (to the north-east) and Iron Age Areas 1–6 running to east and south (bottom left); (c) the PQuota v LOI scatter (bottom right); data scatter according to grid or feature samples, e.g. stoplehull (post hole), Veggrøft (house ditch), ovn (oven), lag (layer), grøft (ditch), Lag_dyrkningslag (cultivation soil layer), Ildsted (hearth), Hus-grophus (pit house), trøkk and vei (paths and roads).

(Continued)

Box 18.4   (Continued)

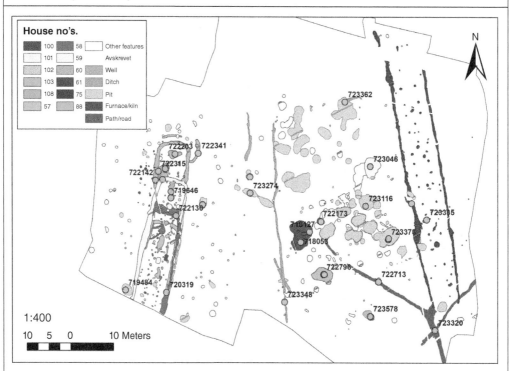

Figure 18.16   Map of features excavated and sampled for soil micromorphology (green circles) in Area 6, with overlapping different house phases on the left (e.g. Houses 58, 59, and 60), separated by a N-S ditch from an area of pits, wells, and pit house (House 100). Pit house 100 could be linked to House 75 by a path that also leads off to the south (map by Torgeir Winther and Marie Ødegaard, Museum of Cultural History, University of Oslo).

excavated features and associated sampling for soil micromorphology (Figure 18.16) and geochemistry; data also includes other features and grid sampling (Figure 18.17). In Area 6, House 75 includes soil micromorphological evidence of byre use (Figure 18.18), with low MS values, small concentrations of CitP and estimated organic matter (LOI), and PQuota ranging from 1.0–2.7. Such geochemical characteristics are consistent with previously studied stabling experiments and archaeological analogues, which lead to a consensus interpretation of animal management at this specific location (Macphail et al., 2004; see also Shahack-Gross, 2011, 2017). Such specific feature findings demonstrating animal management (Macphail et al., 2017) are key to the overall understanding of the Iron Age settlement, which functioned through a mixed farming economy, where the main crop was barley, and where weed plants such as nitrophilous fat hen (*Chenopodium album*) are clearly indicative of manured fields (Gjerpe volume, in prep.).

Box 18.4   (Continued)

Figure 18.17   Map of Area 6, here showing geochemical sampling, often correlated with soil micromorphology sampling (Figure 18.16). Relative amounts of CitP are shown by size of blue circles. There is a coincidence of soil micromorphological identifications of CitP concentrations and byre deposits in Houses 60 (left) and 75 (right), and the well. Small amounts of byre/dung residues and enhanced P/Pquota in Pit house 100 (red; centre) and along the pathway between House 75 and the Pit house 100, were also recorded.

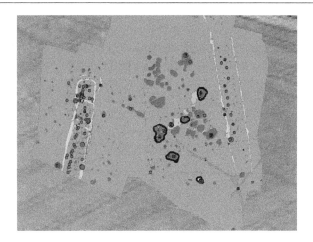

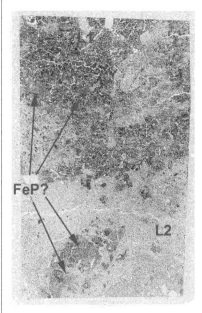

Figure 18.18   Thin section scan of 723335 (Floor 712644, House 75, Area 6). In Layer 2, compaction of fine sands may have caused horizontal cracking; burrowing-down and inwash of humic soil and possibly iron-phosphate staining (FeP?) is also in evidence. Layer 1 is composed of trampled humic muds forming pans, which also likely became stained by secondary iron phosphate from suggested byre use here. This floor layer seems to be refuse from, and location of, a byre (MS = 8–9 χlf 10^{-8} m³ kg⁻¹; 230–240 ppm CitPOI; PQuota 2.1–2.7; 2.7–3.0% LOI compared to lower fills (MS = 10–15 χlf 10^{-8} m³ kg⁻¹; 110–180 ppm CitPOI; PQuota 1–1.6; 0.7–1.1% LOI). Frame width is ~50 mm.

	MS χlf 10^{-8} m³ kg⁻¹	CitPOI (ppm)	PQuota (%)	LOI (%)
Floor	8–9	230–240	2.1–2.7	2.7–3.0
Lower fills	10–15	110–180	1.0–1.6	0.7–1.1

microprobe studies) and closely correlated 48 bulk and 31 pollen analyses; the results were presented chronologically, context-by-context. They are almost completely lost in the archaeological volumes (Bateman et al., 2008; Bowsher et al., 2007), but highlights were published within a thematic chapter (Macphail et al., 2007a).

Soil micromorphological and bulk data have to be presented not only clearly, but also transparently. The presentation of soil micromorphological data, including protocols for data gathering and data "testing", has been discussed earlier (Chapter 17). Ways to present bulk data have also been presented (Chapters 3 and 17) as both tables and diagrams. Sometimes, however, special associations (e.g. field vs. micromorphological findings) may need to be demonstrated through statistics (below 18.5.1). Equally, a site report might include the integration of finds, contexts, descriptions, photographs, altitudes, soils, and geology within a landscape, and here GIS (geographical information systems) becomes a requirement (below 18.5.2). Large numbers of samples and data, particularly images, require that they be archived (below 18.5.3). Currently, there is also the possibility of publishing on the internet, where there is no limit to how much supportive data can be presented; organizations and universities also have designated websites (e.g. https://library. thehumanjourney.net/). Geoarchaeological publications include short and long thematic papers, and large-scale site-based and project-based statements, as tackled by various authors (Holliday et al., 2007).

18.5.1 Statistical Support

Statistical packages are widely available, and used throughout the biological, geographical, and social sciences. Here, we simply provide some examples where statistical analyses of geoarchaeological data have produced some geoarchaeological insights that helped integrate soil micromorphological and bulk analyses, thus making interpretations more robust.

Early Mediaeval London Guildhall (Bowsher et al., 2007) Here, organic matter (LOI), P, magnetic susceptibility (χ), and heavy metals (Cu, Pb, and Zn) were analyzed ($n = 48$) by John Crowther (University of Wales, Lampeter, UK) (see Chapter 17 Laboratory Methods). Pearson product-moment correlation coefficients were used to examine the relationships between these various properties, and analysis of variance (using the Scheffé procedure) was used to compare mean values from groupings of soil microfacies types (SMTs), as identified through soil micromorphology (complemented by microprobe and palynology). An example of one of these SMTs, namely SMT 5a (*well-preserved highly organic stabling refuse*; cf. Brönnimann et al., 2017a; Shahack-Gross, 2011), is given in Table A1.16.6. Analysis of variance was undertaken only on groupings with ≥ 4 samples. In cases where the data for individual properties had a skewness value of ≥ 1.0, a log10 transformation was applied in order to increase the parametricity. Statistical significance was assessed at $p = 0.05$ (i.e. 95% confidence level).

Statistical analyses revealed that, while the majority of the phosphate is present in inorganic forms (Pi) (phosphate-Pi:P ratio, 72.0–92.8%), there is a strong correlation between phosphate-Pi and LOI ($r = 0.705$, $p<0.001$) (Tables 18.1a, b). This suggests that organic sources other than, or in addition to bone, have made a significant contribution to phosphate enrichment; in this case, dung of domestic stock, human excrement, coprolites and cess were the probable sources. The concentrations of three heavy metals (Cu, Pb, and Zn) show highly significant ($p<0.001$) direct correlations. There are also strong direct correlations between phosphate and heavy metal concentrations. Magnetic susceptibility enhancement, however, is much less strongly correlated with the chemical properties, which suggests that the circumstances leading to chemical enrichment are not necessarily the ones associated with burning. It was concluded therefore, that in the early

Table 18.1a   London Guildhall; Summary of analytical data for all samples (n = 48).[*]

Property	n	Mean	Standard deviation	Minimum	Maximum
LOI (%)	45	12.0	9.26	1.20	43.4
pH (1:2.5, water)	40	7.3	0.43	6.1	8.1
Carbonate (est) (%)	47	1.4	2.21	<0.1	>10.0
Phosphate-P_o (mg g^{-1})	48	1.26	0.892	0.190	4.74
Phosphate-P_i (mg g^{-1})	48	7.52	6.75	0.864	48.2
Phosphate-P (mg g^{-1})	48	8.78	7.44	1.20	52.6
Phosphate-P_o:P (%)	48	15.2	4.94	7.2	28.0
Phosphate-P_i:P (%)	48	84.8	4.94	72.0	92.8
χ (10^{-8} m^3 kg^{-1})	48	7.4	36.2	5.5	196
χ_{max} (10^{-8} m^3 kg^{-1})	44	731	370	130	2080
χ_{conv} (%)	44	9.72	5.80	0.26	22.0
Pb (µg g^{-1})	42	345	283	11.4	1280
Zn (µg g^{-1})	42	101	76.0	10.4	468
Cu (µg g^{-1})	42	127	77.0	7.30	333

[*] In certain cases the sample supplied was too small to undertake the full range of analyses. In these cases priority was given to the determination of phosphate and χ.

Medieval levels at the London Guildhall, heavy metal concentrations were linked to organic sources (bone, dung, cess). This is unlike the situation of post-medieval Tower of London (location of Royal Ordinance and Mint - Keevill, 2004; Macphail and Crowther, 2004), where there were links to industry or craft working of alloys, consistent with the archaeological finds recovery of these levels. Other examples of statistical analysis of soil data have been noted already, for example employing the large database from the herding site of Raunds (Northamptonshire, UK) (Macphail, 2011a); see also Figure 18.15 for PCA modeling of chemistry for the Dilling settlement (Vestfold, Norway),

18.5.2   GIS

GIS is a valuable tool for integrating a variety of data sets that originate not only from the geoarchaeologist but the entire research team (see section 16.3.1). It can also make the day-to-day running of a large or long-lived excavation more efficient. Increasingly, as sites are excavated, all finds, features, and samples are located in three dimensions (usually by corrected GPS or Total Station), with additional attribute information, e.g. feature number, trench number, small find number, sample number, etc. As spot dates on finds are produced, archaeological features can be phased. Such GIS "maps" of the site, its contents and surroundings, can then be interrogated to show the distribution patterns of archaeological finds, as well as geoarchaeological data. Moreover, with the increased portability of computers and the use of real-time survey, the data can be interrogated whilst the site is being excavated. This allows for dynamic interpretations and hypothesis generation that can often be tested in the field.

Table 18.1b London Guildhall; Pearson product-moment correlation coefficients (r) for relationships between the various soil properties for all the bulk samples analyzed†

(n = 40−48, depending on combination of variables).

	pH	Carb§	P$_i$§	P$_o$§	P§	P$_i$:P	χ§¶	χ$_{max}$§¶	χ$_{conv}$§¶	Pb§	Zn§	Cu
LOI§	−0.456	n.s.	0.705*	0.822*	0.738*	n.s.	0.324	n.s.	n.s.	0.601*	0.902*	0.569*
pH		0.485	n.s.	−0.337	n.s.	n.s.	n.s.	n.s.	n.s.	n.s.	−0.424	n.s.
Carb§			n.s.	n.s.	n.s.	n.s.	n.s.	n.s.	n.s.	n.s.	n.s.	n.s.
P$_i$§				0.839*	0.997*	0.303	n.s.	0.347	n.s.	0.558*	0.858*	0.643*
P$_o$§					0.881*	n.s.	n.s.	n.s.	n.s.	0.524*	0.847*	0.575*
P§						n.s.	n.s.	0.346	n.s.	0.561*	0.874*	0.643*
P$_i$:P							n.s.	n.s.	n.s.	n.s.	n.s.	n.s.
χ§¶								n.s.	0.720*	0.475	n.s.	n.s.
χ$_{max}$§¶									−0.598*	n.s.	n.s.	n.s.
χ$_{conv}$¶										0.363	n.s.	n.s.
Pb§											0.701*	0.660*
Zn§												0.691*

† Statistical significance: n.s.= not significant (i.e. p=0.05)

* = significant at p<0.001.

§ Indicates log10 transformation applied to the data set.

¶ In the case of the magnetic properties, correlations between the untransformed variables (used in assessing the relative importance of χ$_{conv}$ and χ$_{max}$ in affecting χ are as follows: χ with χ$_{conv}$ r = 0.687 (p<0.001; see Figure 1); and χ with χ$_{max}$ r = −0.134 (n.s.).

Tables 18.1a and18.1b (by kind permission of John Crowther, University of Wales, Lampeter

The production of GIS *Shape Files* allows not only the spatial relationship of mapped features sampling points and sampling grids to be realized, but also and more significantly, the further geospatial processing and analysis of these data. For example, elemental concentrations such as phosphate can be plotted as different size circles (see Box 18.4, Figures 18.16 and 18.17), or interpolated as surfaces; in the case of multiproxy studies, soil micromorphology sequences can be tied into a series of associated bulk samples and mapped quantitative distributions of organic matter (LOI), phosphate, and magnetic susceptibility.

This strategy has been carried out in Scandinavia since ~2000 (Linderholm, 2007, 2010). At the Ørlandet Iron Age settlement study, environmental archaeology reported on macrofossils, the settlement pattern, and how different parts of the site were linked (Figures 18.19–18.21). Geology was important for a number of taphonomic and interpretational reasons, mainly because areas of calcareous shell sand were alkaline, and buffered citric acid extraction of phosphate produced skewed distribution maps (Linderholm et al., 2019). Equally, where long houses were located on shell banks much poorer preservation of macrofossils was recorded and possibly related to more freely draining and oxidizing conditions. Such findings need to be built into any form of interpretation of sites. In Figure 18.19 macrofossil recovery was much higher in House 7 because this one was unusually not situated on a shell bank. In Figure 18.20 wells/waterholes and houses are located, along with sampling locations including soil micromorphology series.

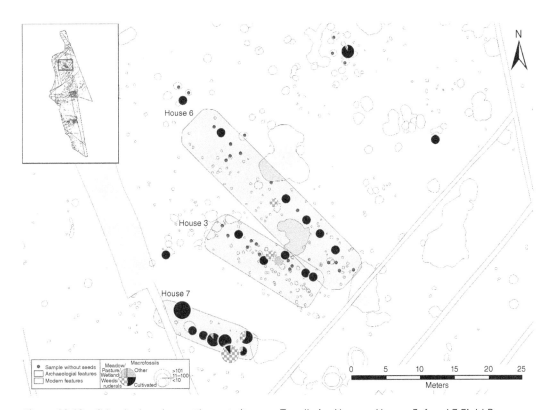

Figure 18.19 Ørlandet Iron Age settlement site, near Trondheim, Norway; Houses 3, 6 and 7, Field B: Relative number and proportion of plant macrofossil remains. Cultivated plants mainly involve barley grain. House 7, with the greatest preservation of macrofossils is not located on a shell sand bank (illustration: Magnar Mojaren Gran; Linderholm et al., 2019; image produced by the kind permission of the NTNU University Museum, Ingrid Ystaard and Magnar Mojaren Gran).

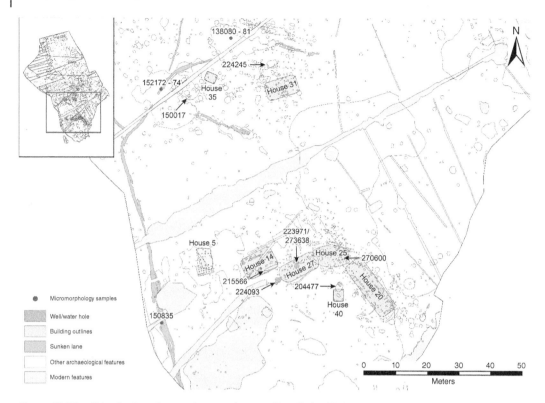

Figure 18.20 Ørlandet Iron Age settlement site, near Trondheim, Norway; excavation areas A (north) and E (south). Buildings, wells/waterholes and a sunken lane-associated bulk sample and micromorphology locations (red dots) are shown (illustration: Magnar Mojaren Gran; Linderholm et al., 2019; image produced by the kind permission of the NTNU University Museum, Ingrid Ystaard and Magnar Mojaren Gran).

Of note is the sunken lane that was used for stock movements from coastal wetland pastures, for example, and aids our understanding of how the village functioned. In the same area of the excavation, bulk sampling allowed the production of phosphate and magnetic susceptibility maps. These were consistent with an interpreted area of artisan and cooking pit activity in the northern portion (MS distribution), and possible sunken lane transportation link between settlement in the south and fields in the northern part (Figures 18.21a and 18.21b); soil micromorphology revealed dung concentrations in the lane deposits including a sample with 300 ppm CitP.

It is essential, however, that such reconstructions are based on relevant data, both at the landscape and the site scale (Courty et al., 1994). Detailed information may have been recorded over a region on soil type, slope type and angle, relief, precipitation, and vegetation, with the intention of relating them to site distribution or land use. The tacit assumption, of course, is that the present configurations of these parameters are close to those of ancient times. Past experiences in the field have shown that site distributions can be truncated or lost by wholesale erosion or burial, thus removing critical information from the database. Workers employing GIS need to be aware of such situations.

At the site level, data may well provide information on artifacts, crops, domestic animals, and wild resources that were utilized during a specific period. As we have emphasized, however, the documentation and reconstruction of past soils is much more problematic. For example, in the case of Ørlandet, post-glacial uplift led to the emergence of larger potential areas of coastal grazing land

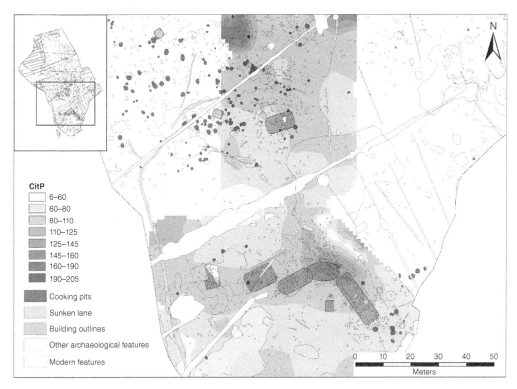

Figure 18.21a   Ørlandet Iron Age settlement site, near Trondheim, Norway; Phosphate data (CitP surface samples) from Excavation areas A (north) and E (south). The sunken lane links southern and northern zones. Note high concentration of CitP in sunken lane (thin section sample 150835 recorded concentrations of dung residues and an example of 300 ppm CitP) in the southern zone and possible transportation route of manure/dung/waste from the house complex (illustration: Magnar Mojaren Gran; Linderholm et al., 2019; image produced by the kind permission of the NTNU University Museum, Ingrid Ystaard and Magnar Mojaren Gran.

during 1,000 years of Iron Age occupation (BC 500 – AD 550). One of the key research topics for uplands in south-west England is recording soil changes associated with Neolithic to Bronze Age activities (a change from brown soils to podzols) with the latter being the modern soil cover (e.g. Balaam et al., 1982; Carey et al., 2020). Erosion, colluviation, alluviation, weathering and podzolization, manuring and plaggen soil or *terra preta* formation, may have transformed the soil cover. The inferred fertility of Bronze Age fields based upon the modern horticultural production of strawberries, did not convince one of the authors at one conference. As geoarchaeologists, therefore, it is our role to make sure that any geological and soil information employed in GIS modeling, is correct.

18.5.3   Archiving

It has already been mentioned that undisturbed, resin-impregnated monolith samples (Chapter 17) form a sample archive, such as that from Fenland studies by Charly French at Cambridge University. A very large number of resin-embedded samples and thin sections, a product of research by Goldberg and Macphail, are also centrally stored at the Institute for Archaeological Science, University of Tübingen, Germany. At the Institute of Archaeology, University College London selected blocks from numerous worldwide projects were reused to aid the creation of a

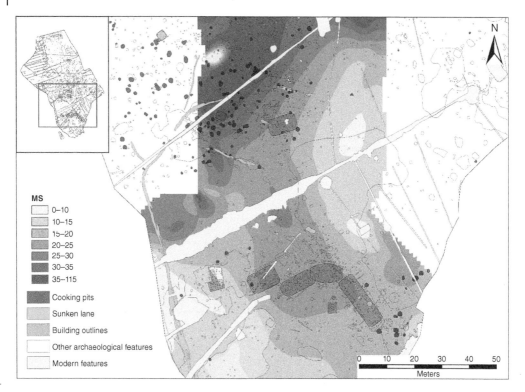

Figure 18.21b As Figure 18.21a; distribution of MS (magnetic susceptibility) data (surface samples). In this case, the greatest activity is recorded in the northern zone and clearly linked to the abundance of cooking pits (Figure 18.20) (illustration: Magnar Mojaren Gran; Linderholm et al., 2019, image produced by the kind permission of the NTNU University Museum, Ingrid Ystaard and Magnar Mojaren Gran).

teaching set. Bulk samples should also be archived. The late Professor G. W. Dimbleby (Institute of Archaeology, UCL) recommended that these be air dried to halt most biological activity that allowed reasonable preservation. It should be remembered, however, that labile soil characteristics such as pH can change on drying, and thus pH measurements are best carried out on samples that still retain their field moisture content. A list of archived samples can be appended to the geoarchaeological report.

Workers normally retain their field notebooks (or xeroxes or scanned versions); if possible, hard copies of datasheets are also stored, although now most information is recorded and archived electronically. As noted earlier, it is common today to produce a digital report that is composed of text, bibliography, tables, figures, and photo-images. Generally, a report is supported by a portable drive/web-archive, where a large number of images and raw data (e.g. field notes, spreadsheets) can be stored. The internet also allows the presentation of whole site reports, their supportive data and images.

Such archiving is particularly important, as some sites are re-opened and re-excavated during the lifetime of an individual researcher or over several generations (Bell et al., 1996; Evans and Limbrey, 1974; Macphail et al., 2003b). It thus behoves any researcher to leave as detailed a "paper trail" as possible in order to facilitate the efforts of later researchers. Personal experience has demonstrated that "retrofitting" one's own research results with those produced only a decade or more earlier at the same site proved to be a difficult task (Holliday et al., 2007).

18.6 Site Interpretation

18.6.1 Introduction

In this book we have repeatedly stressed how to record field and laboratory geoarchaeology data. Less attention has been paid to interpretation, except for illustrative examples and the presentation of type-sites. In the view of one eminent and respected American archaeologist (Professor David Sanger), the easy part is description, the Devil is in the interpretation; and this applies to both *the site* and *the geoarchaeology*. He also suggests that the means to interpret a site takes *something else,* which is hard to teach or acquire. This is not the place, however, to note instances where the temptation has been to forsake interpretations of "difficult situations, materials or sites". Equally, it is bad practice for an interpretation not to be based upon extant data, but rather the interpretation of the data has been fitted or rammed into some kind of preconceived, and often false, notion or model.

However, it is essential to realize that "data does not speak for itself", all data requires an interpretation. This is not necessarily a straightforward process; often data can be equivocal. Geoarchaeological data is after all, just that: data, a list of numbers or a table of descriptions, etc. It is therefore essential to move the discussion beyond the data, to the meaning and the significance of the results. For the interpretations presented, the geoarchaeologist must be clear about *how* they have interpreted their data. In doing so, they need to say how they reached this conclusion, what was the key data used to support this interpretation and the diagnostic features of the data that support this line of reasoning. Likewise, other plausible explanations should equally be considered and again, reasons given that both support alternative arguments and also demonstrate why it is considered a less likely explanation.

It should be clear from previous chapters that the first step must be the correct recording of data, because without this strategy no accurate interpretations are possible. We have also tried to provide examples of what questions need to be asked of a site before the correct recording and analyses can take place. If the archaeologist, through ignorance of geoarchaeology, or the geoarchaeologist, through ignorance of archaeology, do not know what questions to ask during a site investigation, the final interpretation is likely to fall well short of a site's potential. Most geoarchaeologists come from a geoscience, geographical, and/or pedological background, whilst in contrast, geoarchaeology may well be a closed book to many archaeologists. We therefore hope that the thematic studies presented throughout this text will be an aid to both these groups.

18.6.2 Correlating and Integrating Geoarchaeological Data with Those of Other Disciplines

It is important that geoarchaeological data are not collected, analyzed, and interpreted in isolation. Previous good use of complementary geoarchaeological data can be cited (Dimbleby, 1962; Evans, 1972; Haită, 2012; Healy and Harding, 2007; Shillito and Ryan, 2013; Whittle, 2007). Geoarchaeological information is only one part of the whole when studying a site. Holistic archaeological studies involve both cultural and environmental inputs. Geoarchaeology can provide information that is useful to environmental reconstruction and also add to debates concerning cultural activity. It is much more multi-faceted than is often realized within the archaeological literature. It also has a role in theoretical studies. This section seeks to demonstrate how geoarchaeological investigation, if well thought through from the outset, can be successfully integrated with other data in archaeological projects (see Boxes 18.1 and 18.4). This approach first influences the fieldwork and sampling agenda (Chapter 16), laboratory methods applied (Chapter 17), and the way the results are viewed and integrated into the site or project discussion.

18.6.3 Data Gathering

It is crucial that all geoarchaeological sampling is correlated with any other environmental sampling or recovery of artifacts. The exact relationships between the geoarchaeological samples and the archaeological contexts that they are investigating need to be recorded (see section 16.9). This would include the correlation of geoarchaeological findings with those from fossil analyses (Cruise et al., 2009; see also Figure 18.19).

Some examples of the analysis of fossils can be noted. Whole charred cereal grains such as *Triticum* and *Einkorn*, and large pieces of hazelnut shells have been identified in thin section and were reported in Courty et al. (1989). At the Viking Heimdalsjordet settlement (Vestfold, Norway), a charcoal-rich ditch fill was both the focus of a large bulk macrofossil sample and micromorphology; fish were also consumed on site, as recorded in latrine deposits (Figures 18.22a and 18.22b). The identification of wood charcoal to species is more problematic, however, because wood charcoal is only identifiable if sectioned in exactly the right plane, and this does not happen very often (Arene Candide is a notable exception; Macphail et al., 1997). Macrofossil analysis more often relies on flots (or recovered mineralized material) from large (e.g. 2 liter) bulk samples, and in most cases, the exact relationships between important layers and the macrofossils they contain is crucial. Distribution of macrofossil types in long houses has given rise to spatial analyses sometimes linked to patterns of LOI, magnetic susceptibility and phosphate (see Figure 18.19; Engelmark, 1985; Viklund et al., 2013). The worst disputes between geoarchaeologists and other environmentalists often arise from the study of samples from different site locations, so that their individual stories do not tie up. Just as frequently, however, the geoarchaeologist and environmentalists (like the archaeologist) work to different agendas and try to tackle quite dissimilar questions. They may also not wish to fully confront the limitations of their own technique and the ways in which taphonomy and site formation processes affect the extent of recovered biological and geological/pedological materials and information. This is where a useful

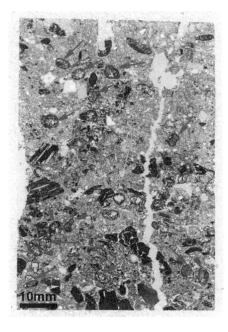

Figure 18.22a Scan of M15196B (Parcel ditch 14031, Section 14750, Layer 6; Heimdalsjordet, Vestfold, Norway). Overall, the fill has an LOI of 11.1%, in part associated with concentration of charred cereal (barley) grains (arrows), recognized as a processed grain store residue, consistent with this coastal trading site's function (Bill and Rødsrud, 2017).

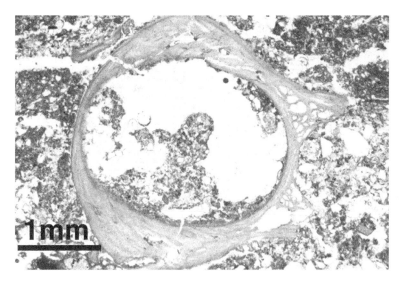

Figure 18.22b   As Figure 18.22a, photomicrograph of M14834B. Fish bone found within cess-stained Layer 4 (SEM/EDS: P = 7.77–8.97%; Fe = 17.8–21.4%). Fishbone is probably a precaudal vertebra of a small gadid (a fish of the cod family) (Rebecca Nicholson, Oxford Archaeology, pers. comm.). PPL.

Figure 18.23a   Field photo of the early Holocene, Neolithic to Medieval peat bog site (arrow) of Lago di Bargone (900 m asl, 9 km from the Mediterranean), which provides both a local and regional environmental record principally based upon palynology, in an area (eastern Apennines) where, because of slope and climate, peat records are rare (Cruise et al., 2009).

dialogue among specialists should take place, as every technique is likely to add some unique information to the consensus interpretation of a site. No one needs to feel defensive at such times. As noted in Box 18.1 if sampling for pollen and macrofossils (including dateable material), geo-chemistry and soil-sediment micromorphology are exactly correlated, a consensus interpretation can be achieved. A similar team approach was applied to Bargone Lake in the Italian Apennines (see Figures 18.23a–18.23c) (Cruise et al., 2009).

Figure 18.23b As Figure 18.23a; photomicrograph of lowermost sediments. During the early Holocene, the site was fully aquatic, and hence the common presence of mid- to fully eutrophic diatoms and diatomaceous silts at around 3.00 m, which together with the palynology are indicative of a lake surrounded by white fir (*Abies*) woodland. PPL, frame width is ~0.27 mm.

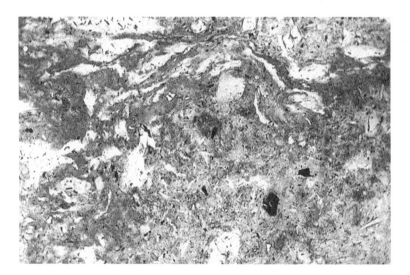

Figure 18.23c As Figure 18.23a, photomicrograph of later Holocene (4350–4040 cal BC) near-shore lake sediments. There has been development of a weakly formed "bog iron" (mean 0.580% Fe; microprobe grid, $n =$ 10), which is also marked by a peak in magnetic susceptibility measured after ignition at 550° C. Iron replacement of some organic materials, suggests fluctuating water tables while pollen recorded a grazed open woodland environment likely contemporary with increased upland pastoral activity during the Chalcolithic and Early Bronze Ages. Such interdisciplinary studies help to answer questions concerning the links, or simple time correlation, between past human activity, vegetation change and climate. PPL, frame width is ~4.62 mm.

If work is to be truly interdisciplinary (Box 18.1), the geoarchaeologist should be aware of their role in a project. Under the umbrella of a project, the geoarchaeologist may take several roles:

- they may be involved as an equal partner amongst a team providing environmental information to the archaeologist/project director, who will develop a consensus interpretational chapter; or

• within the geoarchaeological component interpretations are enhanced by, say, biologists providing data on specific layers or contexts, and here the paleoenvironmentalist plays a supportive role; or
• where, through focused analyses of particular contexts/layers, paleobiological and cultural investigations of full sequences are aided by geoarchaeological information, and here the geoarchaeologist carries out a secondary role.

It should not be forgotten that the geoarchaeological study may well be a team effort, and the individuals/teams providing chemical, magnetic, and soil micromorphological data also need to communicate and discuss the questions that they are addressing, just as much as the geoarchaeologist needs to *connect* with the archaeologist and paleoenvironmentalist.

Finally, too many geoarchaeological publications are site/sondage/theme specific or included in project papers and volumes where geoarchaeological results are simply cherry-picked. Sometimes it is recommended that they should be more holistic. In fact, geoarchaeology can be an equal player in archaeological reporting as shown by the way the morphology of settlements can be reconstructed, with information on how they might have functioned, for example (Macphail et al., 2017a). Suggested investigated settlement components that may have a role to play in settlement reconstruction can be listed as:

• constructions
• tracks and roads
• animal management
• waste disposal
• water management and control
• mortuary features
• specialist domestic, artisan and industrial activity

Some suggested interdisciplinary reconstructions were noted above (see Figures 18.19–18.21) and in Box 18.4. Mortuary features were specifically highlighted in Chapter 15.

18.7 Conclusions

In this chapter we have attempted to demonstrate the way results are reported and published. Both affect the way future site studies are organized and carried out, and the way geoarchaeological information is understood and valued, or not. If geoarchaeology is to be useful and successful, it must be both focused and attentive to the needs of the site and the archaeological and environmental team. Some of the diverse ways of reporting have been discussed above. A field evaluation/ sample assessment differs from a full post-excavation report. Equally, a publication in a journal will contrast strongly with the full archive "document" that can be supplied on a portable memory device or uploaded onto the web. Simply, reports are not articles either in terms of the quantity of data presented or how they are presented. At the reporting stage, we suggest various methods to improve the ways the geoarchaeological information is presented and supplied to the archaeologist. These vary from tables, images, and figures to the use of statistics and integration with GIS. What is emphasized, moreover, is the careful correlation of information with the archaeology and other environmental findings in order to produce both independent and consensus interpretations within a team. This leads to sustainable and robust archaeological reconstructions of both past landscapes and cultural activities.

19

Concluding Remarks and the Geoarchaeological Future

This thorough revision of the 2006 volume – including additions to the chapter list and new content – illustrates that geoarchaeology is an increasingly broadening discipline. Fields of enquiry involve many different kinds of people who are carrying out a vast variety of research both in the field and in the laboratory, some of which was not dreamt of in 2006. The spectrum of specialists and their skills contributing to a wide variety of projects now range from geoscientists (geologists *sensu lato*, geographers, pedologists, and soil chemists) examining landscapes, soils, and deposits, to archaeologists using remote sensing (hyperspectral data sets) as well as soil survey and coring data to infer ages of landscapes and to understand site distributions; at the site and landscape scale these data can be interpreted through GIS and statistical packages. Interestingly, with increased awareness of geoarchaeology, the divisions between geoarchaeologists *sensu stricto* and archaeologists has become increasingly blurred. Furthermore, examples include not only major advances from European and New World sites, but also those from around the globe.

In this revision, we have continued to provide some of the basic characteristics of sediments and soils, which constitute the ultimate stratigraphic framework for archaeological deposits and sites. This is the material that is excavated to expose and capture the archaeological record. The spatiotemporal articulation of these objects and features with deposits and landscapes is covered in the chapter on stratigraphic concepts. In our opinion, an awareness of the concepts *facies and microfacies* is crucial for appreciating the changes in the localized nature of archaeological deposits within a site and its surroundings (Karkanas and Goldberg, 2018b). Such awareness also enables the investigator to link these changes to processes of deposition or formation (usually subsumed under the heading of "site formation processes"). In the first instance, these processes can be of geological origin, and in this light, we have included a discussion of large-scale geological systems, such as those associated with rivers and lakes, as well major environments – coasts and aeolian phenomena. Equally, soils and soil formation are given equal prominence, and our examination of fluvial and wetland systems includes water movement on slopes. One of the most important geological environments for investigation are caves, mainly because millennia-old deposits lost from the surrounding areas can be preserved within them. Moreover, crucial findings of fossil humans and their activities come from caves. In all these chapters we have endeavored to place human activity and site examples as focal points for the reader, caves being a crucial exemplar of this.

Similarly, many deposits in archaeological sites are tied directly to human activities, particularly Late Pleistocene and Holocene ones, where activities associated with hunter-gatherers and more complex societies (e.g. Neolithic, Bronze Age, Roman, Medieval) appear. In the Old World in particular, some deposits owe their very origin to human impact, and theoretically can be traced from

Practical and Theoretical Geoarchaeology, Second Edition. Paul Goldberg and Richard I. Macphail.
© 2022 John Wiley & Sons Ltd. Published 2022 by John Wiley & Sons Ltd.

hillslope clearance and cultivation, and mining disturbance, via colluviation to floodplain sediments. Anthropogenic aspects of deposition and site formation have previously been underplayed in many geoarchaeological publications, particularly those from North America. In view of this paucity, we devoted a relatively large proportion of the book toward the description, analysis, and interpretation of archaeological deposits, and the human activities associated with them. We therefore also include a chapter on human use of materials that allows us to describe archaeological deposits which are solely of human origin (anthropogenic). A case in point has been the study of urban dark earth and the formation of dark earth-like deposits resulting from the processing of waste accumulations.

Techniques such as granulometry may well produce uninterpretable results without having investigated soil micromorphology, for instance. Although much can be, and has been, borrowed from the natural sciences, some archaeological sites and their features have only begun to be understood through experiments designed by archaeologists, the chief examples being studies of buried soils and materials and ethnoarchaeological reconstructions of settlements. An emphasis of ours has been ways to understand better *settlement morphology* – which encompasses both the spatial make-up of a village for example, but also how it functioned – often from only small restricted studies (Macphail et al., 2017). One settlement feature that has been given little emphasis in the past is mortuary archaeology, and we have joined this to forensic geoarchaeological applications.

We have gone into a significant amount of detail describing geoarchaeological field strategies since, like football, geoarchaeological games are won on the practice field; actually, the Battle of Waterloo was won on the playing fields of Eton, or so it is said. No amount of laboratory work can resuscitate incompletely or poorly documented or contextualized field observations, data, or samples. For this reason, we have included many methods such as remote sensing, geotechnical approaches, and deposit modeling that are now carried out in the field (Carey et al., 2018). We have also provided a survey of many of the laboratory techniques that are currently employed in geoarchaeological studies and emphasize that none of these can be employed blindly. Although multi-method approaches are useful, this must not mean that core techniques such as monolith logging and screening, and soil micromorphology, for example, should not be carried out at a sufficiently high level that it cannot provide stand-alone data.

Lastly, we have presented numerous updated practical tips and strategies on how to present geoarchaeological data in written reports, and which data to choose. Field and laboratory information are critical to the unraveling of geoarchaeological problems, but what is included in a report and how it is presented, can make a large difference in how successfully the geoarchaeological results and interpretations are accepted by the geoarchaeologist's audience, clients, and colleagues. Everything must be made clear to archaeologists and other specialist team members of the project, so that the geoarchaeological results can be *usefully* incorporated into the overall understanding of the site and its history. Clear interpretations of settlement/landscape development through time can be achieved when field and laboratory findings are combined properly, using statistical methods when possible. These strategies can be combined with plotting both punctuated information from thin sections and dating linked to chemical and magnetic susceptibility trends across sites employing GIS.

So, what about the future of Geoarchaeology? Before geoarchaeology was even formulated as such in the early part of the 20th century, the major focus on geoarchaeological types of studies was clearly at the landscape or environmental scale, as can be seen by the works of Kirk Bryan, for example, in the American Southwest (see articles in Mandel, 2000a). Emphasis was on reconstructing Quaternary landscape history and paleoenvironments. It still exhibits this tendency, but nevertheless, overall, the situation has changed markedly since the appearance of the first edition of this volume.

Geoarchaeological topics now appear not only in the flagship journal *Geoarchaeology* (www.interscience.wiley.com), but in many international – and more local – journals that deal with Archaeological science, Quaternary studies, Geomorphology, and soils to name a few. In addition, the content shows a very wide variety of topics, including:

- anthropogenic soils in Uruguay
- analysis of Roman building stones in Spain
- sea level fluctuations in the Persian Gulf
- electromagnetic survey at coastal Tel Dor, Israel
- geoarchaeological framework of Late Pleistocene and Early Holocene sites, NW Florida
- lithic raw materials at Early Mesolithic site in Germany
- microanalysis of Middle Stone Age ochre within a sediment block, South Africa
- multitechnique dating of irrigation canals, Sonora, Mexico.

In addition, there has been a marked increase in workshops and meetings on a broad range of geoarchaeological and related topics and themes: micromorphology and phytoliths, hearths/pyrotechnology, urban deposits, through working groups such as *DIG* (Developing International Geoarchaeology) and *Archaeological Soil Micromorphology* (within the various International Soil Science Societies), leading to several special publications (Macphail and Goldberg, 2018a; Nicosia and Stoops, 2017), and, currently, Virtual Soil Micromorphology workshops.

Nevertheless, a few broad subject areas/topics could benefit from additional study. In Europe, archaeological soil micromorphology has developed exponentially since the publication of *Soils and Micromorphology in Archaeology* (Courty et al., 1989). This methodology is now becoming more widespread and almost routine approach across Europe, complementing studies carried out by more traditional methods. Nevertheless, there is a greater need of recording and manipulating data and contextualizing any observations or analyses – GIS is appropriate in this regard – across a broad gamut of scales, from the microstratigraphic, to site, and ultimately to regional scales.

It is heartening that recently in the Mediterranean world, archaeologists and colleagues investigating Late Prehistoric and Classical period sites have shown increased awareness of geoarchaeology in not only understanding how these sites form but also in recognizing surfaces, floors, and decayed construction materials. Yet, given the number of such sites, much more geoarchaeological participation is clearly needed to unravel and discern such processes and features, which are commonly not visible in the field or are poorly documented or verified. Microstratigraphic approaches hold great promise in this regard. Any geoarchaeological review nowadays has also got to consider results coming out of Africa – including ethno-geoarchaeology (Shahack-Gross, 2017) – as well as Asia, where water management, rice cultivation and paleobotanical investigations are closely linked (Rosen et al., 2017; Zhuang, 2018).

North American geoarchaeology still tends to be focused on large- to medium-scale geoarchaeology, with emphasis being given to landscape and soil histories. What appears to be very much lacking is attention to detailed studies of anthropogenic deposits, such as those found in large mounds in the Midwest, or in Pueblos in the southwest. Cultural deposits from Neolithic, Bronze Age, Roman, and Medieval sites have been studied in considerable breadth/depth for more than two decades in the Old World and occur in "historical" sites in North America, from Colonial sites in New England to Spanish Mission sites in the Southwest. Yet the number of geoarchaeological studies has been limited.

We hope that this revised volume will convince archaeologists and specialists alike to incorporate geoarchaeology among other methodologies when excavating a site, from the initial planning stages to the final publications of the results.

Appendix 1

A1.16 (Field Methods)

A1.16.1 Augering

Back strain or even back injury can result from augering. If several days of augering are contemplated, do not forget to pull the auger out using a straight back. Whether working alone or with two people trying to pull the auger out of the ground, one of the best techniques is to grip the handle, then using only the legs "pull" the auger out of the ground. Never just use the back and arms.

A1.17 (Laboratory Methods)

A1.17.1 Bulk Sample Size

Logically, a 10 g sample can be employed for magnetic studies, which are non-destructive, and small sub-samples can then be used for LOI, phosphate, and other extractions.

A1.17.2 Sulfate Problems in Grain Size Analysis

Sometimes the sample may contain SO_4. For example, marine and estuarine deposits and particles settle very quickly leaving a clear supernatant that can be tested for SO_4 content with $BaCl_2$, which forms a white powder (Avery and Bascomb, 1974). SO_4 is removed by drawing off the supernatant, and replacing it with deionized water, a process that may need to be repeated until all the SO_4 is removed.

A1.17.3 Grain Size Analysis

Grain size data is a useful comparison to grain size analysis in thin sections because grains are normally seen as *smaller* than their real size in thin section (Stoops, 2003 14). On the other hand, the presence of ash crystals, clasts, mica, and phytoliths, for example, can be noted in thin section, and in fact clay translocation can be truly identified only through the presence of textural pedofeatures – relict link cappings (e.g. Figure 4.6b).

Practical and Theoretical Geoarchaeology, Second Edition. Paul Goldberg and Richard I. Macphail.
© 2022 John Wiley & Sons Ltd. Published 2022 by John Wiley & Sons Ltd.

A1.17.4   Loss-on-ignition (LOI) and Associated Soil Analyses

Typically, a temperature of 550° C is used for 2 hours in geoarchaeology, although soil scientists are more likely to employ a lower temperature for longer, i.e. 375° C for 6 hours, because some carbonate can be lost above this temperature (Ball, 1964). In fact, the measurement of a sample at both 550° C and 800° C, is believed to yield estimates of organic matter and an approximation of the carbonate content, respectively. Equally, the rubified colors seen after ignition provide a qualitative measure of iron content, because all the iron has become oxidized. Exact amounts of calcium carbonate are calculated from the amount of CO_2 given off from a sample in a calcimeter when HCl is added (Avery and Bascomb, 1974). pH (soil reaction) is measured using a 2.5 proportion of deionized water (H_2O) to soil, although parallel testing is sometimes carried out using $CaCl_2$. In both cases, standard buffer solutions are used (e.g. pH 4.0 and pH 7.0) to calibrate the pH meter – which should also have a thermometer attachment, as temperature also effects pH (Gale and Hoare, 1991: 274). The fresher the soil/sediment sample the better, as drying out and/or microbiological activity in a stored sample, could well transform the pH. Eh (in volts) – level of oxidation-reduction (redox potential) (Levy and Toutain, 1982) and saline content (Avery and Bascomb, 1974: 39) are other characteristics that can be measured and which effect electrical conductivity of deposits, and which are important when attempting to preserve, *in situ*, deposits, biological materials and metal objects for example (French, 2003, Figure 11.3). For example, Tylecote found that acid soils were more aggressive in corroding buried tin-bronzes and copper, while alkaline soils were benign (Tylecote, 1979) – although corrosion still takes place (T. Rehren, pers. comm.). Strangely, peats were also found to be benign, despite their acidity, probably due to the protective action of polyphenols (see also Huisman, 2009).

A1.17.5   Soil Micromorphology – Making Thin Sections from a Long Monolith

Table A1.17.5.1   Example of producing a continuous series of thin sections from a 0.50 m long monolith taken from 10.00–9.50 m OD/height above sea level (asl).

Field monolith sample, e.g. sample [M212] taken at 10.00 m asl	Impregnation in two 160 mm long "resin-proof" containers	Separated blocks ready for cutting into blocks for 75 mm thin sections	Final thin sections (asl)
M212: 10.00–9.50 m (Sub-sampled into 0–150, 150–300, 300–450, 450–500 mm lengths)	**Container A** Sub-sample A1: M212/0–150 mm Sub-sample A2: M212/150–300 mm **Container B** Sub-sample B1: M212/300–450 mm Sub-sample B2: M212/450–500 mm*	Sub-sample A1: M212/0–150 mm Sub-sample A2: M212/150–300 mm Sub-sample B1: M212/300–450 mm Sub-sample B2: M212/450–500 mm	M212a (10.00–9.925 m) M212b (9.925–9.850 m) M212c (9.850–9.775 mm) 212d (9.775–9.700 m) 212f (9.700–9.625 m) 212g (9.625– 9.550 m) 212h (9.550–9.500 m)

*Extra 100 mm long space in container B can be infilled with waste blocks in order not to waste resin.

1.17.6 Soil Micromorphology – Recording and Presenting Soil Micromorphology

Table A1.17.6.1 Suggested outline for presenting soil micromorphological information (and identification of soil microfabric types – SMTs), and any associated data that aid the identification of a microfacies type, for one or multiple thin sections.

Material SMT and MFT	Sample Number	Sampling depth, Soil Micromorphology (SM), *in situ* thin section data (i.e., collected from the thin section itself), and *relevant* associated Bulk Data (BD) including macro- and microfossil data	Context, *brief* comments and discussion, and interpretation
Soil Microfabric Type (SMT)/Microfacies Type (MFT)		**Depth**: Relative sampling depth and /or height above sea level (asl or O.D.) **SM**: • *Heterogeneity*: • *Structure and voids*: • *Coarse mineral* (includes coarse:fine ratio (C:F) and size limit for C:F, e.g. 10 μm): • *Coarse organic*: • *Coarse anthropogenic*: • *Fine fabric* (in addition to color PPL and OIL, and interference colors and autofluorescence; includes information on *related distribution*, and composition of fine material; individual materials can be referred to as soil microfabric types or SMTs): • Pedofeatures (in the order of textural, depletion, crystalline, amorphous and cryptocrystalline, fabric and excrement pedofeatures): ***In situ* data** (from thin section and/or mirror-image polished block): (e.g. Image analysis (IA), microchemistry (microprobe); complementary macro- and microfossil data (identifications of *in situ* charcoal, diatoms, palynomorphs, phytoliths, nematode eggs) **BD**: (e.g. chemistry – loss-on-ignition, phosphate, heavy metals; magnetic susceptibility; grain size; mineralogy (X-Ray diffraction, heavy minerals); diatoms, palynology, phytoliths; macrofossils.	**Context**: (archaeological information – area, context, phase, and preliminary interpretation/ identification) **Comments and discussion**: (Highlighted soil micromorphological data and other analytical information to produce basic 1st and 2nd level identifications/interpretations) **Interpretation**: (Summarized 2nd and 3rd level interpretation)

Table A17.6.2 Freshwater sediment analysis: full presentation of soil micromorphology and complementary data and its brief interpretation for two units at Lower Paleolithic Boxgrove, West Sussex, UK. (Hundreds of Acheulean hand axes found in Unit 4*; along with two human teeth; a human tibia was found just above Unit 4d) (Roberts and Parfitt, 1999; Macphail et al., 2001).

Material (SMT)	Sample Number	Sampling depth, Soil Micromorphology (SM), *in situ* thin section data (i.e. collected from the thin section itself; microprobe and image analysis), and associated Bulk Data (BD)	Context, comments and discussion, and *interpretation*
Sediment Microfabric Type 4d1/1; Microfacies 4d1	M14c/ Unit 4d1	**Relative depth: 270–370 mm** **SM:** *Heterogeneity:* homogeneous *Structure and voids:* massive with dominant fine to medium, coarse meso to macro-channels, showing partial collapse, with polyconcave and closed vughs, and few complex packing voids; coarse voids also include few macro- to mega-vesicles; *Coarse mineral:* C:F, 45:55, dominant coarse silt to very fine sand-size quartz (and chalk clasts), common calcite/aragonite; very few chalk fossils, Cretaceous clay, glauconite, mica, opaques and calcitic root pseudomorph fragments; *Coarse organic:* rare organic staining and amorphous remains of plant tissue lengths; (few calcite root pseudomorphs); *Coarse anthropogenic(?):* 3 mm long bone/scat fragment; *Fine fabric:* cloudy, speckled gray (PPL), very high interference colors (close porphyric, crystallitic b-fabric, XPL), gray to white (OIL); *Pedofeatures: Textural,* very abundant intercalations (interlaced), void infills and ped coatings of gray to pale brown to dark gray calcitic layered impure silt, micrite and brown micritic clay coatings, infilling around collapsed peds; *Crystalline:* few calcite root pseudomorphs; *Fabric:* many fragments of microlaminated textural features; many coarse, 15 mm wide burrows; *Excrement:* rare possible excrements associated with relict roots. **IA:** 8% voids; dominant fine and medium (61.9% – 101–200 μm²), frequent coarse (24.1% – 201–500 μm²) meso voids, few fine (6.5% – 501–1000 μm²) macro voids and very few medium and coarse macro and mega-voids (2.5% – 1000–2000 μm²). **BD:** Sandy silt loam (17.1% clay, 59.9% silt, 22.1% fine sand), 46.2–56.6% CaCO$_3$ (mean 53.5%), 0.458–0.678 phosphate-P mg g^{-1} (mean 0.565 phosphate-P mg g^{-1}), 0.12% org. C., 0.17–0.21% Ext. Fe (mean 0.19%), χ 1.55–1.82 × 10^{-8} m^3 kg^{-1} and 3.54–4.71% $\chi_{conv.}$ ($n = 5$)	**Context:** Quarry 1B, Trench 5; Unit 4d1 **Comments and discussion:** biologically (likely earthworms and roots) worked fresh water pond sediment with later influence of fluvial deposition/meltwater causing collapse of ped (soil structures) and as overlying Unit 8ac is deposited (also causing contemporary textural feature fragmentation), voids are infilled with clay- to silt-size material that is washed in, drastically reducing porosity. **Interpretation:** continuing massive deposition with likely homogenization through biological/freeze-thaw, but with increasing amounts and influence of debris flow sedimentation through time.

Sediment Microfabric Type 4*/1; Microfacies 4*/1

M13c/Unit 4*

Relative depth: 670–730 mm (Upper 4*);

SM:

Heterogeneity: homogeneous

Structure and voids: massive, burrow and channel microstructure; common very fine to fine (0.1 –0.4 mm), frequent medium (0.8 mm) and coarse (1 mm), and few patches of very fine packing pores (15% voids);

Coarse mineral: C:F, 60:40, very dominant very coarse silt-size and very fine sand-size subangular to subrounded quartz (very few feldspars), frequent fine "shell/fossil" fragments; very few rounded glauconite, limonite and other opaques and mica (rare coarse mollusk shell; rare fossils and ostracods and foraminifera); (few)

Coarse organic: (see fine fabric) rare organic staining and amorphous remains of plant tissue lengths; (few calcite root pseudomorphs);

Coarse anthropogenic(?): absent.

Fine fabric: fine speckled, cloudy grayish brown (PPL), moderately high interference colors (dominantly interaggregate related distribution – gefuric, with coated grains to porphyric especially in lowermost 30 mm, XPL), grayish white (OIL) – high amounts of calcitic fine material; rare fine organic and tissue fragments and amorphous staining;

Pedofeatures: *Textural*: many weak, thin (0.5–0.8 mm) chalky clay (micritic) stringers, textural concentrations associated with sub-horizontal divisions (paleosurfaces?) and burrow and channel margins; *Amorphous and cryptocrystalline*: occasional iron staining, especially of root traces; *Fabric*: many 0.5–2 mm and very abundant >2mm size burrows; *Excrement*: *Crystalline*: rare calcitic root pseudomorphs; rare aggregated total excremental fabric?

IA: 15% voids; very dominant fine and medium (78.5% – 101–200 μm²), frequent coarse (20.3% – 201–500 μm²) meso voids, very few fine (0.9% – 501–1000μm²) macro voids and very few medium and coarse macro and mega-voids (0.3% – 1000–2000 μm²).

BD: Sandy silt loam (11.4% clay, 57.3% silt, 30.0% fine sand), 21.1–21.9% CaCO₃ (mean 21.6%), 0.500–0.634 phosphate-P mg g⁻¹ (mean 0.558 phosphate-P mg g⁻¹), 0.05% Org. C., 0.18–0.21% Ext. Fe, χ 2.59–2.73 × 10⁻⁸ m³ kg⁻¹) and 2.54–3.02% χ_conv (n = 3).

Microprobe: 0.14% Na, 0.14% Mg, 1.41% Al, 29.08% Si, 0.03% P, 0.01% S, 0.65% K, 10.54% Ca, 0.02% Ti, 0.01% Mn, and 1.1% Fe (n = 47)

Elemental map: Fe is widespread, with fine (0.1 mm) Fe punctuations (mottling) throughout. Very low P, low Al, and very high Si throughout. High Ca is ubiquitous, and commonly concentrated into laminae.

Lower 4*: 730–760 mm:

SM: compact, massive and channel microstructure; 10–15% voids, dominant fine (0.4 mm) and frequent medium (0.8 mm) channels; Mineral and organic matter as above, but embedded (porphyric; C:F, 45:55), coated and interaggregate related distribution (porphyric, chitonic and gefuric).

BD: 23.0–24.2% CaCO₃ (mean 23.6%), 0.722–804 P mg g⁻¹ (mean 0.763), 0.06% org. C., 0.36–0.37% Ext. Fe, χ 2.91–3.39 × 10⁻⁸ m³ kg⁻¹) and 1.11–1.13% χ_conv. (n = 2).

Field: pale yellow (2.5Y7/4), sharp sloping boundary to Unit 3 below.

Context: Quarry 1B, Trench 5; Unit 4*.

Comments and discussion:

rather open, coarsely burrowed and finely channelled calcitic fine sandy silt loam sediment develops that is slightly more coarse and less calcareous and ferruginous compared to the underlying Unit 3 and basal unit 4* deposits. Also calcitic separations/stringers along possible paleosurfaces and in channels and along burrow margins indicates wash coinciding with biological (fauna and plants) activity. Lower Unit 4*, that little bit more compact, calcareous, and iron stained – similarly calcareous as uppermost Unit 3.

Interpretation:

Appears that primary deposition of Unit 4 was a likely calcareous slurry, forming a compact basal 30 mm layer, with some calcareous sediment washing into voids in Unit 3 – possibly also leading to calcite root pseudomorphs. Upwards, Unit 4* is less densely calcitic and more open, as presumably biological activity by roots and earthworms worked the accumulating sediment that infills above truncated Unit 3.*

Table A17.6.3 Soil analysis: summarized data for soil microfabric types/microfacies types (SMTs) from the mound and buried turf (SMT 3) and buried upper subsoil (SMT 4) from two Neolithic and six Early Bronze Age barrows at Raunds, Northamptonshire, UK (Healy and Harding, 2007; Macphail, 2011a).

Material (SMT)	Sample Number BD = bulk	Sampling depth, Soil Micromorphology (SM), *in situ* thin section data (i.e., collected from the thin section itself), and *relevant associated Bulk Data (BD) including macro- and microfossil data*	Context, comments and discussion, and interpretation
Soil Microfabric Type 3/Soil microfacies Type 3	M8, M35, M36, M39, M40, BD5, 6, 10, 11, 24	e.g: 670–720 mm SM: moderately homogeneous; *Structure and voids*: massive structured, compact; mainly frequent to dominant burrows and chambers, with very few to common channels and vughs (total voids: mean 24% voids, range 10–30% voids; $n = 18$– as above); C:F 65:35; *Coarse mineral*: very poorly sorted; very few gravel-size flint and quartzite; very dominant subangular silt-size to subrounded and rounded medium sand-size quartz; very few opaque minerals and silt-size mica; rare soil clasts; *Fine material*: heavily dotted, blackish, dark reddish brown (PPL), isotic to extremely low birefringence (XPL) (single spaced porphyric, speckled b-fabric); brownish orange with few black inclusions (OIL); many to abundant relict amorphous organic matter; occasional to many fine charcoal and occasional rubified material; *Pedofeatures*: rare to occasional dusty clay, rare to many dark red clay, rare to many pan-like, occasional to many yellowish brown clay and rare to many compound textural features (see Macphail 1999); rare biogenic calcite; rare to occasional blackish iron and manganese nodular impregnations (Fe/Mn). **BD:** sandy loam, with 9–16% clay, 25–31% silt, 56–65% sand; poorly humic 0.2–0.6% Organic C, 2.0–4.6% LOI; mean 330 ppm $P_2O_{5ignited}$, range 230–440 ppm $P_2O_{5ignited}$, 3.4–3.5 P ratio ($n = 3$); mean 1350 ppm P_{nitric}, range 1290–1410 ppm P_{nitric} ($n = 2$)	**Context:** Barrow mound and buried turf. **Comments and discussion**: humic, part-biologically worked soils with anomalous presence of textural pedofeatures (including dark red clay and many pan-like) – not oriented to way-up in turf (also secondary/post-depositional yellow-brown clay). Presence of charcoal, fine burned mineral material and enhanced magnetic susceptibility also contribute to the anthropogenic signature. *Buried topsoil and local turf (of landscape open since earliest Neolithic times) used for barrow construction. Anomalous presence of organic phosphate-rich textural pedofeatures (e.g. red clay) and pans are evidence of soil poaching and a proxy indication of domestic stock concentrations. Post-burial (Saxon) alluvial flooding of site lead to yellow clay inwash.*

| Soil Microfabric Type 4/Soil microfacies type 4 | M9 (M39, M40) BD7, 25, 26 | **SM:** massive structured, with common to dominant burrows and chambers, few to frequent channels and vughs (e.g. total voids: mean 37% voids, range 20–40% voids); C:F 85:15;

Coarse mineral as SMF 3;

Fine mineral: heavily speckled, grayish yellow-brown (PPL), very low interference colors (XPL) (single spaced porphyric, speckled b-fabric); pale yellow orange (OIL); occasional to many relic amorphous organic matter; occasional charcoal and rubified material;

Pedofeatures: rare to many dusty clay, occasional to abundant dark red clay, occasional to abundant pan-like, rare yellowish brown clay and rare to occasional compound textural features (see Key); rare to occasional blackish iron and manganese nodular impregnations (Fe/Mn).

BD: sandy loam to loamy sand, with 6–11% clay, 22–28% silt and 61–72% sand; poorly humic (0.1–0.3% Organic C, 2–3% LOI), non-calcareous (<0.1% $CaCO_3$); 270 ppm $P_2O_{5ignited}$, range 240–330 ppm $P_2O_{5ignited}$, 2.3–2.9 P ratio ($n = 3$).

Microprobe: Textural features

Clay void infill: mean 0.11% P, 1.05% Ca, 0.06% Na, 0.06% S, 7.19% Fe, 0.37% Mn, 0.65% Mg, 16.6% Si, 8.3% Al and 1.56% K ($n = 19$); Coarse capping/pan-like textural feature (excluding large sand grains): mean 0.07% P, 0.48% Ca, 0.19% Na, 0.13% S, 3.29% Fe, 0.03% Mn, 0.21% Mg, 20.6% Si, 3.0% Al and 0.88% K ($n = 6$);dark red clay void coating: mean 0.25% P, 1.03% Ca, 0.38% Na, 0.04% S, 20.6% Fe, 0.12% Mn, 1.09% Mg, 12.2% Si, 7.8% Al and 0.97% K ($n = 9$). | Barrow-buried upper subsoil (Eb or A2 horizon):

As above, with again anomalous presence of textural pedofeatures and microprobe data showing higher concentrations of P here compared to underlying Bt horizon (see Macphail 1999). |

Table A17.6.4 Anthropogenic sediment analysis: summarized data for soil microfabric types/microfacies types (SMTs) from the Early Medieval London Guildhall, UK; data summarized from individual sample descriptions and counts, microprobe bulk chemical and pollen analyses. (Bowsher et al., 2007; Macphail et al., 2007a)

Facies (SMT) (Examples of samples and contexts)	Micromorphology and microprobe	Chemistry	Pollen	Identifications and interpretations
SMT 5a (251) (687) (755) (977) (978) See Figs. 1b, 1d–1i	Very dominantly organic, with (a) intact horizontally layered Poaceae tissues, commonly with abundant phytoliths and long articulated phytoliths; (b) sometimes with dark brown humified plant material, c) intercalated with silt, autofluorescent under blue light, layered Ca, P and K distribution (probe); d) sometimes inclusions of amorphous organic matter with calcium oxalate crystals and Ca/P/K chemistry (probe); commonly limpid brown, speckled brown to dark brown/blackish (PPL), isotic, with rare scatter of high interference colors (isotic or crystallitic, XPL), very dark brown to black (OIL); very abundant plant organs and tissues, with rare pollen, diatoms and ash; wood may occur; very few burrows and excrements.	SMT 5a/b/c: moderate to very high LOI (range, 10.3–43.4%, $n = 13$). Mostly non-calcareous (maximum, 2% carbonate). Clear to strong indications of phosphate (phosphate-P, range = 5.79–20.7 mg g^{-1}), and possible to compelling evidence of heavy metal enrichment (especially Pb and Cu). No χ enhancement in some samples, but possible to strong enhancement in others.	High concentrations of well-preserved, dominantly cereal t. and Poaceae pollen. Samples with >30% grass (see Figure 3) should be viewed as relatively undiluted stabling refuse, with major inputs from animal feed, bedding, and dung. Samples with <30% grass contain very high amounts of cereal t., possibly as special feed ("pig" coprolite association) (Figure 1i). Rare concentrations of woody pollen probably indicative of construction. Weed assemblage probably indicates "local" animal husbandry.	*Well-preserved highly organic stabling refuse* (Periods 10 and 11), with (a) fodder and bedding, (b) dung of herbivores (cattle better preserved than sheep/goat; stabling floor crust and (c) omnivore (pig) dung. These deposits have been very little biologically worked before or after dumping. Also, this material includes "raw," dung being used for lining wattle walls or to support and level (?) floor foundations. (It can be noted that there is also a strong statistical correlation between LOI and phosphate-P, showing phosphate enrichment is due to animal husbandry represented by SMT 5.) (There is also strong evidence of the link between phosphate-P and heavy metal concentrations.)

A1.17.6   Reporting: How to Write a Report – Fieldwork Report Protocol

Introduction to contain:

- name of site and exact geographical location;
- when visited; with whom was the site discussed, and who supplied information – and their affiliations (archaeological company, university); period(s) and type of site (urban, rural, tel);
- what type of exposure (open excavation, backhoe trench, sondage, building cellar);
- why the site was visited/aim of geoarchaeological study (geoarchaeological questions);
- what was done (field evaluation, section description, survey, sampling);
- associated work by other specialists and overall context of the work (ancient erosion, early tillage, urban morphology, human v animal occupation).

Methods

List:

- techniques carried out and authorities employed (for field description, e.g. Munsell Color Chart), sampling, on-site tests (e.g. HCl for carbonate, finger texturing for grain size), coring, survey);
- any specific equipment employed (different augers, geophysical equipment, types, and lengths of monoliths);
- follow-up laboratory techniques, e.g. pH (see Chapter 16 – Lab Tech).

Samples

- Tabulated *list of samples* (type of sample: undisturbed monolith, with measurements, e.g. OD or relative depths).
- Bulk sample – perhaps including size/weight; context number.
- Archaeological character with brief description. Further sampling requirements can be noted in the Discussion, below.

Results

- Present information on the site's location, and known geology, soils and perhaps also modern land use (e.g. using Soil Survey of England and Wales for the UK, USDA guides for the USA);
- systematic presentation of the field information, for example:
  - logically describe the oldest rocks/parent material or earliest stratigraphy first; and
  - present each geological/archaeological section/soil profile in turn – referring to tabulated sample lists and descriptions, section drawings, field drawings, photographs/digital images and any other data (figures and tables) that has been gathered (e.g. pH, HCl carbonate reaction; survey and coring data).

Discussion

- Set the scene by discussing the known information about the geoarchaeology of the site and what the fieldwork has revealed. Perhaps broadly discuss systematically the sequence in terms of oldest to youngest, or by area on the site.
- Discuss the geoarchaeological questions posed by the site – referring to why the site was visited/ aim of geoarchaeological study and what are the broad aims and objectives of the investigation – and what information has been gained from the fieldwork. For example, this is the time to identify which samples have the potential (from a laboratory study) of answering the geoarchaeological questions. It may be necessary here to identify where further samples should be taken.

- Suggest the various techniques that can be employed to answer these questions. Here, a proper literature-supported track record for each technique in providing useful data and interpretations has to be argued. For example, areas of supposed occupation could be isolated by measuring the LOI, P, and magnetic susceptibility of samples taken through a grid survey (see Chapter 16). Or, to seek out possible locations of Paleolithic hearths in caves, soil micromorphology should be combined with microprobe and FTIR analyses to identify ash transformations (see Chapter 10).

Future work and costs
- Here the number of samples and techniques to be applied are listed, based upon the preceding Discussion and the List of Samples.
  - Estimated costs of analyses and reporting (see below) should be given, and this should include the number of working days involved so that the work can scheduled properly.
  - It is also useful here to indicate how long procedures take: e.g. soil micromorphology requires lengthy processing times before thin sections can be made – several months (see Chapter 16). Also, a worker may have a backlog of work and may need to inform the "client" when the work can commence. Reports often have to fit a deadline, so that the archaeologist can write up his report with all contributions present. This is especially true of CRM/mitigation archaeology.

Bibliography
A reasonable list of cited publications; more are normally required for the post-excavation report/ publication (see below).

Lastly, fieldwork may take a single day, and as such the "client" may expect a fieldwork report even the next day (by email) or very soon after, because the geoarchaeological information is required immediately in order to know what important contexts need further sampling, and which areas of the site should be chosen to be further trenched/excavated. It may be that a number of questions were answered on site during the visit, but still these answers need to be reinforced by a thought-out report. This is especially important for short, day-long, days-long trips. During extended periods of fieldwork on site, many "operational" questions can be resolved on site.

Avery, B. W. and Bascomb, C. L. 1974. *Soil Survey Laboratory Techniques*. Soil Survey Technical Monograph, No. 14. Harpenden, Soil Survey of England and Wales.

Ball, D. F. 1964. Loss-on-ignition as an estimate of organic matter and organic carbon in non-calcareous soils. *Journal of Soil Science*, 15: 84–92.

French, C. 2003. *Geoarchaeology in Action. Studies in Soil Micromorphology and Landscape Evolution*. London, Routledge.

Gale, S. J. and Hoare, P. G. 1991. Quaternary Sediments. Petrographic Methods for the Study of Unlithified Rocks. London, Belhaven Press (John Wiley).

Levy, G. and Toutain, F. 1982. Aeration and redox phenomena in soils. In: *Constituents and Properties of Soils* (ed. M. Bonneau and B. Souchier), pp. 355–366. London, Academic Press.

Stoops, G. 2003. *Guidelines for Analysis and description of Soil and Regolith Thin Sections*. Madison, WI, Soil Science Society of America, Inc.

Tylecote, R. F. 1979. The effect of soil conditions on the long term corrosion of buried tin-bronzes and copper. *Journal of Archaeological Science*, 6: 345–368.

References

Abate, N., Elfadaly, A., Masini, N., Lasaponara, R., 2020. Multitemporal 2016–2018 Sentinel-2 Data Enhancement for Landscape Archaeology: The Case Study of the Foggia Province, Southern Italy: *Remote Sensing.* v. 12, no. 1309, p. 1–29.

Acott, T. G., Cruise, G. M., and Macphail, R. I., 1997, Soil micromorphology and high resolution images, *in* Shoba, S., Gerasimova, M., and Miedema, R., eds., *Soil Micromorphology: Diversity, Diagnostics and Dynamics.* Moscow-Wageningen, International Soil Science Society, p. 372–378.

Adderley, W. P., Simpson, I. A., and Davidson, D., 2006, Historic landscape management: a validation of quantitative soil thin-section analyses: *Journal of Archaeological Science*, v. 33, p. 320–334.

Adderley, W. P., Wilson, D. R., Davidson, D., and Simpson, D. D. A., 2018, Anthropogenic features, *in* Stoops, G., Marcelino, V., and Mees, F., eds., *Interpretation of Micromorphological Features of Soils and Regoliths.* Amsterdam, Elsevier, p. 753–777.

Addyman, P. V., Hood, J. S., Kenward, H. K., MacGregor, A., and Williams, D., 1976, Palaeoclimate in urban environmental archaeology at York, England: *World Archaeology*, v. 8, no. 2, p. 220–233.

Adovasio, J. M., Donahue, J., Carlisle, R. C., Cushman, R., Stuckenrath, R., and Wiegman, P., 1984, Meadowcroft Rockshelter and the Pleistocene/Holocene transition in southwestern Pennsylvania, *in* Genoways, H. H., and Dawson, M. R., eds., *Contributions in Quaternary Vertebrate Paleontology, Carnegie Museum of Natural History Special Publication 8*: Pittsburgh, Carnegie Museum of Natural History, p. 347–369.

Adovasio, J. M., Pedler, D. R., Donahue, J., and Stuckenrath, R., 1999, Two decades of debate on Meadowcroft Rockshelter: *North American Archaeologist*, v. 19, no. 4, p. 317–341.

Agapiou, A., Alexakis, D. D., Sarris, A., Hadjimitsis, D. G., 2014. Evaluating the Potentials of Sentinel-2 for Archaeological Perspective. *Remote Sensing.* 6, p. 2176–2194.

AGS, 2018, *Description of anthropogenic materials – a practitioners' guide.* Association of Geotechnical and Geoenvironmental Specialists.

Ahlbrandt, T. S., and Fryberger, S. G., 1982, Eolian Deposits, *in* Scholle, P. A., and Spearing, D. R., eds., *Sandstone Depositional Environments, Memoir 31*: Tulsa, The American Association of Petroleum Geologists, p. 11–47.

Ahler, S. A., 1973, Post-Pleistocene depositional change at Rodgers Shelter, Missouri: *Plains Anthropologist*, v. 18, no. 59, p. 1–26.

Ahler, S. A., 1976, Sedimentary Processes at Rodgers Shelter, *in* Wood, W. R., and McMillan, R. B., eds., *Prehistoric Man and His Environments: A case study in the Ozark Highlands.* New York, Academic Press, p. 123–139.

Akeret, Ö., and Rentzel, P., 2001, Micromorphology and plant macrofossil analysis of cattle dung from the Neolithic lake shore settlement of Arbon Bleiche 3: *Geoarchaeology*, v. 16, no. 6, p. 687–700.

Practical and Theoretical Geoarchaeology, Second Edition. Paul Goldberg and Richard I. Macphail.
© *2022 John Wiley & Sons Ltd. Published 2022 by John Wiley & Sons Ltd.*

Albanese, J., 1974, Geology of the Casper Archeological Site, *in* Frison, G. C., ed., *The Casper site: a Hell Gap bison kill on the High Plains*. New York, Academic Press, p. 266.

Albanese, J. P., and Frison, G. C., 1995, Cultural and landscape change during the middle Holocene, Rocky Mountain area, Wyoming and Montana. Archaeological geology of the Archaic period in North America: *Special Paper – Geological Society of America*, v. 297, p. 1–19.

Albert, R. M., and Weiner, S., 2001, Study of phytoliths in prehistoric ash layers using a quantitative approach, *in* Meunier, J. D., and Colin, F., eds., *Phytoliths, Applications in Earth Science and Human History*: Rotterdam, A. A. Balkema, p. 251–266.

Albert, R. M., Lavi, O., Estroff, L., Weiner, S., Tsatskin, A., Ronen, A., and Lev-Yadun, S., 1999, Mode of occupation of Tabun Cave, Mt Carmel, Israel during the Mousterian Period: A study of the sediments and phytoliths: *Journal of Archaeological Science*, v. 26, no. 10, p. 1249–1260.

Albert, R. M., Weiner, S., Bar-Yosef, O., and Meignen, L., 2000, Phytoliths in the Middle Palaeolithic deposits of Kebara Cave, Mt Carmel, Israel: study of the plant materials used for fuel and other purposes: *Journal of Archaeological Science*, v. 27, no. 10, p. 931–947.

Aldeias, V., 2017, Experimental approaches to archaeological fire features and their behavioral relevance, *in* Proceedings Fire and the Genus Homo, Sintra, Portugal, *Current Anthropology*, v. 58, S191–S205.

Aldeias, V., and Bicho, N., 2016, Embedded Behavior: Human Activities and the Construction of the Mesolithic Shellmound of Cabeco da Amoreira, Muge, Portugal: *Geoarchaeology*, v. 31, no. 6, p. 530–549.

Aldeias, V., Dibble, H. L., Sandgathe, D., Goldberg, P., and McPherron, S. J. P., 2016, How heat alters underlying deposits and implications for archaeological fire features: A controlled experiment: *Journal of Archaeological Science*, v. 67, p. 64–79.

Aldeias, V., Goldberg, P., Sandgathe, D., Berna, F., Dibble, H. L., McPherron, S. P., Turq, A., and Rezek, Z., 2012, Evidence for Neandertal use of fire at Roc de Marsal (France): *Journal of Archaeological Science*, v. 39, no. 7, p. 2414–2423.

Aldeias, V., Gur-Arieh, S., Maria, R., Monteiro, P., and Cura, P., 2016, Shell we cook it? An experimental approach to the microarchaeological record of shellfish roasting: *Archaeological and Anthropological Sciences*, v. 11, p. 389–407.

Alexakis, D., Sarris, A., Astaras, T., Albanakis, K., 2011. Integrated GIS, remote sensing and geomorphologic approaches for the reconstruction of the landscape habitation of Thessaly during the Neolithic period. *Journal of Archaeological Science,* v. 38, 89–100.

Allen, K. M. S., Green, S. W., Zubrow, E. B. W., 1990. *Interpreting space: GIS and archaeology.*

Allen, M. J, 2000, Soils, pollen and lots of snails, *in* Green, M., ed., *A Landscape Revealed. 10,000 Years on a Chalkland Farm*. Stroud, Tempus, p. 36–49.

Allen, M. J., 2005, Beaker Occupation and Development of the Downland Landscape at Ashcombe Bottom, near Lewes, East Sussex: *Sussex Archaeological Collections*, v. 143, p. 7–33.

Allen, M. J., 2007, Prehistoric and medieval environment of Old Town, Eastbourne: *Sussex Archaeological Collections*, v. 145, p. 1–32.

Allen, J. R. L., 1970, *Physical Process of Sedimentation*. London, Allen & Unwin Ltd.

Allen, M. J., 1988, Archaeological and environmental aspects of colluviation in south-east England, *in* Waatteringe, W. G.V., and Robinson, M., eds., *Man-made Soils, BAR International Series 410*. Oxford, British Archaeological Reports, p. 67–92.

Allen, M. J., 1992, Products of erosion and the prehistoric land use of the Wessex Chalk, *in* Bell, M., and Boardman, J., eds., *Past and Present Soil Erosion, Monograph 22*. Oxford, Oxbow, p. 37–52.

Allen, M. J., 1994, *The Land-Use History of the Southern English Chalklands with an Evaluation of the Beaker Period Using Environmental Data: colluvial deposits as environmental and cultural indicators*: Southampton University.

Allen, M. J., 1995, Land molluscs, *in* Wainwright, G. J., and Davies, S. M., eds., *Balksbury Camp, Hampshire: Excavations 1973 and 1981*, Vol. 4. London, English Heritage, p. 92–100.

Allen, M. J., 2008, Late Palaeolithic (Area 3): environmental evidence for the former environment and possible human activity, *in* FitzPatrick, A. P., Powell, A. B., and Allen, M. J., eds., *Archaeological Excavations on the Route of the A27 Westhampnett Bypass, West Sussex, 1992, Wessex Archaeology Report No. 21*: Salisbury, Wessex Archaeology, p. 30–61.

Allen, M. J., 2017, The southern English chalklands: molluscan evidence for the nature of post-glacial woodland cover, *in* Allen, M. J., ed., *Molluscs in Archeology; methods, approaches and applications, Studying Scientific Archaeology 3*. Oxford, Oxbow Books, p. 144–164.

Allen, M. J., and Macphail, R. I., 1987, Micromorphology and magnetic susceptibility studies: their combined role in interpreting archaeological soils and sediments, *in* Fedoroff, N., Bresson, L. M., and Courty, M.-A., eds., *Soil Micromorphology*: Plaisir, Association Française pour l'Étude du Sol, p. 669–676.

Altemuller, H. J., and Van Vliet-Lanoe, B., 1990, Soil thin section fluorescence microscopy, *in* Douglas, L. A., ed., *Soil Micromorphology: A Basic and Applied Science*. Amsterdam, Elsevier, p. 565–579.

Amick, D. S., Mauldin, R. P., and Binford, L. R., 1989, The potential of experiments in lithic technology, Experiments in lithic technology, *BAR International Series 528*, p. 1–14.

Amit, R., Enzel, Y., Crouvi, O., Simhai, O., Matmon, A., Porat, N., McDonald, E., and Gillespie, A. R., 2011, The role of the Nile in initiating a massive dust influx to the Negev late in the middle Pleistocene: *GSA Bulletin*, v. 123, no. 5–6, p. 873–889.

Amit, R., Lekach, J., Ayalon, A., Porat, N., and Grodek, T., 2007, New insight into pedogenic processes in extremely arid environments and their paleoclimatic implications—the Negev Desert, Israel: *Quaternary International*, v. 162, p. 61–75.

Amit, R., Simhai, O., Ayalon, A., Enzel, Y., Matmon, A., Crouvi, O., Porat, N., and McDonald, E., 2011, Transition from arid to hyper-arid environment in the southern Levant deserts as recorded by early Pleistocene cummulic Aridisols: *Quaternary Science Reviews*, v. 30, no. 3–4, p. 312–323.

Ammerman, A., and Polgase, C., 1997, Analyses and descriptions of the obsidian collections from Arene Candide, *in* Maggi, R., ed., *Arene Candide: A Functional and Environmental Assessment of the Holocene Sequence (Excavations Bernabò Brea-Cardini 1940–50). New Series Number 5*: Roma, Memorie dell'Istituto Italiano di Paleontologia Umana, p. 573–592.

Amorosia, A., Brunoa, L., Campo, B., Morelli, A., Rossi, V., Scarponi, D., Hong, W., Bohacs, K. M., and Drexler, T. M., 2017, Global sea-level control on local parasequence architecture from the Holocene record of the Po Plain, Italy: *Marine and Petroleum Geology*, v. 87, p. 99–111.

Ampe, C., and Langohr, R., 1996, Distribution of circular structures and link with the soilscape in the sandy and loamy sandy area of NW Belgium. Fortuitous or a deliberate choice?, *in* Castelletti, L., and Cremaschi, M., eds., *Paleoecology*: Forli, ABACO, p. 59–68.

Amundson, R., 2021, Factors of soil formation in the 21st century: *Geoderma*, v. 391, no. 114960.

Andren, T., 2012, Baltic Sea Basin, since the latest deglaciation, *in* Bengtsson, L., Herschy, R. W., and Fairbridge, R. W., eds., *Encyclopedia of Lakes and Reservoirs*: Dordrecht, Springer, p. 95–102.

Andrews, P., 1990, *Owls, Caves and Fossils. Predation, preservation and accumulation of small mammal bones in caves, with an analysis of the Pleistocene cave faunas from Westbury-sub-Mendip, Somerset, UK*. London, The Natural History Museum, Natural History Museum Publications.

Andrews, P., Cook, J., Currant, A., and Stringer, C., 1999, Westbury Cave. The Natural History Museum Excavations 1976–1984., Bristol, CHERUB (Centre for Human Evolutionary Research at the University of Bristol).

Angelini, I., Artioli, G., and Nicosia, C., 2017, Metals and metalworking residues, *in* Nicosia, C., and Stoops, G., eds., *Archaeological Soil and Sediment Micromorphology*. Chichester, Wiley Blackwell, p. 213–222.

Angelucci, D. E., 2002, The geoarchaeological context, *in* Trinkaus, J. Z. E., ed., *Portrait of the artist as a child: The Gravettian human skeleton from the Abrigo do Lagar Velho, Vol. 22*: Lisboa, Instituto Portugues de Arqueologia, p. 58–91.

Angelucci, D. E., 2008, Geoarchaeological insights from a Roman age incineration feature (ustrinum) at Enconsta de Sant'Ana (Lisbon, Portugal): *Journal of Archaeological Science*, v. 35, no. 9, p. 2624–2633.

Angelucci, D. E., 2017, Lithic artifacts, *in* Nicosia, C., and Stoops, G., eds., *Archaeological Soil and Sediment Micromorphology*. Chichester, Wiley, p. 223–230.

Angelucci, D. E., Boschian, G., Fontanals, M., Pedrotti, A., and Vergès, J. M., 2009, Shepherds and karst: the use of caves and rock-shelters in the Mediterranean region during the Neolithic: *World Archaeology*, v. 41, no. 2, p. 191–214.

Antoine, P. J. A., Catt, J. A., Lautridou, J-P. and Somme, J. 2003. The loess and coversands of northern France and southern England. *Journal of Quaternary Science*, v. 18, p. 309–318.

ApSimon, A. M., Smart, P. L., Macphail, R. I., Scott, K., and Taylor, H., 1992, King Arthur's Cave, Whitchurch, Herefordshire: a reassessment: *Proceedings of the University of Bristol Speleological Society*, v. 19, p. 183–249.

Armour-Chelu, M., and Andrews, P., 1994, Some effects of bioturbation by earthworms (Oligochaeta) on archaeological sites: *Journal of Archaeological Science*, v. 21, p. 433–443.

Arpin, T. L., Mallol, C., and Goldberg, P., 2002, A new method of analyzing and documenting micromorphological thin sections using flatbed scanners: applications in geoarchaeological studies: *Geoarchaeology*, v. 17, p. 305–313.

Arrhenius, O., 1931, Die Bodenanalyse in dienst der Archäologie: *Zeitschrift für Pflanzenerhährung, Düngung und Bodenkunde Teil B*, v. 10, p. 427–439.

Arrhenius, O., 1934, *Fosfathalten i skånska jordar*: Sveriges Geologiska Undersökningar, Ser. C, 393, Årsbok 28, no 3.

Arrhenius, O., 1955, The Iron Age settlements on Gotland and the nature of the soil, Copenhagen, Munksgaard, *Valhagan II*, p. 1053–1064.

Arroyo-Kalin, M., 2012, Slash-burn-and-burn: *Quaternary International*, v. 249, p. 4–18.

Arroyo-Kalin, M., 2017, Amazonian dark earths, *in* Nicosia, C., and Stoops, G., eds., *Archaeological Soil and Sediment Micromorphology*. Chichester, Wiley Blackwell, p. 345–358.

Arroyo-Kalin, M., Neves, E. G., and Woods, W. I., 2008, Anthropogenic Dark Earths of the Central Amazon Region: Remarks on their evolution and polygenetic composition, *in* Woods, W. I., Teixeira, W. G., Lehmann, J., Steiner, C., WinklerPrins, A. M. G. A., and Rebellato, L., eds., *Amazonian Dark Earths: Wim Sombroek's Vision*. New York/Berlin, Springer Science, p. 99–125.

Artioli, G., 2010, *Scientific Methods and Cultural Heritage: An Introduction to the Application of Materials Science to Archaeometry and Conservation Science*. Oxford, Oxford University Press.

Ashley, G. M., 2001, Archaeological sediments in springs and wetlands, *in* Stein, J. K., and Farrand, W. R., eds., *Sediments in Archaeological Context*. Salt Lake City, UT, The University of Utah Press, p. 183–210.

Ashley, G. M., and Driese, S. G., 2000, Paleopedology and paleohydrology of a volcaniclastic paleosol interval: implications for early Pleistocene stratigraphy and paleoclimate record, Olduvai Gorge, Tanzania: *Journal of Sedimentary Research*, v. 70, no. 5, p. 1065–1080.

Ashley, G. M., Ndiema, E. K., Spencer, J. Q. G., Harris, J. W. K., Kiura, P. W., Dibble, L., Du, A., and Lordan, P. T., 2017, Paleoenvironmental Reconstruction of Dongodien, Lake Turkana, Kenya and OSL Dating of Site Occupation During Late Holocene Climate Change: *African Archaeological Review*, v. 34, p. 345–362.

Ashwin, T., and Tester, A., 2014, *A Roman Settlement in the Waveney Valley: Excavations at Scole, 1993-4*. East Anglian Archaeology, Report No. 152, p. 431.

Atkinson, M., and Preston, S. J., 1998, The Late Iron Age and Roman settlement at Elms Farm, Heybridge, Essex, excavations 1993–5: an interim report: *Britannia*, v. XXIX, p. 85–110.

Atkinson, M., and Preston, S. J., 2015, *Heybridge, a Late Iron Age and Roman settlement: excavations at Elms Farm 1993–5.*, East Anglian Archaeology, E-Monograph 29.

Atkinson, R. J. C., 1957, Worms and weathering: *Antiquity*, v. 31, p. 219–233.

Audouze, F., and Enloe, J. G., 1997, High resolution archaeology at Verberie: limits and interpretations: *World Archaeology*, v. 29, no. 2, p. 195–207.

Avery, B. W., 1964, *The Soils and Land-Use of the district around Aylesbury and Hemel Hempstead.* London, Her Majesty's Stationary Office.

Avery, B. W., 1980, *Soil Classification for England and Wales*, Harpenden, Soil Survey of England and Wales.

Avery, B. W., 1990, *Soils of the British Isles*. Wallingford, CAB International.

Avery, B. W., and Bascomb, C. L., 1974, *Soil Survey Laboratory Techniques*, Harpenden, Soil Survey of England and Wales, Soil Survey Technical Monograph.

Avery, G., Cruz-Oribe, K., Goldberg, P., Grine, F., Klein, R. G., Lenardi, M. J., Rink, W. J., Schwarcz, H. P., Thackery, A. I., and Wilson, M. L., 1997, The 1992–1993 Excavations of the Die Kelders Middle and Later Stone Age Cave Site, South Africa: *Journal of Field Archaeology*, v. 24, p. 263–291.

Babel, U., 1975, Micromorphology of soil organic matter, *in* Giesking, J. E., ed., *Soil Components: Organic Components*, Vol. 1. New York, Springer-Verlag, p. 369–473.

Bagnold, R. A., 1941, *The Physics of Blown Sand and Desert Dunes*. London, Methuen and Co.

Bailey, G., 2007, Time perspectives, palimpsests and the archaeology of time: *Journal of Anthropological Archaeology*, v. 26, no. 2, p. 198–223.

Bakels, C. C., 1988, Pollen from plaggen soils in the province of North Brabant, the Netherlands, *in* Groenman-van Waateringe, W., and Robinson, M., eds., *Man-made Soils, International Series 410*. Oxford, British Archaeological Report, p. 55–66.

Bakels, C. C., 1997, The beginnings of manuring in western Europe: *Antiquity*, v. 71, no. 272, p. 442–445.

Bakels, C. C., 2009, The Western European Loess Belt. *Agrarian History, 5300 BC – AD 1000*. New York, Springer.

Baker, R. T., 1976, Changes in the chemical nature of soil phosphate during pedogenesis: *Journal of Soil Science*, v. 27, p. 504–512.

Bal, L., 1982, *Zoological ripening of soils*, Wageningen, Centre for Agricultural Publishing and Documentation, Agricultural Research Report.

Balaam, N., Bell, M., David, A., Levitan, B., Macphail, R. I., Robinson, M., and Scaife, R. G., 1987, Prehistoric and Romano-British sites at Westward Ho!, Devon: archaeological and palaeoenvironmental surveys 1983 and 1984, *in* Balaam, N. D., Levitan, B., and Straker, V., eds., *Studies in Palaeoeconomy and Environment in South West England, British Series 181*. Oxford, British Archaeological Reports, p. 163–264.

Balaam, N., Smith, K., and Wainwright, G. J., 1982, The Shaugh Moor Project: Fourth Report – Environment, context and conclusion: *Proceedings of the Prehistoric Society*, v. 48, p. 203–278.

Balaam, N., Corney, M., Dunn, C., and Porter, H., 1991, The surveys, *in* Sharples, N. M., ed., *Maiden Castle. Excavations and field survey 1985–6, Archaeological Report no 19*. London, English Heritage, p. 37–42.

Balbo, A. L., Madella, M., Vila, A., and Estévez, J., 2010, Micromorphological perspectives on the stratigraphical excavation of shell middens: A first approximation from the ethnohistorical site Tunel VII, Tierra del Fuego (Argentina): *Journal of Archaeological Science*, v. 37, no. 6, p. 1252–1259.

Balbo, A., Madella, M., Godino, I. B., and Álvarez, M., 2011, Shell midden research: An interdisciplinary agenda for the Quaternary and Social Sciences: *Quaternary International*, v. 239, no. 1–2, p. 147–152.

Balek, C. L., 2002, Buried artifacts in stable upland sites and the role of bioturbation: A review: *Geoarchaeology*, v. 17, no. 1, p. 41–51.

Ball, D. F., 1964, Loss-on-ignition as an estimate of organic matter and organic carbon in non-calcareous soils: *Journal of Soil Science*, v. 15, p. 84–92.

Ball, D. F., 1975, Processes of soil degradation: a pedological point of view, *in* Evans, J. G., Limbrey, S., and Cleere, H., eds., *The effect of man on the landscape: the Highland Zone, CBA Research Report No. 11*: Nottingham, The Council for British Archaeology, p. 20–27.

Banerjea, R. Y., Bell, M., Matthews, W., and Brown, A., 2015a, Applications of micromorphology to understanding activity areas and site formation processes in experimental hut floors: *Archaeological and Anthropological Sciences*, v. 7, no. 1, p. 89–112.

Banerjea, R. Y., Fulford, M., Bell, M., Clarke, A., and Matthews, W., 2015b, Using experimental archaeology and micromorphology to reconstruct timber-framed buildings from Roman Silchester: A new approach: *Antiquity*, v. 89, no. 347, p. 1174–1188.

Banerjea, R. Y., Badura, M., Kalējs, U., Cerina, A., Gos, K., Hamilton-Dyer, S., Maltby, E., Seetah, K., and Pluskowski, A., 2016, A multi-proxy, diachronic and spatial perspective on the urban activities within an indigenous community in medieval Riga, Latvia: *Quaternary International*, v. 460, p. 3–21.

Barclay, G. T., 1983, Sites of the third millennium BC to the first millennium AD at North Mains, Strathallan, Perthshire: *Proceedings of the Society of Antiquaries Scotland*, v. 113, p. 122–281.

Barclay, A., Lambrick, G., Moore, J., and Robinson, M., 2003, *Lines in the Landscape. Cursus Monuments in the Upper Thames Valley: excavations at the Drayton and Lechlade Cursuses. Thames Valley Landscapes Monograph Monograph No. 15*. Oxford, Oxford Archaeological Unit.

Barham, A. J., 1995, Methodological approaches to archaeological context recording: X-radiography as an example of a supportive recording, assessment and interpretive technique, *in* Barham, A. J., and Macphail, R. I., eds., *Archaeological Sediments and Soils: Analysis, Interpretation and Management*. London, Institute of Archaeology, p. 145–182.

Barham, A. J., and Macphail, R. I., 1995, *Archaeological Sediments and Soils: Analysis, Interpretation and Management*. London, Institute of Archaeology, University College London, p. 239.

Barker, G., 1985, Prehistoric Farming in Europe. Cambridge, Cambridge University Press.

Bar-Matthews, M., Marean, C. W., Jacobs, Z., Karkanas, P., Fisher, E. C., Herries, A. I. R., Brown, K., Williams, H. M., Bernatchez, J., Ayalon, A., and Nilssen, P. J., 2010, A high resolution and continuous isotopic speleothem record of paleoclimate and paleoenvironment from 90 to 53 ka from Pinnacle Point on the south coast of South Africa: *Quaternary Science Reviews*, v. 29, p. 2131–2145.

Barnes, G. L., 1990, Paddy soils now and then, *World Archaeology*, v. 22, no. 1, p. 1–17.

Barnes, M. A., and Barnes, W. C., 1978, Organic compounds in lake sediments, *in* Lerman, A., ed., *Lakes Chemistry, Geology, Physics*. Berlin, Springer-Verlag, p. 127–152.

Barrat, B. C., 1964, A classification of humus forms and microfabrics in temperate grasslands: *Journal of Soil Science*, v. 15, p. 342–356.

Barton, R. N. E., 1992, *Hengistbury Head, Dorset. Vol. 2: The Late Upper Palaeolithic & Early Mesolithic sites*. Oxford, Oxford University Committee for Archaeology.

Barton, R. N. E., 1997, *English Heritage Book of Stone Age Britain*. London, B.T. Batsford/English Heritage.

Barton, R. N. E., Currant, A. P., Fernandez-Jalvo, Y., Finlayson, J. C., Goldberg, P., Macphail, R., Pettitt, P. B., and Stringer, C. B., 1999, Gibraltar Neanderthals and results of recent excavations in Gorham's, Vanguard and Ibex Caves: *Antiquity*, v. 73, no. 279, p. 13–23.

Barton, R. N. E., Stringer, C., and Finlayson, C., 2012, Gibraltar Neanderthals in Context: A report of the 1995–98 excavations at Gorham's & Vanguards Caves, Gibraltar, Oxford University School of Archaeology: Monograph 75. Oxford, Institute of Archaeology, University of Oxford, p. 328.

Barton, R., Bouzouggar, A., Collcutt, S. N., Marco, Y. C., Clark-Balzan, L., Debenham, N. C., and Morales, J., 2016, Reconsidering the MSA to LSA transition at Taforalt Cave (Morocco) in the light of new multi-proxy dating evidence: *Quaternary International*, v. 413, p. 36–49.

Bartov, Y., Stein, M., Enzel, Y., Agnon, A., and Reches, Z., 2002, Lake levels and sequence stratigraphy of Lake Lisan, the Late Pleistocene precursor of the Dead Sea: *Quaternary Research*, v. 57, p. 9–21.

Bar-Yosef, O. A., M., Mercier, N., Belfer-Cohen, A., Goldberg, P., Housley, R., Laville, H., Meignen, l., Vogel, J. C., Vandermeersch, B., 1996, The dating of the Upper Paleolithic layers in Kebara Cave, Mt. Carmel: *Journal of Archaeological Science*, v. 23, p. 297–306.

Bar-Yosef, O., 1974, Late Quaternary stratigraphy and prehistory in Wadi Fazael, Jordan Valley: *A Preliminary Report: Paléorient*, v. 2, no. 2, p. 415–428.

Bar-Yosef, O., 1994, The Lower Paleolithic of the Near East: *Journal of World Prehistory*, v. 8, p. 211–265.

Bar-Yosef, O., and Goren-Inbar, N., 1993, *The Lithic Assemblages of Ubeidiya- A Lower Palaeolithic site in the Jordan valley*. Jerusalem, The Hebrew University of Jerusalem, Qedem – Monographs of the Institute of Archaeology.

Bar-Yosef, O., and Meignen, L., 2007, *Kebara Cave Mt. Carmel, Israel, American School of Prehistoric Research Bulletin, Volume 49*: Cambridge, Peabody Museum of Archaeology and Ethnology, Harvard University, p. 288.

Bar-Yosef, O., and Phillips, J. L., 1977, Prehistoric Investigations in Gebel Maghara, Northern Sinai, Qedem: Jerusalem, Hebrew University of Jerusalem, p. 269.

Bar-Yosef, O., and Tchernov, E., 1972, *On the Palaeo-ecological History of the Site of 'Ubeidiya*. Jerusalem, The Israel Academy of Sciences and Humanities.

Bar-Yosef, O., and Vandermeersch, B., 2007, Introduction: The framework of the project, *in* Bar-Yosef, O., and Meignen, L., eds., *Kebara Cave, Mt Carmel, Israel – The Middle and Upper Paleolithic Archaeology, Part I, Volume 49*, Cambridge, MA, American School of Prehistoric Research, Peabody Museum, Harvard University, p. 1–22.

Bar-Yosef, O., Arnold, M., Mercier, N., Belfer-Cohen, A., Goldberg, P., Housley, R., Laville, H., Meignen, L., Vogel, J. C., and Vandermeersch, B., 1996, The dating of the Upper Paleolithic Layers in Kebara Cave, Mt. Carmel: *Journal of Archaeological Science*, v. 23, p. 297–306.

Bar-Yosef, O., Vandermeersch, B., Arensburg, B., Belfer-Cohen, A., Goldberg, P., Laville, H., Meignen, L., Rak, Y., Speth, J. D., Tchernov, E., Tillier, A.-M., and Weiner, S., 1992, The Excavations in Kebara Cave, Mt. Carmel: *Current Anthropology*, v. 33, no. 5, p. 497–550.

Bar-Yosef, O., Belfer-Cohen, A., Goldberg, P., Kuhn, S., Meignen, L., Weiner, S., and Vandermeersch, B., 2005, Archaeological background: Hayonim Cave and Meged Rockshelter, *in* Stiner, M., ed., *The Faunas of Hayonim Cave (Israel). A 200,000-Year Record of Paleolithic Diet, Demography and Society*: Cambridge (US), American School of Prehistoric Research, Peabody Museum, Harvard University, p. 17–38.

Bascomb, C. L., 1968, Distribution of pyrophosphate-extractable iron and organic carbon in soils of various groups: *Journal of Soil Science*, v. 19, no. 2, p. 251–268.

Bateman, M. D and Godby, S. P. 2004. Late-Holocene inland dune activity in the UK: a case study from Breckland, East Anglia: *The Holocene*, v. 14, p. 579 – 588.

Bateman, M. D. 1995. Thermoluminescence dating of the British coversand deposits: *Quaternary Science Reviews*, v. 14, p. 791–798.

Bateman, M., Boulter, C., Carr, A., Frederick, C., Peter, D., and Wilder, M., 2007, Preserving the palaeoenvironmental record in drylands: bioturbation and its significance for luminescence-derived chronologies: *Sedimentary Geology*, v. 195, no. 1–2, p. 5–19.

Bateman, N., Cowan, C., and Wroe-Brown, R., 2008, *London's Roman Amphitheatre*: Guildhall Yard, City of London. London, Museum of London Archaeology Service.

Bates, M., and Stafford, E., 2013, *Thames Holocene. A Geoarchaeological Approach to the Investigation of the River Floodplain for High Speed 1, 1994–2003*. Oxford, Oxford Wessex Archaeology.

Bates, M., Pope, M., Shaw, A., Scott, B., and Schwenninger, J.-L., 2013, Late Neanderthal occupation in North-West Europe: rediscovery, investigation and dating of a last glacial sediment sequence at the site of La Cotte de Saint Brelade, Jersey: *Journal of Quaternary Science*, v. 28, no. 7, p. 647–652.

Bates, M. R., Bates, Bates C. R. and Whittaker, J.E. 2007. Mixed Method Approaches to the investigation and mapping of buried Quaternary deposits: Examples from Southern England. *Archaeological Prospection*, 14, p. 104–129.

Bates, M.R., Bates, C.R., 2000. Multidisciplinary approaches to the geoarchaeological evaluation of deeply stratified sedimentary sequences: Examples from Pleistocene and Holocene deposits in Southern England, United Kingdom. *Journal of Archaeological Science*, 27, p. 845–858.

Bates, M. R., Champness, C., Haggart, A., Macphail, R. I., Parfitt, S. A., and Schwenninger, J.-L., 2014, Early Devensian sediments and paleoenvironmental evidence from the excavations at the Royal Oak Portal Paddington, West London, UK: *Proceedings of the Geologists' Association*, v. 125, p. 41–55.

Bauer, E. M., 2017, Two Iron Age Boathouses, *in* Skre, D., ed., *Avaldsnes – A Sea-Kings' Manor in First-Millennium Western Scandinavia*. Berlin, Walter de Gruyter, p. 183–208.

Beach, T. P., Luzzadder-Beach, S., Cook, D., Dunning, N. P., Kennett, D. J., Krause, S., Terry, R., Trein, D., and Valdez, F., 2015, Ancient Maya impacts on the earth's surface: an early anthropocene analog?: *Quaternary Science Reviews*, v. 124, p. 1–30.

Beck, C. C., Feibel, C. S., Wright, J. D., and Mortlock, R. A., 2019, Onset of the African humid period by 13.9 kyr BP at Kabua Gorge, Turkana Basin, Kenya: *The Holocene*, v. 29, p. 1011–1019.

Beckman, G. G., and Smith, K. J., 1974, Micromorphological changes in surface soils following wetting, drying and trampling, *in* Rutherford, G. K., ed., *Soil Microscopy*: Kingston, Ontario, The Limestone Press, p. 832–845.

Begin, Z. B., Nathan, Y., and Ehrlich, A., 1980, Stratigraphy and facies distribution in the Lisan Formation, new evidence from the area south of the Dead Sea, Israel: *Israel Journal of Earth Sciences*, v. 29, p. 182–189.

Behre, K.-E., 1981, The interpretation of anthropogenic indicators in pollen diagrams: *Pollen et Spores*, v. 13, no. 2, p. 225–245.

Beijersbergen, L. T., Bratbak, O. F., and Hufthammer, A. K., 2018, *Korsmyra. Animal Osteological Analyses*: University Museum of Bergen.

Belfer-Cohen, A., and Goldberg, P., 1982, An Upper Paleolithic site in south central Sinai: *Israel Exploration Journal*, v. 32, p. 185–189.

Bell, M., 1981, Seaweed as a prehistoric resource, *in* Brothwell, D., and Dimbleby, G. W., eds., *Environmental Aspects of Coasts and Islands,* International Series 94. Oxford, British Archaeological Reports, p. 117–126.

Bell, M., 1983, Valley sediments as evidence of prehistoric land use on the South Downs: *Proceedings of the Prehistoric Society*, v. 49, p. 118–150.

Bell, M., 1990, *Brean Down Excavations 1983–87*, London, English Heritage.

Bell, M., 1992, The prehistory of soil erosion, *in* Bell, M., and Boardman, J., eds., *Past and Present Soil Erosion,* Monograph 22. Oxford, Oxbow, p. 21–35.

Bell, M., 2007, *Prehistoric Coastal Communities: The Mesolithic in Western Britain*, York, Council for British Archaeology.

Bell, M., 2009, Experimental Archaeology: changing science agendas and perceptual perspectives, *in* Allen, M. J., Sharples, N. M., and O'Connor, T., eds., *Land and People London*, Prehistoric Society, p. 31–46.

Bell, M., and Boardman, J., 1992, *Past and Present Soil Erosion*. Oxford, Oxbow.

Bell, M., Fowler, M. J., and Hillson, S. W., 1996, *The Experimental Earthwork Project, 1960–1992*. York, Council for British Archaeology, Research Report.

Bell, M., Caseldine, A., and Neumann, H., 2000, *Prehistoric Intertidal Archaeology in the Welsh Severn Estuary*. York, Council for British Archaeology.

Bell, M. and Brown, A. 2009. Southern regional overview of Geoarchaeology: Windblown Deposits. Research Department Report 005/2009. English Heritage: Portsmouth.

Bellhouse, R. L., 1982, Soils and archaeology, *in* Alexander, M. J., ed., *North of England Soils Discussion Group, Proceedings*, Vol. 17, p. 41–47.

Bello, S. M., Saladié, P., Cáceres, I., Rodríguez-Hidalgo, A., and Parfitt, S. A., 2015, Upper Palaeolithic ritualistic cannibalism at Gough's Cave (Somerset, UK): The human remains from head to toe: *Journal of Human Evolution*, v. 82, p. 170–189.

Bengtsson, L., 2012, Thermal regime of lakes, *in* Bengtsson, L., Herschy, R. W., and Fairbridge, R. W., eds., *Encyclopedia of Lakes and Reservoirs*: Dordrecht, Springer, p. 798–800.

Berger, J.-F., Salvador, P.-G., Franc, O., Verot-Bourrely, A., and Bravard, J.-P., 2008, La chronologie fluviale postglaciaire du Haut Bassin Rhodanien (The Postglacial fluvial chronology of Upper Rhone basin): *Cahiers de Paléoenvironnement, v. Collection EDYTEM – n° 6*, p. 117–144.

Berge, M.A. and Drahor, M.G. 2011. Electrical Resistivity Tomography Investigations of Multi Layered Archaeological Settlements: Part I – Modelling. *Archaeological Prospection*, 18, p.159–171.

Berggren, M., 1990, *Rygge Bind III. Tiden 1800 til 1980*, Rygge, Rygge sparebank.

Berna, F., Matthews, A. & Weiner, S. 2004. Solubilities of bone mineral from archaeological sites: the recrystallization window. *Journal of Archaeological Science*. v. 31, pp. 867–882.

Berna, F., 2017a, FTIR Microscopy, *in* Nicosia, C., and Stoops, G., eds., *Archaeological Soil and Sediment Micromorphology*. Chichester, Wiley Blackwell, p. 411–415.

Berna, F., and Goldberg, P., 2008, Assessing Paleolithic pyrotechnology and associated hominin behavior in Israel: *Israel Journal of Earth Sciences*, v. 56, p. 107–121.

Berna, F., Behar, A., Shahack-Gross, R., Berg, J., Boaretto, E., Gilboa, A., Sharon, I., Shalev, S., Shilstein, S., Yahalom-Mack, N., Zorn, J. R., and Weiner, S., 2007, Sediments exposed to high temperatures: reconstructing pyrotechnological processes in Late Bronze Age and Iron Age Strata at Tel Dor (Israel): *Journal of Archaeological Science*, v. 34, p. 358–373.

Berna, F., Goldberg, P., Horwitz, L. K., Brink, J., Holt, S., Bamford, M., and Chazan, M., 2012, Microstratigraphic evidence of in situ fire in the Acheulean strata of Wonderwerk Cave, Northern Cape province, South Africa: *PNAS*, v. 109, no. 20, p. 7593–7594.

Bertran, P., 1994, Dégradation des niveaux d'occupation paléolithiques en contexte périglaciaire: Exemples et implications archéologiques: *Paléo*, v. 6, p. 285–302.

Bertran, P., and Texier, J. P., 1990, The recording of pedological, sedimentary and climatic phenomena in colluvial deposits of south-west France. The example of Les Tares sequence, Dordogne, France: L'enregistrement des phenomenes pedo-sedimentaires et climatiques dans les depots colluviaux d'Aquitaine. *L'exemple de la coupe des Tares (Dordogne)*, v. 1, no. 1, p. 77–90.

Bertran, P., and Texier, J.-P., 1999, Facies and microfacies of slope deposits: *Catena*, v. 35, no. 2–4, p. 99–121.

Bertran, P., Caner, L., Langohr, R., Lemée, L., and d'Errico, F., 2008, Continental palaeoenvironments during MIS 2 and 3 in southwestern France: the La Ferrassie rockshelter record: *Quaternary Science Reviews*, v. 27, no. 21, p. 2048–2063.

Bertran, P., Coutard, J.-P., Francou, B., Ozouf, J.-C., and Texier, J.-P., 1994, New data on grèzes bedding and their palaeoclimatic implications, *in* Evans, D. J. A., ed., *Cold Climate Landforms*. Chichester, Wiley & Sons, p. 437–455.

Bertran, P., Hétu, B., Texier, J.-P., and Van Steijn, H., 1997, Fabric characteristics of subaerial slope deposits: *Sedimentology*, v. 44, no. 1, p. 1–16.

Bertran, P., Klaric, L., Lenoble, A., Masson, B., and Vallin, L., 2010, The impact of periglacial processes on Palaeolithic sites: The case of sorted patterned grounds: *Quaternary International*, v. 214, no. 1–2, p. 17–29.

Bertran, P., Todisco, D., Bordes, J.-G., Discamps, E., and Vallin, L., 2019, Perturbation assessment in archaeological sites as part of the taphonomic study: A review of methods used to document the impact of natural processes on site formation and archaeological interpretations: *PALEO. Revue d'archéologie préhistorique*, v. 30, no. 1, p. 52–75.

Bethell, P. H., and Máté, I., 1989, The use of soil phosphate analysis in archaeology: A critique, *in* Henderson, J., ed., *Scientific Analysis in Archaeology*, Monograph No. 19. Oxford, Oxford University Committee, p. 1–29.

Bevan, B. W., and Roosevelt, A. C., 2003, Geophysical Exploration of Guajará, a Prehistoric Earth Mound in Brazil: *Geoarchaeology*, v. 18, p. 287–331.

Beyin, A., Prendergast, M. E., Grillo, K. M., and Wang, H., 2017, New radiocarbon dates for terminal Pleistocene and early Holocene settlements in West Turkana, northern Kenya: *Quaternary Science Reviews*, v. 168, p. 208–215.

Beynon, D. E., and Donahue, J., 1982, The geology and geomorphology of Meadowcroft Rockshelter and the Cross Creek drainage, *in* Carlisle, R. C., and Adovasio, J. M., eds., *Collected papers on the archaeology of Meadowcroft Rockshelter and the Cross Creek drainage*: Pittsburgh, University of Pittsburgh, p. 31–52.

Bibby, J. S., and Mackney, D., 1972, *Land Use Capability Classification*, Harpenden, The Soil Survey, Technical Monograph No. 1.

Biddle, M., Hudson, D., and Heighway, C., 1973, The Future of London's Past: a survey of the archaeological implications of planning and development in the nation's capital: *Rescue: a trust for British Archaeology*, v. Rescue Publication 4.

Biddulph, E., Foreman, S., Stafford, E., Stansbie, D., and Nicholson, R., 2012, *London Gateway. Iron Age and Roman Salt Making in the Thames Estuary; Excavations at Stanford Wharf Nature Reserve, Essex*. Oxford, Oxford Archaeology.

Bierman, P. R., and Gillespie, A. R., 1994, Evidence suggesting that methods of rock-varnish cation-ratio dating are neither comparable nor consistently reliable: *Quaternary Research*, v. 41, no. 1, p. 82–90.

Bill, J., and Daly, A., 2012, The plundering of the ship graves from Oseberg and Gokstad: an example of power politics?: *Antiquity*, v. 86, p. 808–824.

Bill, J., and Rødsrud, C., 2017, Heimdalsjordet – trade, production and communication, *in* Loftsgarden, K., and Glørstad, A. Z., eds., *Viking-Age Transformations: Trade, Craft and Resources in Western Scandinavia*. London, Routledge, p. 212–231.

Binder, D., Brochier, J. E., Duday, H., Helmer, D., Marinval, P., Thiebault, S., and Wattez, J., 1993, L'abri Pendimoun à Castellar (Alpes-Maritimes): nouvelles données sur le complex culturel de la imprimée dans son contexte stratigraphique: *Gallia Préhistoire*, v. 35, p. 177–251.

Bindler, R., Yu, R. L., Hansson, S., Claben, N., and Karlsson, J., 2012, Mining, Metallurgy and the Historical Origin of Mercury Pollution in Lakes and Watercourses in Central Sweden: *Environmental Science and Technology*, v. 46, no. 15, p. 7984–7991.

Binford, L. R., 1981, Behavioral Archaeology and the "Pompeii Premise": *Journal of Anthropological Research*, v. 37, no. 3, p. 195–208.

Bintliff, J., 1992, Erosion in the Mediterranean lands: a reconsideration of pattern, process and methodology, *in* Bell, M., and Boardman, J., eds., *Past and Present Soil Erosion, Monograph 22*. Oxford, Oxbow, p. 125–131.

Birkeland, P. W., 1992, Quaternary soil chronosequences in various environments – extremely arid to humid tropical, *in* Martini, I. P., and Chesworth, W., eds., *Weathering, Soils & Paleosols, Developments in Earth Surface Processes 2*. Amsterdam, Elsevier, p. 261–281.

Birkeland, P. W., 1999, *Soils and Geomorphology*. New York, Oxford University Press.

Blackford, J. J., and Chambers, F. M., 1991, Proxy records of climate from blanket mires: evidence for a Dark Age (1400 BP) climatic deterioration in the British Isles: *The Holocene*, v. 1, p. 63–67.

Blais, J. M., and Kalff, J., 1995, The influence of lake morphometry on sediment focusing: *Limnology and Oceanography*, v. 40, no. 3, p. 582–588.

Blake, M. E., 1947, *Ancient Roman Construction in Italy from the Prehistoric period to Augustus*. Washington, DC, Carnegie Institution of Washington.

Blegen, N., Brown, F. H., Jicha, B. R., Binetti, K. M., Faith, J. T., Ferraro, J. V., Gathogo, P. N., Richardson, J. L., and Tryon, C. A., 2016, The Menengai Tuff: a 36 ka widespread tephra and its chronological relevance to Late Pleistocene human evolution in East Africa: *Quaternary Science Reviews*, v. 152, p. 152–168.

Blenckner, T., 2012, Climate change effects on lakes, *in* Bengtsson, L., Herschy, R. W., and Fairbridge, R. W., eds., *Encyclopedia of Lakes and Reservoirs*: Dordrecht, Springer, p. 165–167.

Bloesch, J., 1995, Mechanisms, measurement and importance of sediment resuspension in lakes: *Marine and Freshwater Research*, v. 46, no. 1, p. 295–304.

Bloesch, J., 2004, Sedimentation and lake sediment formation, *in* O'Sullivan, P., and Reynolds, C. S., eds., *The Lakes Handbook Limnology and Limnetic Ecology*, Vol. 1. Oxford, Wiley Blackwell, p. 197–229.

Bloszies, C., Forman, S. L., and Wright, D. K., 2015, Water level history for Lake Turkana, Kenya in the past 15,000 years and a variable transition from the African Humid Period to Holocene aridity: *Global and Planetary Change*, v. 132, p. 64–76.

Blum, M. D., Abbott, J. T., and Valastro, S., Jr., 1992, Evolution of landscapes on the Double Mountain Fork of the Brazos River, West Texas; implications for preservation and visibility of the archaeological record: *Geoarchaeology*, v. 7, no. 4, p. 339–370.

Blum, M. D., and Valastro, S., Jr., 1992, Quaternary stratigraphy and geoarchaeology of the Colorado and Concho Rivers, West Texas: *Geoarchaeology*, v. 7, no. 5, p. 419–448.

Blume, H.-P., and Leinweber, P., 2004, Plaggen Soils: landscape history, properties, and classification: *Journal of Plant Nutrition and Soil Science*, v. 167, no. 3, p. 319–327.

Blumenschine, R. J., Masao, F. T., Stanistreet, I. G., and Swisher, C. C., 2012, Five decades after Zinjanthropus and *Homo habilis*: Landscape paleoanthropology of Plio-Pleistocene Olduvai Gorge, Tanzania: *Journal of Human Evolution*, v. 63, no. 2, p. 247–437.

Boardman, J., 1992, Current erosion on the South Downs: implications for the past, *in* Bell, M., and Boardman, J., eds., *Past and Present Soil Erosion*. Oxford, Oxbow, p. 9–19.

Boardman, J., and Evans, R., 1997, Soil erosion in Britain; a review, *in* Goudie, A., Alexander, D. E., Gomez, A. B., Slaymaker, H. O., and Trimble, S. W., eds., *The human impact reader; readings and case studies*. Oxford, University Oxford, p. 118–125.

Boaretto, E., Wu, X., Yuan, J., Bar-Yosef, O., Chu, V., Pan, Y., Liu, K., Cohen, D., Jiao, T., Li, S., Gu, H., Goldberg, P., and Weiner, S., 2009, Radiocarbon dating of charcoal and bone collagen associated with early pottery at Yuchanyan Cave, Hunan Province, China: *PNAS* v. 106, no. 24, p. 9595–9600.

Bodziac, W. J., 2000, *Footwear Impression Evidence*, CRC Press.

Bogaard, A., Fraser, R., Heaton, T. H. E., Wallace, M., Vaiglova, P., Charles, M., Jones, G., Evershed, R. P., Styring, A. K., Andersen, N. H., Arbogast, R.-M., Bartosiewicz, L., Gardeisen, A., Kanstrup, M., Maier, U., Marinova, E., Ninov, L., Schäfer, S., and Stephan, E., 2013, Crop manuring and intensive land management by Europe's first farmers: *PNAS* v. 110, no. 31, p. 12589–12594.

Boggs, S., 2001, *Principles of Sedimentology and Stratigraphy*, 3rd edition. Upper Saddle River, NJ, Prentice Hall.

Boggs, S., 2006, *Principles of Sedimentology and Stratigraphy*, 4th edition. Upper Saddle River, NJ, Prentice Hall.

Boggs, S., 2009, *Petrology of Sedimentary Rocks*, 2nd edition. New York, Cambridge University Press.

Boggs, S., 2012, Principles of Sedimentology and *Stratigraphy*, 5th edition. Upper Saddle River, NJ, Pearson Education.

Boggs, S., 2014, *Principles of Sedimentology and Stratigraphy*. Boston, MA, Pearson Education.

Boiffin, J., and Bresson, L. M., 1987, Dynamique de formation des croutes superficielles: apport de l'analyse microscopique, *in* Fedoroff, N., Bresson, L. M., and Courty, M.-A., eds., *Soil Micromorphology*: Plaisir, Association Française pour l'Étude du Sol, p. 393–399.

Boivin, N. L., 1999, Life rhythms and floor sequences: excavating time in rural Rajasthan and Neolithic Çatalhöyük: *World Archaeology*, v. 31, p. 367–388.

Bonde, N., and Christensen, A. E., 1993., Dendrochronological dating of the Viking Age ship burials at Oseberg, Gokstad and Tune, Norway: *Antiquity*, v. 67, p. 575–583.

Bonifay, E., 1956, Les sédiments détritiques grossiers dans le remplissage des grottes – méthode d'étude morphologique et statistique: *L'Anthropologie*, v. 60, p. 447–461.

Bonifay, E., 1975, Stratigraphie du Quaternaire et age des gisements préhistoriques de la zone littorale des Alpes-Maritimes: *Bulletin de la Société Préhistorique française*, v. 72, no. 7, p. 197–208.

Boorman, L., Hazelden, J., and Boorman, M., 2002, New salt marshes for old – salt marsh creation and management, *in* Eurocoast, ed., Littoral 2002, *The Changing Coast*: Porto – Portugal, EUCC, p. 35–45.

Booth, P., Simmonds, A., Boyle, A., Clough, S., Cool, H. E. M., and Poore, D., 2010, *The late Roman cemetery at Lankhills, Winchester: Excavations 2000–2005*. Project Report: Oxford Archaeology.

Borderie, Q., Devos, Y., Nicosia, C., Cammas, C., and Macphail, R. I., 2014a, Dark Earth in the geoarchaeological approach to urban contexts, *in* Arnaud-Fassetta, G., and Carcaud, N., eds., *French Geoarchaeology in the 21st Century*. Paris, CNRS, p. 245–258.

Borderie, Q., Fondrillon, M., Nicosia, C., Devos, Y., and Macphail, R. I., 2014b, Bilan des recherches et nouveaux éclairages sur les terres noires: des processus complexes de stratification aux modalités d'occupation des espaces urbains., *in* Lorans, E., ed., *Archéologie de l'espace urbain – Partie II*. Tours, CTHS, p. 213–223.

Bordes, F., 1954, *Les limons quaternaires du Bassin de la Seine, stratigraphie et archéologie paléolithique*. Paris, Masson, Institut de Paléontologie Humaine, Paris. (Fondation Albert, 1er prince de Monaco). Archives; mémoire 26.

Bordes, F., 1972, *A Tale of Two Caves*. New York, Harper & Row Publishers.

Boschian, G., and Montagnari-Kokelji, E., 2000, Prehistoric shepherds and caves in the Trieste Karst (northeastern Italy): *Geoarchaeology*, v. 15, no. 4, p. 331–371.

Bousman, C. B., Collins, M. B., Goldberg, P., Stafford, T., Guy, J., Baker, B. W., Steele, D. G., Kay, M., Kerr, A., Fredlund, G., Dering, P., Holliday, V., Wilson, D., Gose, W., Dial, S., Takac, P., Balinsky, R., Masson, M., and Powell, J. F., 2002, The Palaeoindian-Archaic transition in North America: new evidence from Texas: *Antiquity*, v. 76, p. 980–990.

Bowler, J. M., 1998, Willandra Lakes revisited: environmental framework for human occupation: *Archaeology in Oceania*, v. 33, no. 3, p. 120–155.

Bowler, J. M., Johnston, H., Olley, J. M., Prescott, J. R., Roberts, R. G., Shawcross, W., and Spooner, N. A., 2003, New ages for human occupation and climatic change at Lake Mungo, Australia: *Nature*, v. 421, no. 6925, p. 837–840.

Bowman, D., 1997, Geomorphology of the Dead Sea western margin, *in* Niemi, T. M., Ben-Avraham, Z., and Gat, J. R., eds., *The Dead Sea, the Lake and Its Setting.* New York, Oxford University Press, p. 217–225.

Bowsher, D., Holder, N., Howell, I., and Dyson, T., 2007, *The London Guildhall: The Archaeology and History of the Guildhall Precinct from the Medieval Period to the 20th Century.* London, Museum of London Archaeological Service, p. 536.

Bradford, J. S. P., and González, A. R., 1960, *Photo Interpretation in Archaeology, Manual of Photographic Interpretation.* Washington, DC, American Society of Photogrammetry, p. 717–733.

Bradley, R. S., 1999, *Paleoclimatology.* San Diego, CA, Academic Press.

Bradley, R. 2007. *The Prehistory of Britain and Ireland.* Cambridge University Press: Cambridge.

Brady, N. C., and Weil, R. R., 2008, *The Nature and Properties of Soils.* New Jersey, Prentice Hall Education.

Brammer, H., 1971, Coatings in seasonally flooded soils: *Geoderma*, v. 6, p. 5–16.

Braun, K., Bar-Matthews, M., Matthews, A., Ayalon, A., Cowling, R. M., Karkanas, P., Fisher, E. C., Dyez, K., Zilberman, T., and Marean, C. W., 2019, Late Pleistocene records of speleothem stable isotopic compositions from Pinnacle Point on the South African south coast: *Quaternary Research*, v. 91, no. 1, p. 265–288.

Breeze, P. S., Groucutt, H. S., Drake, N. A., Louys, J., Scerri, E. M., Armitage, S. J., Zalmout, I. S., Memesh, A. M., Haptari, M. A., and Soubhi, S. A., 2017, Prehistory and palaeoenvironments of the western Nefud Desert, Saudi Arabia: *Archaeological Research in Asia*, v. 10, p. 1–16.

Breuning-Madsen, H., Holst, M. K., and Rasmussen, M., 2001, The chemical environment in a burial mound shortly after construction – an archaeological-pedological experiment: *Journal of Archaeological Science*, v. 28, p. 691–697.

Breuning-Madsen, H., Holst, M. K., Rasmussen, M., and Elberling, B., 2003, Preserved within log coffins before and after barrow construction: *Journal of Archaeological Science*, v. 30, p. 343–350.

Breuning-Madsen, H., Holst, M. K., and Henriksen, P. S., 2012, The hydrology in huge burial mounds built of loamy tills: a case study on the genesis of perched water tables and a well in a Viking Age burial mound in Jelling, Denmark: *Geografisk Tidsskrift-Danish Journal of Geography*, v. 112, p. 40–51.

Brewer, R., 1964, *Fabric and Mineral Analysis of Soils.* New York, Wiley and Sons.

Bridge, J. S., 2003, *Rivers and Floodplains: Forms, Processes, and Sedimentary Record.* Oxford, Blackwell Science.

Bridges, E. M., 1970, *World Soils*, Cambridge, Cambridge University Press.

Bridges, E. M., 1990, *Soil Horizon Designations*, Wageningen, International Soil Reference and Information Centre, Technical paper 19.

Bridgland, D. R., 1999, Analysis of the raised beach gravel deposits at Boxgrove and related sites, *in* Roberts, M. B., and Parfitt, S. A., eds., *Boxgrove. A Middle Pleistocene hominid site at Eartham Quarry, Boxgrove, West Sussex, Archaeological Report 17.* London, English Heritage, p. 100–110.

Brittingham, A., Hren, M. T., Hartman, G., Wilkinson, K. N., Mallol, C., Gasparyan, B., and Adler, D. S., 2019, Geochemical evidence for the control of fire by Middle Palaeolithic Hominins: *Scientific Reports*, v. 9, no. 1, p. 15368.

Brochier, J. E., 1983, Bergeries et feux néolithiques dans le Midi de la France, caractérisation et incidence sur la raisonnement sédimentologique: *Quatar*, v. 33/34, p. 181–193.

Brochier, J. E., 1996, Feuilles ou fumiers? Observations sur le rôle des poussières sphérolitiques dans l'interprétation des dépôts archéologiques Holocènes: *Anthrozoologica*, v. 24, p. 19–30.

Brochier, J. E., Villa, P., and Giacomarra, M., 1992, Shepherds and Sediments: geo-ethnoarchaeology of pastoral sites: *Journal of Anthropological Archaeology*, v. 11, p. 47–102.

Brogioli, G. P., Cremaschi, M., and Gelichi, S., 1988, Orocessi di stratificazione in centri urbani dell'Italia settentrionale: *Archeologia stratigrafica*, v. 1, p. 23–30.

Brönnimann, D., Ismail-Meyer, K., Rentzel, P., Pümpin, C., and Lisá, L., 2017a, Excrements of herbivores, *in* Nicosia, C., and Stoops, G., eds., *Archaeological Soil and Sediment Micromorphology*. Chichester, Wiley Blackwell, p. 55–65.

Brönnimann, D., Pümpin, C., Ismail-Meyer, K., Rentzel, P., and Éguez, N., 2017b, Excrements of omnivores and carnivores, *in* Nicosia, C., and Stoops, G., eds., *Archaeological Soil and Sediment Micromorphology*. Chichester, Wiley Blackwell, p. 67–81.

Brookes, I. A., 2001, Aeolian erosional lineations in the Libyan Desert, Dakhla region, Egypt: *Geomorphology*, v. 39, no. 3–4, p. 189–209.

Brothwell, D., 1986, *The Bogman and the Archaeology of People*. London, British Museum Publications.

Brothwell, D., 1995, *Recent Research on the Lindow Bodies in the Context of Five Years of World Studies Bog Bodies: New Discoveries and New Perspectives*. London, British Museum Press, p. 100–103.

Brown, A. G., 1997, *Alluvial Geoarchaeology*. Cambridge, Cambridge University Press, Cambridge Manuals in Archaeology.

Brown, A., Toms, P., Carey, C., and Rhodes, E., 2013, Geomorphology of the Anthropocene: Time-transgressive discontinuities of human-induced alluviation: *Anthropocene*, v. 1, p. 3–13.

Brown, G., 1988, *Whittington Ave (WIV88)*: Museum of London Archaeological Service.

Brown, G. E., 1990, Testing of concretes, plasters, and stuccos: *Archaeomaterials*, v. 4, p. 185–191.

Brown, L., Stansbie, D., and Webley, L. J., 2009, An Iron Age settlement and post-medieval farmstead at Oxley Park West, Milton Keynes: *Records of Buckinghamshire* v. 49, p. 43–72.

Bruckert, S., 1982, Analysis of the organo-mineral complexes of soils, *in* Bonneau, M., and Souchier, B., eds., *Constituents and Properties of Soils*. London, Academic press, p. 214–237.

Bruins, H. J., and Yaalon, D. H., 1979, Stratigraphy of the Netivot section in the desert loess of the Negev (Israel): *Acta Geologica Hungarica*, v. 22, p. 161–170.

Brunborga, L. A., Julshamna, K., Nortvedta, R., and Frøylanda, L., 2006, Nutritional composition of blubber and meat of hooded seal (*Cystophora cristata*) and harp seal (*Phagophilus groenlandicus*) from Greenland: *Food Chemistry*, v. 96, no. 4, p. 524–531.

Bruxelles, L., Clarke, R. J., Maire, R., Ortega, R., and Stratford, D., 2014, Stratigraphic analysis of the Sterkfontein StW 573 Australopithecus skeleton and implications for its age: *Journal of Human Evolution*, v. 70, p. 36–48.

Bruxelles, L., Stratford, D. J., Maire, R., Pickering, T. R., Heaton, J. L., Beaudet, A., Kuman, K., Crompton, R., Carlson, K. J., and Jashashvili, T., 2019, A multiscale stratigraphic investigation of the context of StW 573 "Little Foot" and Member 2, Sterkfontein Caves, South Africa: *Journal of Human Evolution*, v. 133, p. 78–98.

Bryant, R. G., and Davidson, D. A., 1996, The use of image analysis in the micromorphological study of old cultivated soils: an evaluation based on soils from the island of Papa Stour, Shetland: *Journal of Archaeological Science*, v. 23, no. 6, p. 811–822.

Bryn, H., and Sauvage, R., 2018, *Arkeologisk undersøkelse, Korsmyra, Fræna kommune, Møre og Romsdal*: NTNU Vitenskapsmuseet.

Buckland, P., Östman, S., Wallin, J.-E., Ericson, S., and Linderholm, J., 2017, *Pollen, plant macrofossil and geoarchaeological analyses of profile 11632, Follobanen FO3, Oslo (report for NIKU)*: Environmental Archaeology Laboratory, Umeå University.

Buggle, B., Hambach, U., Glaser, B., Gerasimenko, N., Marković, S., Glaser, I., and Zöller, L., 2009, Stratigraphy, and spatial and temporal paleoclimatic trends in Southeastern/Eastern European loess–paleosol sequences: *Quaternary International*, v. 196, no. 1–2, p. 86–106.

Bukkemoen, G. B., 2013, *Sjøavsatte kulturlag/elvesedimenter. SØRENGA D1A, 234/102, OSLO* (in Norwegian): Kultural History Museum, University of Oslo.

Bull, P. A., and Goldberg, P., 1985, Scanning Electron Microscope Analysis of Sediments from Tabun Cave, Mount Carmel, Israel: *Journal of Archaeological Science*, v. 12, p. 177–185.

Bullock, P., and Murphy, C. P., 1979, Evolution of a paleo-argillic brown earth (Paleudalf) from Oxfordshire, England: *Geoderma*, v. 22, p. 225–252.

Bullock, P., Fedoroff, N., Jongerius, A., Stoops, G., Tursina, T., and Babel, U., 1985, *Handbook for Soil Thin Section Description*. Wolverhampton, Waine Research.

Bunbury, J., 2019, *The Nile and Ancient Egypt: Changing Land- and Waterscapes, from the Neolithic to the Roman Era*. Cambridge, Cambridge University Press.

Büntgen, U., Tegel, W., Nicolussi, K., McCormick, M., Frank, D., Trouet, V., Kaplan, J. O., Herzig, F., Heussner, K.-U., Wanner, H., Luterbacher, J., and Esper, J., 2011, 2500 Years of European Climate Variability and Human: *Science*, v. 331, no. 578, p. 578–582.

Buol, S. W., Hole, F. D., and McCracken, R. J., 1973, *Soil Genesis and Classification*, Ames, The Iowa State University Press.

Burch, M., Treveil, P., and Keene, D., 2011, *The Development of Early Medieval and Later Poultry and Cheapside. Excavations at 1 Poultry and Vicinity, City of London*. London, Museum of London Archaeology.

Burns, A., Pickering, M. D., Green, K. A., Pinder, A. P., Gestsdóttir, H., Usai, M. R., Brothwell, D. R., and Keely, B. J., 2017, Micromorphological and chemical investigation of late-Viking age grave fills at Hofstaðir, Iceland: *Geoderma* v. 306, p. 183–194.

Burrin, P. J., and Scaife, R. G., 1984, Aspects of Holocene valley sedimentation and floodplain development in southern England: *Proceedings of the Geologists' Association*, v. 95, no. 1, p. 81–96.

Butser Ancient Farm, 2009, *Butser Ancient Farm Guide Book*. West Sussex, Butser Ancient Farm.

Butzer, K. W., 1960, Archeology and geology in ancient Egypt: *Science*, v. 132, no. 3440, p. 1617–1624.

Butzer, K. W., 1964, *Environment and Archaeology*. Chicago, IL, Aldine.

Butzer, K. W., 1976, *Geomorphology from the Earth*. New York, Harper & Row.

Butzer, K. W., 1982, *Archaeology as Human Ecology: Method and Theory for a Contextual Approach*. Cambridge, Cambridge University Press.

Butzer, K. W., and Cuerda, J., 1962, Coastal stratigraphy of southern Mallorca and its implications for the Pleistocene chronology of the Mediterranean Sea: *The Journal of Geology*, v. 70, no. 4, p. 398–416.

Byrd, B. F., n.d., *Archaeology around San Elijo Lagoon*: thenaturecollective.org. https://thenaturecollective.org/wp-content/uploads/2019/04/ARCHAEOLOGY-AT-SAN-ELIJO-LAGOON.pdf.

Cabanes, D., Mallol, C., Exposito, I., and Baena, J., 2010, Phytolith evidence for hearths and beds in the late Mousterian occupations of Esquilleu cave (Cantabria, Spain): *Journal of Archaeological Science*, v. 37, no. 11, p. 2947–2957.

Cahen, D., and Moeyersons, J., 1977, Subsurface movements of stone artifacts and their implications for the prehistory of Central Africa: *Nature*, v. 266, no. 5605, p. 812–815.

Cammas, C., 1994, Approche micromorphologique de la stratigraphie urbaine à Lattes: premiers résultats, *Lattara* v. 7: Lattes, A R A L O, p. 181–202.

Cammas, C., 2004, Les "terre noires" urbaines du Nord de la France: première typologie pédo-sédimentaire, *in* Verslype, L., and Brulet, R., eds., *Terres Noires – Dark Earth*:: Louvain-la-Neuve, Université Catholique de Louvain, p. 43–55.

Cammas, C., 2018, Micromorphology of earth building materials: Toward the reconstruction of former technological processes (Protohistoric and Historic Periods): *Quaternary International*, v. 483, p. 160–179.

Cammas, C., David, C., and Guyard, L., 1996a, La question des terre noires dans les sites tardo-antiques et médiéval: le cas du Collège de France (Paris, France), *Proceedings XIII International Congress of Prehistoric and Protohistoric Sciences, Colloquim 14*: Forlì, ABACO, p. 89–93.

Cammas, C., Wattez, J., and Courty, M.-A., 1996b, L'enregistrement sédimentaire des modes d'occupation de l'espace, *in* Castelletti, L., and Cremaschi, M., eds., *Paleoecology; Colloquium 3 of XIII International Congress of Prehistoric and Protohistoric Sciences, Vol. 3*: Forli, ABACO, p. 81–86.

Campana, S., 2017, Drones in archaeology. State-of-the-art and future perspectives: *Archaeological Prospection*, v. 24, no. 4, p. 275–296.

Campbell, G., and Robinson, M., 2007, Environment and Land Use in the Valley Bottom, *in* Healy, F., and Harding, J., eds., *The Raunds Area Project. A Neolithic and Bronze Age Landscape in Northamptonshire Swindon*, English Heritage, p. 18–36.

Campy, M., and Chaline, J., 1993, Missing records and depositional breaks in French late Pleistocene cave sediments: *Quaternary Research*, v. 40, no. 3, p. 318–331.

Canada, C.-N. R., 2004, Fundamentals of Remote Sensing: Ottawa, National Resources Canada.

Cannell, R., 2012, *Archaeological Investigation of the Gokstad Mound 2011*: Museum of Cultural History, University of Oslo.

Cannell, R. J. S., Bill, J., and Macphail, R. I., 2020, Constructing and deconstructing the Gokstad Mound: *Antiquity*, v. 94 (377), p. 1–18.

Cannell, R. J. S., Gustavsen, L., Kristiansen, M., and Nau, E., 2018, Delineating an Unmarked Graveyard by High-Resolution GPR and pXRF Prospection: The Medieval Church Site of Furulund in Norway: *Journal of Computer Applications in Archaeology*, v. 1, no. 1, p. 1–18.

Canti, M. G., 2017, Coal, *in* Nicosia, C., and Stoops, G., eds., *Archaeological Soil and Sediment Micromorphology*. Chichester, Wiley Blackwell, p. 143–145.

Canti, M., 1995, A mixed approach to geoarchaeological analysis, *in* Barham, A. J., and Macphail, R. I., eds., *Archaeological Sediments and Soils: Analysis, Interpretation and Management*. London, Institute of Archaeology, p. 183–190.

Canti, M., 1999, The production and preservation of faecal spherulites: animals, environment and taphonomy: *Journal of Archaeological Science*, v. 26, p. 251–258.

Canti, M., 2017, Burnt carbonates, *in* Nicosia, C., and Stoops, G., eds., *Archaeological Soil and Sediment Micromorphology*. Chichester, Wiley Blackwell p. 181–188.

Canti, M., and Brochier, J. E., 2017, Plant ash, *in* Nicosia, C., and Stoops, G., eds., *Archaeological Soil and Sediment Micromorphology*. Chichester, Wiley Blackwell, p. 147–153.

Canti, M., Carter, S., Davidson, D., and Limbrey, S., 2006, Problems of unscientific method and approach in "Archaeological soil and pollen analysis of experimental floor deposits; with special reference to Butser Ancient Farm, Hampshire, UK" by R. I. Macphail, G. M. Cruise, M. Allen, J. Linderholm and P. Reynolds: *Journal of Archaeological Science*, v. 33, p. 295–298.

Carey, C. J., and Juleff, G., 2013, Geochemical survey and metalworking: a case study from Exmoor, southwest Britain, *in* Humphries, J., and Rehren, T., eds., *The World of Iron*. London, Archetype, p. 383–392.

Carey, C. J. Howard, A.J., Jackson, R. and Brown, A.G. 2017. Using geoarchaeological deposit modelling as a framework for archaeological evaluation and mitigation in alluvial environments. *Journal of Archaeological Science: Reports*, 11, p. 658–673

Carey, C. J., White, H., Macphail, R., Bray, L. S. and Scaife, R. 2021. Analysis of a brown earth palaeosol and derived sediments associated with a Mesolithic pit, a Late Neolithic – Early Bronze Age burnt mound and an Early Bronze Age burnt mound on Exmoor, UK. *Journal of Archaeological Science: Reports*, 35, 102675.

Carey, C., and Hunnisett, K., 2019, *Holwell Hut Circle and Reave Report 1: Geoarchaeological analysis of the pre reave and roundhouse deposit sequences*. Brighton: University of Brighton.

Carey, C., Howard, A. J., Jackson, R. and Brown, A., 2017, Using geoarchaeological deposit modelling as a framework for archaeological evaluation and mitigation in alluvial environments: *Journal of Archaeological Science: Reports*, v. 11, p. 658–673.

Carey, C., Howard, A. J., Knight, D., Corcoran, J., and Heathcote, J. E., 2018, *Deposit Modelling and Archaeology*. Exeter, Short Run Press. www.brighton.ac.uk/_pdf/research/set-groups/deposit-modelling-and-archaeology-volume.pdf.

Carey, C., Wickstead, H. J., Juleff, G., Anderson, J. C., and Barber, M. J., 2014, Geochemical survey and metalworking: analysis of chemical residues derived from experimental non-ferrous metallurgical processes in a reconstructed roundhouse: *Journal of Archaeological Science*, v. 49, p. 383–397.

Carey, C., White, H., Macphail, R., Bray, L., Scaife, R., Coyle McClung, L., and Macleod, A., 2020, Analysis of prehistoric brown earth paleosols under the podzol soils of Exmoor, UK: *Geoarchaeology*, v. 35, no. 5, p. 772–799.

Carpentier, F., 2015, *Minoans under the microscope. Archaeological soil micromorphology at the Cretan Bronze Age site of Sissi*: KU Leuven, 336 p.

Carpentier, F., and Vandermeulen, B., 2016, High-resolution photography for soil micromorphology slide documentation: *Geoarchaeology*, v. 31, no. 6, p. 603–607.

Carrión, J. S., Ochando, J., Fernández, S., Blasco, R., Rosell, J., Munuera, M., Amorós, G., Martín-Lerma, I., Finlayson, S., and Giles, F., 2018, Last Neanderthals in the warmest refugium of Europe: Palynological data from Vanguard Cave: *Review of Palaeobotany and Palynology*, v. 259, p. 63–80.

Carruthers, W. J., 2000, Mineralised plant remains, *in* Lawson, A. J., ed., *Potterne 1982–5. Animal Husbandry in Later Prehistoric Wiltshire*, Wessex Archaeology Report No. 17. Salisbury, Wessex Archaeology, p. 72–84.

Carter, M. R., 1993, *Soil Sampling and Methods of Analysis*. London, Lewis Publishers.

Carter, S., 1987, *The Reconstruction of land-Snail Death Assemblages*: University of London.

Carter, S., 1998, The use of peat and other organic sediments as fuel in northern Scotland: identifications derived from soil thin sections, *in* Coles, C. M. M. a. G., ed., *Life on the Edge: Human Settlement and Marginality*, Monograph 100. Oxford, Oxbow, p. 99–104.

Carter, S. P., 1990, The stratification and taphonomy of shells in calcareous soils: implications for landsnail analysis in archaeology: *Journal of Archaeological Science*, v. 17, p. 495–507.

Carter, S., and Davidson, D., 1998, An evaluation of the contribution of soil micromorphology to the study of ancient arable cultivation: *Geoarchaeology*, v. 13, p. 535–547.

Carver, M. O. H., 1987, The nature of urban deposits, *in* Schofield, J., and Leech, R., eds., *Urban Archaeology in Britain, CBA Research Report 61*: York, Council for British Archaeology, p. 9–26.

Carver, M. O. H., 1998, Sutton Hoo. *Burial ground of Kings?* London, British Museum Press.

Carver, R. E., 1971, *Procedures in Sedimentary Petrology*. New York, Wiley-Interscience, p. 653.

Catt, J. A. 1978. The contribution of loess to soils in lowland Britain, *in* S. Limbrey, and J. G. Evans (eds.), *The Effects of Man of the Landscape: The Lowland Zone*. York, Council for British Archaeology Research Report, p. 12–20.

Catt, J. A. E., 1990, Paleopedology manual: *Quaternary International*, v. 6, p. 1–95.

Catt, J. A., 1986, Soils and Quaternary Geology. *A Handbook for Field Scientists*, Oxford, Clarendon Press.

Catt, J. A., 1999, Particle size distribution and mineralogy of the deposits, *in* Roberts, M. B., and Parfitt, S. A., eds., *Boxgrove. A Middle Pleistocene hominid site at Eartham Quarry, Boxgrove, West Sussex*, Archaeological Report 17. London, English Heritage, p. 111–118.

Catt, J. A., and Bronger, A., 1998, Reconstruction and Climatic Implications of Paleosols, *Catena Special Issue*, Vol. 34 (1–2). Amsterdam, Elsevier, p. 1–207.

Challis, K., Forlin, P., Kincey, M., 2011. A Generic Toolkit for the Visualization of Archaeological Features on Airborne LiDAR Elevation Data. Archaeological Prospection, 18, p. 279–289.

Challis, K., Howard, A.J., Moscrop, D., Gearey, B., Smith, D., Carey, C., Thompson, A., 2006. Using airborne lidar intensity to predict the organic preservation of waterlogged deposits, in, S Campana and M Forte (eds), *From Space to Place 2nd International Conference of Remote Sensing*. BAR S1568. p. 93–98.

Challis, K., Kincey, M., Howard, A.J., 2009. Airborne remote sensing of valley floor geoarchaeology using Daedalus ATM and CASI. *Archaeological Prospection*, 16, p.17–33.

Champness, C., Teague, S., and Ford, B., 2012, Holocene Environmental Change and Roman Floodplain Management at Pilgrim's School, Cathedral Close, Winchester, Hampshire: *Proc. Hampshire Field Club Archaeol. Soc.*, v. 67, no. 1, p. 25–68.

Chen, F., Lasaponara, R., Masini, N., 2017. An overview of satellite synthetic aperture radar remote sensing in archaeology: From site detection to monitoring. *Journal of Cultural Heritage*, 23, p. 5–11. https://doi.org/10.1016/j.culher.2015.05.003

Chen, F. H., Zhang, J., Liu, J., Cao, X., Hou, J., Zhu, L., Xu, X., Liu, X., Wang, M., Wu, D., Huang, L., Zeng, T., Zhang, S., Huang, W., Zhang, X., and Huang, L., 2020, Climate change, vegetation history, and landscape responses on the Tibetan Plateau during the Holocene: a comprehensive review: *Quaternary Science Reviews*, v. 243, p. 106444.

Chorley, R. J., Schumm, S. A., and Sugden, D. E., 1984, *Geomorphology*. London, Methuen.

Ciezar, P., Gonzalez, V., Pieters, M., Rodet-Belarbi, I., and Van-Ossel, P., 1994, In suburbano – new data on the immediate surroundings of Roman and early medieval Paris, *in* Hall, A. R., and Kenward, H. K., eds., *Urban-Rural Connexions: Perspectives from Environmental Archaeology*. Oxford, Oxford: Oxbow Books, p. 137–146.

Clark, A., 2000, *Seeing beneath the soil: prospecting methods in archaeology*. New edition. London, Routledge.

Clark, J. L., and Ligouis, B., 2010, Burned bone in the Howieson's Poort and post-Howieson's Poort Middle Stone Age deposits at Sibudu (South Africa): Behavioral and taphonomic implications: *Journal of Archaeological Science*, v. 37, no. 10, p. 2650–2661.

Clark, K., 2000, Architect's specification: building analysis and conservation, *in* Roskams, S., ed., *Interpreting Stratigraphy, International Series 910*. Oxford, British Archaeological Reports, p. 17–24.

Clayden, B., and Hollis, J. M., 1984, *Criteria for Differentiating Soil Series*, Harpenden, Soil Survey of England and Wales, Soil Survey Technical Monograph.

Cohen, A., Campisano, C., Arrowsmith, R., et al., 2016, The Hominin Sites and Paleolakes Drilling Project: inferring the environmental context of human evolution from eastern African rift lake deposits: *Scientific Drilling*, v. 21, p. 1–16.

Colcutt, S. N., 1999, Structural sedimentology at Boxgrove, *in* Roberts, M., and Parfitt, S. A., eds., *Boxgrove. A Middle Pleistocene Hominid Site at Eartham Quarry, Boxgrove, West Sussex.*, Archaeological Report 17. London, English Heritage, p. 42–99.

Collins, M. B., and Mear, C. E., 1998, The site and its setting, in Collins, M. B., ed., *Wilson-Leonard: An 11,000-year Archaeological Record of Hunter-Gatherers in Central Texas, Studies in Archaeology 31*. Austin, TX, Texas Archaeological Research Laboratory, p. 5–31.

Collins, M. B., Bousman, C. B., Goldberg, P., Takac, P. R., Guy, J. C., Lanata, J. L., Stafford, T., Jr., and Holliday, V. T., 1993, The Paleoindian Sequence at the Wilson-Leonard Site, Texas: *Current Research in the Pleistocene*, v. 10, p. 10–12.

Collins, M. E., Carter, B. J., Gladfelter, B. G., and Southard, R. J., 1995, *Pedological Perspectives in Archaeological Research, SSSA Special Publication Number 44*: Madison, Soil Science Society of America, Inc.

Collins, M. B., Bailey, G. L., Bousman, C. B., Dial, S. W., Goldberg, P., Guy, J., Holliday, V. T., Mear, C., and Takac, P. R., 1998, *Wilson-Leonard An 11,000-year Archeological Record of Hunter-Gatherers in*

Central Texas, Vol. I: Introduction, Background, and Syntheses, Index of Texas Archaeology: Open Access Gray Literature from the Lone Star State, no. 1, article 24.

Collinson, J. D. and Thompson, D. B. 1989, Sedimentary Structures, 2nd edition, Chapman & Hall, London.

Collinson, J., and Mountney, N., 2019, *Sedimentary Structures.* Edinburgh, Dunedin Academic Press Ltd.

Conard, N. J., and Bolus, M., 2003, Radiocarbon dating the appearance of modern humans and timing of cultural innovations in Europe: new results and new challenges: *Journal of Human Evolution,* v. 44, p. 331–371.

Connell, B., Gray Jones, A., Redfern, R., and Walker, D., 2012 *A Bioarchaeological Study of Medieval Burials on the Site of St Mary Spital: Excavations at Spitalfields Market, London E1, 1991–2007.* London, Museum of London Archaeology.

Conolly, J. and Lake, M., 2006. *Geographical Information systems in Archaeology.* Cambridge Manuals in Archaeology.

Conry, M. J., 1971, Irish Plaggen soils, their distribution, origin and properties: *Journal of Soil Science,* v. 22, p. 401–416.

Conway, J., 1983, An investigation of soil phosphorus distribution within occupation deposits from a Romano-British hut group: *Journal of Archaeological Science,* v. 10, p. 117–128.

Conyers, L. B., 1995, The use of ground-penetrating radar to map the buried structures and landscape of the Ceren Site, El Salvador: *Geoarchaeology,* v. 10, no. 4, p. 275–299.

Conyers, L. B., and Goodman, D., 1997, *Ground-penetrating Radar: An Introduction for Archaeologists,* Walnut Creek, CA, AltaMira Press.

Cook, S. R., Banerjea, R. Y., Marchall, L. J., Fulford, M., Clarke, A., and van Zwieten, C., 2010, Concentrations of copper, zinc and lead as indicators of hearth usage at the Roman own of Calleva Atrebatum (Silchester, Hampshire, UK): *Journal of Archaeological Science,* v. 37, p. 871–879.

Cook, S. R., Clarke, A. S., and Fulford, M. G., 2005, Soil geochemistry and detection of early Roman precious metal and copper alloy working in the Roman town of Calleva Atrebatum (Silchester, Hampshire, UK). *Journal of Archaeological Science,* v. 32, p. 805–812.

Cooke, R., Warren, A., and Goudie, A., 1993, *Desert Geomorphology.* London, UCL Press.

Cooper, L. P., Davis, W., et al., 2017, Making and Breaking Microliths: A Middle Mesolithic Site at Asfordby, Leicestershire: *Proceedings of the Prehistoric Society,* v. 83, p. 43–96.

Corbett, W. M. 1973. *Breckland Forest Soils.* Harpenden, Soil Survey of England and Wales.

Cornwall, I. W., 1953, *Soils for the Archaeologist.* New York, The Macmillan Company.

Coronato, A., Fanning, P., Salemme, M., Oría, J., Pickard, J., and Ponce, J. F., 2011, Aeolian sequence and the archaeological record in the Fuegian steppe, Argentina: *Quaternary International,* v. 245, no. 1, p. 122–135.

Corrêa, G. R., Schaefer, C. E., and Gilkes, R. J., 2013, Phosphate location and reaction in an archaeoanthrosol on shell-mound in the Lakes Region, Rio de Janeiro State, Brazil: *Quaternary International,* v. 315, p. 16–23.

Costamagno, S., Théry-Parisot, I., Brugal, J.-P., and Guibert, R., 2005, Taphonomic consequences of the use of bones as fuel. Experimental data and archaeological applications: Biosphere to Lithosphere: *New Studies in Vertebrate Taphonomy.* Oxbow Books. Oxford, p. 51–62.

Couchoud, I., 2003, Processus géologiques de formation du site moustérien du Roc de Marsal (Dordogne, France): *Paleo,* v. 15, p. 51–68.

Courty, M.-A., 1984, Formation et évolution des accumulations cendreuses: *approche micromorphologique, Actes du Colloque Interrégional sur le Néolithique, 1981*: Le Puy, p. 341–353.

Courty, M.-A., 1989, Analyse microscopique des sédiments du remplissage de la grotte de Vaufrey (Dordogne), *in* Rigaud, J.-P., ed., *La Grotte de Vaufrey*, Mémoire de la Societé Préhistorique Française, p. 183–209.

Courty, M.-A., 2001, Microfacies analysis assisting archaeological stratigraphy, *in* P. Goldberg, Holliday, V. T., and Ferring, C. R., eds., *Earth Sciences and Archaeology*. New York, Kluwer, p. 205–239.

Courty, M.-A., 2012, Ancestral processing of exceptional organo-mineral materials: microfacies and multi-analytical study, *in* Poch, R. M., Casamitjana, M., and Francis, M. L., eds., *Proceedings of the 14th International Working Meeting on Soil Micromorphology – Lleida 8–14 July 2012*: Lleida, Universitat of Lleida and International Union of Soil Sciences, p. 321–325.

Courty, M.-A., and Fedoroff, N., 1982, Micromorphology of a Holocene dwelling, Proceedings Nordic Archaeometry, *PACT* 7, p. 257–277.

Courty, M.-A., and Fedoroff, N., 1985, Micromorphology of recent and buried soils in a semiarid region of Northwestern India: *Geoderma*, v. 35, no. 4, p. 287–332.

Courty, M.-A., and Nørnberg, P., 1985, Comparison between buried uncultivated and cultivated Iron Age soils on the west coast of Jutland, Denmark, *in* Edgren, T., and Jungner, H., eds., *Proceedings of the Third Nordic Conference on the Application of Scientific Methods in Archaeology*: Helsinki, The Finnish Antiquarian Society, p. 57–69.

Courty, M.-A., and Vallverdú, J., 2001, The microstratigraphic record of abrupt climate changes in cave sediments of the Western Mediterranean: *Geoarchaeology*, v. 16, no. 5, p. 467–500.

Courty, M.-A., Goldberg, P., and Macphail, R. I., 1989, *Soils and Micromorphology in Archaeology*, 1st edition. Cambridge, Cambridge University Press, Cambridge Manuals in Archaeology.

Courty, M.-A., Macphail, R. I., and Wattez, J., 1991, Soil micromorphological indicators of pastoralism with special reference to Arene Candide, Fianle Ligure, Italy: *Rivista di Studi Liguri*, v. **LVII**, p. 127–150.

Courty, M.-A., Goldberg, P., and Macphail, R. I., 1994, Ancient people – lifestyles and cultural patterns, *Transactions of the 15th World Congress of Soil Science, International Society of Soil Science, Mexico, Vol. 6a*: Acapulco, International Society of Soil Science, p. 250–269.

Courty, M.-A., Fedoroff, N., Jones, M. K., and McGlade, J., 1994, Environmental dynamics, *in* van der Leeuw, S. E., ed., *Temporalities and Desertification in the Vera Basin, Southeast Spain, Archaeomedes Project, Vol. 2*: Brussels, p. 19–84.

Courty, M.-A., Cricsi, A., Fedoroff, N., Greenwood, P., Grice, K., Mermoux, M., Smith, D. C., and Thiemens, M., 2008, Regional manifestation of the widespread disruption of soil-landscapes by the 4 kyr BP impact-linked dust-event using pedo-sedimentary micro-fabrics, *in* Kapur, S., Memut, A., and Stoops, G., eds., *New Trends in Soil Micromorphology*. New York, Springer, p. 211–236.

Courty, M.-A., Carbonell, E., Poch, J. V., and Banerjea, R., 2012, Microstratigraphic and multi-analytical evidence for advanced Neanderthal pyrotechnology at Abric Romani (Capellades, Spain): *Quaternary International*, v. 247, p. 294–313.

Cowan, C., 2003, *Urban development in north-west Roman Southwark: Excavations 1974–90*, Monograph 16. London, MOLAS, p. 209.

Cowgill, J., 2003, The iron production industry and its extensive demand upon woodland resources: a case study from Creeton Quarry, Lincolnshire, *in* Murphy, P., and Wiltshire, P. E. J., eds., *The Environmental Archaeology of Industry, Symposia of the Association for Environmental Archaeology No. 20*. Oxford, Oxbow, p. 48–57.

Cowley, D., Moriarty, C., Geddes, G., Brown, G., Wade, T. and Nichol, C. 2017. UAVs in Context: Archaeological Airborne Recording in a National Body of Survey and Record. *Drones*, 2, (1), 2. https://doi.org/10.3390/drones2010002

Crampton, C. B., 1963, The development and morphology of iron pan podzols in Mid and South Wales: *Journal of Soil Science*, v. 14, p. 282–302.

Creekmore, A. 2010. The structure of Upper Mesopotamian cities: Insight from fluxgate gradiometer survey at Kazane Höyük, southeastern Turkey. *Archaeological Prospection*, 17, p. 73–88.

Cremaschi, M., 1987, Paleosols and Vetusols in the Central Po Plain (Northern Italy). *A Study in Quaternary Geology and Soil Development*: Milano, Unicopli.

Cremaschi, M., and Nicosia, C., 2010, Corso Porta Reno, Ferrara (Northern Italy); a study in the formation processes of urban deposits, *Il Quaternario – Italian Journal of Quaternary Sciences*, v. 23, no. 2, p. 395–408.

Cremaschi, M., and Nicosia, C., 2012, Sub-Boreal aggradation along the Apennine margin of the Central Po Plain: geomorphological and geoarchaeological aspects: Géomorphologie: *Relief, Processus, Environnement*, v. 18, no. 2, p. 155–174.

Cremaschi, M., Di Lernia, S., and Trombino, L., 1996, From taming to pastoralism in a drying environment. Site formation processes in the shelter of the Tadrat Massif (Libya, Central Sahara), *in* Castelletti, L., and Cremaschi, M., eds., *Paleoecology, Proceedings of Int. Union of Prehistoric and Protohistoric Sciences*: Forlì, ABACO, p. 87–106.

Cremaschi, M., Trombino, L., and Zerboni, A., 2018, Palaeosols and Relict Soils: A Systematic Review, *in* Stoops, G., Marcelino, V., and Mees, F., eds., *Interpretation of Micromorphological Features of Soils and Regoliths*. Amsterdam, Elsevier, p. 863–894.

Cremeens, D. L., 1995, Pedogenesis of Cotiga Mound, A 2100-year-old woodland mound in Southwest West Virginia: *Soil Science Society of America Journal*, v. 59, p. 1377–1388.

Cremeens, D. L., Landers, D. B., and Frankenberg, S. R., 1997, Geomorphic setting and stratigraphy of Cotiga Mound, Mingo County, West Virginia: *Geoarchaeology*, v. 12, no. 5, p. 459–477.

Crombé, P., 1993, Tree-fall features on Final Palaeolithic and Mesolithic sites situated on sandy soils: how to deal with it: *Helenium*, v. XXXIII, no. I, p. 50–66.

Crombé, P., Langohr, R., and Louwagie, G., 2015, Mesolithic hearth-pits: fact or fantasy? A reassessment based on the evidence from the sites of Doel and Verrebroek (Belgium): *Journal of Archaeological Science*, v. 61, p. 158–171.

Crouvi, O., Amit, R., Enzel, Y., Porat, N., and Sandler, A., 2008, Sand dunes as a major proximal dust source for late Pleistocene loess in the Negev Desert, Israel: *Quaternary Research*, v. 70, no. 2, p. 275–282.

Crouvi, O., Amit, R., Enzel, Y., and Gillespie, A. R., 2010, Active sand seas and the formation of desert loess: *Quaternary Science Reviews*, v. 29, no. 17–18, p. 2087–2098.

Crouvi, O., Amit, R., Porat, N., Gillespie, A. R., McDonald, E. V., and Enzel, Y., 2009, Significance of primary hilltop loess in reconstructing dust chronology, accretion rates, and sources: an example from the Negev Desert, Israel: *Journal of Geophysical Research: Earth Surface*, v. 114, no. F2.

Crowther, J., 1996a, Phosphate migration around buried bones, *in* Bell, M., Fowler, P. J., and Hillson, S. W., eds., *The Experimental Earthwork Project 1960–1992, Vol. 100*: York, Council for British Archaeology, p. 195–196.

Crowther, J., 1996b, Report on sediments from Building 5, Old Market Street "86", *in* A. G. Marvell, ed., *Excavations at Usk 1986–1988* (Brittania XXVII, 51–110): Britannia, v. XXVII, p. 92–99.

Crowther, J., 1996c, Soil chemistry, *in* Bell, M., Fowler, P. J., and Hillson, S. W., eds., *The Experimental Earthwork Project 1960–1992*, Research Report 100: York, Council for British Archaeology, p. 107–118.

Crowther, J., 1997, Soil phosphate surveys: critical approaches to sampling, analysis and interpretation: *Archaeological Prospection*, v. 4, p. 93–102.

Crowther, J., 2000, Phosphate and magnetic susceptibility studies, *in* Bell, M., Caseldine, A., and Neumann, H., eds., *Prehistoric Intertidal Archaeology in the Welsh Severn Estuary*, Vol. 120. York, CBA Research report, p. 57–58 (and CD).

Crowther, J., 2003, Potential magnetic susceptibility and fractional conversion studies of archaeological soils and sediments: *Archaeometry*, v. 45, no. 4, p. 685–701.

Crowther, J., 2007, Chemical and magnetic properties of soils and pit fills, *in* Whittle, A., ed., *The Early Neolithic on the Great Hungarian Plain: investigations of the Körös culture site of Ecsegfalva 23, Co. Békés*, Vol. I: Budapest, Institute of Archaeology, p. 227–254.

Crowther, J., and Barker, P., 1995, Magnetic susceptibility: distinguishing anthropogenic effects from the natural: *Archaeological Prospection*, v. 2, p. 207–215.

Crowther, J., Macphail, R. I., and Cruise, G. M., 1996, Short-term burial change in a humic rendzina, Overton Down Experimental Earthwork, Wiltshire, England: *Geoarchaeology*, v. 11, no. 2, p. 95–117.

Cruise, G. M., 1990, Holocene peat initiation in the Ligurian Apennines, northern Italy: *Review of Palaeobotany and Palynology*, v. 63, p. 173–182.

Cruise, G. M., and Macphail, R. I., 2000, Microstratigraphical Signatures of Experimental Rural Occupation Deposits and Archaeological Sites, *in* Roskams, S., ed., *Interpreting Stratigraphy*, Vol. 9. York, University of York, p. 183–191.

Cruise, G. M., Macphail, R. I., Linderholm, J., Maggi, R., and Marshall, P. D., 2009, Lago di Bargone, Liguria, N. Italy: a reconstruction of Holocene environmental and land-use history: *The Holocene*, v. 19, no. 7, p. 987–1003.

Currant, A. P., 1991, A Late Glacial interstadial mammal fauna from Gough's Cave, Somerset, England, *in* Barton, N., Roberts, A. J., and Roe, D. A., eds., *The Late Glacial in North-west Europe; human adaptation and environmental change at the end of the Pleistocene*. London, Council for British Archaeology, p. 48–50.

Dabkowski, J., 2014, High potential of calcareous tufas for integrative multidisciplinary studies and prospects for archaeology in Europe: *Journal of Archaeological Science*, v. 52, p. 72–83.

Dalan, R. A., and Bevan, B. W., 2002, Geophysical indicators of culturally emplaced soils and sediments: *Geoarchaeology*, v. 17, no. 8, p. 779–810.

Dalrymple, J. B., Blong, R. J., and Conacher, A. J., 1968, A hypothetical nine unit landsurface model: *Zeitschift für Geomorphologie*, v. 12, p. 60–76.

Dalsgaard, K. and Odgaard, B. V. 2001. Dating sequences of buried horizons of podzols developed in wind-blown sand at Ulfborg, Western Jutland: *Quaternary International*, v. 78, p. 53–60.

Dalwood, H., and Edwards, R., 2004, *Excavations at Deansway, Worcester, 1988–89: Romano-British small town to late medieval city*, CBA Research Report No 139. York, Council for British Archaeology, p. 605.

Dammers, K., and Joergensen, R. G., 1996, Progressive loss of Carbon and Nitrogen from simulated daub on heating: *Journal of Archaeological Science*, v. 23, p. 639–648.

Danin, A., and Ganor, E., 1991, Trapping of airborne dust by mosses in the Negev Desert, Israel: *Earth Surface Processes and Landforms*, v. 16, no. 2, p. 153–162.

Darmark, K., In prep., *Håkonshellaveien lok 1 -kulturlagslokal med senmesolitiska bostadslämningar. Askeladden id 138737, Alvøy 130/29, Bergen kommune, Vestland.* Utgravingsrapport (unpublished excavation report från Fornminneseksjonen). Bergen, University of Bergen.

Darvill, T. 2010. *Prehistoric Britain*. London, Routledge.

Darwin, C., 1888, *The Formation of Vegetable Mould through the Action of Worms, with Observations on their Habits*. London, Murray.

Davidson, D. A., and Shackley, M., 1976, *Geoarchaeology: Earth Science and the Past*. London, Duckworth.

Davidson, D. A., Carter, S., Boag, B., Long, D., Tipping, R., and Tyler, A., 1999, Analysis of pollen in soils: processes of incorporation and redistribution of pollen in five soil profile types: *Soil Biology & Chemistry*, v. 31, p. 643–653.

Davis, O., 2012. *Processing and Working with LiDA RData in ArcGIS : A Practical Guide for Archaeologists*. Royal Commission on the Ancient and Historic Monuments of Wales.

de Chardin, P. T., and Young, C.-C., 1929, Preliminary report on the Chou Kou Tien fossiliferous deposit: *Bulletin of Geological Society of China*, v. 8, p. 173–202.

De Coninck, F., 1980, Major mechanisms in formation of spodic horizons: *Geoderma*, v. 24, p. 101–128.

de la Torre, I., Albert, R. M., Macphail, R., McHenry, L. J., Pante, M. C., Rodríguez-Cintas, A., Stanistreet, I. G., and Stollhofen, H., 2018, The contexts and early Acheulean archaeology of the EF-HR paleo-landscape (Olduvai Gorge, Tanzania): *Journal of Human Evolution*, v. 120, p. 274–297.

De Smedt, P., Van Meirvenne, M., Meerschman, E., Saey, T., Bats, M., Court-Picon, M., De Reu, J., Zwertvaegher, A., Antrop, M., Bourgeois, J., De Maeyer, P., Finke, P.A., Verniers, J. and Crombé, P., 2011. Reconstructing palaeochannel morphology with a mobile multicoil electromagnetic induction sensor. *Geomorphology,* 130, p. 136–141. https://doi.org/10.1016/j.geomorph.2011.03.009

Deák, J., Gebhardt, A., Lewis, H. A., Usai, M. R., and Lee, H., 2017, Soils disturbed by vegetation clearance and tillage, *in* Nicosia, C., and Stoops, G., eds., *Archaeological Soil and Sediment Micromorphology*. Chichester, Wiley Blackwell, p. 233–264.

Delporte, H., Delibrias, G., Delpech, F., Donard, E., Heim, J. L., Laville, H., Marquet, J. C., Mourer-Chauvire, C., Paquereau, M. M., and Tuffreau, A., 1984, *Le grand abri de La Ferrassie. Fouilles 1968–1973*. Paris, Editions du Laboratoire de Paléontologie Humaine et de Préhistoire, mémoire 7.

Delvigne, J., and Stoops, G., 1990, Morphology of mineral weathering and neoformation I, *in* Douglas, L. A., ed., *Soil Micromorphology: a basic and applied science*. Amsterdam, Elsevier, p. 471–482.

Dennell, R., 2013, The Nihewan Basin of North China in the Early Pleistocene: continuous and flourishing, or discontinuous, infrequent and ephemeral occupation?: *Quaternary International*, v. 295, p. 223–236.

Denny, C. S., and Goodlett, J. C., 1956, Microrelief resulting from fallen trees, *in* Denny, C. S., ed., *Surficial Geology and Geomorphology of Potter County*, Vol. 288, US Geological Survey professional Paper, p. 59–65.

Derevianko, A., Shunkov, M., Agadjanian, A., Baryshnikov, G., Malaeva, E., Ulianov, V., Kulik, N., Postnov, A., and Anoikin, A., 2003, *Paleoenvironment and Paleolithic human occupation of Gorny Altai: Subsistence and adaptation in the vicinity of Denisova Cave*. Institute of Archaeology and Ethnography SB RAS Press, Novosibirsk.

Derevyanko, A. P., 2015, Human origins: New discoveries, interpretations, and hypotheses: *Herald of the Russian Academy of Sciences*, v. 85, no. 5, p. 381–391.

Devos, Y., 2018, Dark Earth in Brussels (Belgium). A geoarchaeological study [Doctoral Thesis]: Vrije Universiteit Brussel.

Devos, Y., Vrydaghs, L., A., D., and Fechner, K., 2009, An archaeopedological and phytolitarian study of the "Dark Earth" on the Site of Rue de Dinant (Brussels, Belgium): *Catena*, v. 78, no. 3, p. 270–284.

Devos, Y., Nicosia, C., Vrydaghs, L., and Modrie, S., 2013, Studying urban stratigraphy: Dark Earth and a microstratified sequence on the site of the Court of Hoogstraeten (Brussels, Belgium). Integrating archaeopedology and phytolith analysis: *Quaternary International*, v. 315, p. 147–166.

Devos, Y., Nicosia, C., Vrydaghs, L., Speleers, L., van der Valk, J., Marinova, E., Claes, B., Albert, R. M., Esteban, I., Ball, T. B., Court-Picon, M., and Degraeve, A., 2016, An integrated study of Dark Earth from the alluvial valley of the Senne river (Brussels, Belgium): *Quaternary International*, v. 460, p. 175–197.

Devos, Y., Nicosia, C., and Wouters, B., 2020a, Urban geoarchaeology in Belgium: Experiences and innovations: *Geoarchaeology*, v. 35, no. 1, p. 27–41.

Devos, Y., Hodson, M. J., and Vrydaghs, L., 2020b, Auto-fluorescent phytoliths: A new method for detecting heating and fire: *Environmental Archaeology*, v. 26, no. 4, p. 388–405.

Devoy, R. J. N., 1982, Analysis of the geological evidence for the Holocene sea-level movements in southeast England: *Proceedings of the Geological Association*, v. 93, no. 1, p. 65–90.

Dibble, H. L., Abodolahzadeh, A., Aldeias, V., Goldberg, P., McPherron, S. P., and Sandgathe, D. M., 2017, How did hominins adapt to Ice Age Europe without fire?: *Current Anthropology*, v. 58, no. S16, p. S278–S287.

Dibble, H. L., Aldeias, V., Goldberg, P., McPherron, S. P., Sandgathe, D., and Steele, T. E., 2015, A critical look at evidence from La Chapelle-aux-Saints supporting an intentional Neandertal burial: *Journal of Archaeological Science*, v. 53, p. 649–657.

Dibble, H. L., and Rezek, Z., 2009, Introducing a new experimental design for controlled studies of flake formation: results for exterior platform angle, platform depth, angle of blow, velocity, and force: *Journal of Archaeological Science*, v. 36, no. 9, p. 1945–1954.

Dibble, H., McPherron, S. P., Goldberg, P., and Sandgathe, D., 2018a, *The Middle Paleolithic Site of Pech de l'Azé IV, Cave and Karst Systems of the World*: Basel, Springer International.

Dibble, H. L., Sandgathe, D., Goldberg, P., McPherron, S., and Aldeias, V., 2018b, Were Western European Neandertals able to make fire?: *Journal of Paleolithic Archaeology*, published online.

Dick, W. A., and Tabatabai, M. A., 1977, An alkaline oxidation method for the determination of total phosphorus in soils: *Journal of the Soil Science Society of America*, v. 41, p. 511–514.

Dimbleby, G. W., 1962, *The Development of British Heathlands and their Soils*. Oxford, Clarendon Press.

Dimbleby, G. W., 1976, Climate, soil and man: *Philosophical Transactions of the Royal Society, London, Ser. B*, no. 275, p. 197–208.

Dimbleby, G. W., 1985, *The Palynology of Archaeological Sites*. London, Academic Press.

Dimbleby, G. W., and Gill, J. M., 1955, The occurrence of podzols under deciduous woodland in the New Forest: *Forestry*, v. 28, p. 95–106.

Dinies, M., Plessen, B., Neef, R., and Kürschner, H., 2015, When the desert was green: Grassland expansion during the early Holocene in northwestern Arabia: *Quaternary International*, v. 382, p. 293–302.

Dinn, J. and Roseff, R., 1992, Alluvium and archaeology in the Herefordshire valleys, *in* Alluvial Archaeology in Britain: Proceeding of a conference sponsored by the RMC Group plc 2–5 January 1991. London, British Museum, p. 141–155.

Discamps, E., Muth, X., Gravina, B., Lacrampe-Cuyaubère, F., Chadelle, J.-P., Faivre, J.-P., and Maureille, B., 2016, Photogrammetry as a tool for integrating archival data in archaeological fieldwork: Examples from the Middle Palaeolithic sites of Combe-Grenal, Le Moustier, and Regourdou: *Journal of Archaeological Science: Reports*, v. 8, p. 268–276.

Dodonov, A. E., 1995, Geoarchaeology of Palaeolithic sites in loesses of Tajikistan (Central Asia), *in* Johnson, E., ed., *Ancient Peoples and Landscapes*: Lubbock, Museum of Texas Tech University, p. 127–136.

Dodonov, A. E., and Baiguzina, L. L., 1995, Loess stratigraphy of Central Asia; palaeoclimatic and palaeoenvironmental aspects. Aeolian sediments in the Quaternary record: *Quaternary Science Reviews*, v. 14, no. 7–8, p. 707–720.

Dodonov, A. E., Sadchikova, T. A., Sedov, S. N., Simakova, A. N. and Zhou, L. P. 2006. Multidisciplinary approach for paleoenvironmental reconstruction in loess-paleosol studies of the Darai Kalon section, Southern Tajikistan: *Quaternary International*, v. 152–153, p. 48–58.

Doerschner, N., Fitzsimmons, K. E., Blasco, R., Finlayson, G., Rodríguez-Vidal, J., Rosell, J., Hublin, J.-J., and Finlayson, C., 2019, Chronology of the Late Pleistocene archaeological sequence at Vanguard Cave, Gibraltar: Insights from quartz single and multiple grain luminescence dating: *Quaternary International*, v. 501, p. 289–302.

Donahue, J., and Adovasio, J. M., 1990, Evolution of sandstone rockshelters in eastern North America; A geoarchaeological perspective, *in* Lasca, N. P., and Donahue, J., eds., *Archaeological Geology of North America, Centennial Volume No. 4*: Boulder, Geological Society of America, p. 231–251.

Doneus, M., Verhoeven, G., Atzberger, C., Wess, M. and Rusˇ, M. 2014. New ways to extract archaeological information from hyperspectral pixels. *Journal of Archaeological Science*, 52, pp. 84–96. https://doi.org/10.1016/j.jas.2014.08.023

Doran, G. H., 2002, *Windover: Multidisciplinary Investigations of an Early Archaic Florida Cemetery.* Gainesville, FL, University of Florida Press.

Dorn, R. I., 1983, Cation-ratio dating; a new rock varnish age-determination technique: *Quaternary Research (New York)*, v. 20, no. 1, p. 49–73.

Dorn, R. I., 1991, Rock varnish: *American Scientist*, v. 79, no. 6, p. 542–553.

Dorn, R. I., and Oberlander, T. M., 1982, Rock varnish: *Progress in Physical Geography: Earth and Environment*, v. 6, no. 3, p. 317–367.

Douglas, L. A., 1990, *Soil Micromorphology: A Basic and Applied Science, Proceedings of the VIIIth International Working Meeting of Soil Micromorphology, San Antonio, Texas – July 1988*. Amsterdam, Elsevier.

Drescher, H. E., Harms, U., and Huschenbeth, E., 1977, Organochlorines and heavy metals in the harbour seal *Phoca vitulina* from the German North Sea Coast: *Marine Biology*, v. 1, p. 99–106.

Dreslerová, D., Hajnalová, M., Trubač, J., Chuman, T., Kočár, P., Kunzová, E., and Šefrna, L., 2021, Maintaining soil productivity as the key factor in European prehistoric and Medieval farming: *Journal of Archaeological Science: Reports*, v. 35.

Drewett, P. L., 1989, Anthropogenic soil erosion in prehistoric Sussex: excavations at West Heath and Ferring, 1984: *Sussex Archaeological Collections*, v. 127, p. 11–29.

Driessen, P., Deckers, J., Spaargaren, O., and Nachtergaele, F., 2001, *Lecture Notes on the Major Soils of the World*, FAO.

Driskell, B. N., 1994, Stratigraphy and chronology at Dust Cave: *Journal of Alabama Archaeology*, v. 40, no. 1, p. 17–34.

Driskell, B. N., 1996, Stratified Late Pleistocene and Early Holocene deposits at Dust Cave, northwestern Alabama: Paleoindian and Early Archaic Southeast, p. 315–330.

Duchaufour, P., 1982, *Pedology*. London, Allen and Unwin.

Dunn, R. K., and Mazzullo, S. J., 1993, Holocene paleocoastal reconstruction and its relationship to Marco Gonzalez, Ambergris Caye, Belize: *Journal of Field Archaeology* v. 20, no. 2, p. 121–131.

Dunning, N. P., Beach, T. P., and Luzzadder-Beach, S., 2012, Kax and Kol: Collapse and resilience in lowland Maya civilization: *PNAS*, v. 109, p. 3652–3657.

Dunning, N., Rue, D. J., Beach, T., Covich, A., and Traverse, A., 1998, Human-environment interactions in a tropical watershed: the paleoecology of Laguna Tamarindito, El Petén, Guatemala: *Journal of Field Archaeology*, v. 25, no. 2, p. 139–151.

Durand, N., Monger, H. C., Canti, M. G., and Verrecchia, E. P., 2018, Calcium Carbonate Features, *in* Stoops, G., Marcelino, V., and Mees, F., eds., *Interpretation of Micromorphological Features of Soils and Regoliths*. Amsterdam, Elsevier, p. 205–258.

Duval, M., Grün, R., Parés, J. M., Martín-Francés, L., Campaña, I., Rosell, J., Shao, Q., Arsuaga, J. L., Carbonell, E., and de Castro, J. M. B., 2018, The first direct ESR dating of a hominin tooth from Atapuerca Gran Dolina TD-6 (Spain) supports the antiquity of Homo antecessor: *Quaternary Geochronology*, v. 47, p. 120–137.

Edwards, L. E., and Owen, D. E., 2005, North American Commission on Stratigraphic Nomenclature: *AAPG Bulletin*, v. 89, no. 11, p. 1547–1591.

Eidt, R., 1973, A rapid chemical field test for archaeological site surveying: *American Antiquity*, v. 38, no. 2, p. 206–210.

Eidt, R. C., 1977, Detection and Examination of Anthrosols by Phosphate Analysis: *Science*, v. 197, p. 1327–1333.

Eidt, R. C., 1984, *Advances in Abandoned Settlement Analysis: Application to prehistoric Anthrosols in Colombia, South America*, Milwaukee, The Center for Latin America, University of Wisconsin-Milwaukee.

Engelmark, R., 1985, Carbonized seeds in postholes – a reflection of human activity, *in* Edgren, T., and Jungner, H., eds., *Proceedings of the Third Nordic Conference on the Application of Scientific Methods in Archaeology*. 57–69. Helsinki, The Finnish Antiquarian Society, p. 205–209.

Engelmark, R., 1992, A review of farming economy in South Scania based on botanical evidence, *in* Larsson, L., et al., eds., *The Archaeology of the Cultural Landscape: Field Work and Research in a South Swedish Region*. Stockholm, Almqvist & Wiksell International, p. 369–376.

Engelmark, R., and Linderholm, J., 1996, Prehistoric land management and cultivation. A soil chemical study, *in* Mejdahl, V., and Siemen, P., eds., *Proceedings from the 6th Nordic Conference on the Application of Scientific Methods in Archaeology, Esbjerg 1993*, Arkaeologiske Rapporter Number 1. Esbjerg, Esbjerg Museum, p. 315–322.

Engelmark, R., and Linderholm, J., 2008, *Miljöarkeologi Människa och Landskap – en komplicerad dynamik. Projektet Öresundsförbindelsen. (Environmental Archaeology. Man and Landscape – a dynamic interrelation. The Öresund Fixed Link Project)*. Malmö, Kulturmilö.

Engelmark, R., and Viklund, K., 1986, Järnålders-jordbruk in Norrland – Teori och praktik: *Populär Arkeologi*, v. 4, no. 2, p. 22–24.

Entwhistle, J. A., Abrahams, P. W., and Dodgshon, R. A., 1998, Multi-element analysis of soils from Scottish Historical sites. Interpreting land-use history through physical and geochemical analysis of soil: *Journal of Archaeological Science*, v. 25, p. 53–68.

Entwistle, J., Abrahams, P. W., and Dodgshon, R. A., 2000, The geoarchaeological significance and spatial variability of a range of physical and chemical soil properties from a former habitation site, Isle of Skye: *Journal of Archaeological Science*, v. 27, no. 4, p. 287–303.

Enzel, Y., Kadan, G. & Eyal, Y. 2000. Holocene earthquakes inferred from a fan-delta sequence in the Dead Sea graben. Quaternary Research. v. 53, pp. 34–48.

Enzel, Y., Amit, R., Crouvi, O., and Porat, N., 2010, Abrasion-derived sediments under intensified winds at the latest Pleistocene leading edge of the advancing Sinai–Negev erg: *Quaternary Research*, v. 74, no. 1, p. 121–131.

Enzel, Y., and Bar-Yosef, O., 2017, *Quaternary of the Levant: Environments, Climate Change, and Humans*. Cambridge, Cambridge University Press.

Enzel, Y., Bookman, R., Sharon, D., Gvirtzman, H., Dayan, U., Ziv, B., and Stein, M., 2003, Late Holocene climates of the Near East deduced from Dead Sea level variations and modern regional winter rainfall: *Quaternary Research*, v. 60, p. 263–273.

Ervynck, A., Laleman, C., Stoops, G., Bastiaens, J., Deforce, K., De Groote, K., Demiddele, H., Defender, K., Hendrix, V., Langohr, R., Lievois, D., Louwyer, S., Meersschaert, L., Schelvis, J., Ge. Stoops, Gu. Stoops, Van Neer, W., Van Strydonck, M., and Verbruggen, C., 1999, Het "Zwarte

Laag"-project. Ophoggingslaag, straatvuil, baggerspecie, stort of compothoop? Datering, herkomst en betekenis van de "Zwarte Laag" in gent (0.-VI.). *Archaeologia Mediavalis Chronique/Kroniek*, v. 22, p. 64–66.

Evans, D. L., Vis, B. N., Dunning, N. P., Graham, G., and Isendahl, C., 2021, Buried solutions: How Maya urban life substantiates soil connectivity: *Geoderma*, v. 387, 114925.

Evans, D., Pottier, C., Fletcher, R., Hensley, S., Tapley, I., Milne, A., and Barbetti, M., 2007, A comprehensive archaeological map of the world's largest pre-industrial settlement complex at Angkor, Cambodia: *PNAS*, v. 104, no. 36, p. 14277–14282.

Evans, E. E. 1957. *Irish Folk Ways*. London: Routledge and Kegan Paul.

Evans, J. G., 1971, Habitat changes on the calcareous soils of Britain: the impact of Neolithic man, *in* Simpson, D. D. A., ed., *Economy and Settlement in Neolithic and Early Bronze Age Britain and Europe*: Leicester, Leicester University Press, p. 27–74.

Evans, J. G., 1972, *Land Snails in Archaeology*. London, Seminar Press.

Evans, J. G., 1975, *The Environment of Early Man in the British Isles*. London, Paul Elek.

Evans, J. G., 1999, *Land & Archaeology*, Stroud, Tempus, Histories of Human Environment in the British Isles.

Evans, J. G., and Limbrey, S., 1974, The experimental earthwork on Morden Bog, Wareham, Dorset, England: 1963–1972: *Proceedings of the Prehistoric Society*, v. 40, p. 170–202.

Evans, R., 1992, Erosion in England and Wales – the present the key to the past, *in* Bell, M., and Boardman, J., eds., *Past and Present Soil Erosion, Monograph 22*. Oxford, Oxbow, p. 53–66.

Evans, R., and Jones, R. J. A., 1977, Crop marks and soils at two archaeological sites in Britain: *Journal of Archaeological Science*, v. 4, p. 63–76.

Evershed, R. P., 2008, Organic residue analysis in archaeology: the archaeological biomarker revolution: *Archaeometry*, v. 50, no. 6, p. 895–924.

Evershed, R. P., Bethell, P. H., and Walsh, N. J., 1997, 5ß-stigmastanol and related 5ß-stanols as biomarkers of manuring: analysis of modern experimental material and assessment of the archaeological potential: *Journal of Archaeological Science*, v. 24, p. 485–495.

Eyre, S. R., 1968, *Vegetation and Soils: a World Picture*. London, Edward Arnold (Publishers) Ltd.

Fa, D. A., and Sheader, M., 2000, Zonation patterns and fossilization potential of the rocky-shore biota along the Atlantic-Mediterranean interface: a possible framework for environmental reconstruction, *in* Finlayson, C., Finlayson, G., and Fa, D., eds., *Gibraltar during the Quaternary*, Vol. 1: Gibraltar, Gibraltar Government Heritage Publications Monographs, p. 237–251.

Faegri, K., and Iverson, J., 1989, *Textbook of Pollen Analysis*, 4th edition. Chichester, John Wiley & Sons.

Fairbridge, R. W., and Bengtsson, L., 2012, Europe, lakes review, *in* Bengtsson, L., Herschy, R. W., and Fairbridge, R. W., eds., *Encyclopedia of Lakes and Reservoirs*: Dordrecht, Springer, p. 249–258.

Faivre, J.-P., Discamps, E., Gravina, B., Turq, A., Guadelli, J.-L., and Lenoir, M., 2014, The contribution of lithic production systems to the interpretation of Mousterian industrial variability in southwestern France: The example of Combe-Grenal (Dordogne, France): *Quaternary International*, v. 350, p. 227–240.

Falck, T., and Gundersen, J., 2007, *Bebyggelsesplan for Sørengutstikkeren. Plan for arkeologisk overvåking og beredskap (Built-in plan for the Søreng Outlet. (Plan for archaeological monitoring and emergency preparedness)*: Norsk Sjøfartsmuseum.

FAO, 2003, *World Soil Resources*: www.fao.org/ag/agI/agII/wrb/soilres.stm#down.

FAO, 2015, *World reference base for soil resources 2014 (2015 update). International soil classification system for naming soils and creating legends for soil maps*, Vienna, International Union of Soil Science, World Soil Resources Reports.

FAO-Unesco, 1988, *Soil Map of the World*: FAO.

Farrand, W. R., 1975, Sediment analysis of a prehistoric rockshelter: the Abri Pataud: *Quaternary Research (New York)*, v. 5, no. 1, p. 1–26.

Farrand, W. R., 1984, Stratigraphic classification: Living within the law: *Quarterly Review of Archaeology*, v. 5, no. 1, p. 1–5.

Farrand, W. R., 1990, Origins of Quaternary-Pleistocene-Holocene Stratigraphic Terminology, Special Paper 242, *in* Laporte, L. F., ed., *Establishment of a Geologic Framework for Paleoanthropology*: Boulder, Geological Society of America, p. 15–22.

Farres, P. J., Wood, S. J., and Seeliger, S., 1992, A conceptual model of soil deposition and its implications for environmental reconstruction, *in* Bell, M., and Boardman, J., eds., *Past and Present Soil Erosion, Monograph 22*. Oxford, Oxbow, p. 217–226.

Faul, M. L., and Smith, R. T., 1980, Phosphate analysis and three possible Dark Age ecclesiastical sites in Yorkshire: *Landscape History*, v. 2, p. 21–38.

Fechner, K., Baes, R., Louwagie, G., and Gebhardt, A., 2014, Relic Holocene buried colluvial and alluvial deposition in the basins of the Scheldt, the Meuse, the Seine and the Rhine (Belgium, Luxembourg and Northern France). A prospective state of research in rescue excavations avec la collaboration de Deschodt L., Bécu B., Schartz E., *in*: Meylemans E., Poesen J., In't Ven I., eds., 2014. *The Archaeology of Erosion, the Erosion of Archaeology. Conference Brussels, 28–30 April 2008*. Relicta Monographien 9 (VIOE, Brussels): 147–190.

Fechner, K., Langohr, R., and Devos, Y., 2004, Archaeopedological checklist: Proposal for a simplified version for the routine archaeological recording of Holocene rural and urban sites in North-Western Europe, *in* Carver, G., ed., *Digging in the Dirt. Excavation in the new millenium*, International Series 1256. Oxford, British Archaeological Reports, p. 241–254.

Feder, K. L., 1997, Data Preservation: Recording and Collecting, *in* Hester, T. R., Shafer, H. J., and Feder, K. L., eds., *Field Methods in Archaeology*: Mountain View, CA, Mayfield Publishing Co, p. 113–142.

Fedoroff, N., and Goldberg, P., 1982, Comparative micromorphology of two late Pleistocene palaeosols (in the Paris basin): *Catena*, v. 9, p. 227–251.

Fedoroff, N., Courty, M.-A., and Guo, Z., 2018, Palaeosols and Relict Soils, *in* Stoops, G., Marcelino, V., and Mees, F., eds., *Interpretation of Micromorphological Features of Soils and Regoliths*. Amsterdam, Elsevier, p. 821–662.

Feibel, C. S., 2001, Archaeological sediments in lake margin environments, *in* Stein, J. K., and Farrand, W. R., eds., *Sediments in Archaeological Context*. Salt Lake City, UT, The University of Utah Press, p. 127–148.

Fern, C. J. R., 2015, *Before Sutton Hoo: the Prehistoric Remains and Early Anglo-Saxon Cemetery at Tranmer House*, Bromeswell, Suffolk, East Anglian Archaeology.

Ferraris, M., 1997, Ochre remains, *in* Maggi, R., ed., *Arene Candide: a Functional and Environmental Assessment of the Holocene Sequence (Excavations Bernarbo' Brea-Cardini 1940–50)*. Roma, Istituto Italiano di paleontologia Umana, p. 593–598.

Ferring, C. R., 1986, Rates of fluvial sedimentation: implications for archaeological variability: *Geoarchaeology*, v. 1, no. 3, p. 259–274.

Ferring, C. R., 1990, *Late Quaternary geology and geoarchaeology of the upper Trinity River drainage basin, Texas*, Geological Society of America, Annual Meeting Field trip, Dallas, p. 81.

Ferring, C. R., 1992, Alluvial pedology and geoarchaeological research, *in* Holliday, V. T., ed., *Soils in Archaeology: Landscape and Human Occupation*. Washington, DC, Smithsonian Institution Press, p. 1–39.

Ferring, C. R., 1995, Middle Holocene environments, geology and archaeology in the Southern Plains, *in* Bettis, E. A., III, ed., *Archaeological geology of the Archaic period in North America, Special Paper 297*: Boulder, CO, The Geological Society of America, p. 21–35.

Ferring, C. R., 2001, Geoarchaeology in alluvial landscapes, *in* Goldberg, P., Holliday, V. T., and Ferring, C. R., eds., *Earth Sciences and Archaeology*. New York, Kluwer-Plenum, p. 77–106.

Ferring, R., Oms, O., Agustí, J., Berna, F., Nioradze, M., Shelia, T., Tappen, M., Vekua, A., Zhvania, D., and Lordkipanidze, D., 2011, Earliest human occupations at Dmanisi (Georgian Caucasus) dated to 1.85–1.78 Ma: *PNAS*, v. 108, no. 26, p. 10432–10436.

Findlay, D. C., Colborne, G. J. N., Cope, D. W., Harrod, T. R., Hogan, D. V., and Staines, S. J., 1984, *Soils and their use in South West England*, Harpenden, Lawes Agricultural Trust, Soil Survey of England and Wales.

Finlayson, C., Pacheco, F. G., Rodríguez-Vidal, J., Fa, D. A., López, J. M. G., Pérez, A. S., Finlayson, G., Allue, E., Preysler, J. B., and Cáceres, I., 2006, Late survival of Neanderthals at the southernmost extreme of Europe: *Nature*, v. 443, no. 7113, p. 850–853.

Finlayson, C., Fa, D. A., Espejo, F. J., Carrión, J. S., Finlayson, G., Pacheco, F. G., Vidal, J. R., Stringer, C., and Ruiz, F. M., 2008, Gorham's Cave, Gibraltar—the persistence of a Neanderthal population: *Quaternary International*, v. 181, no. 1, p. 64–71.

Fisher, R. V., and Schmincke, H.-U., 2012, *Pyroclastic Rocks*. Berlin, Springer-Verlag.

FitzPatrick, E. A., 1956, An indurated soil horizon formed by permafrost: *Journal of Soil Science*, v. 7, p. 248–254.

FitzPatrick, E. A., 1984, *Micromorphology of Soils*. London, Chapman and Hall.

FitzPatrick, E. A., 1993, *Soil Microscopy and Micromorphology*. Chichester, John Wiley & Sons.

Fitzsimmons, K. E., Stern, N., and Murray-Wallace, C. V., 2014, Depositional history and archaeology of the central Lake Mungo lunette, Willandra Lakes, southeast Australia: *Journal of Archaeological Science*, v. 41, p. 349–364.

Fladmark, K. R., 1982, An Introduction to the Prehistory of British Columbia: *Canadian Journal of Archaeology*, v. 6, p. 95–156.

Fleming, A. 2008. *The Dartmoor Reaves*, 2nd ed. Oxford, Oxbow.

Flügel, E., 2009, *Microfacies of Carbonate Rocks: analysis, interpretation and application*, 2nd edition. Berlin; New York, Springer, xxiii.

Folk, R. L., 1974, *Petrology of Sedimentary Rocks*. Austin, TX, Hemphill Publishing Co.

Fondrillon, M., 2007, *La formation du sol urbain: étude archéologique des terres noires à Tours (4e-12e siècle)*: Université François Rabelais Tours.

Fondrillon, M., Devos, Y., Graz, Y., Laurent, C., Macphail, R., and Vrydaghs, L., 2009, Processus d'urbanisation au Moyen-Age: recherches interdisciplinaires sur les terres noires à Bruxelles et à Tours (5e-14e siècle). *XVIIe colloque du GMPCA, Archéométrie: Ressources, sociétés, biodiversité*, Montpellier, France; May 2009.

Ford, T. D., and Cullingford, C. H. D., 1976, *The Science of Speleology*. London, Academic Press.

Ford, B. M., and Teague, S., 2011, *Winchester – a City in the Making, Oxford Archaeology Monograph No 12*. Oxford, Oxford Archaeology, p. 402.

Ford, T. D., and Pedley, H. M., 1996, A review of tufa and travertine deposits of the world: *Earth Science Reviews*, v. 41, p. 117–175.

Ford, D., and Williams, P. D., 2007, *Karst Hydrogeology and Geomorphology*. Chichester, John Wiley & Sons.

Forget, M. C. L., and Shahack-Gross, R., 2016, How long does it take to burn down an ancient Near Eastern city? The study of experimentally heated mud-bricks: *Antiquity*, v. 90 no. 353 p. 1213–1225.

Fowler, M. J., 2002, Satellite Remote Sensing and Archaeology: a Comparative Study of Satellite Imagery of the Environs of Figsbury Ring, Wiltshire: *Archaeological Prospection*, v. 9, p. 55–69.

Fowler, P. J., and Evans, J. G., 1967, Plough-marks, lynchets and early fields: *Antiquity*, v. 41, p. 289–301.

Fox, C. A., 1985, Micromorphological characterisation of histosols, *in* Douglas, L. A., and Thompson, R., eds., *Soil Micromorphology and Soil Classification, Special Publication Number 15*: Madison, Wisconsin, Soil Science Society of America, p. 85–104.

Frahm, E., 2019, Scanning Electron Microscopy (SEM): Applications in Archaeology, *in* Basu, A., ed., *Encyclopedia of Global Archaeology*, Springer nature Switzerland, p. 6487–6495.

Francus, P., 2004, *Image analysis, sediments and paleoenvironments*. Dordrecht; Boston, Kluwer Academic Publishers, Developments in paleoenvironmental research; v. 7, xviii.

Frechen, M., Dermann, B., Boenigk, W., and Ronen, A., 2001, Luminescence chronology of aeolianites from the section at Givat Olga, coastal plain of Israel: *Quaternary Science Reviews*, v. 20, no. 5–9, p. 805–809.

Frechen, M., Neber, A., Tsatskin, A., Boenigk, W., and Ronen, A., 2004, Chronology of Pleistocene sedimentary cycles in the Carmel Coastal Plain of Israel: *Quaternary International*, v. 121, p. 41–52.

Frederick, C., 2001, Evaluating causality of landscape change: examples from alluviation, *in* Goldberg, P., Holliday, V. T., and Reid Ferring, C., eds., *Earth Sciences and Archaeology*. New York, Kluwer, p. 55–76.

Frederick, C. D., Bateman, M. D., and Rogers, R., 2002, Evidence for eolian deposition in the Sandy Uplands of East Texas and the implications for archaeological site integrity: *Geoarchaeology*, v. 17, p. 191–217.

French, C. A. I., 2003, *Geoarchaeology in Action: Studies in Soil Micromorphology and Landscape Evolution*. London, Routledge.

French, C., 2015, *A Handbook of Geoarchaeological Approaches for Investigating Landscapes and Settlement Sites*. Studying Scientific Archaeology I. Oxford, Oxbow.

French, C., 2016, Colluvial settings, *in* Gilbert, A. S., ed., *Encyclopedia of Geoarchaeology*, Springer, p. 157–170.

French, C., and Milek, K., 2012, The geoarchaeological evidence, *in* Tipper, J., ed., *Experimental Archaeology and Fire: the investigation of a burnt reconstruction at West Stow Anglo-Saxon village*, East Anglian Archaeology 146: Bury St Edmunds, Archaeological Service, Suffolk County Council, p. 77–89.

French, C., and Whitelaw, T. M., 1999, Soil erosion, agricultural terracing and site formation processes at Markiani, Amorgos, Greece: the micromorphological perspective: *Geoarchaeology*, v. 14, no. 2, p. 151–189.

French, C., Sulas, F., and Petrie, C., 2017, Expanding the research parameters of geoarchaeology: *Archaeological and Anthropological Sciences*, v. 9, p. 1613–1626.

Friesem, D. E., Boaretto, E., Eliyahu-Behar, A., and Shahack-Gross, R., 2011, Degradation of mud brick houses in an arid environment: a geoarchaeological model: *Journal of Archaeological Science*, v. 38, no. 5, p. 1135–1147.

Friesem, D. E., Tsartsidou, G., Karkanas, P., and Shahack-Gross, R., 2014a, Where are the roofs? A geo-ethnoarchaeological study of mud brick structures and their collapse processes, focusing on the identification of roofs: *Archaeological and Anthropological Sciences*, v. 6, p. 73–92.

Friesem, D. E., Karkanas, P., Tsartsidou, G., and Shahack-Gross, R., 2014b, Sedimentary processes involved in mud brick degradation in temperate environments: a micromorphological approach in an ethnoarchaeological context in northern Greece: *Journal of Archaeological Science*, v. 41, p. 556–567.

Friesem, D. E., Wattez, J., and Onfray, M., 2017, Earth construction materials, *in* Nicosia, C., and Stoops, G., eds., Archaeological Soil and Sediment Micromorphology. Chichester, Wiley, p. 99–110.

Friesem, D. E., Abadi, I., Shaham, D., and Grosman, L., 2019, Lime plaster cover of the dead 12,000 years ago–new evidence for the origins of lime plaster technology: *Evolutionary Human Sciences*, v. 1.

Frison, G. C., 1974, *The Casper site: a Hell Gap bison kill on the High Plains*. New York, Academic Press, Studies in Archeology.

Fritz, W. J., and Moore, J. N., 1988, *Basics of Physical Stratigraphy and Sedimentology*. New York, Wiley.

Frumkin, A., 2001, The Cave of the Letters sediments; indication of an early phase of the Dead Sea depression?: *Journal of Geology*, v. 109, no. 1, p. 79–90.

Frumkin, A., 2013, Salt karst, *in* Shroder, J. F., ed., *Treatise on Geomorphology*, Vol. 6. San Diego, CA, Academic Press, p. 407–424.

Frumkin, A., Carmi, I., Magaritz, M., Zak, I., and Anonymous, 1991, The Holocene climatic record in the salt caves of Mount Sedom, Israel. Program and abstracts of the 14th international radiocarbon conference: *14th international radiocarbon conference*, v. 33, no. 2, p. 197–198.

Frumkin, A., Bar-Matthews, M., and Vaks, A., 2008, Paleoenvironment of Jawa basalt plateau, Jordan, inferred from calcite speleothems from a lava tube: *Quaternary Research*, v. 70, no. 3, p. 358–367.

Frumkin, A., Shimron, A., and Rosenbaum, J., 2003, Radiometric dating of the Siloam Tunnel, Jerusalem: *Nature*, v. 425, p. 169–171.

Fuchs, M., 2007, An assessment of human versus climatic impacts on Holocene soil erosion in NE Peloponnese, Greece: *Quaternary Research*, v. 67, no. 3, p. 349–356.

Fuchs, M., 2011, *Ausgrabungsprojekt E-18 / Norwegen* (Unpublisert rapport, top. ark., Kulturhistorisk museum, Oslo): Department of Geomorphology, Bayreuth University.

Fuchs, M., and Lang, A., 2001, OSL dating of coarse-grain fluvial quartz using single-aliquot protocols on sediments from NE Peloponnese, Greece: *Quaternary Science Reviews* v. 20 p. 783–787.

Fuchs, M., and Lang, A., 2009, Luminiscence dating of hillslope deposits – a review: *Geomorphology*, v. 109, no. 1–2, p. 17–26.

Fulford, M., and Wallace-Hadrill, A., 1995–6, The House of Amarantus at Pompeii (I, 9, 11–12): an interim report on survey and excavations in 1995–96: *Revista di Studi Pompeiani*, v. VII, p. 77–113.

Fulford, M., and Wallace-Hadrill, A., 1998, Unpeeling Pompeii (Archaeology, chronology, dating): *Antiquity*, v. 72, no. 275, p. 128–145.

Fuller, D. Q., Qin, L., Zheng, Y. F., Zhao, Z. J., Chen, X. G., Hosoya, L. A., and Sun, G. P., 2009, The domestication process and domestication rate in rice: spikelet bases from the Lower Yangtze: *Science*, v. 323, p. 1607–1610.

Fyfe, R. M., Brown, A. G. and Coles, B. J. 2003. Mesolithic to Bronze Age vegetation change and human activity in the Exe Valley, Devon, UK. *Proceedings of the Prehistoric Society*, 69, p. 161–181.

Gabunia, L., Vekua, A., Lordkipanidze, D., Swisher, C. C., III, Ferring, R., Justus, A., Nioradze, M., Tvalchrelidze, M., Antón, S. C., Bosinski, G., Jöris, O., de Lumley, M.-A., Majsuradze, G., and Mouskhelishvili, A., 2000, Earliest Pleistocene hominid cranial remains from Dmanisi, Republic of Georgia: Taxonomy, geological setting, and age: *Science*, v. 288, no. 5468, p. 1019–1025.

Gade, M., Kohlus, J. and Kost, C. 2017. SAR imaging of archaeological sites on intertidal flats in the German Wadden Sea. *Geosciences*, 7, p. 1–14.

Gaffney, C., and Gater, J., 2003, *Revealing the Buried Past: Geophysics for Archaeologists*. Stroud, Gloucestershire, Tempus.

Gaffney, V, Neubauer, W, Garwood, P, Gaffney, C., Löcker, K., Bates, R., De Smedt, P., Baldwin, E., Chapman, H., Hinterleitner, A., Wallner, M., Nau, E., Filzwieser, R., Kainz, J., Trausmuth, T., Schneidhofer, P., Zotti G., Lugmayer, A., Trinks,I. and Corkum, T. 2018. Durrington walls and the Stonehenge Hidden Landscape Project 2010–2016. *Archaeological Prospection*, 25, p. 255– 269.

Gage, M. D., 1999, Vacuum chambers for use in resin impregnation of core samples: *Geoarchaeology*, v. 14, no. 3, p. 307–311.

Gage, M. D., 2000, *Ground-penetrating Radar and Core Sampling at the Moundville Site*. [Masters Thesis]: University of Alabama.

Gage, M. D., and Jones, V. S., 1999, Ground penetrating radar: *Journal of Alabama Archaeology*, v. 45, no. 1, p. 49–61.

Gale, S. J., and Hoare, P. G., 2011, *Quaternary Sediments. Petrographic methods for the study of unlithified rocks*. Caldwell. New Jersey, Blackburn Press.

Galili, E., Benjamin, J., Hershkovitz, I., Weinstein-Evron, M., Zohar, I., Eshed, V., Cvikel, D., Melamed, J., Kahanov, Y., and Bergeron, J., 2017, *Atlit-Yam: A unique 9000 year old prehistoric village submerged off the Carmel Coast, Israel–The SPLASHCOS Field School, 2011, Under the Sea: Archaeology and Palaeolandscapes of the Continental Shelf*. Springer, p. 85–102.

Galinié, H., 2000, *Terres Noires – 1*, Documents. Sciences de la Ville, Vol. 6: Tours, La Maison des Sciences de la Ville, p. 119.

Galinié, H., 2004, L'expression terres noires, un concept d'attente, *in* Verslype, L., ed., *Terres Noires Dark Earth*: Louvain-La-Neuve, Université Catholique de Louvain, p. 1–11.

Galinié, H., Lorans, E., Macphail, R. I., Seigne, J., Fondrillon, M., Laurent, A., and Moreau, A., 2007, Chapter 53. La fouille du square Prosper-Mérimée. The excavation in Prosper-Mérimée Square, *in* Galinié, H., ed., *Tours, antique et médiéval. Lieux de vie Temps de la ville, 30th Supplément: spécial de la collection Recherches sur Tours*. Tours, Revue Archéologique du Centre de la France (FERACF), p. 171–180.

Galway-Witham, J., Cole, J. and Stringer, C. 2019. Aspects of human physical and behavioural evolution during the last 1 million years: *Journal of Quaternary Science*, v. 34 no. 6, p. 355–378.

Gao, X., Zhang, S., Zhang, Y., and Chen, F., 2017, Evidence of hominin use and maintenance of fire at Zhoukoudian: *Current Anthropology*, v. 58, no. S16, p. S267–S277.

Garcia-Suarez, A., Portillo, M., and Matthews, W., 2020, Early animal management strategies during the Neolithic of the Konya Plain, Central Anatolia: integrating micromorphological and microfossil evidence: *Environmental Archaeology: the Journal of Human Palaeoecology*, v. 25, no. 2, p. 208–226.

Gardiner, J., Allen, M. J., Lewis, J. S. C., Wright, J., and Macphail, R. I., 2016, *A Long Blade site at Underdown Lane, Herne Bay, Kent, and a Model for Habitat Use in the British Early Postglacial*. Maidstone, Kent Archaeological Society.

Garfinkel, Y., 1987, Burnt lime products and social implications in the pre-pottery Neolithic B villages of the Near East: *Paléorient*, v. 13, no. 1, p. 69–76.

Garrison, E.G. 2003, *Techniques in Archaeological Geology*, Springer, Berlin, Heidelberg, New York.

Garrison, E. 2016, *Techniques in Archaeological Geology*, 2nd edition, Springer-Verlag, Berlin, Heidelberg.

Garrod, D., and Bate, D. M. A., 1937, *The Stone Age of Mount Carmel*, Vol. 1. Oxford, Clarendon Press.

Gasche, H., and Tunca, O., 1983, Guide to archaeostratigraphic classification and terminology: Definitions and principles: *Journal of Field Archaeology*, v. 10, p. 325–335.

Gaylarde, C. C., Rodríguez, C. H., Navarro-Noya, Y. E., and Ortega-Morales, B. O., 2012, Microbial Biofilms on the Sandstone Monuments of the Angkor Wat Complex, Cambodia: *Current Microbiology*, v. 64, no. 2, p. 85–92.

Gé, T., Courty, M.-A., Matthews, W., and Wattez, J., 1993, Sedimentary formation processes of occupation surfaces, *in* Goldberg, P., Nash, D. T., and Petraglia, M. D., eds., *Formation Processes in Archaeological Contexts, Monographs in World Archaeology No. 17*: Madison, Wisconsin, Prehistory Press, p. 149–163.

Gebhardt, A., 1990, *Evolution du Paleopaysage Agricole dans Le Nord-Ouest de la France: apport de la micromorphologie*: L'Université de Rennes I.

Gebhardt, A., 1992, Micromorphological analysis of soil structural modification caused by different cultivation implements, *in* Anderson, P. C., ed., *Prehistoire de l'Agriculture: nouvelles approaches experimentales et ethnographiques, Monographie de CRA No. 6*: Paris, Centre Nationale de la Recherche Scientifique, p. 373–392.

Gebhardt, A., 1993, Micromorphological evidence of soil deterioration since the mid-Holocene at archaeological sites in Brittany, France: *The Holocene*, v. 3, no. 4, p. 331–341.

Gebhardt, A., 1995, Soil micromorphological data from traditional and experimental agriculture, *in* Barham, A. J., and Macphail, R. I., eds., *Archaeological Sediments and Soils: Analysis, Interpretation and management*. London, Institute of Archaeology, p. 25–40.

Gebhardt, A., 1997, "Dark Earth": Some results in Rescue Archaeological Context in France, *in* Macphail, R. I., and Acott, T., eds., *unpublished Bulletin 1 of the Archaeological Soil Micromorphology Working Group, Vol. 1995–1997*: Greenwich, University of Greenwich, p. 45–47.

Gebhardt, A., 2007, Impact of charcoal production activities on soil profiles: the micromorphological point of view: *ARCHAEOSCIENCES, revue d'archéométrie*, v. 31, p. 127–136.

Gebhardt, A., and Langohr, R., 1999, Micromorphological study of construction materials and living floors in the medieval motte of Werken (West Flanders, Belgium): *Geoarchaeology*, v. 14, no. 7, p. 595–620.

Gebhardt, A., and Langohr, R., 2015, Traces de roulage ou de labour ? Le diagnostic micromorphologique: *ArchéoSciences*, v. 39, p. 31–38.

Gebhardt, A., and Marguerie, D., 2006, les sols, leur couvert végétale et leur utilisation au Néolithique, *in* Le Roux, C.-T., ed., Monuments mégalithiques à Locmariaquer Lwe Long Tumulus d'Er Grah dans son Environnement, Gallia Préhistoire sup XXXVIII, CNRS Editions, p. 14–23.

Gebhardt, A., Fechner, K., and Occhietti, S., 2014, Grandes phases de pédogenèse, d'érosion et d'anthropisation des sols au cours de la seconde moitié de l'Holocène en Lorraine (France): *ArchéoSciences*, v. 38, no. 1, p. 7–29.

Gerson, R., and Amit, R., 1987, Rates and modes of dust accretion and deposition in an arid region; the Negev, Israel, *in Proceedings Special Scientific Meeting of the Geological Society of London: Desert Sediments, Ancient and Modern*. London, Vol. 35, p. 157–169.

Gifford, D. P., 1978, Ethnoarchaeological observations of natural processes affecting cultural materials, *in* Gould, R. (Ed.), Explorations in Etnoarchaeology, University of New México Press, Albuquerque, pp. 77–102.

Gifford, D. P., 1985, The third dimension in site structure: an experiment in trampling and vertical dispersal: *American Antiquity*, v. 50, p. 803–818.

Gilbert, A. S., Goldberg, P., Holliday, V. T., Mandel, R. D., and Sternberg, R. S., 2017, *Encyclopedia of Geoarchaeology*: Dordrecht, Springer Netherlands.

Gilbertson, D. D., 1995, Studies of lithostratigraphy and lithofacies: a selective review of research developments in the last decade and their applications to geoarchaeology, *in* Barham, A. J., and Macphail, R. I., eds., *Archaeological Sediments and Soils: analysis, interpretation and management*. London, Institute of Archaeology, p. 99–145.

Gillieson, D., 2009, *Caves: Processes, Development and Management*. Oxford, John Wiley & Sons.

Gillings, M., Haciguzeller, P. and Lock, G. 2020. *Archaeological Spatial Analysis*. Routledge: Abingdon.

Gimingham, C. H., 1975, *Ecology of Heathlands*. London, Chapman and Hall Ltd.

Giosan, L., Clift, P., Macklin, M. G., Fuller, D. Q., Constantinescu, S., Durcan, J. A., Stevesn, T., Duller, G., Tabrez, A. R., Gangal, K., Adhikari, R., Alizai, A., Flip, F., Van Laningham, S., and Syvitski, J., 2012, Fluvial landscapes of the Harappan civilization: *PNAS*, v. 109, p. 10138–10139.

Gischler, E., and Hudson, H. J., 2004, Holocene development of the Belize Barrier Reef: *Sedimentary Geology*, v. 164, p. 223–236.

Giuntoli, S., 1994, *The Golden Book of Pompeii, Herculaneum*, Mt. Vesuvius, Florence, Casa Editrice Bonechi.

Gladfelter, B. G., 1992, Soil properties of sediments in Wadi Feiran, Sinai: A geoarchaeological interpretation, *in* Holliday, V. T., ed., *Soils in Archaeology: Landscape Evolution and Human Occupation*. Washington, DC, Smithsonian Institution Press, p. 169–192.

Gladfelter, B. G., 2001, Archaeological sediments in humid alluvial environments, *in* Stein, J. K., and Farrand, W. R., eds., *Sediments in Archaeological Context*. Salt Lake City, UT, The University of Utah Press, p. 93–125.

Glaser, B., and Woods, W. I., 2004, *Amazonian Dark Earths: Exploration in Space and Time*. New York, Springer.

Glob, P. V., 1969, *The Bog People; Iron Age Man Preserved*. Ithaca, NY, Cornell University Press.

Goder-Goldberger, M., Crouvi, O., Caracuta, V., Horwitz, L. K., Neumann, F. H., Porat, N., Scott, L., Shavit, R., Jacoby-Glass, Y., and Zilberman, T., 2020, The Middle to Upper Paleolithic transition in the southern Levant: New insights from the late Middle Paleolithic site of Far'ah II, Israel: *Quaternary Science Reviews*, v. 237, p. 106304.

Goldberg, P. 1969, Analyses of sediments of Jerf 'Ajla and Yabrud rockshelters Syria, In: Ters, M. (ed) *Union Internationale pour L'Etude du Quaternaire, VIII Congres INQUA*, p. 747–754, Paris.

Goldberg, P., 1973, *Sedimentology, Stratigraphy and Paleoclimatology of et-Tabun Cave, Mount Carmel, Israel*, [Doctoral dissertation]: The University of Michigan.

Goldberg, P., 1977a, Late Quaternary Stratigraphy of Gebel Maghara, *in* Bar-Yosef, O., and Phillips, J. L., eds., *Prehistoric Investigations in Gebel Maghara, Northern Sinai, Qedem, Vol. 7*: Jerusalem, Hebrew University, p. 11–31.

Goldberg, P., 1977b, Nahal Aqev (D35) Stratigraphy and Environment of Deposition, *in* Marks, A. E., ed., *Prehistory and Paleoenvironments in the Central Negev*, Vol. 2, p. 56–60.

Goldberg, P., 1979a, Micromorphology of Pech-de-l'Azé II sediments: *Journal of Archaeological Science*, v. 6, p. 1–31.

Goldberg, P., 1979b, Micromorphology of sediments from Hayonim Cave, Israel: *Catena*, v. 6, no. 2, p. 167–181.

Goldberg, P., 1979c, Geology of Late Bronze Age mudbrick from Tel Lachish: *Tel Aviv, Journal of the Tel Aviv Institute of Archaeology*, v. 6, p. 60–71.

Goldberg, P., 1980, Micromorphology in archaeology and prehistory: *Paléorient*, v. 6, p. 159–164.

Goldberg, P., 1981, Applications of micromorphology in archaeology, *in* Bullock, P., and Murphy, C. P., eds., *Soil Micromorphology. Vol. 1. Techniques and Applications*: Berkhamsted, A B Academic Publishers, p. 139–150.

Goldberg, P., 1983, The Geology of Boker Tachtit, Boker, and Their Surroundings, *in* Marks, A. E., ed., *Prehistory and Paleoenvironments in the Central Negev, Israel*, Vol. III: Dallas, Department of Anthropology, Institute for the Study of Earth and Man, Southern Methodist University, p. 39–62.

Goldberg, P., 1984, Late Quaternary history of Qadesh Barnea Northeastern Sinai: *Zeitschrift für Geomorphologie N.F.*, v. 28 (2), p. 193–217.

Goldberg, P., 1986, Late Quaternary environmental history of the Southern Levant: *Geoarchaeology*, v. 1, no. 3, p. 225–244.

Goldberg, P., 1987, The Geology and Stratigraphy of Shiqmim, *in* Levy, T. E., ed., *Shiqmim 1: Studies Concerning Chalcolithic Societies in the Northern Negev Desert, Israel (1982–1984)*. Oxford, B.A.R., p. 35–43 and p. 435–444.

Goldberg, P., 1988, The Archaeologist as viewed by the Geologist: *Biblical Archaeologist*, December, p. 197–202.

Goldberg, P., 1994, Interpreting Late Quaternary continental sequences in Israel, *in* Bar-Yosef, O., and Kra, R. S., eds., *Late Quaternary Chronology and Paleoclimates of the Eastern Mediterranean*. Tucson, AZ, Radiocarbon, p. 89–102.

Goldberg, P., 1996, Micromorphological Analysis of Selected Samples from the Alamo, *in* Meissner, B., ed., *Alamo restoration and Conservation Project, Archaeological Survey Report 245*: San Antonio, Centre for Archaeological Research, The University of Texas at San Antonio, p. 115–120.

Goldberg, P., 2000a, Micromorphological aspects of site formation at Keatley Creek, *in* Hayden, B., ed., *The Ancient Past of Keatley Creek*: Simon Fraser University. Burnaby, B.C., Archaeology Press, p. 79–95.

Goldberg, P., 2000b, Micromorphology and site formation at Die Kelders Cave 1, South Africa: *Journal of Human Evolution*, v. 38, p. 43–90.

Goldberg, P., 2001a, Geoarchaeology: *Geotimes*, v. 46, no. 7, p. 38–39.

Goldberg, P., 2001b, Some micromorphological aspects of prehistoric cave deposits: *Cahiers d'archéologie du CELAT*, v. 10, série archéometrie 1, p. 161–175.

Goldberg, P., and Aldeias, V., 2018, Why does (archaeological) micromorphology have such little traction in (geo)archaeology?: *Archaeological and Anthropological Sciences*, v. 10, p. 269–278.

Goldberg, P., and Arpin, T., 1999, Micromorphological analysis of sediments from Meadowcroft Rockshelter, Pennsylvania: *Implications for radiocarbon dating: Journal of Field Archaeology*, v. 26, no. 3, p. 325–342.

Goldberg, P. and Bar-Yosef, O. 1995. Sedimentary environments of prehistoric sites in Israel and the Southern Levant, in, Johnson, E., ed., *Ancient Peoples and Landscapes*. Lubbock, TX, Museum of Texas Tech University, p. 29–49.

Goldberg, P., and Bar-Yosef, O., 1998, Site formation processes in Kebara and Hayonim Caves and their significance in Levantine prehistoric caves, *in* Akazawa, T., Aoki, K., and Bar-Yosef, O., eds., *Neandertals and Modern Humans in Western Asia*. New York, Plenum, p. 107–125.

Goldberg, P., and Bar-Yosef, O., 2019, Cave dwellers in Southwest Asia – Chapter 25, *in* White, W. B., Culver, D. C., and Pipan, T., eds., *Encyclopedia of Caves*, 3rd edition. San Diego, CA, Academic Press, p. 218–222.

Goldberg, P., and Berna, F., 2010, Micromorphology and context: *Quaternary International*, v. 214, no. 1–2, p. 56–62.

Goldberg, P., and Brimer, B., 1983, Late Pleistocene Geomorphic Surfaces and Environmental History of Avdat/Havarim Area, Nahal Zin, *in* Marks, A. E., ed., *Prehistory and Paleoenvironments in the Central Negev, Israel*, Vol. III: Dallas, Department of Anthropology, Institute for the Study of Earth and Man, Southern Methodist University, p. 1–13.

Goldberg, P., and Bull, P. A., 1982, Scanning electron microscope analysis of sediments from Tabun Cave, Mount Carmel, Israel: *Eleventh international congress on sedimentology*, v. 11, p. 143–144.

Goldberg, P., and Holliday, V. T., 1998, Geology and stratigraphy of the Wilson-Leonard site, *in* Collins, M. B., ed., *Wilson-Leonard, An 11,000-year Archeological Record of Hunter-Gatherers in Central Texas, Studies in Archaeology*, Vol. 31. Austin, TX, Archeological Research Laboratory, University of Texas at Austin, p. 77–121.

Goldberg, P., and Macphail, R. I., 2000, Micromorphology of Sediments from Gibraltar Caves: Some Preliminary Results from Gorham's Cave and Vanguard Cave, *in* Finlayson, C., Finlayson, G., and Fa, D., eds., *Gibraltar during the Quaternary: the southernmost part of Europe in the last two million years*: Gibraltar, Gibraltar Government Heritage Publications, p. 93–108.

Goldberg, P., and Macphail, R. I., 2003, Short contributions: strategies and techniques in collecting micromorphology samples: Geoarchaeology: *an International Journal*, v. 18, no. 5, p. 571–578.

Goldberg, P., and Macphail, R. I., 2012, Gorham's Cave sediment micromorphology, *in* Barton, R. N. E., Stringer, C., and Finlayson, C., eds., *Neanderthals in Context. A report of the 1995–1998 excavations at Gorham's and Vanguard Caves, Gibraltar, Monograph 75*. Oxford, Oxford University School of Archaeology, p. 50–61; Appendix 52: 314–321; Color figures at: www.arch.ox.ac.uk/gibraltar.

Goldberg, P., and Macphail, R. I., 2016, Microstratigraphy, *in* Gilbert, A. S., ed., *Encyclopedia of Geoarchaeology*: Dordrecht, Springer Scientific, p. 532–537.

Goldberg, P., McPherron, S.J., Dibble, H.L., Sandgathe, D.M., 2018. Stratigraphy, Deposits, and Site Formation, in: Dibble, H.L., McPherron, S.J., Goldberg, P., Sandgathe, D.M. (Eds.), *The Middle Paleolithic Site of Pech de l'Azé IV*, Springer, pp. 21–74.

Goldberg, P., and Nathan, Y., 1975, The Phosphate Minerology of et-Tabun Cave, Mount Carmel, Israel: *Mineralogical Magazine*, v. 40, p. 253–258.

Goldberg, P., and Rosen, A., 1987, Early Holocene Paeloenvironments, *in* Levy, T. E., ed., *Shiqmim 1: Studies Concerning Chalcolithic Societies in the Northern Negev Desert, Israel (1982–1984)*. Oxford, B.A.R., p. 23–33 and 434.

Goldberg, P., and Sherwood, S. C., 1994, Micromorphology of Dust Cave sediments: some preliminary results: *Journal of Alabama Archaeology*, v. 40, no. 1, p. 57–65.

Goldberg, P., and Sherwood, S. C., 2006, Deciphering human prehistory through the geoarcheological study of cave sediments: *Evolutionary Anthropology*, v. 15, no. 1, p. 20–36.

Goldberg, P., and Whitbread, I., 1993, Micromorphological study of a Bedouin tent floor, *in* Goldberg, P., Nash, D. T., and Petraglia, M. D., eds., *Formation Processes in Archaeological Context, Monographs in World Archaeology No. 17*: Madison, Prehistory Press, p. 165–188.

Goldberg, P., Gould, B., Killebrew, A., and Yellin, J., 1986, The provenance of Late Bronze Age ceramics from Deir el Balah: comparison of the results of neutron activation analysis and thin section analysis, *in* Olin, J. S., and Blackman, M. J., eds., *Proceedings of 24th International Archaeometry Symposium (Washington DC 1984)*, p. 341–351.

Goldberg, P., Lev-Yadun, S., and Bar-Yosef, O., 1994, Petrographic thin sections of archaeological sediments: a new method for palaeobotanical studies: *Geoarchaeology*, v. 9, no. 3, p. 243–257.

Goldberg, P., Weiner, S., Bar-Yosef, O., Xu, Q., and Liu, J., 2001, Site formation processes at Zhoukoudian, China: *Journal of Human Evolution*, v. 41, p. 483–530.

Goldberg, P., Schiegl, S., Meligne, K., Dayton, C., and Conard, N. J., 2003, Micromorphology and site formation at Hohle Fels Cave, Swabian Jura, Germany: *Eiszeitalter und Gegenwart*, v. 53, p. 1–25.

Goldberg, P., Laville, H., Meignen, L., and Bar-Yosef, O., 2007, Stratigraphy and geoarchaeological history of Kebara Cave, Mount Carmel, *in* Bar-Yosef, O., and Meignen, L., eds., *Kebara Cave Mt. Carmel, Israel: The Middle and Upper Paleolithic Archaeology, Part 2*. Cambridge, American School of Prehistoric Research Bulletin 49. Peabody Museum of Archaeology and Ethnology Harvard University, p. 49–89.

Goldberg, P., Berna, F., and Macphail, R. I., 2009a, Comment on "DNA from Pre-Clovis Human Coprolites in Oregon, North America": *Science*, v. 325, p. 148-c.

Goldberg, P., Miller, C. E., Schiegl, S., Berna, F., Ligouis, B., Conard, N. J., and Wadley, L., 2009b, Bedding, hearths, and site maintenance in the Middle Stone Age of Sibudu Cave, KwaZulu-Natal, South Africa: *Archaeological and Anthropological Sciences*, v. 1, p. 95–122.

Goldberg, P., Dibble, H., Berna, F., Sandgathe, D., McPherron, S., and Turq, A., 2012, New evidence on Neandertal use of fire: Examples from Roc de Marsal and Pech de l'Azé IV: *Quaternary International*, v. 247, no. 1, p. 325–340.

Goldberg, P., Berna, F., and Chazan, M., 2015, Deposition and Diagenesis in the Earlier Stone Age of Wonderwerk Cave, Excavation 1, South Africa: *African Archaeological Review*, v. 32, p. 613–643.

Goldberg, P., Aldeias, V., Dibble, H., McPherron, S., Sandgathe, D., and Turq, A., 2017a, Testing the Roc de Marsal Neandertal "Burial" with Geoarchaeology: *Archaeological and Anthropological Sciences*, v. 9, p. 1005–1015.

Goldberg, P., Miller, C. E., and Mentzer, S. M., 2017b, Recognizing fire in the Paleolithic archaeological record: *Current Anthropology* v. 58, no. S16, p. S175–S190.

Goldberg, P., McPherron, S. J., Dibble, H. L., and Sandgathe, D. M., 2018, Stratigraphy, deposits, and site formation, *in* Dibble, H. L., McPherron, S. J., Goldberg, P., and Sandgathe, D. M., eds., *The Middle Paleolithic Site of Pech de l'Azé IV*, Springer, p. 21–74.

Goldberg, P., Conard, N. J., and Miller, C. E., 2019, Geißenklösterle Stratigraphy and Micromorphology, *in* Conard, N. J., Bolus, M., and Münzel, S., eds., *Geißenklösterle: Chronostratigraphie, Paläoumwelt und Subsistenz im Mittel- und Jungpaläolithikum der Schwäbishen Alb*: Tübingen, Kerns Verlag, p. 25–61.

Goldman-Finn, N. S., 1994, Dust Cave in regional context: *Journal of Alabama Archaeology*, v. 40, no. 1, p. 212–231.

Goldsmith, Y., Stein, M., and Enzel, Y., 2014, From dust to varnish: Geochemical constraints on rock varnish formation in the Negev Desert, Israel: *Geochimica et Cosmochimica Acta*, v. 126, p. 97–111.

Gollwitzer, M., 2012, Brønner, graver og bosetningsspor fra bronsealder til middelalder på Hesby (lok. 13), *in* Mjærum, A., and Gjerpe, L.-E., eds., *Dyrking, bosetninger og graver i Stokke og Sandefjord, Vol. 1*: Bergen, Fagbokforlaget, p. 107–215.

Gómez, M. J. B., and di Monti, J. C. G., 2009, La minería del lapis specularis y su relación con las ciudades romanas de Segóbriga, Ercávica y Valeria, *in Proceedings La ciudad romana de Valeria (Cuenca)*, Universidad de Castilla-La Mancha, p. 211–226.

Goodfriend, G. A., 1990, Rainfall in the Negev Desert during the middle Holocene, based on ^{13}C of organic matter in land snail shells: *Quaternary Research (New York)*, v. 34, no. 2, p. 186–197.

Goodman, D. and Piro, S. 2013. *GPR remote sensing in archaeology*. Springer: Berlin.

Goren, Y., and Goldberg, P., 1991, Petrographic thin sections and the development of Neolithic plaster production in Northern Israel: *Journal of Field Archaeology*, v. 18, p. 131–138.

Goren-Inbar, N., Alperson-Afil, N., Sharon, G., and Herzlinger, G., 2018, *The Site of Gesher Benot Ya'aqov, The Acheulian Site of Gesher Benot Ya'aqov Vol. IV*. Amsterdam, Springer, p. 21–41.

Goring-Morris, A. N., 1987, *At the Edge: Terminal Pleistocene Hunter-Gatherers in the Negev and Sinai*. Oxford, BAR International Series.

Goring-Morris, N., 1993, From foraging to herding in the Negev and Sinai: the Early to Late Neolithic transition: *Paléorient*, v. 19, no. 1, p. 65–89.

Goring-Morris, A. N., and Goldberg, P., 1990, Late Quaternary dune incursions in the Southern Levant: archaeology, chronology and paleoenvironments: *Quaternary International*, v. 5, p. 115–137.

Goring-Morris, N., and Belfer-Cohen, A., 1997, The articulation of cultural processes and late Quaternary environmental changes in Cisjordan: *Paléorient*, v. 23, no. 2, p. 71–93.

Gose, W., 2000, Paleomagnetic studies of burned rocks: *Journal of Archaeological Science*, v. 27, p. 409–421.

Goudie, A. S., 2008, The history and nature of wind erosion in deserts: *Annual Review of Earth and Planetary Sciences* v. 36, p. 97–119.

Goudie, A. S., and Middleton, N. J., 2001, Saharan dust storms; nature and consequences: *Earth-Science Reviews*, v. 56, no. 1–4, p. 179–204.

Graham, E., 1998, Metaphor and metamorphism: some thoughts on environmental metahistory, *in* Balée, W., ed., *Advances in Historical Ecology*: New York, Columbia University Press.

Graham, E., 2006, A Neotropical Framework for Terra Preta, *in* Balée, W., and Erickson, C. L., eds., *Time and Complexity in Historical Ecology*. New York, Columbia University Press, p. 57–86.

Graham, E., and Pendergast, D. M., 1989, Excavations at the Marco Gonzalez Site, Abergris Cay, Belize, 1986: *Journal of Field Archaeology*, v. 16, p. 1–16.

Graham, E., Macphail, R. I., Crowther, J., Turner, S., Stegemann, J., Arroyo-Kalin, M., Duncan, H., Austin, P., Whittet, R., and Rosique, C., 2016, Past and future earth: archaeology and soil studies on Ambergris Caye, Belize: *Archaeology International*, v. 19, p. 97–108.

Graham, E., Macphail, R. I., Turner, S., Crowther, J., Stegemann, J., Arroyo-Kalin, M., Duncan, L., Whittet, R., Rosique, C., and Austin, P., 2017, The Marco Gonzalez Maya site, Ambergris Caye, Belize: Assessing the impact of human activities by examining diachronic processes at the local scale, *Quaternary International*, Vol. 437 B, p. 115–142.

Graham, I. D. G., and Scollar, I., 1976, Limitations on magnetic prospection in archaeology imposed by soil properties. *Archaeo-Physika*, v. 6, p. 1–124.

Griffin, J.W. 1974, *Investigations in Russell Cave*: Russell Cave National Monument, Alabama, National Park Service.

Grimes, W. F., 1968, *The Excavations of Roman and Medieval London*. London, Routledge and Kegan Paul.

Grøn, O., and Kuznetsov, O., 2004, What is a hunter-gatherer settlement? An ethno-archaeological and interdisciplinary approach, *Section 7 The Mesolithic. Acts of the XIVth UISPP Congress, University of Liège, Belgium, 2–8 September 2001*, BAR International Series 1302. Oxford, British Archaeological Reports, p. 47–53.

Grøn, O., Aurdal, L., Christensen, F., Solberg, R., Macphail, R., Lous, J., and Loska, A., 2005, *Locating invisible cultural heritage sites in agricultural fields. Development of methods for satellite monitoring of cultural heritage sites – report 2004*: Norwegian Directorate for Cultural Heritage.

Grøn, O., Boldreel, L. O., Smith, M. F., Joy, S., Boumda, R. T., Mäder, A., Bleicher, N., Madsen, B., Cvikel, D., Nilsson, B., et al. 2021, Acoustic mapping of submerged Stone Age sites – A HALD approach: *Remote Sensing*, v. 13, no. 3, p. 445.

Guccione, M. J., Sierzchula, M. C., Lafferty, R. H., and Kelley, D., 1998, Site Preservation along an active meandering and avulsing river: the Red River, Arkansas: *Geoarchaeology*, v. 13, no. 5, p. 475–500.

Guélat, M., and Federici-Schenardi, M., 1999, Develier-Courtételle (Jura) L'histoire d'une cabane en fosse reconstituée grâce à la micromorphologie: *Helvetica Archaeologica*, v. 118/119, p. 58–63.

Guérin, G., Frouin, M., Talamo, S., Aldeias, V., Bruxelles, L., Chiotti, L., Dibble, H. L., Goldberg, P., Hublin, J.-J., Jain, M., Lahaye, C., Madelaine, S., Maureille, B., McPherron, S. J. P., Mercier, N., Murray, A. S., Sandgathe, D., Steele, T. E., Thomsen, K. J., and Turq, A., 2015, A multi-method luminescence dating of the Palaeolithic sequence of La Ferrassie based on new excavations adjacent to the La Ferrassie 1 and 2 skeletons: *Journal of Archaeological Science*, v. 58, p. 147–166.

Guillet, B., 1982, Study of turnover of soil organic matter using radio-isotopes, *in* Bonneau, M., and Souchier, B., eds., *Constituents and Properties of Soils*. London, Academic Press, p. 238–257.

Gundersen, I. M., 2016, Gård og utmark i Gudbrandsdalen. *Arkeologiske undersøkelser i Fron 2011–2012*. Kristiansand, Portal forlag.

Gur-Arieh, S., Shahack-Gross, R., Maeir, A. M., Lehmann, G., Hitchcock, L. A., and Boaretto, E., 2014, The taphonomy and preservation of wood and dung ashes found in archaeological cooking installations: case studies from Iron Age Israel: *Journal of Archaeological Science*, v. 46, p. 50–67.

Gustavs, S., 1998, Spätkaiserzeitliche Baubefunde von Klein Köris, Lkr. Dahme-Spreewald, in Henning, J., and Leube, A., eds., *Haus und Hof im östlichen Germanien (Berlin, 1994), Volume Universtätsforschungen zur prähistorischen Archäologie Band 50 and Schriften zur Archäologie der germanischen und slawischen Frühgeschichte Band 2*. Bonn, Dr Rudolf Habelt GmbH, p. 40–66.

Gutiérrez-Rodríguez, M., Goldberg, P., Peinado, F. J. M., Schattner, T., Martini, W., Orfila, M., and Acero, C. B., 2019, Melting, bathing and melting again. Urban transformation processes of the Roman city of Munigua: the public thermae: *Archaeological and Anthropological Sciences*, v. 11, no. 1, p. 51–67.

Gutiérrez-Rodríguez, M., Toscano, M., and Goldberg, P., 2018, High-resolution dynamic illustrations in soil micromorphology: A proposal for presenting and sharing primary research data in publication: *Journal of Archaeological Science: Reports*, v. 20, p. 565–575.

Gutiérrez-Zugasti, I., Andersen, S. H., Araújo, A. C., Dupont, C., Milner, N., and Monge-Soares Antonio M, A. M., 2011, Shell midden research in Atlantic Europe: State of the art, research problems and perspectives for the future: *Quaternary International*, v. 239, no. 1–2, p. 70–85.

Gyoung-Ah Lee, G. W. C., Li Liu, and Xingcan Chen, 2007, Plants and people from the Early Neolithic to Shang periods in North China: *PNAS*, v. 104, no. 3, p. 1087–1092.

Haaland, M. M., Czechowski, M., Carpentier, F., Lejay, M., and Vandermeulen, B., 2019, Documenting archaeological thin sections in high-resolution: A comparison of methods and discussion of applications: *Geoarchaeology*, v. 34, no. 1, p. 100–114.

Haaland, M. M., Friesem, D. E., Miller, C. E., and Henshilwood, C. S., 2017, Heat-induced alteration of glauconitic minerals in the Middle Stone Age levels of Blombos Cave, South Africa: Implications for evaluating site structure and burning events: *Journal of Archaeological Science*, v. 86, p. 81–100.

Haaland, M. M., Miller, C. E., Unhammer, O. F., Reynard, J. P., van Niekerk, K. L., Ligouis, B., Mentzer, S. M., and Henshilwood, C. S., 2020, Geoarchaeological investigation of occupation deposits in Blombos Cave in South Africa indicate changes in Site use and settlement dynamics in the southern Cape during MIS 5b-4: *Quaternary Research*, v. 100, p. 170–223.

Haesaerts, P., and Teyssandier, N., The early Upper Paleolithic occupations of Willendorf II (Lower Austria): a contribution to the chronostratigraphic and cultural context of the beginning of the Upper Paleolithic in Central Europe, *in The Chronology of the Aurignacia and of the Transitional Technocomplexes: dating, stratigraphies, cultural implications. Proceedings of Symposium 6.1 of the XIVth Congress of the UISPP, Université of Liège, Belgium, 2–8 September, 2001,* 2003, Vol. **33**, p. 33–51.

Haesaerts, P., Borziak, I., Vhirica, V., Koulakovsa, L., and van der Plicht, J., 2003, The East Carpathian loess record: a reference for the Middle and Late Pleniglaical stratigraphy in Central Europe: *Quaternaire (Paris)*, v. 14, no. 3, p. 163–188.

Haesaerts, P., Mestdagh, H., and Bosquet, D., 1999, The sequence of Remicourt (Hesbaye, Belgium): new insights on the ped- and chronostratigraphy of the Rocourt Soil: *Geologica Belgica*, v. 2/3–4, p. 5–27.

Hageman, J. B., and Bennett, D. A., 2003, Construction of digital elevation models for archaeological applications, *in* Wescot, K. L., and Brandon, R. J., eds., *Practical Applications of GIS for Archaeologists: A Predictive Modelling Toolkit*, p. 113–127.

Haită, C., 2003, Micromorphology. Inhabited space disposition and uses. Analysis of an occupation zone placed outside the dwellings, *in* Popovici, D., ed., *Archaeological Pluridisciplinary Researches at Borduşani-Popină, Pluridisciplinary Researches series VI*: Bucharest, National Museum of Romanian History, Ialomita County Museum, p. 51–74.

Haită, C., 2012, *Sedimentologie şi Micromorfologie. Aplicaţii în Arheologie.* Târgovişte, Cetatea de Scaun.

Hajic, E. R., Mandel, R. D., Ray, J. H., and Lopinot, N. H., 2007, Geoarchaeology of stratified Paleoindian deposits at the Big Eddy site, southwest Missouri, USA: *Geoarchaeology*, v. 22, no. 8, p. 891–934.

Hakanson, L., 1977, The influence of wind, fetch and water depth on the distribution of sediments in Lake Vanern, Sweden: *Canadian Journal of Earth Science*, v. 14, p. 397–412.

Hakanson, L., 2012, Sedimentation processes in lakes, *in* Bengtsson, L., Herschy, R. W., and Fairbridge, R. W., eds., *Encyclopedia of Lakes and Reservoirs*: Dordrecht, Springer, p. 701–710.

Hakanson, L., and Jansson, M., 1983, *Principles of Lake Sedimentology*. Berlin, Springer-Verlag.

Hamblin, W. K., 1996, *Atlas of stereoscopic aerial photographs and remote sensing imagery of North America*, Glenview, IL, Crystal Productions.

Hamilton, D, Marshall, P, Roberts, HM, Bronk Ramsey, C and Cook, G, 2007d Appendix 1 Gwithian: scientific dating in Nowakowski JA et al, Return to Gwithian: shifting the sands of time: *Cornish Archaeology*, v. 46, p. 61–70.

Hamilton-Taylor, J., and Davison, W., 1995, Redox-driven cycling of trace elements in lakes, *in* Lerman, A., Imboden, D. M., and Gat, J. R., eds., *Physics and Chemistry of Lakes*. Berlin, Springer, p. 217–263.

Hannon, N., 2018. Airborne Laser Scanning and Lidar. *Encyclopaedia of Archaeological Sciences*, p. 1–3.

Harden, J. W., 1982, A quantitative index of soil development from field descriptions. Examples from a chronosequence in central California: *Geoderma*, v. 28, p. 1–28.

Harding, J., and Healy, F., 2011, *The Raunds Area Project. A Neolithic and Bronze Age Landscape in Northamptonshire, Vol 2. Supplementary Studies*. Swindon, English Heritage, p. 978.

Harding, P., Ellis, C., and Grant, M. J., 2014, Late Upper Palaeolithic Farndon Fields, p. 12–70.

Harris, E.C. 1979. The laws of archaeological stratigraphy. *World Archaeology*. v. 11, pp. 111–117.

Harris, E.C. 1989, Principles of Archaeological Stratigraphy, Academic Press, London.

Harrower, M., McCorriston, J., and Oches, E. A., 2002, Mapping the Roots of Agriculture in Southern Arabia: the Application of Satellite Remote Sensing, Global Positioning System and Geographic Information System Technologies: *Archaeological Prospection*, v. 9, p. 35–42.

Hartman, G., Hovers, E., Hublin, J.-J., and Richards, M., 2015, Isotopic evidence for Last Glacial climatic impacts on Neanderthal gazelle hunting territories at Amud Cave, Israel: *Journal of Human Evolution*, v. 84, p. 71–82.

Harward, C., Holder, N., Phillpotts, C., and Thomas, C., 2019, *The medieval priory and hospital of St Mary Spital and the Bishopsgate suburb: excavations at Spitalfields Market, London E1, 1991–2007*. London, MOLA.

Hassan, F., 1995, Late Quaternary geology and geomorphology of the area of the vicinity of Ras en Naqb, *in* Henry, D. O., ed., *Prehistoric Cultural Ecology and Evolution: insights from Southern Jordan*. New York, Plenum, p. 23–41.

Hastings, C. M., and Moseley, M. E., 1975, The adobe of Huaca del Sol and Huaca de la Luna: *American Antiquity*, v. 40, p. 196–203.

Haughton, C., and Powlesland, D., 1999, West Heslerton. *The Anglian Cemetery*. London, English Heritage.

Haury, E. W., 1950, The geology and fossil vertebrates of Ventana Cave, *Archaeology and Stratigraphy of Ventanna Cave*, p. 75–126.

Hawley, N., Wang, X., Brownawell, B., and Flood, R., 1996, Resuspension of bottom sediments in Lake Ontario during the unstratified period, 1992–1993: *Journal of Great Lakes Research*, v. 22, p. 707–721.

Hayden, B., 2004, *The ancient past of Keatley Creek. Vol. III: Excavations*: Burnaby, BC, Archaeology Press, Simon Fraser University.

Haynes, C. V., Jr., 1991, Geoarchaeological and paleohydrological evidence for a Clovis-Age drought in North America and its bearing on extinction: *Journal of Quaternary Research*, v. 35, no. 3, p. 438–450.

Haynes, C. V., Jr., 1995, Geochronology of paleoenvironmental change, Clovis type site, Blackwater Draw, New Mexico: *Geoarchaeology*, v. 10, no. 5, p. 317–388.

Hazelden, J., Sturdy, R. G., and Loveland, P. J., 1987, Saline soils in North Kent, *in* Jarvis, M. G., ed., *SEESOIL, Vol. 4*: Bedford, South East Soils Discussion Group, p. 2–23.

Healy, F., and Harding, J., 2007, The Raunds Area Project. *A Neolithic and Bronze Age Landscape in Northamptonshire*. Swindon, English Heritage, p. 324.

Heathcote, A. J., Filstrup, C. T., and Downing, J. A., 2013, Watershed Sediment Losses to Lakes Accelerating Despite Agricultural Soil Conservation Efforts: *PLoS One*, v. 8, no. 1, p. e53554.

Heathcote, J. L., 2002, *An Investigation of the Pedosedimentary Characteristics of Deposits Associated with Managed Livestock*, [Doctoral Dissertation]: University College London.

Heimdahl, J., 2005, *Urbanised nature in the past. Site formation and environmental development in two Swedish towns, AD 1200–1800*, Stockholm, Stockholm University.

Helms, J. G., McGill, S. F., and Rockwell, T. K., 2003, Calibrated, late Quaternary age indices using clast rubification and soil development on alluvial surfaces in Pilot Knob Valley, Mojave Desert, southeastern California: *Quaternary Research*, v. 60, p. 377–393.

Henning, J., 1996, *Landwirtschaft der Franken, Die Franken – Wegbereiter Europas. Vor 1500 Jahren*: König Chlodwig und seine Erben: Mainz, p. 174–185.

Henning, J., 2009, Revolution or Relapse? Technology, Agriculture and Early Medieval Archaeology in Germanic Central Europe, *in* Ausenda, G., Delogu, P., and Wickham, C., eds., *The Langobards Before the Frankish Conquest; an ethnographic perspective, Studies in Historical Archaeoethnology 8*: San Marino, The Boydell Press, p. 151–175.

Henning, J., and Macphail, R. I., 2004, Das karolingische Oppidum Büraburg: Archälogische und mikromorphologische. Stedien zur Funktion einer frümittelalterlichen Bergbefestigung in Nordhessen (The Carolingian times *oppidum* Büraburg: archaeological and soil investigations on the function of an early medieval hillfort in North Hesse), *in* Hänsel, B., ed., *Parerga Praehistorica. Jubiläumsschrift zur Prähistorischen Archäologie 15 Jahre UPA, Band 100*. Bonn, Verlag Dr Rudolf habelt GmbH, p. 221–252.

Henning, J., McCormick, M., and Fischer, T., 2012, Decempagi at the end of antiquity and the fate of the Roman road system in eastern Gaul, in: (ed.), *in* Bidwell, P., ed., *Proceedings of the XXIst International Limes (Roman Frontiers) Congress, 2009 at Newcastle upon Tyne*. Oxford, British Archaeological Reports.

Henshilwood, C. S., d'Errico, F., and Watts, I., 2009, Engraved ochres from the Middle Stone Age levels at Blombos Cave, South Africa: *Journal of Human Evolution*, v. 57, no. 1, p. 27–47.

Henshilwood, C. S., Sealy, J. C., Yates, R., Cruz-Uribe, K., Goldberg, P., Grine, F. E., Klein, R. G., Poggenpoel, C., Niekerk, K. v., and Watts, I., 2001, Blombos Cave, Southern Cape, South Africa: Preliminary Report on the 1992–1999 Excavations of the Middle Stone Age Levels: *Journal of Archaeological Science*, v. 28, p. 421–448.

Herdendor, C. E., 1990, Distribution of the World's Large Lakes, *in* Tilzer, M. M., and Serruya, C., eds., *Large Lakes, Ecological Structure and Function*. Berlin, Springer-Verlag, p. 3–38.

Herschy, R. W., 2012a, Great Lakes, North America, *in* Bengtsson, L., Herschy, R. W., and Fairbridge, R. W., eds., *Encyclopedia of Lakes and Reservoirs*: Dordrecht, Springer, p. 303–314.

Herschy, R.W. 2012b, Trophiclake classification, in Bengtsson, L., Herschy, R. W., and Fairbridge, R. W., eds., *Encyclopedia of Lakes and Reservoirs*: Dordrecht, Springer, p. 813.

Hershkovitz, I., Weber, G. W., Quam, R., Duval, M., Grün, R., Kinsley, L., Ayalon, A., Bar-Matthews, M., Valladas, H., and Mercier, N., 2018, The earliest modern humans outside Africa: *Science*, v. 359, no. 6374, p. 456–459.

Herz, N., and Garrison, E., 1998, *Geological Methods for Archaeology*. New York, Oxford University Press.

Hester, T. R., Shafer, H. J., and Feder, K. L., 1997, *Field Methods in Archaeology*, 7th edition. Mountain View, CA, Mayfield Publishing Company.

Hewitson, C., Ramsey, E., Shaw, M., Hislop, M., and Cuttler, R., 2010, *Archaeological Investigations at Old Hall Street, Wolverhampton, 2000-2007*. Birmingham, Birmingham Archaeology.

Hicks, A., 2015, Medieval Town and Augustinian Friary Settlement c1325-1700. *Canterbury Whitefriars Excavations 1999-2004*. Canterbury, Canterbury Archaeological Trust.

Hicks, A., and Houlistan, M., 2018, Within the walls: the developing town c AD 750-1325. *Canterbury Whitefriars Excavations 1999-2004*. Canterbury, Canterbury Archaeological Trust.

Hill, C. L., 2009, Stratigraphy and sedimentology at Bir Sahara, Egypt: Environments, climate change and the Middle Paleolithic: *Catena*, v. 78, no. 3, p. 250-259.

Hill, C., and Fortí, P., 1997, *Cave Minerals of the World*, 2nd edition. Huntsville, AL, National Speleological Society.

Hill, D., 1998, *Cave Minerals of the World*. Huntsville, AL, National Speleological Society.

Hilton, J., 1985, A conceptual framework for predicting the occurrence of sediment focusing and sediment redistribution in small lakes: *Limnology and Oceanography*, v. 30, no. 6, p. 1131-1143.

Hilton, J., Lishman, J. P., and Allen, P. V., 1986, The dominant processes of sediment distribution and focusing in a small, eutrophic, monomictic lake: *Limnology and Oceanography*, v. 31, no. 1, p. 125-133.

Hilton, M. R., 2002, *Evaluating Site Formation Processes at a Higher Resolution: An Archaeological Case Study in Alaska Using Micromorphology and Experimental Techniques*, [Doctoral Dissertation]: University of California.

Historic England, 2015, *Geoarchaeology. Using earth sciences to understand the archaeological record*, Historic England.

Historic England, 2016a, *Historic England Research*: Historic England Research, no. 3, p. 5.

Historic England, 2016b, *Preserving Archaeological Remains*. London, Historic England.

Historic England, 2020, *Deposit Modelling and Archaeology. Guidance for Mapping Buried Deposits*. London, Historic England.

Hjulström, F., 1939, Transportation of detritus by moving water, *in* Trask, P. D., ed., *Recent Marine Sediments: A symposium*: Tulsa, American Association of Petroleum Geologists, p. 5-31.

Hodge, C., A. H., Burton, R. G. O., Corbett, W. M., Evans, R., George, H., Heaven, F. W., Robson, J. D., and Seale, R. S., 1983, *Soils of England and Wales, Sheet 4 Eastern England*, Southampton, Ordnance Survey, Soils of England and Wales.

Hodgson, J. M., 1997, *Soil Survey Field Handbook*, Silsoe, Soil Survey and Land Research Centre.

Hoffecker, J. F., Holliday, V. T., Anikovich, M. V., Dudin, A. E., Platonova, N. I., Popov, V. V., Levkovskaya, G. M., Kuz'mina, I. E., Syromyatnikova, E. V., Burova, N. D., Goldberg, P., Macphail, R. I., Forman, S. L., Carter, B. J., and Crawford, L. J., 2016, Kostenki 1 and the early Upper Paleolithic of Eastern Europe: *Journal of Archaeological Science: Reports*, v. 5, p. 307-326.

Hoffmann, D. L., Standish, C. D., García-Diez, M., Pettitt, P. B., Milton, J. A., Zilhão, J., Alcolea-González, J. J., Cantalejo-Duarte, P., Collado, H., and De Balbín, R., 2018, U-Th dating of carbonate crusts reveals Neandertal origin of Iberian cave art: *Science*, v. 359, no. 6378, p. 912-915.

Höfle, B. and Pfeifer, N. 2007. Correction of laser scanning intensity data: Data and model-driven approaches. *ISPRS Journal of Photogrammetry and Remote Sensing*, 62, p. 415-433.

Holden, J. 2017. *An Introduction to Physical Geography and the Environment*. Pearson: Harlow.

Holden, T. G., 1995, *The Last Meals of the Lindow Bog Men, Bog Bodies: New Discoveries and New Perspectives*. London, British Museum Press, p. 76-82.

Holliday, V. T., 1990, Pedology in archaeology, *in* Lasca, N. P., and Donahue, J., eds., *Archaeological Geology of North America, Centennial Special Vol. 4*, Boulder, CA, Geological Society of America, p. 525-540.

Holliday, V. T., 1992, *Soils in Archaeology. Landscape Evolution and Human Occupation.* Washington, DC, Smithsonian Institution Press.

Holliday, V. T., 1997, Paleoindian Geoarchaeology of the Southern High Plains. *Texas Archaeology and Ethnohistory Series.* Austin, TX, University of Texas Press.

Holliday, V. T., 2004, *Soils in Archaeological Research.* Oxford, New York, Oxford University Press.

Holliday, V. T., and Gartner, W. G., 2007, Methods of soil P analysis in archaeology: *Journal of Archaeological Science*, v. 34, p. 301–333.

Holliday, V. T., and Miller, D. S., 2013, The Clovis landscape: *Paleoamerican odyssey*, p. 221–245.

Holliday, V. T., Hoffecker, J., Goldberg, P., Macphail, R. I., Forman, S., Anikovich, M. V., and Sinitsyn, A. A., 2007, Geoarchaeology of the Kostenki-Borshchevo sites, Don River Valley, Russia: *Geoarchaeology*, v. 22, no. 2, p. 181–228.

Holliday, V., and Johnson, E., 1989, Lubbock Lake: Late Quaternary cultural and environmental change on the Southern High Plain, USA: *Journal of Quaternary Science*, v. 4, no. 2, p. 145–165.

Holtedahl, H., 1975, The geology of Hardangerfjord, West Norway: *Norges Geol. Undersøkelse*, v. **323**, p. 1–87.

Homsey, L. K., and Capo, R. C., 2006, Integrating geochemistry and micromorphology to interpret feature use at Dust Cave, a Paleo-Indian through middle-archaic site in Northwest Alabama: *Geoarchaeology*, v. 21, no. 3, p. 237–269.

Homsey, L., and Sherwood, S., 2010, Interpretation of prepared clay surfaces at Dust Cave, Alabama: *Ethnoarchaeology*, v. 2, no. 1, p. 73–98.

Hopkins, D. W., Wiltshire, P. E. J., and Turner, B. D., 2000, Microbial characteristics of soils from graves: an investigation at the interface of soil microbiology and forensic science: *Applied Soil Ecology*, v. 14, p. 283–288.

Horowitz, A., 1979, *The Quaternary of Israel.* New York, Academic Press.

Horrocks, M., Coulson, S. A., and Walsh, K. A. J., 1999, Forensic palynology: Variation in the pollen content of soil on shoes and shoeprints in soil: *Journal of Forensic Science*, v. 44, no. 1, p. 119–122.

Horwitz, L. K., and Goldberg, P., 1989, A Study of Pleistocene and Holocene hyaena coprolites: *Journal of Archaeological Science*, v. 16, p. 71–94.

Hovers, E., 2009, *The lithic assemblages of Qafzeh Cave.* Oxford, Oxford University Press, Human Evolution series.

Hovers, E., Rak, Y., Lavi, R., and Kimbel, W., 1995, Hominid remains from the Amud Cave in the context of the Levantine Middle Paleolithic: *Paléorient*, v. 21, no. 2, p. 47–61.

Howard, A. J., Macklin, M. G., and Passmore, D. G., 2003, *Alluvial Archaeology in Europe*: Lisse, Balkema Publishers.

Hu, Y. F., Zhou, B., Lu, H. Y., Zhang, J. P., Min, S. Y., Dai, M. Z., Xu, S. Y., Yang, Q., and Zheng, H. B., 2020, Abundance and morphology of charcoal in sediments provide no evidence of massive slash-and-burn agriculture during the Neolithic Kuahuqiao culture, China: *PLoS One*, v. 15, no. 8, p. e0237592.

Huckleberry, G., 2001, Archaeological sediments in dryland alluvial environments, *in* Stein, J. K., and Farrand, W. R., eds., *Sediments in Archaeological Context*. Salt Lake City, UT, The University of Utah Press, p. 67–92.

Huckleberry, G., and Fadem, C., 2007, Environmental change recorded in sediments from the Marmes rockshelter archaeological site, southeastern Washington state, USA: *Quaternary Research*, v. 67, p. 21–32.

Hughes, P. D., 2010, Geomorphology and Quaternary stratigraphy: The roles of morpho-, litho-, and allostratigraphy: *Geomorphology*, v. 123, no. 3, p. 189–199.

Hughes, P. J., and Lampert, R. J., 1977, Occupational disturbance and types of archaeological deposit: *Journal of Archaeological Science*, v. 4, p. 135–140.

Huisman, D. J., 2009, *Degradation of Archaeological Remains*. Den Haag, Sdu Uitgevers b.v., p. 245.

Huisman, D. J., and Milek, K., 2017, Turf as constructional material, *in* Nicosia, C., and Stoops, G., eds., *Archaeological Soil and Sediment Micromorphology*. Chichester, Blackwell Wiley, p. 113–119.

Humphrey, J. D., and Ferring, C. R., 1994, Stable isotopic evidence for latest Pleistocene and Holocene climatic change in North-Central Texas: *Quaternary Research*, v. 41, p. 200–213.

Hunter, J., Roberts, C., and Martin, A., 1997, *Studies in Crime: An Introduction to Forensic Archaeology*. London, Routledge.

Hunter, K., 2012, Plant macrofossils, *in* Biddulph, E., Foreman, S., Stafford, E., Stansbie, D., and Nicholson, R., eds., *London Gateway. Iron Age and Roman salt making in the Thames Estuary; Excavations at Stanford Wharf Nature Reserve, Essex, Oxford Archaeology Monograph No. 18*. Oxford, Oxford Archaeology, p. 85.

Iceland, H. B., and Goldberg, P., 1999, Late Terminal Classic Maya pottery in Northern Belize; a petrographic analysis of sherd samples from Colha and Kichpanha: *Journal of Archaeological Science*, v. 26, p. 951–966.

Imberger, J., 1998, *Physical Processes in Lakes and Oceans*. Washington, DC, American Geophysical Union.

Imboden, D. M., 2004, The motion of lake waters, *in* O'Sullivan, P., and Reynolds, C. S., eds., *The Lakes Handbook Limnology and Limnetic Ecology*, Vol. 1. Oxford, Wiley Blackwell, p. 115–152.

Imeson, A. C., and Jungerius, P. D., 1976, Aggregate stability and colluviation in the Luxembourg Ardennes: an experimental and micromorphological study: *Earth Surface Processes*, v. 1, p. 259–271.

Imeson, A. C., Kwaad, F. J. P. M., and Mucher, H. J., 1980, Hillslope processes and deposits in forested areas of Luxembourg, *in* Cullingford, R. A., Davidson, D. A., and Lewin, J., eds., *Timescales in Geomorphology*. Chichester, John Wiley and Sons, p. 31–42.

Ingram, S. E., and Hunt, R. C., 2015, Traditional Arid lands Agriculture. *Understanding the Past for the Future*. Tucson, AZ, University of Arizona Press.

Institute of Archaeology of Chinese Academy of Social Sciences and Banpo Museum of Xi'an, 1963, *Xi'an Banpo: Excavation of a Prehistoric Village*. Beijing, Cultural Relics Press.

Institute of Archaeology of Chinese Academy of Social Sciences and Yi-Luo River Joint Archaeology Team, 2019, *Pre-Qin Period Sites in the Middle and Eastern Parts of the Luoyang Basin: Report of the Systematic Regional Survey Carried Out Between 1997–2007*. Beijing, Science Press.

Isendahl, C., 2012., Agro-urban landscapes: the example of Maya lowland cities: *Antiquity*, v. 86, no. 334, p. 1112–1125.

Ismail-Meyer, K., 2017, Plant remains, *in* Nicosia, C., and Stoops, G., eds., *Archaeological Soil and Sediment Micromorphology*. Chichester, Wiley Blackwell, p. 131–136.

Ismail-Meyer, K., and Rentzel, P., 2004, Mikromorphologische Untersuchung der Schichtabfolge, *in* Jacomet, S., Leuzinger, U., and Schibler, J., eds., *Die jungsteinzeitliche Seeufersiedlung Arbon/ Bleiche 3. Umwelt und Wirtschaft*, Archäologie im Thurgau/Band 12: Kanton Thurgau, Departement für Erziehung und Kultur des Kantons Thurgau, p. 66–80.

Ismail-Meyer, K., Rentzel, P., and Wiemann, P., 2013, Neolithic Lakeshore Settlements in Switzerland: new insights on site formation processes from micromorphology: *Geoarchaeology*, v. 28, p. 317–339.

Ismail-Meyer, K., Stolt, M. H., and Lindbo, D. L., 2018, Soil organic matter, *in* Stoops, G., Marcelino, V., and Mees, F., eds., *Interpretation of Micromorphological Features of Soils and Regoliths*. Amsterdam, Elsevier, p. 471–512.

Ismail-Meyer, K., Vach, W., and Rentzel, P., 2020, Do still waters run deep? Formation processes of natural and anthropogenic deposits in the Neolithic wetland site Zug-Riedmatt (Switzerland): *Geoarchaeology*, v. 35, no. 6, p. 921–951.

Itkin, D., Crouvi, O., Monger, H. C., Shaanan, U., and Goldfus, H., 2018, Pedology of archaeological soils in tells of the Judean foothills, Israel: *Catena*, v. 168, p. 47–61.

Jackson, L. J., Ellis, C., Morgan, A. V., and McAndrews, J. H., 2000, Glacial lake levels and Eastern Great Lakes Palaeo-Indians: *Geoarchaeology*, v. 15, no. 5, p. 415–440.

Jackson, M. L., Levelt, T. W. M., Syers, J. K., Rex, R. W., Clayton, R. N., Clayton, G. D., Sherman, G. D., and Uehara, G., 1971, Geomorphological relationships of tropospherically-derived quartz in soils on the Hawaiian Islands: *Soil Science Society of America Proceedings*, v. 35, p. 515–525.

Jackson, R. and Miller, D., 2011, *Wellington Quarry, Herefordshire (1986–96): Investigations of a landscape in the Lower Lugg Valley*. Oxford: Oxbow Books.

Jacob, J. S., 1995, Archaeological pedology in the Maya lowlands, *in* Collins, M. E., Carter, B. J., Gladfelter, B. G., and Southard, R. J., eds., *Pedological Perspectives in Archaeological Research, SSSA Special Publication Number 44*, Madison, Soil Science Society of America, Inc., p. 51–80.

Jacobi, R., undated, *The Late Pleistocene Archaeology of Somerset*.

Jacobs, P. M., Konen, M. E., and Curry, B. B., 2009, Pedogenesis of a catena of the Farmdale–Sangamon Geosol complex in the north central United States: Palaeogeography, Palaeoclimatology, *Palaeoecology*, v. 282, no. 1, p. 119–132.

Jacobs, Z., 2008, Luminescence chronologies for coastal and marine sediments: *Boreas*, v. 37, no. 4, p. 508–535.

Jacobs, Z., Duller, G. A. T., Wintle, A. G., and Henshilwood, C. S., 2006, Extending the chronology of deposits at Blombos Cave, South Africa, back to 140 ka using optical dating of single and multiple grains of quartz: *Journal of Human Evolution* v. 51 p. 255–273.

Jacobs, Z., Li, B., Shunkov, M. V., Kozlikin, M. B., Bolikhovskaya, N. S., Agadjanian, A. K., Uliyanov, V. A., Vasiliev, S. K., O'Gorman, K., Derevianko, A. P., and Roberts, R. G., 2019, Timing of archaic hominin occupation of Denisova Cave in southern Siberia: *Nature*, v. 565, no. 7741, p. 594–599.

Jacobs, Z., Wintle, A. G., and Duller, G. A. T., 2003, Optical dating of dune sand from Blombos Cave, South Africa, I: multiple grain data: *Journal of Human Evolution*, v. 44, no. 5, p. 599.

Jambrina-Enríquez, M., Herrera-Herrera, A. V., Rodríguez de Vera, C., Leierer, L., Connolly, R., and Mallol, C., 2019, n-Alkyl nitriles and compound-specific carbon isotope analysis of lipid combustion residues from Neanderthal and experimental hearths: Identifying sources of organic compounds and combustion temperatures: *Quaternary Science Reviews*, v. 222, p. 105899.

James, P., 1999, Soil variability in the area of an archaeological site near Sparta, Greece: *Journal of Archaeological Science*, v. 26, p. 1273–1288.

James, P., Chester, D., and Duncan, A., 2000, Volcanic soils: their nature and significance for archaeology: *Geological Society*. London, Special Publications, v. 171, no. 1, p. 317–338.

Jankowski, N. R., Jacobs, Z., and Goldberg, P., 2015, Optical dating and soil micromorphology at MacCauley's Beach, New South Wales, Australia: *Earth Surface Processes and Landforms*, v. 40, p. 229–242.

Jankowski, N. R., Stern, N., Lachlan, T. J., and Jacobs, Z., 2020, A high-resolution late Quaternary depositional history and chronology for the southern portion of the Lake Mungo lunette, semi-arid Australia: *Quaternary Science Reviews*, v. 233, p. 106224.

Jarvis, R. A., Allison, J. W., Bendelow, J. W., Bradley, R. I., Carroll, D. M., Furness, R. R., Kilgour, I. N. L., King, S. J., and Matthews, B., 1983, *Soils of England and Wales. Sheet 1 Northern England*. Ordnance Survey.

Jarvis, M. G., Allen, R. H., Fordham, S. J., Hazleden, J., Moffat, A. J., and Sturdy, R. G., 1983, *Soils of England and Wales. Sheet 6. South East England*. Ordnance Survey.

Jarvis, M. G., Allen, R. H., Fordham, S. J., Hazleden, J., Moffat, A. J., and Sturdy, R. G., 1984, *Soils and Their Use in South-East England*, Harpenden, Soil Survey of England and Wales.

Jarvis, K. E., Wilson, H. E., and James, S. L., 2004, Assessing element variability in small soil samples taken during forensic investigation, *in* Pye, K., and Croft, D. J., eds., *Forensic Geoscience: Principles, Techniques and Applications., Special Publications 232.* London, Geological Society of London, p. 171–182.

Jaubert, J., Verheyden, S., Genty, D., Soulier, M., Cheng, H., Blamart, D., Burlet, C., Camus, H., Delaby, S., and Deldicque, D., 2016, Early Neanderthal constructions deep in Bruniquel Cave in southwestern France: *Nature*, v. 534, no. 7605, p. 111–114.

Jedoui, Y., Kallel, N., Fontugne, M., Ben Ismail, H., Rabet, A. M., and Montacer, M., 1998, A high relative sea-level stand in the middle Holocene of southeastern Tunisia: *Marine Geology*, v. 147, no. 1–4, p. 123–130.

Jelinek, A. J., 1982, The Tabun Cave and Paleolithic man in the Levant: *Science*, v. 216, p. 1369–1375.

Jelinek, A. J., Farrand, W. R., Haas, G., Horowitz, A., and Goldberg, P., 1973, New excavations at the Tabun Cave, Mount Carmel, Israel, 1967–1972: a preliminary report: *Paléorient*, v. 1/2, p. 151–183.

Jenkins, D., 1994, Interpretation of interglacial cave sediments from a hominid site in North Wales: translocation of Ca-Fe-phosphates, *in* Rinrose-Vaose, A., ed., *Proceedings of IX International Working-Meeting on Soil Micromorphology, Townsville, Australia, July 1992.* Amsterdam, Elsevier, p. 293–305.

Jennings, J. D., 1957, Danger cave, Salt Lake City, University of Utah Anthropological Papers 27.

Jenny, H., 1941, *Factors of Soil Formation*, New York, McGraw-Hill.

Jerardino, A., and Marean, C. W., 2010, Shellfish gathering, marine paleoecology and modern human behavior: perspectives from cave PP13B, Pinnacle Point, South Africa: *Journal of Human Evolution*, v. 59, no. 3–4, p. 412–424.

Jewell, A. M., Drake, N., Crocker, A. J., Bakker, N. L., Kunkelova, T., Bristow, C. S., Cooper, M. J., Milton, J. A., Breeze, P. S., and Wilson, P. A., 2021, Three North African dust source areas and their geochemical fingerprint: *Earth and Planetary Science Letters*, v. 554, p. 116645.

Jia, L., 1999, *Chronicle of Zhoukoudian (1927–1937)*: Shanghai, Shanghai Scientific and Technical Publishers, p. 152.

Jia, L., and Huang, W., 1990, *The Story of Peking Man.* Beijing, Hong Kong, Foreign Languages Press, Oxford University Press.

Jing, Z., Rapp, G., Jr., and Gao, T., 1995, Holocene landscape evolution and its impact on the Neolithic and Bronze Age sites in the Shangqui area, northern China: *Geoarchaeology*, v. 10, p. 481–513.

Joffe, J., 1949, *Pedology.* New Brunswic, New Jersey, Pedology Publications.

Johnson, D. L., 2002, Darwin would be proud: bioturbation, dynamic denudation, and the power of theory in science: *Geoarchaeology*, v. 17, no. 1, p. 7–40.

Johnson, T. C., 1984, Sedimentation in large lakes: *Annual Review of Earth and Planetary Sciences*, v. 12, p. 179–204.

Jones, A. K. G., 1985, Trichurid ova in archaeological deposits: their value as indicators of ancient faeces, *in* Fieller, N. R. J., Gilbertson, D. D., and Ralph, N. G. A., eds., *Palaeoenvironmental Investigations: Research Design, Methods and Data Analysis*, Vol. 258. Oxford, British Archaeological Reports International Series, p. 105–115.

Jones, B. F., and Bowser, C. J., 1978, The mineralogy and related chemistry of lake sediments, *in* Lerman, A., ed., *Lakes: Chemistry, Geology, Physics.* Berlin, Springer-Verlag, p. 179–235.

Jones, R. J. A., and Evans, R., 1975, Soil and crop marks in the recognition of archaeological sites by aerial photography, *in* Wilson, D. R., ed., *Aerial Reconnaissance for Archaeology*, Vol. 12. Oxford, Council for British Archaeology Research Reports, p. 1–12.

Jongerius, A., 1983, The role of micromorphology in agricultural research, *in* Bullock, P., and Murphy, C. P., eds., *Soil Micromorphology, Vol. 1: Techniques and Applications*: Berkhamsted, A B Academic Publishers, p. 111–138.

Judson, S., 1961, Perspectives: archaeology and the natural sciences: *American Scientist*, v. 49, no. 3, p. 410–414.

Judson, S., 1963, Erosion and deposition of Italian stream valleys during historic time: *Science*, v. 140, no. 3569, p. 898–899.

Junge, A., Lomax, J., Shahack-Gross, R., Finkelstein, I., and Fuchs, M., 2018, Chronology of an ancient water reservoir and the history of human activity in the Negev Highlands, Israel: *Geoarchaeology*, v. 33, p. 695–707.

Kaiser, K., Barthelmes, A., Pap, S. C., Hilgers, A., Janke, W., Kühn, P., and Theuerkauf, M., 2006, A Lateglacial palaeosol cover in the Altdarss area, southern Baltic Sea coast (northeast Germany): investigations on pedology, geochronology and botany: *Netherlands Journal of Geosciences*, v. 85, no. 3, p. 197–220.

Kapur, S., Sakarya, N., and FitzPatrick, A. P., 1999, Mineralogy and micromorphology of Chalcolithic and Early Bronze Age Ikiztepe ceramics: *Geoarchaeology*, v. 7, p. 327–337.

Karami, F., Balci, N., and Guven, B., 2019, A modeling approach for calcium carbonate precipitation in a hypersaline environment: A case study from a shallow, alkaline lake: *Ecological Complexity*, v. 39, p. 100774.

Karkanas, P., 2007, Identification of lime plaster in prehistory using petrographic methods: a review and reconsideration of the data on the basis of experimental and case studies: *Geoarchaeology*, v. 22, no. 7, p. 775–796.

Karkanas, P., 2019, Microscopic deformation structures in archaeological contexts: *Geoarchaeology*, v. 34, no. 1, p. 15–29.

Karkanas, P., and Efstratiou, N., 2009, Floor sequences in Neolithic Makri, Greece: Micromorphology reveals cycles of renovation: *Antiquity*, v. 83, no. 322, p. 955–967.

Karkanas, P. and Goldberg, P., 2010. Site formation processes at Pinnacle Point Cave 13B (Mossel Bay, Western Cape Province, South Africa): resolving stratigraphic and depositional complexities with micromorphology: *Journal of Human Evolution*, v. 59, no. 3–4, p. 256–273.

Karkanas, P., and Goldberg, P., 2018a, Phosphatic features, *in* Stoops, G., Marcelino, V., and Mees, F., eds., *Interpretation of Micromorphological Features of Soils and Regoliths*. Amsterdam, Elsevier, p. 323–346.

Karkanas, K., and Goldberg, P., 2018b, Reconstructing Archaeological Sites. *Understanding the Geoarchaeological Matrix*. Chichester, Wiley Blackwell.

Karkanas, P., and Van de Moortel, A., 2014, Micromorphological analysis of sediments at the Bronze Age site of Mitrou, central Greece: patterns of floor construction and maintenance: *Journal of Archaeological Science*, v. 43, p. 198–213.

Karkanas, P., Kyparissi-Apostolika, N., Bar-Yosef, O., and Weiner, S., 1999, Mineral assemblages in Theopetra, Greece: A framework for understanding diagenesis in a prehistoric cave: *Journal of Archaeological Science*, v. 26, p. 1171–1180.

Karkanas, P., Bar-Yosef, O., Goldberg, P., and Weiner, S., 2000, Diagenesis in prehistoric caves: the use of minerals that form in situ to assess the completeness of the archaeological record: *Journal of Archaeological Science*, v. 27, p. 915–929.

Karkanas, P., Rigaud, J.-P., Simek, J. F., Albert, R. M., and Weiner, S., 2002, Ash bones and guano: a study of the minerals and phytoliths in the sediments of Grotte XVI, Dordogne, France: *Journal of Archaeological Science*, v. 29, no. 7, p. 721–732.

Karkanas, P., Koumouzelis, M., Kozlowski, J., K., Sitlivy, V., Sobczyk, K., Berna, F., and Weiner, S., 2004, The earliest evidence for clay hearths: Aurignacian features in Klisoura Cave 1, southern Greece: *Antiquity*, v. 78, p. 513–525.

Karkanas, P., Shahack-Gross, R., Ayalon, A., Bar-Matthews, M., Barkai, R., Frumkin, A., Gopher, A., and Stiner, M. C., 2007, Evidence for habitual use of fire at the end of the Lower Paleolithic: Site-formation processes at Qesem Cave, Israel: *Journal of Human Evolution*, v. 53, no. 2, p. 197–212.

Karkanas, P., Pavlopoulos, K., Kouli, K., Ntinou, M., Tsartsidou, G., Facorellis, Y., and Tsourou, T., 2011, Palaeoenvironments and site formation processes at the Neolithic lakeside settlement of Dispilio, Kastoria, Northern Greece: *Geoarchaeology*, v. 26, no. 1, p. 83–117.

Karkanas, P., Brown, K. S., Fisher, E. C., Jacobs, Z., and Marean, C. W., 2015, Interpreting human behavior from depositional rates and combustion features through the study of sedimentary microfacies at site Pinnacle Point 5-6, South Africa: *Journal of Human Evolution*, v. 85, p. 1–21.

Karkanas, P., Tourloukis, V., Thompson, N., Giusti, D., Panagopoulou, E., and Harvati, K., 2018, Sedimentology and micromorphology of the Lower Palaeolithic lakeshore site Marathousa 1, Megalopolis basin, Greece: *Quaternary International*, v. 497, p. 123–136.

Karkanas, P., Berna, F., Fallu, D., and Gauß, W., 2019, Microstratigraphic and mineralogical study of a Late Bronze Age updraft pottery kiln, Kolonna site, Aegina Island, Greece: *Archaeological and Anthropological Sciences*, v. 11, no. 10, p. 5763–5780.

Karkanas, P., Marean, C., Bar-Matthews, M., Jacobs, Z., Fisher, E., and Braun, K., 2020, Cave life histories of non-anthropogenic sediments help us understand associated archaeological contexts: *Quaternary Research*, v. 99, p. 270–289.

Keay, S.J., Parcak, S.H. and Strutt, K.D. 2014. High-resolution space and ground-based remote sensing and implications for landscape archaeology: The case from Portus, Italy. *Journal of Archaeological Science*, 52, p. 277–292.

Keddy, P. A., 2010, *Wetland Ecology Principles and Conservation*, 2nd edition. Cambridge, Cambridge University Press.

Keeley, H. C. M., Hudson, G. E., and Evans, J., 1977, Trace element contents of human bones in various states of preservation: *Journal of Archaeological Science*, v. 4, p. 19–24.

Keevill, G. D., 1995, Processes of collapse in Romano-British Buildings: a review of the evidence, *in* Shepherd, L., ed., *Interpreting Stratigraphy 5*. Bawdeswell, Interpreting Stratigraphy 5, p. 26–37.

Keevill, G., 2004, *The Tower of London Moat: Archaeological Excavations 1995–9 (Historic Royal Palaces Monograph)*. Oxford, Oxford Archaeological Unit for English Heritage.

Kellerman, A. M., Dittmar, T., Kothawala, D., and Tranvik, L. J., 2014, Chemodiversity of dissolved organic matter in lakes driven by climate and hydrology: *Nature Communications*, v. 5, p. 3804.

Kelley, J. T., Belknap, D. F., and Claesson, S., 2010, Drowned coastal deposits with associated archeological remains from a sea-level "slowstand": northwestern Gulf of Maine, USA: *Geology*, v. 38, p. 695–698.

Kelly, R. L., and Thomas, D. H., 2017, *Archaeology*, 7th edition. New York, Wadsworth Cengage Learning.

Kelts, K., and Hsu, K. J., 1978, Freshwater carbonate sedimentation, *in* Lerman, A., ed., *Lakes: Chemistry, Geology, Physics*. Berlin, Springer-Verlag, p. 295–323.

Kemp, R. A., 1985a, The Valley Farm Soil in southern East Anglia, *in* Boardman, J., ed., *Soils and Quaternary Landscape Evolution*. Chichester, John Wiley and Sons, p. 179–196.

Kemp, R. A., 1985b, The decalcified Lower Loam at Swanscombe, Kent: a buried Quaternary soil: *Proceedings of the Geological Association*, v. 96, p. 343–354.

Kemp, R. A., 1986, Pre-Flandrian Quaternary soils and pedogenic processes in Britain, *in* Wright, V. P., ed., *Paleosols. Their Recognition and Interpretation*. Oxford, Blackwell Scientific Publications, p. 242–262.

Kemp, R. A., 1999, Soil micromorphology as a technique for reconstructing palaeoenvironmental change, *in* Singhvi, A. K., and Derbyshire, E., eds., *Paleoenvironmental Reconstruction in Arid Lands*. New Delhi, Oxford and IBH Publishing Co. PVT. LTD., p. 41–71.

Kemp, R., Whiteman, C., and Rose, J., 1993, Palaeoenvironmental and stratigraphic significance of the Valley Farm and Barham Soils in Eastern England: *Quaternary Science Reviews*, v. 12, no. 10, p. 833–848.

Kemp, R. A., Jerz, H., Grottemthaler, W., and Preece, R. C., 1994, Pedosedimentary fabrics of soils within loess and colluvium in southern England and southern Germany, *in* Ringrose-Voase, A. J., and Humphreys, G. S., eds., *Soil Micromorphology: studies in management and genesis*. Amsterdam, Elesevier, p. 207–219.

Kemp, R. A., McDaniel, P. A., and Busacca, A. J., 1998, Genesis and relationship of macromorphology and micromorphology to contemporary hydrological conditions of a welded Argixeroll from the Palouse in Idaho: *Geoderma*, v. 83, p. 309–329.

Kennett, D. J., and Hodell, D. A., 2017, AD 750–1100 Climate change and critical transitions in Classic Maya sociopolitical networks, *in* Weiss, H., ed., *Megadrought and Collapse: From Early Agriculture to Angkor*. Oxford, Oxford Scholarship Online, p. 1–26.

Kenward, H. F., and Hall, A. R., 1997, Enhancing bioarchaeological interpretation using indicator groups: stable manure as a paradigm: *Journal of Archaeological Science*, v. 24, p. 663–673.

Kenward, H. K., and Hall, A. R., 1995, *Biological Evidence from Anglo-Scandinavian Deposits at 16–22 Coppergate*, York, York Archaeological Trust.

Kerr, P. K., 1995, Phosphate imprinting within Mound A at the Huntsville site, *in* Collins, M. E., Carter, B. J., Gladfelter, B. G., and Southard, R. J., eds., *Pedological Perspectives in Archaeological Research, SSSA Special Publication Number 44*: Madison, Soil Science Society of America, p. 133–149.

Khokhlova, O., Sverchkova, A., Myakshina, T., and Kalmykov, A., In review, A geoarchaeological study of the large Essentuksky 1 kurgan (dated to the second quarter of the 4th millennium BC) in Ciscaucasia, Russia: *Geoarchaeology*.

Khokhlova, O., Sverchkova, A., Myakshinaa, T., Makeev, A., and Tregub, T., 2021, Environmental trends during the Bronze Age recorded in paleosols buried under a big kurgan in the steppes of the Ponto-Caspian area: *Quaternary International*, v. 583, p. 83–93.

Kidder, T. R., 2002, Mapping Poverty Point: *American Antiquity*, v. 67, no. 1, p. 89–101.

Kidder, T. R., and Liu, H., 2017, Bridging theoretical gaps in geoarchaeology: Archaeology, geoarchaeology, and history in the Yellow River valley, China: *Archaeological and Anthropological Sciences*, v. 9, no. 8, p. 1585–1602.

Kidder, T. R., Liu, H. W., and Li, M. L., 2012a, Sanyangzhuang: early farming and a Han settlement preserved beneath Yellow River flood deposits: *Antiquity*, v. 86, no. 331, p. 30–47.

Kidder, T. R., Liu, H., Xu, Q., and Li, M., 2012b, The alluvial geoarchaeology of the Sanyangzhuang Site on the Yellow River floodplain, Henan Province, China: *Geoarchaeology*, v. 27, no. 4, p. 324–343.

Kile-Vesik, J., and Orvik, K., 2020, *Gravhauger. Voldskogen, 51/43, Rygge, Østfold*: University of Oslo.

King, R. B., Baillie, I. C., Abell, T. M. B., Dunsmore, J. R., Gray, D. A., Pratt, J. H., Versey, H. R., Wright, A. C. S., and Zisman, S. A., 1992, *Land Resource Assessment of Northern Belize*. Vols 1 and 2, Kent, U.K., Natural Resources Institute Bulletin.

Kingery, W. D., Vandiver, P. B., and Prickett, M., 1988, The beginnings of pyrotechnology, part II;: production and use of lime and gypsum plaster in the pre-pottery Neolithic Near-East: *Journal of Field Archaeology*, v. 15, p. 219–244.

Klein, C., 1986, *Fluctuations of the Level of the Dead Sea and Climatic Fluctuations during Historical Times*, [Doctoral Dissertation]: Hebrew University of Jerusalem.

Klingebiel, A. A., and Montgomery, P. H., 1961, *Land Capability Classification*. Washington, USDA, Soil Conservation Service Agricultural Handbook No. 210.

Kobe, F., Bezrukova, E. V., Leipe, C., Shchetnikov, A. A., Goslar, T., Wagner, M., Kostrova, S. S., and Tarasov, P. E., 2020, Holocene vegetation and climate history in Baikal Siberia reconstructed from pollen records and its implications for archaeology: *Archaeological Research in Asia*, v. 23, p. 100209.

Kolka, R. A., and Thompson, J. A., 2006, Wetland geomorphology, soils, and formative processes, *in* Batzer, P. L., and Sharitz, R., eds., *Ecology of Freshwater and Estuarine Wetlands*: Los Angeles, University of California Press, p. 7–42.

Kooistra, M. J., 1978, *Soil Development in Recent Marine Sediments of the Intertidal Zone in the Oosterschelde – the Netherlands: a Soil Micromorphological Approach*. Wageningen, Soil Survey Institute, Soil Survey Papers.

Kooistra, M. J., and Pulleman, M. M., 2018, Features related to faunal activity, *in* Stoops, G., Marcelino, V., and Mees, F., eds., *Interpretation of Micromorphological Features of Soils and Regoliths*, 2nd edition. Amsterdam, Elsevier, p. 447–469.

Kovda, I., and Mermut, A. R., 2018, Vertic features, *in* Stoops, G., Marcelino, V., and Mees, F., eds., *Interpretation of Micromorphological features in Soils and Regoliths*. Amsterdam, Elsevier, p. 605–632.

Krahtopoulou, A., and Frederick, C., 2008, The stratigraphic implications of long-term terrace agriculture in dynamic landscapes: Polycyclic terracing from Kythera Island, Greece: *Geoarchaeology*, v. 23, no. 4, p. 550–585.

Krinsley, D. H., and Doornkamp, J. C., 2011, *Atlas of Quartz Sand Surface Textures*. Cambridge, Cambridge University Press.

Kristiansen, S. M., Ljungberg, T. E., Christiansen, T. T., Dalsgaard, K., Haue, N., Greve, M. H., and Nielsen, B. H., 2020, Meadow, marsh and lagoon: Late Holocene coastal changes and human-environment interactions in northern Denmark: *Boreas*, 50, 279–293.

Kruger, R. P., 2015, A burning question or, some half-baked ideas: patterns of sintered daub creation and dispersal in a modern wattle and daub structure and their Implications for archaeological interpretation: *Journal of Archaeological Method and Theory*, v. 22, p. 883–912.

Krumbein, W. C., and Pettijohn, F. J., 1938, *Manual of Sedimentary Petrography*. New York, Appleton-Century-Crofts, Inc.

Krumbein, W. C., and Sloss, L. L., 1963, *Stratigraphy and Sedimentation*. San Francisco, CA, W.H. Freeman.

Kubiëna, W. L., 1938, *Micropedology*. Ames, IA, Collegiate Press.

Kubiëna, W. L., 1953, *The Soils of Europe*. London, Thomas Murby.

Kubiëna, W. L., 1970, *Micromorphological Features of Soil Geography*. New Brunswick, NJ, Rutgers University Press, xxiii.

Kühn, P., Aguilar, J., Miedema, R., and Bronnikova, M., 2018, Textural pedofeatures and related horizons, *in* Stoops, G., Marcelino, V., and Mees, F., eds., *Interpretation of Micromorphological Features of Soils and Regoliths*, 2nd edition. Amsterdam, Elsevier, p. 377–424.

Kuhn, S. L., Stiner, M. C., Güleç, E., Ozer, I., Yılmaz, H., Baykara, I., Açıkkol, A., Goldberg, P., Martınez Molina, K., Unay, E., and Suata-Alpaslan, F., 2009, The early Upper Paleolithic occupations at Uçagızlı Cave (Hatay, Turkey): *Journal of Human Evolution*, v. 56, p. 87–113.

Kukla, G. J., 1977, Pleistocene Land – Sea Correlations I. Europe: *Earth-Science Reviews*, v. 13, p. 307–374.

Kvamme, K. L., 2001, Current practices in archaeogeophysics, *in* Goldberg, P., Holliday, V. T., and Ferring, C. R., eds., *Earth Sciences and Archaeology*. New York, Kluwer Academic/Plenum, p. 353–384.

Kwaad, F. J. P. M., and Mücher, H. J., 1979, The formation and evolution of colluvium on arable land in northern Luxembourg: *Geoderma*, v. 22, no. 2, p. 173–192.

Lambrecht, G., de Vera, C. R., Jambrina-Enríquez, M., Crevecoeur, I., Gonzalez-Urquijo, J., Lazuen, T., Monnier, G., Pajović, G., Tostevin, G., and Mallol, C., 2021, Characterisation of charred organic matter in micromorphological thin sections by means of Raman spectroscopy: *Archaeological and Anthropological Sciences*, v. 13, no. 1, p. 1–15.

Lambrick, G., 1992, Alluvial archaeology of the Holocene in the Upper Thames Basin 1971–1991: a review, *in* Needham, S., and Macklin, M. G., eds., *Alluvial Archaeology in Britain*. Oxford, Oxbow, p. 209–228.

Landeschi, G., 2019, Rethinking GIS, three-dimensionality and space perception in archaeology: *World Archaeology*, v. 51, no. 1, p. 17–32.

Landuydt, C. J., 1990, Micromorphology of iron minerals from bog ores of the Belgian Campine Area, *in* Douglas, L. A., ed., *Soil Micromorphology: a Basic and Applied Science*. Amsterdam, Elsevier, p. 289–301.

Langohr, R., 1991, Soil characteristics of the Motte of Werken (West Flanders, Belgium), *in* Tauber, J., ed., *Methoden und Perspektiven der Archaologie des Mittelalters, Tagungsberichte zum interdisziplinaren Kolloquium, September 1989, Liestal, Switzerland*, 209–223: Heft, Berichte aus der Arbeit des Amtes fur Museen und Archaologie des Kantons Baseeland.

Langohr, R., 1993, Types of tree windthrow, their impact on the environment and their importance for the understanding of archaeological excavation data: *Helenium*, v. XXXIII, no. 1, p. 36–49.

Langohr, R. 2019. The edaphic factor. Settlement of the first farmers in the Belgian loess belt, *in* Deák J., Ampe C., and Hinsch Mikkelsen J., eds., *Soils as Records of Past and Present. From Soil Surveys to Archaeological Sites: Research Strategies for Interpreting Soil Characteristics. Raakvlak, Archaeology, Monuments and Landscapes of Bruges and Hinterland, Belgium*.

Lang-Yona, N., Maier, S., Macholdt, D. S., Müller-Germann, I., Yordanova, P., Rodriguez-Caballero, E., Jochum, K. P., Al-Amri, A., Andreae, M. O., and Fröhlich-Nowoisky, J., 2018, Insights into microbial involvement in desert varnish formation retrieved from metagenomic analysis: *Environmental Microbiology Reports*, v. 10, no. 3, p. 264–271.

Larbey, C., Mentzer, S. M., Ligouis, B., Wurz, S., and Jones, M. K., 2019, Cooked starchy food in hearths ca. 120 kya and 65 kya (MIS 5e and MIS 4) from Klasies River Cave, South Africa: *Journal of Human Evolution*, v. 131, p. 210–227.

Lasaponara, R., Masini, N., 2011. Satellite remote sensing in archaeology: Past, present and future perspectives. *Journal of Archaeological Science*, 38 (9), p. 1995, 2496.

Lasaponara, R. and Masini N. 2012. *Satellite Remote Sensing: A New Tool for Archaeology*. Springer: London.

Lathrap, D. W., 1968, Aboriginal Occupation and Changes in River Channel on the Central Ucayali, Peru: *American Antiquity*, v. 33, no. 1, p. 62–79.

Lattman, L. H., and Ray, R. G., 1965, *Aerial Photographs in Field Geology*. New York, Holt, Rinehart and Winston.

Laurent, A., and Fondrillon, M., 2010, Mesurer la ville par l'évaluation et la caractérisation du sol urbain: l'exemple de Tours: *Revue archéologique du Centre de la France*, v. 49, p. 307–343.

Lautridou, J. P., and Ozouf, J. C., 1982, Experimental frost shattering; 15 years of research at the Centre de geomorphologie du CNRS: *Progress in Physical Geography*, v. 6, no. 2, p. 215–232.

Laville, H., 1964, *Recherches sédimentologiques sur la paéoclimatologie du Wurm recent en Périgord*: l'Université de Bordeaux, p. 4–19.

Laville, H., and Goldberg, P., 1989, The collapse of the Mousterian sedimentary regime and the beginning of the Upper Paleolithic at Kebara Cave, Mount Carmel, *in* Bar-Yosef, O., and Vandermeersch, B., eds., *Investigations in South Levantine Prehistory/Préhistoire du Sud-Levant*, Oxford, British Archaeological Reports, International Series, v. 197, p. 75–96.

Laville, H., and Tuffreau, A., 1984, Les dépôts du grand abri de la Ferrassie: stratigraphie, signification climatique et chronologie: *Le Grand Abri de La Ferrassie, Fouilles 1968–1973*, p. 22–50.

Laville, H., Rigaud, J.-P., and Sackett, J., 1980, *Rock Shelters of the Perigord*. New York, Academic Press.

Lawson, A. J., 2000, *Potterne 1982–5: Animal Husbandry in Later Prehistoric Wiltshire*, Salisbury, Wessex Archaeology.

Le Tensorer, J.-M., 1972, Analyse chimique des remplissages quaternaires, méthode et premiers résultats, interprétation paléoclimatique: *Bulletin de l'Association française pour l'étude du quaternaire*, v. 9, no. 3, p. 155–169.

Lechtman, H. N., and Hobbs, L. W., 1983, Roman concrete and Roman architectural revolution: *Journal of Archaeological Science*, v. 13, p. 81–128.

Leckebusch, J. 2003. Ground-penetrating radar: A modern three-dimensional prospection method. *Archaeological Prospection,* 10, p.213–240. https://doi.org/10.1002/arp.211

Lee, C. S., Qi, S. H., Zhang, G., Luo, C. L., Zhao, L. Y. L., and Li, X. D., 2008, Seven thousand years of records on the mining and utilization of metals from lake sediments in Central China: *Environmental Science and Technology*, v. 42, no. 13, p. 4732–4738.

Lee, H., French, C., and Macphail, R. I., 2014, Microscopic examination of ancient and modern irrigated paddy soils in South Korea, with special reference to the formation of silty clay concentration features: *Geoarchaeology*, v. 29, p. 326–348.

Legros, J. P., 1992, Soils of Alpine Mountains, *in* Martini, I. P., and Chesworth, W., eds., *Weathering, Soils & Paleosols, Developments in Earth Surface Processes 2*. Amsterdam, Elsevier.

Lehmann, G., Kern, D., Glaser, B., and Woods, W. I., 2004, *Amazonian dark Earths: Origins, Properties, Management*, New York, Kluwer Academic Publishers.

Lehmkuhl, F., Nett, J., Pötter, S., Schulte, P., Sprafke, T., Jary, Z., Antoine, P., Wacha, L., Wolf, D., Zerboni, A., Hošek, J., Marković, S. B., Obreht, I., Sümegi, P., Veres, D., Zeeden, C., Boemke, B., Schaubert, V., Viehweger, J., and Hambach, U., 2020, Loess landscapes of Europe–Mapping, geomorphology, and zonal differentiation: *Earth-Science Reviews*, v. 215, p. 103496.

Leierer, L., Carrancho Alonso, Á., Pérez, L., Herrejón Lagunilla, Á., Herrera-Herrera, A. V., Connolly, R., Jambrina-Enríquez, M., Hernández Gómez, C. M., Galván, B., and Mallol, C., 2020, It's getting hot in here – Microcontextual study of a potential pit hearth at the Middle Paleolithic site of El Salt, Spain: *Journal of Archaeological Science*, v. 123, p. 105237.

Leierer, L., Jambrina-Enríquez, M., Herrera-Herrera, A. V., Connolly, R., Hernández, C. M., Galvan, B., and Mallol, C., 2019, Insights into the timing, intensity and natural setting of Neanderthal occupation from the geoarchaeological study of combustion structures: A micromorphological and biomarker investigation of El Salt, unit Xb, Alcoy, Spain: *PLoS One*, v. 14, no. 4.

Leigh, D. S., 1998, Evaluating artifact burial by eolian versus bioturbation processes, South Carolina Sandhills, USA: *Geoarchaeology*, v. 13, no. 3, p. 309–330.

Leigh, D. S., 2001, Buried artifacts in sandy soils; techniques for evaluating pedoturbation versus sedimentation, *in* Goldberg, P., Holliday, V. T., and Ferring, C. R., eds., *Earth Sciences and Archaeology*. New York, Kluwer-Plenum, p. 269–293.

Lelong, R., and Souchier, B., 1982, Ecological significance of the weathering complex: relative importance of general and local factors, *in* Bonneau, M., and Souchier, B., eds., *Constituents and Properties of Soils*. London, Academic Press, p. 82–108.

Leone, A., 1998, Change or No Change? Revised perceptions of urban transformation in late Antiquity, *TRAC 98 (Proceedings of the Eighth Annual Theoretical Roman Archaeology Conference (Leicester 1998)*. Oxford, p. 121–130.

Leonov, A., Anikushkin, M., Bobkov, A., Rys, I., Kozlikin, M., Shunkov, M., Derevianko, A., and Baturin, Y., 2014, Development of a Virtual 3d Model of Denisova Cave in the Altai Mountains1: *Archaeology, Ethnology and Anthropology of Eurasia*, v. 42, no. 3, p. 14–20.

Leonova, N., Nesmeyanov, S., Vinogradova, E. and Voeykova, O. 2015. Upper Paleolithic subsistence practices in the southern Russian Plain: paleolandscapes and settlement system of Kamennaya Balka sites: *Quaternary International*, v. 355, p. 175–187.

Leroi-Gourhan, A., 1984, *Pincevent: campement magdalénien de chasseurs de rennes*. Paris, Ministère de la culture Direction du Patrimoine Sous-direction de l'archéologie: Impr. nationale, Guides archaéologiques de la France, 3.

Leroi-Gourhan, A., and Brézillon, M. N., 1972, *Fouilles de Pincevent. Essai d'analyse éthnographique d'un habitat magdalénien (la section 36)*. Paris, Centre National de la Recherche Scientifique, Supplément à "Gallia Préhistoire".

Leshchinskiy, S. V., Zenin, V. N., and Bukharova, O. V., 2021, The Volchia Griva mammoth site as a key area for geoarchaeological research of human movements in the Late Paleolithic of the West Siberian Plain: *Quaternary International*, v. 587–588, p. 368–383.

Levine, M. N., Joyce, A. A., and Goldberg, P., 2004, Earthen Mound Construction at Río Viejo on the Pacific Coast of Oaxaca, Mexico, *in Proceedings Poster presentation at the 69th Annual Meeting of the Society for American Archaeology*, Montreal.

Levy, T. E., 1995, Cult, metallurgy and rank societies – Chalcolithic Period (ca. 4500–3500 BCE), *in* Levy, T. E., ed., *The Archaeology of Society in the Holy Land*. London, Leicester University Press, p. 226–244.

Lewis, H. A. 2012. *Investigating Ancient Tillage. An Experimental and Soil Micromorphological Study*. Oxford, British Archaeological Reports 2388.

Lewis, J. S. C., and Rackham, J., 2011, Three Ways Wharf, Uxbridge. *A Lateglacial and Early Holocene hunter-gatherer site in the Colne Valley*. London, Museum of London.

Lewis, J. S., Wiltshire, P., and Macphail, R. I., 1992, A Late Devensian/Early Flandrian site at Three Ways Wharf, Uxbridge: environmental implications, *in* Needham, S., and Macklin, M. G., eds., *Alluvial Archaeology in Britain*, Monograph 27. Oxford, Oxbow, p. 235–248.

Lewis, J., Leivers, M., Brown, L., Smith, A., Cramp, K., Mepham, L., and Phillpotts, C., 2010, Landscape Evolution in the Middle Thames, Valley. *Heathrow Terminal 5 Excavation Vol. 2*, Framework Archaeology Monograph 13: Oxford/Salisbury, Framework Archaeology.

Li, Z. T., Xu, Q. H., Zhang, S. R., Hun, L. Y., Li, M. Y., Xie, F., Wang, F. G., and Liu, L. Q., 2014, Study on stratigraphic age, climate changes and environment background of Houjiayao Site in Nihewan Basin: *Quaternary International*, v. 349, p. 42–48.

Ligouis, B., 2017, Reflected light, *in* Nicosia, C., and Stoops, G., eds., *Archaeological Soil and Sediment Micromorphology*. Chichester, Wiley Blackwell, p. 461–470.

Lillesand, T. M., and Kiefer, R. W., 1994, *Remote Sensing and Image Interpretation*, 3rd edition. New York, Wiley & Sons, xvi.

Limbrey, S., 1975, *Soil Science and Archaeology*. London, Academic Press.

Limbrey, S., 1990, Edaphic opportunism? A discussion of soil factors in relation to the beginnings of plant husbandry in south-west Asia, *in* Thomas, K., ed., *Soils and Early Agriculture, World Archaeology 22, 1*. London, Routledge, p. 45–52.

Limbrey, S., 1992, Micromorphological studies of buried soils and alluvial deposits in a Wiltshire river valley, *in* Needham, S., and Macklin, M. G., eds., *Alluvial Archaeology in Britain*. Oxford, Oxbow, p. 53–64.

Linderholm, J., 2003, Soil chemical surveying: a path to a different understanding of prehistoric sites and societies in Northern Sweden, *in* Boschian, G., ed., *Second International Conference on Soils and Archaeology, Pisa, 12th-15th May, 2003. Extended Abstracts: Pisa, Dipartimento di Scienze Archeologiche, Università di Pisa*, p. 114–119.

Linderholm, J., 2007, Soil chemical surveying: a path to a deeper understanding of prehistoric sites and societies in Sweden: *Geoarchaeology*, v. 22, no. 4, p. 417–438.

Linderholm, J., 2010, Soil prospection: chemical and magnetic susceptibility attributes from off-site and intra-site perspectives. A case study from a late Mesolithic-early Neolithic dwelling in northern Sweden, *in* Linderholm, J., ed., *The Soil as a Source Material in Archaeology. Theoretical Considerations and Pragmatic Applications*, Archaeology and Environment 25: Umeå, Umeå University, p. 1–43.

Linderholm, J., and Geladi, P., 2015, Classification of archaeological soil and sediment samples using near infrared techniques: *NIR News*, v. 23, no. 7, p. 6–9.

Linderholm, J., and Lundberg, E., 1994, Chemical characterisation of various archaeological soil samples using main and trace elements determined by inductively coupled plasma atomic emission spectrometry: *Journal of Archaeological Science*, v. 21, p. 303–314.

Linderholm, J., Macphail, R. I., Bill, J., Bukkemoen, G. B., Ericson, S., Östman, S., and Engelmark, R., Submitted, The Sørenga D1A borehole site, Oslo Harbour, Norway: a multi-analytical geoarchaeological and palaeoenvironmental approach. , in Barnett, C., and Walker, T., eds., Environment, Archaeology and Landscape - Papers in Honour of Professor Martin Bell, Oxford, Archaeopress, p. 65–76.

Linderholm, J., Macphail, R. I., Östman, S., and Eriksson, S., 2020, *Sammanslagna rapporter, RAÄ 165, Norum sn*, Bohuslän: University of Umeå.

Linderholm, J., Macphail, R., Buckland, P., Ostman, S., Eriksson, S., Wallen, J.-E., and Engelmark, R., 2019, Ørlandet Iron Age Settlement Pattern Development: Geoarchaeology (geochemistry and soil micromorphology) and Plant Macrofossils, *in* Ystgaard, I., ed., *Human-environment Interaction during the Iron Age and Early Medieval Period in Vik. Archaeology at Ørland air base*: Oslo, Cappelen Damm Akademisk, p. 107–134.

Lisá, L., Kočár, P., Bajer, A., Kočárová, R., Syrová, Z., Syrový, J., Porubčanová, M., Lisý, P., Peška, M., and Ježková, M., 2020, The Floor – a voice of human lifeways. *A geo-ethnographical study of historical and recent floors at Dolní Němči Mill, Czech Republic Archaeological and Anthropological Sciences*, v. 12, no. 115.

Littmann, E. R., 1958, Ancient Mesoamerican mortars, plasters, and stuccos: the composition and origin of sascab: *American Antiquity*, v. XXIV, no. 2.

Littmann, T., 1991, Recent African dust deposition in West Germany – sediment characteristics and climatological aspects: *CATENA Supplement*, v. 20, p. 57–73.

Liu, B., Wang, N. Y., Chen, M. H., Wu, X. H., Mo, D. W., Liu, J. G., Xu, S. J., and Zhuang, Y., 2017, Earliest hydraulic enterprise in China, 5,100 years ago: *PNAS*, v. 114, no. 52, p. 13637–13642.

Liu, L., Chen, X., and Jiang, L., 2004, A study of Neolithic water buffalo remains from Zhejiang, China: *Bulletin of the Indo-Pacific Prehistory Association*, v. 24 (Indo-Pacific Prehistory: the Taipei Papers, Vol. 2), p. 113–120.

Liu, L., Chen, X., Lee, Y. K., Wright, H., and Rosen, A., 2002–2004, Settlement patterns and developing of social complexity in the Yiluo Region, north China: *Journal of Field Archaeology*, v. 29, no. 1 and 2, p. 75–100.

Liversage, D., Munro M. A. R, Courty M. A, and Nørnberg, P., 1987, Studies of a buried Iron Age field: *Acta Archaeologica*, v. 1, p. 55–84.

Livingstone, I., and Warren, A., 1996, *Aeolian Geomorphology; an introduction*, Harlow, UK, Addison Wesley Longman.

Loffler, H., 2004, The origin of lake basins, *in* O'Sullivan, P., and Reynolds, C. S., eds., *The Lakes Handbook: Limnology and Limnetic Ecology*, Vol. 1. Oxford, Wiley Blackwell, p. 8–60.

Longa, E. D., Corso, M. D., Vicenzutto, D., Nicosia, C., and Cupito, M., 2019, The Bronze Age settlement of Fondo Paviani (Italy) in its territory. Hydrography, settlement distribution, environment and in-site analysis: *Journal of Archaeological Science*: Reports, v. 28, p. 102018.

Longworth, G., Becker, L. W., Thompson, R., Oldfield, F., Dearing, J. A., and Rummery, T. A., 1979, Mössbauer and magnetic studies of secondary iron oxides in soils: *Journal of Soil Science*, v. 30, p. 93–110.

Love, S., 2012, The geoarchaeology of mudbricks in architecture: A methodological study from Çatalhöyük, Turkey: *Geoarchaeology*, v. 27, no. 2, p. 140–156.

Lu, P., Lu, J. Q., Zhuang, Y., Chen, P. P., Wang, H., Tian, Y., Mo, D. W., Xu, J. J., Gu, W. F., Hu, Y. Y., Wei, Q. L., Yan, L. J., Xia, W., and Zhai, H. G., 2020, Evolution of Holocene alluvial landscapes in the northeastern Songshan Region, Central China: *Chronology, models and socio-economic impact: Catena*, v. 197, p. 104956.

Luo, L., Wang, X., Guo, H., Lasaponara, R., Zong, X., Masini, N., Wang, G., Shi, P., Khatteli, H., Chen, F., Tariq, S., Shao, J., Bachagha, N., Yang, R. and Yao, Y. 2019. Airborne and spaceborne remote sensing for archaeological and cultural heritage applications: A review of the century (1907–2017). *Remote Sensing of Environment*, 232, 111280

Lutz, H. J., and Griswold, F. S., 1939, The influence of tree roots on soil morphology: *American Journal of Science*, v. 237, p. 389–400.

Luzzadder-Beach, S., Beach, T. P., and Dunning, N. P., 2012, Wetland fields as mirrors of drought and the Maya abandonment: *PNAS*, v. 109, p. 3646–3651.

Luzzadder-Beach, S., Beach, T. P., Hutson, S., and Krause, S., 2016, Sky-earth, lake-sea: climate and water in Maya history and landscape: *Antiquity*, v. 90, p. 426–442.

MacEachern, S., 2012, The Holocene history of the southern Lake Chad Basin: archaeological, linguistic and genetic evidence: *African Archaeological Review*, v. **29**, p. 253–271.

Mackney, D., 1961, A podzol development sequence in oakwoods and heath in central England: *Journal of Soil Science*, v. 12, p. 23–40.

Macnab, J. W., 1965, British strip lynchets: *Antiquity*, v. 39, p. 279–290.

Macphail, R. I., 1981, Soil and botanical studies of the "Dark Earth", *in* Jones, M., and Dimbleby, G. W., eds., *The Environment of Man: the Iron Age to the Anglo-Saxon Period, British Series 87*. Oxford, British Archaeological Reports, p. 309–331.

Macphail, R. I., 1983, The micromorphology of dark earth from Gloucester, London and Norwich: an analysis of urban anthropogenic deposits from the Late Roman to Early Medieval periods in England, *in* Bullock, P., and Murphy, C. P., eds., *Soil Micromorphology, Vol. 1: Techniques and Applications*: Berkhamsted, A B Academic Publishers, p. 245–252.

Macphail, R. I., 1986, Paleosols in archaeology: their role in understanding Flandrian pedogenesis, *in* Wright, V. P., ed., *Paleosols. Their Recognition and Interpretation*. Oxford, Blackwell Scientific Publications, p. 263–290.

Macphail, R. I., 1987a, A review of soil science in archaeology in England, *in* Keeley, H. C. M., ed., *Environmental Archaeology: A Regional Review Vol. II*, Occasional paper No. 1. London, Historic Buildings & Monuments Commission for England, p. 332–379.

Macphail, R. I. 1987b, *Soil Report on the Cairn and Field System at Chysauster, Penzance, Cornwall*. English Heritage, Ancient Monuments Laboratory Report 111/87.

Macphail, R. I., 1990a, *Soil report on Carn Brea, Redruth, Cornwall, with some reference to similar sites in Brittany*, France: Ancient Monuments Laboratory, English Heritage, 55/90.

Macphail, R. I., 1990b, The soils, *in* Saville, A., ed., *Hazleton North, Gloucestershire, 1979–82: The Excavation of a Neolithic Long Cairn of the Cotswold-Severn Group, Archaeological Report no. 13*. London, English Heritage, p. 223–226.

Macphail, R. I., 1990c, Soil history and micromorphology, *in* Bell, M., ed., *Brean Down Excavations 1983–1987, Archaeological Report No. 15*. London, English Heritage, p. 187–196.

Macphail, R. I., 1990d, Soil micromorphological evidence of the impact of ancient agriculture, *Transactions of the 14th International Congress of Soil Science, Kyoto, Japan, August 1990, Vol. VII*: Kyoto, International Society of Soil Science, p. 264–269.

Macphail, R. I., 1991, The archaeological soils and sediments, *in* Sharples, N. M., ed., *Maiden Castle: Excavations and field survey 1985–6, Archaeological Report no 19*. London, English Heritage, p. 106–118.

Macphail, R. I., 1992a, Soil micromorphological evidence of ancient soil erosion, *in* Bell, M., and Boardmand, J., eds., *Past and Present Soil Erosion, Monograph 22*. Oxford, Oxbow, p. 197–216.

Macphail, R. I., 1992b, Late Devensian and Holocene soil formation, *in* Barton, R. N. E., ed., *Hengistbury Head, Dorset. Vol. 2: The Late Upper Palaeolithic & Early Mesolithic sites, Monograph No. 34*. Oxford, Oxford University Committee for Archaeology, p. 44–51.

Macphail, R. I., 1994a, The reworking of urban stratigraphy by human and natural processes, *in* Hall, A. R., and Kenward, H. K., eds., *Urban-Rural Connexions: Perspectives from Environmental Archaeology, Monograph 47*. Oxford, Oxbow, p. 13–43.

Macphail, R. I., 1994b, Soil micromorphological investigations in archaeology, with special reference to drowned coastal sites in Essex, *in* Cook, H. F., and Favis-Mortlock, D. T., eds., *SEESOIL*, Vol. 10. Wye, South East Soils Discussion Group, p. 13–28.

Macphail, R. I., 1995, *Report on the Soils at Westhampnett Bypass, West Sussex: with Special Reference to the Micromorphology of the Late-glacial Soil and Marl at Area 3*. Salisbury, Wessex Archaeology.

Macphail, R. I., 1998, A reply to Carter and Davidson's "An evaluation of the contribution of soil micromorphology to the study of ancient arable agriculture": *Geoarchaeology*, v. 13, no. 6, p. 549–564.

Macphail, R. I., 1999a, Sediment micromorphology, *in* Roberts, M. B., and Parfitt, S. A., eds., *Boxgrove, A Middle Pleistocene Hominid Site at Eartham Quarry, Boxgrove, West Sussex, Archaeological Reports no. 17*. London, English Heritage, p. 118–148.

Macphail, R. I., 1999b, *Eton Rowing Lake Neolithic midden deposits: a soil micromorphological and chemical study*: Oxford Archaeological Unit.

Macphail, R. I., 2000, Soils and microstratigraphy: a soil micromorphological and micro-chemical approach, *in* Lawson, A. J., ed., *Potterne 1982–5: Animal Husbandry in Later Prehistoric Wiltshire, Archaeology Report No. 17*. Salisbury, Wessex Archaeology, p. 47–70.

Macphail, R. I., 2003a, Industrial activities – some suggested microstratigraphic signatures: ochre, building materials and iron-working, *in* Wiltshire, P. E. J., and Murphy, P., eds., *The Environmental Archaeology of Industry, AEA Symposia No. 20*. Oxford, Oxbow, p. 94–106.

Macphail, R. I., 2003b, Soil microstratigraphy: a micromorphological and chemical approach, *in* Cowan, C., ed., *Urban development in north-west Roman Southwark Excavations 1974–90, Monograph 16*. London, MoLAS, p. 89–105.

Macphail, R. I., 2004, *Land at Underdown Lane, Eddington, Herne Bay, Kent: Soil Micromorphology*. Salisbury, Wessex Archaeology.

Macphail, R. I., 2005, Soil micromorphology and chemistry, *in* Shelley, A., ed., *Dragon Hall, King Street, Norwich: Excavation and Survey of a Late Medieval Merchant's Trading Complex, Report No. 112*: Norwich, East Anglian Archaeology, p. 175–178.

Macphail, R. I., 2006, *Hyena Den, Wookey Hole, Somerset: Soil Micromorphology (report for Roger Jacobi, Natural History/British Museum)*. London, Institute of Archaeology, University College London.

Macphail, R. I., 2009, Marine inundation and archaeological sites: first results from the partial flooding of Wallasea Island, Essex, UK., 2009, Antiquity Project Gallery; http://antiquity.ac.uk/projgall/macphail/.

Macphail, R. I., 2010, *Chongokni, Hantan River, Imjin River Valley, Korea (2009): soil micromorphology*. Seoul, Seoul National University.

Macphail, R. I., 2011a, Soils and sediments, *in* Harding, J., and Healy, F., eds., *The Raunds Area Project. A Neolithic and Bronze Age Landscape in Northamptonshire. Vol. 2 Supplementary Studies*: Swindon, English Heritage, p. 737–838.

Macphail, R. I., 2011b, CD Table 11 Micromorphology – summarised soil data and interpretation; CD Table 12 Micromorphology – facies types (soil microfabric types and associated data), *in* Burch, M.,

Treveil, P., and Keene, D., eds., *The Development of Early Medieval and Later Poultry and Cheapside: Excavations at 1 Poultry and Vicinity, City of London, MOLA Monograph 38*. London, Museum of London.

Macphail, R. I., 2011c, Micromorphological analysis of road construction sediments, *in* An engineered Iron Age road, associated Roman use (Margary Route 64), and Bronze Age activity recorded at Sharpstone Hill, 2009, Malim, T. and Hayes, L: *Transactions of the Shropshire and Historical Society*, v. 85, p. 53–55.

Macphail, R. I., 2011d, Soil micromorphology, *in* Fulford, M., and Rippon, S., eds., *Pevensey Castle, Sussex. Excavations in the Roman fort and Medieval keep, 1993–95, Wessex Archaeology Report No. 26*: Salisbury, Wessex Archaeology and University of Reading, p. 109–121.

Macphail, R. I., 2012, *Gokstad Mound, Sandefjord, Vestfold, Norway: Soil Micromorphology (Report for Museum of Cultural History, University of Oslo)*: Institute of Archaeology, University College London.

Macphail, R. I., 2013, Site formation processes in archaeology: soil and sediment micromorphology. Proceedings of the 14th IWMSM Session 5, Lleida, Spain, July 2012 (Guest Editorial): *Quaternary International*, v. 315, p. 1–2.

Macphail, R. I., 2014, Reconstructing past land use from dark earth: examples from England and France, *in* Lorans, E., and Rodier, X., eds., *Archéologie de l'espace urbain – Partie II*, Vol. 21/17: Tours, CTHS, p. 251–261.

Macphail, R. I., 2016, *Follobaneprosjektet: Follobanen FO3 Arkeologigropa soil micromorphology (including SEM/EDS) (report for NIKU Norsk institutt for kulturminneforskning)*: Institute of Archaeology, University College London.

Macphail, R. I., 2016b, Privies and latrines, *in* Gilbert, A. S., ed., *Encyclopedia of Geoarchaeology*: Dordrecht, Springer Scientific, p. 682–687.

Macphail, R. I., 2016c, House pits and grubenhausen, *in* Gilbert, A. S., ed., *Encyclopedia of Geoarchaeology*: Dordrecht, Springer Scientific, p. 425–432.

Macphail, R. I., 2018, *Søndre gate, Trondheim (church and pre-church contexts), Norway; Soil Micromorphology (including SEM/EDS)* (report for NIKU): Institute of Archaeology, University College London.

Macphail, R. I., 2019a, *Holwell (Bronze Age Roundhouse), Dartmoor; Soil Micromorphology including SEM/EDS – Report 1*. Unpublished report for the Dartmoor Holwell project.

Macphail, R. I., 2019b, Sediment micromorphology, *in* Barton, R. N. E., Bouzouggar, A., Colcutt, S. N., and Humphrey, L. T., eds., *Cemeteries and Sedentism in Later Stone Age of NW Africa: Excavations at Grottes des Pigeons, Taforalt, Morocco, Vol. 147*: Mainz, Römisch-Germanisches Zentralmuseum Leibniz-Forschungsinstitut für Archäologie, p. 117–142.

Macphail, R. I., 2019c, *Sotrasambandet Project, Sites 11 and 13, Bildøy, Norway: soil micromorphology including SEM/EDS analyses (report for Bergen University)*: Institute of Archaeology, University College London.

Macphail, R. I., 2019d, Archaeological Soil Micromorphology Working Group, *in* Basu, A., ed., *Encyclopedia of Global Archaeology*, Springer Nature Switzerland.

Macphail, R. I., and Courty, M.-A., 1985, Interpretation and significance of urban deposits, *in* Edgren, T., and Jungner, H., eds., *Proceedings of the Third Nordic Conference on the Application of Scientific Methods in Archaeology*: Helsinki, The Finnish Antiquarian Society, p. 71–83.

Macphail, R. I., and Crowther, J., 2002a, *Canterbury CW12 (Cycling Centre – Roman rampart site): soil micromorphology and chemistry*. Canterbury, Canterbury Archaeological Trust.

Macphail, R. I., and Crowther, J., 2002b, *Battlesbury, Hampshire: soil micromorphology and chemistry (W4896)*. Salisbury, Wessex Archaeology.

Macphail, R. I., and Crowther, J., 2004, Tower of London Moat: sediment micromorphology, particle size, chemistry and magnetic properties, *in* Keevil, G., ed., *Tower of London Moat Excavation, Historic Royal Palaces Monograph 1*. Oxford, Oxford Archaeology, p. 41–43, 48–50, 78–79, 82–83, 155, 183–186, 202–204 and p. 271–284.

Macphail, R. I., and Crowther, J., 2006a, *The soil micromorphology, phosphate and magnetic susceptibility from White Horse Stone, Aylesford, Kent (WHS98) (report for Oxford Archaeology), CTRL Specialist Archive Report Series*. London, Institute of Archaeology, University College London, p. 1–48.

Macphail, R. I., and Crowther, J., 2006b, *Spitalfields: microstratigraphy (soil micromorphology, microprobe, chemistry and magnetic susceptibility) (Report for Museum of London Archaeological Service)*: Institute of Archaeology, University College London.

Macphail, R. I., and Crowther, J., 2007a, Soil micromorphology, chemistry and magnetic susceptibility studies at Huizui (Yiluo region, Henan Province, northern China), with special focus on a typical Yangshao floor sequence: *Bulletin of the Indo-Pacific Prehistory Association*, v. 27, p. 93–113.

Macphail, R. I., and Crowther, J., 2007b, *Freeschool Lane and Vine Street, Leicester: soil micromorphology, chemistry and magnetic susceptibility*: Leicester University Archaeological Service.

Macphail, R. I., and Crowther, J., 2008, Illustrations from soil micromorphology and complementary investigations, *in* Thiemeyer, H., ed., *Archaeological Soil Micromorphology – Contributions to the Archaeological Soil Micromorphology Working Group Meeting 3rd to 5th April 2008, Vol. D30*: Frankfurt A.M, Frankfurter Geowiss. Arb., p. 81–87.

Macphail, R. I., and Crowther, J., 2010, *Terminal 5: soil micromorphology, chemistry, magnetic susceptibility and particle size analyses (report for Framework Archaeology)*. Oxford, Framework Archaeology, Vol. 2, Section 19.

Macphail, R. I. and Crowther, J., 2011, Experimental pig husbandry: soil studies from West Stow Anglo-Saxon Village, Suffolk, UK, Antiquity Project Gallery, Vol. **85**, 330, Antiquity (http://antiquity. ac.uk/projgall/macphail330/).

Macphail, R. I., Crowther, J., and Berna, F., 2012, *Stanford Wharf, Essex (London Compensation Sites A and B; COMPA 09): Soil Micromorphology, Microchemistry, Chemistry, Magnetic Susceptibility and FTIR. (Report for Oxford Archaeology)*: Institute of Archaeology, University College London.

Macphail, R. I., and Crowther, J., 2015a, *Olduvai Gorge (HWKEE, MNK SKULL and MNK T5), Tanzania: Soil Micromorphology and Chemistry*. London, Institute of Archaeology, University College London.

Macphail, R. I., and Crowther, J., 2015b, *Thameslink (BVK11) 11–15 Borough High Street, London Borough of Southwark: Soil Micromorphology, Particle Size, Chemistry and Magnetic Susceptibility (Report for Oxford Archaeology)*: Institute of Archaeology, University College London.

Macphail, R. I., and Crowther, J., 2016, *Merv, Turkmenistan: analysis and interpretation of sample sequences 027 and 028 (soil micromorphology, chemistry and magnetic susceptibility)*. Institute of Archaeology, University College London.

Macphail, R. I., and Cruise, G. M., 2000, Soil micromorphology on the Mesolithic site, *in* Bell, M., Caseldine, A., and Neumann, H., eds., *Prehistoric Intertidal Archaeology in the Welsh Severn Estuary, Research Report 120*. York, Council for British Archaeology, p. 55–57 and CD-ROM.

Macphail, R. I., and Cruise, G. M., 2001, The soil micromorphologist as team player: a multianalytical approach to the study of European microstratigraphy, *in* Goldberg, P., Holliday, V., and Ferring, R., eds., *Earth Science and Archaeology*. New York, Kluwer Academic/Plenum Publishers, p. 241–267.

Macphail, R. I., and Goldberg, P., 1990, The micromorphology of tree subsoil hollows: their significance to soil science and archaeology, *in* Douglas, L. A., ed., *Soil-Micromorphology: a Basic and Applied Science, Developments in Soil Science 19*. Amsterdam, Elsevier, p. 425–429.

Macphail, R. I., and Goldberg, P., 1995, Recent advances in micromorphological interpretations of soils and sediments from archaeological sites, *in* Barham, A. J., and Macphail, R. I., eds., *Archaeological Sediments and Soils: Analysis, Interpretation and Management*. London, Institute of Archaeology, p. 1–24e.

Macphail, R. I., and Goldberg, P., 1999, The soil micromorphological investigation of Westbury Cave, *in* Andrews, P., Cook, J., Currant, A., and Stringer, C., eds., *Westbury Cave. The Natural History Museum Excavations 1976–1984*: Bristol, CHERUB (Centre for Human Evolutionary Research at the University of Bristol), p. 59–86.

Macphail, R. I., and Goldberg, P., 2000, Geoarchaeological investigation of sediments from Gorham's and Vanguard Caves, Gibraltar: Microstratigraphical (soil micromorphological and chemical) signatures, *in* Stringer, C. B., Barton, R. N. E., and Finlayson, J. C., eds., *Neanderthals on the Edge*. Oxford, Oxbow Books, p. 183–200.

Macphail, R. I., and Goldberg, P., 2003, Gough's Cave, Cheddar, Somerset: Microstratigraphy of the Late Pleistocene/earliest Holocene -sediments: *Bulletin of Natural History Museum London (Geology)*. v. 58 (supplement), p. 51–58.

Macphail, R. I., and Goldberg, P., 2018a, *Applied Soils and Micromorphology in Archaeology*. Cambridge, Cambridge University Press.

Macphail, R. I., and Goldberg, P., 2018b, Archaeological materials, *in* Stoops, G., Marcelino, V., and Mees, F., eds., *Interpretation of Micromorphological Features of Soils and Regoliths*. Amsterdam, Elsevier, p. 779–819.

Macphail, R. I., and Linderholm, J., 2004a, Neolithic land use in south-east England: a brief review of the soil evidence, *in* Cotton, J., and Field, D., eds., *Towards a New Stone Age*: York, CBA, p. 29–37.

Macphail, R. I., and Linderholm, J., 2004b, "Dark earth": recent studies of "dark earth" and "dark earth-like" microstratigraphy in England, *in* Verslype, L., and Brulet, R., eds., *Terres Noire; Dark Earth*: Louvain-la-Neuve, Université Catholique de Louvain, p. 35–42.

Macphail, R. I., and Linderholm, J., 2017, Avaldsnes: Scientific Analyses – Microstratigraphy (soil micromorphology and microchemistry, soil chemistry and magnetic susceptibility), *in* Skre, D., ed., *Avaldsnes – A Sea-King's Manor in First-Millennium Western Scandinavia, Band 104*. Berlin, De Gruyter, p. 379–420.

Macphail, R. I., and McAvoy, J. M., 2008, A micromorphological analysis of stratigraphic integrity and site formation at Cactus Hill, an Early Paleoindian and hypothesized pre-Clovis occupation in south-central Virginia, USA: *Geoarchaeology*, v. 23, no. 5, p. 675–694.

Macphail, R. I., Romans, J. C. C., and Robertson, L., 1987, The application of micromorphology to the understanding of Holocene soil development in the British Isles; with special reference to cultivation, *in* Fedoroff, N., Bresson, L. M., and Courty, M.-A., eds., *Soil Micromorphology*: Plaisir, Association Française pour l'Étude du Sol, p. 669–676.

Macphail, R. I., Courty, M.-A., and Gebhardt, A., 1990a, Soil micromorphological evidence of early agriculture in north-west Europe: *World Archaeology*, v. 22, no. 1, p. 53–69.

Macphail, R. I., Hather, J., Hillson, S. W., and Maggi, R., 1994c, The Upper Pleistocene deposits at Arene Candide: soil micromorphology of some samples from the Cardini 1940–42 excavations: *Quaternaria Nova*, v. IV, p. 79–100.

Macphail, R. I., Crowther, J., and Cruise, G. M., 1995, The soils, *in* Thurley, S., ed., *The King's Privy Garden at Hampton Court Palace 1689–1995*. London, Apollo, p. 116–118.

Macphail, R. I., Courty, M.-A., Hather, J., and Wattez, J., 1997, The soil micromorphological evidence of domestic occupation and stabling activities, *in* Maggi, R., ed., *Arene Candide: a Functional and Environmental Assessment of the Holocene Sequence (Excavations Bernabò Brea-Cardini 1940–50)*: Roma, Memorie dell'Istituto Italiano di Paleontologia Umana, p. 53–88.

Macphail, R. I., Cruise, G. M., Mellalieu, S. J., and Niblett, R., 1998, Micromorphological interpretation of a "Turf-filled" funerary shaft at St. Albans, United Kingdom: *Geoarchaeology*, v. 13, no. 6, p. 617–644.

Macphail, R. I., Crowther, J., and Cruise, G. M., 1999, *King Arthur's Cave: Soils of the Allerød Palaeosol (report for Oxford University)*. London, Institute of Archaeology, University College London.

Macphail, R. I., Cruise, G. M., Engelmark, R., and Linderholm, J., 2000, Integrating soil micromorphology and rapid chemical survey methods: new developments in reconstructing past rural settlement and landscape organization, *in* Roskams, S., ed., *Interpreting Stratigraphy*, Vol. 9. York, University of York, p. 71–80.

Macphail, R. I., Acott, T. G., and Crowther, J., 2001, *Boxgrove: Sediment microstratigraphy (soil micromorphology, image analysis and chemistry)*. London, Institute of Archaeology.

Macphail, R. I., Galinié, H., and Verhaeghe, F., 2003a, A future for dark earth?: *Antiquity*, v. 77, no. 296, p. 349–358.

Macphail, R. I., Crowther, J., Acott, T. G., Bell, M. G., and Cruise, G. M., 2003b, The experimental earthwork at Wareham, Dorset after 33 years: changes to the buried LFH and Ah horizon: *Journal of Archaeological Science*, v. 30, p. 77–93.

Macphail, R. I., Cruise, G. M., Allen, M. J., Linderholm, J., and Reynolds, P., 2004, Archaeological soil and pollen analysis of experimental floor deposits; with special reference to Butser Ancient Farm, Hampshire, UK: *Journal of Archaeological Science*, v. 31, p. 175–191.

Macphail, R. I., Cruise, G. M., Allen, M. J., and Linderholm, J., 2006, A rebuttal of the views expressed in "Problems of unscientific method and approach in Archaeological soil and pollen analysis of experimental floor deposits; with special reference to Butser Ancient Farm, Hampshire, UK by R. I. Macphail, G. M. Cruise, M. Allen, J. Linderholm and P. Reynolds" by Matthew Canti, Stephen Carter, Donald Davidson and Susan Limbrey: *Journal of Archaeological Science*, v. 33, p. 299–305.

Macphail, R. I., Crowther, J., and Cruise, G. M., 2007, Micromorphology and post-Roman town research: the examples of London and Magdeburg, *in* Henning, J., ed., *Post-Roman Towns and Trade in Europe, Byzantium and the Near-East. New Methods of Structural, Comparative and Scientific Methods in Archaeology*. Berlin, Walter de Gruyter & Co. KG, p. 303–317.

Macphail, R. I., Crowther, J., and Cruise, G. M., 2007b, Microstratigraphy: soil micromorphology, chemistry and pollen, *in* Bowsher, D., Dyson, T., Holder, N., and Howell, I., eds., *The London Guildhall. An archaeological history of a neighbourhood from early medieval to modern times, MoLAS Monograph 36*. London, Museum of London Archaeological Service, p. 18, 25–16, 35, 39, 55–16, 57, 59, 76, 90, 97, 98, 134, 154–135, p. 428–430.

Macphail, R. I., Crowther, J., and Cruise, G. M., 2008, Microstratigraphy: soil micromorphology, chemistry and pollen, *in* Bateman, N., Cowan, C., and Wroe-Brown, R., eds., *The Roman amphitheatre of London and its surroundings*. London, Museum of London Archaeological Service, p. 160–164.

Macphail, R. I., Crowther, J., and Cruise, G. M., 2009, *Pilgrims' School, Winchester: Soil micromorphology, pollen, chemistry and magnetic susceptibility (Report for Oxford Archaeology)*: Institute of Archaeology, University College London.

Macphail, R. I., Allen, M. J., Crowther, J., Cruise, G. M., and Whittaker, J. E., 2010, Marine inundation: Effects on archaeological features, materials, sediments and soils: *Quaternary International*, v. 214, no. 1–2, p. 44–55.

Macphail, R. I., Crowther, J., and Macphail, G. M., 2011, Soil micromorphology, chemistry and magnetic susceptibility, *in* Ford, B. M., and Teague, S., eds., *Winchester – A City in the Making. Archaeological investigations between 2002 and 2007 on the sites of Northgate House, Staple Gardens and the former Winchester Library, Jewry St., Oxford Archaeology Monograph No 12*. Oxford, Oxford Archaeology, p. 376, CD Part 373.317 (Soil Micromorphology Report.pdf).

Macphail, R. I., Crowther, J., and Berna, F., 2012a, Soil micromorphology, microchemistry, chemistry, magnetic susceptibility and FTIR, *in* Biddulph, E., Foreman, S., Stafford, E., Stansbie, D., and Nicholson, R., eds., *London Gateway*. Iron Age and Roman salt making in the Thames Estuary; Excavations at Stanford Wharf Nature Reserve, Essex (http://library.thehumanjourney.net/909), Oxford Archaeology Monograph No. 18. Oxford, Oxford Archaeology, p. 193.

Macphail, R. I., Goldberg, P., and Barton, R. N. E., 2012b, Vanguard Cave sediments and soil micromorphology, *in* Barton, R. N. E., Stringer, C., and Finlayson, C., eds., *Neanderthals in Context. A report of the 1995–1998 excavations at Gorham's and Vanguard Caves, Gibraltar, Monograph 75.* Oxford, Oxford University School of Archaeology, p. 193–210; color figures at: www.arch.ox.ac.uk/gibraltar.

Macphail, R. I., Bill, J., Cannell, R., Linderholm, J., and Rødsrud, C. L., 2013a, Integrated microstratigraphic investigations of coastal archaeological soils and sediments in Norway: the Gokstad ship burial mound and its environs including the Viking harbour settlement of Heimdaljordet, Vestfold: *Quaternary International*, no. 315, p. 131–146.

Macphail, R. I., Crowther, J., and Wenbann-Smith, F. F., 2013b, Soil micromorphology, loss-on-ignition, phosphate-P concentrations and magnetic susceptibility analyses (and Appendix 7, Thin Section Data), *in* Wenbann-Smith, F. F., ed., *The Ebbsfleet Elephant Excavations at Southfleet Road, Swanscombe in advance of High Speed 1, 2003 –4, Oxford Archaeology Monograph*. Oxford, Oxford Archaeology, p. 111–137; p. 511–521.

Macphail, R. I., Cruise, G. M., and Linderholm, J., 2014, Soil micromorphology and chemistry, *in* Ashwin, T., and Tester, A., eds., *A Roman Settlement in the Waveney Valley: Excavations at Scole, 1993–4*, East Anglian Archaeology Report No. 152, p. Chapter 9, Section VI, 422–431 (CD).

Macphail, R. I., Linderholm, J., and Crowther, J., 2015, *Alna River Sediments Oslo: Investigation of human impact in estuary sediments from Oslo. Sørenga D1A, borehole site: Sediment micromorphology and particle size analysis. Report for Cultural History Museum, University of Oslo.* London, Institute of Archaeology, University College London.

Macphail, R. I., Cruise, G. M., Courty, M.-A., Crowther, J., and Linderholm, J., 2016, 27. E6 Gudbrandsdalen Valley Project (Brandrud, Fryasletta, Grytting and Øybrekka), Oppland, Norway: soil micromorphology (with selected microchemistry, bulk soil chemistry, carbon polymer, particle size and pollen analyses), *in* Gundersen, I. M., ed., *Gård og utmark i Gudbrandsdalen. Arkeologiske undersøkelser i Fron 2011–2012.* Kristiansand, Portal forlag, p. 304–317.

Macphail, R. I., Bill, J., Crowther, J., Haită, C., Linderholm, J., Popovici, D., and Rødsrud, C. L., 2017a, European ancient settlements – a guide to their composition and morphology based on soil micromorphology and associated geoarchaeological techniques; introducing the contrasting sites of Chalcolithic Borduşani-Popină, Borcea River, Romania and the Viking Age Heimdaljordet, Vestfold, Norway: *Quaternary International*, v. 460, p. 30–47.

Macphail, R. I., Graham, E., Crowther, J., and Turner, S., 2017b, Marco Gonzalez, Ambergris Caye, Belize: A geoarchaeological record of ground raising associated with surface soil formation and the presence of a Dark Earth: *Journal of Archaeological Science*, v. 77, p. 35–51.

Macphail, R. I., Linderholm, J., and Eriksson, S., 2017c, *Korsmyra 1, Bud, Møre og Romsdal, Norway: soil micromorphology (including SEM/EDS), chemistry and magnetic susceptibility studies (report for Cultural History Museum, University of Oslo).* London, Institute of Archaeology, University College London.

Macphail, R. I., Linderholm, J., and Eriksson, S., 2017d, *Åker gård 7/201 (Nye Åker E6), Hamark, Hedmark, Norway: soil micromorphology (including SEM/EDS), chemistry and magnetic susceptibility studies (report for Cultural history Museum, University of Oslo).* London, Institute of Archaeology, University of London.

Macphail, R. I., Linderholm, J., Erikson, S., and Hristov, C., 2020a, *Gjellestad Ship Mound, 28/1, Halden k., Østfold, Norway; Soil Micromorphology (including SEM/EDS), Chemistry and Magnetic Susceptibility (Report for Cultural History Museum, University of Oslo)*: Institute of Archaeology, University College London.

Macphail, R. I., Linderholm, J., Eriksson, S., and Hristov, C., 2020b, *KBM 4286 – Råduspladsen Nord (City Hall Square, Copenhagen, Denmark); Soil micromorphology, magnetic susceptibility, phosphate and geochemical elemental analyses (XRF) (report for Copenhagen Museum)*: Institute of Archaeology, University College London.

Macphail, R. I., Linderholm, J., Gjerpe, L.-E., Buckland, P., Eriksson, S., and Hristov, K., in press, Dobbeltspor Dilling in Rygge, Østfold, Norway: Geoarchaeology and Morphology of a Mixed Farming Settlement *in* Gjerpe, L.-E., ed., *Dilling – en landsby fra førromersk jernalder (Dilling – a pre-roman Iron Age village)*.

Macphail, R. I., Linderholm, J., and Gjerpe, L.-E., in review, Speculations on farming development during the early Iron Age (500 BC–550 AD) of southern Norway focusing on the Dobbeltspor Dilling Project, *in* Sulas, F., Lewis, H. A., and Arroyo-Kalin, M., eds., *Inspired Geoarchaeologies*.

Macumber, P. G., Head, M. J., Chivas, A. R., and De Deckker, P., 1991, Implications of the Wadi al-Hammeh sequences for the terminal drying of Lake Lisan, Jordan: Palaeogeography, Palaeoclimatology, *Palaeoecology*, v. 84, no. 1–4, p. 163–173.

Madella, M., Jones, M. K., Goldberg, P., Goren, Y., and Hovers, E., 2002, Exploitation of plant resources by Neanderthals in Amud Cave (Israel): the evidence from phytolith studies: *Journal of Archaeological Science*, v. 29, no. 7, p. 703–719.

Magaritz, M., Goodfriend, G. A., Berger, W. H., and Labeyrie, L. D., 1987, Movement of the desert boundary in the Levant from latest Pleistocene to early Holocene, *in* Berger, W. H. and Labeyrie, L. D., eds, *Abrupt Climatic Change; Evidence and Implications, Nato Science Series C*, v. 216. Dordrecht, Springer, p. 173–183.

Magaritz, M., Kaufman, A., and Yaalon, D. H., 1981, Calcium carbonate nodules in soils: $^{18}O/^{16}O$ and $^{13}C/^{12}C$ ratios and ^{14}C Contents: *Geoderma*, v. 25, p. 157–172.

Maggi, R., 1997, *Arene Candide: a Functional and Environmental Assessment of the Holocene Sequence (Excavations Bernabò Brea-Cardini 1940–50)*, Roma, Memorie dell'Istituto Italiano di Paleontologia Umana.

Maher, L. A., 2004, *The Epipalaeolithic in Context: Palaeolandscapes and Prehistoric Occupation of Wadi Ziqlab, Northern Jordan*, [Doctoral Dissertation]: University of Toronto.

Malim, T., and Hayes, L., 2011, An engineered Iron Age road, associated Roman use (Margary Route 64), and Bronze Age activity recorded at Sharpstone Hill, 2009: *Transactions of the Shropshire Archaeological and Historical Society*, v. 85, p. 7–80.

Malinowski, R., 1979, Concretes and mortars in ancient aqueducts: *Concrete International*, v. 1, p. 66–76.

Mallol, C., 2004, *Micromorphological Observations from the Archaeological Sediments of 'Ubeidiya (Israel), Dmanisi (Georgia) and Gran Dolina-Td10 (Spain) for the Reconstruction of Hominid Occupation Contexts*, [Doctoral Dissertation]: Harvard University, 277 p.

Mallol, C., and Goldberg, P., 2017, Cave and Rock Shelter Sediments, *in* Nicosia, C., and Stoops, G., eds., *Archaeological Soil and Sediment Micromorphology*. Oxford, Wiley Blackwell, p. 359–381.

Mallol, C., and Henry, A., 2017, Ethnoarchaeology of Paleolithic fire: methodological considerations: *Current Anthropology*, v. 58, no. S16, p. S217–S229.

Mallol, C., Marlowe, F. W., Wood, B. M., and Porter, C. C., 2007, Earth, wind, and fire: ethnoarchaeological signals of Hadza fires: *Journal of Archaeological Science*, v. 34, p. 2035–2052.

Mallol, C., Cabanes, D., and Baena, J., 2010, Microstratigraphy and diagenesis at the upper Pleistocene site of Esquilleu Cave (Cantabria, Spain): *Quaternary International*, v. 214, no. 1–2, p. 70–81.

Mallol, C., Hernández, C. M., Cabanes, D., Machado, J., Sistiaga, A., Pérez, L., and Galván, B., 2013a, Human actions performed on simple combustion structures: An experimental approach to the study of Middle Palaeolithic fire: *Quaternary International*, v. 315, p. 3–15.

Mallol, C., Hernández, C. M., Cabanes, D., Sistiaga, A., Machado, J., Rodríguez, Á., Pérez, L., and Galván, B., 2013b, The black layer of Middle Palaeolithic combustion structures. Interpretation and archaeostratigraphic implications: *Journal of Archaeological Science*, v. 40, no. 5, p. 2515–2537.

Mallol, C. M., Mentzer, S. M., and Miller, C. E., 2017, Combustion features, *in* Nicosia, C., and Stoops, G., eds., *Encyclopedia of Archaeological Soil and Sediment Micromorphology*. Oxford, Wiley Blackwell, p. 299–330.

Maltby, E., and Caseldine, C. J., 1982, Prehistoric soil and vegetation development on Bodmin Moor, Southwestern England: *Nature*, v. 297, p. 397–400.

Malvern Instruments, 2007, *Mastersizer 2000 manual*. Malvern, Malvern Instruments.

Mandel, R. D., 1992, Soils and Holocene landscape evolution in Central and Southwestern Kansas: Implications for archaeological research, *in* Holliday, V. T., ed., *Soils in Archaeology: Landscape evolution and human occupation*. Washington, DC, Smithsonian Institution Press, p. 41–100.

Mandel, R., 1995, Geomorphic controls of the Archaic record in the Central Plains of the United States, *in* Bettis, E. A., III, ed., *Archaeological Geology of the Archaic period in North America; Special Paper 297*: Boulder, CO, The Geological Society of America, p. 37–66.

Mandel, R. D., 2000a, *Geoarchaeology in the Great Plains*. Norman, OK, University of Oklahoma Press.

Mandel, R. D., 2000b, The Past, Present, and Future, *in* Mandel, R. D., ed., *Geoarchaeology in the Great Plains*. Norman, OK, University of Oklahoma Press, p. 286–295.

Mandel, R., 2008, Buried Paleoindian-age landscapes in stream valleys of the Central Plains, USA: *Geomorphology*, v. 101, no. 1–2, p. 342–361.

Mandel, R., and Bettis, E. A., III, 2001, Use and analysis of soils by archaeologists and geoscientists; a North American perspective, *in* Goldberg, P., Holliday, V. T., and Ferring, C. R., eds., *Earth Sciences and Archaeology*. New York, Kluwer Academic/Plenum Publishers, p. 173–204.

Mandel, R. D., Arpin, T., and Goldberg, P., 2003, *Stratigraphy, Lithology, and Pedology of the South Wall at the Hopeton Earthworks, South-Central Ohio*: Kansas Geological Survey Open File Report 2003-46, p. 49.

Mandel, R. D., Thoms, A. V., Nordt, L. C., and Jacob, J. S., 2018, Geoarchaeology and paleoecology of the deeply stratified Richard Beene site, Medina River valley, South-Central Texas, USA: *Quaternary International*, v. 463, p. 176–197.

Mania, D., and Mania, U., 2005, The natural and socio-cultural environment of *Homo erectus* at Bilzingsleben, Germany, *in* Gamble, C., and Porr, M., eds., *The Hominid Individual in Context. Archaeological Investigations of Lower and Middle Palaeolithic Landscapes, Locales and Artifacts*. London, Routledge, p. 98–114.

Mannino, M. A., and Thomas, K., 2004–2006, New radiocarbon dates for hunter gatherers and early farmers in Sicily: *Accordia Research Papers*, v. 10, p. 13–33.

Mannino, M. A., Thomas, K. D., Leng, M. J., Piperno, M., Tusa, S., and Tagliacozzo, A., 2007, Marine resources in the Mesolithic and Neolithic at the Grotta dell'Uzzo (Sicily): evidence from isotope analyses of marine shells: *Archaeometry*, v. 49, no. 1, p. 117–133.

Marcelino, V., Schaefer, C. E., and Stoops, G., 2018, Oxic and related Materials, *in* Stoops, G., Marcelino, V., and Mees, F., eds., *Interpretation of Micromorphological Features of Soils and Regoliths*. Amsterdam, Elsevier, p. 663–690.

Marcolongo, B., and Mantovani, F., 1997, *Photogeology: Remote Sensing Applications in Earth Science*, Enfield, NH, Science Publishers Inc.

Marean, C. W., 2014, The origins and significance of coastal resource use in Africa and Western Eurasia: *Journal of Human Evolution*, v. 77, p. 17–40.

Marean, C. W., Goldberg, P., Avery, G., Grine, F. E., and Klein, R. G., 2000, Middle Stone Age Stratigraphy and Excavations at Die Kelders Cave 1 (Western Cape Province, South Africa): the 1992, 1993, and 1995 Field Seasons: *Journal of Human Evolution*, v. 38, p. 7–42.

Marean, C. W., Bar-Matthews, M., Bernatchez, J., Fisher, E., Goldberg, P., Herries, A. I. R., Jacobs, Z., Jerardino, A., Karkanas, P., Minichillo, T., Nilssen, P. J., Thompson, E., Watts, I., and Williams, H. M., 2007, Early human use of marine resources and pigment in South Africa during the Middle Pleistocene: *Nature*, 449, p. 905–908.

Marean, C. W., Bar-Matthews, M., Fisher, E., Goldberg, P., Herries, A., Karkanas, P., Nilssen, P. J., and Thompson, E., 2010, The stratigraphy of the Middle Stone Age sediments at Pinnacle Point Cave 13B (Mossel Bay, Western Cape Province, South Africa): *Journal of Human Evolution*, v. 59, no. 3–4, p. 234–255.

Marković, S. B., Hambach, U, Stevens, T., Kukla, G. J., Heller, F., McCoy, W. D., Oches, E. A., Buggle, B. and Zöller, L. 2011. The last million years recorded at the Stari Slankamen (Northern Serbia) loess-palaeosol sequence: revised chronostratigraphy and long-term environmental trends: *Quaternary Science Reviews*, v. 30, no. 9–10, p. 1142–1154.

Markovič, S. B., Korač, M., Mrđič, N., Buylaert, J. P., Thiel, C., McLaren, S. J., Stevens, T., Tomič, N., Petič, N., Jovanovič, M., Vasiljević, D. A., Sümegi, P. Gavrilov, M. B., and Obreht, I. 2014. Palaeoenvironment and geoconservation of mammoths from the Nosak loess–palaeosol sequence (Drmno, northeastern Serbia): Initial results and perspectives: *Quaternary International*, v. 334–335, p. 34–39.

Marks, A. E., 1983. The sites of Boker Tachtit and Boker: a brief introduction, *in* Marks, A. E., ed., *Prehistory and Paleoenvironments in the Central Negev, Israel, III*. Dallas, TX, Department of Anthropology, Institute for the Study of Earth and Man, Southern Methodist University, p. 15–37.

Martínez-Pabello, P. U., Sedov, S., Solleiro-Rebolledo, E., Solé, J., Pi-Puig, T., Alcántara-Hernández, R. J., Lebedeva, M., Shishkov, V., and Villalobos, C., 2021, Rock varnish in La Proveedora/Sonora in the context of desert geobiological processes and landscape evolution: *Journal of South American Earth Sciences*, v. 105, p. 102959.

Maslin, S. P., 2015, The taphonomy and micromorphology of sunken-featured buildings from Lyminge, Kent: A comparative mixed-method analysis: *Environmental Archaeology*, v. 20, no. 2, p. 202–220.

Mason, J. A., and Kuzila, M. S., 2000, Episodic Holocene loess deposition in central Nebraska: *Quaternary International*, v. 67, no. p. 119–131.

Matarazzo, T., 2015, *The Early Bronze Age Village of Afragola in Southern Italy*. Oxford, Archaeopress.

Matarazzo, T., Berna, F., and Goldberg, P., 2010, Occupation surfaces sealed by the Avellino eruption of Vesuvius at the Early Bronze Age Village of Afragola in Southern Italy: A micromorphological analysis: *Geoarchaeology*, v. 25, no. 4, p. 437–466.

Mateu, M., Bergadà, M. M., and Garcia i Rubert, D., 2013, Manufacturing technical differences employing raw earth at the protohistoric site of Saint Jaume (Alcanar, Tarragona, Spain): construction and furniture elements: *Quaternary International*, v. 315, p. 76–86.

Matsui, A., Hiraya, R., Mijaji, A., and Macphail, R. I., 1996, Availability of soil micromorphology in archaeology in Japan (in Japanese): *Archaeology Society of Japan*, v. 38, p. 149–152.

Matthews, W., 1995, Micromorphological characterisation and interpretation of occupation deposits and microstratigraphic sequences at Abu Salabikh, Southern Iraq, *in* Barham, A. J., and Macphail,

R. I., eds., *Archaeological Sediments and Soils: Analysis, Interpretation and Management*. London, Institute of Archaeology, p. 41–74.

Matthews, W., 2005, Micromorphological and microstratigraphic traces of uses and concept of space, *in* Hodder, I., ed., *Inhabiting Çatalhöyük: Reports from the 1995–99 Seasons, BIAA Monograph 40*. Cambridge, McDonald Institute, p. 355–398.

Matthews, W., 2010, Geoarchaeology and taphonomy of plant remains and microarchaeological residues in early urban environments in the Ancient Near East: *Quaternary International*, v. 214, no. 1–2, p. 98–113.

Matthews, W., French, C. A. I., Lawrence, T., Cutler, D. F., and Jones, M. K., 1996, Multiple surfaces: the micromorphology, *in* Hodder, I., ed., *On the Surface: Çatalhöyük 1993–1995*. Cambridge, The McDonald Institute for Research and British Institute of Archaeology at Ankara, p. 301–342.

Matthews, W., French, C. A. I., Lawrence, T., Cutler, D. F., and Jones, M. K., 1997, Microstratigraphic traces of site formation processes and human activities: *World Archaeology*, v. 29, no. 2, p. 281–308.

Matthews, W., Hastorf, C. A., and Ergenekon, B., 2000, Ethnoarchaeology: studies in local villages aimed at understanding aspects of the Neolithic site, *in* Hodder, I., ed., *Towards Reflexive Method in Archaeology: the Example at Çatalhöyük*. Cambridge, McDonald Institute for Archaeological Research and British Institute of Archaeology at Ankara., p. 177–188.

Matthews, R., Matthews, W., Raheem, K. R., and Richardson, A., 2020, *The Early Neolithic of the Eastern Fertile Crescent. Excavations at Bestansur and Shimshara, Iraqi Kurdistan. Central Zagros Archaeological Project, 2*. Oxford, Oxbow.

May, D. W., and Holen, S. R., 2014, Early Holocene alluvial stratigraphy, chronology, and Paleoindian/ Early Archaic geoarchaeology in the Loup River Basin, Nebraska, U.S.A: *Quaternary International*, v. 342, p. 73–90.

McAuliffe, J. R., 2000, Desert Soils, *in* Philipps, S. J., and Comus, P. W., eds., *A Natural History of the Sonoran Desert*: Tucson, AZ, Arizona-Sonora Desert Museum Press, p. 87–104.

McBratney, A. B., Moran, C. J., Stewart, J. B., Cattle, S. R., and Koppi, A. J., 1992, Modifications to a method of rapid assessment of soil macropore structure by image analysis: *Geoderma*, v. 53, p. 255–274.

McBratney, A. B., Bishop, T. F. A., and Teliatnikov, I. S., 2000, Two soil profile reconstruction techniques: *Geoderma*, v. 97, no. 3–4, p. 209–221.

McCann, J. M., Woods, W. I., and Meyer, D. W., 2001, Organic matter and anthrosols in Amazonia: Interpreting the Amerindian legacy, *in* Rees, R. M., Ball, B. C., Campbell, C. D., and Watson, C. A., eds., *Sustainable Management of Soil Organic Matter*. Wallingford, CAB International, p. 180–189.

McDonnell, G., ed., 2012. *North Lincolnshire Coversands Research Project*. Report for the ALSF scheme. Bradford, University of Bradford.

McIntyre, D. S., 1958a, Permeability measurements of soil crusts formed by raindrop impact: *Soil Science*, v. 85, p. 185–189.

McIntyre, D. S., 1958b, Soil splash and the formation of surface crusts by raindrop impact: *Soil Science*, v. 85, p. 261–266.

McKeague, A., Macdougall, J. I., and Miles, N. M., 1973, Micromorphological, physical and mineralogical properties of a catena of soils from Prince Edward Island, in relation to their classification and genesis: *Canadian Journal of Soil Science*, v. 53, p. 281–295.

McKeague, J. A., 1985, Clay skins and argillic horizons, *in* Bullock, P. a. M., C. P., ed., *Soil Micromorphology, Vol. 2: Soil Genesis*: Berkhamsted, AB Academic Publishers, p. 367–388.

McKee, E. D., 1979, Introduction to a study of global sand seas, *in* McKee, E. D., ed., *A study of global sand seas*; U. S. Geological Survey Professional Paper, Vol. 1052, p. 1–19.

McOmish, D., Field, D., and Brown, G., 2010, The Late Bronze Age and Early Iron Age Midden Site at East Chisenbury: *Wiltshire Archaeological and Natural History Magazine*, v. 103, p. 35–101.

McPherron, S. P., Braun, D. R., Dogandžić, T., Archer, W., Desta, D., and Lin, S. C., 2014, An experimental assessment of the influences on edge damage to lithic artifacts: a consideration of edge angle, substrate grain size, raw material properties, and exposed face: *Journal of Archaeological Science*, v. 49, p. 70–82.

Mees, F., 2018, Authigenic silicate minerals – sepiolite-palygorskite, zeolites and sodium silicates, *in* Stoops, G., Marcelino, V., and Mees, F., eds., *Interpretation of Micromorphological Features of Soils and Regoliths*. Amsterdam, Elsevier, p. 177–203.

Mees, F., and Stoops, G., 2018, Sulphidic and sulphuric materials, *in* Stoops, G., Marcelino, V., and Mees, F., eds., *Interpretation of Micromorphological Features of Soils and Regoliths*. Amsterdam, Elsevier, p. 347–376.

Mees, F., and Tursina, T. V., 2018, Salt minerals in saline soils and salt crusts, *in* Stoops, G., Marcelino, V., and Mees, F., eds., *Interpretation of Micromorphological Features of Soils and Regoliths*. Amsterdam, Elsevier, p. 289–321.

Meignen, L., Bar-Yosef, O., and Goldberg, P., 1989, Les structures de combustion moustériennes de la grotte de Kébara (Mont Carmel, Israel): *Nature et Fonction des Foyers Préhistoriques, Mémoires du Musée de Préhistoire d'Ille de France*, v. 2, p. 141–146.

Meignen, L., Goldberg, P., and Bar-Yosef, O., 2007, The Hearths at Kebara Cave and Their Role in Site Formation Processes, *in* Bar-Yosef, O., and Meignen, L., eds., *Kebara Cave Mt. Carmel, Israel. The Middle and Upper Paleolithic Archaeology. Part I*. Cambridge, Massachusetts, Peabody Museum of Archaeology and Ethnology Harvard University, p. 91–122.

Meignen, L., Goldberg, P., Albert, R. M., and Bar-Yosef, O., 2009, Structures de combustion, choix des combustibles et degré de mobilité des groupes dans le Paléolithique moyen du Proche-Orient: exemples des grottes de Kébara et d'Hayonim (Israël), *in* Théry-Parisot, I., Costamagno, S., and Henry, A., eds., *Gestion des combustibles au Paléolithique et Mésolithique: nouveaux outils, nouvelles interprétations/Fuel management during the Paleolithic and Mesolithic period: new tools, new interpretations, Vol. 1914*. Oxford, Archeopress, p. 111–118.

Meignen, L., Goldberg, P., and Bar-Yosef, O., 2017, Together in the field: interdisciplinary work in Kebara and Hayonim caves (Israel): *Archaeological and Anthropological Sciences*, v. 9, no. 8, p. 1603–1612.

Mekel, J. F. M., 1978, *The use of aerial photographs and other images in geological mapping, Enschede*, The Netherlands, International Institute for Aerial Survey and Earth Sciences (ITC), ITC textbook.

Menotti, F., and O'Sullivan, A., 2012, *The Oxford Handbook of Wetland Archaeology*. Oxford, Oxford University Press.

Mentzer, S. M., 2014, Microarchaeological approaches to the identification and interpretation of combustion features in prehistoric archaeological sites: *Journal of Archaeological Method and Theory*, v. 21, no. 3, p. 616–668.

Mentzer, S. M., 2016, Hearths and Combustion Features, *in* Gilbert, A. S., ed., *Encyclopedia of Geoarchaeology*: Dordrecht, Springer, p. 411–424.

Mentzer, S. M., 2017a, Rockshelter settings, *in* Gilbert, A. S., ed., *Encyclopedia of Geoarchaeology*: Dordrecht, Springer, p. 725–743.

Mentzer, S. M., 2017b, Micro XRF, *in* Nicosia, C., and Stoops, G., eds., *Archaeological Soil and Sediment Micromorphology*. Chichester, Wiley Blackwell, p. 431–440.

Mentzer, S. M., and Quade, J., 2013, Compositional and isotopic analytical methods in archaeological micromorphology: *Geoarchaeology*, v. 28, p. 87–97.

Mentzer, S. M., Romano, D. G., and Voyatzis, M. E., 2017, Micromorphological contributions to the study of the ritual behavior at the ash altar to Zeus on Mt. Lykaion, Greece: *Archaeological and Anthropological Sciences*, v. 9, p. 1017–1043.

Mercer, R., 1981, Excavations at Carn Brea: *Cornish Archaeology*, v. 20, p. 259–263.

Mercier, N., and Valladas, H., 2003, Reassessment of TL age estimates of burnt flints from the Paleolithic site of Tabun Cave, Israel: *Journal of Human Evolution*, v. 45, p. 401–409.

Mercier, N., Valladas, H., Valladas, G., and Reyss, J.-L., 1995, TL dates of burnt flints from Jelinek's excavations at Tabun and their implications: *Journal of Archaeological Science*, v. 22, p. 495–509.

Mercier, N., Valladas, H., Froget, L., Joron, J.-L., Reyss, J.-L., Weiner, S., Goldberg, P., Meignen, L., Bar-Yosef, O., and Belfer-Cohen, A., 2007, Hayonim Cave: a TL-based chronology for this Levantine Mousterian sequence: *Journal of Archaeological Science*, v. 34, no. 7, p. 1064–1077.

Mertens, G., and Elsen, J., 2006, Use of computer assisted image analysis for the determination of the grain-size distribution of sands used in mortars: *Cement and Concrete Research*, v. 36, no. 8, p. 1453–1459.

Middleton, W. D., and Douglas-Price, T., 1996, Identification of activity areas by multi-element characterisation of sediments from modern and archaeological house floors using inductively coupled plasma-atomic emission spectroscopy: *Journal of Archaeological Science*, v. 23, p. 673–687.

Miedema, R., Jongmans, A. G., and Slager, S., 1974, Micromorphological observations on pyrite and its oxidation products in four Holocene alluvial soils in the Netherlands, *in* Rutherford, G. K., ed., *Soil Microscopy*. Kingston, Ontario, The Limestone Press, p. 772–794.

Mighall, T. M., Timberlake, S., Foster, I. D. L., Krupp, E., and Singh, S., 2009, Ancient copper and lead pollution records from a raised bog complex in Central Wales, UK: *Journal of Archaeological Science*, v. 36, p. 1504–1515.

Mikkelsen, J. H., Langohr, R., Boas, N. A., and Macphail, R. I., 2006, Land use and environmental degradation in Bronze Age settlements, Eastern Jutland, Denmark, *in* Engelmark, R., and Linderholm, J., eds., *Proceedings from the 8th Nordic Conference on the Application of Scientific Methods in Archaeology in Umeå 2001, Archaeology and Environment 21*: Ümea, Ümea University, p. 81–92.

Mikkelsen, J. H., Langohr, R. and Macphail, R. I. 2007. Soilscape and land-use evolution related to drift sand movements since the Bronze Age in Eastern Jutland, Denmark. *Geoarchaeology*, v. 22, no. 2, p. 155–180.

Mikkelsen, J. H., Langohr, R., Vanwesenbeeck, V., Bourgeois, I., and De Clercq, W., 2019, The Byre's Tale. Farming nutrient-poor cover sands at the edge of the Roman Empire (NW-Belgium), *in* Deak, J., Ampe, C., and Mikkelsen, J. H., eds., *Soils as Records of Past and Present From soil surveys to archaeological sites: research strategies for interpreting soil characteristics. Proceedings of the Geoarchaeological Meeting Bruges (Belgium), 6 & 7*: Bruges, Raakvlak, p. 64–85.

Milek, K., 2006, *Houses and Households in Early Icelandic Society: Geoarchaeology and the Interpretation of Social Space*, [Doctoral Dissertation]: University of Cambridge.

Milek, K. B., 2012, Floor formation processes and the interpretation of site activity areas: an ethnoarchaeological study of turf buildings at Thverá, northeast Iceland: *Journal of Anthropological Archaeology*, v. 31, no. 2, p. 119–137.

Milek, K., and French, C., 2007, Soils and sediments in the settlement and harbour at Kaupang, *in* Skre, D., ed., *Kaupang in Skiringssal*: Aarhus, Aarhus University Press, p. 321–360.

Milek, K. B., and Roberts, H. M., 2013, Integrated geoarchaeological methods for the determination of site activity areas: a study of a Viking Age house in Reykjavik, Iceland: *Journal of Archaeological Science*, v. 40, no. 4, p. 1845–1865.

Miller, C. E., 2011, Deposits as Artifacts: *Mitteilungen der Tübinger Verein zur Förderung der Ur- und Frühgeschichtlichen Archäologie* v. 12, p. 91–107.

Miller, C. E., 2015, *A Tale of Two Swabian Caves: Geoarchaeological investigations at Hohle Fels and Geißenklösterle*, Tübingen, Kerns Verlag, Tübingen Publications in Prehistory.

Miller, C. E., Conard, N. J., Goldberg, P., and Berna, F., 2009, Dumping, sweeping and trampling: experimental micromorphological analysis of anthropogenically modified combustion features, *in* Théry-Parisot I., Chabal L., Costamagno S. eds., *The Taphonomy of Burned Organic Residues and Combustion Features in Archaeological Contexts, Proceedings of the Round Table, May 27–29 2008, P@lethnology*, v. **2**, p. 25–37.

Miller, C. E., Goldberg, P., and Berna, F., 2013, Geoarchaeological investigations at Diepkloof Rock Shelter, Western Cape, South Africa: *Journal of Archaeological Science*, v. 40, no. 9, p. 3432–3452.

Miller, C. E., Mentzer, S. M., Berthold, C., Leach, P., Ligouis, B., Tribolo, C., Parkington, J., and Porraz, G., 2016, Site-formation processes at Elands Bay Cave, South Africa: *Southern African Humanities*, v. 29, no. 1, p. 69–128.

Milner, N., Conneller, C., Taylor, B., and Schadla-Hall, R. T., 2012, *The Story of Star Carr*, York, Council for British Archaeology.

Minnis, P. E., and Whalen, M. E., 2015, Ecology and food economy, *in* Minnis, P. E., and Whalen, M. E., eds., *Ancient Paquimé and the Casas Grandes World, American Studies in Anthropology*: Tucson, AZ, University of Arizona Press, p. 41–57.

Misarti, N., Fuinney, B. P., and Maschner, H., 2011, Reconstructing site organisation in the eastern Aleutian Islands, Alaska using multi-element chemical analysis of soils: *Journal of Archaeological Science*, v. 38, p. 1441–1455.

Miskovsky, J.-C., 1969, Sédimentologie des couches supérieures de la grotte du Lazaret: *Mémoire de la Société Préhistorique Française*, v. tome 7, p. 25–51.

Mizoguchi, K., 2013, *The Archaeology of Japan: from the Earliest Rice Farming Villages to the Rise of the State*. Cambridge, Cambridge University Press.

Mjærum, A., 2012, Dyrkningsspor og fegate fra eldre jernalder på Hørdalen (lok. 51), *in* Gjerpe, L.-E., and Mjærum, A., eds., *E18-prosjektet Gulli-Langåker. Jordbruksbosetning og graver i Tønsberg og Stokke, Bind 2*: Bergen, Fagbokforlaget, p. 187–256.

Mjærum, A., 2020, The emergence of mixed farming in eastern Norway: *Agricultural History Review* v. 68, no. 1, p. 1–21.

Montufo, A. M., 1997, The use of satellite imagery and digital image processing in landscape archaeology. A Case Study from the Island of Mallorca, Spain: *Geoarchaeology*, v. 12, p. 71–85.

Mora, P., Mora, L., and Philipott, P., 1984, *Conservation of wall paintings*. London, Butterworths.

Morisawa, M., 1985, *Rivers*. New York, Longman.

Morley, M. W., Goldberg, P., Sutikna, T., Tocheri, M. W., Prinsloo, L. C., Jatmiko, Saptomo, E. W., Wasisto, S., and Roberts, R. G., 2017, Initial micromorphological results from Liang Bua, Flores (Indonesia): Site formation processes and hominin activities at the type locality of Homo floresiensis: *Journal of Archaeological Science*, v. 77, p. 125–142.

Morley, M. W., Goldberg, P., Uliyanov, V. A., Kozlikin, M. B., Shunkov, M. V., Derevianko, A. P., Jacobs, Z., and Roberts, R. G., 2019, Hominin and animal activities in the microstratigraphic record from Denisova Cave (Altai Mountains, Russia): *Scientific Reports*, v. 9, no. 1, p. 13785.

Morris, E. H., 1944, Adobe bricks in a pre-Spanish wall near Aztec, New Mexico: *American Antiquity*, v. 9, p. 434–438.

Moussa, N. M., McKenzie, H. G., Bazaliiskii, V. I., Goriunova, O. I., Bamforth, F., and Weber, A. W., 2021, Insights into Lake Baikal's ancient populations based on genetic evidence from the Early Neolithic Shamanka II and Early Bronze Age Kurma XI cemeteries: *Archaeological Research in Asia*, v. 25, p. 100238.

Mücher, H. J., 1997, The response of a soil ecosystem to mowing and sod removal: a micromorphological study in The Netherlands, *in* Shoba, S., Gerasimova, M., and Miedema, R., eds., *Soil Micromorphology: Studies on Soil Diversity, Diagnostics and Dynamics*: Moscow-Wagenigen, International Soil Science Society, p. 271–281.

Mücher, H. J., Slotboom, R. T., and ten Veen, W. J., 1990, Palynology and micromorphology of a man-made soil. A reconstruction of the agricultural history since Late-medieval times of the Posteles in the Netherlands: *Catena*, v. 17, p. 55–67.

Mücher, H. J., van Steijn, H., and Kwaad, F. J. P. M., 2018, Colluvial and mass wasting deposits, *in* Stoops, G., Marcelino, V., and Mees, F., eds., *Interpretation of Micromorphological Features of Soils and Regoliths* (2nd Edition). Amsterdam, Elsevier, p. 21–36.

Muhs, D. R., and Bettis, E., 2003, Quaternary loess-paleosol sequences as examples of climate-driven sedimentary extremes, *in* Chan, M. A., and Archer, A. W., eds., *Extreme depositional environments: Mega end members in geologic time: Special Papers, Vol. 370*: Boulder, Geological Society of America, p. 53–74.

Muhs, D. R., Cattle, S. R., Crouvi, O., Rousseau, D.-D., Sun, J., and Zárate, M. A., 2014, Loess Records, *in* Knippertz, P., and Stuut, J.-B. W., eds., *Mineral Dust: A Key Player in the Earth System*: Dordrecht, Springer Science, p. 411–441.

Müller, W., and Pasda, C., 2011, Site formation and faunal remains of the Middle Pleistocene site Bilzingsleben: *Quartär*, v. 58, p. 25–49.

Munro, N. R., Ibrahim, M. A. M., Abuzied, H. and el-Hassan, B. 2012. Aeolian sand landforms in parts of Sudan and Nubia. Origins and impacts on past and present land use. Sudan & Nubia: *The Sudan Archaeological Research Society*, v. 16, p. 140–154.

Murphy, P. 1984. Environmental Archaeology in East Anglia, *in* Keeley, H. C. M., ed., *Environmental Archaeology, A Regional Review*. Occasional Paper No. 6. London, Directorate of Ancient Monuments and Historic Buildings.

Murphy, P., and Fryer, V. 1999. The plant macrofossils, *in* Niblett, R., ed., *The Excavation of a Ceremonial Site at Folly Lane, Verulamium*. London, Society for the Promotion of Roman Studies.

Murphy, M. A., and Salvador, A., 1998, International stratigraphic guide – an abridged version: *Episodes*, v. 22, no. 4, p. 255–271.

Murphy, M. A., and Salvador, A. E., 2020, Stratigraphic Guide, Chapter 3. Definitions and Procedures, https://stratigraphy.org/guide/defs, 2020, International Committee on Stratigraphy.

Murphy, C. P., Bullock, P., and Turner, R. H., 1977, The measurement and characterisation of voids in thin sections by image analysis. Part 1. Principles and techniques: *Journal of Soil Science*, v. 29, p 498–508.

Myhre, B., 2004, Agriculture, landscape and society ca. 4000 BC–AD 800, *in* Almås, R., ed., *Norwegian Agricultural History*: Trondheim, Tapir Academic Press, p. 14–77.

Nadel, D., and Nadler, M., 2006, Ohalo I- Shaldag Beach: a final report on an Epipaleolithic-Neolithic workshop-site in the Sea of Galilee: *Journal of the Israel Prehistoric Society*, v. 36, p. 39–98.

Nadel, D., Tsatskin, A., Belmaker, M., Boaretto, E., Kislev, M., Mienis, H., Rabinovich, R., Simchoni, O., Simmons, T., Weiss, E., and Zohar, I., 2004, On the shore of a fluctuating lake: environmental evidence from Ohalo II (19 500 BP): *Israel Journal of Earth Sciences*, v. 53, no. 3–4, p. 207–224.

Napieralski, J., Barr, I., Kamp, U., Kervyn, M., 2013. *Remote Sensing and GIScience in Geomorphological Mapping*. Academic Press, San Francisco.

Nash, D. J., Ciborowski, J. R., Ullyott, J. S., Parker Pearson, M., Darvill, T, Greaney, S., Maniatis, G. and Whitaker, K. 2020. Origins of the sarsen megaliths at Stonehenge. *Science Advances*, 6 (31), P. 1–8.

National Cultural Heritage Administration, 2009, *Fieldwork Protocol*. Beijing, Cultural Relics Press.

Naudinot, N., Bourdier, C., Laforge, M. Paris, C., Bellot-Gurlet, L., Beyries, S., Thery-Parisot, I., and Le Goffic, M., 2017, Divergence in the evolution of Paleolithic symbolic and technological systems: The shining bull and engraved tablets of Rocher de l'Impératrice, *PLoS One*, v. 12, no. 3.

Needham, S., and Macklin, M. G., 1992, *Alluvial Archaeology in Britain*. Oxford, Oxbow.

Neev, D., Bakler, N., and Emory, K. O., 1987, *Mediterranean Coasts of Israel and Sinai*. New York, Taylor & Francis.

Neogi, S., French, C., Durcan, J. A., Singh, R. N., and Petrie, C. A., 2020, Geoarchaeological insights into the location of Indus settlements on the plains of northwest India: *Quaternary Research*, v. 94, p. 137–155.

Nesse, W. D., 2014, *Introduction to Optical Mineralogy*, 4th edition. New York, Oxford University Press.

Neubauer, W., Eder-Hinterleitner, A., Seren, S. and Melichar, P. 2002. Georadar in the Roman civil town Carnuntum, Austria: Anapproach for archaeological interpretation of GPR data. *Archaeological Prospection*, 9, p. 135–156.

Newcomer, M. H., and Sieveking, G. D. G., 1980, Experimental flake scatter patterns: a new interpretive technique: *Journal of Field Archaeology*, v. 7, p. 343–352.

Newell, R. R., 1980, Mesolithic dwelling structures: fact and fantasy: *Veröffentlichungen des Museums für Ur- und Frühgeschichte Potsdam*, v. Band 14/15, p. 235–284.

Nials, F. L., 2011, *An Early Agricultural period irrigation system in the Tucson basin, USA.*, 76th Meeting of the Society for American Archaeology: Sacramento, California.

Nian, X. M., Zhou, L. P., and Yuan, B. Y., 2013, Optically Stimulated Luminescence dating of terrestrial sediments in the Nihewan Basin and its implications for the evolution of ancient Nihewan Lake (in Chinese with English abstract): *Quaternary Sciences*, v. 33, p. 403–414.

Niblett, R. 1999. *The Excavation of a Ceremonial Site at Folly Lane, Verulamium. Vol. 14, Britannia Monograph Series*. London, Society for the Promotion of Roman Studies.

Nichols, G., 1999, *Sedimentology & Stratigraphy*. Oxford, Blackwell Science.

Nichols, G., 2013, *Sedimentology and Stratigraphy*, 2nd edition. Chichester, Wiley-Blackwell.

Nicolaysen, N., 1882, *Langskibet fra Gokstad ved Sandefjord*, Kristiania, Cammermeyer.

Nicosia, C., and Stoops, G., 2017, *Archaeological Soil and Sediment Micromorphology*. Oxford, John Wiley & Sons.

Nicosia, C., Langohr, R., Mees, F., Arnoldus-Huyzendveld, A., Bruttini, J., and Cantini, F., 2012, Archaeo-pedological study of medieval Dark Earth from the Uffizi gallery complex in Florence (Italy). *Geoarchaeology*, v. 27, p. 105–122.

Nicosia, C., Devos, Y., and Macphail, R. I., 2017, European "Dark Earth", *in* Nicosia, C., and Stoops, G., eds., *Archaeological Soil and Sediment Micromorphology*: Chichester, Wiley Blackwell, p. 331–344.

Nielsen, A. F., 1991, Trampling the archaeological record: an experimental study: *American Antiquity*, v. 96, p. 483–503.

Niemi, T. M., 1997, Fluctuations of Late Pleistocene Lake Lisan in the Dead Sea Rift, *in* Niemi, T. M., Ben-Avraham, Z., and Gat, J. R., eds., *The Dead Sea: The Lake and its Setting*. New York, Oxford University Press, p. 226–235.

Nigst, P. R., Haesaerts, P., Damblon, F., Frank-Fellner, C., Mallol, C., Viola, B., Götzinger, M., Niven, L., Trnka, G., and Hublin, J.-J., 2014, Early modern human settlement of Europe north of the Alps occurred 43,500 years ago in a cold steppe-type environment: *PNAS*, v. 111, no. 40, p. 14394–14399.

Nigst, P. R., Viola, T. B., Haesaerts, P., Blockley, S., Damblon, F., Frank, C., Fuchs, M., Götzinger, M., Hambach, U., and Mallol, C., 2008, New research on the Aurignacian of Central Europe: A first note on the 2006 fieldwork at Willendorf II: *Quartär*, v. 55, p. 9–15.

Nikon, 2021, Polarized Light Microscopy – https://www.microscopyu.com/techniques/polarized-light, MicroscopyU, Nikon.

Nordt, L. C., 1995, Geoarchaeological investigations of Henson Creek; a low-order tributary in central Texas: *Geoarchaeology*, v. 10, no. 3, p. 205–221.

Nordt, L. C., 2004, Late Quaternary alluvial stratigraphy of a low-order tributary in central Texas, USA and its response to climate and sediment supply: *Quaternary Research*, v. 62, p. 289–300.

Norman, P., and Reader, F. W., 1912, Further discoveries relating to Roman London, 1906–12: *Archaeologia*, v. LXIII, p. 257–344.

Nowakowski, J. A. 2007. *Excavations of a Bronze Age landscape and a post- Roman industrial settlement 1953–1961, Gwithian, Cornwall. Assessment of key datasets (2005–2006) Vol. I.* Report for the ALSF scheme. Cornwall County Council: Truro.

Nowakowski, J. A., 2009. Living in the Sands – Bronze Age Gwithian, Cornwall, Revisited, in Allen, J.M, Sharples, N., and O'Conner, T., eds., *Land and People. Papers in memory of John G Evans*, Prehistoric Society Research Paper 2, p. 115–125.

Nowakowski, J. A., Quinnell, H., Sturgess, J., Thomas, C. and Thorpe, C. 2007. Return to Gwithian: shifting the sands of time: *Cornish Archaeology*, v. 46, p. 13–76.

Nunn, P. D., 2000, Environmental catastrophe in the Pacific islands around A.D. 1300: *Geoarchaeology*, v. 15, no. 7, p. 715–740.

Obreht, I., Hambach, U., Veres, D., Zeeden, C., Bösken, J., Stevens, T., Marković, S. B., Klasen, N., Brill, D., Burow, C. and Lehmkuhl, F. 2017. Shift of large-scale atmospheric systems over Europe during late MIS 3 and implications for Modern Human dispersal. *Scientific Reports*, v. 7, p. 1–10.

O'Connor, S., Barham, A., Aplin, K., and Maloney, T., 2017, Cave stratigraphies and cave breccias: implications for sediment accumulation and removal models and interpreting the record of human occupation: *Journal of Archaeological Science*, v. 77, p. 143–159.

Odah,H., Abdallatif, T.F., El-Hemaly, I.A. and El-All, E.A. 2005. Gradiometer survey to locate the ancient remains distributed to the northeast of the Zoser Pyramid, Saqqara, Giza, Egypt: *Archaeological Prospection*, v. 12, p. 61–68.

Oldfield, F., and Crowther, J., 2007, Establishing fire incidence in temperate soils using magnetic measurements: *PALAEO*, v. 249, p. 362–369.

Opitz, R., Herrmann, J., 2018. Recent trends and long-standing problems in archaeological remote sensing: *Journal of Computer Applications in Archaeology*, v. 1, p. 19–41.

O'Sullivan, P., and Reynolds, C. S., 2004, *The Lakes Handbook Limnology and Limnetic Ecology*, Vol. 1. Oxford, Wiley Blackwell.

Øye, I., 2009, Settlement patterns and field systems in medieval Norway: *Landscape History*, v. 30, no. 2, p. 37–54.

Pan, B. T., Su, H., Liu, X. F., Hu, X. F., Zhou, T., Hu, C. S., and Li, J. J., 2007, River terraces of the yellow river and their genesis in eastern Lanzhou basin during last l.2 Ma (in Chinese with English abstract): *Quaternary Sciences*, v. 27, p. 172–180.

Panda, D., Subramanian, V., and Panigrahy, R. C., 1995, Geochemical fractionation of heavy metals in Chilika lake (east coast of India) – a tropical coastal lagoon: *Environmental Geology*, v. 26, no. 4, p. 199–210.

Pape, J. C., 1970, Plaggen soils in the Netherlands: *Geoderma*, v. 4, p. 229–255.

Parcak, S. H. 2009. *Satellite Remote Sensing for Archaeology*. Routledge: London.

Parés, J. M., Álvarez, C., Sier, M., Moreno, D., Duval, M., Woodhead, J., Ortega, A., Campaña, I., Rosell, J., and de Castro, J. B., 2018, Chronology of the cave interior sediments at Gran Dolina archaeological site, Atapuerca (Spain): *Quaternary Science Reviews*, v. 186, p. 1–16.

Parés, J. M., Campaña, I., Duval, M., Sier, M. J., Ortega, A. I., López, G. I., and Rosell, J., 2020, Comparing depositional modes of cave sediments using magnetic anisotropy: *Journal of Archaeological Science*, v. 123, p. 105241.

Parker Pearson, M., Bevins, R., Ixer, R., Pollard, J., Richards, C., Welham, K., Chan, B., Edinborough, K., Hamilton, D., Macphail, R., Schlee, D., Simmons, E., and Smith, M., 2015, Craig Rhos-y-felin: a Welsh bluestone megalith quarry for Stonehenge: *Antiquity*, v. 89 348, p. 1331–1352.

Parker Pearson, M., Pollard, A. M., Richards, C., Welham, K., Kinnaird, T., Shaw, D., Simmonds, E., Stanford, A., Bevins, R., Ixer, R., Rylatt, J., and Edinborough, K., 2021, The original Stonehenge? A dismantled stone circle in the Preseli Hills of west Wales: *Antiquity*, v. 95, no. 379, p. 85–103.

Parks, D. A. and Rendell, H. M. 1992. Thermoluminescence dating and geochemistry of loessic deposits in southeast England: *Journal of Quaternary Science*, v. 7, p. 99–109.

Parkyn, A. 2010. A survey in the park: Methodological and practical problems associated with geophysical investigation in a late Victorian municipal park: *Archaeological Prospection*, 17, p. 161–174.

Parnell, J. J., Terry, R. E., and Golden, C., 2001, Using in-field phosphate testing to rapidly identify middens at Piedras Negras, Guatemala: *Geoarchaeology*, v. 16, no. 8, p. 855–873.

Parssinen, M. H., Salo, J. S., and Rasanen, M. E., 1996, River floodplain relocations and the abandonment of Aborigine settlements in the Upper Amazon Basin: a historical case study of San Miguel de Cunibos at the Middle Ucayali River: *Geoarchaeology*, v. 11, no. 4, p. 345–359.

Patania, I., Goldberg, P., Cohen, D. J., Yuan, J., Wu, X., and Bar-Yosef, O., 2020, Micromorphological and FTIR analysis of the Upper Paleolithic early pottery site of Yuchanyan cave, Hunan, South China: *Geoarchaeology*, v. 35, no. 2, p. 143–163.

Payton, R. W., 1983, The micromorphology of some fragipans and related horizons in British soils with particular reference to their consistence characteristics, *in* Bullock, P., and Murphy, C. P., eds., *Soil Micromorphology, Vol. 1: Techniques and Applications*: Berkhamsted, A B Academic Publishers, p. 317–336.

Peacock, D. P. S., 1982, *Pottery in the Roman World: An Ethnoarchaeological Approach*. London, Longmans.

Pearsall, W. H., 1952, The pH of natural soils and its ecological significance: *Journal of Soil Science*, v. 3, no. 1, p. 41–51.

Pei, S., Li, X. L., Liu, D. C., Ma, N., and Peng, F., 2009, Preliminary study on the living environment of hominids at the Donggutuo site, Nihewan Basin: *Chinese Science Bulletin (D)*, v. 54, p. 3896–3904.

Pendergast, D. M., and Graham, E., 1987, No site too small: the ROM's Marco Gonzalez excavations in Belize: *Rotunda*, v. 20, no. 1, p. 34–40.

Pérez-García, L., Sánchez-Palencia, F. J., and Torres-Ruiz, J., 2000, Tertiary and Quaternary alluvial gold deposits of Northwest Spain and Roman mining (NW of Duero and Bierzo Basins): *Journal of Geochemical Exploration*, v. 71, no. 2, p. 225–240.

Pérez-Juez, A., 2011, Excavaciones en la Casa 2 del yacimiento de Torre d'en Galmés, Alaior: propuestas para el hábitat talayótico: *Jornades d'Arqueologiade les Illes Balears*, v. III, p. 119–129.

Pérez-Juez, A., 2013, Talayotic Culture, *in* Bagnall, R. S., Brodersen, K., Champion, C. B., Erskine, A., and Huebner, S. R., eds., *The Encyclopedia of Ancient History*. Oxford, Blackwell Publishing, p. 6518–6520.

Pérez-Juez, A., 2014, Spain: Archaeological Heritage Management, *in* Smith, C., ed., *Encyclopedia of Global Archaeology*. New York, NY, Springer.

Pérez-Juez, A., and Goldberg, P., 2018, Evidence of Quarrying at the Iron Age Site of Torre d'en Galmés, Menorca, Spain: Boletín Geológico y Minero (Instituto Geológico y Minero de España) v. 129, p. 353–370.

Peterken, G. F., 1996, *Natural Woodland. Ecology and Conservation in Northern Regions*. Cambridge, Cambridge University Press.

Petrie, C. A., Singh, R. N., Bates, J., Dixit, Y., French, C., Hodell, D. A., Jones, P., Lancelotti, C., Lynam, F., Neogi, S., Pandey, A. K., Pawar, V., Redhouse, D. I., and Singh, D., 2017, Adaptation to variable environments, resilience to climate change: investigating land, water and settlement in Indus Northwest India: *Current Anthropology*, v. 58, p. 1–30.

Pettijohn, F. J., 1975, *Sedimentary Rocks*, 3rd edition. New York, Harper & Row, Publishers.

Pettitt, P., 1997, High resolution Neanderthals? Interpreting Middle Palaeolithic intrasite spatial data: *World Archaeology*, v. 29, no. 2, p. 208–224.

Peyrony, D., 1934, La Ferrassie. *Moustérien, Périgordien, Aurignacien*. Paris, Editions Leroux, Préhistoire.

Phillips, J. L., 1988, The Upper Paleolithic of the Wadi Feiran, southern Sinai: *Paleorient*, v. 14, p. 183–200.

Phillips, S. J., and Comus, P. W., 2000, A *Natural History of the Sonoran Desert*. Tucson, AZ, Arizona-Sonora Desert Museum, p. 628.

Picard, L., and Baida, U., 1966, *Geological Report on the Lower Pleistocene of the 'Ubeidiya Excavations*. Jerusalem, Israel Academy of Sciences.

Pickering, M. D., Ghislandi, S., Usai, M. R., Wilson, C., Connelly, P., Brothwell, D., and Keely, B. J., 2018, Signatures of degraded body tissues and environmental conditions in grave soils from a Roman and an Anglo-Scandinavian age burial from Hungate, York: *Journal of Archaeological Science*, v. 99, p. 87–98.

Piggott, S. 1965. *Ancient Europe*. Edinburgh, Edinburgh University Press.

Pirrie, D., Butcher, A. R., Power, M. R., Gottlieb, P., and Miller, G. L., 2004, Rapid quantitative mineral and phase analysis using an automated scanning electron microscope (QemScan); potential applications in forensic science, *in* Pye, K., and Croft, D. J., eds., *Forensic Geoscience: Principles, Techniques and Applications, Special Publications 232*. London, The Geological Society of London, p. 123–136.

Pirrie, D., Rollinson, G. K., Power, M. R., and Webb, J. A., 2013, Automated forensic soil mineral analysis; testing the potential of lithotyping: *Geological Society of London*, v. Special Publications, no. 384, p. 47–64.

Poch, R. M., Artieda, O., and Lebedeva, M., 2018, Gypsic features, *in* Stoops, G., Marcelino, V., and Mees, F., eds., *Interpretation of Micromorphological Features of Soils and Regoliths*. Amsterdam, Elsevier, p. 259–287.

Pohl, M. D., Bloom, P. R., and Pope, K. O., 1990, Interpretation of wetland farming in northern Belize: excavations at San Antonio Rio Hondo, *in* Pohl, M. D., ed., *Ancient Maya wetland agriculture. Excavations on Albion Island, Northern Belize*. Boulder, CO, Westview, p. 187–254.

Polo-Díaz, A., Eguíluz, M. A., Ruiz, M., Pérez, S., Mújika, J., Albert, R., and Eraso, J. F., 2016, Management of residues and natural resources at San Cristóbal rock-shelter: Contribution to the characterisation of Chalcolithic agropastoral groups in the Iberian Peninsula: *Quaternary International*, v. 414, p. 202–225.

Polunin, O., and Smythies, B. E., 1973, *Flowers of South-West Europe. A Field Guide*. London, Oxford University Press.

Ponomarenko, D., and Ponomarenko, E., 2019, Describing krotovinas: a contribution to methodology and interpretation: *Quaternary International*, v. 502, p. 238–245.

Ponting, M., 2004, The scanning electron microscope and the archaeologist: *Physics Education*, v. 39, no. 2, p. 166–170.

Pope, K. O., and Dahlin, B. H., 1989, Ancient Maya wetland agriculture: new insights from ecological and remote sensing research: *Journal of Field Archaeology*, v. 16, no. 1, p. 87–106.

Pope, M. I., Parfitt, S. A., and Roberts, M. B., 2020, *The Horse Butchery Site: A High-resolution Record of Lower Palaeolithic Hominin Behviour at Boxgrove, UK, SpoilHeap Monograph 23*: Brighton, SpoilHeap Press, p. 157.

Popovici, D., Haită, C., Bălășescu, A., Radu, V., Vlad, F., and Tomescu, I., 2003, *Archaeological Pluridisciplinary Researches at Borduşani-Popin*ă, Bucharest, National Museum of Romanian History, Ialomita County Museum.

Portillo, M., Dudgeon, K., Allistone, G., Raeuf Aziz, K., and Matthews, W., 2020, The taphonomy of plant and livestock dung microfossils: an ethnoarchaeological and experimental approach: *Environmental Archaeology*, v. 26, p. 439–454.

Potts, R., Behrensmeyer, A. K., and Ditchfield, P., 1999, Paleolandscape variation and Early Pleistocene hominid activities: members 1 and 7, Olorgesailie Formation, Kenya: *Journal of Human Evolution*, v. 37, p. 747–788.

Powlesand, D., Lyall, J., Hopkinson, G., Donoghue, D., Beck, M., Harte, A. and Stott, D. 2006. Beneath the sand – Remote sensing, archaeology, aggregates and sustainability: a case study from Heslerton, the Vale of Pickering, North Yorkshire, UK: *Archaeological Prospection*, v. 13, p. 291–299.

Preece, R. C., 1992, Episodes of erosion and stability since the late-glacial: the evidence from dry valleys in Kent, *in* Bell, M., and Boardman, J., eds., *Past and present Soil Erosion, Volume Monograph 22*. Oxford, Oxbow, p. 175–183.

Preece, R. C., and Bridgland, D. R., 1998, *Late Quaternary Environmental Change in North-West Europe: Excavations at Holywell Coombe, South-east England*. London, Chapman & Hall.

Preece, R. C., Kemp, R. A., and Hutchinson, J. N., 1995, A Late-glacial colluvial sequence at Watcombe Bottom, Ventnor, Isle of Wight, England: *Journal of Quaternary Science*, v. 10, no. 2, p. 107–121.

Prendergast, M. E., and Beyin, A., 2018, Fishing in a fluctuating landscape: terminal Pleistocene and early Holocene subsistence strategies in the Lake Turkana Basin, Kenya: *Quaternary International*, v. 471, p. 203–221.

Price, T. D., and Feinman, G. M., 2005, *Images of the Past*, 4th edition. New York, McGraw-Hill.

Pringle, J. K., Ruffell, A., Jervis, J. R., Donnelly, L., McKinley, J., Hansen, J., and Harrison, M., 2012, The use of geoscience methods for terrestrial forensic searches: *Earth-Science Reviews*, v. 114, no. 1–2, p. 108–123.

Pringle, J. K., Stimpson, I. G., Wisniewski, K. D., Heaton, V., Davenward, B., Mirosch, N., Spencer, F., and Jervis, J. R., 2021, Geophysical monitoring of simulated clandestine burials of murder victims to aid forensic investigators: *Geology Today*, v. 37, no. 2, p. 63–65.

Proudfoot, V. B., 1976, The analysis and interpretation of soil phosphorous in archaeological contexts, *in* Davidson, D. A., and Shackley, M. L., eds., *Geo-archaeology: Earth Science and the Past*. London, Duckworth, p. 93–113.

Provencher, L., and Dubois, J.-M. M., 2012, Lake shore nomenclature, *in* Bengtsson, L., Herschy, R. W., and Fairbridge, R. W., eds., *Encyclopedia of Lakes and Reservoirs*: Dordrecht, Springer, p. 463–467.

Pümpin, C., Le Bailly, M., and Pichler, S., 2017, Ova of intestinal parasites, *in* Nicosia, C., and Stoops, G., eds., *Archaeological Soil and Sediment Micromorphology*. Chichester, Wiley Blackwell, p. 91–98.

Pye, E., 2000/2001, Wall painting in the Roman empire: colour, design and technology: *Archaeology International*, p. 24–27.

Pye, K., 1987, *Aeolian Dust and Dust Deposits*. London, Academic Press.

Pye, K., 1995, The nature, origin and accumulation of loess: *Quaternary Science Reviews*, v. 14, p. 653–666.

Qin, Z., Storozum, M., Zhao, H., Liu, H. W., Fu, K., and Kidder, T. R., 2019, Cereals, soils and iron at Sanyangzhuang: Western Han agricultural production in the Central Plains: *Antiquity*, v. 93, no. 369, p. 685–701.

Quinn, P. S., 2013, *Ceramic Petrography. The Interpretation of Archaeological Pottery and Related Artifacts in Thin Section*. Oxford, Archeopress.

Rabenhorst, M. C., Wilding, L. P., and Girdner, C. L., 1984, Airborne dusts in the Edwards Plateau region of Texas: *Soil Society of America Journal*, v. 48, no. p. 621–627.

Rabinovich, R., and Tchernov, E., 1995, Chronological, paleoecological and taphonomical aspects of the Middle Paleolithic site of Qafzeh, Israel, *in* Buitenhuis, H., and Uerpmann, H.-P., eds., *Archaeozoology of the Near East II*: Leiden, Backhuys Publishers, p. 5–44.

Radley, J. and Simms, C. 1967. Wind erosion in East Yorkshire: *Nature*, v. 216, p. 20–22.

Radu, V., 2003, Zooarchaeology. VI.3. Several data about fish and fishing importance in the palaeoeconomy of the Gumelniţa A2 community from Borduşani – Popină, *in* Popovici, D., Haită, C., Bălăşescu, A., Radu, V., Vlad, F., and Tomescu, I., eds., *Archaeological pluridisciplinary researches at Borduşani – Popină.*, *Pluridisciplinary researches series, VI*,: Târgovişte, National Museum Library, Editura Cetatea de Scaun, p. 75–94.

Raghavan, H., Gaillard, C., and Rajaguru, S. N., 1991, Genesis of calcretes from the calc-pan site of Singi Talav near Didwana, Rajasthan, India: a micromorphological approach: *Geoarchaeology*, v. 6, no. 2, p. 151–168.

Ramsay, P. J., and Cooper, J. A. G., 2002, Late Quaternary sea-level change in South Africa: *Quaternary Research*, v. 57, no. 1, p. 82–90.

Ranov, V., 1995, The "Loessic Palaeolithic" in South Tadjikistan, Central Asia: its industries, chronology and correlation: *Quaternary Science Reviews*, v. 14, p. 731–745.

Rapp, A., and Nihlén, T., 1991, Desert dust-storms and loess deposits in North Africa and South Europe: *Catena Supplement*, v. 20, p. 43–55.

Rapp, G. R. J., and Hill, C., 1998, *Geoarchaeology*. New Haven, CT, Yale University Press.

Rapp, G., Jr., 1975, The archaeological field staff: the geologist: *Journal of Field Archaeology*, v. 2, p. 229–237.

Rasmussen, P., 1993, Analysis of goat/sheep faeces from Egolzwil 3, Switzerland: evidence for branch and twig foddering of livestock in the Neolithic: *Journal of Archaeological Science*, v. 20, p. 479–502.

Ray, R. G., 1960, *Aerial Photographs in Geologic Interpretation and Mapping*. Washington, DC, U.S. Govt. Print. Off., Geological Survey Professional Paper 373.

Rea, D. K., Owen, R. M., and Meyers, P. A., 1981, Sedimentary processes in the Great Lakes: *Reviews of Geophysics and Space Physics*, v. 19, p. 635–648.

Rech, J. A., Quintero, L. A., Wilke, P. J. and Winer, E. R. 2007. The lower paleolithic landscape of 'Ayoun Qedim, al-Jafr Basin, Jordan: *Geoarchaeology*, v. 22, p. 261–275.

Reineck, H. E., and Singh, I. B., 1986, *Depositional Sedimentary Environments*. Berlin, Springer-Verlag.

Reitz, E. J., and Wing, E. S., 1999, *Zooarchaeology*. Cambridge, Cambridge University Press, Cambridge Manuals in Archaeology.

Rellini, I., Firpo, M., Martino, G., Riel-Salvatore, J., and Maggi, R., 2013, Climate and environmental changes recognized by micromorphology in Palaeolithic deposits at Arene Candide (Liguria, Italy): *Quaternary International*, v. 315 Site formation processes in archaeology, p. 42–55.

Rendu, W., Beauval, C., Crevecoeur, I., Bayle, P., Balzeau, A., Bismuth, T., Bourguignon, L., Delfour, G., Faivre, J.-P., and Lacrampe-Cuyaubère, F., 2014, Evidence supporting an intentional Neandertal burial at La Chapelle-aux-Saints: *PNAS*, v. 111, no. 1, p. 81–86.

Rendu, W., Beauval, C., Crevecoeur, I., Bayle, P., Balzeau, A., Bismuth, T., Bourguignon, L., Delfour, G., Faivre, J.-P., Lacrampe-Cuyaubère, F., Muth, X., Pasty, S., Semal, P., Tavormina, C., Todisco, D., Turq, A., and Maureille, B., 2016, Let the dead speak. . .comments on Dibble et al.'s reply to "Evidence supporting an intentional burial at La Chapelle-aux-Saints": *Journal of Archaeological Science*, v. 69, p. 12–20.

Renfrew, C., 1976, Archaeology and the earth sciences, *in* Davidson, D. A., and Shakley, M. L., eds., *Geoarchaeology: Earth Science and the Past*. London, Duckworth, p. 1–5.

Renfrew, C., and Bahn, P., 2001, *Archaeology: Theories, Methods and Practice*. London, Thames and Hudson.

Rentzel, P., 1997, Geoarchäologische beobachtungen an der Römischen Wasserleitung von Leistal nach Augst, *in* Ewald, J., Hartman, M., and Rentzel, P., eds., *Die Römischen Wasserleitung von Leistal nach Augst*: Liestal, Schweiz, Berichte aus Archäologie und Kantonmuseum Baselland, p. 37–62.

Rentzel, P., 1998, Musselwhite Grubenstrukturen aus spätlatènezeitlichen Fundstelle Basel-Gasfabrik: *Geoarchäologische interpretation der Grubenfüllungen, Jahresbericht der archäologischen Bodenforschung des Kantons Basel-Stadt 1995*: Basel, p. 35–79.

Rentzel, P., 2009, The arena floor of the amphitheater of Augst-Nine towers. Geo-archaeological investigations, *in* Hufschmid, T., ed., *Amphitheatrum in Provincia et Italia. Architecture and use Roman amphitheater of Augusta Raurica to Puteoli* (with contributions by Ph. Rentzel, N. Frésard and ME Fox), Vol. 43. Augst, Research in Augst, p. 569–578.

Rentzel, P., and Narten, G.-B., 2000, *Zur Entstehung von Gehniveaus in sandig-lehmigen Ablagerungen – Experimente und archäologische Befunde (Activity surfaces in sandy-loamy deposits – experiments and archaeological examples), Jahresbericht 1999*: Basel, Archäologische Bodenforschung des Kantons Basel-Stadt, p. 107–127.

Rentzel, P., Nicosia, C., Gebhardt, A., Brönnimann, D., Pümpin, C., and Ismail-Meyer, K., 2017, Trampling, poaching and the effects of traffic, *in* Nicosia, C., and Stoops, G., eds., *Archaeological Soil and Sediment Micromorphology*. Chichester, Wiley Blackwell, p. 281–298.

Retallack, G. J., 1996, *Soils of the Past*, 2nd edition. Oxford, Blackwell Science.

Retallack, G. J., 2001, *Soils of the Past: An Introduction to Paleopedology*, 2nd edition. Oxford, Wiley.

Reynolds, P., 1979, Iron Age Farm. *The Butser Experiment*. London, British Museum Publications Ltd.

Reynolds, P., 1981, Deadstock and livestock, *in* Mercer, R., ed., *Farming Practice in British Prehistory*: Edinburgh, Edinburgh University Press, p. 97–122.

Reynolds, P., 1987, *Ancient Farming*, Aylesbury, Shire Publications Ltd, Shire Archaeology.

Reynolds, P., 1995, The life and death of a post-hole, *in* Shepherd, E., ed., *Interpreting Stratigraphy 5, Proceedings of a Conference held at Norwich Castle Museum on 16th June 1994 and supported by the Norfolk Archaeological Unit*, p. 21–25.

Reynolds, P., and Shaw, C., 2000, Butser Ancient Farm. *The Open Air Laboratory for Archaeology*, Waterlooville, Butser Ancient Farm.

Rhines, P. B., 1998, Circulation, convection, and mixing in rotating stratified basins with sloping topography, *in* Imberger, J., ed., *Physical Processes in Lakes and Oceans*. Washington, DC, American Geophysical Union, p. 435–451.

Rhodes, S. E., Ziegler, R., Starkovich, B. M., and Conard, N. J., 2018, Small mammal taxonomy, taphonomy, and the paleoenvironmental record during the Middle and Upper Paleolithic at Geißenklösterle Cave (Ach Valley, southwestern Germany): *Quaternary Science Reviews*, v. 185, p. 199–221.

Richter, D., Grün, R., Joannes-Boyau, R., Steele, T. E., Amani, F., Rué, M., Fernandes, P., Raynal, J.-P., Geraads, D., and Ben-Ncer, A., 2017, The age of the hominin fossils from Jebel Irhoud, Morocco, and the origins of the Middle Stone Age: *Nature*, v. 546, no. 7657, p. 293–296.

Rick, T. C., 2002, Eolian processes, ground cover, and the archaeology of coastal dunes: a taphonomic case study from San Miguel Island, California, U.S.A: *Geoarchaeology*, v. 17, no. 8, p. 811–833.

Ridgeway, V., Taylor, J., and Biddulph, E., 2019, *A Bath House, Settlement and Industry on Roman Southwark's North Island. Excavations along Thameslink Borough Viaduct and at London Bridge Station, Thameslink Monograph Series No. 1*. London, PCA and Oxford Archaeology (OAPCA).

Riley, D. N., 1987, *Air Photography and Archaeology*, Philadelphia, University of Pennsylvania Press.

Ringberg, B., 1994, The Swedish Varve Chronology: *Pact*, v. 41-I.2, p. 25–34.

Ringberg, B., and Erlstrom, M., 1999 Micromorphology and petrography of Late Weichselian glaciolacustrine varves in southeastern Sweden: *Catena* v. 35 p. 147–177.

Ringrose-Voase, A. J., 1990, One-dimensional image analysis of soil structure; I, Principles: *Journal of Soil Science*, v. 41, no. 3, p. 499–512.

Rink, W. J., and Schwarcz, H. P., 2005, ESR and uranium series dating of teeth from the lower Paleolithic site of Gesher Benot Ya'aqov, Israel: confirmation of paleomagnetic age indications: *Geoarchaeology*, v. 20, no. 1, p. 57–66.

Rink, W., Bartoll, J., Goldberg, P., and Ronen, A., 2003, ESR dating of archaeologically relevant authigenic terrestrial apatite veins from Tabun Cave, Israel: *Journal of Archaeological Science*, v. 30, p. 1127–1138.

Rink, W. J., Schwarcza, H. P., Ronenb, A. and Tsatskin, A. 2004. Confirmation of a near 400 ka age for the Yabrudian industry at Tabun Cave, Israel: *Journal of Archaeological Science*, v. 31, p. 15–20.

Ritter, D. F., Kochel, R. C., and Miller, J. R., 2011, *Process Geomorphology*, 5th edition. Long Grove, IL, Waveland Press.

Roberts, D., Valdez-Tullett, A., Marshall, P., Last, J., Oswald, A., Barclay, A., Dunbar, E., Forward, A., Law, M., Linford, N., Linford, P., López-Dóriga, I., Manning, A., Payne, A., Pelling, R., Powell, A., Reimer, P., Russell, M., Small, F., Soutar, S., Vallender, J., Worley, F.,BIshop, B. 2018. Recent investigations at two long barrows and reflections on their context in the Stonehenge World Heritage site and environs. *Internet Archaeolology*, 47.

Roberts, M. B., and Parfitt, S. A., 1999, Boxgrove. *A Middle Pleistocene Hominid Site at Eartham Quarry, Boxgrove, West Sussex*. London, English Heritage.

Roberts, M. B., and Pope, M. I., 2018, *The Boxgrove Wider Area Project: Mapping the Early Middle Pleistocene Deposits of the Slindon Formation Across the Coastal Plain of West Sussex and Eastern Hampshire*. Brighton, SpoilHeap Press.

Roberts, M. B., Stringer, C. B., and Parfitt, S. A., 1994, A hominid tibia from Middle Pleistocene sediments at Boxgrove, UK: *Nature*, v. 369, p. 311–313.

Robinson, D., and Rasmussen, P., 1989, Botanical investigations at the Neolithic lake village at Weier, N. E. Switzerland: leaf hay and cereals as animal fodder, *in* Milles, A., Williams, D., and Gardner, N., eds., *The Beginnings of Agriculture, International Series 496*. Oxford, British Archaeological Reports, p. 137–148.

Robinson, M., 1991, The Neolithic and Late Bronze Age insect assemblages, *in* Needham, S., ed., *Excavation and Salvage at Runnymede Bridge, 1978*. London, British Museum, p. 277–326.

Robinson, M., 1992, *Environment, archaeology and alluvium on the river gravels of the South Midlands floodplains, Alluvial Archaeology in Britain*. Oxford, Oxbow, p. 197–208.

Rodríguez-Cintas, Á., and Cabanes, D., 2017, Phytolith and FTIR studies applied to combustion structures: the case of the Middle Paleolithic site of El Salt (Alcoy, Alicante): *Quaternary International*, v. 431, Part A, p. 16–26.

Rodríguez-Vidal, J., d'Errico, F., Pacheco, F. G., Blasco, R., Rosell, J., Jennings, R. P., Queffelec, A., Finlayson, G., Fa, D. A., and López, J. M. G., 2014, A rock engraving made by Neanderthals in Gibraltar: *PNAS*, v. 111, no. 37, p. 13301–13306.

Rødsrud, C. L., 2020, Burial mounds, ard marks, and memory: a case study from the early Iron Age at Bamble, Telemark, Norway: *European Journal of Archaeology*, v. **23**, p. 207–226.

Rødsrud, C., 2007, Graver og Bosetningsspor på Bjørnstad (Lokalitet 44), *in* Bårdseth, G. A., ed., *Hus, gard og graver langs E6 i Sarpsborg kommune. E6-prosjeket Østfold, Varia, Band 2*: Oslo, Kulturhistorisk Museum Fornminneseksjonen, p. 91–181.

Rodwell, J. S., 1992, *British Plant Communities. Vol. 3. Grasslands and Montane Communities*. Cambridge, Cambridge University Press.

Rogers, M. J., Harris, J. W. K., and Feibel, C. S., 1994, Changing patterns of land use by Plio-Pleistocene hominins in the Lake Turkana Basin: *Journal of Human Evolution*, v. 27, p. 139–158.

Rognon, P., Coudé-Gaussen, G., Fedoroff, N., and Goldberg, P., 1987, Micromorphology of loess in the northern Negev (Israel), *in* Fedoroff, N., Bresson, L. M., and Courty, M.-A., eds., *Micromorphologie des Sols – Soil Micromorphology*: Plaisir, Association Française pour l'Étude du Sol, p. 631–638.

Rognon, P., Coudé-Gaussen, G., Fedoroff, N., and Goldberg, P., 1987, Micromorphology of loess in the Northern Negev (Israel), *in* Fedoroff, N., Bresson, L. M., and Courty, M.-A., eds., *Micromorphologie De Sols – Soil Micromorphology*: Plaisir, Association Française pour l'Étude du Sol, p. 631–638.

Rollefson, G. O., 1990, The uses of plaster at Neolithic 'Ain Ghazal Jordan: *Archeomaterials*, v. 4, p. 33–54.

Romans, J. C. C., and Robertson, L., 1974, Some aspects of the genesis of alpine and upland soils in the British Isles, *in* Rutherford, G. K., ed., *Soil Microscopy*. Kingston, Ontario, The Limestone Press, p. 498–510.

Romans, J. C. C., and Robertson, L., 1975a, Soils and archaeology in Scotland, *in* Evans, J. G., Limbrey, S., and Cleere, H., eds., *The Effect of Man on the Landscape: the Highland Zone, Research Report No. 11*: York, The Council for British Archaeology, p. 37–39.

Romans, J. C. C., and Robertson, L., 1975b, Some genetic characteristics of the freely drained soils of the Ettrick Association in east Scotland: *Geoderma*, v. 14, p. 297–317.

Romans, J. C. C., and Robertson, L., 1983a, An account of the soils at North Mains, *in* "Sites of the third millenium BC to the first millenium AD at North Mains, Strathallan, Perthshire", by Barclay, G. J: *Proceedings of the Society of Antiquities Scotland*, v. 113, p. 260–269.

Romans, J. C. C., and Robertson, L., 1983b, The environment of north Britain: soils, *in* Chapman, J. C., and Mytum, H. C., eds., *Settlement in North Britain 1000 BC to AD 1000*, Vol. 118. Oxford, British archaeological reports, British Series, p. 55–80.

Romans, J. C. C., and Robertson, L., 1983c, The general effects of early agriculture on soil, *in* Maxwell, G. S., ed., *The Impact of Aerial Reconnaissance on Archaeology, Research Report No. 49*. London, Council for British Archaeology, p. 136–141.

Ringrose-Voase, A. J., and Bullock, P., 1984, The automatic recognition and measurement of soil pore types by image analysis and computer programs: *Journal of Soil Science*, v. 35, p. 673–684.

Röpke, A., and Dietl, C., 2017, Burnt soils and sediments, *in* Nicosia, C., and Stoops, G., eds., *Archaeological Soil and Sediment Micromorphology*. Chichester, Wiley Blackwell, p. 173–179.

Rosas, A., Huguet, R., Pérez-González, A., Carbonell, E., De Castro, J. M. B., Vallverdú, J., Van Der Made, J., Allué, E., García, N., Martínez-Pérez, R., Rodríguez, J., Sala, R., Saladie, P., Benito, A., Martínez-Maza, C., Bastir, M., Sánchez, A., and Parés, J. M., 2006, The "Sima del Elefante" cave site at Atapuerca (Spain): *Estudios Geologicos*, v. 62, no. 1, p. 327–348.

Rösch, M., Ehrmann, O., Herrmann, L., et al., 2004, Slash-and-burn experiments to reconstruct Late Neolithic shifting cultivation: *International Forest Fire News*, v. 30, p. 70–74.

Rose, J., Boardman, J., Kemp, R. A., and Whitman, C. A., 1985, Palaeosols and the interpretation of the British Quaternary stratigraphy, *in* Richards, K. S., Arnett, R. R., and Ellis, S., eds., *Geomorphology and Soils*. London, George Allen & Unwin, p. 348–375.

Rosen, A. M., 1986, Cities of Clay. *The Geoarchaeology of Tells*. Chicago, IL, University Press of Chicago.

Rosen, A. M., 1991, "BA" guide to artifacts: microartifacts and the study of ancient societies: *The Biblical Archaeologist*, v. 54, no. 2, p. 97–103.

Rosen, A. M., 2007, Civilizing Climate. *Social Responses to Climate Change in the Ancient Near East*, Lanham, Altamira Press.

Rosen, A., 1986, Environmental change and settlement at Tel Lachish, Israel: *BASOR*, v. 266, p. 45–58.

Rosen, A., 2007, The role of environmental change in the development of complex societies in China: a study from the Huizui site: *Bulletin of the Indo-Pacific Prehistory Association*, v. 27, p. 39–48.

Rosen, A., 2008, The impact of environmental change and human land use on alluvial valleys in the Loess Plateau of China during the Middle Holocene: *Geomorphology*, v. 101, p. 298–307.

Rosen, A., Macphail, R., Liu, L., Chen, X., and Weisskopf, A., 2017, Rising social complexity, agricultural intensification, and the earliest rice paddies on the Loess Plateau of northern China: *Quaternary International*, Vol. 437B, p. 50–59.

Rosen, S. A., 1988, Notes on the origins of pastoral nomadism: a case study from the Negev and Sinai: *Current Anthropology*, v. 29, no. 3, p. 498–506.

Rosen, S., Savinetsky, A., Plakht, Y., Kisseleva, N., Khassanov, B., Pereladov, A., and Haiman, M., 2005, Dung in the desert: preliminary results of the Negev Holocene Ecology Project: *Current Anthropology*, v. 46, no. 2, p. 317–326.

Roskin, J., 2012a, Evidence for widespread episodic Late Pleistocene and late Holocene dune-dammed lakes in the northeast Sinai (Egypt) and northwest Negev Desert, Israel: *Quaternary International*, v. 279–280, p. 414.

Roskin, J., 2012b, Palaeoclimate interpretation of the Late Pleistocene and late Holocene dune encroachments into the northwest Negev, Israel: *Quaternary International*, v. 279–280, p. 414.

Roskin, J., Porat, N., Tsoar, H., Blumberg, D. G., and Zander, A. M., 2011a, Age, origin and climatic controls on vegetated linear dunes in the northwestern Negev Desert (Israel): *Ninth international conference on Luminescence and electron spin resonance dating LED99*, v. 30, no. 13–14, p. 1649–1674.

Roskin, J., Tsoar, H., Porat, N., and Blumberg, D. G., 2011b, Palaeoclimate interpretations of Late Pleistocene vegetated linear dune mobilization episodes: evidence from the northwestern Negev dunefield, Israel: *Quaternary Science Reviews*, v. 30, no. 23–24, p. 3364–3380.

Roskin, J., Katra, I., Porat, N., and Zilberman, E., 2013a, Evolution of Middle to Late Pleistocene sandy calcareous paleosols underlying the northwestern Negev Desert Dunefield (Israel): *Palaeogeography, Palaeoclimatology, Palaeoecology*, v. 387, p. 134–152.

Roskin, J., Katra, I., and Blumberg, D. G., 2013b, Late Holocene dune mobilizations in the northwestern Negev dunefield, Israel: a response to combined anthropogenic activity and short-term intensified windiness: *Quaternary International*, v. 303, p. 10–23.

Roskin, J., Bookman, R., Friesem, D. E., and Vardi, J., 2017, A late Pleistocene linear dune dam record of aeolian-fluvial dynamics at the fringes of the northwestern Negev dunefield: *Sedimentary Geology*, v. 353, p. 76–95.

Rothe, M., Kleeberg, A., and Hupfer, M., 2016, The occurrence, identification and environmental relevance of vivianite in waterlogged soils and aquatic sediments: *Earth-Science Reviews*, v. 158, p. 51–64.

Rothwell, R. G., and Rack, F. R., 2006, New techniques in sediment core analysis: an introduction, *in* Rothwell, R. G., ed., *New Techniques in Sediment Core Analysis, Special Publications 267*. London, Geological Society of London, p. 1–29.

Rowan, L. C., and Mars, J. C., 2003, Lithologic mapping in the Mountain Pass, California area using Advanced Spaceborne Thermal Emission and Reflection Radiometer (ASTER) data: *Remote Sensing of Environment*, v. 84, p. 350–366.

Rowell, D. L., 1994, *Soil Science: Methods and Applications*. New York, John Wiley & Sons.

Rowsome, P., 2000, Heart of the City. *Roman, Medieval and Modern London Revealed by Archaeology at 1 Poultry*. London, Museum of London Archaeology Service.

Rozo, M. G., Nogueira, A. C., and Truckenbrodt, W., 2012, The anastomosing pattern and the extensively distributed scroll bars in the middle Amazon River: *Earth Surface Processes and Landforms*, v. 37, no. 14, p. 1471–1488.

Runia, L. T., 1988, So-called secondary podzolisation in barrows, *in* Groenman-van Waateringe, W., and Robinson, M., eds., *Man-made Soils*, International Series 410. Oxford, British Archaeological Reports, p. 129–142.

Salquea, M., Radi, G., Tagliacozzo, A., Pino Uriac, B., Wolfram, S., Hohled, I., Stäuble, H., Hofmann, D., Whittleg, A., Pechtl, J., Schade-Lindigi, S., Eisenhauer, U., and Evershed, R. P., 2012, New insights into the Early Neolithic economy and management of animals in Southern and Central Europe revealed using lipid residue analyses of pottery vessels: *Anthropozoologica* v. 47, no. 2, p. 45–62.

Salway, P., 1993, *The Oxford Illustrated History of Roman Britain*. Oxford, Oxford University Press.

Sandgathe, D. M., 2017, Identifying and describing pattern and process in the evolution of hominin use of fire: *Current Anthropology*, v. 58, no. S16, p. S360–S370.

Sandgathe, D. M., Dibble, H. L., Goldberg, P., McPherron, S. P., Turq, A., Niven, L., and Hodgkins, J., 2011, On the role of fire in Neandertal adaptations in Western Europe: evidence from Pech de l'Azé and Roc de Marsal, France: *PaleoAnthropology* p. 216–242.

Sandor, J. A., 1992, Long term effects of prehistoric agriculture on soils: examples from New Mexico and Peru, *in* Holliday, V. T., ed., *Soils in Archaeology: Landscape Evolution and Human Occupation*. Washington, DC, Smithsonian Institution Press, p. 217–245.

Sandor, J. A., and Homburg, J. A., 2015, Agricultural soils of the prehistoric southwest: unknown unknowns, *in* Ingram, S. E., and Hunt, R. C., eds., *Traditional Arid lands Agriculture. Understanding the Past for the Future*: Tucson, AZ, University of Arizona Press, p. 54–88.

Sandor, J. A., Huckleberry, G., Hayashida, F. M., Parcero-Oubiña, C., Salazar, D., Troncoso, A., and Ferro-Vázquez, C., 2021, Soils in ancient irrigated agricultural terraces in the Atacama Desert, Chile: *Geoarchaeology*, p. 1–24.

Sanger, D., Kelley, A. R., and Almquist, H., 2003, Geoarchaeological and cultural interpretations in the lower Penobscot valley, Maine, *in* Hart, J., and Cremeens, D., eds., *Geoarchaeology of Landscapes in the Glaciated Northeast*: Albany, NY. New York State Museum, Bulletin 497, p. 135–150.

Sankey, D., 1998, Cathedrals, granaries and urban vitality in late Roman London, *in* Watson, B., ed., Roman London. *Recent Archaeological Work, Supplementary Series no. 24*: Portsmouth, Rhode Island, Journal of Roman Archaeology, p. 78–82.

Santschi, P., Hohener, P., Benoit, G., and Brink, M. B., 1990, Chemical processes at the sediment-water interface: *Marine Chemistry*, v. 30, p. 269–315.

Sara, T. R., Macphail, R., Goldberg, P., and Larson, B., 2006, *Archaeological Context and Geoarchaeological Study, Camp Lemonier, Djibouti, Africa*: Geo-Marine Inc, Miscellaneous Reports of Investigation Number 349.

Sarmast, M., Farpoor, M. H., and Boroujeni, I. E., 2017, Soil and desert varnish development as indicators of landform evolution in central Iranian deserts: *Catena*, v. 149, p. 98–109.

Savage, S.H., Levy, T.E. and Jones, I.W. 2012. Prospects and problems in the use of hyperspectral imagery for archaeological remote sensing: a case study from the Faynan copper mining district, Jordan. *Journal of Archaeological Science*, 39, p. 407–420.

Saville, A., 1990, Hazleton North. *The Excavation of a Neolithic Long Cairn of the Cotswold-Severn Group*. London, English Heritage.

Scaife, R. G., 1995, Pollen Analysis: Intertidal sites (Blackwater Sites, 3, 18 and 28; Crouch Sites 9 and 8), *in* Wilkinson, T. J., and Murphy, P. L., eds., *The Archaeology of the Essex Coast, Vol. I: The Hullbridge Survey, Report No. 71*. Chelmsford, East Anglian Archaeology, p. 43–51.

Scaife, R. G., and Macphail, R. I., 1983, The post-Devensian development of heathland soils and vegetation, *in* Burnham, P., ed., *Soils of the Heathlands and Chalklands*, Vol. 1. Wye, South-East Soils Discussion Group, p. 70–99.

Scarborough, V. L., and Gallopin, G. G., 1991, A water storage adaptation in the Maya Lowlands: *Science*, v. 251, p. 658–662.

Scerri, E. M. L., Breeze, P. S., Parton, A., Groucutt, H. S., White, T. S., Stimpson, C., Clark-Balzan, L., Jennings, R., Alsharekh, A., and Petraglia, M. D., 2015, Middle to Late Pleistocene human habitation in the western Nefud Desert, Saudi Arabia: *Quaternary International*, v. 382, p. 200–214.

Scharlotta, I., 2018, Differentiating mobility and migration in middle Holocene Cis-Baikal, Siberia: *Journal of Archaeological Science: Reports*, v. 17, p. 919–931.

Schiegl, S. G., Goldberg, P., Bar-Yosef, O., and Weiner, S., 1996, Ash deposits in Hayonim and Kebara Caves, Israel: macroscopic, microscopic and mineralogical observations, and their archaeological implications: *Journal of Archaeological Science*, v. 23, p. 763–781.

Schiegl, S., Goldberg, P., Pfretzschner, H.-U., and Conard, N. J., 2003, Paleolithic burnt bone horizons from the Swabian Jura: distinguishing between in situ fire places and dumping areas: *Geoarchaeology*, v. 18, no. 5, p. 541–565.

Schiffer, M. B., 1987, *Formation Processes in the Archaeological Record*. Albuquerque, University of New Mexico Press.

Schlezinger, D. R., and Howes, B. L., 2000, Organic phosphorous and elemental ratios as indicators of prehistoric human occupation: *Journal of Archaeological Science*, v. 27, p. 479–492.

Schmid, E., 1958, Höhlenforschung und Sedimentanalyse: *Schriften des Institutes für Ur-und Frügeshicthe des Schweitz*, v. 13.

Schmid, E., 1963, Cave sediments and prehistory, *in* Brothwell, D., ed., *Science in Archaeology*, p. 123–138.

Schmidt ,A. 2013. *Earth Resistance for Archaeologists, Geophysical Methods for Archaeology*. Altamira Press

Schmidt, A., Linford, P., Linford, N., David, A., Gaffney, C., Sarris, A. and Fassbinder, J. 2015. *EAC Guidelines for the use of geophysics in archaeology: questions to ask and points to consider, EacGuidelin 2*. https://www.europae-archaeologiae-consilium.org/eac-guidelines

Schoch, R. M., 1989, *Stratigraphy: Principles and Methods*. New York, Van Nostrand Reinhold.

Schoeneberger, P., Wysocki, D., and Benham, E., 2012, Soil Survey Staff. 2012. Field Book for Describing and Sampling Soils, Version 3.0: *Natural Resources Conservation Service, National Soil Survey Center, Lincoln, NE*, v. 36.

Schofield, D., and Hall, D. M., 1985, A method to measure the susceptibility of pasture soils to poaching by cattle: *Soil Use and Management*, v. 1, p. 134–138.

Schofield, T, Carter, T, Jackson, N. and Moir, R. 2020. Integrating geophysical survey and excavation at the Freston Early Neolithic causewayed enclosure, Suffolk (UK): *Archaeological Prospection*, 28 (1), p. 1–13

Schreiner, A., 1997, Einführing in die Quartärgeologie, 2. Auflage, Stuttgart, E. Schweizerbart'sche Verlagsbuchhandlung.

Schuldenrein, J., 1995, Geochemistry, phosphate fractionation, and the detection of activity areas at prehistoric North American sites, *in* Collins, M. E., Carter, B. J., Gladfelter, B. G., and Southard, R. J., eds., *Pedological Perspectives in Archaeological Research, SSSA Special Publication Number 44*: Madison, Soil Science Society of America, p. 107–132.

Schuldenrein, J., 2007, A reassessment of the Holocene stratigraphy of the Wadi Hasa Terrace and Hasa formation, Jordan: *Geoarchaeology*, v. 22, no. 6, p. 559–588.

Schuldenrein, J., and Clark, G. A., 1994, Landscape and prehistoric chronology of west-central Jordan: *Geoarchaeology*, v. 9, no. 1, p. 31–55.

Schuldenrein, J., and Clark, G. A., 2001, Prehistoric landscapes and settlement geography along the Wadi Hasa, West-Central Jordan. Part I: geoarchaeology, human palaeoecology and ethnographic modelling: *Environmental Archaeology*, v. 6, p. 23–38.

Schuldenrein, J., and Goldberg, P., 1981, Late Quaternary paleoenvironments and prehistoric site distributions in the Lower Jordan Valley: a preliminary report: *Paléorient*, v. 7, p. 57–81.

Schuldenrein, J., Wright, R. P., Mughal, M. R., and Khan, M. A., 2004, Landscapes, soils, and mound histories of the Upper Indus Valley, Pakistan: new insights on the Holocene environments near ancient Harappa: *Journal of Archaeological Science*, v. 31, p. 777–797.

Schulz, E., Biester, H., Bogenrieder, A., Eckmeier, E., Ehrmann, O., Gerlach, R., Hall, M., Hartktopf-Fröder, C., Herrmann, L., Kury, B., Rösch, M., and Schier, W., 2011, How long will it take? Regeneration of vegetation and soil after clearing, burning and cultivation. The Forchtenberg-Experiment, *in* Sauer, D., Jahn, R., and Stahr, K., eds., *Landscape & Soils through Time*: Hohenheim, IUSS, p. 112–113.

Schuster, M., and Nutz, A., 2017, Lacustrine wave-dominated clastic shorelines: modern to ancient littoral landforms and deposits from the Lake Turkana Basin (East African Rift System, Kenya: *Journal of Paleolimnology*, v. 59, p. 221–243.

Segerström, U., 1991, Soil pollen analysis – an application for tracing ancient arable patches: *Journal of Archaeological Science*, v. 18, p. 165–175.

Shahack-Gross, R., 2011, Herbivorous livestock dung: formation, taphonomy, methods For identification, and archaeological significance: *Journal of Archaeological Science*, v. 38, p. 205–218.

Shahack-Gross, R., 2017, Animal gathering enclosures, *in* Nicosia, C., and Stoops, G., eds., *Archaeological Soil and Sediment Micromorphology*. Chichester, Wiley Blackwell, p. 265–280.

Shahack-Gross, R., Marshall, F., and Weiner, S., 2003, Geo-ethnoarchaeology of pastoral sites: the identification of livestock enclosures in abandoned Maasai settlements: *Journal of Archaeological Science*, v. 30, p. 439–459.

Shahack-Gross, R., Berna, F., Karkanas, P., and Weiner, S., 2004a, Bat guano and preservation of archaeological remains in cave sites: *Journal of Archaeological Science*, v. 31, no. 9, p. 1259–1272.

Shahack-Gross, R., Marshall, F., Ryan, K., and Weiner, S., 2004b, Reconstruction of spatial organisation in abandoned Maasai settlements: implications for site structure in Pastoral Neolithic of East Africa: *Journal of Archaeological Science*, v. 31, p. 1395–1411.

Shahack-Gross, R., Albert, R.-M., Gilboa, A., Nagar-Hilman, O., Sharon, I., and Weiner, S., 2005, Geoarchaeology in an urban context: the uses of space in a Phoenician monumental building at Tel Dor (Israel): *Journal of Archaeological Science*, v. 32, no. 9, p. 1417–1431.

Shahack-Gross, R., Ayalon, A., Paul Goldberg, Goren, Y., Ofek, B., Rabinovich, R., and Hovers, E., 2008a, Formation processes of cemented features in karstic cave sites revealed using stable oxygen and carbon isotopic analyses: a case study at Middle Paleolithic Amud Cave, Israel: *Geoarchaeology*, v. 23, no. 1, p. 43–62.

Shahack-Gross, R., Simons, A., and Ambrose, S. H., 2008b, Identification of pastoral sites using stable nitrogen and carbon isotopes from bulk sediment samples: a case study in modern and archaeological pastoral settlements in Kenya: *Journal of Archaeological Science*, v. 35, p. 983–990.

Shahack-Gross, R., Gafri, M., and Finkelstein, I., 2009, Identifying threshing floors in the archaeological record: a test case at Iron Age Tel Megiddo, Israel: *Journal of Field Archaeology*, v. 34, no. 2, p. 171–184.

Shahack-Gross, R., Berna, F., Karkanas, P., Lemorini, C., Gopher, A., and Barkai, R., 2014, Evidence for the repeated use of a central hearth at Middle Pleistocene (300 ky ago) Qesem Cave, Israel: *Journal of Archaeological Science*, v. 44, p. 12–21.

Shang, Y., Kaakinen, A., Beets, C. J., and Prins, M. A., 2018, Aeolian silt transport processes as fingerprinted by dynamic image analysis of the grain size and shape characteristics of Chinese loess and Red Clay deposits: *Sedimentary Geology*, v. 375, p. 36–48.

Sharer, R. J., and Ashmore, W., 2003, *Archaeology, Discovering our past*. New York, McGraw-Hill.

Sharples, N. M., 1991, Maiden Castle. *Excavations and field survey 1985–6, Archaeological Report no 19.* London, English Heritage, p. 284.

Sheets, P. D., 1992, *The Ceren Site: A Prehistoric Village Buried by Volcanic Ash in Central America.* Fort Worth, TX, Harcourt Brace College Publishers, Case Studies in Archaeology Series.

Sheets, P. D., 2002, *Before the Volcano Erupted: The Ancient Ceren Village in Central America.* Austin, TX, University of Texas Press.

Shelley, A., 2005, *Dragon Hall, King Street, Norwich: Excavation and Survey of a Late Medieval Merchant's Trading Complex, Report No. 112*: Norwich, East Anglian Archaeology, p. 206.

Shepard, A. O., 1985, *Ceramics for the Archaeologist*, 5th edition. Washington DC, Carnegie Institution of Washington.

Sherwood, S. C., 2001, *The Geoarchaeology of Dust Cave: A Late Paleoindian Through Middle Archaic Site in the Middle Tennessee River Valley*, [Doctoral Dissertation]: University of Tennessee.

Sherwood, S. C., and Chapman, J., 2005, The identification and potential significance of early Holocene prepared clay surfaces: examples from Dust Cave and Icehouse Bottom: *Southeastern Archaeology*, v. 24, no. 1, p. 70–82.

Sherwood, S. C., and Goldberg, P., 2001, A geoarchaeological framework for the study of karstic cave sites in the eastern woodlands: *Midcontinental Journal of Archaeology*, v. 26, no. 2, p. 145–167.

Sherwood, S. C., and Kidder, T. R., 2011, The DaVincis of dirt: geoarchaeological perspectives on Native American mound building in the Mississippi River basin: *Journal of Anthropological Archaeology*, v. 30, no. 1, p. 69–87.

Sherwood, S. C., Simek, J. F., and Polhemus, R. R., 1995, Artifact size and spatial process; macro- and microartifacts in a Mississippian house: *Geoarchaeology*, v. 10, no. 6, p. 429–455.

Sherwood, S., 2001, Microartifacts, *in* Goldberg, P., Holliday, V. T., and Ferring, C. R., eds., *Earth Sciences and Archaeology*. New York, Kluwer Academic/Plenum Publishers, p. 327–352.

Sherwood, S., 2008, Increasing the resolution of cave archaeology, *in* Dye, D. H., ed., *Cave Archaeology of the Eastern Woodlands. Essays in Honor of Patty Jo Watson*: Knoxville, The University of Tennessee Press, p. 27–48.

Sherwood, S., and Kidder, T. R., 2011, The DaVincis of dirt: Geoarchaeological perspectives on Native American mound building in the Mississippi River basin: *Journal of Anthropological Archaeology*, v. 30, p. 69–87.

Shillito, L. M., and Ryan, P., 2013, Surfaces and streets: phytoliths, micromorphology and changing use of space at Neolithic Çatalhöyük (Turkey). *Antiquity*, v. 87, p. 684–700.

Shimelmitz, R., Kuhn, S. L., Jelinek, A. J., Ronen, A., Clark, A. E., and Weinstein-Evron, M., 2014, "Fire at will": the emergence of habitual fire use 350,000 years ago: *Journal of Human Evolution*, v. 77, p. 196–203.

Shimelmitz, R., Weinstein-Evron, M., Ronen, A., Kuhn, S.L., 2016. The Lower to Middle Paleolithic transition and the diversification of Levallois technology in the Southern Levant: evidence from Tabun Cave, Israel: *Quaternary International* 409, Part B, 23–40.

Shunkov, M., Kulik, N., Kozlikin, M., Sokol, E., Miroshnichenko, L., and Ulianov, V., 2018, The phosphates of Pleistocene–Holocene sediments of the eastern gallery of Denisova Cave: *Doklady Earth Sciences*, v. 478, no. 1, p. 46–50.

Siart, C., Forbinger, M. and Bubenzer, O. 2018. Digital geoarchaeology: bridging the gap between archaeology, geosciences and computer sciences, *in*, Siart, C., Forbinger, M., Bubenzer, O. (Eds.), *Digital Geoarchaeology: New Techniques for Interdisciplinary Human Environmental Research.* Springer, pp. 1–10.

Sidell, E. J., 2000, Dark earth and obscured stratigraphy, *in* Huntley, J. P., and Stallibrass, S., eds., *Taphonomy and Interpretation, Symposia of the Association for Environmental Archaeology No. 14.* Oxford, Oxbow, p. 35–42.

Sidell, E. J., 2003, The London Thames: a decade of research into the river and its floodplain, *in* Howard, A. J., Macklin, M. G., and Passmore, D. G., eds., *Alluvial Archaeology in Europe*: Lisse, A. A. Balkema Publishers, p. 133–143.

Sigurðardòttir, S., 2008, *Building with Turf*, Skagafjörður Heritage Museum.

Sigurgeirsson, M. A., Hauptfleisch, U., Newton, A., and Einarsson, A., 2013, Dating of the Viking Age Landnám tephra sequence in Lake Myvatn sediment, North Iceland: *Journal of the North Atlantic*, v. 21, p. 1–11.

Simmonds, A., Wenbann-Smith, F. F., Bates, M., Powell, K., Sykes, D., Devaney, R., Stansbie, D., and Score, D., 2011, *Excavations in North-West Kent 2005–2007. One hundred thousand years of human activity in and around the Darent Valley*. Oxford, Oxford Archaeology Ltd.

Simmons, I. G., 1975, The ecological setting of Mesolithic man in the highland zone, *in* Evans, J. G., Limbrey, S., and Cleere, H., eds., *The Effect of Man on the Landscape: The Highland Zone, Research Report 11*: York, CBA, p. 57–63.

Simpson, D., Lehouck, A., Van Meirvenne, M., Bourgeois, J., Thoen, E. and Vervloet, J. 2008. Geoarchaeological prospection of a medieval manor in the Dutch polders using an electromagnetic induction sensor in combination with soil augerings: *Geoarchaeology* 23, p. 305–319.

Simpson, I. A., Perdikaris, S., Cook, G., Campbell, J. L., and Teesdale, W. J., 2000, Cultural sediment analyses and transitions in early fishing activity at Langenesvæet, Vesterålen, Northern Norway: *Geoarchaeology*, v. 15, no. 8, p. 743–763.

Simpson, I., Barrett, J. H., and Milek, K., 2005, Interpreting the Viking Age to mediaeval period Transition in Norse Orkney through cultural soil and sediment analyses: *Geoarchaeology*, v. 20, p. 355–377.

Singer, B. S., 2014, A Quaternary geomagnetic instability time scale: *Quaternary Geochronology*, v. 21, p. 29–52.

Singer, R., and Wymer, J., 1982, *The Middle Stone Age at Klasies River Mouth in South Africa*, Chicago, The University of Chicago Press.

Sivitskis, A.J., Harrower, M.J., David-Cuny, H., Dumitru, I.A., Nathan, S., Wiig, F., Viete, D.R., Lewis, K.W., Taylor, A.K., Dollarhide, E.N., Zaitchik, B., Al-Jabri, S., Livi, K.J.T. and Braun, A. 2018. Hyperspectral satellite imagery detection of ancient raw material sources: Soft-stone vessel production at Aqir al-Shamoos (Oman): *Archaeological Prospection*, 25 (4), p. 363–374.

Sjöberg, A., 1976, Phosphate analysis of anthropic soils: *Journal of Field Archaeology*, v. 3, no. 4, p. 447–454.

Skaarup, J., and Grøn, O., 2004, *Møllegabet II: A submerged Mesolithic settlement in southern Denmark*. Oxford, Archaeopress.

Skre, D., 2007, *Kaupang in Skiringssal*: Aarhus, Aarhus University Press.

Skre, D., 2017, *Avaldsnes – A Sea-Kings' Manor in First-Millenium Western Scandinavia*. Berlin, de Gruyter.

Slon, V., Hopfe, C., Weiß, C. L., Mafessoni, F., de la Rasilla, M., Lalueza-Fox, C., Rosas, A., Soressi, M., Knul, M. V., and Miller, R., 2017, Neandertal and Denisovan DNA from Pleistocene sediments: *Science*, v. 356, no. 6338, p. 605–608.

Slon, V., Mafessoni, F., Vernot, B., de Filippo, C., Grote, S., Viola, B., Hajdinjak, M., Peyrégne, S., Nagel, S., and Brown, S., 2018, The genome of the offspring of a Neanderthal mother and a Denisovan father: *Nature*, v. 561, no. 7721, p. 113–116.

Slotten, V., Lentz, D., and Sheets, P., 2020, Landscape management and polyculture in the ancient gardens and fields at Joya de Cerén, El Salvador: *Journal of Anthropological Archaeology*, v. 59, p. 101191.

Smart, P., and Tovey, N. K., 1981, *Electron microscopy of soils and sediments: examples*. Oxford/New York, Clarendon Press; Oxford University Press, viii.

Smart, P., and Tovey, N. K., 1982, *Electron Microscopy of Soils and Sediments: Techniques*. Oxford/ New York, Clarendon Press, p. xiii.

Smith, G., Macphail, R. I., Mays, S. A., Nowakowski, J., Rose, P., Scaife, R. G., Sharpe, A., Tomalin, D. J., and Williams, D. F., 1996, Archaeology and environment of a Bronze Age Cairn and Romano-British field system at Chysauster, Gulval, near Penzance, Cornwall: *Proceedings of the Prehistoric Society*, v. 62, p. 167–219.

Smith, K., Coppen, J., Wainwright, G. J. and Beckett, S. 1981. The Shaugh Moor Project: third report-settlement and environmental investigations. *Proceedings of the Prehistoric Society*, 47, p. 205–273.

Smith, M. F., 2020, Geoarchaeological investigations at the Ryan-Harley Paleoindian site, Florida (8JE1004): Implications for human settlement of the Wacissa River Basin during the Younger Dryas: *Geoarchaeology*, v. 35, no. 4, p. 451–466.

Smith, N. J. H., 1980, Anthrosols and human carrying capacity in Amazonia: *Annuals of the Association of American Geographers*, v. 70, p. 553–566.

Smyth, M. P., Dunning, N. P., and Dore, C. D., 1995, Interpreting prehistoric settlement patterns: lessons from the Maya center of Sayil, Yucatan: *Journal of Field Archaeology*, v. 22, no. 3, p. 321–347.

Sneh, A., 1983, Redeposited Loess from the Quaternary Besor Basin, Israel: *Israel Journal of Earth Sciences*, v. 32, p. 63–69.

Soil Conservation Service, 1994, *Keys to Soil Taxonomy*. Washington DC, U.S. Department of Agriculture.

Soil Survey Staff, 2014, *Keys to Soil Taxonomy*, 12th edition. Washington, DC, U. S. Department of Agriculture-Natural Resources Conservation Service.

Sombroek, W. G., 1966, Amazon Soils. *A Reconnaissance of the Soils of the Brazilian Amazon Region*. Wageningen, Centre for Agricultural Publication and Documentation.

Sombroek, W. G., Kern, D., Rodrigues, T., Cravio, M., Jarbas, T., Woods, W. I., and Glaser, B., 2002, Terra Preta and Terra Mulata: pre-Columbian Amazon kitchen middens and agricultural fields, their sustainability and their replication, *17th World Congress of Soil Science*, Symposium 18, 1935: Bankok, Thailand, p. 1–9.

Sordoillet, D., Weller, O., Rouge, N., Buatier, M., and Sizun, J. P., 2018, Earliest salt working in the world: from excavation to microscopy at the prehistoric sites of Țolici and Lunca (Romania). *Journal of Archaeological Science*, v. 89, p. 46–55.

Sorensen, A. C., 2017, On the relationship between climate and Neandertal fire use during the Last Glacial in south-west France: *Quaternary International*, v. 436, p. 114–128.

Sørensen, R., Henningsmoen, K. E., Høeg, H., Stabell, B., and Bukholm, K. M., 2007, Geology, soils, vegetation and sea levels in the Kaupang area, *in* Skre, D., ed., *Kaupang in Skiringssal*: Aarhus, Denmark, Aarhus University Press.

Soressi, M., Jones, H. L., Rink, W. J., Maureille, B., and Tillier, A.-M., 2007, The Pech-de-l'Azé I Neandertal child: ESR, uranium series, and AMS 14C dating of its MTA type B context: *Journal of Human Evolution*, v. 52, p. 455–466.

Soressi, M., McPherron, S. P., Lenoire, M., Dogandzic, T., Goldberg, P., Jacobs, Z., Maigrot, Y., Martisius, N. L., Miller, C. E., Rendu, W., Richards, M., Skinner, M. M., Steele, T. E., Talamo, S., and Texier, J.-P., 2013, Neandertals made the first specialized bone tools in Europe: *PNAS*, v. 110, no. 35, p. 14186–14190.

Soressi, M., Rendu, W., Texier, J.-P., Daulny, É. C. L., D'Errico, F., Laroulandie, V., Maureille, B., Niclot, M., Schwortz, S., and Tillier, A.-M., 2008, Pech-de-l'Azé I (Dordogne, France): nouveau regard sur un gisement moustérien de tradition acheuléenne connu depuis le XIXe siècle: Les sociétés Paléolithiques d'un grand Sud-Ouest: nouveaux gisements, nouvelles méthodes, nouveaux résultats.-*Actes des journées décentralisées de la SPF des 24–25 novembre 2006*, p. 95–132.

Spataro, M., 2002, *The First Farming Communities of the Adriatic: Pottery Production and Circulation in the Early and Middle Neolithic*, Trieste, Società per la Preistoria e Protostoria della Regione Friuli-Venezia Giulla.

Speth, J. D., and Tchernov, E., 2007, The Middle Paleolithic occupations at Kebara Cave: a faunal perspective, *in* Bar-Yosef, O., and Meignen, L., eds., *Kebara Cave Mt. Carmel, Israel: The Middle and Upper Paleolithic Archaeology.* Cambridge, American School of Prehistoric Research Bulletin 49, p. 165–260.

Speth, J. D., Meignen, L., Bar-Yosef, O., and Goldberg, P., 2012, Spatial organization of Middle Paleolithic occupation X in Kebara Cave (Israel): concentrations of animal bones: *Quaternary International*, v. 247, no. 1, p. 85–102.

Spoor, F., Leakey, M. G., Gathogo, P. N., Brown, F. H., Anton, S. C., McDougall, I., Kiarie, C., Manthi, F. K., and Leakey, L. N., 2007, Implications of new early Homo fossils from Ileret, east of Lake Turkana, Kenya: *Nature*, v. 448, p. 688–691.

Stabel, H. H., 1985, Mechanisms controlling the sedimentation sequence of various elements in prealpine lakes, *in* Stumm, W., ed., *Chemical Processes in Lakes.* New York, Wiley, p. 143–167.

Stafford, E., and Teague, S., 2016, *From Blackfriars to Bankside. Medieval and later riverfront archaeology along the route of Thameslink, Central London.* London, PCA and Oxford Archaeology (OAPCA).

Stafseth, T., 2021, *Rådhuspladsen Nord, KBM 4286* (Unpublished excavation report): Museum of Copenhagen.

Stahlschmidt, M. C., Miller, C. E., Ligouis, B., Goldberg, P., Berna, F., Urban, B., and Conard, N. J., 2015a, The depositional environments of Schöningen 13 II-4 and their archaeological implications: *Journal of Human Evolution*, v. 89, p. 71–91.

Stahlschmidt, M. C., Miller, C. E., Ligouis, B., Hambach, U., Goldberg, P., Berna, F., Richter, D., Urban, B., Serangeli, J., and Conard, N. J., 2015b, On the evidence for human use and control of fire at Schöningen: *Journal of Human Evolution*, v. 89, p. 181–201.

Stanley, D. J., Chen, Z., and Song, J., 1999, Inundation, sea-level rise and transition from Neolithic to Bronze Age cultures, Yangtze delta, China: *Geoarchaeology*, v. 14, no. 1, p. 15–26.

Stead, I. M., Bourke, J. B., and Brothwell, D. R., 1986, *Lindow Man: The Body in the Bog.* [London]; Ithaca, NY, Published for the Trustees of the British Museum by British Museum Publications; Cornell University Press.

Stein, J. K., 1987, Deposits for archaeologists: *Advances in Archaeological Method and Theory*, v. 11, p. 337–393.

Stein, J. K., 1990, Archaeological stratigraphy, *in* Lasca, N. P., and Donahue, J., eds., *Archaeological Geology of North America, Centennial Special Vol. 4*, Boulder, CA, Geological Society of America, p. 513–523.

Stein, J. K., 1992, *Deciphering a Shell Midden.* San Diego, CA, Academic Press, xix.

Stein, J. K., and Linse, A. R., 1993, *Effects of Scale on Archaeological and Geoscientific Perspectives.* Boulder, CO, Geological Society of America, Special paper.

Stephens, E. P., 1956, The uprooting of trees: a forest process: *Soil Science Society of America Proceedings*, v. 20, p. 113–116.

Stern, N., Porch, N., and McDougall, I., 2002, FxJj43: a window into a 1.5 million-year-old palaeolandscape in the Okote Member of the Koobi Fora Formation, Northern Kenya: *Geoarchaeology*, v. 17, no. 4, p. 349–392.

Stewart, D. J., 1999, Formation processes affecting submerged archaeological sites: an overview: *Geoarchaeology*, v. 14, no. 6, p. 565–587.

Stiner, M. C., 2005, *The Faunas of Hayonim Cave (Israel). A 200,000-Year Record of Paleolithic Diet, Demography and Society.*, Cambridge, Peabody Museum, Harvard University, American School of Prehistoric Research, Bulletin 48.

Stiner, M., Kuhn, S. L., Surovell, T. A., Goldberg, P., Meignen, L., Weiner, S., and Bar-Yosef, O., 2001, Bone preservation in Hayonim Cave (Israel): a macroscopic and mineralogical study: *Journal of Archaeological Science*, v. 28, p. 643–659.

Stiner, M., Kuhn, S., Surovell, T., Goldberg, P., Margaris, A., Meignen, L., Weiner, S., and Bar-Yosef, O., 2005, Bone, ash, and shell preservation in Hayonim Cave, *in* Stiner, M., ed., *The Faunas of Hayonim Cave (Israel). A 200,000-Year Record of Paleolithic Diet, Demography and Society, Vol. 48*: Cambridge (US), American School of Prehistoric Research, Peabody Museum, Harvard University, p. 59–79.

Stoops, G., 1984, Petrographic study of mortar and plaster samples, *in* Devreker, J., and Waelkens, M., eds., "Les Fouilles de la Rijksuniversiteit te Gent a Pessiononte 1967–73" : *Dissertationes Archaeologicae Gardenses*, v. XXII, p. 164–170.

Stoops, G., 2003, *Guidelines for Analysis and Description of Soil and Regolith Thin Sections*, Madison, Wisconsin, Soil Science Society of America, Inc.

Stoops, G., 2017a, Laterite as construction material, *in* Nicosia, C., and Stoops, G., eds., *Archaeological Soil and Sediment Micromorphology*. Chichester, Wiley Blackwell, p. 111–112.

Stoops, G., 2017b, Fluorescence microscopy, *in* Nicosia, C., and Stoops, G., eds., *Archaeological Soil and Sediment Micromorphology*. Chichester, Wiley Blackwell, p. 393–397.

Stoops, G., 2020, Guidelines for Analysis and Description of Soil and Regolith Thin Sections, 2nd edition. Chichester: Wiley.

Stoops, G., and Marcelino, V., 2018, lateritic and bauxitic materials, *in* Stoops, G., Marcelino, V., and Mees, F., eds., *Interpretation of Micromorphological Features of Soils and Regoliths*. Amsterdam, Elsevier, p. 691–720.

Stoops, G., and Schaefer, C. E. G. R., 2018, Pedoplasmation: formation of soil material, *in* Stoops, G., Marcelino, V., and Mees, F., eds., *Interpretation of Micromorphological Features in Soils and Regoliths*. Amsterdam, Elsevier, p. 59–71.

Stoops, G., Canti, M., and Kapur, S., 2017a, Calcareous Mortars, Plasters and Floors, *in* Nicosia, C., and Stoops, G., eds., *Archaeological Soil and Sediment Micromorphology*. Chichester, Wiley, p. 189–200.

Stoops, G., Marcelino, V., and Mees, F., 2018, *Interpretation of Micromorphological Features of Soils and Regoliths*, 2nd edition. Amsterdam, Elsevier, p. 1000.

Stoops, G., Tsatskin, A., and Canti, M. G., 2017b, Gypsic mortars and plasters, *in* Nicosia, C., and Stoops, G., eds., *Archaeological Soil and sediment Micromorphology*. Chichester, Wiley Blackwell, p. 201–204.

Stoops, G., Sedov, S., and Shoba, S., 2018b, *Regolith and Soils on Volcanic Ash, Interpretation of Micromorphological Features of Soils and Regoliths*. Amsterdam, Elsevier, p. 721–751.

Straulino, L., Sedov, S., Michelet, D., and Balanzario, S., 2013, Weathering of carbonated materials in ancient Maya constructions (Río Bec and Dzibanché): limestone and stucco deterioration patterns: *Quaternary International*, v. 315, p. 87–100.

Stringer, C. B., Trinkhaus, E., Roberts, M. B., Macphail, R. I., and Parfitt, S. A., 1998, The Middle Pleistocene human tibia from Boxgrove: *Journal of Human Evolution*, v. 34, p. 509–547.

Stringer, C. B., Barton, R. N. E., Currant, J. C., Finlayson, J. C., Goldberg, P., Macphail, R., and Pettit, P. B., 1999, Gibraltar Palaeolithic revisited: new excavations at Gorham's and Vanguard Caves, *in* Davies, W., and Charles, R., eds., *Dorothy Garrod and the Progress of the Palaeolithic (Studies in the Prehistoric Archaeology of the Near East and Europe)*. Oxford, Oxbow Books, p. 84–96.

Stringer, C., 2000, The Gough's Cave human fossils: an introduction: *Bulletin of the Natural History Museum. Geology Series*, v. 56, no. 2, p. 135–139.

Stumm, W., 2004, Chemical processes regulating the composition of lake waters, *in* O'Sullivan, P., and Reynolds, C. S., eds., *The Lakes Handbook: Limnology and Limnetic Ecology*, Vol. 1. Oxford, Wiley Blackwell, p. 79–106.

Sukopp, H., Blume, H.-P., and Kunick, W., 1979, The soil, flora and vegetation of Berlin's waste lands, *in* Laurie, I. C., ed., *Nature in Cities*. Chichester, John Wiley, p. 115–132.

Sulas, F., 2018, Traditions of water in the highlands of the northern Horn of Africa, *in* Sulas, F., and Pikirayi, I., eds., *Water and Societies*, Routledge, p. 155–175.

Summerfield, M. A., 1991, *Global Geomorphology: An Introduction to the Study of Landforms*. New York, John Wiley & Sons.

Sunaga, K., Sakagami, K., and Seki, T., 2003, Chemical and physical properties of buried cultivated soils of Edo Periods under volcanic mudflow deposits of 1783 eruption of Asama Volcano (Tables and abstract in English): *Pedologist*, v. 47, no. 1, p. 14–28.

Sutikna, T., Tocheri, M. W., Faith, J. T., Jatmiko, Due Awe, R., Meijer, H. J. M., Wahyu Saptomo, E., and Roberts, R. G., 2018, The spatio-temporal distribution of archaeological and faunal finds at Liang Bua (Flores, Indonesia) in light of the revised chronology for Homo floresiensis: *Journal of Human Evolution*, v. 124, p. 52–74.

Sutikna, T., Tocheri, M. W., Morwood, M. J., Saptomo, E. W., Jatmiko, Awe, R. D., Wasisto, S., Westaway, K. E., Aubert, M., Li, B., Zhao, J.-x., Storey, M., Alloway, B. V., Morley, M. W., Meijer, H. J. M., van den Bergh, G. D., Grün, R., Dosseto, A., Brumm, A., Jungers, W. L., and Roberts, R. G., 2016, Revised stratigraphy and chronology for Homo floresiensis at Liang Bua in Indonesia: *Nature*, v. 532, p. 366.

Talamo, S., Aldeias, V., Goldberg, P., Chiotti, L., Dibble, H. L., Guérin, G., Hublin, J.-J., Madelaine, S., Maria, R., Sandgathe, D., Steele, T. E., Turq, A., and McPherron, S. J. P., 2020, The new ^{14}C chronology for the Palaeolithic site of La Ferrassie, France. The disappearance of Neandertals and the arrival of *Homo sapiens* in France: *Journal of Quaternary Science*, v. 35, no. 7, p. 961–973.

Tan, K. H., 1984, *Andosols*. New York, Van Nostrannd Reinhold Company.

Tang, Y., Jia, J., and Xie, X., 2003, Records of magnetic properties in Quaternary loess and its paleoclimatic significance: *Quaternary International*, v. 108, p. 33–50.

Tankard, A. J., 1976, The stratigraphy of a coastal cave and its palaeoclimatic significance: *Palaeoecology of Africa*, v. 9, p. 151–159.

Tankard, A. J., and Schweitzer, F. R., 1974, The geology of Die Kelders Cave and environs: a palaeoenvironmental study: *South African Journal of Science*, v. 70, p. 365–369.

Tankard, A. J., and Schweitzer, F. R., 1976, Textural analysis of cave sediments: Die Kelders, Cape Province, South Africa, *in* Davidson, D. A., and Shackley, M. L., eds., *Geoarchaeology: Earth Science and the Past*. London, Duckworth, p. 289–316.

Tansley, A. G., 1939, *The British Islands and their Vegetation*. Cambridge, Cambridge University Press.

Tapete, D., Banks, V., Jones, L., Kirkham, M. and Garton, D. 2017. Contextualising archaeological models with geological, airborne and terrestrial LiDAR data: The Ice Age landscape in Farndon Fields, Nottinghamshire, UK: *Journal of Archaeological Science*, v. 81, p. 31–48.

Tapete, D. and Cigna, F. 2017.Trends and perspectives of space-borne SAR remote sensing for archaeological landscape and cultural heritage applications. *Journal of Archaeological Science: Reports*, 14, pp. 716–726.

Taxel, I., Sivan, D., Bookman, R., and Roskin, J., 2018, An Early Islamic inter-settlement agroecosystem in the coastal sand of the Yavneh Dunefield, Israel: *Journal of Field Archaeology*, v. 43, no. 7, p. 551–569.

Taylor, G. H., Teichmüller, M., A., D., Diessel, C. F. K., Littke, R., and Robert, P., 1998, *Organic Petrology*. Berlin-Stuttgart, Gebrüder Bornträger.

Tchernov, E., 1987, The age of 'Ubeidiya Formation, and early Pleistocene hominid site in the Jordan Valley, Israel: *Israel Journal of Earth Sciences* v. 36, p. 3–30.

Terrible, F., Wright, R., and FitzPatrick, E. A., 1997, Image analysis in soil micromorphology: from univariate approach to multivariate solution, *in* Shoba, S., Gerasimova, M., and Miedema, R., eds., *Soil Micromorphology: Studies on Soil Diversity, Diagnostics, Dynamics*: Moscow – Wageningen, International Society of Soil Science, p. 397–417.

Texier, J.-P., 2006, La Ferrassie, *in* Texier, J.-P., Kervazo, B., Lenoble, A., and Nespoulet, R., eds., *Sédimentogenèse de Sites Préhistoriques Classiques du Périgord, Édition Numérique I 2006*: Les Eyziies, Pôle International de la Préhistoire, p. 23–30.

Texier, J.-P., 2009, *Histoire géologique de sites préhistoriques classiques du Périgord une vision actualisée la Micoque, la grotte Vaufrey, le Pech de l'Azé I et II, la Ferrassie, l'abri Castanet, le Flageolet, Laugerie Haute*. Paris, Édition du Comité des travaux historiques et scientifiques, Collection Documents préhistoriques, v. n° 25.

Texier, J.-P., Agsous, S., Kervazo, B., Lenoble, A., and Nespoulet, R., 2006, *Sédimentogenèse de sites préhistoriques classiques du Périgord*, Les Eyzies, Pôle International de la Préhistoire.

Théry-Parisot, I., Costamagno, S., Brugal, J. P., Fosse, O., and Guibert, R., 2005, The use of bone as fuel during the Palaeolithic, experimental study of bone combustible properties, *in* Mulville, J., and Outram, A., eds., *The Zooarchaeology of Fats, Oils, Milk and Dairying*, Oxbow Book, p. 50–59.

Thieme, H., 1997, Lower Palaeolithic hunting spears from Germany: *Nature*, v. 385, p. 807–810.

Thirly, M., Galbois, J., and Schmitt, J.-M., 2006, Unusual phosphate concretions related to groundwater flow in a continental environment: *Journal of Sedimentary Research*, v. 76, p. 866–877.

Thomas, D. H., 1998, *Archaeology*, 3rd edition. Fort Worth, TX, Harcourt Brace College Publishers.

Thomas, D. S. G., 1989, Aeolian sand deposits, *in* Thomas, D. S. G., ed., *Arid Zone Geomorphology*. New York, Halsted Press (John Wiley & Sons), p. 232–261.

Thorndycroft, V. R., Pirrie, D., and Brown, A. G., 1999, Tracing the record of early alluvial tin mining on Dartmoor, UK., *in* Pollard, A. M., ed., *Geoarchaeology: Exploration, Environments, Resources*. London, Geological Society Special Publications, p. 91–102.

Thorson, R. M., 1990, Archaeological Geology: *Geotimes*, v. 35, p. 32–33.

Thorson, R. M., and Hamilton, T. D., 1977, Geology of the Dry Creek site: a stratified early man site in interior Alaska: *Quaternary Research*, v. 7, p. 172–188.

Thurley, S., 1995, *The King's Privy Garden at Hampton Court Palace 1689–1995*. London, Apollo.

Tite, M. S., 1972, The influence of geology on the magnetic susceptibility of soils on archaeological sites: *Archaeometry*, v. 14, p. 229–236.

Tite, M. S., and Mullins, C. E., 1971, Enhancement of magnetic susceptibility of soils on archaeological sites: *Archaeometry*, v. 13, p. 209–219.

Toffolo, M. B., Brink, J. S., van Huyssteen, C., and Berna, F., 2017, A microstratigraphic reevaluation of the Florisbad spring site, Free State Province, South Africa: formation processes and paleoenvironment: *Geoarchaeology*, v. 32, no. 4, p. 456–478.

Tøssebro, C., Åstveit, L. I., Macphail, R. I., et al., in prep, Seasonality, use and reuse of two Late Mesolithic dwellings from Western Norway. *Journal of Field Archaeology*.

Touchart, L., 2012, Baikal, Lake, *in* Bengtsson, L., Herschy, R. W., and Fairbridge, R. W., eds., *Encyclopedia of Lakes and Reservoirs*: Dordrecht, Springer, p. 83–90.

Tovey, N. K., and Sokolov, V. N., 1997, Image analysis applications in soil micromorphology, *in* Shoba, S., Gerasimova, M., and Miedema, R., eds., *Soil Micromorphology: Studies on Soil Diversity, Diagnostics, Dynamics*: Moscow – Wageningen, International Society of Soil Science, p. 345–356.

Trinks, I., Hinterleitner, A., Neubauer, W., Nau, E., Löcker, K., Wallner, M., Gabler, M., Filzwieser, R., Wilding, J., Schiel, H., Jansa, V., Schneidhofer, P., Trausmuth, T., Sandici, V., Ruß, D.,Flöry, S., Kainz, J., Kucera, M., Vonkilch, A., Tencer, T., Gustavsen, L., Kristiansen, M., Bye-Johansen, L.M., Tonning, C., Zitz, T., Paasche, K., Gansum, T. and Seren, S. 2018. Large-area high-resolution ground-penetrating radar measurements for archaeological prospection. *Archaeological Prospection*, 25, pp. 171–195.

Tsatskin, A., Weinstein-Evron, M., Ronen, A., Polishook, B.E., 1992. *Preliminary Paleoenvironmental Studies of the Lowest Layers (G and F) of Tabun Cave, Mt. Carmel*. Israel Geological Society annual meeting, p.153.

Tsatskin, A., and Nadel, D., 2003, Formation processes at the Ohalo II submerged prehistoric campsite, Israel, inferred from soil micromorphology and magnetic susceptibility studies: *Geoarchaeology*, v. 18, no. 4, p. 409–432.

Tsoar, H., and Goodfriend, G. A., 1994, Chronology and palaeoenvironmental interpretation of Holocene aeolian sands at the inland edge of the Sinai-Negev erg: *The Holocene*, v. 4, no. 3, p. 244–250.

Tucker, M. E., 1988, *Techniques in Sedimentology*. Oxford, Blackwell Scientific Publications.

Tucker, M. E., 2001, *Sedimentary Petrology: An Introduction to the Origin of Sedimentary Rocks*. Oxford, Blackwell Science.

Tucker, M. E., 2009, *Sedimentary Petrology: An Introduction to the Origin of Sedimentary Rocks*, John Wiley & Sons.

Tundisi, J. E., and Matsumura Tundisi, T., 2012, *Limnology*. London, Taylor & Francis.

Turner, B. D., 1987, Forensic entomology: insects against crime: *Science Progress*, v. 71, p. 169–180.

Turner, B. D., and Wiltshire, P., 1999, Experimental validation of forensic evidence: a study of the decomposition of buried pigs in a heavy clay soil: *Forensic Science International*, v. 101, p. 113–112.

U.S. Geological Survey Geologic Names Committee, 2018, *Divisions of geologic time-major chronostratigraphic and geochronologic units*, Fact Sheet 20183054.

Uchida, E., Maeda, N., and Nakagawa, T., 1999, The laterites of the Anghor monuments, Cambodia. The grouping of the monuments on the basis of laterites: *Journal of Mineralogy Petrology and Economic Geology*, v. 94, p. 162–175.

Újvári, G., Varga, A., Raucsik, B., and Kovács, J., 2014, The Paks loess-paleosol sequence: a record of chemical weathering and provenance for the last 800 ka in the mid-Carpathian Basin: *Quaternary International*, v. 319, p. 22–37.

Usai, M. R., Pickering, M. D., Wilson, C. A., Keely, B. J., and Brothwell, D. R., 2014, "Interred with their bones": soil micromorphology and chemistry in the study of human remains: *Antiquity Journal*, v. 88, no. 339, p. 12.

USDA, 2014, *Keys to Soil Taxonomy*. Washington DC, Department of Agriculture (USDA) Natural Resources Conservation Service (NRCS).

Valentin, C., 1983, Effects of grazing and trampling around recently drilled water holes on the soil deterioration in the Sahelian zone: *Soil Erosion and Conservation* (Soil Conservation Society of America), no. p. 51–65.

Valentin, C., 1991, Surface crusting in two alluvial soils of northern Niger: *Geoderma*, v. 48, p. 201–222.

Valladas, H., Geneste, J. M., Joron, J. L., and Chadelle, J. P., 1986, Thermoluminescence dating of Le Moustier (Dordogne, France): *Nature*, v. 322, no. 6078, p. 452–454.

Valladas, H., Reyss, J. L., Joron, J. L., Valladas, G., Bar-Yosef, O., and Vandermeersch, B., 1988, Thermoluminescence dating of Mousterian "Proto-Cro-Magnon" remains from Israel and the origin of modern man: *Nature*, v. 331, no. 6157, p. 614–616.

Vallverdú, J., Allué, E., Bischoff, J. L., Cáceres, I., Carbonell, E., Cebrià, A., García-Antón, D., Huguet, R., Ibáñez, N., Martínez, K., Pastó, I., Rosell, J., Saladié, P., and Vaquero, M., 2005, Short human occupations in the Middle Palaeolithic level i of the Abric Romaní rock-shelter (Capellades, Barcelona, Spain): *Journal of Human Evolution*, v. 48 p. 157–174.

Vallverdú, J., Courty, M.-A., Carbonell, E., Canals, A., and Burjachs, F., 2001, Les sédiments d'Homo Antecessor de Gran Dolina, (Sierra de Atapuerca, Burgos, Espagne). Interprétation micromorphologique des processus de formation et enregistrement paléoenvironnemental des sédiments: *L'Anthropologie*, v. 105, p. 45–69.

van Andel, T. H., Zanger, E., and Demitrack, S., 1990, Land use and soil erosion in prehistoric and historic Greece: *Journal of Field Archaeology*, v. 17, p. 379–396.

van de Westeringh, W., 1988, Man-made soils in the Netherlands. especially in sandy areas ("Plaggen Soils"), *in* Groenman-van Waateringe, W., and Robinson, M., eds., *Man-made Soils, International Series 410*. Oxford, British Archaeological Reports, p. 5–19.

Van der Drift, J., 1964, Soil fauna and soil profile in some inland-dune habitats, *in* Jongerius, A., ed., *Soil Micromorphology*. Amsterdam, Elsevier, p. 69–81.

van der Lubbe, H. J. L., Krause-Nehring, J., Junginger, A., Garcin, Y., Joordens, J. C. A., Davies, G. R., Beck, C. C., Feibel, C. S., Johnson, T. C., and Vonhof, H. B., 2017, Gradual or abrupt? Changes in water source of Lake Turkana (Kenya) during the African Humid Period inferred from Sr isotope ratios: *Quaternary Science Reviews*, v. 174, p. 1–12.

van der Meer, J. J. M., 1982, *The Fribourg Area, Switzerland. A study in Quaternay geology and soil development*. Amsterdam, University of Amsterdam.

Van Keuren, S., and Roos, C. I., 2013, Geoarchaeological evidence for ritual closure of a kiva at Fourmile Ruin, Arizona: *Journal of Archaeological Science*, v. 40, no. 1, p. 615–625.

van Metre, P. C., and Fuller, C. C., 2009, Dual-core mass-balance approach for evaluating mercury and ^{210}Pb atmospheric fallout and focusing to lakes: *Environmental Science and Technology*, v. 43, no. 1, p. 26–32.

Van Nest, J., 2002, The good earthworm: How natural processes preserve upland Archaic archaeological sites of western Illinois, U.S.A: *Geoarchaeology*, v. 17, no. 1, p. 53–90.

Van Nest, J., Charles, D. K., Buikstra, J. E., and Asch, D. L., 2001, Sod blocks in Illinois Hopewell Mounds: *American Antiquity*, v. 66, p. 633–650.

Van Ranst, E., Wilson, M. A., and Righi, D., 2018, Spodic materials, *in* Stoops, G., Marcelino, V., and Mees, F., eds., *Interpretation of Micromorphological Features of Soils and Regoliths*. Amsterdam, Elsevier, p. 633–662.

Van Vliet, B., 1980, Approche des conditions physico-chimiques favorisant l'autofluorescence des minéraux argileux: *Pedologie*, v. 30, p. 369–390.

Van Vliet, B., 1982, Structures and microstructures associées à la formation de glace de ségrégation: leurs conséquences, *Proceedings of the Fourth Canadian Permafrost Conference*: Ottawa, p. 116–122.

Van Vliet, B., Faivre, P., Andreux, F., Robin, A. M., and Portal, J. M., 1983, Behaviour of some organic components in blue and ultra-violet light: application to the micromorphology of podzols, andosols and planosols, *in* Bullock, P., and Murphy, C. P., eds., *Soil Micromorphology, Vol. 1: Techniques and Applications*: Berkhamsted, A B Academic Publishers, p. 91–99.

Van Vliet-Lanoë, B., 1985, Frost effects in soils, *in* Boardman, J., ed., *Soils and Quaternary Landscape Evolution*. Chichester, John Wiley & Sons, p. 117–158.

van Vliet-Lanoe, B., 1986, Micromorphology, *in* Callow, P., and Cornford, J. M., eds., *La Cotte de St. Brelade*: Norwich, Geo Books, p. 1961–1978.

Van Vliet-Lanoë, B., 1998, Frost and soils: implications for paleosols, paleoclimates and stratigraphy: *Catena*, v. 34, no. 1–2, p. 157–183.

Van Vliet-Lanoë, B., and Fox, C. A., 2018, Frost action, *in* Stoops, G., Marcelino, V., and Mees, F., eds., *Interpretation of Micromorphological Features of Soils and Regoliths*, 2nd edition. Amsterdam, Elsevier, p. 575–603.

Van Vliet-Lanoë, B., Helluin, M., Pellerin, J., and Valadas, B., 1992, Soil erosion in Western Europe: from the last interglacial to the present, *in* Bell, M., and Boardman, J., eds., *Past and Present Soil Erosion, Monograph 22*. Oxford, Oxbow, p. 101–114.

Vandermeersch, B., 1981, Les hommes fossiles de Qafzeh (Israël). Paris, Editions du CNRS, *Cahiers de Paléontologie (Paléoanthropologie)*.

Vandermeersch, B., and Bar-Yosef, O., 2019, The Paleolithic Burials at Qafzeh Cave, Israel: *PALEO*. Revue d'archéologie préhistorique, v. 30, no. 1, p. 256–275.

Veneman, P. I. M., Jacke, P. V., and Bodine, S. M., 1984, Soil formation as affected by pit and mound relief in Massachusetts, USA: *Geoderma*, v. 33, p. 89–99.

Vepraskas, M. J., Lindbo, D. L., and Stolt, M. H., 2018, Redoximorphic Features, *in* Stoops, G., Marcelino, V., and Mees, F., eds., *Interpretation of Micromorphological Features of Soils and Regoliths*. Amsterdam, Elsevier, p. 425–445.

Verdonck, L., Launaro, A., Vermeulen, F. and Millett, M. 2020. Ground-penetrating radar survey at Falerii Novi: a new approach to the study of Roman cities. *Antiquity*, v. 94, p. 705–723.

Verhegge, J, Storme, A, Cruz, F. and Crombé, P. 2021. Cone penetration testing for extensive mapping of deeply buried late glacial cover sand landscape paleotopography. *Geoarchaeology*, v. 36, p. 130–148.

Viklund, K., 1998, *Cereals, Weeds and Crop Processing in Iron Age Sweden. Methodological and interpretive aspects of archaeobotanical evidence*, Umea, Archaeology and Environment 14, 192 p.

Viklund, K., Engelmark, R., and Linderholm, J., 1998, *Fåhus från bronsålder till idag*, Skrifter om skogs- och lantbrukshistoria 12: Lund, Nordiska Museet.

Viklund, K., Linderholm, J., and Macphail, R. I., 2013, Integrated Palaeoenvironmental Study: Micro- and Macrofossil Analysis and Geoarchaeology (soil chemistry, magnetic susceptibility and micromorphology), *in* Gjerpe, L.-E., ed., *E18-prosjektet Gulli-Langåker. Oppsummering og arkeometriske analyser, Vol. Bind 3*: Bergen, Fagbokforlaget, p. 25–83.

Villagran, X. S., Giannini, P. C. F., and DeBlasis, P., 2009, Archaeofacies analysis: using depositional attributes to identify anthropogenic processes of deposition in a monumental shell mound of Santa Catarina State, southern Brazil: *Geoarchaeology*, v. 24, no. 3, p. 311–335.

Villagran, X. S., Balbo, A. L., Madella, M., Vila, A., and Estevez, J., 2011a, Experimental micromorphology in Tierra del Fuego (Argentina): building a reference collection for the study of shell middens in cold climates: *Journal of Archaeological Science*, v. 38, no. 3, p. 588–604.

Villagran, X. S., Klokler, D., Peixoto, S., Deblasis, P., and Giannini, P. C. F., 2011b, Building coastal landscapes: zooarchaeology and geoarchaeology of Brazilian shell mounds: *Journal of Island and Coastal Archaeology*, v. 6, no. 2, p. 211–234.

Villagran, X. S., Schaefer, C. E. G. R., and Ligouis, B., 2013, Living in the cold: geoarchaeology of sealing sites from Byers Peninsula (Livingston Island, Antarctica): *Quaternary International*, v. 315 – Site formation processes in archaeology, p. 184–199.

Villagran, X. S., Strauss, A., Miller, C., Ligouis, B., and Oliveira, R., 2017a, Buried in ashes: site formation processes at Lapa do Santo rockshelter, east-central Brazil: *Journal of Archaeological Science*, v. 77, p. 10–34.

Villagran, X. S., Huisman, D. J., Mentzer, S. M., Miller, C. E., and Jans, M. M., 2017b, Bone and other skeletal tissue, *in* Nicosia, C., and Stoops, G., eds., *Archaeological Soil and sediment Micromorphology*. Chichester, Wiley Blackwell, p. 11–38.

Villaseñor, I., and Graham, E., 2010, The use of volcanic materials for the manufacture of pozzolanic plasters in the Maya lowlands: a preliminary report: *Journal of Archaeological Science*, v. 37, p. 1339–1347.

Vincent, P. J., Lord, T. C., Telfer, M. W. and Wilson, P. 2010. Early Holocene loessic colluviation in northwest England: new evidence for the 8.2 ka event in the terrestrial record?: *BOREAS*, v. **40**, p. 105–115.

Vint, J. M., and Nials, F. L., 2015, *The Anthropogenic Landscape of Las Capas, an Early Agricultural Irrigation Community in Southern Arizona (AP-50)*. Tucson, AZ, Archaeology Southwest, p. 294.

Vissac, C., 2002, *Les Terres Raportées dans les Jardins du XVIe au XIXe Siècles Caracterisations de l'Impact Anthropique à Differentes Echèlles d'Organisation du Sol*: [Doctoral Thesis], Angers, 237 p.

Vita-Finzi, C., 1969, *The Mediterranean Valleys*. Cambridge, Cambridge University Press.

Vita-Finzi, C., 1973, *Recent Earth History*. London, Macmillan.

Vita-Finzi, C., and Stringer, C., 2007, The setting of the Mt. Carmel caves reassessed: *Quaternary Science Reviews*, v. 26, no. 3, p. 436–440.

Vittori Antisari, L., Cremonini, S., Desantis, P., Calastri, C., and Vianello, G., 2013, Chemical characterisation of anthro-technosols from Bronze to Middle Age in Bologna (Italy). *Journal of Archaeological Science*, v. 40, p. 3660–3671.

Vos, K., Vandenberghe, N., and Elsen, J., 2014, Surface textural analysis of quartz grains by scanning electron microscopy (SEM): from sample preparation to environmental interpretation: *Earth-Science Reviews*, v. 128, p. 93–104.

Vött, A., Lang, F., Brückner, H., Gaki-Papanastassiou, K., Maroukian, H., Papanastassiou, D., Giannikos, A., Hadler, H., Handl, M., and Ntageretzis, K., 2011, Sedimentological and geoarchaeological evidence of multiple tsunamigenic imprint on the Bay of Palairos-Pogonia (Akarnania, NW Greece): *Quaternary International*, v. 242, no. 1, p. 213–239.

Vrydaghs, L., and Devos, Y., 2018, Visibility, preservation and colour: a descriptive system for the study of opal phytoliths in (archaeological) soil and sediment thin section: *Environmental Archaeology*, v. 25, no. 2, p. 170–177.

Wacha, L. and Frechen, M. 2011. The geochronology of the "Gorjanović loess section" in Vukovar, Croatia: *Quaternary International*, v. 240, no. 1–2, p. 87–99.

Wadley, L., Sievers, C., Bamford, M., Goldberg, P., Berna, F., and Miller, C., 2011, Middle Stone Age bedding construction and settlement patterns at Sibudu, South Africa: *Science*, v. 334, p. 1388–1391.

Wagstaff, M., 1992, Agricultural terraces: the Vailikos Valley, Cyprus, *in* Bell, M., and Boardman, J., eds., *Past and Present Soil Erosion, Monograph 22*. Oxford, Oxbow, p. 155–161.

Walker, R. G., and Cant, D. J., 1984, Sandy Fluvial Systems, *in* Walker, R. G., ed., *Facies Models: Geoscience Canada Reprint Series 1*, Geological Association of Canada, p. 71–89.

Walker, T. M. 2018. The Gwithian landscape. *Mollusks and archaeology on the Cornish sand dunes*. Oxford, Archaeopress.

Wallwork, J. A., 1983, *Earthworm Biology*. London, Edward Arnold, Studies in Biology.

Wang, P., Jiang, H. C., Yuan, D. Y., Liu, X. W., and Zhang, B., 2010, Optically stimulated luminescence dating of sediments from the Yellow River terraces in Lanzhou: Tectonic and climatic implications: *Quaternary Geochronology*, v. 5, no. 2–3, p. 181–186.

Wang, Z., Zhuang, C., Saito, Y., Chen, J., Zhan, Q., and Wang, X., 2012, Holocene sea-level change and coastal environmental response on the southern Yangtze delta plain, China: implications for the rise of Neolithic culture: *Quaternary Science Reviews*, v. 35, p. 51–62.

Wang, N. Y., Dong, C. W., Xu, H. G., and Zhuang, Y., 2020, Letting the stones speak: An interdisciplinary survey of stone collection, transportation and construction at Liangzhu City, prehistoric Lower Yangtze River: *Geoarchaeology*, v. **35**, p. 625–643.

Waters Rist, A. L., Lieverse, A. R., Novikov, A. G., Goriunova, O. I., Kaharinskii, A. A., and McKenzie, H. G., 2021, Spatial and temporal differences in Late Neolithic Serovo to Early Bronze Age Glazkovo forager diet in Lake Baikal's Little Sea Microregion, *Siberia Archaeological Research in Asia*, v. 25, p. 100235.

Waters, M. R., 1986, *The Geoarchaeology of Whitewater Draw, Arizona*. Tucson, AZ, University of Arizona Press.

Waters, M. R., 1992, *Principles of Geoarchaeology: A North American Perspective*. Tucson, AZ, University of Arizona Press.

Waters, M. R., 2000, Alluvial stratigraphy and geoarchaeology in the American Southwest: *Geoarchaeology*: v. 15, p. 537–557.

Waters, M. R., and Nordt, L. C., 1995, Late Quaternary floodplain history of the Brazos River in east-central Texas: *Quaternary Research* , v. 43, no. 3, p. 311–319.

Waters, M. R., and Ravesloot, J. C., 2000, Late Quaternary geology of the Middle Gila River, Gila River Indian Reservation, Arizona: *Quaternary Research (New York)*, v. 54, no. 1, p. 49–57.

Waters, M. R., Forman, S. L., Jennings, T. A., Nordt, L. C., Driese, S. G., Feinberg, J. M., Keene, J. L., Halligan, J., Lindquist, A., and Pierson, J., 2011, The Buttermilk Creek complex and the origins of Clovis at the Debra L. Friedkin site, Texas: *Science*, v. 331, no. 6024, p. 1599–1603.

Watson, B., 1998, A brief history of archaeological exploration in Roman London, *in* Watson, B., ed., *Roman London. Recent Archaeological Work*, Supplementary series No. 24: Portsmouth, Rhode Island, Journal of Roman Archaeology, p. 13–22.

Watson, A., 1989a, Desert crusts and varnishes, *in* Thomas, D. S. G., ed., *Arid Zone Geomorphology*. London, Belhaven Press, p. 25–55.

Watson, A., 1989b, Windflow characteristics and aeolian entrainment, *in* Thomas, D. S. G., ed., *Arid Zone Geomorphology*. London, Belknap Press, p. 209–231.

Wattez, J., Courty, M.-A., and Macphail, R. I., 1990, Burnt organo-mineral deposits related to animal and human activities in prehistoric caves, *in* Douglas, L. A., ed., *Soil Micromorphology: A Basic and Applied Science, Developments in Soil Science 19*. Amsterdam, Elsevier, p. 431–439.

Wauchope, R., 1938, *Modern Maya Houses, a Study of the Archaeological Significance*. Washington, DC, Carnegie Institution of Washington.

Weber, A. W., and McKenzie, H. G., 2003, Prehistoric Foragers of the Cis-Baikal, Siberia. *Proceedings of the First Conference of the Baikal Archaeological Project, Northern Hunter-Gatherers Research Series*: Edmonton, The University of Alberta Press.

Weber, A. W., Link, D. W., and Katzenberg, M. A., 2002, Hunter-Gatherer Culture Change and Continuity in the Middle Holocene of the Cis-Baikal, Siberia: *Journal of Anthropological Archaeology*, v. 21, p. 230–299.

Wegener, O., 2009, Soil micromorphological investigations on trampling floors in pit houses (Grubenhäuser) of the deserted medieval town Marsleben (Saxony-Anhalt), *in* Thiemeyer, H., ed., *Archaeological Soil Micromorphology - Contributions to the Archaeological Soil Micromorphology Working Group Meeting 3rd to 5th April 2008*, Volume D30. Frankfurt A M, Frankfurter Geowiss. Arb., p. 133–141.

Weiner, S. Q., X.; Goldberg, P.; Liu, J.; Bar-Yosef, O., 1998, Evidence for the use of fire at Zhoukoudian, China: *Science*, v. 281, p. 251–253.

Weiner, S., 2010, *Microarchaeology: Beyond the Visible Archaeological Record*. Cambridge, New York, Cambridge University Press.

Weiner, S., and Goldberg, P., 1990, On-site Fourier transform infrared spectrometry at an archaeological excavation: *Spectroscopy*, v. 5, p. 47–50.

Weiner, S., Goldberg, P., and Bar-Yosef, O., 1993, Bone preservation in Kebara Cave, Israel using on-site Fourier transform infrared spectrometry: *Journal of Archaeological Science*, v. 20, p. 613–627.

Weiner, S., Schiegl, S., Goldberg, P., and Bar-Yosef, O., 1995, Mineral assemblages in Kebara and Hayonim Caves, Israel: Excavation strategies, bone preservation, and wood ash remnants: *Israel Journal of Chemistry*, v. 35, p. 143–154.

Weiner, S., Bar-Yosef, O., Goldberg, P., Xu, Q., and Liu, J., 2000, Evidence for the use of fire at Zhoukoudian: *Acta Anthropologica Sinica*, Supplement to vol. 19, p. 218–223.

Weiner, S., Goldberg, P., and Bar-Yosef, O., 2002, Three-dimensional distribution of minerals in the sediments of Hayonim Cave, Israel: Diagenetic processes and archaeological implications: *Journal of Archaeological Science*, v. 29, p. 1289–1308.

Weiner, S., Berna, F., Cohen-Ofri, I., Shahack-Gross, R., Albert, R. M., Karkanas, P., Meignen, L., and Bar-Yosef, O., 2007, Mineral distributions in Kebara Cave: diagenesis and its effect on the archaeological record, *in* Bar- Yosef, O., and Meignen, L., eds., *Kebara Cave, Mt. Carmel, Israel: The*

Middle and Upper Paleolithic Archaeology. Cambridge, American School of Prehistoric Research Bulletin 49, Peabody Museum of Archaeology and Ethnology, Harvard University, p. 131–146.

Weissbrod, L., and Weinstein-Evron, M., 2020, Climate variability in early expansions of Homo sapiens in light of the new record of micromammals in Misliya Cave, Israel: *Journal of Human Evolution,* v. 139, p. 102741.

Wenban-Smith, F. F., 2013, *The Ebbsfleet Elephant. Excavations at Southfleet Road, Swanscombe in advance of High Speed 1, 2003–4.* Oxford, Oxford Archaeology, Oxford Archaeology Monograph No. 20.

Wenban-Smith, F. F., Allen, P., Bates, M. R., Parfitt, S. A., Preece, R. C., Stewart, J. R., Turner, C., and Whittaker, J. E., 2006, The Clactonian elephant butchery site at Southfleet Road, Ebbsfleet, UK: *Journal of Quaternary Science,* v. 21, no. 5, p. 471–483.

Wenban-Smith, F., Bates, M., Allen, P., Bates, R., and Hutchinson, J. N., 2013, Geology, stratigraphy and site phasing, *in* Wenban-Smith, F., ed., *The Ebbsfleet Elephant Excavations at Southfleet Road, Swanscombe in advance of High Speed 1, 2003–4.* Oxford, Oxford Archaeology Monograph, p. 57–110.

Wendorf, F., and Schild, R., 1980, *Prehistory of the Eastern Sahara.* New York, Academic Press.

Wendorf, F., Schild, R., and Close, A. E., 1993, *Egypt during the Last Interglacial: The Middle Paleolithic of Bir Tarfawi and Bir Sahara East.* New York, Plenum Press.

Wendorf, F., Schild, R., Close, A. E., Schwarcz, H. P., Miller, G. H., Grün, R., Bluszcz, A., Stokes, S., Morawska, L., Huxtable, J., Lundberg, J., Hill, C. L., and McKinney, C., 1994, A Chronology for the Middle and Late Pleistocene wet episodes in the Eastern Sahara, *in* Bar-Yosef, O., and Kra, R. S., eds., *Late Quaternary Chronology and Paleoclimates of the Eastern Mediterranean.* Tucson, AZ, Radiocarbon, p. 147–168.

Wentworth, C. K., 1922, A scale of grade and class terms for clastic sediments: *The Journal of Geology,* v. 30, no. 5, p. 377–392.

West, L. T., Bradford, J. M., and Norton, L. D., 1990, Crust morphology and infiltrability in surface soils from the southeast and midwest U.S.A., *in* Douglas, L. A., ed., *Soil Micromorphology: A Basic and Applied Science.* Amsterdam, Elsevier, p. 107–113.

West, S. E., 1985, West Stow: the Anglo-Saxon village: *East Anglian Archaeology,* v. 24, no. 1 and 2.

Westaway, K. E., Roberts, R. G., Sutikna, T., Morwood, M. J., Drysdale, R., Zhao, J., and Chivas, A. R., 2009a, The evolving landscape and climate of western Flores: an environmental context for the archaeological site of Liang Bua: *Journal of Human Evolution,* v. 57, no. 5, p. 450–464.

Westaway, K. E., Sutikna, T., Saptomo, W. E., Jatmiko, Morwood, M. J., Roberts, R. G., and Hobbs, D. R., 2009b, Reconstructing the geomorphic history of Liang Bua, Flores, Indonesia: a stratigraphic interpretation of the occupational environment: *Journal of Human Evolution,* v. 57, no. 5, p. 465–483.

Wetzel, A., and Unverricht, D., 2020, Sediment dynamics of estuarine Holocene incised-valley fill deposits recorded by Siphonichnus (ancient Red River, Gulf of Tonkin). *Palaeogeography, Palaeoclimatology, Palaeoecology,* Vol. 560, 110041.

Wheatley, D., and Gillings, M., 2002, *Spatial Technology and Archaeology: The Archaeological Applications of GIS.* London, Taylor & Francis.

White, D. 2013. LIDAR, point clouds, and their archaeological applications, in D Comer and M Harrower(Eds.), Mapping *Archaeological Landscapes from Space.* Springer: New York, p. 175–186.

White, J. A., Schulting, R. J., Lythe, A., Hommel, P., Bronk Ramsey, C., Moiseyev, V., Khartanovich, V., and Weber, A. W., 2020, Integrated stable isotopic and radiocarbon analyses of Neolithic and bronze age hunter-gatherers from the Little Sea and Upper Lena micro- regions, Cis-Baikal, Siberia: *Journal of Archaeological Science,* v. 119, p. 105161.

White, R. E., 2013, *Principles and Practice of Soil Science: The Soil as a Natural Resource,* 4th edition. Chichester, Wiley-Blackwell.

White, W. B., and Culver, D. C., 2012, *Encyclopedia of Caves*, 2nd edition. San Diego, CA, Academic Press, p. 966.

White, W. B., Culver, D., and Pipan, T., 2019, *Encyclopedia of Caves*, 3rd edition. San Diego, CA, Academic Press, p. 1250.

Whittle, A., 2007, *The Early Neolithic on the Great Hungarian Plain: investigations of the Körös culture site of Ecsegfalva 23, Co. Békés*: Budapest, Institute of Archaeology, p. 809.

Whittle, A., Rouse, A. J., and Evans, J. G., 1993, A Neolithic downland monument in its environment: excavations at the Easton Down Long Barrow, Bishops Canning, North Wiltshire: *Proceedings of the Prehistoric Society*, v. 59, p. 197–239.

Whittlesey, J. H., Myers, J. W., and Allen, C. C., 1977, The Whittlesey Foundation 1976 Field Season: *Journal of Field Archaeology*, v. 4, no. 2, p. 181–196.

Wieder, M., and Adan-Bayewitz, D., 2002, Soil parent materials and the pottery of Roman Galilee: a comparative study: *Geoarchaeology*, v. 17, p. 393–415.

Wieder, M., and Gvirtzman, G., 1999, Micromorphological indications on the nature of the late Quaternary Paleosols in the southern coastal plain of Israel: *Catena*, v. 35, no. 2–4, p. 219–237.

Wiedner, K., Schneeweiß, J., Dippold, M. A., and Glaser, B., 2015, Anthropogenic Dark Earth in Northern Germany – The Nordic Analogue to terra preta de Índio in Amazonia: *Catena*, v. 132, p. 114–125.

Wilkinson, T. J., 1990, Soil development and early land use in the Jazira region, Upper Mesopotamia, *in* Thomas, K., ed., *Soils and Early Agriculture, World Archaeology 22, No. 1*. London, Routledge, p. 87–103.

Wilkinson, T. J., 1997, Holocene environments of the High Plateau, Yemen. Recent geoarchaeological investigations: *Geoarchaeology*, v. 12, no. 8, p. 833–864.

Wilkinson, T. J., 2003, *Archaeological Landscapes of the Near East*. Tucson, AZ, The University of Arizona Press.

Wilkinson, T. J., and Murphy, P. L., 1995, *The Archaeology of the Essex Coast, Volume I: The Hullbridge Survey*. Chelmsford, Essex County Council, East Anglian Archaeology report No. 71.

Wilkinson, T. J., and Murphy, P., 1986, Archaeological survey of an intertidal zone: the submerged landscape of the Essex coast, England: *Journal of Field Archaeology*, v. 13, p. 177–194.

Wilkinson, T. J., Murphy, P. L., Brown, N., and Heppell, E., 2012, *The Archaeology of the Essex Coast, Vol. II. Excavations at the Prehistoric Site of The Stumble, East Anglian Archaeology 142*. Chelmsford, Essex County Council.

Williams, T., 2007, The city of Sultan Kala, Merv, Turkmenistan: communities, neighbourhoods and urban planning from the eighth to the thirteenth century, *in* Bennison, A. K., and Gascoigne, A., eds., *Cities in the Pre-modern Islamic world: The Urban Impact of Religion, State and Society*. London, Routledge, p. 42–62.

Williams, T., 2018, Flowing into the city: Approaches to water management in the early Islamic city of Sultan Kala, Turkmenistan, *in* Altaweel, M., and Zhuang, Y., eds., *Water Societies and Technologies from the Past and Present*. London, UCL Press, open access, p. 157–179.

Williams, A. J., Pagliai, M., and Stoops, G., 2018a, Physical and biological surface crusts and seals, *in* Stoops, G., Marcelino, V., and Mees, F., eds., *Interpretation of Micromorphological Features in Soils and Regoliths*. Amsterdam, Elsevier, p. 539–574.

Williams, T. J., Collins, M. B., Rodrigues, K., Rink, W. J., Velchoff, N., Keen-Zebert, A., Gilmer, A., Frederick, C. D., Ayala, S. J., and Prewitt, E. R., 2018b, Evidence of an early projectile point technology in North America at the Gault Site, Texas, USA: *Science advances*, v. 4, no. 7, eaar5954, 1–7.

Wilson, H. E., 2004, *Chemical and Textural Analysis of Soils and Sediments in Forensic Investigations*: Royal Holloway University of London.

Wilson, C. A., 2017, Electron Probe X-ray Microanalysis (SEM-EPMA) Techniques, *in* Nicosia, C., and Stoops, G., eds., *Archaeological Soil and Sediment Micromorphology*. Chichester, Wiley Blackwell, p. 451–460.

Wilson, C. A., Davidson, D. A., and Cresser, M. S., 2008, Multi-element soil analysis: an assessment of its potential as an aid to archaeological interpretation: *Journal of Archaeological Science*, v. 35, p. 412–424.

Wiltshire, P. E. J., 2002, Environmental Profiling and Forensic Palynology: Background and potential value to the criminal investigator, Scientific Support for Crime Scenes that involve human remains, *British Association for Human Identification*, p. 14–21.

Wiltshire, P., 2019, *Traces*. London, 535 Bonnier Books.

Wiltshire, P. E. J., Edwards, K. J., and Bond, S., 1994, Microbially-derived metallic sulphide spherules, pollen, and the waterlogging of archaeological sites: *Proceedings of the American Association of Sedimentary Palynologists*, v. 29, p. 207–221.

Winter, T. C., 2004, The hydrology of lakes, *in* O'Sullivan, P., and Reynolds, C. S., eds., *The Lakes Handbook Limnology and Limnetic Ecology*, Vol. 1. Oxford, Wiley Blackwell, p. 61–78.

Wiseman, J., El-Baz, F., 2007. *Remote Sensing in Archaeology*. New York, Springer.

Woods, W. I., and McCann, J. M., 1999, The Anthropogenic origin and persistence of Amazonian Dark Earths: *Conference of Latin Americanist Geographers*, v. 25, p. 7–14.

Woodward, J. C., and Goldberg, P., 2001, The sedimentary records in Mediterranean rockshelters and caves: archives of environmental change: *Geoarchaeology*, v. 16, no. 4, p. 465–466.

Wouters, B., Devos, Y., Milek, K., Vrydaghs, L., Bartholomieux, B., Tys, D., Moolhuizen, C., and van Asch, N., 2016, Medieval markets: a soil micromorphological and archaeobotanical study of urban stratigraphy of Lier (Belgium). *Quaternary International*, v. 460, p. 48–64.

WRB, 2006, *World Reference Base for Soil Resources*. 2nd edition. Rome, FAO.

WRB, 2015, *World Reference Base for Soil Resources 2014 (Update 2015)*, Rome, FAO.

Wright, D. K., Forman, S. L., Kiura, P. W., Bloszies, C., and Beyin, A., 2015, Lakeside view: sociocultural responses to changing water levels of Lake Turkana, Kenya: *African Archaeological Review*, v. 32, p. 335–367.

Wright, D. K., Thompson, J. C., Schilt, F., Cohen, A., Choi, J. H., Mercader, J., Nightingale, S., Miller, C. E., Mentzer, S. M., Walde, D., and Welling, M., 2017, Approaches to Middle Stone Age landscape archaeology in tropical Africa: *Journal of Archaeological Science*, v. 77, p. 64–77.

Wright, V. P., 1986, Paleosols. *Their Recognition and Interpretation*. Oxford, Blackwell Scientific Publications.

Wright, V. P., 1992, Problems in detecting environmental changes in Pre-Quaternary paleosols, *Seesoil*, v. **8**. Wye, South East Soils Discussion Group, p. 5–12.

Wrighton, F. E., 1951, "Plant ecology at Cripplegate, 1950", in "City bombed sites survey": *The London Naturalist*, v. 30, p. 73–79.

Wroth, K., Cabanes, D., Marston, J. M., Aldeias, V., Sandgathe, D., Turq, A., Goldberg, P., and Dibble, H. L., 2019, Neanderthal plant use and pyrotechnology: phytolith analysis from Roc de Marsal, France: *Archaeological and Anthropological Sciences* v. **11**, p. 4325–4346.

Wurz, S., Bentsen, S. E., Reynard, J., Van Pletzen-Vos, L., Brenner, M., Mentzer, S., Pickering, R., and Green, H., 2018, Connections, culture and environments around 100 000 years ago at Klasies River main site: *Quaternary International*, v. 495, p. 102–115.

Xia, Z. K., 1992, The study of the change of ancient lakeshore in the Datong-Yangyuan Basin (in Chinese with English abstract): *Geographical Research*, v. 11, p. 52–59.

Xia, Z. K., Deng, H., and Wu, H. L., 2000, Geomorphologic Background of the Prehistoric Cultural Evolution in the Xar Moron River Basin, Inner Mongolia (in Chinese with English abstract): *Acta Geographica Sinica*, v. 55, no. 3, p. 329–336.

Xu, Q., Tian, M., Li, L., and Liu, J., 1996, A brief introduction to the Quaternary geology and paleoanthropology of the Zhoukoudian site, Beijing, *in* Deng, N., ed., *Field Trip Guide; Vol. 6, Beijing and Its Adjacent Areas*. Beijing, China, Geological Publishing House, p. T203.201–T203.210.

Yaalon, D. H., and Dan, J., 1974, Accumulation and distribution of loess-derived deposits in the semi-desert and desert fringe areas of Israel: *Zeitschrift für Geomorphologie Supplement Band*, v. 20, p. 91–105.

Yaalon, D. H., and Ganor, E., 1975, The influence of dust on soils during the Quaternary: *Soil Science*, v. 116, p. 146–155.

Yaalon, D., and Kalmar, D., 1972, Vertical movement in an undisturbed soil: continuous measurement of swelling and shrinkage with a sensitive apparatus: *Geoderma*, v. 8, no. 4, p. 231–240.

Yaalon, D. H., and Kalmar, D., 1978, Dynamics of cracking and swelling clay soils: Displacement of skeletal grains, optimum depth of slickensides, and rate of intra-pedonic turbation: *Earth Surface Processes*, v. 3, p. 31–42.

Yi, S., 1988, Quaternary geology and paleoecology of hominid occupation of Imjin Basin: *The Korean Journal of Quaternary Research*, v. 2, no. 1, p. 25–30.

Yi, S., 2005, New data on the formation of the basalt plain in the Imjin River Basin: *Journal of the Korean Geomorphological Association*, v. 12, no. 3, p. 21–38.

Yi, S., 2015, Hand Axes in the Imjin River Basin, Korea – Implications for Late Pleistocene Hominin Evolution in East Asia, *in* Kaifu, Y., Izuho, M., Goebel, T., Sato, H., and Ono, A., eds., *Emergence and Diversity of Modern Human Behavior in Paleolithic Asia*: College Station, Texas A&M University Press, p. 259–269.

Yoo, Y., 2007, Long-Term Changes in the Organization of Lithic Technology: A Case Study from the Imjin-Hantan River Area, Korea, [Doctoral Dissertation]: McGill University.

Yule, B., 1990, The "dark earth" and late Roman London: *Antiquity*, v. 64, p. 620–628.

Yuretich, R., Melles, M., Sarata, B., and Grobe, H., 1999, Clay minerals in the sediments of Lake Baikal; a useful climate proxy: *Journal of Sedimentary Research*, v. 69, no. 3, p. 588–596.

Zachariae, G., 1964, Welche bedeutung haben enchytraeen im waldboden?, *in* Jongerius, A., ed., *Soil Micromorphology*. Amsterdam, Elsevier, p. 57–67.

Zangger, E., 1992, Neolithic to present soil erosion in Greece, *in* Bell, M., and Boardman, J., eds., *Past and Present Soil Erosion, Monograph 22*. Oxford, Oxbow, p. 133–147.

Zarzycka, S. E., Surovell, T. A., Mackie, M. E., Pelton, S. R., Kelly, R. L., Goldberg, P., Dewey, J., and Kent, M., 2019, Long-distance transport of red ocher by Clovis foragers: *Journal of Archaeological Science*: Reports, v. 25, p. 519–529.

Zauyah, S., Schaefer, C. E. G. R., and Simas, F. N. B., 2018, Saprolites, *in* Stoops, G., Marcelino, V., and Mees, F., eds., *Interpretation of Micromorphological features of Soils and Regoliths*. Amsterdam, Elsevier, p. 37–57.

Zeeden, C., and Hambach, U. F., 2021, Magnetic susceptibility properties of loess from the Willendorf archaeological site: Implications for the syn/post-depositional interpretation of magnetic fabric: *Frontiers in Earth Science*, v. 8, p. 1–11.

Zeist, W. v., and Bottema, S., 1982, Vegetational history of the eastern Mediterranean and the Near East during the last 20,000 years, *in* Bintliff, J. L., and Zeist, W. v., eds., *Palaeoclimates, Palaeoenvironments and Human Communities in the Eastern Mediterranean Region in Later Prehistory Part ii, International Series 133 ii*. Oxford, British Archaeological Reports, p. 277–321.

Zeuner, F. E., 1946, *Dating the Past. An Introduction to Geochronology*. London, Methuen & Co. Ltd.

Zeuner, F. E., 1953, The chronology of the Mousterian at Gorham's Cave Gibraltar: *The Prehistoric Society*, v. 19, no. 2, p. 180–188.

Zeuner, F. E., 1958, *Dating the Past: An Introduction to Geochronology*, 4th edition. London, Methuen & Co.

Zeuner, F. E., 1959, *The Pleistocene Period*. London, Hutchinson Scientific & Technical.

Zhang, J. F., Qiu, W. L., Wang, X. Q., Hu, G., Li, R. Q., and Zhou, L. P., 2010, Optical dating of a hyperconcentrated flow deposit on a Yellow River terrace in Hukou, Shaanxi, China: *Quaternary Geochronology*, v. 5, no. 2–3, p. 194–199.

Zhang, J. F., Wang, X. Q., Qiu, W. L., Shelach, G., Hu, G., Fu, X., Zhuang, M. G., and Zhou, L. P., 2011, The paleolithic site of Longwangchan in the middle Yellow River, China: chronology, paleoenvironment and implications: *Journal of Archaeological Science*, v. 38, no. 7, p. 1537–1550.

Zhang, L. T., and Yang, X. L., 2012, Chinese lakes, *in* Bengtsson, L., Herschy, R. W., and Fairbridge, R. W., eds., *Encyclopedia of Lakes and Reservoirs*: Dordrecht, Springer, p. 156–160.

Zhang, S. S., Wang, P. F., Wang, C., and Hou, J., 2015, Sediment resuspension under action of wind in Taihu Lake, China: *International Journal of Sediment Research*, v. 30, no. 1, p. 48–62.

Zhang, S., Zhang, J. F., Zhao, H., Liu, X. J., and Chen, F. H., 2020, Spatiotemporal complexity of the "Greatest Lake Period" in the Tibetan Plateau: *Science Bulletin*, v. 65, p. 1317–1319.

Zheng, Y., Sun, G., Qin, L., Li, C., Wu, X., and Chen, X., 2009, Rice fields and modes of rice cultivation between 5000 and 2500 BC in east China: *Journal of Archaeological Science*, v. 36, p. 2609–2616.

Zhuang, Y., 2017, State and irrigation: arch(a)eological and textual evidence of water management in late Bronze Age China: *WIREs-Water*, v. 4, no. 4, p. e1217.

Zhuang, Y., 2018, Paddy fields, water management and agricultural development in the prehistoric Taihu Lake region and the Ningshao Plain, *in* Zhuang, Y., and Altaweel, M., eds., *Water Societies and Technologies from the Past and Present*. London, UCL Press, open access, p. 87–108.

Zhuang, Y., and Kidder, T. R., 2014, Archaeology of Anthropocene in the Yellow River region, China, 8000–2000 cal. BP: *The Holocene*, v. 24, no. 11, p. 1602–1623.

Zhuang, Y., Bao, W., and French, C., 2013, River floodplain aggradation history and culture activities: geoarchaeological investigation at the Yuezhuang site, Lower Yellow River, China: *Quaternary International*, v. 315, p. 101–115.

Zhuang, Y., Ding, P., and French, C., 2014, Water management and agricultural intensification of rice farming at the late-Neolithic site of Maoshan, Lower Yangtze River, China: *The Holocene*, v. 24, no. 5, p. 531–545.

Zilberman, E., 1993, The late Pleistocene sequence of the northwestern Negev flood plains; a key to reconstructing the paleoclimate of southern Israel in the last glacial: *Israel Journal of Earth Sciences*, v. 41, no. 2–4, p. 155–167.

Zolitschka, B., Francus, P., Ojala, A. E. K., and Schimmelmann, A., 2015, Varves in lake sediments – a review: *Quaternary Science Reviews*, v. 117, p. 1–41.

Zong, Y., 2004, Mid-Holocene sea-level highstand along the Southeast Coast of China: *Quaternary International*, v. 117, p. 55–67.

Zong, Y., Chen, Z., Innes, J. B., Chen, C., Wang, Z., and Wang, H., 2007, Fire and flood management of coastal swamp enabled first rice paddy cultivation in east China: *Nature*, v. 449, p. 459–463.

ZPICRA (Zhejiang Provincial Institute of Cultural Relics and Archaeology), 2011, Summary of archaeological work at the Tianluoshan site: phase I (2004–2008) (in Chinese), *in* School of Archaeology and Museology of Peking University and ZPICRA (Zhejiang Provincial Institute of Cultural Relics and Archaeology), eds., *Synthetic Studies of the Ecological Remains from the Tianluoshan Site*. Beijing, Cultural Relics Press, p. 7–39.

ZPICRA (Zhejiang Provincial Institute of Cultural Relics and Archaeology), and Museum of Xiaoshan., 2004, *Kuahuqiao: An Excavation Report*. Beijing, Cultural Relics Press.

ZPICRA (Zhejiang Provincial Institute of Cultural Relics and Archaeology) and Museum of Haining, 2015, *Xiaodouli: An Excavation Report*. Beijing, Cultural Relics Press.

Index

Practical and Theoretical Geoarchaeology, Second Edition. Paul Goldberg and Richard I. Macphail.
© 2022 John Wiley & Sons Ltd. Published 2022 by John Wiley & Sons Ltd.

Archaeological, Geological, and Chronological Periods and Cultures

Sites and place names

Practical and Theoretical Geoarchaeology, Second Edition. Paul Goldberg and Richard I. Macphail.
© 2022 John Wiley & Sons Ltd. Published 2022 by John Wiley & Sons Ltd.